HARMFUL CHEMICAL SUBSTANCES
Volume 1: Elements in Groups I–IV of the Periodic Table and their Inorganic Compounds

Ellis Horwood and Prentice Hall
are pleased to announce their collaboration in a new imprint whose list will encompass outstanding works by world-class chemists, aimed at professionals in research, industry and academia. It is intended that the list will become a by-word for quality, and the range of disciplines in chemical science to be covered are:

ANALYTICAL CHEMISTRY
ORGANIC CHEMISTRY
INORGANIC CHEMISTRY
PHYSICAL CHEMISTRY
POLYMER SCIENCE & TECHNOLOGY
ENVIRONMENTAL CHEMISTRY
CHEMICAL COMPUTING & INFORMATION SYSTEMS
BIOCHEMISTRY
BIOTECHNOLOGY

HARMFUL CHEMICAL SUBSTANCES
Volume 1: Elements in Groups I–IV of the Periodic Table and their Inorganic Compounds

Editors:
V. A. FILOV, Sc.D. (Biol.)
Head, Department of Chemistry, Biology and Toxicology,
Institute of Oncology, and President, Institute for the Study of
Xenobiotics, St Petersburg, Russia

B.A. IVIN, Sc.D. (Chem.)
Professor, Corresponding Member of the Russian Academy of
Natural Sciences, and Head of Department of Organic Chemistry,
Chemical-Pharmaceutical Institute, St. Petersburg, Russia

A.L. Bandman, M.D.
Formerly Senior Researcher, Library of the U.S.S.R. Academy of
Sciences, Leningrad, Russia

ELLIS HORWOOD
NEW YORK   LONDON   TORONTO   SYDNEY   TOKYO   SINGAPORE

First published in 1993 by
ELLIS HORWOOD LIMITED
Market Cross House, Cooper Street,
Chichester, West Sussex, PO19 1EB, England

A division of
Simon & Schuster International Group
A Paramount Communications Company

First published in Russian as **Vrednye khimichesklye veschestra. Neorganicheskiye soyedineniga
elementor 1–IV grup,** the first of a series edited by V.A. Filov.
© 1988 Khimiya, St Petersburg.

Translated into English by V.E. Tatarchenko

English Edition © Ellis Horwood Limited, 1993

Printed and bound in Great Britain
by Bookcraft, Midsomer Norton

British Library Cataloguing in Publication Data

A catalogue record for this book is available from the British Library

ISBN 0–13–383373–9

Library of Congress Cataloging-in-Publication Data

Available from the publisher

List of contributors

A. L. Bandman, Library of the USSR Academy of Sciences (now retired)

L. S. Dubeikovskaya, Institute of Occupational Health, Leningrad

V. A. Filov, Institute of Oncology, Leningrad (editor of the *Harmful Chemical Substances* series)

G. A. Gudzovsky, Medical Institute, Stavropol

B. A. Ivin, Chemical-Pharmaceutical Institute, Leningrad

B. A. Katsnelson, Medical Research Center for Prophylaxis and Health Protection of Industrial Workers, Sverdlovsk

M. N. Korshun, Institute of Occupational Hygiene, Kiev

Yu. A. Krotov, I. P. Pavlov Medical Institute, Leningrad

M. I. Mikheyev, Institute for Advanced Medical Studies, Leningrad

N. A. Minkina, Research Institute of Occupational Health, Leningrad

A. V. Moskvin, Chemical-Pharmaceutical Institute, Leningrad

V. Ya. Rusin, Pedagogical Institute, Yaroslavl

V. V. Semenova, Medical Institute of Sanitation and Hygiene, Leningrad

I. M. Trakhtenberg, Institute of Occupational Hygiene, Kiev

Table of contents

Preface to the English-language edition

This edition is a translation of a revised text of the first book in the multivolume series of books being published in Russian by the Leningrad (now St. Peterburg) branch of Khimiya Publishers under the general title *Harmful Chemical Substances* [Vrednye Khimicheskiye Veshchestva].

The series, which is expected to be completed by the end of this decade, will comprise 12 to 15 volumes and cover the vast majority of substances that do or can, under certain circumstances, act as environmental pollutants or toxic agents. So far, five volumes have been published, viz. *Elements in Groups I—IV of the periodic Table and Their Inorganic Compounds* (1988); *Elements in Groups V—VIII of the Periodic Table and Their Inorganic Compounds* (1989); *Radioactive Substances* (1990); *Hydrocarbons and Halogenated Hydrocarbons* (1990); and *Nitrogen-Containing Organic Compounds* (1992). In the USSR, all five were sold out almost instantaneously, although over 30 000 copies of each have been issued. This is an indication of the great demand for information on chemical pollutants and toxic substances among professionals of various categories. The present series is designed to meet this demand.

The next two volumes, which are now in preparation, will be *Oxygen- and Halogen-Containing Organic Compounds* and *Naturally Occurring Organic Compounds*.

The Russian text of the first volume has been revised and updated specifically for the English-language edition. In particular, the material applicable Russia alone has been deleted from the section *Measures to control exposure* and threshold limit values adopted in the USA have been included in the section *Hygienic standards*; the Appendix has also been reworked.

I hope that this and other books in the series, when translated and published in English, will be found useful by readers in the United Kingdom, the USA, and other countries. My hope is strengthened by the fact that the contributors have drawn heavily upon Russian sources of information that are often inaccessible to English-speaking people because of the language barrier.

In conclusion, the contributors and I wish to express deep gratitude to V. E. Tatarchenko for his generous and competent assistance in revising the Russian text and for his translation of it into English.

V. FILOV
Leningrad, 1992

Note. The geographical names and administrative terms relating to the former USSR are those that were in use at the time the book was being written.

Preface to the 1988 Russian edition

This serial publication has direct relevance to the challenging problem of protecting the man's environment — a problem that has been generated by man himself and will undoubtedly remain one of growing immediate concern in future. The impacts of man on his environment and his health have in many respects become more substantial than those of Nature. This is particularly true of the emissions of chemical agents into the environment from man-made sources, which have far exceeded in scope and intensity those from natural sources; the latter are usually either of low intensity or localized in time and space (these sources are well exemplified by volcanoes). Already in the early stages of civilization, man frequently experienced adverse effects of various chemical substances during the processing of mineral or organic raw materials and when using the finished products. In fact, deterioration of environmental quality began when people first collected into villages and utilized fire, and it has existed as a serious problem since then but particularly under the ever-increasing impacts of exponentially growing population and of industrialization. In recent decades, with the rapid development of science and technology, environmental contamination of air, water, soil, and food has become a real threat to the continued existence of many plant and animal communities and even species, and may ultimately threaten the very survival of the human race.

The scientific and technological revolution has been closely associated with the phenomenon of urbanization, a hallmark of which is the ever accelerating expansion of cities and urban agglomerations. Thus, the number of cities with a population of more than one million had increased from only 4 one hundred or so years ago, to 12 at the outset of the present century, to 75 by 1950, and to as many as 141 by 1960, and the number reaching the million mark has more than doubled since then. Large urban agglomerations are being formed at a rapid pace. One example is the agglomeration with a population of about 40 million and an area of more than 150 000 km^2 on the Atlantic coast of the USA. Another is the extremely large agglomeration, with a population of no less than 60 million, under formation on the Pacific coast of Japan through the confluence of Tokyo, Yokohama, Kyoto, Nagoya, Osaka, and Kobe. It is anticipated that three megalopolises will appear in the USA by the end of this century — Sansan (San Francisco–San Diego), Chipits (Chicago–Pittsburg), and Boswash (Boston–Washington), with populations of 20, 40, and 80 million, respectively.

Cities (especially large ones) discharge, not unlike volcanoes, enormous quantities of oxides of carbon, nitrogen, and sulfur, ammonium, phenols, hydrocarbons, various fumes, and a host of other noxious agents. There is a relationship between the size of cities and the concentration of pollutants therein. In cities with a population over 500 000, for example, atmospheric pollution is 1.5–2 times greater than in those with a population of about 100 000. It has been estimated that increasing the number of motor vehicles from 10 000 to 50 000 entails a 2- to 4-fold rise in the carbon monoxide content of the urban air. Small urban centers tend to remain relatively clean, although towns and settlements located near mines as well as those developing on the basis of oil-extracting, oil-processing, metallurgical, chemical or some other industries are often very polluted.

Pollution from man-made sources may involve all environmental media, including air, water, and soil, and it frequently extends over large areas. Typical examples are acid rains, pollution of seas by oil spilt from tankers, and releases of radioactive materials from nuclear power plants as a result of equipment failure or accidents. In many situations, the pollution picture is fairly complex, since adverse physical factors (noise, heat, electromagnetic radiation, etc.) are added to the chemical components of pollution. Chemical pollutants include synthetic substances that do not occur naturally; one of these is the notorious dioxin. They can react with each other to form compounds with unknown or unpredictable toxic characteristics.

It is commonly believed that the rural environment is much less polluted than the urban or periurban. Indeed, concentrations of many toxic products in the air over rural areas are, on average, only 10 times as high as over the oceans, while being 150 as high over urban areas. Yet agricultural activities, too, are associated with increasing utilization of chemicals such as fertilizers and particularly pesticides.

Environmental pollution is inevitable when chemicals are used in warfare; the classic example is the Vietnam war.

If the adverse effects of chemical pollutants on man and on the biosphere as a whole are to be effectively prevented, it is necessary to know the man-made sources of pollution, physical and chemical properties of pollutants, and how they act not only on humans and animals, but also on other forms of life, both aquatic and terrestrial, as well as on biocenoses. Moreover, information is needed on toxic and acceptable concentrations of pollutants and methods of their detection and measurement, on measures to be taken in the event of exposure to pollutants, and on many other aspects relating to pollution control and environmental sanitation. The pertinent knowledge and information are scattered throughout the literature, and there is an urgent need to gather them in comprehensive handbooks that could be used by any specialist concerned with environmental protection, adverse health effects of chemicals, or other related problem areas. No such handbooks have been published so far, although there is a prototype — the book *Vrednye veshchestva v promyshlennosti* [Harmful substances in industry]. This invaluable reference work has contributed greatly to the protection of workers in various industries (including agricultural workers) from adverse chemical agents. It therefore seems worthwhile to dwell briefly on the history of its publication.

The first edition of this book, written by Nikolai V. Lazarev with Pyotr I. Astrakhantsev as the coauthor, appeared in two volumes in the mid-1930s. It summarized the factual material accumulated in the field of toxicology (primarily industrial toxicology) by that time and satisfied the requirements of the then young and rapidly developing industry in the USSR. The second, considerably enlarged and updated edition, prepared under Lazarev's guidance by a team of authors who worked in the Laboratory of Industrial Toxicology of the Leningrad Institute of Industrial Hygiene and Occupational Diseases, appeared in 1951 and the third, stereotype edition, in 1954. The fourth edition, thoroughly revised and brought up to date under the editorship of N. V. Lazarev, was issued in 1963; in 1965 and 1971, the book was reprinted as the fifth and sixth editions, while 1969 saw the publication of an additional, third volume that included information on new industrially important compounds and pesticides. The most recent, seventh edition appeared in three revised and updated volumes in 1975–77, with two close pupils of Lazarev — E. N. Levina and I. D. Gadaskina — acting as the editors; in 1985 a fourth volume was added.

The handbook *Harmful substances in industry* has been extensively used in East European countries. Between 1954 and 1964 it was published in the Polish, Czech, Roumanian, and Chinese languages.

Today, however, it is obviously not sufficient to confine oneself to considering pollutants exclusively from the angle of industry where many comprehensive measures have been implemented to protect the health of workers from harmful effects of chemical pollutants. It was therefore decided to launch a series of handbooks that could by justly called *'Harmful substances in the environment'*.

Although following the general principles laid down in Lazarev's handbooks, this multivolume series is much broader in scope and will deal, in particular, with many compounds not previously covered—for example certain groups of drugs. It is planned to issue no less than 10 volumes in the series, at least one per year beginning with 1988; for each, a team of authors who are leading specialists in the respective fields, will be formed.

The authors and editors devote all these volumes to the memory of Nikolai V. Lazarev, the founder of industrial toxicology in our country.

INFORMATION AND DOCUMENTATION ON CHEMICAL POLLUTANTS OF THE ENVIRONMENT

It may be useful to consider briefly the information and documentation pertaining to chemical environmental pollutants, their toxic effects, and related matters. Since 1986, the Library of the USSR Academy of Sciences in Leningrad has been regularly publishing the bibliographic index *Vrednye veshchestva v okruzhayushchei srede* (Harmful substances in the environment) that alerts to the literature concerned with man-made pollution of air, water, soil, plants, and animals, including the hazards and toxic effects of pollutants, their environmental sources and levels, physicochemical and hygienic characteristics, transfer and transformation in the biosphere, impacts on organisms in the environment and on ecosystems, methods of detection and analysis in various media, and acceptable levels of exposure.

Most of the relevant secondary information is published by the principal information agency of the country—the All-Union Institute for Scientific and Technological Information (VINITI), mainly in its abstract journals *Toksikologiya* (Toxicology) and *Okhrana prirody i vosproizvodstvo prirodnykh resursov* (Environmental protection and resource management). About 95% of all items appearing annually in the latter publication are abstracts of articles from approximately 1700 journals and other serial publications (of which some 70% are in languages other than Russian), about 4% are abstracts or summaries of books, while the rest cover patents, standards, manuscripts deposited in libraries, and other miscellaneous materials. *Toxicology* issues monthly about 600 abstracts of works, most of which are concerned with problems of ecological toxicology or with toxicologic characteristics of individual chemicals that are actual or potential pollutants.

Individual aspects relating to environmental pollution and its implications are covered in many other discipline-oriented abstract journals produced by VINITI, including those devoted to biology, biochemistry, geophysics, transport, and power engineering—to mention only a few. *Biologiya* (Biological abstracts), for example, contains abstracts from journal articles and other published materials concerned with pollution of the environment with wastewaters and other products of human activity, alterations in vegetation under the influence of various anthropogenic factors, the impact of these on animals, etc. *Biokhimiya* (Biochemical abstracts) publishes, for instance, information about the impact of toxic environmental contaminants on biochemical processes in plants and animals. *Geofizika* (Geophysical abstracts) contains abstracts dealing with alterations in the chemical composition of waters brought about by anthropogenic influences and with methods and instrumentation for measuring the pollution of water, soil, and air. *Zhivotnovodstvo* (Animal husbandry abstracts) contains information about effects of environmental pollutants on agricultural animals. Information on biologic effects of environmental pollutants generated by internal combustion engines and by metallurgical processes can be found, respectively, in the abstract journals *Dvigateli vnutrennego sgoraniya* (Internal combustion engines) and *Metallurgiya* (Metallurgy). To facilitate quick location of abstracts in the various journals, a classified subject index of these abstracts appears each year in the first issue of the above-mentioned journal *Environmental protection and resource management.*

VINITI also issues, on a regular basis, cumulative information in the form of collections of abstracts on particular subjects; one example is *Sistemy, pribory, i metody kontrolya kachestva okruzhayushchei sredy* (Systems, instruments, and methods for environmental quality monitoring) — a monthly publication containing up to 200 abstracts pertaining to this area. Relevant information can also be found in a number of other publications by that Institute, including *Okhrana i uluchsheniye gorodskoi sredy* (Protection and improvement of the urban environment) and *Tekhnologicheskiye aspekty okhrany okruzhayushchei sredy* (Environmental engineering). Important components of the information services offered by VINITI are its publications under the general title *Itogi nauki i tekhniki* (Latest developments in science and technology), each of which reviews recent process in a particular field (e.g. toxicology).

Important sources of current awareness information are various bibliographic indexes (retrospective, current, cumulative, universal, departmental, subject-oriented, etc.) issued by major libraries of the country (e.g. V. I. Lenin State Library, State Public Library of Science and Technology, Central Scientific Agricultural Library, and Central Scientific Medical Library in Moscow and State Public Library in Leningrad), libraries and institutes of the USSR Academy of Sciences and of Academies of Sciences of the Union Republics, by departments of scientific and technical information in various ministèrs, institutes, etc.

Of the large international reference information systems, mention should be made of the International Referral System for Sources of Environmental Information (INFOTERRA) setup in Nairobi, Kenya, by the United Nations

Environment Programme (UNEP) to link up users with the sources of information they need.

Another international system is the International Register of Potentially Toxic Chemicals (IRPTC) operating in Geneva under the United Nations Environment Programme and publishing, among other things, reviews of Soviet literature concerning the toxicity of, and hazards posed by, individual chemicals and environmental contaminants. The IRPTC Bulletin, published three times a year, contains summaries of new data on chemicals.

Mention should also be made of the series of documents under the general title *Environmental Health Criteria,* being published by the World Health Organization (WHO) under the joint sponsorship of UNEP, ILO (International Labour Organisation), and WHO. The main purpose of these criteria documents is to compile, review, and evaluate available information on the biologic effects of pollutants and other environmental factors that may influence human health, and to provide a scientific basis for decisions aimed at protecting man from the adverse consequences of exposure to such environmental factors, both in the occupational and the general environment.

In addition to the international information systems mentioned above, abstracts, current awareness information, articles, and reviews bearing on problems of environmental pollution and related aspects are contained in a large number of periodicals; the more important of those published in English are the following: *Biological Abstracts, Chemical Abstracts, Ecological Abstracts, Environmental Abstracts, Pollution Abstracts, Current Contents (Agricultural Biology, Environmental Sciences, Life Sciences), Environmental Periodicals Bibliography, Index Medicus, Sciences Citation Index; Annual Review of Ecology and Systematics, Archives of Enviromental Contamination and Toxicology, Archives of Environmental Health, Archives of Industrial Hygiene and Toxicology, Bulletin of Enviromental Contamination and Toxicology, Ecotoxicology, Environmental Health Perspectives, Environmental Research, Environmental Sciences, International Journal of Environmental Studies, Journal of Agriculture and Food Chemistry, Journal of Environmental Pathology and Toxicology, Toxicology, Toxicology and Applied Pharmacology, Toxicology Letters, Toxic Substances Journal.*

For example, the *Environmental Abstracts* journal, issued monthly in the USA, derives its information from about 3000 journals published in the USA and other countries, governmental reports, proceedings of symposia and other meetings, etc. A similar volume of data input is used in *Pollution Abstracts* which is published bimonthly in the USA and contains information relating to the pollution and condition of air and drinking and sea water, problems of waste management and pesticide application, etc. Brief information on pertinent articles published in journals outside the USSR appears regularly in the indexes *Zarubezhnye nauchnye i tekhnicheskiye zhurnaly* (Foreign scientific and technical journals) issued by the USSR Library of Science and Technology.

ORGANIZATION OF MATERIAL IN THIS SERIAL PUBLICATION

The first two volumes of the series are devoted to elements in groups I–IV (Volume 1) and V–VIII (Volume 2) of the periodic table and their inorganic compounds. The subsequent volumes will cover radioactive substances and, in so far as possible, the organic compounds that can act as environmental pollutants and/or toxic agents.

For individual elements or compounds, brief information is given on their physical and chemical properties, natural occurrence and levels in environmental media and in organisms, methods of production, uses, and man-made sources of emission into the environment. (Further details on properties of the elements and compounds dealt with in the main text can be found in the Appendix.) These sections are followed by those presenting more detailed data on their toxic effects in experimental mammals and in man on single (acute), repeated (short-term), and chronic (long-term) exposure; for many substances, information on environmental transport and transformation is also provided, as are data on their toxicity to plants, insects, and aquatic organisms. A section on the absorption, distribution, and elimination in animals and man of the element or compound(s) concerned is included in a vast majority of chapters. There is also a separate section presenting, often in tabular form, safe exposure limits (hygienic standards) adopted in the USSR for the substances under discussion. Methods for detecting and quantitative determination of the particular pollutants in environmental media and biologic materials are then briefly covered. The last two sections are devoted to measures for preventing hazardous exposures and (in many chapters) to first-aid measures to be taken in cases of poisoning.

Each chapter concludes with a list of cited literature. In addition, at the end of the book there is a list of publications of a more general nature (predominantly books) that are each cited in a number of chapters. These publications as well as those listed within chapters can serve as a convenient source for those seeking more detailed information. Many references have been omitted in instances where good bibliographies can be found in the cited publications.

Each chapter has been written mainly by one or two authors (their names are appear in the table of contents), with contributions from other authors where appropriate. In the first volume of the series, for example, Professor B. A. Ivin and Dr. A. V. Moskvin are the authors or coauthors of the sections *Identity and physicochemical properties of the element, Production*, and *Uses;* A. V. Moskvin has also compiled the Appendix. Information on hygienic standards for all chapters has been provided by Yu. A. Krotov and V. V. Semenova, while material for the section on exposure control measures has been contributed by L. S. Dubeikovskaya.

The editor is much indebted to E. N. Levina, M.D., and I. D. Gadaskina, M.D., for the very useful discussions on various aspects of this multivolume publication, and to Academician S. N. Golikov for his generous assistance in its organization as well as for his unflagging attention and friendly advice given during the preparation of all materials. Special mention should be made of the large role played by Professor V. Ya. Rusin in generating the very idea of launching this project. Finally, the important contributions of Dr E. A. Pyaivinen to the preparation of the manuscripts are gratefully acknowledged.

V. FILOV,
Leningrad,
January, 1987

Notes on the structure and use of this book

The elements together with their compounds are dealt with sequentially by groups and subgroups of the periodic table in the order of increasing atomic numbers of the elements within the respective subgroups.

Each element and its compounds are allotted a separate chapter with the exception of rare earths (which are all dealt with in a single chapter) and carbon and silicon (whose compounds are considered in separate chapters). Although thorium and uranium and their compounds are covered in Volume 3 of the series, devoted to radioactive substances,[*] these actinides are also included in this volume in view of their chemical toxicity in addition to the radiation hazards they present.

The main text of each chapter (again with the exception of those referred to above) is preceded by an alphabetical list of inorganic compounds for which toxicity data are presented, as are their common synonyms or related terms, including minerals (abbreviated 'min.'). Throughout, the systematic names of compounds recommended by the International Union of Pure and Applied Chemistry (IUPAC) are used, with the exception of chemicals for which the older terminology is still widely employed, at least in the USSR.

In the vast majority of chapters, the main text is divided into the following sections:
- Identity and physicochemical properties of the element
- Natural occurrence and environmental levels
- Production
- Uses
- Man-made sources of emission into the environment

---

[*] *Radioaktivnye veshchestva* (Radioactive substances), published in Russian by Khimiya, Leningrad, in 1990 (463 pages).

* Toxicity
* Absorption, distribution, and elimination
* Hygienic standards
* Methods of determination
* Measures to control exposure

In addition, a section outlining first-aid measures to be taken in poisoning is also included in many chapters. Each chapter concludes with a list of cited literature[**].

In the section *Identity and physicochemical properties of the element*, some physical and chemical properties of the element concerned are indicated. A more extensive coverage of these properties, both of the element itself and of its compounds, will be found in the Appendix. In it, the material is arranged in the same way as in the main text, i.e. the elements appear in the order of growing atomic numbers and the compounds are arranged in the alphabetic order of their systematic names.

The sections *Production* and *Uses* list the more common methods of production and the main uses of the element and its inorganic compounds.

The section *Natural occurrence and environmental levels* contains information on the occurrence and levels of the element in components of the geosphere and biosphere (excluding animal tissues and organs and human foods), and also data, where available, about its migration and transformation in nature. In some chapters, the following two symbols are used in this section:

K_b =coefficient of biologic absorption of the element, defined as the ratio of its concentration in plant ash to that in the continental granite layer; and

K_w =coefficient of water migration of the element, defined as the ratio of its concentration in the dry residue of water to that in the continental granite layer.

The most extensive section is, naturally enough, that on toxicity. It is divided into several subsections that deal with the effects produced by the element and its compounds in laboratory animals and in man on acute, subacute, and chronic exposures and also, where data are available, with their effects on plants, fish, and other organisms.

The section *Hygienic standards* presents, as a rule in tabular form, safe exposure limits set for the element and its compounds in the USSR. Threshold limit values adopted in the USA are also given. In various chapters the following accepted abbreviated designations of Soviet 'hygienic standards' are used in this section:

[**] Where no reference to a published source appears in the main text, this usually means that the material has been taken from the handbook *Vrednye veshchestva v promyshlennosti* (Harmful Substances in Industry), Vol. 3 (edited by N. V. Lazarev & I. D. Gadaskina), published in Russian by Khimiya, Leningrad, 1977 (608 pages).

MAC   = maximum allowable concentration.

MAC_{ad} = average daily maximum allowable concentration of a substance in the atmosphere of residential areas.

MAC_{hm} = highest momentary allowable concentration of a substance in the atmosphere of residential areas, i.e. the concentration that should not be exceeded for any length of time.

MAC_s = maximum allowable concentration of a substance in the cultivated layer of soil.

MAC_w = maximum allowable concentration of a substance in bodies of water for 'sanitary-domestic uses' (i.e. those used for water-supply, public, and/or recreational purposes).

MAC_{wf} = maximum allowable concentration of a substance in bodies of water used for breeding and/or catching fish.

MAC_{wz} = maximum allowable concentration of a substance in the air of the working zone, i.e. the concentration daily exposure to which for 8 hours (with the exception of non-working days) or during another period but not more than 41 hours per week, throughout the entire working life, will not cause any disease or deviation from a normal state of health detectable by currently available methods of investigation, either during the work itself or in the long term, in the present and subsequent generations.

TSEL   = tentative safe exposure level: a temporary hygienic standard specifying the highest acceptable concentration of a substance in air or in water bodies.

$TSEL_{aa}$ = tentative safe exposure level of a substance in the atmospheric air of residential areas.

For comparison, the USA standards TLV-TWA, TLV-C, and TLV-STEL are also included in this section.

TLV-TWA = threshold limit value — time weighted average: the time weighted average concentration for a normal 8-hour working day and a 40-hour working week, to which nearly all workers may be repeatedly exposed day after day, without adverse effect.

TLV-C   = concentration that should not be exceeded even instantaneously.

TLV-STEL = threshold limit value — short term exposure limit: the concentration to which workers can be exposed continuously for a short period of time without suffering from (1) irritation, (2) chronic or irreversible tissue change, or (3) narcosis of sufficient degree to increase the likelihood of accidental injury, impair self-rescue or materially reduce work efficiency, provided that the daily TLV-TWA is not exceeded.

The meaning of the 'hazard classes' referred to in the *Hygienic standards* and some other sections is explained in Table 1.

Table 1. Classification of chemicals by hazard in the USSR [*]

Parameter	Hazard Class			
	1	2	3	4
Maximum allowable concentration (MAC)				
in the air of working zone (mg/m^3)	<0.1	0.1—1.0	1.1—10.0	>10.0
Median lethal dose (LD_{50}) with oral (usually by gavage) administration (mg/kg)	<15	15—150	151—5000	>5000
Median lethal dose (LD_{50}) with dermal application (mg/kg)	<100	100—500	500—2500	>2500
Median lethal concentration (LC_{50}) in air (mg/m^3)	<500	500—5000	5001—50000	>50000
Coefficient of potential poisoning by inhalation [*]	>300	300—30	29—3	<3
Acute action zone [**]	<6.0	6.0—18.0	18.1—54.0	>54.0
Chronic action zone [***]	>10.0	10.0—5.0	4.9—2.5	<2.5

Notes:

[*] Class 1—extremely dangerous substances; class 2—highly dangerous substances; class 3—moderately dangerous substances; class 4—slightly dangerous substances.

[**] The coefficient of potential poisoning by inhalation is the ratio of the maximum attainable concentration of a given substance in the air at 20°C to the median lethal concentration of that substance for mice.

[***] The acute action zone is the ratio of the median lethal concentration of a given substance to the acute action threshold for that substance. The chronic action zone is the ratio of the acute action threshold for a given substance to the chronic action threshold (i.e., the lowest concentration that produces adverse effects in a long-term study with 4-hour exposure daily 5 times per week for a period of at least 4 months) for that substance.

It seems worthwhile to define also the meaning of the following six terms used in this book:

Cumulation coefficient: the ratio of the summated median lethal dose (ΣLD_{50}) given in the course of an experiment with repeated exposure to the single-exposure LD_{50};

Harmful substance (chemical): a substance exposure to which at particular doses or concentrations may cause disease or deviation from the normal state of health detectable by currently available methods of investigation during the period of exposure or in the long term, in this or subsequent generations;

LC_{50} = median lethal concentration: the concentration of a substance required to kill 50% of the test animals by the inhalation route within a specified period. It is usually expressed in mg/m^3 of air (sometimes in mg/L of air);

LD_{50} = median lethal dose: the dose of a substance required to kill 50% of the test animals, usually within 2 weeks, after a single administration by a particular route (oral, intravenous, intraperitoneal, etc.). It is usually expressed in mg/kg animal body weight. As a rule, administration via the oral route to laboratory animals is by gavage;

Summated threshold index: an indicator of the ability of the central nervous system (CNS) to summate subthreshold impulses;

Threshold dose (concentration): the minimal dose (concentration) at the lower limit of the toxicity range (zone) of a substance (i.e. the dose or concentration at which an effect just begins to occur).

Hydrogen and its compounds

Hydrogen peroxide (perhydrol); deuterium oxide (heavy water)

IDENTITY AND PHYSICOCHEMICAL PROPERTIES OF THE ELEMENT

Hydrogen (H) is a nonmetallic element in Group I of the periodic table, with the atomic number 1. It consists of two stable isotopes: the light hydrogen, or protium (^{1}H), and the heavy hydrogen (^{2}H), or deuterium (D). Naturally occurring hydrogen compounds contain, on average, 1 deuterium atom per 6800 hydrogen atoms. There exists superheavy hydrogen (^{3}H), or tritium (T); in nature, the latter is produced by the action of neutrons from cosmic rays on atmospheric nitrogen.

Under ordinary conditions, hydrogen exists as diatomic molecules (H_2). Its oxidation state is +1 in most compounds and -1 in metal hydrides. It is a reducing agent, but usually shows low activity in ordinary circumstances. When in mixture with O_2, it explodes at temperatures above 550°C with the formation of water. Explosive limits range from 4% to 94% for hydrogen-oxygen mixtures and 4% to 74% for hydrogen-air mixtures. When mixed with O_2 in a 2:1 ratio by volume, it forms a detonating gas. Hydrogen reduces metals from their oxides. It forms hydrogen halides with halogens, ammonia with nitrogen (at elevated temperatures and pressures with a catalyst), hydrogen sulfide with sulfur (at 600°C), and methane with pure carbon (at high temperatures). It gives hydrides with alkaline, alkaline–earth, and many rare–earth metals as well as with some other metals. See also Appendix.

NATURAL OCCURRENCE AND ENVIRONMENTAL LEVELS

Hydrogen makes up 0.15% of the earth's crust by weight, or 17% in terms of the number of atoms. Water contains 10.72% of hydrogen by weight, and hydrogen and O_2 account together for 96.54% of the total mass of seawater. Free hydrogen constitutes 5×10^{-6} of the total volume (or 3.48×10^{-6}%) of the atmosphere up to a height of 90 km; volcanic gases may contain as much as 50% hydrogen by volume. Hydrogen makes up about 10.0% of the substance of plants and 9.7% of that of animal organisms [Venchikov].

In nature, hydrogen is involved in forming compounds such as, for example, ammonia (NH_3), hydrogen sulfide (H_2S), sulfuric acid (H_2SO_4), and, at high temperatures, hydrides (LiH, CaH_2) as well as compounds with halogens and other elements. Hydrogen has key roles in biochemical processes. The H^+–Na^+ exchange across cell membranes is essential for the regulation of pH within cells and the control of cell growth, proliferation, and volume; for the transepithelial absorption and secretion of Na, H^+, chloride ions, and organic anions; and for metabolic reactions of the cell. H^+–Na^+ exchange has also been implicated in a number of pathophysiologic processes, including those associated with acid-base imbalance, essential hypertension, organ hypertrophy, and malignant growth (for review, see Mahnensmith & Aronson). The numerous intermolecular hydrogen bonds determine the type and specificity of enzymatic and other reactions in living organisms. As a constituent of water, hydrogen plays critical roles in bio-atmospheric cycles and bioenergetic processes. Soil-inhabiting hydrogen bacteria exist that are able to oxidize hydrogen and use the resultant energy for taking up carbon from CO_2. The bacterial biomass has an average hydrogen content of about 7400 mg/100 g dry weight [Vadkovskaya & Lukashev].

The naturally produced **tritium** (T) makes up 4×10^{-15}% of the total number of hydrogen atoms in the atmosphere. Tritium artificially produced in nuclear power plants settles in soil at rates of 0.007–0.07 cm/s in summer and less than 0.0005 cm/s in winter, with the bulk of tritium being converted to its oxides. Tritium concentrations in soils vary with the composition, porosity, and moisture and organic matter contents of the soil. From 50% to 90% of the sedimented tritium is absorbed by the upper (2 cm) soil layer. In the soil, tritium undergoes microbiological oxidation at rates that depend on the physicochemical properties of the soil [Dunstall *et al.*]. According to Karavayeva *et al.*, the vertical movement and distribution of HTO in the soil as well as HTO uptake by plants depend on the rate ratio of ascending and descending flows of soil moisture which in turn vary with the degree of soil irrigation. These authors have shown experimentally that HTO concentrations in the soil moisture, averaged for individual soil layers, decline with increasing strength of HTO bonding with the solid phase of the soil. A method for calculating tritium dispersion in the environment is described in Bell *et al.*

Hydrogen peroxide (H_2O_2), which is produced in the air naturally (the mechanism of production is considered in Heikes, Heikes *et al.*, and Zika &

Saltzman), decomposes in the atmosphere during daytime under the action of solar radiation, with the formation of reactive HO radicals that undergo partial recombination in night-time. In a south-eastern area of the Ontario province (Canada), the mean 5-minute concentration of H_2O_2 in the air was calculated to be 2.9 ppb by volume [Slemr et al.]. H_2O_2 concentrations in rainwater, measured inside clouds at mountain tops in October, ranged from 0.3 to 72.0 mmol [Penkett].

PRODUCTION

Hydrogen is produced industrially from coke-oven gases formed during coal coking and from gases generated in oil refining by removal of all components of the gaseous mixture except hydrogen through deep freezing (these components then liquefy more readily than hydrogen). It is also made by catalytically reacting water vapor with natural fuel gases (mainly gaseous hydrocarbons) that form in the earth's crust. Moreover, it is often produced from water gas or steam–air gas and from water by electrolysis.

Hydrogen peroxide can be obtained by oxidation of sulfuric acid or an acid solution of ammonium hyposulfate, or by hydrolysis of peroxydisulfuric acid, but the main modern method is by the autooxidation of an alkyl anthraquinol.

Deuterium oxide is made by electrolysis of pure water, by fractional distillation of liquid hydrogen followed by burning off of the deuterium, or by water distillation [Gamburg et al.].

USES

Hydrogen is used in the chemical industry for the manufacture of ammonia, methanol and other alcohols, and of various products that are obtained from hydrogen and carbon monoxide by synthesis. It is also used for the hydrogenation of solid and liquid fuels; for the hydrorefining of oil products, fats, coals, and tars; in metal welding and metal cutting; in processes of microbiological synthesis; in special thermometers and in electrodes. In biologic research and medical practice it is utilized in studies of tissue blood flows [Brill et al.; Gozhenko et al.] and metabolic processes [Rosado]; it is also employed in gastroenterology [Gasbarrini et al.; Castello & Gerbino].

The hydrogen isotopes deuterium and tritium have found widespread use in the nuclear industry. **Deuterium oxide** is used as a neutron moderator and heat-transfer agent in nuclear reactors and serves as a source of deuterons (D^+) for nuclear and thermonuclear reactions; it is also employed for research purposes.

Hydrogen peroxide is used for disinfection and sterilization (it has a broad spectrum of antimicrobial actions, possesses sporicidal activity, is odorless and resistant to cold); in the canning and brewing industries; in rocket fuels; in the chemical industry as an oxidant of vat dyes and for the production of peroxide compounds; as a bleaching agent; in surgery as a hemostatic [Shah]; and in bio-organic chemistry [Sverdlov & Kalinina].

MAN-MADE SOURCES OF HYDROGEN PEROXIDE EMISSION INTO THE ENVIRONMENT

Two major anthropogenic sources of emission into the environment are aqueous hydrogen peroxide solutions such as 30% solutions prepared for technical use (known as perhydrol) and 1–3% solutions used in medicine. Stabilizers such as sodium pyrophosphate, salicylic acid, or orthophosphoric acid are usually added to perhydrol to prolong its storage life. During disinfection operations, the concentration of hydrogen peroxide aerosol may reach 200–400 ml/m^3 [Lyarsky & Tsetlin; Lyarsky *et al.*]. Hydrogen peroxide concentrations of up to 700 mg/m^3 developed in the indoor air of workplaces where this compound was produced electrochemically [Diterikhs & Nevskaya].

TOXICITY

Effects of hydrogen peroxide
The threshold concentration of H_2O_2 required to impart a metallic taste to water has been reported to be 87 mg/L [Antonova *et al.*].

Bacteria
It has been shown on a bacterial model that 0.01- and 0.1-molar hydrogen peroxide solutions, while displaying low antibacterial activity themselves, synergistically enhance the bactericidal effect of ionizing radiation [Vasilieva & Samoilenko].

Fish
Hydrogen peroxide was reported to be toxic to fish in a concentration of 0.25 mg/ml water. Fish exposed to this concentration had a blue skin, white, and extremely slimy gills, and ulcers on both the skin and gills [Metelev *et al.*].

Animals and man

Acute exposure

Animals. LD$_{50}$ and LC$_{50}$ values of hydrogen peroxide for mice and rats are given in Table 1.

For rats, threshold concentrations were 60 mg/m^3 with 4-hour exposure by inhalation and 110 mg/m^3 with application to the skin. The inhalation exposure resulted in elevated NAD diaphorase activity in bronchial epithelium, while the dermal exposure brought about moderate skin hyperemia and thickening caused by decomposition of the H_2O_2 that had penetrated under the skin, with the resultant formation of oxygen bubbles.

Toxic doses with intravenous injection were as follows: 1.0–1.5 cm^3 of 2–3% solutions for horses, 1.0–2.0 cm^3 of a 3% solution for dogs, and 2.0 cm^3 of 0.4–0.5% solutions for rabbits. After such a dose, the horse was unable to stand and remained dyspneic for 20 to 60 min; convulsions of short duration and tachycardia were noted. The abnormal state persisted for up to 2.5 h, after which time the horse stood up spontaneously and defecated. In the dogs, dyspnea, tachycardia, strong intestinal peristalsis, and defecation were observed. The rabbits showed marked dyspnea, excitation, convulsions, and slowed and weakened breathing; some of them died [Petrov].

Table 1. Median lethal doses and concentrations of hydrogen peroxide

Species	Route of administration	Dose or concentration
Mouse	Oral	2000–2538 mg/kg
	Intraperitoneal	880 mg/kg
Rat	Oral	4050 mg/kg
	Subcutaneous	620 mg/kg
	Inhalation	2000 mg/m^3
	Dermal	4060 mg/m^3

Sources: Kondrashov; Lyarsky *et al.*

Skin application of hydrogen peroxide in high concentration caused subcutaneous emphysema in mice [Lyarsky *et al.*]. Mice exposed to a 28% solution (20 mg/cm^2) developed emphysema that varied in severity from one animal to another; after the application of a 10% solution (7 mg/cm^2), emphysema was less pronounced, but signs of general intoxication (excitation/inhibition, ataxia, tremor, paresis of the extremities, dyspnea) were in evidence within 5 to 10 min post-application. Skin applications of 15–20% solutions to rabbits led to their death after 10 min in the presence of gas emboli.

Concentrated solutions cause severe changes (ulcers and hemorrhages) on mucous membranes of the eyes and gastrointestinal tract. Vapors and aerosols have irritant effects on ocular and nasopharyngeal mucous membranes and may cause corneal opacity.

Man. Cases of accidental or intentional poisoning resulting from the ingestion of perhydrol in amounts ranging from a single swallow to 500 ml have been described [Blokhas; Polyansky]. Nearly all victims experienced pain in the throat, esophagus, and abdomen and discharged bloody froth from the mouth followed by bloody vomits (sometimes bloody stools) and a state of shock; death ensued several hours to days post-ingestion as a result of massive gas embolism. Autopsy revealed profound necrotic inflammatory changes in the esophagus and stomach, edema in the larynx, signs of confluent serous desquamative pneumonia, necrobiotic areas in the liver, toxic edema and swelling of the brain substance, and other manifestations of local irritant and systemic toxic actions of perhydrol which itself was indetectable in the cadavers because of its rapid decomposition.

As found by Kondrashov in studies on volunteers, threshold concentrations of H_2O_2 vapor for isolated cutaneous exposures of 60, 30, 15, and 5 min in duration were 80, 110, 140, and 180 mg/m^3, respectively. When volunteers were exposed to low concentrations of H_2O_2 vapor by inhalation, the degree of respiratory tract irritation mainly depended on the concentration rather than the duration of exposure; the threshold concentration was 10 mg/m^3 and the no-observed-effect level, 5 mg/m^3. The threshold concentration of H_2O_2 vapor for skin irritation was determined at 20 mg/m^3. Van der Zee *et al.* described damaging effects of H_2O_2 on human red cell membranes. Takasu *et al.* concluded from their tissue-culture studies that H_2O_2 generation in the body is regulated by the level of free cytoplasmic calcium.

Chronic exposure

Animals. Exposure of rats to H_2O_2 vapor at 10.0 mg/m^3 for 5 h/day 5 days per week during 4 months resulted in elevated peroxidase activity of the blood at 2 and 3 months of exposure and in inhibited succinate dehydrogenase, monoamine oxidase, acid phosphatase, and esterase activities and elevated acid phosphatase activity in the lungs at 4 months; the concentration of 1.0 mg/m^3 produced no demonstrable effects. When shaved rat skin was exposed to H_2O_2 vapor using the same scheme as above, elevations of monoamine oxidase, succinate dehydrogenase, NAD diaphorase, and lactate dehydrogenase activities in the epidermis were recorded at the end of the experiment. The concentrations of 1.0 mg/m^3 and 0.1 mg/m^3 were described as threshold and ineffective, respectively.

Slowed gains in body weight, elevated peroxidase activity in the blood, depressed catalase activity in the liver, reticulocytosis, and albuminuria were observed in rats administered H_2O_2 by gavage in concentrations equal to 1/5 or 1/10 of the LD$_{50}$. The percentage of normally ovulating females decreased, as did sperm motility in males. When H_2O_2 was given to rats or rabbits in drinking water for 6 months, the lowest effective dose was 0.05 mg/kg body weight (daily intake of 1.0 mg H_2O_2/L water). Adverse effects of ingested aqueous H_2O_2 solutions on the reproductive function and offspring of rats were also observed [Antonova].

Some mice of the C57BL/6J strain administered 0.4% or 0.1% H_2O_2 solutions in drinking water from the age of 8 weeks to the age of 108 weeks developed gastric and duodenal erosions or duodenal adenocarcinoma [IRPTC Bulletin].

Man. Persons occupationally exposed to H_2O_2 concentrations of 10 to 700 mg/m^3 complained of lacrimation, hoarseness, cough, a pricking sensation in the nostrils, and a metallic taste in the mouth. Symptoms of systemic intoxication (headache, increased fatiguability, and drowsiness) were also noted [Diterikhs & Nevskaya].

Effects of deuterium oxide (heavy water)

Heavy water (D_2O) is absorbed from the gastrointestinal tract at a fast rate [Pulyaevsky]. The heavy water that has entered the circulation rapidly reaches a state of equilibrium with both extracellular and intracellular water, although its levels differ in different tissues. After perfusion of the abdominal cavity with a D_2O-containing Ringer solution in rats, D_2O levels in the liver, heart, and brain were higher than in the blood, which can be explained by the ability of deuterium to enter not only the aqueous phase but also structural elements of organs. With long-term multiple exposures, the rates of deuterium incorporation into cellular structures and its concentrations in organs were much higher than those observed after a short single exposure [Pulyaevsky].

Experiments in which mice and rats were given various doses of D_2O in drinking water showed the following [Pulyaevsky]. When 15–20% of the body water was replaced by D_2O, rats became hyperexcitable. Replacement of 20% to 25% of the body water resulted in more strongly marked hypersensitivity, and many of the rats developed convulsions. In mice, elevated metabolic rates, fever, and increased pilomotor activity were observed, while skin lesions, ulcers on the paws and snout, and necrotic changes in the tail were noted in rats. Animals in which 30% of the body water had been replaced by heavy water, refused to eat, were rapidly losing weight, and became comatose. D_2O concentrations in the body water between 30% and 35% were fatal for some animals. Mice and rats whose body water contained 50% of D_2O all died within 4 to 5 weeks. None of the animals developed tolerance to heavy water.

When D_2O solutions were administered parenterally to mice, symptoms of intoxication were similar to those seen in animals given D_2O in drinking water, but were evolving much more rapidly. The average survival times of mice given 40% and 75% D_2O solutions were 60 and 11 days, respectively. Injection of water containing 50% or more D_2O led to erythropenia in addition to other signs and symptoms. Heavy water elevated the level of nitrogenous protein and decreased those of plasma protein, creatinine, and amino acids, while increasing the content of lactic acid and inorganic phosphate. In mice given a 99.5% D_2O solution daily in a dose of 100 g/kg, decreased metabolic rates were recorded on the 5th day when the body fluids contained 30–33% of D_2O. These mice showed signs of nervous system disease and died on the 7th day. D_2O caused hypertrophy of the adrenals and altered endocrine metabolism. On post-mortem examination, the liver and adrenals had increased weights, the level of hepatic DNA was elevated, while the adrenals showed cortical hyperplasia and alterations in the medullary tissue; the weight of the spleen was decreased [Pulyaevsky].

As pointed out by Pulyaevsky, the relevant data in the literature mainly concern the rapidly metabolized D_2O fraction contained in the aqueous medium of the body; no information regarding the metabolism of the deuterium bound to organic molecules could be found. Animal studies suggest that the D_2O concentration in body fluids that does not exceed tolerance limits of the organism is 10%. It has been estimated that about 5 liters of D_2O will have to be ingested to attain such a concentration in the human body (which will take at least 4 days at a normal level of consumption of water and other liquids). Since such a situation does not occur under ordinary circumstances, D_2O is thought not to present a real hazard to human health.

HYGIENIC STANDARDS

Kondrashov considers that the maximal permissible level of hydrogen peroxide on the skin is between 0.8 and 1.0 mg/dm^2 for exposures of short duration; he recommends that the maximal permissible level of H_2O_2 vapor in the air of workplaces be set at 0.3 mg/m^3. In the USA, the TLV-TWA for hydrogen peroxide has been set at 1.4 mg/m^3.

METHODS OF DETERMINATION

Chromatographic determination of **hydrogen** in air is based on the use of a gas-absorption or gas-liquid chromatograph equipped with an argon ionization detector (detection limit 1 μg hydrogen) or with a katharometer and flame ionization detector (*Guidelines for the Determination of Noxious Substances in Air*); the determination of hydrogen in expired air by gas chromatography has been used for diagnostic purposes in intestinal disturbances (sensitivity 20 ppm) [Castello & Gerbino; Rosado]. Chemiluminescent methods for determining **hydrogen peroxide** in air rely on its reaction with luminol in the presence of a catalyst (detection limit 0.05 mg/m^3) [Peregud & Gorelik] or on the chemiluminescence of lucigenin in the presence of H_2O_2 in an alkaline medium [Tovmasian *et al.*]. A colorimetric method for measuring H_2O_2 in air is based on the formation of pertitanic acid as a result of the peroxide reacting with titanium sulfate (the reported range of measurable concentrations is 10–90 μg/5 ml) [Peregud]. Hydrogen peroxide can also be measured in air by absorption spectroscopy using a diode laser [Slemr *et al.*] or by fluorimetry [Lazrus *et al.*]. A procedure for measuring H_2O_2 in water has been described [Madsen & Kromis]. Chemiluminescent assays for H_2O_2 in biological media are based on the catalytic oxidation of luciferin (N-N'-dimethylbiacridinium dinitrate) by the peroxide in an alkaline medium (the detection limit is 8×10^{-7} g/ml) [Galstian & Uloyan]. **Deuterium** in water can be measured by infrared spectrophotometry (with it, the amount of water in the body can be determined); the sensitivity was approximately 300 ppm [Shakar].

MEASURES TO CONTROL EXPOSURE AT THE WORKPLACE

These measures depend on the specific methods used for the production of hydrogen or its compounds. In factories where hydrogen is derived from a gaseous mixture in the process of coal coking or oil refining, the primary consideration in health risk assessment is ambient air pollution by concomitant components and destruction products of the raw material. Particular care needs to be taken to ensure that the transport lines work properly and that the equipment is hermetically sealed and in good repair. A warning system should be installed to signal any gas emissions into the air that exceed the permissible level. Preventive measures to be taken in the production and transportation of hydrogen peroxide are detailed in the paper by Diterikhs & Nevskaya. Given that hydrogen and other gases produce irritant effects, a set of medical measures to control them should be considered. In prophylactoria at the factories, procedures by which the respiratory function can be restored and the irritation of the nasopharyngeal mucosa can be eliminated should be available. Workers engaged in the production or use of hydrogen and its compounds should undergo medical examinations at regular intervals.

For personal protection, workers should be supplied with appropriate overalls, gloves (made of polyvinylchloride, polyethylene or some other suitable plastic material), and goggles or face masks made of transparent polymeric material.

FIRST AID IN POISONING

A person who has ingested perhydrol should be immediately given large quantities of water to drink and be induced to vomit. Analgesics and antishock therapy with the use of papaverine and atropine should then be administered if necessary, followed by delivery to a hospital by ambulance if the poisoning is severe.

REFERENCES

Antonova, V. I. (1981) *Gig. San.*, No. 1, 72–73 (in Russian).
Antonova, V. I. *et al.* (1974) *Gig. San.*, No. 10, 22–23 (in Russian).
Bell, R. P. *et al.* (1985) *Fusion Technol.*, 8, Part 2: 2582–2586.
Blokhas, T. V. (1968) *Sudebno-Meditsinskaya Ekspertiza*, No. 1, 54 (in Russian).
Brill, G. E. *et al.* (1983) *Patol. Fiziol. Eksper. Ter.*, No. 5, 83–85 (in Russian).

Castello, G. & Gerbino, T. C. (1987) *J. Chromatography*, **416**, 119—124.
Diterikhs, D. D. & Nevskaya, A. I. (1967) In: *Gigiyena truda v khimicheskoi promyshlennosti* [Occupational Health in the Chemical Industry]. Moscow, pp. 386—403 (in Russian).
Dunstall, T. G. *et al.* (1985) *Fusion Technol.*, **8**, Part 2: 2551—2556.
Galstian, G. & Uloyan, S. M. (1978) *Gig. San.*, No. 5, 75—76 (in Russian).
Gamburg, D. Yu. *et al.* (1989) *Vodorod* [Hydrogen]. Khimiya. Moscow (in Russian).
Gasbarrini, G. *et al.* (1986) In: G. Dobrillo *et al.* (eds), *Problems and Controversies in Gastroenterology.* Raven Press, New York, p. 123.
Gozhenko, A. I. *et al.* (1985) *Fiziol. Zhurnal SSSR*, No. 5, 673—875 (in Russian).
Guidelines for the Determination of Noxious Substances in Air (1979) [Metodicheskiye ukazaniya po metodam opredeleniya vrednykh veshchestv v vozdukhe], No. 15. Publication of the USSR Ministry of Health, Moscow (in Russian).
Heikes, B. G. (1984) *Atmos. Environ.*, **18**, 1433—1445.
Heikes, B. G. *et al.* (1982) *J. Geophys. Res.*, **87**, 3045—3051.
IRPTC Bulletin (1981) [Bulletin of the International Register of Potentially Toxic Compounds], Vol. **4**, No. 1 (Russian edition).
Karavayeva, E. N. *et al.* (1985) *Ekologiya*, No. 3, 85—87 (in Russian).
Kondrashov, V. A. (1977) *Gig. Truda Prof. Zabol.*, No. 10, 22—24 (in Russian).
Lazrus, A. L. *et al.* (1986) *Anal. Chem.*, **58**, 594—597.
Lyarsky, P. P. *et al.* (1983) *Gig. San.*, No. 6, 28—31 (in Russian).
Lyarsky, P. P. & Tsetlin, V. M. (1981) *Dezinfektsiya aerozolyami* [Disinfection with Aerosols]. Moscow (in Russian).
Madsen, B. C. & Kromis, M. S. (1984) *Anal. Chem.*, **56**, 2849—2850.
Mahnensmith, R. L. & Aronson, P. S. (1985) *Circ. Res.*, **56**, 773—788.
Metelev, V. V. *et al.* (1971) *Vodnaya toksikologiya* [Aquatic Toxicology]. Kolos, Moscow (in Russian).
Penkett, S. A. (1986) *Nature* (London), **319**, 624.
Peregud, E. A. (1976) *Khimicheskiy analiz vozdukha (novye i usovershenstvovannye metody)* [Chemical Analysis of Air (New and Improved Methods)]. Khimiya, Leningrad (in Russian).
Peregud, E. A. & Gorelik, D. O. (1981) *Instrumentalnye metody kontrolya zagryazneniya atmosfery* [Instrumental Methods of Atmospheric Pollution Monitoring]. Khimiya, Leningrad (in Russian).
Petrov, V. P. (1940) *Farmakol. Toksikol.*, No. 5, 61—66 (in Russian).
Polyansky, A. D. (1970) In: *Aktualnye voprosy sudebnoi meditsiny i patologicheskoi anatomii* [Current Topics in Forensic Medicine and Pathologic Anatomy]. Tallinn, pp. 210—212 (in Russian).
Pulyaevsky, A. G. (1986) *Gig. San.*, No. 3, 63—66 (in Russian).
Rosado, J. L. (1984) *Clin. Chem.*, **30**, 1838—1842.
Shah, J. S. (1984) *Anesthesiology*, **61**, 631—632.
Shakar, J. J. (1988) *Anal. Chem.*, **58**, 1460—1461.
Slemr, F. *et al.* (1986) *J. Geophys. Res.*, **91D**, 5371—5378.
Sverdlov, E. A. & Kalinina, N. F. (1983) *Bioorgan. Khim.*, No. 9, 1696—1698 (in Russian).
Takasu, N. *et al.* (1987) *Biochem. Biophys. Res. Commun.*, **148**, 1572—1577.
Tovmasian, A. P. *et al.* (1979) *Gig. Truda Prof. Zabol.*, No. 3, 58—59 (in Russian).
Vadkovskaya, I. K. & Lukashev, K. P. (1977) *Geokhimicheskie osnovy okhrany biosfery* [Protection of the Biosphere: Geochemical Principles]. Nauka i Tekhnika, Minsk (in Russian).
Van der Zee, J. *et al.* (1985) *Biochim. Biophys. Acta*, **818**, 38—44.
Vasilieva, E. I. & Samoilenko, I. I. (1978) *Radiobiologiya*, 25, 309—313 (in Russian).
Venchikov, A. I. (1982) *Printsipy lechebnogo primeneniya mikroelementov v kachestve biotikov* [Principles for the Therapeutic Use of Trace Elements as Biotics]. Ylym, Ashkhabad (in Russian).
Zika, R. G. & Saltzman, E. S. (1982) *Geophys. Res. Lett.*, **9**, 231—234.

Lithium and its compounds

Lithium bromide; l. carbonate; l. chloride; l. dihydrogen phosphate
(l. orthophosphate); l. hydride; l. metaniobate; l. metaphosphate;
l. metatantalate; l. nitrate; l. nitrate trihydrate; l. sulfate

IDENTITY AND PHYSICOCHEMICAL PROPERTIES OF THE ELEMENT

Lithium (Li) is an alkali metal in group I of the periodic table, with the atomic number 3. Its natural isotopes are ^{6}Li (7.52%) and ^{7}Li (92.48%).

Lithium exists in two modifications, α and β. It is ductile and viscous, and can be easily drawn out into a thread or cut with a knife. It is the lightest of all metals.

Lithium is highly reactive. In air it becomes coated with a dark gray deposit of the oxide Li_2O and nitride Li_3N. Its oxidation state in compounds is +1.

In reacting with oxygen, lithium burns with a red flame to the oxide Li_2O. It reacts with water less vigorously than do the other alkali metals; in the reaction, the hydroxide LiOH is produced and hydrogen is released. It reacts vigorously with mineral acids to form the respective salts. In aqueous solutions, the lithium cation is more solvated and less mobile than the cations of other alkali metals. Lithium combines directly with halogens to form halides. On heating with sulfur, hydrogen, nitrogen, carbon, and silicon it forms the sulfide Li_2S_4, hydride LiH, nitride Li_3N, carbide Li_2C_2, and silicide Li_6Si_2. On dissolution of lithium in liquid ammonia, lithium amide is produced. Lithium does not react with phosphorus directly. It imparts viscosity or hardness to metals when alloyed to them. It forms intermetallic compounds with aluminum, zinc, magnesium, cadmium, mercury, thallium, lead, bismuth, and tin. See also Appendix.

NATURAL OCCURRENCE AND ENVIRONMENTAL LEVELS

Lithium occurs in ores, mainly in minerals. It tends to concentrate in acid magmatic rocks and residual aluminosilicates. Its major minerals are spodumene $LiAlSi_2O_6$ (contains 6–7% Li_2O), petalite $LiAlSi_4O_{10}$ (up to 4.9% Li_2O), amblygonite $(Li,Na)AlPO_4(F,OH)$ (8–10% Li_2O), and lithium micas such as zinnwaldite (3.0–3.5% Li_2O) and lepidolite (4–6% Li_2O). The clarke of lithium is estimated at $(20–32) \times 10^{-4}\%$ in the earth's crust and at $30 \times 10^{-4}\%$ in the granite layer of the continental crust [Dobrovolsky].

The oceans contain an estimated 232 900 million tonnes of dissolved lithium, at average concentrations of 170 µg/L in water and $4.9 \times 10^{-4}\%$ in total salts [Dobrovolsky]; its residence time in seawater is estimated to be 2.3×10^6 years [Henderson]. In river water, its average concentration is 2.0 µg/L, and its calculated global runoff from rivers amounts to 74 000 tonnes per year ($K_w = 0.57$) [Dobrovolsky].

The abundance of lithium in surface soils is rather uniform. Its mean concentrations range from 12 mg/kg in light organic soils to 98 mg/kg in alluvial soils. Relatively high lithium levels occur in solonchak, chestnut, and prairie soils. Water-soluble forms of lithium in the soil profile reach up to about 5% of its total soil content [Kabata-Pendias & Pendias].

Continental vegetation contains an estimated 3.8 million tonnes of lithium, at average wet weight, dry weight, and ash weight concentrations of $0.6 \times 10^{-4}\%$, $1.5 \times 10^{-4}\%$, and $30 \times 10^{-4}\%$, respectively. The amount of lithium taken up annually through the increment in phytomass throughout the global land area is estimated to be 258.8 thousand tonnes, or 1.73 kg/km^2 ($K_b = 1.00$) [Dobrovolsky].

Mean lithium contents in individual plant families in the USSR were calculated to be (mg/kg dry weight) 0.10 for Polygonaceae, 0.24 for Gramineae, 0.67 for Leguminosae, and 2.9 for Rosaceae. In the USA, the values ranged 0.01–3.1 for Leguminosae, 0.01–31 for Solanaceae, and 0.07–1.5 for Gramineae, while in New Zealand they ranged 0.02–13 for Gramineae and 0.03–143 for Leguminosae. Some plant species of the Solanaceae family are capable of accumulating the metal in excess of 1000 mg/kg. In vegetables, mean lithium contents were reported to be (mg/kg dry weight) 0.05 in corn (grains), 0.06 in onion (bulbs), 0.2 in carrot (roots), 0.3 in lettuce (leaves), 0.5 in cabbage (leaves), 0.8 in corn ears and stover, 6.2 in chard (leaves), and 6.6 in celery (leaves); orange fruits had a mean content of 0.2 mg/kg [Kabata-Pendias & Pendias]. The content of lithium in animals is estimated at $n \times 10^{-4}\%$ by weight [Venchikov]. (*See also* Birch; Ribas.)

PRODUCTION

Metallic **lithium** is obtained by electrolysis of a mixture of fused lithium chloride and potassium at 400–460°C, followed by removal of mechanical inclusions and of impurities (K, Na, Mg, Ca, Al, Si, Fe) through remelting and refining at reduced pressure. Metallothermic methods for lithium production also exist.

Lithium bromide is made by the direct union of lithium with bromine or by reaction of lithium carbonate with hydrogen bromide. **L. carbonate** is produced by passage of carbon monoxide into a lithium hydroxide solution or by the action of potassium or sodium carbonate on solutions of lithium salts (usually sulfate) near their boiling points. **L. chloride** can be obtained by reacting lithium or its hydroxide with chlorine or by dissolving lithium carbonate or hydroxide in hydrochloric acid. **L. hydride** is made by direct union of lithium with hydrogen at 680–700°C or by reduction of lithium oxide or hydroxide with aluminum or magnesium at a moderate hydrogen pressure. **L. metaniobate** is prepared by fusing niobium(V) oxide with lithium oxide, by treatment of niobium(V) hydroxide with lithium hydroxide solution, or by decomposing lithium ore concentrates. For the preparation of **l. metatantalate**, tantalum(V) oxide is fused with lithium hydroxide. **L. nitrate** is made by reaction of lithium carbonate or hydroxide with dilute nitric acid, followed by evaporation of the solution and heating of the residue to 200°C in vacuum. **L. nitrate trihydrate** is separated from aqueous solutions at temperatures below 30°C. **L. sulfate** is obtained by reacting lithium carbonate with sulfuric acid followed by heating of the resultant lithium sulfate hydrate at 500°C.

USES

Lithium finds widespread use in nonferrous metallurgy and in the aircraft industry in the form of various alloys (with Mg, Pb, Cu, Ag, or Al) which are plastic, hard, light, and corrosion-resistant. In the nuclear power industry, lithium is used in the production of tritium, in control rods for nuclear reactors, as a heat-transfer agent in uranium reactors, and as a solvent of uranium and thorium. In the silicate industry, its minerals spodumene and lepidolite are used in the production of various materials, for example glass with enhanced resistance to cracking and breakage. In the rubber industry, it is used in a dispersed form to promote polymerization processes. In ferrous metallurgy it is employed for the deoxidation and modification of many alloys, and lithium batteries.

Lithium and its compounds are used in pyrotechnics, the chemical, pharmaceutical, and textile industries, and in medicine for the treatment of certain mental disorders.

Lithium salts such as the bromide and chloride are very hygroscopic and are used for the conditioning and drying of air. **Lithium carbonate** finds use in the production of most other lithium compounds, in pyrotechnics, as a catalyst in the manufacture of plastics, in ferrous metallurgy for desulfurizing steel, as a flux in aluminum production, and in the production of ceramics and glasses. **Lithium chloride** is used for the production of lithium by electrolysis, in dry storage batteries, as a flux in the smelting of metals and the welding of magnesium, aluminum, and light alloys. **Lithium hydride** is used in the synthesis of diborane, as a reducing agent in organic synthesis; and as a convenient source of hydrogen for filling aerostats and some types of equipment designed for rescue operations in the open sea. **Lithium nitrate** is used in pyrotechnics and as a stabilizer of liquid ammonia. From **lithium sulfate**, the heads of ultrasonic flaw detectors are made. The isotope ^{6}Li is used as a thermonuclear fuel both in civilian and military applications [Ribas].

MAN-MADE SOURCES OF EMISSION INTO THE ENVIRONMENT

It has been estimated that the annual amount of lithium released worldwide into the environment from coal burning exceeds 20 times that entering the oceans and seas with river discharges. The lithium contained in industrial effluents used for irrigating farmlands is accumulated by the soil and plants, and the consumption of vegetables grown on such soils may result in chronic poisoning [Grushko, 1979; Lukashev et al.].

In a factory where compounds of lithium were made, its concentrations in the air ranged 0.68 to 0.80 mg/m³ at centrifuges during lithium loading and unloading operations, 0.37 to 0.42 mg/m³ over baths when lithium salt solutions were drained from them, and 0.01 to 0.46 mg/m³ elsewhere in the factory [Gostinsky & Krasnopevtseva, 1978]. In a synthetic fiber plant, lithium concentrations between 24 and 40 mg/m³ were common in the air sampled near cylinder dryers during the charging and discharging of lithium chloride [Smirnova].

TOXICITY

Plants

Crop plants grown on soils irrigated with water containing **lithium chloride** in concentrations of 1.2 to 4.0 mg/L showed symptoms of poisoning; toxic concentrations of lithium sulfate ranged from 2.0 to 5.0 mg/L [Grushko, 1972].

Citrus trees are highly susceptible to high soil levels of **lithium**. Moderate to severe toxic effects were evident in these plants at lithium concentrations in leaves as low as 4 to 40 mg/kg. In corn plants growing in lithium-enriched soils, damaged root tips, necrotic spots in the interveinal tissues of leaves, and some other signs of injury were observed. The uptake of lithium by plants is inhibited by calcium, so that its toxicity can be reduced by adding lime to the soil [Kabata-Pendias & Pendias].

Aquatic organisms

Toxic concentrations of **lithium chloride,** calculated as the metal ion, were 100 mg/L for freshwater fish after 24 h of exposure and 7.2 mg/L (lake water) or 16 mg/L (river water) for daphnids after 48 h of exposure at 23°C [Grushko, 1979]. For young brook trout, lithium chloride was toxic at 10.6 g/L after 16 h of exposure. Its threshold concentration for goldfish was 3.4 g/L [Metelev *et al.*].

Animals and man

Acute exposure

Animals

LD_{50} values of lithium compounds are given in Table 1.

The clinical picture of acute poisoning with **lithium** was characterized by general inhibition, strongly depressed responses by external stimuli, diarrhea with bloody and mucous stools, convulsions during the first few hours of observation followed later by paralysis, respiratory disturbances, and arrhythmia. The mucous membranes and skin became cyanotic. Death occurred within 24 hours; its onset was accelerated by physical activity.

Animals administered **lithium chloride, lithium dihydrogen phosphate,** or **lithium metaphosphate** parenterally in a lethal or sublethal dose exhibited motor excitation, abnormal gait, and leg twitching during the first few minutes post-administration; later these manifestations of poisoning were succeeded by general inhibition, adynamia, involuntary urination and defecation, convulsions, and respiratory distress.

Mice, rats, rabbits, and guinea pigs poisoned with **lithium chloride** by various routes in doses causing partial mortality, developed ulcers in the gastrointestinal tract, which were more pronounced after oral administration. The gastrointestinal effects of **lithium bromide** and **lithium carbamate** were less severe, while **lithium sulfate** caused small erosions in the stomach, but only after oral administration and in doses that exceeded the LD_{50}. **Lithium metaniobate** and **lithium metatantalate** did not alter the external appearance or behavior of the animals. Slight to moderate fibrogenic activity of these compounds was noted in rats after intratracheal administration. Animals given **lithium carbonate** showed slowed weight gains.

Acute studies of several lithium compounds have shown that the major determinants of toxicity are indeed the lithium cations rather than anions. Lithium salts inhibited oxygen consumption in mice, rats, and guinea-pigs as well as oxidative phosphorylation in rat brain mitochondria. Their effects on energy metabolism correlated directly with the concentration of lithium ions in tissues and did not depend on the nature of the anionic component of the salt.

Table 1. Median lethal doses of lithium compounds

Compound	Animal species	Route of administration	LD_{50} (mg/kg)
Bromide	Mouse	Oral	2353
		Subcutaneous	1680
Carbonate	Mouse	Oral	531
		Subcutaneous	413
		Intraperitoneal	360
	Rat	Oral	553
	Rabbit	Oral	404
	Frog	Subcutaneous	452
Chloride	Mouse	Oral	1165
		Subcutaneous	970
		Intraperitoneal	680
	Rat	Oral	1530
		Intraperitoneal	925
	Rabbit	Oral	775
		Subcutaneous	700
	Guinea-pig	Intraperitoneal	624
Metaniobate	Mouse	Oral	8000
	Rat	Oral	8000
Metaphosphate	Mouse	Oral	1200
	Rat	Oral	1370
Metatantalate	Mouse	Oral	750
	Rat	Oral	750
Nitrate trihydrate	Mouse	Oral	1200
	Rat	Oral	1400
Sulfate	Mouse	Oral	1190
		Subcutaneous	953

Lithium poisoning stimulated glycogenolysis and inhibited glycogenesis. Poisoned animals showed inhibited aldolase activity, elevated lactate dehydrogenase activity, decreased pyruvic acid levels, and increased levels of lactic acid. The blood concentrations of cholesterol and β-lipoproteins were also elevated. Electrolyte balance in the blood, internal organs, and central nervous system structures was altered. Pyridine nucleotides were present in increased concentrations in the blood, whereas their levels in the liver were below normal because of the damage caused by lithium ions to nicotinamide enzymes in tissues.

Cardiac and renal functions can also be impaired by lithium. Its cardiotoxicity was due to disturbances of tissue metabolism in the myocardium and was manifested in bradycardia and electrocardiographic abnormalities at blood concentrations above 1.0 mEq/L. The magnitude of electrocardiographic changes was directly related to the lithium dose, and it also correlated with the extent of sodium and potassium imbalance in the myocardium and the blood.

The toxicity of lithium salts is associated with the hormonal status and sex of the host; thus, it was more strongly marked in animals with depressed pituitary or adrenal functions. Both lithium and its salts are capable of producing cumulative effects.

Man

Clinical manifestations of acute lithium intoxication have been observed in cases of intentional poisoning and in patients treated with excessive doses of lithium preparations for mental disorders [Birch; Ribas]. As regards occupational settings, Gostinsky & Krasnopevtseva [1977] concluded that the risk of acute poisoning with lithium by dermal absorption in workers using lithium or its compounds was negligible, although, as these authors have demonstrated, the Li^+ ion can penetrate through human skin.

Acutely poisoned persons complained of general weakness, somnolence, loss of appetite, thirst, and dryness in the mouth; nausea, vomiting, and profuse diarrhea also occurred in some cases. Other symptoms included tremor of the lips, mandible, and hands, hyperreflexia, vertigo, dysarthria, and visual impairment. Epileptic seizures, convulsions, and mental deterioration also occurred in some cases of severe poisoning, as did coma and death.

Toxic manifestations in patients receiving lithium preparations for therapeutic purposes included arterial hypotension, arrythmia, proteinuria, edema, polyuria, and polydipsia [Berman; Brakhnova; Grushko, 1979].

In persons exposed to **lithium hydride** by inhalation at 0.2–0.5 mg/m^3, hyperemia of the skin and lacrimation often occurred, followed in some cases by persistent conjunctivitis and perforation of the nasal septum. Lithium hydride solutions caused skin burns in concentrations ⩾0.01 mg/L.

Chronic exposure

Animals

Dogs given **lithium chloride** in the diet at 150 mg/kg body weight for several months, lost weight and appetite, became lethargic, weak, and ataxic, and showed increased erythrocyte counts and hemoglobin levels. With increasing length of exposure, deterioration of conditioned reflexes, twitching of individual muscle groups, tremor, and motor incoordination appeared and progressed, and the animals eventually died in a state of coma. Histopathologic examination revealed degenerative lesions in internal organs as well as changes in the nervous system where cells in the brainstem, thalamus/hypothalamus region, medulla oblongata, and lateral horns of the spinal cord were predominantly affected [Neretin].

In another long-term study, where lithium chloride was applied to the tail skin of rats at 3.7 mg/cm^2, body-weight loss, lack of grooming, dullness of the wool, and diminished urinary output were observed, as were alterations in orienting responses and increases in the time required by the animal to regain the capacity for rectilinear movement after hexenal (hexobarbital sodium) administration. Notable biochemical changes included reduced alanine and asparagine transaminase activities. Histopathologic examination demonstrated perivascular edema and swollen nerve cells in the brain and degenerative changes in convoluted tubules of the kidneys. The dose of 1.2 mg/cm^2 was considered as the threshold level and that of 0.005 mg/cm^2, as the no-observed-effect level. Lithium chloride was classified as moderately hazardous with regard to dermal toxicity.

Man
Workers exposed to **lithium chloride** aerosols (40–60 mg/g^3 of air) for several years suffered from dermatitis that affected the face and/or hands, but no cases of chronic systemic intoxication were detected [Smirnova].

Effects of dermal application
Lithium nitrate trihydrate applied to the skin causes petechial eruptions, while **lithium dihydrogen phosphate** makes the skin dry and may cause its desquamation. Lithium compounds such as the carbonate, nitrate, metaniobate, and metatantalate have been found not to pass through the skin barrier [Popova *et al.*; Spiridonova & Shabalina].

A high capacity for penetration through intact skin is possessed by **lithium chloride**, and its effects with skin application have been studied in detail by Gostinsky & Krasnopevtseva and Smirnova. In Smirnova's experiments, lithium chloride was applied to shaved areas of rabbit's skin as concentrated aqueous solution or in powder form; the powder was dried at 100°C and applied either alone or in vegetable oil or petrolatum. The powder and ointments applied at 20 mg/cm^2 led to dermatitis in 4 days; the skin exposed to the solution became swollen and was seen to have cracks and fine crusts 24 h post-application. The threshold dose of lithium chloride applied by the dermal route was 1.2 mg/cm^2; at this dose level, significant changes in the summated threshold index and orienting responses were observed.

ABSORPTION, DISTRIBUTION, AND ELIMINATION
After parenteral administration of lithium chloride to mice at 320 or 400 mg/kg, guinea-pigs at 230 mg/kg, and rats at 200 mg/kg, lithium reached its highest concentrations in 60 min both in the blood and in all organs where they were measured. With oral administration of this compound by gavage to mice at

400 mg/kg, peak lithium concentrations occurred after 6 h in muscles and after 1 h in the blood and other organs. In terms of accumulation of lithium ions, tissues and biological fluids of mice, guinea-pigs, rats, and rabbits given a non-lethal lithium dose by any route (parenteral or oral), can be arranged in a decreasing order as follows: thyroid, heart, bile, lung, blood, saliva, adrenals, spleen, skeletal muscle, liver, bone, brain, erythrocytes, eyeball. The time required for lithium ions to be completely absorbed from the gastrointestinal tract is about 8 h. Lithium does not bind to plasma proteins, but can cross the blood-brain barrier and its concentration in the cerebrospinal fluid may reach 40% of that in the plasma. Its level in the saliva may be severalfold higher than in the plasma. Lithium ions were found in the milk of nursing mothers given lithium preparations for therapeutic purposes. Animal experiments have shown that lithium is retained in the pituitary where its concentration directly depends on the administered dose. Lithium ions also penetrate the placental barrier and accumulate in the embryo.

During therapy with lithium drugs, plasma lithium should preferably be measured between the 8th and 12th hours after the last dose and should not exceed values around 1.5 mEq/L (5–11 µg/ml). Approximately 95% of the administered dose is excreted in urine, 4% in sweat, and 1% in feces. In humans, lithium levels were reported to range 0.13–0.27 µg/L in lymph nodes, 0.05–0.07 µg in lungs, 3–5 ng/g in brain, 2–4 ng/g in testis, and 4–8 ng/g in blood [Berman]. A discussion of lithium toxicokinetics can be found in Jaeger *et al.*

HYGIENIC STANDARDS

Exposure limits adopted for lithium and its chloride in the USSR are given in Table 2. In the USA, a TLV-TWA of 0.025 mg/m^3 has been set for lithium hydride.

Table 2. Exposure limits for lithium and lithium chloride

	Workroom air	Contamination of hands	Atmospheric air	Water sources	
	MAC_{wz} (mg/m^3)	MPL[a] (mg/cm^2)	$TSEL_{aa}$ (mg/m^3)	MAC_w (mg/L)	Hazard class
Lithium	—	—	—	0.03[b]	2
Lithium chloride	0.5	0.05	0.02	—	1

[a] Maximum permissible level.
[b] Based on sanitary/toxicological criteria.

METHODS OF DETERMINATION

In *air*: photometry based on the reaction of lithium with nitroanthranilazo (sensitivity 0.3 μg in the assayed volume) [Peregud]. In *natural* and *waste waters, air, snow, soil,* and *plant materials*: flame emission spectrometry (sensitivity 1 μg/m³ for air and 0.01 mg/ml for aqueous media) [Loseva *et al.*; Lurie; Novikov *et al.*; Sopach *et al.*]. In *blood serum*: spectrometric methods [Kim & Gracy; Moynier; Wheeling & Christian]. In *foods*: flame atomic emission spectrophotometry (detection limit 0.0013 mg/kg for a 10-g sample) [Ewans & Read]. Procedures for determining lithium in biological materials by atomic absorption spectrometry are described in *Toxic Metals and Their Analysis* [Berman].

The problems associated with biological monitoring of lithium alloys are discussed by Bencze *et al.*

MEASURES TO CONTROL EXPOSURE

At the workplace, the provision of healthy working conditions for those handling lithium or its compounds is a difficult task because of high lithium toxicity. All processes involving the production or use of this element and its compounds should proceed in sealed equipment that completely excludes manual operations. In workplaces where lithium is used, closed well-ventilated boxes with built-in gloves for handling the metal must be installed. Since lithium can be absorbed by the equipment, clothing, and skin, good housekeeping and personal hygiene are essential. The workrooms should be cleaned up at least once per shift using wet methods or vacuum cleaners. Wash stands with a cold and hot water supply should be available in the work areas. There should be a ban on taking food and keeping personal effects in these areas. Showering after work is mandatory. Rest rooms with adequate relaxation facilities should be provided at the factory. The rapid and complete restoration of physiologic functions in the workers can be promoted by a rational regimen of work and rest combined with the appropriate diet.

All workers must undergo preplacement and periodic medical examinations.

For *personal protection*, respirators should be worn where the risk of exposure to the dust of lithium, lithium compounds, or their breakdown products is high; the working clothes should meet the standards laid down for the particular operations, and appropriate barrier creams or rubber gloves should be used for hand protection.

The main measures to be implemented for protection of the *general environment* include utilization or recycling of valuable by-products and raw materials to the greatest possible extent; effective purification of gaseous and particulate emissions from plants producing lithium compounds; and monitoring the use of lithium-containing industrial wastewaters in agriculture to ensure that the prescribed rates of their application are not exceeded and the input/output balance of lithium in soils is not upset.

FIRST AID IN POISONING

After the ingestion of a lithium salt, vomiting should be induced and/or the stomach washed thoroughly, followed by oral administration of activated carbon and an isotonic sodium sulfate solution. The victim should drink large quantities of liquid. Hospitalization is necessary if his condition becomes worse.

REFERENCES

Bencze, K. *et al.* (1991) *Sci. Tot. Env.*, **101**, 83—90.
Berman, E. (1980) *Toxic Metals and their Analysis*. Heyden, London.
Birch, N. J. (1988) Chapter 32 *Lithium* in: Seiler, H. *et al.* (eds) *Handbook on Toxicity of Inorganic Compounds*. Marcel Dekker, New York, pp. 383—393.
Brakhnova, I. T. (1971) *Toksichnost poroshkov metallov i ikh soyedineniy* [Toxicity of Metals and their Compounds in Powder Form]. Naukova Dumka, Kiev (in Russian).
Dobrovolsky, V. V. (1983) *Geografiya mikroelementov. Globalnoe rasseyanie* [Geography of Trace Elements. Global Dispersion]. Mysl, Moscow (in Russian).
Ewans, W. H. & Read, J. I. (1985) *Analyst*, **110**, 619—623.
Gostinsky, V. D. & Krasnopevtseva, G. B. (1977) In: *Kozhnyi put postupleniya promyshlennykh yadov v organism i ego profilaktika* [Absorption of Industrial Poisons by the Cutaneous Route and its Prevention]. Moscow, pp. 22—27 (in Russian).
Gostinsky, V. D. & Krasnopevtseva, G. B. (1978) In: *Aktualnye voprosy ekologicheskoi toksikologii* [Current Topics in Ecological Toxicology]. Ivanovo, pp. 74—75 (in Russian).
Grushko, Ya. M. (1972) *Yadovitye metally i ikh neorganicheskiye soyedineniya v promyshlennykh stochnykh vodakh* [Poisonous Metals and their Inorganic Compounds in Industrial Waste Waters]. Meditsina, Moscow (in Russian).
Grushko, Ya. M.(1979) *Vrednye neorganicheskie soyedineniya v promyshlennykh stochnykh vodakh* [Harmful Inorganic Compounds in Industrial Waste Waters]. Khimiya, Leningrad (in Russian).
Henderson, P. (1982) *Inorganic Geochemistry*. Pergamon Press, Oxford.
Jaeger, A. *et al.* (1985—86) *Clin. Toxicol.*, **23**, 501—517.
Kabata-Pendias, A. & Pendias, H. (1986) *Trace Elements in Soils and Plants*. CRC Press, Boca Raton (Fl.).
Khakimov, Kh. Kh. & Tatarskaya, A. Z. (1985) *Periodicheskaya sistema i biologicheskaya rol elementov* [The Periodic Table and the Biological Role of Elements]. Meditsina, Tashkent (in Russian).
Kim, W. & Gracy, C. (1984) *Anal. Lett.*, **B17**, 217—227.
Loseva, A. F. *et al.* (1984) *Spektralnye metody opredeleniya mikroelementov v obyektakh biosfery* [Spectral Methods for Measuring Trace Elements in the Biosphere]. Publication of Rostov University, Rostov (in Russian).
Lukashev, K. I. *et al.* (1984) *Chelovek i priroda: Geokhimichekiye i ekologicheskiye aspecty ratsionalnogo prirodopolzovaniya* [Man and Nature: Geochemical and Ecological Aspects of Environmental Management]. Nauka i Tekhnika, Minsk (in Russian).
Lurie, Yu. Yu. (1984) *Analiticheskaya khimiya promyshlennykh stochnykh vod* [Analytical Chemistry of Industrial Waste Waters]. Khimiya, Moscow (in Russian).
Metelev, V. V. *et al.* (1971) *Vodnaya toksikologiya* [Aquatic Toxicology]. Kolos, Moscow (in Russian).
Moynier, J. (1986) *Path. Biol.*, **34**, 51—56.

Neretin, V. Ya. (1959) In: *Materialy k toksikologii nekotorykh soyedineniy litiya* [Toxicology of Selected Lithium Compounds]. Moscow (in Russian).

Novikov, Yu. V. *et al.* (1981) *Metody opredeleniya vrednykh veshchestv v vode vodoyemov* [Methods for Measuring Noxious Substances in Water Bodies]. Meditsina, Moscow (in Russian).

Peregud, E. A. (1976) *Khimicheskiy analiz vozdukha (novye i usovershenstvovannye metody)* [Chemical Analysis of Air (new and improved methods)]. Khimiya, Leningrad (in Russian).

Popova, O. Ya. *et al.* (1978) In: *Aktualnye problemy gigiyeny truda* [Current Topics in Occupational Hygiene]. Moscow, pp. 28–32 (in Russian).

Ribas, B. (1991) Chapter II 17 *Lithium* in: Merian, E. (ed.) *Metals and Their Compounds in the Environment—Occurrence, Analysis and Biological Relevance.* VCH, New York, pp. 1015–1024.

Smirnova, E. S. (1983) In: *Aktualnye voprosy gigiyeny truda i profpatologii v nekotorykh otraslyakh khimicheskoi promyshlennosti* [Current Problems of Occupational Health in the Chemical Industry]. Moscow, pp. 33–39 (in Russian).

Sopach, E. D. *et al.* (1976) *Gig. San.*, No. 9, 58–60 (in Russian).

Spiridonova, V. S. & Shabalina, L. P. (1979) *Gig. Truda Prof. Zabol.*, No. 9, 49–50 (in Russian).

Vasilenko, V. N. *et al.* (1985) *Monitoring zagryazneniya snezhnogo pokrova* [Monitoring the Pollution of the Snow Cover]. Gidrometeoizdat, Leningrad (in Russian).

Venchikov, A. I. (1982) *Printsipy lechebnogo primeneniya mikroelementov v kachestve biotikov* [Principles for the Therapeutic Use of Trace Elements as Biotics]. Ylym, Ashkhabad (in Russian).

Wheeling, K. & Christian, G. D. (1984) *Anal. Lett.*, B17, 217–227.

Sodium and its compounds

Sodium bromide; **s. carbonate** (soda ash); **s. carbonate decahydrate** (crystalline soda, washing soda); **s. chlorate; s. chloride** (common salt; halite [rock salt] min.); **s. chlorite trihydrate; s. fluoride; s. hexafluoroaluminate** (cryolite [min.]); **s. hexahydroxostannate(IV)** (s. stannate trihydrate); **s. hydrogen carbonate** (s. acid carbonate, s. bicarbonate, baking soda, soda); **s. hydroxide** (caustic soda); **s. iodide; s. metasilicate** (s. silicate); **s. orthophosphate** (s. phosphate); **s. perborate tetrahydrate; s. sulfate** (thenardite(α) [min.]); **s. sulfite; s. thiosulfate pentahydrate** (s. hyposulfate, antichlor); **s. triphosphate** (s. tripolyphosphate)

IDENTITY AND PHYSICOCHEMICAL PROPERTIES
OF THE ELEMENT

Sodium (Na) is an alkali metal in group I of the periodic table, with the atomic number 11. It has one naturally occurring isotope, ^{23}Na.

Sodium is a ductile soft metal that can be readily cut with a knife. Its oxidation state in compounds is +1. It is highly active chemically. It reacts violently with water forming sodium hydroxide and releasing hydrogen. It readily oxidizes in air to form the oxide Na_2O and the peroxide Na_2O_2. At 200°C it reacts with hydrogen to produce the hydride NaH. It forms the amide $NaNH_2$ with ammonia and halides with the halogens (it burns in an atmosphere of fluorine or chlorine). It reduces many metals to their salts. It reacts with sulfur at room temperature. At 800–900°C sodium vapor combines with carbon to produce the carbide Na_2C_2. Sodium forms alloys with many metals. See also Appendix.

NATURAL OCCURRENCE AND ENVIRONMENTAL LEVELS

A total of 222 minerals of sodium are known. Among the more important sources of sodium and its compounds are halite (rock salt) NaCl, Chile saltpeter $NaNO_3$, thenardite Na_2SO_4, mirabilite $Na_2SO_4 \cdot 10H_2O$, and trona $NaH(CO_3)_2 \cdot 2H_2O$.

Average abundances of sodium are (in %) : 2.5–2.6 in the earth's crust, 0.63 in soils (0.33 in sandstones, 0.66 in shales and clays), 1.06 in seawater, 0.22 in plants, 0.1 in animals. In seawater, its principal species is Na^+; its residence time is 6.8×10^7 years, and its average concentrations are 10.5–11.0 mg/g (in sea salt, they reach up to 290 mg/g). Its average abundance in ferromanganese oxide deposits of the world's ocean is 1.94%. In river water, its mean content is 6300 µg/L. For bacterial biomass, levels up to 460 mg/100 g dry weight have been reported. Its reported mean concentrations in vegetables are (mg%) 65 in carrot, 37 in potato, and 25 in radish. Coal and petroleum contained sodium at 2000 µg/g and 2 µg/g, respectively [Henderson; Lukashev et al.; Ostromogilsky et al.; Venchikov].

Sodium concentrations ranged from 2.7 to 12.0 ng/m^3 (mean, 3.2 ng/m^3) in air samples taken at the South Pole [Zoller et al.], 0.9–5.1 mg/L in snow water samples from the environs of Moscow [Vasilenko et al.], and 0.76–3.9 mg/L in those from the South and North Poles [Karlsson et al.]. In the state of Mississippi, USA, sodium was present at levels up to 11 µg/g in river water and 120 µg/g in well water [Choi].

Sodium ions are constituents of extracellular and intracellular fluids of various plants and animals. In animals, sodium is among the major elements involved in mineral metabolism, in the maintenance of osmotic pressure and acid-base balance, and in nerve impulse conduction. See also Birch.

PRODUCTION

Sodium can be obtained by electrolysis of its fused hydroxide or by reduction of its salts with carbon or other reducing agents at temperatures above their melting points.

Sodium bromide is made by the action of bromine on sodium hydroxide solutions in the presence of a reducing agent (formic acid, ammonia) or by exchange reaction between iron(II) bromide and s. carbonate or between calcium bromide and s. sulfate. **S. carbonate** is made by reacting a s. chloride solution with ammonia and carbon(IV) oxide followed by calcination of the s. hydrocarbonate precipitate; or by calcination of s. sulfate with carbon and calcium carbonate (limestone). **S. carbonate decahydrate** crystallizes from aqueous solutions of s. carbonate at temperatures below 32°C. **S. chlorate** is made by electrolysis of s. chloride solution or by chlorination of s. hydroxide, carbonate, or hydrocarbonate solution. **S. chloride** is recovered from deposits of the mineral halite (rock salt) and from seawater or salt lakes. **S. chlorite** is obtained by exchange reaction between solutions of either barium chlorite and s. sulfate, calcium chlorite and s. carbonate, or zinc chlorite and s. hydroxide, or by reaction of chlorine(IV) oxide with s. peroxide or with s. hydroxide and hydrogen peroxide. **S. fluoride** occurs as the mineral villiaumite and is a constituent of cryolite and other minerals; it can be obtained by sintering calcium fluoride with s. carbonate and silicon oxide, by decomposing s. hexafluorosilicate, or by dissolving s. carbonate or hydroxide in hydrofluoric acid. **S. hexafluoroaluminate** is the major constituent of cryolite and can be obtained from hot fluorine-enriched aqueous solutions of this mineral; industrially, it is made by neutralization of hydrofluoric acid with s. carbonate and aluminium hydroxide or by carbonization of a mixture of s. fluoride and s. aluminate solutions. **S. hydrogen carbonate** is made by saturating s. carbonate solutions with carbon(IV) oxide at 75°C. **S. hydroxide** is obtained by electrolysis of s. chloride solutions or by reacting a hot s. carbonate solution with calcium hydroxide. **S. iodide** is produced in exchange reaction between Fe_3I_8 and s. carbonate. **S. nitrite** is made by absorbing nitrogen oxides with aqueous alkaline solutions followed by their evaporation. **S. orthophosphate** is obtained by neutralization of orthophosphoric acid with s. hydroxide. **S. perborate** is prepared by treatment of s. metaborate with hydrogen peroxide or of orthoboric acid with s. peroxide. **S. sulfate** is the basic constituent of the mineral mirabilite (Glauber's salt). Anhydrous s. sulfate is obtained by melting mirabilite, by evaporating aqueous solutions of s. sulfate, or by reacting s. chloride with sulfuric acid at 500–550°C. Anhydrous **s. sulfite** is made by evaporation of concentrated aqueous solutions of s. sulfite at 95–100°C or by dehydration of s. sulfite heptahydrate; it also forms as a by-product in the production of phenol from benzenesulfonic acid by alkaline fusion. **S. thiosulfate** is made by dissolving sulfur in a hot s. sulfite solution or by reacting s. hydrosulfide with s. hydrosulfite; it is also obtained as a by-product in the production of s. hydrosulfite, sulfur dyes, and thiocarbanilide and in the removal of sulfur from industrial gases. **S. triphosphate** is produced by heating a solid mixture of the hydrous and dihydrous sodium orthophosphates in the 2:1 molar ratio.

USES

Metallic **sodium** is used as a heat-transfer agent in airplane engines, machines for pressure-die casting, nuclear reactors (in the form of alloys with potassium), and in various chemical processes; in sodium-vapor lamps; in metallurgy, especially as a reducing agent and for strengthening alloys; in making triethyl lead; and in the chemical industry.

Sodium bromide is used in medicine and photography. **S. carbonate** finds use in the manufacture of glass, aluminum, soap, s. hydroxide and s. hydrogen carbonate, detergents, various salts and paints, in petroleum refining, for desulfurizing cast iron, washing wool, laundering, etc. **S. chlorate** serves as a herbicide and defoliant and as an oxidizing agent in the manufacture of chlorine(IV) oxide and s. perchlorate. **S. chloride** (common salt) has a variety of applications, including its use as the raw material in the manufacture of chlorine and of s. hydroxide, carbonate, and sulfate. **S. chlorite** is employed in the production of chlorine(IV) oxide, for the decontamination and deodorization of water, and as a bleaching agent in the textile and paper and pulp industries. **S. fluoride** is used in the chemical, glass, and cement industries, in metallurgy (for electrolytic production of aluminum, beryllium, etc.), in the manufacture of casein glues and agents for rust removal, as a preservative for wood, meat, and butter, in the fluoridation of drinking water, and in medicine in the treatment of dental caries, osteoporosis and otosclerosis.

S. hexafluoroaluminate is used in the electrochemical production of aluminum as an aluminum oxide-dissolving electrolyte, in the manufacture of aluminum alloys, glass, enamels, and for other purposes. **S. hydrogen carbonate** is employed in bread baking, food industries, medicine, and in fire extinguishers. **S. hydroxide** is used in making artificial fibers, soap, aluminum, and paints, in the paper and pulp industry, in petroleum refining, and in the textile industry as a bleaching and mercerizing agent. **S. iodide** is used in medicine as a source of iodine and as an expectorant. **S. nitrite** is used in the manufacture of dyes and iodine, in the food industry, and in medicine as an antidote and a pain-relieving agent. **S. orthophosphate** is a detergent. **S. perborate** is a component of synthetic cleansing agents and is used in medicine as an antiseptic. **S. sulfate** is used in the glass, soap, sulfate cellulose, textile, and leather goods industries, in nonferrous metallurgy, in medicine and veterinary as an antidote, cathartic, and laxative, and as a source of s. silicate and s. sulfite. **S. sulfite** is employed in photography, the pharmaceutical industry, in the manufacture of artificial fibers, and in medicine. **S. thiosulfate** is used in photography and as a reagent in analytical chemistry. **S. triphosphate** serves as an inorganic base for synthetic cleansing agents. (See also Birch.)

MAN-MADE SOURCES OF EMISSION INTO THE ENVIRONMENT

It has been reported that about 4000 µg of sodium enters the ambient air in the fly ash per gram of burned coal, and that urban industrial aerosols contain this element at an average concentration of 8200 µg/g. Some coals (e.g. those from Donbas coal mines in the USSR) contain elevated sodium levels, and the burning of such coals results in considerable surface corrosion of the boilers and in high sodium levels in the emitted gases [Loseva *et al.*]. Sodium concentrations in emissions from iron and steel works ranged 0.31 to 1.1% and the amounts entering soils from these emissions, 3.7 to 9.4 mg/kg soil [Garmash].

Substantial quantities of sodium and its compounds are contained in municipal wastewaters and in effluents from a wide range of industrial activities (chemical, petrochemical, textile, leather-goods, soap, synthetic rubber, canning, and other industries). Such effluents, discharged into waterways, contained up to 200 g of sodium chloride per liter [Vinogradov & Beletsky]. Very high in sodium are also effluents from the processing of potassium and magnesium ores; when discharged into a fresh water body, they increased its sodium content to hundreds or even thousands of milligrams per liter.

Industrial effluents as well as domestic sewage are significant sources of sodium in soils from which it may enter food chains. In one study, carrot plants harvested from a soil fertilized with industrial and domestic sewage sludge containing 175–185 mg Na/kg, had an average sodium concentration of 76 mg% as compared to 65 mg% in the control plants [Leshchenko et al.]. Large quantities of sodium enter soil in areas where sodium chloride is used to remove ice from highways. (The amount of NaCl employed for this purpose rose from 0.5 million tonnes in 1947 to 9 million tonnes in 1970.) In areas where sodium is used to soften natural waters, its levels in drinking water were also elevated [Craun].

TOXICITY

Aquatic organisms

In the Black Sea, maximum permissible sodium concentrations for aquatic organisms of various systematic groups or community complexes (microorganisms, phytoplankton, zooplankton, phytobenthos, zoobenthos, fish) vary from 1.3 to 7.5 g/L. Vinogradov & Beletsky have concluded from their studies on the distribution patterns of various types of fauna in the Black, Azov, and Caspain Seas that the salinity of sea areas with a total salt content in water of less than 13% should not be increased further by sodium chloride-containing effluents; they recommend that areas with higher salinity values be selected for their discharge, indicating that no change in the type of fauna will ensue following a local 3% increase in the salinity of such areas by effluents high in this salt.

In fish, toxic sodium concentrations cause neuromuscular paralysis and damage to the branchial epithelium; a characteristic sign of poisoning is dark coloration of the fish body. Lethal concentrations of several sodium salts for fish are shown in Table 1. Crustaceans exposed to sodium at 680 mg/L showed reduced reproductive capacity [Metelev et al.]

Animals and man

Lethal doses of sodium compounds for animals

LD_{50} values of five sodium compounds for laboratory mammals are given in Table 2.

Table 1. Lethal concentrations of sodium salts for fish

	Fish	LC (g/L)	Survival time (h)
Bromide	Carp	25–27	≤24*
Carbonate decahydrate	Carp	0.25	≤24*
Chlorate	Roach & Carp	17–18	≤24
Chloride	Roach & Tench	13	≤24
Hydrogen carbonate	Various fish	10	≤48*
Iodide	Roach	10	≤24
Metasilicate	Trout	3.2	≤48*
Nitrite	Various fish	10	≤2
Phosphate	Various fish	4	≤24*
Sulfate	Crucian Carp	14.8	≤24
Thiosulfate	Various fish	30	≤4.5

* Only some of the fish died within 24 or 48 h.
Source: Metelev *et al.*

Table 2. Median lethal doses of sodium compounds

Compound	Animal species	Route of administration	LD_{50} (mg/kg)
Carbonate	Mouse	Subcutaneous	2210
	Rat	Oral	4200
Chloride	Mouse	Subcutaneous	3150
		Intraperitoneal	2900
	Rat	Oral	3000
		Intraperitoneal	2600
Chlorate	Mouse	Oral	3600
	Rat	Oral	6500
	Guinea-pig	Oral	6100
Chlorite	Mouse	Oral	350
	Rat	Oral	350
	Guinea-pig	Oral	300
Hydrogen carbonate	Mouse	Oral	3360
	Rat	Oral	9940

Sources: Babyan *et al.*; Bergquist; Liublina & Dvorkin.

Sodium

Animals
In rats exposed to sodium and potassium for 6 months in their drinking water containing these elements in concentrations of 75 and 7.5 mg/L or 100 and 10 mg/L, respectively, a number of changes were observed, including erythropenia and leukopenia, decreased hemoglobin levels, lowered activities of alanine and aspartate transaminases in the blood serum, and elevated levels of ascorbic acid in the adrenals and liver and of urea and creatinine in the urine. The relative weights of their hearts and stomachs were decreased. On histopathologic examination, mucosal edema and hyperemia, hemorrhagic areas, and increased numbers of mitotic cells in the liver and small gut were noted, as was focal dystrophic degeneration of muscle cells in the myocardium [Omelyanets *et al.*]. In another study by these authors, rats maintained for 3 months on drinking water containing 150 mg Na/L and 15 mg K/L, showed increased levels of catecholamines, longer periods of iodine excretion by the thyroid, reduced intensity of tissue respiration in the liver and brain, and lowered succinate dehydrogenase and lactate dehydrogenase activities; these changes were all long-lasting. Other changes included reduced acetylcholinesterase activity and glycogen level in hepatic tissue, increased duration of Hexenal (hexobarbital sodium)-induced sleep, elevated sodium levels in the blood serum and skeletal muscle, and greatly increased latencies of reflex action.

Sodium carbonate

Animals
Acute and chronic studies in mice and rats have shown that the primary target for sodium carbonate on inhalation exposure is the upper respiratory tract. The threshold concentration of washing soda for irritancy was 100 mg/m^3, as calculated after measuring a set of parameters characterizing respiratory function; after 4-h exposure to this concentration, the respiratory frequency was altered and the odor threshold was increased. With oral administration by gavage, the toxicity of sodium carbonate to rats and mice was extremely low, the LD_{50} values being 5.6 and 6.6 g/kg body weight, respectively [Kamaldinova *et al.*].

Rats exposed to sodium carbonate aerosol at 16 mg/m^3 for periods up to 4 months, exhibited reduced odor perception, diminished breathing frequency at the beginning and at the end of the exposure period, altered ratio of sodium and potassium ions, increased pH, augmented oxygen consumption, and blood pressure elevation. Hematologic effects included altered protein spectrum in the serum and reduced phagocytic activity of neutrophils. Pathologic examination demonstrated inflammatory changes in the nasal mucosa similar to those seen in chronic rhinitis; bronchial dilatation; and areas of emphysema and inflammation in the lungs. After exposure of rats to a sodium carbonate concentration of

2.2 mg/m^3 over the same period as above, no changes in their condition or behavior were observed.

In rabbits, sodium carbonate placed into the conjunctival sac in an amount of 50 mg, caused moderate conjunctivitis accompanied by profuse lacrimation and strong scleral hyperemia; the conjunctivitis lasted for 7 days.

Man

Longshoremen employed at a seaport in the loading of washing soda from ships to tank cars were found to be exposed to a fine dust of this compound for 65 to 90% of the working time, at concentrations that were at times as high as $311-345 \text{ mg/m}^3$. Medical check-ups of these dock laborers, most of whom were aged under 40 years, showed burns, erosions, or eczemas on the skin as well as deviations, perforations, and other defects of the nasal septum in many of them (nasal defects were present even in some of those employed for a short time). An increased incidence of exacerbations of chronic rhinitis, pharyngitis, conjunctivitis, and gastrointestinal disorders was also noted. The incidence of disabling upper respiratory tract and gastrointestinal diseases was 1.5 times higher than in a control group of longshoremen not exposed to the dust. The threshold concentration for irritant effects with 1-minute exposure by inhalation was 40 mg/m^3 [Kamaldinova *et al.*].

Sodium chloride

Animals

Teratogenic effects of sodium chloride in mice have been reported [Nishimura & Miyamoto].

Man

Ingestion of sodium chloride in an amount close to the lethal dose produced cyanosis, tachycardia, blood pressure elevation, and tension in the anterior abdominal wall on the left; the patient was diagnosed as having acute pancreatitis [Danilenko].

The toxicity of sodium chloride on long-term ingestion in excessive amounts through food or water is often manifested in hypertension. Severe hypertensive disease accompanied by frequent vascular crises and unresponsive to hypotensive agents was widespread in a village whose residents were consuming drinking water high in this salt (concentrations up to 2100 mg/L were recorded) throughout their lives [Fatula]; the disease, which usually developed by age of 25–28 years, was about four times more common than in a control village. Moreover, considerably increased prevalence rates of pulmonary tuberculosis and emphysema, chronic bronchitis, peripheral nervous system disease, and disorders of the liver and gall-bladder were noted during the 17-year observation period of that study. (The role of sodium in the genesis of hypertension is discussed in Luft & Ganten.)

In a group of workers exposed to a dust of NaCl (76.8%) and KCl (22.8%) composed of particles up to 4.8 μm in diameter, chronic bronchitis was observed four times as frequently as in the control group. Also, the sense of smell was impaired or absent in some of the exposed workers.

Sodium chlorite

Animals
Oral administration of this compound to mice and rats in a very large dose (2000–5000 mg/kg) ·rapidly led to twitching of the limbs and gross cyanosis of the paws, tail, and ears; the animals all died 5 to 10 min post-administration. Single oral doses of 600–1200 mg/kg produced similar effects, but deaths occurred after 30 to 45 min. After an oral dose of 300 to 400 mg/kg, clinical manifestations of poisoning became evident in 2 to $2\frac{1}{2}$ h, while deaths began to be recorded in 3 h.

Repeated oral administration of sodium chlorite to rats in doses equal to 1/10 or 1/50 of the LD_{50} for a period of 2 months stimulated erythropoiesis and thrombocytopoiesis, caused methemoglobinemia, impaired liver function, and reduced blood levels of urea; phasic changes in phagocytic activity were also recorded, as were increases in the summated threshold index. Histopathologic examination of rats given 1/10 of the LD_{50} demonstrated focal necrobiosis of the liver parenchyma; thickened interalveolar septa, hyperplastic peribronchial tissue, and focal emphysema in the lungs; and dystrophic changes in the cytoplasm of epithelial cells in the convoluted renal tubules. In addition, changes indicative of oxygen deficiency were found in the central nervous system. With the lower dose (1/50 of the LD_{50}), morphologic changes were much less marked.

Sodium chlorite is not a cumulative poison; thus, the cumulation coefficient, calculated for rats given this compound at 1/5 of the LD_{50} orally for 20 days, was 18.6.

In a chronic oral toxicity study, in which sodium chlorite was administered to rats at 0.01, 0.1, or 1.0 mg/kg for 6 months, animals on the 1.0 mg/kg dose showed elevated alkaline phosphatase activity, phasic changes in acetylcholinesterase activity in the blood, and impaired excretory function of the liver; other hematologic findings were reduced levels of sulfhydryl groups, inhibited bactericidal activity of the serum, and depressed phagocytic activity of neutrophils. Sulfhydryl groups (both total and free) were reduced in liver homogenates, as were ascorbic acid levels in the adrenals in the absence of their hypertrophy. Moreover, shortened times of sperm motility were recorded. Histopathologic examination demonstrated hepatic dystrophy, hyperplasia of peribronchial lymphoid tissue, focal myocardial dystrophy, granular degeneration of convoluted renal tubules, hypertrophy and hyperplasia of reticular cells in the spleen, and activation of vascular endothelium in the brain. The dose of 0.1 mg/kg was much less damaging while the lowest dose did not cause any detectable changes.

Sodium hydrogen carbonate

Animals
The threshold for acute inhalation toxicity of sodium hydrogen carbonate in rats was 74 mg/m^3, as estimated from changes in plasma potassium, serum chlorides, total protein, and some other biochemical parameters [Babayan et al.].

These authors also found that long-term inhalation exposure of rats to this compound at concentrations of about 77 mg/m^3 reduced plasma and serum levels of potassium, albumins, and alanine transaminase and caused elevations of sodium levels in the erythrocytes, myocardium, and lungs, of total lipids and cholesterol in the blood serum, of chlorides and creatinine in the urine, and of creatine kinase and adenosine triphosphate in the myocardium. The concentration of 15 mg/m^3 was described as being close to the threshold level; the cumulation coefficient was calculated to be 12.9.

Sodium hydrogen carbonate was slightly irritating to the conjunctiva of rats. No absorption through the skin was observed [Babayan et al.].

Sodium hydroxide

Animals
Sodium hydroxide exerts very pronounced local effects, causing severe chemical burns in the skin and mucous membranes. In rabbits, uniform application of a 25% solution to their shaved skin for 2–3 min resulted in strongly marked systemic intoxication, manifested in lethargy, greatly decreased food intake, rapid weight loss, and fever. On the fifth day, extensive necrosis of all skin layers accompanied by acute inflammation was observed. Signs of necrosis and acute inflammation were still present on day 45 post-application, although young granulation tissue and new capillaries were already seen to be forming on day 15. On day 60, development of scar tissue was observed, but complete epithelialization occurred later (after day 75 in some animals).

Man
Sodium hydroxide mist or fumes contacting the eyes may cause severe conjunctival edema and hyperemia, corneal cloudiness, and lesions of the iris. On entering the eye in the form of solution or dust, the hydroxide produces a chemical burn that varies in severity depending on the degree to which the cornea and conjunctiva of the eyeball have been damaged, and may lead to blindness. Skin contact can also result in severe lesions extending deep into the dermis and accompanied by colliquative necrosis. Ulcers that healed very slowly

with scar formation have also been described. Clinical manifestations of oral poisoning include burns of the oral, esophageal, and gastric mucous membranes, hypersalivation, nausea and vomiting (often with blood), severe pains in the mouth and in the retrosternal and abdominal regions, and impossibility of swallowing; collapse may occur. Kidney and liver damage and pulmonary edema have been reported. Death may follow from shock within 24 h after ingestion [Tarakhovsky et al.]. Lethal oral doses ranged from 10 to 20 mg [Mogosh].

ABSORPTION, DISTRIBUTION, AND ELIMINATION IN MAN

The main sources of exposure to sodium are drinking water and food. In the USA, sodium concentrations in drinking-water supplies provided by 2100 surveyed water systems that served more than a half of the country's population were found to range from 0.4 to 1900 mg/L, although values higher than 20 and 250 mg/L were recorded for 42% and 5% of the systems, respectively; the daily sodium intake by an adult person was estimated to vary from 1600 to 9600 mg [Graun; Luft & Ganten].

In the body, most of the sodium is found in extracellular fluid; the exchangeable sodium present in the average adult person totals 3890 mEq, of which 2450 mEq is contained in the 17.5 L of the extracellular fluid and 1440 mEq in the 30.3 L of the internal environment of the cells. Sodium is found in all tissues; its average levels are (in mEq/kg): bones, 78; teeth, 208; muscles, 31; heart, 80; lungs, 108; brain, 74; liver, 83; kidneys, 76; erythrocytes, 35; blood serum, 140; cerebrospinal fluid, 130; lymph, 135.

The sodium ingested in food rapidly distributes in the body. Its absorption begins in the stomach but proceeds for the most part in the small intestine. The bulk of ingested sodium is eliminated in the feces [Birch].

HYGIENIC STANDARDS

Exposure limits adopted for sodium and its compounds in the USSR are given in Table 3. In the USA, TLV-C values of 0.3 and 2.0 mg/m^3 have been set for sodium azide and sodium hydroxide, respectively, and TLV-TWA values of 0.05 mg/m^3 for the fluoroacetate and 5.0 mg/m^3 for the bisulfite, metabisulfite and pyrophosphate.

Table 3. Exposure limits for sodium and its compounds

	Workroom air		Atmospheric air			Water sources	
	MAC_{wz} (mg/m^3)	Aggregative state	MAC_{hm} (mg/m^3)	MAC_{ad} (mg/m^3)	$TSEL_{aa}$ (mg/m^3)	MAC_w (mg/L)	Hazard class
Sodium	—	—	—	—	—	200.0[a]	2
Sodium							
carbonate	5.0	Aerosol	—	—	0.04	—	3
chlorate	5.0	Aerosol	—	—	—	20.0[b]	3
chloride	5.0	Aerosol	—	—	0.15	—	3
chlorite	1.0	Aerosol	—	—	—	—	3
fluoride	0.2	Vapor/Aerosol	0.03	0.01	—	—	2
hexafluoro-aluminate	0.5	Aerosol	0.2	0.03	—	—	2
hexahydroxo-stannate	—	—	—	—	0.05	—	—
hydroxide	0.5	Aerosol	—	—	0.01	—	2
nitrite	—	—	—	—	0.005	—	—
perborate	1.0	Aerosol	—	—	0.02	—	2
silicate (as SiO$_2$)	—	—	—	—	—	30.0[a]	2
sulfate	10.0	Aerosol	—	—	0.1	—	4
sulfide	0.2	Aerosol	—	—	—	—	2
sulfite	—	—	—	—	0.1	—	—
thiocyanate	10.0	Aerosol	—	—	—	—	4
triphosphate	—	—	—	—	0.5	—	—

[a] Based on sanitary/toxicological criteria.
[b] Based on organoleptic criteria.

METHODS OF DETERMINATION

The determination of caustic alkalies present in the air as aerosols is based on the ability of acid-base indicators to alter their color according to the pH of the medium. The detection limit of colorimetric determinations has been put at 1 µg for a 5-ml sample [Demchenko & Guz; Vilner]. A densitometric method has a detection limit of 4 µg in the solution volume analyzed [*Specifications for Methods of Determining Noxious Substances in Air.*]. A method for iodometric determination of sodium chlorite in air with a detection limit of 0.08 mg/m^3 is described in *Guidelines for the Determination of Noxious Substances in Air.* Sodium in natural waters can be determined by ionic chromatography, for example with the method described by Basta & Tabatabai (detection limit 0.05 mg/L). Sodium in industrial and domestic wastewaters can be measured by flame emission spectrometry (detection limit <0.1 mg/L) or by a gravimetric method in which it precipitates in the form of sodium zinc uranylacetate [Lurie]. Procedures for determining sodium in natural waters, soil, plant materials, and animal tissues by flame photometry have been reviewed by Poluektov. A method for sodium determination in blood serum using ion-selective electrodes has been described by Dobrolyubova. Electrometric detrminations of sodium in foods using a sodium ion-selective electrode had reported sensitivities of 790 mg/100 g for butter and 653 mg/100 g for cheese [Florence] and 0.7–1.1% for bacon, ham, salami, and sausage [Fulton *et al.*]. Use of atomic absorption spectrometry to measure sodium in powdered milk has been described [Guardia *et al.*].

MEASURES TO CONTROL EXPOSURE

At the workplace, the basic technical measures include segregation of the dust-generating processes and hermetic enclosure and venting of the equipment. Efforts should be made to exclude spillage of raw materials during their transportation. Wet methods or vacuum cleaners must be used for cleaning the polluted areas.

Personal protective equipment should be provided to workers handling sodium and its compounds for respiratory, skin and eye protection. Wearing a respirator and chemical safety goggles may be necessary. Showering at the end of the shift and washing of contaminated hands without delay should be insisted upon.

Pollution of the *general environment* should be abated by reducing atmospheric emissions of sodium and its compounds from fuel combustion processes and by introducing effective methods for the treatment of industrial and domestic wastewaters. Also, suitable substitutes for sodium chloride should be sought to keep highways free from ice. Where sodium chlorate is used as a herbicide or defoliant, arrangements should be made to prevent its excessive accumulation in the soil and thus avoid the contamination of food, water, and atmospheric air by this compound.

FIRST AID IN POISONING

In case of eye contamination with caustic soda (sodium hydroxide), the eyes should be immediately flushed with jets of water for 15–20 min, after which a 2% novocain [procaine] solution (or a 0.5% dicaine [synonym: tetracaine] solution) and a 0.25% solution of levomycetin [synonym: chloramphenicol] or another antibiotic should be dropped into the eyes, followed by the introduction of sterile mineral or peach oil into the conjunctival sac. The patient may have to wear dark glasses for some time. Admission to a specialized medical institution may be necessary. In the event of upper airway irritation, the victim should be moved to fresh air, gargle his throat and nose with water, drink warm milk or mineral water (e.g. Borzhomi water), take codeine, and, if there is severe cough, have mustard plasters applied to the chest and back; oxygen therapy may be required. Skin areas contaminated with sodium hydroxide must be immediately washed with copious amounts of water, followed by the application of dressings impregnated with a 0.1% solution of rivanol (2-ethoxy-6,9-diaminoacridine lactate) or a 0.02% furaciline [synonym: nitrofurazone] solution. Treatment for skin burns depends on their severity. A person who has ingested caustic soda should have his stomach washed out immediately with water containing milk or egg whites (if the washing is not possible or has failed, he should drink large quantities of water and then be made to vomit), followed by ingestion of beaten-up egg whites, milk, and/or vegetable oil. Antishock therapy and admission to a hospital may be necessary.

REFERENCES

Babayan, E. A. *et al.* (1989) *Gig. Truda Prof. Zabol.*, No. **5**, 30−32 (in Russian).
Basta, N. T. & Tabatabai, M.A. (1985) *J. Environ. Qual.*, **14**, 450−455.
Bergqvist, U. (1983) *New metals*. University of Stockholm (Institute of Physics).
Birch, N. J. (1988) Chapter 56 *Sodium* in: Seiler, H. *et al.* (eds) *Handbook on Toxicity of Inorganic Compounds*. Marcel Dekker, New York, pp. 625−629.
Choi, J. H. (1984) *Environ. Conserv.*, **11**, 171−173.
Craun, G. F. (1984) In: Bitton, G. & Gerba, Ch. (eds) *Ground-water Pollution Microbiology*. Wiley, New York, pp. 135−179.
Danilenko, V. S. (1989) *Farmakol. Toksikol.*, Kiev, No. **24**, 87−103 (in Russian).
Demchenko, M. P. & Guz, E.I. (1975) *Gig. San.*, No. **2**, 94−95 (in Russian).
Dobrolyubova, B. A. (1984) *Gig. San.*, No. **8**, 68−69 (in Russian).
Fatula, M. P. (1977) *Gig. San.*, No. **2**, 7−11 (in Russian).
Florence, E. (1986) *Analyst*, **111**, 571−573.
Fulton, B. A. *et al.* (1984) *Anal. Chem.*, **56**, 2919−2920.
Garmash, G. A. (1983) In: *Khimiya v selskom khoziaistve* [Chemistry in Agriculture]. No. 10, pp. 45−48 (in Russian).
Guardia, M. *et al.* (1986) *Analyst*, **111**, 1375−1377.
Guidelines for the determination of noxious substances in air (1979) [Metodicheskiye ukazaniya po metodam opredeleniya vrednykh veshchestv v vozdukhe], No. 15. Publication of the USSR Ministry of Health, Moscow (in Russian).
Henderson, P. (1982) *Inorganic Geochemistry*. Pergamon Press, Oxford.
Kamaldinova, M. M. *et al.* (1976) *Gig. Truda Prof. Zabol.*, No. **11**, 55−57 (in Russian).
Karlsson, V. *et al.* (1985) *Chemoshere*, **14**, 1127−1131.
Leshchenko, P. D. *et al.* (1972) *Vopr. Pitaniya*, No. **1**, 81−85 (in Russian).
Linnik, P. N. & Nabivanets, B. I. (1986) *Formy migratsii metallov v presnykh poverkhnostnykh vodakh* [Migration of Metals in Fresh Surface Waters]. Gidrometeoizdat, Leningrad (in Russian).
Liublina, E. I. & Dvorkin, E. A. (1983) In: *Gigiyenicheskaya toksicologiya metallov* [Hygienic Toxicology of Metals]. Meditsina, Moscow, pp. 25−29 (in Russian).
Loseva, A. F. *et al.* (1984) *Spektalnye metody opredeleniya mikroelementov v obyektakh biosfery* [Spectral Methods for Measuring Trace Elements in the Biosphere]. Publication of Rostov University, Rostov (in Russian).
Lurie, Yu. Yu. (1984) *Analiticheskaya khimiya promyshlennykh stochnykh vod* [Analytical Chemistry of Industrial Waste Waters]. Khimiya, Moscow (in Russian).

Luft, F. & Ganten, D. (1985) *Klin. Wschr.*, 63, 788–792.

Metelev, V. V. *et al.* (1971) *Vodnaya toksikologiya* [Aquatic Toxicology]. Kolos, Moscow (in Russian).

Mogosh, G. (1984) *Ostrye otravleniya* [Acute Poisonings] (Russian translation from Roumanian published in Bucharest).

Nishimura, H. & Miyamoto, S. (1969) *Acta Anat.* (Basel), 74, 121–124.

Omelyanets, N. I. *et al.* (1984) In: *Gigiyena naselennykh mest* [Hygiene in Populated Areas]. Kiev, No. 23, pp. 72–77 (in Russian).

Ostromogilsky, A. Kh. *et al.* (1981) *Mikroelementy v atmosfere fonovykh rayonov sushi i okeana* [Trace Elements in Background Land and Sea Areas]. Obninsk (in Russian).

Poluektov, N. S. (1967) *Metody analiza po fotometrii plamenem* [Methods of Analysis by Flame Photometry]. Moscow (in Russian).

Specifications for Methods of Determining Noxious Substances in Air (1974) [Tekhnicheskiye usloviya na metody opredeleniya vrednykh veshchestv v vozdukhe], No. 10. Publication of the Ministry of the Navy, Moscow (in Russian).

Tarakhovsky, M. L. *et al.* (1982) *Lecheniye ostrykh otravleniy* [Treatment of Acute Intoxications]. Zdorovye, Kiev (in Russian).

Vasilenko, V. N. *et al.* (1985) *Monitoring zagryazneniya snezhnogo pokrova* [Monitoring the Pollution of the Snow Cover]. Gidrometeoizdat, Leningrad (in Russian).

Venchikov, A. I. (1985) *Printsipy lechebnogo primeneniya mikroelementov v kachestve biotikov* [Principles for the Therapeutic Use of Trace Elements as Biotics]. Ylym, Ashkhabad (in Russian).

Vilner, T. A. (1972) *Gig. San.*, No. 9, 71–72 (in Russian).

Vinogradov, A. K. & Beletsky, V. I. (1985) *Gidrobiologicheskiy Zh.*, No. 5, 44–48 (in Russian).

Zoller, W. H. *et al.* (1974) *Science*, 183, 198–200.

Potassium and its compounds

Potassium bromide; **p. carbonate** (potash); **p. chloride** (sylvinite [min.]);
p. fluoride; **p. hydroxide** (caustic potash, caustic potash lye, potassium hydrate);
p. iodide; **p. nitrate** (niter, saltpeter); **p. orthophosphate**; **p. peroxodisulfate**
(p. persulfate); **p.–magnesium sulfate hexahydrate** (schoenite [min.]); **p. sulfate**
(arcanite [min.]).

IDENTITY AND PHYSICOCHEMICAL PROPERTIES OF THE ELEMENT

Potassium (K) is an alkali metal in group I of the periodic table, with the atomic number 19. Its natural isotopes are ^{39}K (93.08%), ^{41}K (6.91%), and ^{40}K (0.01%), the latter being radioactive (half life $= 1.32 \times 10^9$ years).

Potassium is a soft metal that can be easily cut with a knife and can be molded and rolled in the cold state; it retains its plasticity at low temperatures.

In compounds, potassium is present in the oxidation state +1. It is highly active chemically. It rapidly oxidizes in air to the oxide K_2O and peroxide K_2O_2; it ignites when heated. At 200°C it reacts with hydrogen to form the hydride KH. It reacts vigorously with halogens to form the halides. On heating it combines with sulfur, selenium, tellurium, and carbon. It forms a variety of compounds with metals and reacts as a reducing agent with many compounds, reducing the oxides of nitrogen, carbon, boron, aluminum, silicon, mercury, silver, nickel, lead, cobalt, tin, copper, chromium, bismuth, and titanium. In reacting with water and acids, it forms salts and displaces hydrogen. See also Appendix.

NATURAL OCCURRENCE AND ENVIRONMENTAL LEVELS

The major potassium minerals are sylvine (KCl, 52.4% K), carnallite (KCL $\cdot$ MnCl$_2$ $\cdot$ 6H$_2$O, 35.8% K), kainite (KMg(SO$_4$)Cl $\cdot$ 3H$_2$O, 14% K), and langbeinite (K$_2$Mg$_2$(SO$_4$)$_3$, 18.8% K). Potassium is present in feldspars and micas.

Average abundances of potassium are 2.5% in the earth's crust, 1.36% in soils (1.1% in sandstones; 2.28% in shales and clays), 0.038% in seawater, 1.5 $\times$ 10^{-4}% in river water, 0.3% in plants, and 0.27% in animals.

In seawater, the principal species of potassium is K$^+$ and its residence time is 7 $\times$ 10^6 years; its average abundance in ferromanganese oxide deposits of the world's ocean is 0.64% [Henderson].

Like sodium, calcium, and magnesium, potassium is a major ion (macrocomponent) of natural waters and forms compounds readily soluble in aqueous media. Its principal natural sources in fresh surface waters are igneous rocks and their chemical degradation products. Potassium is readily released from rocks by weathering processes and enters water bodies in large quantities as a result of the dissolution of its compounds containing carbon, chlorine, or sulfur. In fresh waters, potassium is taken up by organisms and absorbed by clays; in the oceans, it is also taken up by organisms and is absorbed by bottom muds. Because of their good solubility, many naturally occurring compounds of macrocomponents migrate in natural surface waters predominantly in the form of dissolved ions, whereas the bulk of potassium is present in suspended matter. In natural waters, 80% of K$^+$ ions occur as free uncomplexed ions, but they greatly decrease in amount in the presence of organic ligands. Potassium, like sodium, is poorly hydrolyzed in such waters. Potassium migration on the earth's surface is slight [Linnik & Nabivanets].

In bacterial biomass, potassium levels up to 11.5 g/100 g dry weight have been reported; its reported percentage contents in plants are as follows (average ash weight values): algae, 5–18.2; mosses, 8.0; lichens, up to 10.0; horsetails, 11.2; fungi, 28.4; ferns, 35.4 [Lukashev et al.; Vadkovskaya & Lukashev; Venchikov].

Potassium concentrations ranged from 0.14 to 0.5 ng/m^3 (mean, 0.3 ng/m^3) in air samples taken at the South Pole [Zoller et al.], 0.4–1.0 mg/L in snow water samples from the environs of Moscow [Vasilenko et al.], and 0.031–0.319 mg/L in those from the South and North Poles [Karlsson et al.].

In animals and humans, potassium plays major roles in the generation of bioelectric potentials, in the maintenance of osmotic pressure, in carbohydrate metabolism, and in protein synthesis.

PRODUCTION

Potassium is commonly obtained by exchange reaction of metallic sodium with potassium hydroxide at 380–440°C or with potassium chloride at 760–800°C.

Potassium carbonate is produced by saturating a potassium hydroxide solution or magnesium carbonate suspension with carbon dioxide in a potassium chloride solution, or as a by-product in the processing of nepheline to obtain alumina. **P. chloride** is extracted from sylvinite or carnallite by treatment with water or lye. **P. fluoride** is ·made principally by reaction of p. carbonate with a stoichiometric amount of hydrofluoric acid, by fusing fluorite with p. carbonate or p. hydroxide, or by decomposition of p. hexafluorosilicate through heating with p. carbonate. **P. hydroxide** is made by electrolysis of p. chloride solutions. **P.-magnesium sulfate** is produced in the processing of kainite–langbeinite ore. **P. nitrate** is produced by exchange reaction between sodium nitrate and p. chloride or by the action of nitric acid or nitrogen oxides on p. carbonate or p. chloride. **P. orthophosphate** is made by neutralization of orthophosphoric acid with p. hydroxide. **P. sulfate** is produced by exchange reaction between p. chloride and magnesium sulfate or sulfuric acid or by calcination of langbeinite with carbon.

USES

Potassium is chiefly used to obtain its peroxide for use in oxygen regeneration; as a catalyst in the manufacture of some synthetic rubbers; for drying gases and making them free of oxygen; and, when alloyed with sodium, as a heat-transfer agent in nuclear reactors, and as a reducer in the production of titanium.

Potassium carbonate is used in the glass and soap industries and as a starting material in the production of other potassium compounds, including potash fertilizers. **P. chloride** finds use as a fertilizer and a raw material for the production of p. hydroxide and p. salts. **P. fluoride** is used as a salt additive to cryolite in the preparation of aluminum by electrolysis; as an electrolyte in the production of fluorine; in the manufacture of wallboard compounds and of fluxes for soldering and welding; and as a fluorinating agent in organic synthesis. **P. hydroxide** is employed in making liquid soaps and in the production of potassium salts. **P.-magnesium sulfate** is used as a fertilizer. **P. nitrate** is used in pyrotechnics, in curing meat, in making glasses, and as a fertilizer. **P. orthophosphate** serves as an absorbent of hydrogen sulfide and industrial gases. **P. sulfate** is used as a fertilizer and in the manufacture of alums and other salts.

MAN-MADE SOURCES OF EMISSION INTO THE ENVIRONMENT

An important anthropogenic source of potassium in soil may be solid emissions from metallurgical works. In one study, the amount of potassium entering the soil with such emissions within a distance of 1 to 3 km from an iron and steel plant was found to be between 4.3 and 11.8 mg/kg of soil per year (the control soil contained 0.8 mg/kg) [Garmash]. Another substantial source of potassium deposition on land may be exhausts from motor vehicles. For example, 10 to 30 mg of potassium was deposited per m^2 of land area per month within 50 m of a highway with dense traffic [Berinia].

As reported by Shipilov, plants processing potassium–magnesium ores each produced tens of thousands of cubic meters of salt residues and clayey slurries that contained, among other substances, 11 to 19% of potassium chloride. In the production of this compound, 45 to 94 kg of spent water were found to be discharged together with other wastes from the plants per tonne of processed ore. In that author's view, such water, by percolating along with melt and storm waters through the accumulated wastes, can readily reach both artesian waters and open water bodies (such as lakes or rivers) together with the alkaline elements they have entrained during the percolation process. River water samples collected at various distances downstream from potash factories contained potassium in amounts running into hundreds or thousands of mg/L, whereas those collected upstream contained, on average, only 3 mg/L.

As found by Barymova & Chernyshov, industrial effluents, notably those from factories manufacturing heat-resistant magnetic tapes, can add substantial quantities of potassium sulfate to water bodies used as sources of municipal and domestic water supplies. These authors also showed that decomposing municipal and domestic wastes together with the water flowing from asphalted roads were responsible for the high concentrations of potassium present in surface run-offs from urban territories with inadequate waste-disposal facilities and heavy motor traffic.

Measurements of potassium wash-out from lands carried out over a number of years in the central wooded steppe zone of the USSR gave the following average annual ranges (in kg/km^2): 69–104 from mowed or grazed (unplowed) lands, 62–485 from fall-plowed or fall-sown lands, and 171–1590 from urban lands occupied by old or new apartment buildings. Radish and carrot plants harvested in an area irrigated with industrial and domestic wastewaters at a rate of 300 tonnes per hectare had average potassium concentrations of 277.2 and 263.7 mg%, respectively, in their roots, as compared with 226.8 and 173.2 mg% in plants from a control area [Barymova & Chernyshov].

TOXICITY

Aquatic organisms

Potassium chloride produced signs of toxicity in perch at a concentration of 5 g/L after 1.5 h of exposure; at 10 g/L it was toxic to perch and whitefish on 18-hour exposure. The nitrate at 5.0 g/L and sulfate at 4.3 g/L were lethal to various (unspecified) fish after 15 to 27-h exposure, but appeared harmless at 0.9 and 1.3 g/L, respectively, on 50-hour exposure, while the **orthophosphate** was toxic at 1.0 g/L after 24 h of exposure. Crustaceans exposed to potassium at 53 mg/L showed reduced reproductive capacity [Metelev *et al.*].

Potassium bromide at 7 g/L and **potassium iodide** at 5 g/L each caused 100% mortality in rainbow trout by day 4 of exposure. The poisoned fish did not exhibit any signs of excitation, poorly responded to tactile stimuli, lay in a state of narcosis on the bottom of the aquarium, and were dying imperceptibly. The LD_{50} of the bromide and iodide for juvenile rainbow trout was 3.2 g/L on 5-day exposure at 6–9°C [Goreva]. In a 90-day experiment, the bromide at 12.5 mg/L and the iodide at 25 mg/L were found by that author not to have adverse effects on rainbow trout as judged by survival rates, body-weight gains, and hematologic parameters. Nor did these compounds impart any extraneous odor or taste to the flesh of these fish (either raw or boiled) in concentrations up to 1600 mg/L. Rainbow trout exposed to their toxic concentrations exhibited hyperchromic anemia, abnormally large erythrocytes, altered metabolic rates, and reduced bodily responsiveness.

Animals and man

Acute exposure. For rats, the oral LD of **potassium fluoride** was 245 mg/kg, and that of the **hydroxide**, 365 mg/kg [Bergqvist]. With intraperitoneal administration, the LD_{50} of **potassium chloride** was 770 mg/kg for rats and 820 mg/kg for mice.

In man, the manifestations of acute poisoning by **potassium hydroxide** are essentially the same as those described for sodium hydroxide (q.v.). The mechanism of the resultant necrotic process involves the action of OH^- ions (produced through dissociation of KOH or NaOH molecules), the formation of soluble alkaline proteinates, and the denaturation of protein molecules. The lethal oral dose of potassium (as well as sodium) hydroxide for man is between 10 and 20 mg; the concentration of 3 mg/m^3 was recommended as the permissible inhalation exposure level [Mogosh].

Potassium carbonate

Animals

In an acute toxicity study, the oral LD_{50}s for mice and rats were 2570 ± 142 mg/kg and 2980 ± 140 mg/kg, respectively, while the threshold of acute toxicity for rats on inhalation exposure was 54 mg/m^3 as calculated from changes in the summated threshold index, rectal temperature, and respiratory rate [Babayan et al.].

In rats chronically exposed to potassium carbonate aerosol in a concentration of about 40 mg/m^3, leukocytosis, increased erythrocyte counts, elevated levels of creatinine in urine and of sodium in plasma, reduced activity of total serum lactate dehydrogenase, hematuria, and elevated sodium and potassium levels in urine were recorded, as were bradycardia, increased minute volume of blood, and several electrocardiographic and rheographic changes. The concentration of 4.7 mg/m^3 was taken as the threshold level. The coefficient of cumulation was calculated to be 8.32. The irritant effect of potassium carbonate on the conjunctiva was described as slight. Neither dermal absorption nor local dermal effects were observed [Babayan et al.].

Chlorine-free potassic fertilizers

The most important compounds in this group, widely used in agriculture as fertilizers, are **potassium sulfate** and **potassium–magnesium sulfate**. In fertilizer factories, high concentrations of their airborne dust may be generated during various operations (drying, granulation, weighing, bagging, transportation) and also in premises where finished products are stored. In one factory, for instance, dust levels up to 470 mg/m^3 were recorded. Such high levels may be due to discontinuity of the production process, inadequate enclosure of the apparatus, conveyors, or loading/unloading appliances, the lack of an effective exhaust ventilation system in the drying department, and the difficulty of applying wet methods of dust control because the dust is highly hygroscopic.

Airborne dusts of potassium sulfate and potassium–magnesium sulfate consist of fairly small particles (60% are less than 4.4 μm in diameter) that can readily penetrate deep into the lungs and may lead to systemic intoxication given the good solubility of these compounds.

Acute toxicity studies in animals

For mice, rats, and guinea-pigs, the LD_{50} of potassium sulfate was about 6.6 g/kg body weight with oral administration by gavage and 1.1–1.25 g/kg with intraperitoneal injection; the respective LD_{50} values of potassium-magnesium sulfate were similar. No sex or age differences, nor appreciable species differences, in the sensitivity to these compounds were detected. After a lethal oral or intraperitoneal dose, a short period of excitation was succeeded by general inhibition, muscular weakness and atony, motor incoordination, convulsions, and respiratory distress; deaths occurred within 2 h. Post-mortem examination showed that the poisoning had upset the ionic/electrolyte balance in the interior organs. With inhalation exposure to potassium sulfate or potassium-magnesium sulfate, the thresholds for acute toxicity were at the level of 200 mg/m^3, while those for irritant effects on the respiratory tract (as estimated from changes in respiratory frequency and in cellular responses of the lungs and upper airways) ranged between 60 and 70 mg/m^3. The cumulative potential of these compounds was described as very low. In rabbits and guinea-pigs, they both were only slightly irritant to the skin and mucous membranes. No absorption through the skin was reported [Gnatiuk].

Chronic toxicity studies in animals
When rats were chronically exposed by inhalation to potassium sulfate at 5.5 and 15.0 mg/m^3 or to potassium–magnesium sulfate in the same concentrations (4 h/day, 5 days/week, for 4 months), both compounds were found to be polytropic in their action, especially at the higher concentration. The effects included increased motor activity; leukocytosis (in the case of potassium–magnesium sulfate); changes in the absorptive capacity of the liver, kidneys, and spleen; decreases in the relative weights of these organs and of the adrenals; an initial increase in total dehydrogenase activity followed by its subsequent decrease; augmented cardiac automatism; elevated levels of lipids and hydroxyproline in the lung tissue; disordered carbohydrate, lipid, and protein metabolism; and altered permeability of the histohematic barriers. In animals exposed to the lower (5.5 mg/m^3) concentration of either compound, these manifestations of poisoning were decreasing or disappearing in the course of the 4-month exposure period to be indetectable thereafter; the observed changes were therefore interpreted as adaptive responses. With the higher (15 mg/m^3) concentration, the abnormalities were long-lasting, some of them (such as decreased total dehydrogenase activity) progressed with time, and some persisted during the post-exposure observation period. Irritation of the upper airways and lungs and alteration in the immune status of the animals were also noted [Gnatiuk].

   In another chronic study reported by the same author, potassium sulfate administered to rats orally by gavage at 620 mg/kg body weight produced predominantly cardiovascular effects, notably blood pressure falls, electrocardiographic changes, bradycardia, and arrhythmia. Histropathologic examination demonstrated cardiac hypertrophy, structural damage to cardiac muscle cells (manifested in vacuolar and granular dystrophy and in necrosis of individual cells or of their groups), thickened vessel walls, connective-tissue proliferation, and clear evidence of impaired myocardial circulation. Biochemical changes in the myocardium included reductions in creatine phosphatase and succinic dehydrogenase activities and in RNA and glycogen levels together with increases in the activities of intracellular enzymes such as alkaline and acid phosphatases, cytochrome oxidase, and transaminases. In rats concurrently exposed to potassium sulfate *and* high temperature, functional and structural changes similar to those listed above occurred with a much lower dose of this compound (124 mg/kg).

Effects of occupational exposure in man
Health examinations of workers employed in the manufacture of chlorine-free potassic fertilizers showed substantially increased overall morbidity rates in terms of both the number of sick individuals and the number of days off for sickness, the rates being higher by factors of 1.36 and 1.6, respectively, than in a group of matched controls. Among workers exposed to the dust of these fertilizers, similarly increased rates of morbidity due to respiratory disease were recorded. A rising trend in the incidence of diseases, including those of the respiratory tract, was observed during the 4 year study period. The authors concluded that potassic fertilizers may impair the health of workers by acting on the respiratory tract directly as well as by reducing the bodily resistance to disease in general [Khodykina; Rumiantsev *et al.*]

Potassium persulfate

Oral toxicity studies in animals
The oral LD_{50} of potassium persulfate was reported to be 1320 ± 150 mg/kg for mice and 925 ± 100 mg/kg for rats. Autopsy demonstrated vascular congestion in the brain, liver, and spleen [Baturina *et al.*].

Daily oral administration of potassium persulfate to rats in a dose equal to 1/500 of the LD_{50} over a period of 1.5 months led to decreases in hemoglobin levels, erythrocyte counts, and creatine phosphokinase activity in the blood, and to an initial rise in serum catalase and peroxidase levels, followed by a marked decline. Histopathologic examination showed the presence of moderately hypertrophic hepatocytes, hypertrophic or hyperplastic microphagocytes, and dilated veins in the liver; swollen and pyknotic nuclei in Malpighian corpuscles in the kidney; and hypertrophic lymphoid follicles in the spleen. On histochemical study, intensified nucleic acid metabolism and elevated succinic dehydrogenase (SDH) and monoamine oxidase (MAO) activities were recorded in the thyroid, indicating, in the authors' view, enhanced activity of this gland and, hence, a high affinity of the compound for it. In rats dosed with potassium persulfate at 1/100 or 1/50 of the LD_{50} over the same period as above, SDH activity was inhibited in the liver, while in the kidney this activity as well as that of lactate dehydrogenase (LDH) were elevated in the proximal and distal convoluted tubules and depressed in the renal corpuscles and medullary substance; in addition, reduced LDH, SDH, and MAO activities occurred in the spleen and brain. Increased embryonal mortality was also observed in rats given the persulfate at 1/100 of the LD_{50} value [Baturina *et al.*].

Dermal effects in animals and man

Potassium persulfate caused dermatitis when applied to the intact skin of rabbits and guinea-pigs. In workers handling this compound, skin cracks and flaking that disappeared after the cessation of contact were observed on the hands and arms [Baturina *et al.*].

ABSORPTION, DISTRIBUTION, AND ELIMINATION

Potassium enters the body in food and water. It is the major intracellular cation. [Birch]. It has been estimated that the adult human body contains 4000 to 9000 mEq, or 160 to 250 g, of this element, of which only about 2% is present in extracellular fluids (interstitial fluid and blood plasma). The daily requirement is 2–3 g in an adult and 12–16 mg/kg body-weight in a child. The distribution in the body is as follows (in mEq/kg, average values): bones, 15; teeth, 17; muscles, 100; heart, 64; lungs, 38; brain, 84; liver, 55; kidneys, 45; erythrocytes, 150; blood serum, 4.5; cerebrospinal fluid, 2.3; lymph, 2.2. Potassium turnover in the body occurs at extremely high rates (per minute, 3.3–4% of the potassium was found to be exchanged in the brain and 8–10.7% in the retina). Elimination from the body is mainly by way of the urine (93–95%) [Krokhalev].

According to Bergqvist, the average adult human body contains 140 g of potassium, with 62% of this amount being in the muscles and 11% in the skeleton.

As suggested by experiments on rats [Brooks *et al.*], a high dietary intake of potassium may result in its increased plasma concentration, an elevated plasma level of vasopressin, and a consequent rise in systolic blood pressure, with concomitant increases in urine output and vasopressin excretion in the urine.

HYGIENIC STANDARDS

Exposure limits adopted for potassium compounds in the USSR are set out in Table 1.

METHODS OF DETERMINATION

Potassium hydroxide (caustic potash) present as aerosol in the air can be determined by a densitometric method based on the capacity of acid-base indicators to change color according to the pH of the medium (sensitivity 4 μg in the assayed volume) [*Specifications for Methods of Determining Noxious Substances in Air*]. For measuring potassium in soil, aqueous media, and biologic material, flame photometry can be used [Poluektov]. In wastewaters it can be measured by flame emission spectrometry (sensitivity 0.1 mg/L) or by titrimetry [Lurie] and in natural waters, by ion-exchange chromatography (detection limit

0.1 mg/L) [Basta & Tabatabai]. For its determination in blood plasma and serum, a method using ion-selective electrodes has been proposed [Dobrolyubova], as has been a flame-photometric technique for measuring it in plant samples during mass screening surveys [Sudakov et al.].

Table 1. Exposure limits for potassium compounds

	Workroom air		Atmospheric air	
	MAC_{wz} (mg/m^3)	Aggregative state	$TSEL_{aa}$ (mg/m^3)	Hazard class
Potassium carbonate	2.0	Aerosol	0.05	3
chloride	5.0	Aerosol	0.1	3
hydroxide	2.0*	—	—	—
nitrate	5.0	Aerosol	—	3
silicofluoride (as F)	0.2	Vapor/Aerosol	—	2
sulfate	10.0	Aerosol	—	3
Potassium–magnesium sulfate	5.0	Aerosol	—	3

* This is also the threshold limit value (TWA) adopted in the USA.

MEASURES TO CONTROL EXPOSURE

At the workplace
The control measures recommended for sodium and its compounds also apply to potassium.

In factories manufacturing chlorine-free potassic fertilizers, special attention should be paid to dust control. The most radical technical measures include the use of hermetically sealed apparatus, conveyors, and loading/unloading devices, the installation of an integrated exhaust system in the drying departments, and the mechanization of operations involved in the transportation, weighing, and bagging of raw materials and finished products.

Appropriate *personal protection* devices (respirators, goggles, clothing, footwear, gloves) should be used by workers exposed to potassium and its compounds.

Protection of the general environment
Where potassium compounds are used as fertilizers, their accumulation in the soil should be monitored so that the application rates established for specific crops in particular climatic and geographic zones are not exceeded.

FIRST AID IN POISONING

First aid measures to be taken in poisoning with potassium hydroxide are the same as those described for sodium hydroxide.

REFERENCES

Babayan, E. A. *et al.* (1989) *Gig. Truda Prof. Zabol.*, No. **5**, 30–32 (in Russian).

Barymova, N. A. & Chernyshov, E. P. (1982) In: *Vzaimodeistvie khozyaistva i prirody v gorodskikh promyshlennykh geotekhsistemakh* [Interaction of Economic Activity and Nature in Urban Industrial Geotechnic Systems]. Moscow, pp. 31–45 (in Russian).

Basta, N. T. & Tabatabai, M. A. (1985) *J. Environ. Qual.*, **14**, 450–455.

Baturina, T. S. *et al.* (1984) In: *Gigiyena naselennykh mest* [Hygiene in Populated Areas]. Kiev, No. 23, pp. 32–35 (in Russian).

Bergqvist, U. (1983) *New metals.* University of Stockholm (Institute of Physics).

Berinia, D. Zh. *et al.* (1980) In: *Zagryazneniye prirodnoi sredy vybrosami avtotransporta* [Environmental Pollution by Automobile Exhausts]. Riga, pp. 16–27 (in Russian).

Birch, N. J. & Karim, A. R. (1988) Chapter 46 *Potassium* in: Seiler, H. *et al.* (eds) *Handbook on Toxicity of Inorganic Compounds.* Marcel Dekker, New York, pp. 543–547.

Brooks, D. P. *et al.* (1986) *Experientia*, **42**, 1012–1014.

Dobrolyubova, B. A. (1984) *Gig. San.*, No. **8**, 68–69 (in Russian).

Garmash, G. A. (1983) In: *Khimiya v selskom khoziaystve* [Chemistry in Agriculture]. No. 10, pp. 45–48 (in Russian).

Gnatiuk, M. S. (1989) In: *Gigiyena okruzhaiushchei sredy. Tezisy dokladov respublikanskoi nauchnoi konferentssi* [Environmental Hygiene. Abstracts of Papers Presented to a Workshop]. Kiev, pp. 68–69 (in Russian).

Goreva, V. A. (1984) *Gidrobiologicheskiy Zh.*, No. **5**, 50–54 (in Russian).

Karlsson, V. *et al.* (1985) *Chemosphere*, **14**, 1127–1131.

Khodykina, T. M. (1984) *Gig. Truda Prof. Zabol.*, No. **7**, 47–49 (in Russian).

Krokhalev, A. A. (1972) *Vodnyi i elektrolitnyi obmen (Ostrye rasstroistva)* [Acute Disturbances of Water and Electrolyte Metabolism]. Moscow (in Russian).

Linnik, P. N. & Nabivanets, B. I. (1986) *Formy migratsii metallov v presnykh poverkhnostnykh vodakh* [Migration of Metals in Fresh Surface Waters]. Gidrometeoizdat, Leningrad (in Russian).

Lukashev, K. I. *et al.* (1984) *Chelovek i priroda: Geokhimicheskiye i ekologicheskiye aspekty ratsionalnogo prirodopolzovaniya* [Man and Nature: Geochemical and Ecological Aspects of Environmental Management]. Nauka i Tekhnika, Minsk (in Russian).

Lurie, Yu. Yu. (1984) *Analiticheskaya khimiya promyshlennykh stochnykh vod* [Analytical Chemistry of Industrial Waste Waters]. Khimiya, Moscow (in Russian).

Metelev, V. V. *et al.* (1971) *Vodnaya toksikologiya* [Aquatic Toxicology]. Kolos, Moscow (in Russian).

Mogosh, G. (1984) *Ostrye otravleniya* [Acute Poisonings]. (Russian translation from Roumanian published in Bucharest).

Poluektov, N. S. (1967) *Metody analiza po fotometrii plameni* [Methods of Analysis Using Flame Photometry]. Khimiya, Moscow (in Russian).

Rumiantsev, G. I. *et al.* (1985) *Gig. San.*, No. **8**, 14–18 (in Russian).

Shipilov, A. L. (1980) *Gig. San.*, No. **12**, 63–64 (in Russian).

Specifications for methods of determining noxious substances in air (1974) [Tekhnicheskiye usloviya na metody opredeleniya vrednykh veshchestv v vozdukhe], No. 10, Publication of the Ministry of the Navy, Moscow (in Russian).

Sudakov, V. D. *et al.* (1985) In: *Khimiya v selskom khoziaistve* [Chemistry in Agriculture]. No. 2, pp. 66–67 (in Russian).

Vadkovskaya, I. K. & Lukashev, K. P. (1977) *Geokhimicheskie osnovy okhrany biosfery* [Protection of the Biosphere: Geochemical Principles]. Nauka i Tekhnika, Minsk (in Russian).

Vasilenko, V. N. *et al.* (1985) *Monitoring zagryazneniya snezhnogo pokrova* [Monitoring the Pollution of the Snow Cover]. Gidrometeoizdat, Leningrad (in Russian).

Venchikov, A. I. (1982) *Printsipy lechebnogo primeneniya mikroelementov v kachestve biotikov* [Principles for the Therapeutic Use of Trace Elements as Biotics]. Ylym, Ashkhabad (in Russian).

Zoller, W. H. *et al.* (1974) *Science*, **183**, 198−200.

Rubidium and its compounds

Rubidium chloride; r. hydroxide; r. iodide; r. orthoarsenate; r. sulfate

IDENTITY AND PHYSICOCHEMICAL PROPERTIES
OF THE ELEMENT

Rubidium (Rb) is an alkali metal in group I of the periodic table, with the atomic number 37. It has two naturally occurring isotopes: ^{85}Rb (72.15%) and the β-radioactive ^{87}Rb (27.85%) with a half-life of 5×10^{10} years.

Rubidium is a soft viscous metal with a pasty consistency at ordinary temperatures. It is highly reactive. Its oxidation state in compounds is +1. It ignites is reacting with oxygen to form the peroxide Rb_2O_2 and superoxide RbO_2 and with halogens to form the halides. Its reaction with water gives the hydroxide RbOH with the evolution of hydrogen which inflames. It reacts explosively with sulfur to form the sulfide Rb_2S and with phosphorus to form the phosphide RbP_5. It gives the hydride RbH with hydrogen at high temperatures and pressures and the nitride Rb_3N with nitrogen in a field of silent electric charge. It forms the amide $RbNH_2$ with liquid ammonia, the carbide Rb_2C_2 with carbon, and the silicide RbSi with silicon. At temperatures above 300°C it destroys glass, reducing silicon from the SiO_2 and silicates. It reacts violently with acids, forming the respective salts and releasing hydrogen. It forms alloys with other alkali metals and alkaline-earth metals as well as with mercury, antimony, bismuth, and gold. See also Appendix.

NATURAL OCCURRENCE AND ENVIRONMENTAL LEVELS

Rubidium does not form minerals of its own. It is found as an isomorphic impurity in minerals of potassium and cesium and is contained in granitoid rocks and pegmatites. Its salts are present in many minerals. The clarke of rubidium is estimated at $(90-150) \times 10^{-4}\%$ in the earth's crust and at $180 \times 10^{-4}\%$ in the granite layer of the continental crust [Dobrovolsky].

In seawater, rubidium occurs at an average concentration of 0.12 mg/L, mainly as Rb^+. The oceans contain an estimated 41.1 million tonnes of the metal, and its residence time there is estimated to be between 4 and 5 million years. Accumulations of nonbiogenic particles (clay muds) in the oceans contain rubidium at an average level of $110 \times 10^{-4}\%$, while those of biogenic particles (carbonate muds) contain $10 \times 10^{-4}\%$. The common range of its concentrations in surface river waters is 0.6–1.1 µg/L. In soils, rubidium levels range from 5×10^{-3} to $1 \times 10^{-2}\%$ (averaging $2.7 \times 10^{-2}\%$ in sandstones and $2.0 \times 10^{-2}\%$ in shales and clays). Its estimated total content in the continental vegetation is 12.5 million tonnes, the average levels in plants being $(2-5) \times 10^{-4}\%$ on a fresh weight basis and $5.0 \times 10^{-4}\%$ on a dry weight basis; its contents in marine algae varied from 0.61 to 2.4 ppm. In atmospheric precipitation, its concentrations average 0.15 µg/L in solution and 0.24 µg/L in suspension; the average amount deposited on land annually from the atmosphere was calculated to be 0.21 ng/m^2. Its contents in coal and crude oil averaged 14.4 mg/kg and 0.15 mg/kg, respectively [Dobrovolsky; Henderson; Lukashev et al.; Venchikov].

Rubidium contents in food and forage plants were as follows (mg/kg dry weight): cereal grains, 4; corn grains, 3; onion bulbs, 1; lettuce leaves, 14; cabbage leaves, 12; bean pods, 51; soya-bean seeds, 220; apple fruits, 50; avocado fruits, 20; clover tops, 44; alfalfa tops, 98; forage grass tops, 130 [Kabata-Pendias & Pendias].

PRODUCTION

Metallic rubidium is commonly produced by chemical reduction of rubidium chloride in vacuum at 700–800°C. Rubidium salts are often obtained as by-products in the making of lithium, magnesium, or potassium salts. The mining and production of rubidium compounds have considerably increased in recent years.

USES

Rubidium and its compounds are mainly used in cathodes for photocells, as additives to the gas medium of neon or argon lamps, in various special alloys, and as catalysts in organic syntheses; some compounds have uses in the manufacture of semiconductors, ferro- and piezoelectrics, in electron optics, and in electrical equipment.

Rubidium is also used in pharmacological and cytological studies [Hunter *et al.*; Marakhova *et al.*; Verennikov *et al.*], in the treatment of transitory ischemia, and in radiation medicine [Davie *et al.*; Shea *et al.*; Yano].

MAN-MADE SOURCES OF EMISSION INTO THE ENVIRONMENT

In factory premises where rubidium compounds were produced, particulate matter suspended in the air contained the metal and its salts in concentrations of up to 25.4 mg/m^3 as a result of their escape in water droplets or vapor from the surface of heated solutions. Much higher rubidium concentrations (up to 96.4 mg/m^3) were registered in the workroom air during the leaching, crystallization, and evaporation of its salts and especially when various pieces of the equipment were washed down.

Other sources of rubidium in the environment are coal dust (where rubidium content was reported to range 9–22 µg/kg [Manchuk & Ryabov]) and fly ash from coal-fired electric power stations. Clover plants from an area where fly ash from such a station was deposited, contained significantly higher amounts of rubidium than did those harvested in a control area, and guinea pigs fed the contaminated clover (which made up 45% of their diet) for 90 days had elevated rubidium levels in their kidneys, livers, and muscles as compared to control animals [Stoewsand *et al.*]. A potential source of environmental pollution with rubidium is automobile exhaust [Szymczyk *et al.*].

For monitoring seawater pollution by rubidium, mollusks such as mussels may be used, since they are good concentrators of various elements including rubidium. The bodies and shells of Black Sea mussels residing in uncontaminated media contained 0.47 to 19.13 mg Rb/kg [Morozova *et al.*].

TOXICITY

Rubidium chloride and **rubidium sulfate** at concentrations ≤1 mg/L and **rubidium hydroxide** at concentrations ≤0.05 mg/L did not alter the odor, color, or turbidity of water. Rubidium salts may decrease the biochemical oxygen demand (BOD). The highest concentration without effect on the BOD in water was 0.1 mg/L for the chloride and sulfate and 0.05 mg/L for the hydroxide. These three compounds were stable in water if present in low concentrations; in dissolved form, the chloride and sulfate were found to persist 15 days in concentrations of ≤5 mg/L and the hydroxide, in concentrations of ≤1 mg/L [Romantsova *et al.*].

Aquatic organisms

The lowest concentration of **rubidium chloride** toxic to *Scenedesmus* algae and *Microregma* protozoans was 14 mg/L [Grushko].

Animals and man

Acute exposure

Animals

Lethal doses of three rubidium compounds for mice and rats are given in Table 1.

Table 1. Median lethal doses of rubidium compounds

Compound	Animal species	Route of administration	LD_{50} (mg/kg)
Chloride	Rat	Oral	4440
	Rat	Intraperitoneal	1200
	Mouse	Oral	3800
	Mouse	Intraperitoneal	1160
Hydroxide	Rat	Oral	586
	Rat	Oral	992
	Mouse	Oral	900
Iodide	Rat	Oral	4708

Animals lethally poisoned with **rubidium chloride** were lethargic and inhibited, crowded together in the cage, had ruffled coats, and breathed frequently; they died on the second day in the presence of tetanic convulsions. Poisoning with the **hydroxide** was first manifested in restlessness, followed by a state of lethargy, paresis of the hind limbs, and death. On pathologic examination, gross vascular abnormalities were present in the interior organs; the lungs were inflated and had hemorrhages in places, the liver was mace-colored and flabby, the gut was atonic, and the stomach was distended, had thin walls with necrotic areas, and appeared hyperemic and edematous in the pyloric region.

Man

Persons poisoned by **rubidium chloride** displayed hyperactivity, aggressiveness, and anxiety; acute poisoning may be fatal [Meltzer *et al.*; Stolk].

Repeated exposure

Animals

In the course of repeated oral administration of **rubidium chloride** in single doses of 44.4 mg/kg (a dose level equal to 1/100 of the LD_{50}), rats showed reductions in cholinesterase and lactate dehydrogenase activities and sulfhydryl levels in the blood, erythropenia, and leukocytosis along with increases in the summated threshold index [Romantsova et al.]. At the end of that 3-month study, reduced cholinesterase and succinate dehydrogenase activities and depressed sulfhydryl levels in hepatic tissue were   recorded. In addition, this compound displayed gonadotoxic activity, as was indicated by significant decreases in testicular size, the number of spermatozoa, their osmotic and acidic resistance, and the time of their motility. A mutagenic effect was also noted (increased number of abnormal mitoses in bone marrow cells). Histopathologic examination showed dystrophic and necrobiotic changes in the kidneys, spleen, myocardium, and brain.

Rubidium salts are highly cumulative substances. The cumulation coefficient has been determined at 1.7 for the **chloride** and at 2.2 for the **hydroxide**. The chloride has been reported to inhibit all components of exploratory behavior in animals [G. Syme & L. Syme].

Chronic exposure

Animals

When rats were exposed by inhalation to **rubidium hydroxide** at 10 mg/m^3 or to **rubidium chloride** at 30 mg/m^3 for 4 h daily up to 4 months, both compounds caused systemic intoxication, but the hydroxide, unlike the chloride, produced leukocytosis and did not adversely affect the protein-forming function of the liver [Khosid].

Rats exposed to **rubidium sulfate** through inhalation at 21.7 mg/m^3 for 4 h/day, 5 days/week over 4 months experienced disturbances in energy and cholesterol metabolism that persisted during 9 months of the recovery period [Akinfieva].

In rats ingesting **rubidium** in their drinking water at a level of 0.4 mg/kg body weight for a period on 7 months, elevated lactate dehydrogenase activity, increased sulfhydryls, and reduced cholinesterase activity were recorded in the blood along with a strongly marked tendency to increased sulfhydryl levels and succinate dehydrogenase activity in hepatic tissues. The element also displayed considerable neurotoxicity, manifested in general excitation and aggressiveness of the animals, and in persistently increased values of the summated threshold index. The main sites of its accumulation were found to be the liver, spleen, brain, and erythrocytes. Pathologic examination demonstrated chronic gastritis and some evidence of dystrophy in the liver and kidneys. In addition, embryo- and fetotoxic effects were observed, including increased total and preimplantation embryonal mortality, decreased placental size, reduced numbers of normally developing embryos (various developmental defects were noted), and decreases in the fetal survival rate and in the number of pups born alive per female. An increased frequency of chromosomal aberrations (polyploidy) in bone marrow cells was also recorded. The dose of 0.004 mg Rb/kg was described as the no-effect-level with regard to embryotoxicity, mutagenicity, and general toxicity [Romantsova et al.]. The authors proposed that the maximum permissible level of rubidium for water (MAC$_w$) be set at 0.1 mg/L on the basis of toxicological criteria of harmfulness, and that this element be assigned to hazard class 2.

Man

Workers aged 21 to 40 years who had been occupationally exposed to **rubidium** for periods of 2 to 10 years complained of increased excitability and sweating, rapid onset of fatigue, bad sleep, frequent headaches, numbness in the fingertips, and dyspepsia. On examination, the more common findings were a neurasthenic syndrome (in the presence of vasoneurosis), impaired renal and gastrointestinal functions, bradycardia, lowered arterial blood pressure, electrocardiographic abnormalities, and increased humidity and reduced electric resistance of the skin.

Effects of local application

Experiments with dermal application of various rubidium compounds to rabbits, rats, and guinea-pigs have shown [Akinfieva & Khamidulina] that some of them can produce not only severe local effects but also systemic intoxication by being absorbed through the skin. The application of **rubidium hydroxide** at 20 mg/cm^2 of skin resulted in the emergence of hemorrhagic areas, followed 30 to 40 min later by the onset of necrotic changes; after 4 h of exposure, a continuous area of gross necrosis with a dense crust had formed at the application site. Two to three weeks later, a coarse scar tissue was seen to have formed at that site. The hydroxide produced marked corrosive effects when applied to ocular mucous membranes of rabbits. Animals exposed to **rubidium sulfate** or **orthoarsenate** for 10 days through daily dermal application showed a slowing down of body-weight gains beginning with the 5th day; later, abnormal behavioral reactions were also noted (including the escape response), as were increases in the summated threshold index and in the blood content of sulfhydryl groups. The systemic effects of the sulfate were more pronounced. Sensitizing activity of **rubidium hydroxide** solutions was observed following intradermal administration.

ABSORPTION, DISTRIBUTION, AND ELIMINATION

Rubidium enters the human body in food or in inspired air. An adult human contains 320 to 650 mg of the element, with 32% of this amount being present in the bones and 26% in the muscles. Rubidium readily crosses the erythrocyte membrane. Its average blood levels have been estimated at 0.00032% in men and 0.00028% in women; most of the rubidium in the blood occurs combined with potassium in the intracellular fluids. In animals rubidium salts were absorbed from the gastrointestinal tract almost completely, and rubidium levels in the liver, muscle, and kidney ranged from 100 to 200 ppm. Augmented rubidium toxicity in the presence of potassium deficiency has been noted [Vassalo et al.; Whanger; Yano]. Detailed experimental data on gastric, biliary, and renal excretion of rubidium have been presented by Schäfer & Forth.

METHODS OF DETERMINATION

Some of the methods proposed for the determination of rubidium are the following. In *minerals*: flame photometry [Poluektov]; in *soil*: spectral analysis [Loseva et al.]; in *water*: flame photometry (sensitivity 0.008 mg/L) [Novikov et al.]; in *food*: flame atomic emission spectrophotometry (detection limit 0.19 mg/kg for a 10-g sample) [Ewans & Read]; in *human plasma* and *erythrocytes*: graphite-furnace atomic absorption spectrophotometry (sensitivity 29 nmol/L for plasma and 12 nmol/L for erythrocytes; detection limit 24 nmol/L for plasma and 4.8 nmol/L for erythrocytes) [Halls et al.]. The latter method has been reported to be approximately 30-fold more sensitive than previously described techniques involving atomic absorption spectrophotometry.

MEASURES TO CONTROL EXPOSURE AT THE WORKPLACE

All processes involving the production or use of rubidium or its compounds should be mechanized to the maximum possible extent. The crushing and grinding equipment should be remotely controlled. Safe working conditions can be assured by completely enclosing all sources of dust. All workers engaged in rubidium production or application must undergo health examinations. Arrangements should be made for medical treatment of respiratory abnormalities that may be caused by dust exposure. It is advisable that a prophylactorium be set up at the plant to offer health-promoting, preventive, and therapeutic procedures for the workers.

For *personal protection,* the use of respirators in work areas containing airborne dust is mandatory, as is the use of gas masks in those where chlorine may be emitted; goggles and gloves should be worn for eye and hand protection.

REFERENCES

Akinfieva, T. A. (1989) *Gig. Truda Prof. Zabol.*, No. 3, 16—20 (in Russian).
Akinfieva, T. A. & Khamidulina, Kh. Kh. (1986) *Gig. Truda Prof. Zabol.*, No. 3, 50—53 (in Russian).

Bergqvist, U. (1983) *New metals.* University of Stockholm (Institute of Physics).

Davie, R. J. *et al.* (1988) Chapter 50 *Rubidium* in: Seiler, H. *et al.* (eds) *Handbook on Toxicity of Inorganic Compounds.* Marcel Dekker, New York, pp. 567–570.

Dobrovolsky, V. V. (1983) *Geografiya mikroelementov. Globalnoe rasseyanie* [Geography of Trace Elements. Global Dispersion]. Mysl, Moscow (in Russian).

Ewans, W. H. & Read, J. I. (1985) *Analyst,* 110, 619–623.

Grushko, Ya. M. (1979) *Vrednye neorganicheskiye soyedineniya v promyshlennykh stochnykh vodakh* [Harmful Inorganic Compounds in Industrial Waste Waters]. Khimiya, Leningrad (in Russian).

Hallis, K. F. *et al.* (1985) *Clin. Chem.,* 31, 274–276.

Henderson, P. (1982) *Inorganic Geochemistry.* Pergamon Press, Oxford.

Hunter, D. D. *et al.* (1985) *Anal. Biochem.,* 149, 392–398.

Kabata-Pendias, A. & Pendias, H. (1986) *Trace Elements in Soils and Plants.* CRC Press, Boca Raton (Fl.).

Khosid, G. M. (1965) In: *Khimicheskiye factory vneshnei sredy i ikh gigiyenicheskoye znacheniye* [Environmental Chemicals and their Health Effects]. Moscow, pp. 39–41 (in Russian).

Loseva, A. F. *et al.* (1984) *Spectralnye metody opredeleniya mikroelementov v obyektakh biosfery* [Spectral Methods for Measuring Trace Elements in the Biosphere]. Publication of Rostov University, Rostov (in Russian).

Lukashev, K. I. *et al.* (1984) *Chelovek i priroda: Geokhimicheskiye i ekologicheskiye aspekty ratsionalnogo prirodopolzovaniya* [Man and Nature: Geochemical and Geological Aspects of Environmental Management]. Nauka i Tekhnika, Minsk (in Russian).

Manchuk, V. A. & Ryabov, N. A. (1983) In: *Metally. Gigiyenicheskiye aspekty otsenki i ozdorovleniya okruzhayushchei sredy* [Metals: Hygienic Aspects of Evaluation and Environmental Health Promotion]. Moscow, pp. 94–199 (in Russian).

Marakhova, I. I. *et al.* (1987) *Tsitologiya,* No. 2, 202–207 (in Russian).

Meltzer, H. L. *et al.* (1969) *Nature,* 223, 321–322.

Morozova, N. P. *et al.* (1984) In: *Biokhimicheskiye i toksikologicheskie issledovaniya zagriazneniya vodoyemov* [Biochemical and Toxicological Studies on Pollution of Water Bodies]. Moscow, pp. 21–28 (in Russian).

Novikov, Yu. V. *et al.* (1981) *Metody opredeleniya vrednykh veschestv v vode vodoyemov* [Methods for Measuring Noxious Substances in Water Bodies]. Meditsina, Moscow (in Russian).

Poluektov, N. S. (1967) *Metody analiza po fotometrii plameni* [Methods of Analysis Using Flame Photometry]. Khimiya, Moscow (in Russian).

Romantsova, O. A. *et al.* (1988) *Gig. San.,* No. 5, 76–78 (in Russian).

Schäfer, S. G. & Forth, W. (1980) In: *Mechanisms of Toxicity and Hazard Evaluation.* Elsevier, Amsterdam, pp. 547–550.

Shea, M. J. *et al.* (1987) *J. Nucl. Med.,* 28, 89–97.

Stoewsand, G. S. *et al.* (1976) In: *Proceedings of the 10th International Congress on Nutrition, Kyoto, 1975.* Kyoto, 1976, p. 724.

Stolk, J. M. (1970) *Science,* 168, 501–503.

Syme, G. Y. & Syme, L. A. (1979) *Psychopharmacology,* 61, 227–229.

Szymczyk, S. *et al.* (1981) *Nucl. Instrum. Meth,* 181, 281–284.

Vassalo, M. *et al.* (1985) *Life Sci.,* 37, 835–840.

Venchikov, A. I. (1982) *Printsipy lechebnogo primeneniya mikroelementov v kachestve biotikov* [Principles for the Therapeutic Use of Trace Elements as Biotics]. Ylym, Ashkhabad (in Russian).

Vereninov, A. A. *et al.* (1985) *Tsitologiya,* No. 2, 1359–1366 (in Russian).

Whanger, P. D. (1982) In: Hatchcock, J. N. (ed.) *Nutritional Toxicology.* Academic Press, New York, Vol. 1, pp. 163–208.

Yano, Y. (1987) *Int. J. Rad. Appl. Instrum.* (A), 38, 205–211.

Cesium and its compounds

Cesium arsenide; c. bromide; c. carbonate; c. chloride; c. dihydrogen
orthoarsenate; c. hydroxide; c. iodide; c. nitrate; c. sulfate

IDENTITY AND PHYSICOCHEMICAL PROPERTIES
OF THE ELEMENT

Cesium (Cs) is an alkali metal in group I of the periodic table, with the atomic
number 55. Naturally occurring cesium consists of one isotope, ^{133}Cs.

Cesium is ductile and very soft, and has a pasty consistency at ordinary
temperatures. It is highly reactive. Its oxidation state in compounds is +1. It
ignites in reacting with oxygen to form the peroxide Cs_2O_2 and the superoxide
CsO_2. It reacts violently with water (even in the ice form) with the evolution of
hydrogen (which ignites) and the formation of cesium hydroxide CsOH. It
reacts, igniting, with halogens to form the halides. It reacts explosively with
sulfur to form the sulfide Cs_2S and with phosphorus to give the phosphide
Cs_2P_5. It gives the hydride CsH with hydrogen at high temperatures and
pressures and the nitride Cs_3N with nitrogen in a field of silent electric charge.
It forms the amide $CsNH_2$ with liquid ammonia and the silicide CsSi with silicon
in an argon atmosphere at 600°C. At temperatures above 300°C it destroys glass,
reducing silicon from the SiO_2 and silicates. It reacts violently with acids,
forming the respective salts and releasing hydrogen. It forms alloys with other
alkali metals as well as with alkaline-earth metals and with mercury, gold,
antimony, and bismuth. See also Appendix.

NATURAL OCCURRENCE AND ENVIRONMENTAL LEVELS

Two rare minerals of cesium are known — pollucite $(Cs,Na)[AlSi_2O_6] \cdot nH_2O$ and avogardite $(C,Cs)BF_4$. Cesium occurs as an impurity in beryl, carnallite, and volcanic glass. Its clarke is estimated at $(3.0-3.7) \times 10^{-4}\%$ in the earth's crust and at $3.8 \times 10^{-4}\%$ in the granite layer of the continental crust [Dobrovolsky].

In seawater, cesium is contained at an average concentration of 4×10^{-4} mg/L, mainly as Cs^+, its residence time being 6×10^5 years [Henderson]. Its average concentrations in surface river waters are $0.03-0.04$ µg/L and reach 0.1 µg/L in the Upper Volga region where its atmospheric precipitation amounted to 0.36 kg/m^2 per year, the average concentration in precipitation being 0.04 µg/L. The continental vegetation contains an estimated 0.38 million tonnes of cesium, at an average concentration of $0.06 \times 10^{-4}\%$ on a fresh weight basis and $3 \times 10^{-4}\%$ on an ash weight basis. Its average concentrations in organisms are (in µg/g dry weight) $0.01-0.1$ in marine algae, $0.05-0.2$ in terrestrial plants, $0.067-0.503$ in arthropods, 0.04 in amphibians, and 0.05 in mammals. In soils, its levels average $5 \times 10^{-4}\%$ $(1.0 \times 10^{-3}$ in shales and clays, $7 \times 10^{-5}\%$ in sandstones) [Dobrovolsky; Vadkovskaya & Lukashev].

Tea leaves commonly contained cesium at levels of 0.5 to 1.0 mg/kg ash weight and desert plants, at 0.03 to 0.4 mg/kg dry weight (with a mean value of 0.1 mg/kg). Lettuce, barley, and flax accumulated the metal mainly in roots (highest value, 0.32 mg/kg dry weight) and in old leaves (highest value, 0.16 mg/kg dry weight). Its wet weight concentrations ranged from 0.2 to 3.3 µg/kg in vegetables and from 0.1 to 2.9 µg/kg in fruits [Kabata-Pendias & Pendias].

PRODUCTION

Cesium is obtained directly from pollucite by thermal reduction in vacuum or by concentration of pollucite ore followed by decomposition, precipitation, crystallization, and refining of the concentrate. It is also recovered by extraction and sorption techniques from underground waters obtained during petroleum production.

USES

Cesium and its compounds are mainly used in electron tubes, photoelectric cells, photomultipliers, thermionic converters, ferroelectric and piezoelectric crystals, in the manufacture of storage batteries, semiconductors, and certain instruments, and in petroleum cracking. [See also Davie & Coleman.]

MAN-MADE SOURCES OF EMISSION INTO THE ENVIRONMENT

Cesium compounds can enter the working environment as condensation or disintegration aerosols and as fumes from processes in which these compounds are produced or used. The air of workplaces is likely to contain particularly high concentrations of cesium hydroxide and cesium chloride (tens of milligrams per m^3). Significant sources of cesium in the general atmosphere may be coal dust and stack emissions from coal-burning factories: in coal dust, levels up to 3 $\mu g/g$ were recorded [Manchuk & Ryabov]. Cesium deposited by the Chernobyl disaster will continue to be recycled in organic and low-potassium soils [Coughtrey et al.].

TOXICITY TO ANIMALS AND MAN

Acute exposure

Animals

LD_{50} values of several cesium salts for mice, rats, and frogs are given in Table 1.

For cats and rabbits, LD_{50} values of the **carbonate, chloride, nitrate,** and **sulfate** with intravenous administration were in the ranges 640–660 mg/kg and 830–970 mg/kg, respectively. The clinical picture of acute poisoning was marked by excitation, convulsions, and respiratory distress, followed by death. The surviving animals appeared inhibited, but did show reflex reactions to conventional stimuli.

The oral LD_{50} of **cesium dihydrogen orthoarsenate** was determined at 116 mg/kg for mice and 345 mg/kg for rats [Tarasenko & Lemeshevskaya]. As indicated by these authors, the acute toxicity of this compound is due not only to the cation but also, in large measure, to the anion. In rats, the threshold for acute toxicity on 4-h inhalation exposure to the dihydrogen orthoarsenate was 5.2 mg/m^3. Its cumulation coefficient was calculated to be 0.99. The oral LD_{50} of **cesium hydroxide** was 759 mg/kg for mice and 570 mg/kg for rats. In rats, the thresholds for acute toxicity and for irritant effects on inhalation exposure to this compound were 42.4 mg/m^3 and 20.5 mg/m^3, respectively. Its cumulation coefficient was determined at 1.2.

Man

The threshold concentration of **cesium hydroxide** producing irritant effects on inhalation was determined at 5.02 mg/m^3.

Table 1. Median lethal doses of cesium salts

Compound	Animal species	Route of administration	LD_{50} (mg/kg)
Bromide and iodide	Rat	Oral	1400
Carbonate	Mouse/rat	Oral	2170/2333
		Subcutaneous	2300/2200
		Intraperitoneal	1000/810
		Intravenous	940/877
	Frog	Oral	220
		Subcutaneous	135
Chloride	Mouse/rat	Oral	3000/2004
		Subcutaneous	2750/2590
		Intraperitoneal	1500/1316
		Intravenous	1230/1075
Nitrate	Mouse/rat	Oral	2300/2390
		Subcutaneous	2200/2830
		Intraperitoneal	1450/1100
		Intravenous	1300/1210
	Frog	Oral	232
		Subcutaneous	120
Sulfate	Mouse/rat	Oral	3180/2830
		Subcutaneous	2800/2919
		Intraperitoneal	1790/1700
		Intravenous	1780/1220
	Frog	Oral	535
		Subcutaneous	145

Source: Bruk.

Chronic exposure

Animals
In a 4-month inhalation toxicity study, rats exposed to **cesium dihydrogen orthoarsenate** by inhalation at 4.96 mg/m^3 for 4 h daily 5 days per week showed strongly marked signs of systemic poisoning, such as reduced weight gains, leukocytosis, hypoalbuminuria, lowered hippuric acid levels, and increased values of the summated threshold index. In the blood, reduced pH, elevated levels of phospholipids, β-lipoproteins, and cholesterol, hyperkalemia, and increased lactic acid levels were noted; serum levels of uric acid and creatinine were also raised. Pathologic examination revealed dystrophic changes in the kidneys, heart, lungs, and testes. The concentration of 0.33 mg/m^3 was found to

be close to the threshold level for chronic toxicity. Inhalation exposure of rats to cesium dihydrogen orthoarsenate at 4.6 mg/m³ or 0.43 mg/m³ by inhalation throughout the period of pregnancy resulted in effects described as embryotropic and gonadotropic [Silayev & Lemeshevskaya; Tarasenko & Lemeshevskaya].

During and after 4 months of inhalation exposure to **cesium hydroxide** at 10 mg/m³, rats exhibited clear evidence of systemic poisoning including reticulocytosis, leukocytosis, abnormal ratios of total protein to its fractions in blood serum, altered acetylcholinesterase activity, and variability of blood sugar levels. The concentration of 1.0 mg/m³ was interpreted as the no-effect level.

Man

Among workers engaged in cesium production, abnormalities of cardiac activity, dystrophic changes in the upper respiratory tract, and a neurasthenic syndrome with signs of vasoneurosis were noted; leached-out epithelium and pigments were detected in the urine.

Effects of local application

Animals

In rabbits, application of a 4% **cesium hydroxide** solution to the skin resulted in edema and reddening of the affected skin area, followed by the emergence of a sharply delimited dense scab that was not connected to subjacent tissues; 1 month after detachment of the scab, a scar formed at the application site. When a 0.4% cesium hydroxide solution was placed into the conjunctival sac of a rabbit, purulent inflammation, corneal clouding, and ulcerous lesions in the lower eyelid conjunctiva were observed [Khosid].

Cesium dihydrogen orthoarsenate can produce local skin effects as well as systemic poisoning by being absorbed through the skin. When a 20% ointment of this compound was applied to the skin of rats, guinea pigs, and rabbits, small ulcers appeared at the application site in about 15 days; all animals died by day 20. Autopsy demonstrated dystrophic changes and vascular lesions. The threshold dose for cutaneous effects with 4-h exposure was 500 mg/kg.

ABSORPTION, DISTRIBUTION, AND ELIMINATION

Cesium enters the body through the respiratory tract and, to some extent, via the digestive tract. By virtue of their good solubilities, cesium and its compounds are well absorbed by these routes and will cause systemic poisoning if the dose is sufficiently high. Absorption from the intestine is almost 100%, and cesium and its compounds are also readily absorbed from subcutaneous tissues, lungs, muscles, and wound surfaces. Cesium readily passes the placental and other biological barriers. Its distribution in organs and tissues is fairly uniform, with the exception of muscles. From 44% to 65% of the administered dose was found to be deposited in muscles and 13% in bones. The total human body burden of cesium averages 1.25 mg. Its concentrations in mammalian blood ranged up to 2.8 μg/L. From 40 to 70% of cesium is excreted in urine and 10 to 15 times less in feces [Tarasenko & Lemeshevskaya].

HYGIENIC STANDARDS

In the USSR, occupational exposure limits (MAC$_{wz}$) have been set for cesium arsenide at 0.03 mg/m^3 (aerosol, hazard class 1) and cesium hydroxide at 0.3 mg/m^3 (aerosol, hazard class 2). In the USA, the TLV-TWA for the hydroxide is 2.0 mg/m^3.

METHODS OF CESIUM DETERMINATION

Flame photometry has been used for determining cesium in atmospheric air, in natural, industrial, and waste waters, and in snow, soil, and plants (sensitivity 0.001 mg/m^3 for air and 0.02 mg/L for aqueous media); nephelometric determination using calcium hexacyanoferrate(II) has been described as an ancillary procedure (sensitivity 0.05 mg/m^3 for air and 1.0 mg/L for aqueous media) [Novikov et al.; Sopach et al.]. The use of emission spectral analysis for measuring cesium in plants [Loseva et al.] and of flame atomic absorption spectrometry for its measurement in animal tissues [Koli et al.] has been reported.

MEASURES TO CONTROL EXPOSURE AT THE WORKPLACE

All processes involved in the production and use of cesium and its compounds should be mechanized to the greatest possible extent. For separating the final products, air classifiers hermetically linked to discharge openings of the roasting furnaces and to the storage bunkers should be used. Powders should be fed to and discharged from blenders by means of enclosed auger conveyors hermetically connected to these. The prepared powders should be weighed and packaged by automatic machines in a specially designated area.

All workers at the factory should be instructed about and trained in the safe handling of cesium and its compounds. They are subject to medical examination before employment and at regular intervals during the period of employment.

For *personal protection*, appropriate protective devices must be used to prevent eye and skin contact with cesium and its compounds. Respirators must be worn if the workroom air is polluted by aerosols.

REFERENCES

Bruk, M. M. (1964) In: *Farmakologiya i toksikologiya* [Pharmacology and Toxicology], Vol. 1, Kiev, pp. 200–205 (in Russian).

Coughtrey, P. J. *et al.* (1989) *Envir. Poll.*, **62**, 281–315.

Davie, R. J. & Coleman, I. P. L. (1988) Chapter 17 *Cesium* in: Seiler, H. *et al.* (eds) *Handbook on Toxicity of Inorganic Compounds*. Marcel Dekker, New York, pp 217–221.

Dobrovolsky, V. V. (1983) *Geografiya mikroelementov. Globalnoe rasseyanie* [Georgaphy of Trace Elements. Global Dispersion]. Mysl, Moscow (in Russian).

Henderson, P. (1982) *Inorganic Geochemistry*. Pergamon Press, Oxford.

Kabata-Pendias, A. & Pendias, H. (1986) *Trace Elements in Soils and Plants*. CRC Press, Boca Raton (Fl.).

Koli, A. K. *et al.* (1985) *Environ. Int.*, **11**, 23–24.

Khosid, G. M. (1965) In: *Khimicheskie faktory vneshnei sredy i ikh gigiyenicheskoe znachenie* [Environmental Chemicals and their Health Effects]. Moscow, pp. 39–41 (in Russian).

Loseva, A. F. *et al.* (1984) *Spektralnye metody opredeleniya mikroelementov v obiektakh biosfery* [Spectral Methods for Determining Trace Elements in the Biosphere]. Rostov-on-Don (in Russian).

Manchuk, V. A. & Ryabov, N. A. (1983) In: *Metally. Gigiyenicheskie aspekty otsenki i ozdorovleniya okruzhayushchei sredy* [Metals: Hygienic Aspects of Evaluation and Environmental Health Promotion]. Moscow, pp. 194–199 (in Russian).

Novikov, Yu. V. *et al.* (1981) *Metody opredeleniya vrednykh veshchestv v vode vodoyemov* [Methods for Measuring Noxious Substances in Water Bodies]. Meditsina, Moscow (in Russian).

Silayev, A. A. & Lemeshevskaya, E. P. (1980) *Gig. Truda Prof. Zabol.*, No. 9, 46–48 (in Russian).

Sopach, E. D. *et al.* (1976) *Gig. San.*, No. 9, 58–60 (in Russian).

Tarasenko, N. Yu. & Lemeshevskaya, E. P. (1978) *Vestn. Akad. Med. Nauk SSSR*, No. 8, 10–18 (in Russian).

Vadkovskaya, I. K. & Lukashev, K. P. (1977) *Geokhimicheskie osnovy okhrany biosfery* [Protection of the Biosphere: Geochimical Principles]. Nauka i Tekhnika, Minsk (in Russian).

Copper and its compounds

Copper(I) (cuprous) compounds: **chloride** (nantokite(α) [min.]); **iodide** (marshite [min.]); **oxide** (cuprite [min.]); **sulfide** (chalcosine(α,β) [min.]).

Copper(II) (cupric) compounds: **acetate; carbonate; carbonate hydroxide** {copper(II) sodium carbonate–sodium hydroxide (1/1); basic cupric carbonate; malachite [min.]; Cupronil [a mixture containing 35% of copper(II) carbonate hydroxide]}; **chloride; chloride hydroxide hydrate** {copper(II) chloride–copper(II) hydroxide–water (1/3/1); basic cupric chloride hydrate; Cuprozan [a mixture containing 37.5% of copper(II) chloride hydroxide]}; **hydroxide; nitrate trihydrate; oxide** (tenorite [min.]); **sulfate; sulfate hydroxide** (basic cupric sulfate); **sulfate pentahydrate** (blue vitriol; Bordeaux mixture); **sulfide** (covelline [min.])

IDENTITY AND PHYSICOCHEMICAL PROPERTIES
OF THE ELEMENT

Copper (Cu) is a metal in group I of the periodic table, with the atomic number 29. Its natural isotopes are ^{63}Cu (69.1%) and ^{65}Cu (30.9%).

Copper is highly ductile and malleable and an exellent conductor of heat and electricity, being second to silver in the latter respect.

In compounds, copper exists in the oxidation state +1 (copper(I), or cuprous, compounds), +2 (copper(II), or cupric, compounds), and rarely +3 (copper(III) compounds). Its chemical activity is not high. It remains virtually unchanged in dry air, but in the presence of moisture and carbon dioxide, its surface becomes coated with a deposit of copper(II) carbonate hydroxide. On heating in air at temperatures between 200 and 375°C, it oxidizes to the black copper(II) oxide CuO; at higher temperatures, a double-layer scale appears on its surface, the outer layer being copper(II) oxide and the inner layer, red copper(I) oxide Cu_2O.

In the presence of moisture copper reacts with chlorine in the cold to form copper(II) chloride $CuCl_2$, and it also readily combines with the other halogens. It reacts with sulfur and selenium to give the sulfide CuS and selenide CuSe. It does not combine with hydrogen, nitrogen, or carbon even at high temperatures. Strongly heated copper reacts with ammonia to form the copper(I) nitride Cu_3N and with N_2O or NO to form the copper(I) oxide Cu_2O or copper(II) oxide CuO. Copper carbides are produced by reaction of acetylene with ammonia solutions of copper salts. Copper does not react with non-oxidizing acids. It reacts with nitric acid to give copper(II) nitrate $Cu(NO_3)_2$ and nitrogen oxides. On reaction with hot sulfuric acid, it gives copper(II) sulfate $CuSO_4$ and sulfur dioxide SO_2. Copper(I) salts are virtually insoluble in water, whereas copper(II) salts are readily soluble.

Copper forms numerous complex compounds. For copper(I), the characteristic coordination numbers are 2, 3, and 4, while for copper(II) they are 4 and 6.

When exposed to a moist atmosphere, copper and its alloys form a stable fine surface film protecting them from further corrosion. In the initial period of atmospheric corrosion, copper oxides and sulfides of brown color are produced that can form a black layer on the surface; after several years of exposure, a greenish coating known as patina appears, which is a basic copper sulfate or, in marine environments, a basic copper chloride. These compounds are both highly resistant to further atmospheric influences. See also Appendix.

NATURAL OCCURRENCE AND ENVIRONMENTAL LEVELS

Copper is widely distributed in all continents and is present in most living organisms. Among copper-bearing minerals, sulfides, sulfates, phosphates, and chlorides predominate. Copper is also found in carbonates and oxides as well as in the native state. Native copper is usually found in the areas of oxidation of certain copper-sulfide deposits, most often in association with malachite $CuCO_3 \cdot (CuOH)_2$ (contains ~57% Cu) or azurite $Cu_3[(OH)CO_3]_2$ (65–69% Cu). It is also encountered in volcanic and sedimentary rocks. The global reserves of copper in ores are estimated at 465 million tonnes.

The clarke of copper has been estimated at $(47–55) \times 10^{-4}\%$ in the earth's crust and at $22 \times 10^{-4}\%$ in the granite layer of the continental crust. The oceans contain an estimated total of 5×10^9 tonnes of copper; annually, they receive up to 3.0×10^5 tonnes from rivers and deposit about 4.0×10^4 tonnes. Average copper concentrations are about 0.9 μg/L in seawater and 7 μg/L in river and lake waters. River water contains the metal in suspended matter, whereas seawater contains it in dissolved form. The copper deposited in seawater becomes a constituent of muds and nodules. Ferromanganese oxide deposits of the Pacific take up an estimated 32 000 tonnes of copper per year and contain it at an average concentration of $5300 \times 10^{-4}\%$ [Dobrovolsky; Vadkovskaya & Lukashev].

An important role in copper migration in the hydrosphere is played by hydrobionts such as plankton, zoobenthos, and phytobenthos. For example, some plankton species are capable of concentrating copper by a factor of 90 000 [Mullins]. Mussels living near or at the sea bottom filter up to 250 m^3 of water/m^2 of bottom area per day and transfer various metals, including copper, to the bottom sediment as well as accumulating them in their bodies. The bodies and shells of mussels inhabiting the Black Sea contained copper in concentrations of 1.96 to 3.06 mg/kg wet weight; algae in the Pacific contained it at levels up to 5 μg/g dry weight and those in the North Sea, at 2.5 to 67.0 mg/kg dry weight. In the Gulf of Riga, average annual concentrations in hydrobionts were as follows (mg Cu/kg dry weight): 1.0 in plankton, 40.0 in benthic crustaceans, 0.8 in algae, 8.0 in mollusks, 0.9 in fish.

The most common primary minerals of copper are simple and complex sulfides which are readily soluble in weathering processes and release copper ions, especially in acid environments. Copper cations in soils and sediments exhibit a great capacity for chemical interaction with their mineral and organic components. Mean copper levels in uncontaminated soils range from 6 to 60 mg/kg, being highest in ferrallitic soils and lowest in sandy and organic ones. The most common characteristic of copper distribution in soil profiles is its accumulation in the upper soil horizons where its predominant mobile form is believed to be the 2+ cation. Copper is abundant in soil solutions of all soil types. Its concentration in soil solutions obtained by various techniques from different soils varied from 0.003 to 0.135 mg/L [Alloway; Kabata-Pendias & Pendias].

Most of the copper in soil (>90%) occurs as complexes with organic compounds. In the various layers of a fully developed soil profile, copper migrates in ionic, colloidal, and suspended states, and in the process of this migration the proportion of copper in the soluble ionic form decreases with an increase of that in the colloidal state which is an intermediate form in the transition to the suspended state. Accumulation of the metal migrating in this state takes place at a depth of 1.8–2.0 m, in the lower part of the alluvial soil layer. Migration rates vary markedly with the pH and redox potential of the soil medium. In an acid soil, copper is more mobile by 1 or 2 orders of magnitude than in a neutral or alkaline soil. Well-drained soils with low pH values are low in copper [Bolshakov et al.; Bergqvist & Sundbom].

On average, cereal crops remove from soil about 20 to 30 g of copper per hectare of soil per year, while forest biomass removes about 40 g. Copper applied in a soil remains there for a long time; for example, its concentrations were still elevated in pasture herbage 10 years after the application of copper-containing fertilizer. Repeated application may result in copper accumulation to concentrations toxic for some crops [Bergqvist & Sundbom; Bolshakov et al.; Kabata-Pendias & Pendias].

Copper is one of the biocatalysts essential to plant life. Continental vegetation contains an estimated total of 25 million tonnes of copper, at average fresh weight, dry weight, and ash weight concentrations of $3.2 \times 10^{-4}\%$, $10.0 \times 10^{-4}\%$, and $200 \times 10^{-4}\%$, respectively [Dobrovolsky]. Its dry weight concentrations range from 8.4 to 24.0 mg/kg in freshwater plants and average 14.0 mg/kg in terrestrial plants. In the USSR, barley, oat, and winter wheat grains had mean copper concentrations of 5.2, 3.2, and 5.1 mg/kg dry weight, respectively. The corresponding figure for barley in Norway is 5.5 and in Great Britain, 4.3; in the USA, rye and barley grains contained 4–8 and 4–15 mg/kg [Kabata-Pendias & Pendias]. The mean copper content in grapes and apricots was calculated to be 3.3 ± 1.3 and 2.4 ± 1.2 mg/kg, respectively. Fir, spruce fir, oak, birch, and mountain ash (Sorbus aucuparia) trees contained copper at 1 to 12 mg/kg dry weight, while mosses and lichens contained 5–9 mg/kg. The entire biomass of the Earth contains an estimated 200 million tonnes of this metal [Kovalsky].

Copper migrating in the atmosphere occurs in the form of solid aerosol particles of 0.25–25.0 μm in diameter. Atmospheric air samples collected over different countries or regions contained copper in the following concentrations (ng/m^3): Japan, 11–200; Norway, 2.5; West Germany, 8–4900; Shetland Islands, 20; North, Central, and South America, 3–153, 7–100, and 30–180, respectively; the South Pole, 0.03–0.06 [Kabata-Pendias & Pendias].

PRODUCTION

Copper is recovered from its concentrated sulfide, oxide, or carbonate ores by pyro- or hydrometallurgical processes.

Copper(II) chloride is made by chloridizing roasting of copper sulfides with sodium chloride at 550–600°C (the exit gases contain HCl, Cl_2, SO_2, SO_3, and arsenic compounds) or by dissolving Cu(II) oxide and Cu(II) carbonate in hydrochloric acid. Cu(II) hydroxide is produced by reaction of Cu(II) salts with alkali; Cu(II) carbonate hydroxide, by precipitation from Cu(II) sulfate with sodium carbonate; and Cu(II) chloride hydroxide, by precipitation from Cu(II) chloride solution with an aqueous solution of calcium hydroxide or calcium carbonate. Cu(II) nitrate is obtained by dissolving copper or its oxide CuO in a 25% nitric acid solution. Cu(II) oxide is prepared by roasting copper glance (Cu_2S) at 700°C or by calcining the reaction products of Cu(II) salts and alkalies; Cu(I) oxide is made by calcination of copper in limited air supply, by the action of alkali and a reducing agent (such as hydrazine, hydroxylamine, or grape sugar) on a Cu(II) sulfate solution, or by electrolysis. **Cu(II) sulfate** is made by the action of sulfuric acid on copper, its oxide (CuO), or pyrite cinders or by sulfidizing roasting of copper glance or other sulfide ores. **Cu(II) sulfide** is obtained as a precipitate by passing hydrogen sulfide through solutions of Cu(II) salts; it also occurs naturally as covellite.

USES

Copper has a wide diversity of uses in the electrical, refrigeration, and heavy industries, in powder metallurgy, and in communications. Much of the copper produced is utilized in various alloys (as bronze, brass, and copper-nickel alloys) that find extensive appication as construction, antifriction, corrosion-resistant, and heat- or electricity-conducting materials in the instrument-making, machine- and ship-building, automobile, and aircraft industries. New crystalline compounds containing Cu^{2+}, Ca^{2+}, O^{2-} and other bivalent elements seem to allow "warm" superconduction [Scheinberg]. Copper metal is also used as a pigment. Livestock and poultry feeds are not infrequently supplemented with copper to promote growth or provide antibiotic activity. (See Sarkar for other uses.)

Copper(I) chloride is used as a catalyst and an absorbent of carbon monoxide; **Cu(II) chloride** in mineral paints, as a mordant in dyeing and printing, and in pyrotechnics; **Cu(II) nitrate** in cotton printing, in making enamels, and in pyrotechnics; **Cu(I)** and **Cu(II) oxides** as pigments and in the production of glasses and enamels; **Cu(II) sulfate** as a wood preservative, in mineral paints, in dyeing and printing, and in electroplating. **Cu(II) oxide** and **Cu(II) sulfate** are also used in making some types of artificial fibers. **Cu(II) sulfate pentahydrate** finds use in medicine.

Many copper compounds are employed as pesticides either alone (as Cu(I) oxide and Cu(II) sulfate) or in mixtures. Examples of the latter are the pesticide Cupronil which contains 35% **Cu(II) carbonate hydroxide** and 15% zinc *N,N*-dimethyldithiocarbamate; the fungicide Cuprozan containing 37.5% **Cu(II) chloride hydroxide** and 15% zineb (zinc ethylene-*bis*-dithiocarbamate); Bordeaux mixture—a fungicide prepared by reacting **Cu(II) sulfate** with hydrated lime; and Burgundy mixture—a fungicide similar to Bordeaux mixture but containing sodium carbonate instead of lime.

Copper compounds have catalytic properties in many chemical reactions and can activate formation of chlorinated dioxins and dibenzofurans during incineration processes [Scheinberg].

MAN-MADE SOURCES OF EMISSION INTO
THE ENVIRONMENT

The major man-made sources of environmental copper are nonferrous metal industries (stack emissions, liquid and solid wastes), transport, copper-containing fertilizers and pesticides, welding and galvanization processes, and the combustion of hydrocarbon fuels in various industries.

On a global basis, the annual anthropogenic environmental releases of copper were estimated to be 56 000 tonnes with atmospheric emissions, 77 000 tonnes with solid and liquid wastes, and 94 000 tonnes with fertilizers; the chemical industries were calculated to be responsible for the discharge of up to 155 thousand tonnes of copper to the earth's surface per year. Although coal contained on average about 15 mg Cu/kg, some coal dusts contained as much as 340 mg/kg. The total amount of copper emitted annually into the atmosphere as a result of coal and oil combustion was estimated at about 2100 tonnes [Manchuk & Ryabov].

In one urban area, copper concentrations up to 1.3 mg/m^3 were recorded in the ambient air within several kilometers of a copper smelting and refining works as compared to an average of only 0.09 mg/m^3 in that area as a whole [Sidorenko]. The 2 cm soil layer in the immediate vicinity of the works contained 1.1% of copper, mainly in water-soluble and hence readily migrating forms. Average copper contents in potato, carrot, and tomato plants grown at distances of 1–3 km from the works on a soil that had copper levels of 0.3 to 0.7 mg/kg were found to be 3.02, 4.97, and 11.08 mg/kg dry weight, respectively, as compared to 1.93, 2.96, and 4.05 in those grown at a distance of about 25 km where the soil contained only 0.05 mg Cu/kg. Copper-polluted soils lose their orderly structure and their total porosity decreases, with a consequent impairment of the water permeability and of the air and water regimes.

At a distance of 50 m from a highway with heavy automobile traffic in the Baltic region of the USSR, soil pollution with copper occurred at a rate of 60–140 μg/m^2 per month, and dandelion plants collected there had mean dry-weight copper levels of about 1.7 mg/kg in leaves and 1.3 mg/kg in roots; in an area 25 m from the highway, where the soil pollution was much more intensive, the respective values were 8.2 and 13.2 mg/kg [Berinia].

Five kilometers away from a heavy metal-processing factory, the foliage of many trees underwent necrosis; on analysis, it was found to contain copper at 16–17 mg/kg dry weight (against 6–8 mg/kg in control samples) [Ginoyan].

Considerable quantities of copper are introduced into soils in fertilizers, pesticides, irrigation waters, and sewage. Excessive copper levels in soil reduce the activity and accelerate the decomposition of phosphatases and ureases. In forest soils containing more than 100 mg Cu/kg, the mineralization of phosphorus and nitrogen occurs at greatly diminished rates.

In one region of Moldavia in the USSR, the copper content of the top (0–30 cm) soil layer increased from 7–8 mg/kg to 14 mg/kg shortly after the treatment of fruit trees with Bordeaux mixture and to 18 mg/kg after spraying them with a 0.8% suspension of Cuprozan (fungicidal formulation). One month later, the copper level in the top soil decreased, suggesting that some of the copper had migrated to the underlying soil layers and could subsequently reach water sources. Indeed, copper concentrations were found to have increased, during one vegetation period, from 1.1 to 2.03 mg/L in a pond located on sandy loam soil in that region and as much as 10-fold in well water sampled there. As a result of soil saturation with copper, its concentrations in apples ranged from 4.16 to 5.10 mg/kg as compared to around 1.0 mg/kg in control samples. The application of blue vitriol at rates of 0.5 to 2.0 mg/kg soil resulted in copper migration to barley and onion plants in amounts that positively correlated with copper levels in the soil [Merenyuk & Medzhibovskaya].

In seawater, the sources of copper are river discharges, deposition from the air, atmospheric precipitation, and human activities at sea. In the Black Sea, for instance, its average concentration was 30.2 µg/L [Ryabinin & Lazareva], as compared with the estimated average of 0.9 µg/L for the world's ocean as a whole. In the early 1980s, according to Dmitriev & Tarasova, the average concentration in the uppermost (10 µm) layer of the oceans attained 235 µg/L.

Effluents from the metallurgical, machine-building, chemical/pharmaceutical, and some other industries contained up to 500 mg Cu/L [Grushko, 1972]. It has been established that the copper content should not exceed 2 mg/L in waters used for long-term irrigation of agricultural crops and 5 mg/L in those used for short-term irrigation of crops with low copper sensitivity. Copper concentrations in excess of 1.9 mg/L inhibited sludge fermentation in wastewater-treatment plants. The threshold copper level in wastewater, defined as the concentration at which the treatment efficiency is reduced by 5%, has been found to be 1.9 mg/L. Contamination of drinking water with copper and other metals from corrosion of distribution pipes has been described [Alam & Sadiq].

Valuable bioindicators of environmental pollution by copper compounds are certain birds since they concentrate much copper in their plumage (e.g. values of 3.6 and 14 mg/kg dry weight were reported for jays and grouse, respectively [Dobrovolskaya]). Good bioindicators of atmospheric pollution with copper may be domestic dogs whose coats can accumulate considerable amounts of heavy metals, including copper (in one study, levels up to 33 µg Cu/g were detected) [Ginoyan]. Effective copper concentrators and thus potential bioindicators of water pollution are blue-green algae and mollusks; some species of the latter accumulated up to 6.8 mg of copper per kg dry weight from water that contained it at concentrations of 3.5×10^{-4} mg/L [Zakharova & Udelnova; Zhulidov et al.]. As bioindicators, polychete worms may also be useful, for which the LC_{50} of copper was determined to be between 0.2 and 0.25 mg/L after 28 days of exposure [Reich et al.].

TOXICITY

Soil microflora

Copper and its compounds may be very toxic to soil microorganisms. After treatment of plants with Cuprozan or Bordeaux mixture, total microbial counts in the soil decreased by 21 to 72%; these two fungicides proved to be particularly toxic to spore-bearing organisms whose numbers declined by up to 92%. In sandy loam soils containing copper at 3 mg/kg or more, a markedly reduced activity of nitrifying bacteria was noted [Merenyuk & Medzhibovkaya]. Some microorganisms isolated from soil or activated sludge were capable of accumulating copper in appreciable amounts [Kovalsky]. Mehra found that *Bacillus* spp. grew better in growth media containing up to 3 μg Cu/g than in those with no added copper.

Plants

For most plants, copper levels in soil ranging from 6 to 15 mg/kg are regarded as too low, those between 15 and 60 mg/kg as normal, and those above 60 mg/kg as excessive. Excess copper in the soil is conducive to chlorosis in plants. For barley and clover, toxic copper levels in soil at which the yield or height of plants was reduced by 5–10% were 20 ppm and 18 ppm, respectively. The threshold concentrations of copper in plants (its lowest values at which signs of toxicity were absent) were 10 mg/kg for cereal crops and 32 mg/kg for leguminous crops on a dry weight basis. In water used to irrigate agricultural crops, the lowest copper concentrations leading to manifestations of toxicity were 0.1 mg/L for orange and mandarine seedlings, 0.17–0.20 mg/L for sugar beet, tomato, and barley plants, and 0.5 mg/L for flax plants [Goncharuk].

The major signs of toxicity in common crop plants are dark green leaves followed by chlorosis; thick, short, or barbed-wire-like roots; and depressed tillering. Plants growing in soils contaminated with copper from geological, industrial, or agricultural sources may die; only copper-tolerant individuals will survive [Scheinberg]. Copper-sensitive crops include various cereals and legumes, spinach, citrus seedlings, and gladioluses [Kabata-Pendias & Pendias]. Effects of copper on plant growth, photosynthesis, and chlorophyll *a* synthesis have been described by Wang & Chang.

Soil invertebrates

For earthworms, toxic copper concentrations in soil lie in the range from 100–120 ppm upward (dry weight); the 98-hour LC_{50} was 181 ppm in a soil low in organic matter and 2760 ppm in a peat soil. Scheinberg reported that 100 mg Cu/kg soil decreased earthworm cocoon production but not hatchability.

Aquatic organisms

Metallic copper is moderately toxic to fish, whereas soluble copper salts such as the **chlorides** and **nitrates** may be very toxic, even in low concentrations (0.01–0.02 mg/L). The reversibility of toxic effects is slight; in particular, poisoned fishes failed to survive when transferred to clean water at the narcosis stage. In soft water the toxicity of copper salts is considerably higher than in hard water where some of the copper forms relatively nontoxic carbonates. Of particular importance from a toxicological viewpoint is **copper(II) sulfate** which is used as an effective algicide. When exposed to this compound, perches died after 24–48 h at its concentration of 0.25 mg/L and after 1.5–5.5 h at 2 mg/L; still more sensitive were whitefish which died at 0.1 mg/L [Ilyaletdinov; Metelev *et al.*]. Extremely sensitive to the sulfate were leeches (they died at 0.08 mg/L), whereas insect larvae were fairly resistant (mosquito larvae, e.g., survived for several days at concentrations of 25 to 250 mg/L). Protozoans did not survive for more than 24 h at 0.5 mg/L while tolerating the exposure to 0.1 mg/L for at least 48 h without any ill-effects. **Copper(II) chloride** was lethal to loach after 3 to 7 h of exposure to 0.019 mg/L. **Copper(II) nitrate** killed stickleback at 0.02 mg/L [Grushko, 1979]. For rainbow trout (*Salmo gairdneri*), the 96 hour LC_{50} of copper was 520 µg/L in hard water and as low as 20 µg/L in soft water [Dixon & Sprague].

For marine organisms, the following toxic and threshold concentrations of copper have been reported [Patin] :

	Toxic (mg/L)			Threshold (mg/L)
Unicellular algae	10^{-2}	to	1	10^{-3} to 10^{-2}
Macrophytes	10^{-2}	to	10^2	10^{-2}
Protozoans	10^{-2}	to	10^2	–
Crustaceans	10^{-3}	to	10^2	10^{-2}
Mollusks	10^{-2}	to	10	10^{-2}
Worms	10^{-1}	to	10	10^{-3} to 10^{-1}
Fish	10^{-2}	to	1	10^{-2} to 10^{-1}

Several studies have looked at the importance of synergistic effects of copper and other elements such as cadmium and zinc on aquatic organisms [Dive *et al.*; Scheinberg].

Animals and man

General considerations

Copper is a highly toxic metal that can cause acute poisoning and has a broad spectrum of actions with diverse clinical manifestations [Roshchin]. The most critical factor in the mechanism of toxicity is the ability of copper ions to block sulfhydryl groups in proteins, especially enzymes. The high degree of hepatotoxicity shown by copper and its compounds is associated with their ability to increase the permeability of mitochondrial membranes and with the tendency of copper to lodge in hepatic cell lysosomes. Acute copper intoxication may be accompanied by strongly marked hemolysis [Helman *et al.*]. Ruminants susceptible to copper toxicity show symptoms of hepatitis and hemolytic anemia [Scheinberg]. The damage caused by copper to blood-forming organs has been attributed to its direct action on young cells, in particular the inhibition of enzymatic processes responsible for the proliferation and differentiation of these cells, and also to the impairment of erythrocyte membrane permeability; the blood changes may be manifested as a hemolytic anemia in which the Coombs test is negative [Ringenberg *et al.*]. Intoxication with copper compounds may involve autoimmune reactions and impairment of monoamine metabolism [Lychagin *et al.*].

In experiments with sea-urchin embryos and with cells of murine transplantable Ehrlich's tumor, embryotoxic and cytotoxic effects of copper sulfate were demonstrated and believed to have resulted from the inhibition of protein synthesis by copper [Berisha *et al.*].

Acute/single exposure

Animals

LD_{50} values for copper and its compounds administrated by the oral (gavage) and intraperitoneal routes are shown in Table 1.

With intraperitoneal injection into mice, the threshold dose of **copper(II) oxide** for effects on body weight, the weights of internal organs, and blood sulfhydryls was determined at 45.4 mg/kg.

A single inhalation exposure to a dust of pure **electrolytic copper** in concentrations of 800 to 900 mg/m^3 caused excitation in rabbits followed by their death; the dust was highly irritating to the upper respiratory tract mucosa and the conjunctivae. When rats were caused to inhale copper dust at 100 mg/m^3 for 5 h/day, more than half of the animals were dead after 7 days [Brakhnova]. At autopsy of rats that had been inhaling **copper(II) oxide** aerosol, its particles were seen to have extensively penetrated into the alveolocapillary membrane (blood-air barrier) to invade the ultrastructures of alveolar cells [Bandura].

Dusts of **copper sulfide ore** were lethal to rats on 4-hour inhalation exposure at 217 mg/m^3 [Izmerov *et al.*]. In dogs, a short-term inhalation exposure to **copper(II) carbonate** dust led to the development of pulmonary edema.

Intratracheal administration of **pure copper** dust in an amount of 20 mg caused 100% mortality in rats within 24 h; the introduction of 2 mg by this route on two occasions separated by a 2-week interval also resulted in the death of all animals. **Copper(II) oxide** administered by the same pathway in an amount of 50 mg was lethal to only 50% of the rats. As shown by pathologic examinations, the animals developed bronchitis, purulent catarrhal inflammation of the lungs, and eventually diffuse sclerosis of the interstitial tissue and lesions of pulmonary vessels; changes were also noted in the liver. Similar effects were produced in the lungs of rats by intratracheally administered dusts from copper ores, a copper ore-concentrating factory, and a copper smelter [Tsoi], and also by malachite dust. Copper dust was found to intensify the silicotic process by specifically damaging pulmonary vessels and stimulating the inflammatory response of lung tissue [Dzhangozina & Slutsky]. Malachite dust was very toxic, killing 20% of rats after intratracheal insufflation in a relatively small amount (10 mg). A single administration of copper smelter dust by this route led to severe changes in the liver parenchyma several days later; after 3 months, necrobiotic foci were noted in the cytoplasm of hepatic cells and necrosis in Kupffer's cells.

Table 1. Median lethal doses of copper and its compounds

	Animal	Route of administration	LD_{50} (mg/kg)
Copper	Mouse	Intraperitoneal	0.07
Copper(II) carbonate hydroxide	Rat	Oral	159
Copper(II) chloride	Mouse	Intraperitoneal	6.7–9.4
	Rat	Oral	140
	Rat	Intraperitoneal	3.7
Copper(II) chloride hydroxide	Mouse	Oral	470
	Rat	Oral	812
	Cock	Oral	1263
Copper(II) nitrate	Rat	Oral	940
Copper(I) oxide	Rat	Oral	470
Copper(II) oxide	Mouse	Intraperitoneal	273
	Rat	Oral	470
Copper(II) sulfate pentahydrate	Mouse	Oral	43
	Rat	Oral	300–520
	Cock	Oral	693
Copper sulfide ore	Mouse	Intraperitoneal	2015

Mice receiving subcutaneous testosterone injections developed teratomas after an inratesticular administration of 0.05 ml of an emulsion composed of peach oil and an equal amount of 10% **copper(II) sulfate** solution in 1.5% formalin [Bresler].

A subcutaneous injection of **copper(II) sulfate** at 50 mg/kg or **copper(II) chloride** at 12.5 mg/kg stimulated the growth of Guérin's carcinoma in mice exposed to high ambient temperatures (35–40°C) [Savitsky & Golovan].

Copper orthoarsenate hydrate, a by-product of copper smelting, caused 50% mortality among mice in the oral dose of 157 mg/kg and the intraperitoneal dose of 28 mg/kg. The poisoning was marked by dyspnea, tremor of the limbs, hypothermia, and diarrhea; autopsy revealed dystrophy of parenchymal organs. Multiple (19 times) intragastric administration of the latter dose led to hypochromic anemia, impaired thermoregulation, and gastroenterocolitis [Davydova *et al.*].

In cocks, **copper(II) sulfate pentahydrate** and **copper(II) chloride hydroxide** given orally in single lethal doses produced testicular atrophy with a sharp reduction of spermatogenesis [Shivanandappa *et al.*].

Man

Copper is a gastrointestinal-tract irritant, irritating the nerve endings in the stomach and initiating the vomiting reflex. Emetic doses of copper salts are 0.2–0.5 g; doses of 1–2 g have caused severe and sometimes fatal poisoning. Ingestion of **copper sulfate** or **basic copper acetate** was immediately followed by nausea, vomiting, gastric pain, and diarrhea. Hemoglobin soon appeared in the plasma and urine, and jaundice and anemia developed in many cases; decreased erythrocyte resistance and sometimes sulfhemoglobinemia and bilirubinemia were observed. In severe cases, death occurred in the presence of acute renal failure; autopsy showed hemorrhages in the gastric and intestinal mucosas, focal or diffuse necrosis in the liver and kidneys, and nephrosis [Chand; Csasta *et al.*; Deodhar & Deshpande; Scheinberg].

Inhalation of dusts, fumes, and mists of **copper** or its **salts** may cause acute irritation in the upper respiratory tract, and congestion of the nasal and pharyngeal mucous membranes. Copper fumes may cause salivation, nausea, vomiting, gastric pain, diarrhea, hemorrhagic gastritis, and cramps in the calves and even muscular rigor and prostration.

Workers exposed to **copper** and **copper(II) oxide** dusts in concentrations of 0.22 to 14 mg/m^3 (generated by such processes as the welding and cutting of copper articles, the rolling of copper ingots, or arc-furnace smelting of copper scrap) experienced a sweet taste in the mouth and irritation in the upper airways after 1–2 h of exposure. A few hours later, headache, weakness (especially in the legs), reddening of the throat and conjunctivae, nausea, and muscle pain supervened; in addition, vomiting, malaise, chills, and temperature rises to 38–39°C occurred in some cases. After 24 h, the temperature tended to return to normal, but other manifestations of poisoning (weakness, headache, dizziness, rapid pulse, lymphocytosis) usually remained for some time. Rather similar clinical pictures were presented by workers who had been inhaling **copper(II) carbonate** dust while grinding this compound (additionally, nasal hemorrhage, elevated blood bilirubin, and jaundice were noted), as well as by those who had been spraying **copper(II) oxide** or removing residues of copper compounds from apparatus.

In agricultural workers, severe shaking chills, high-grade fever ($\geqslant 39°C$), profuse sweating, a feeling of prostration, aching muscle pains, headache, irritation of the pharyngeal and laryngeal mucosas, and cough productive of green sputum (both during and after the fever) developed a few hours after the dry treatment of grain seed with a fungicide containing the **carbonate** and **sulfate** of copper(II) and some ($\sim 0.005\%$) arsenic. In some cases, chills during evening hours occurred as the only symptom preceding the major attack that came on later [Brodsky *et al.*].

Colicky abdominal pains and sometimes nasal bleeding were experienced by workers engaged in spraying vineyards with a copper-containing pesticide such as **blue vitriol, Bordeaux mixture, copper(II) sulfate hydroxide,** or **copper(II) chloride hydroxide** (aerosol concentrations in the air varied from 1 to 130 mg/m^3) [Krasnyuk].

The pathogenesis of 'copper fever' has at its basis necrosis of the alveolar endothelium with subsequent absorption of the resultant protein denaturation products. The fever usually subsides within 3 or 4 days. Rather rapid habituation to the exposure is common. A feature of copper fever, distinguishing it from other metal fume fevers such as zinc chills is that it affects the gastrointestinal tract. Gastrointestinal disorders may also result from inhalation of **bronze** dust or of **copper(II) hydroxide** in the form of small droplets.

Cough, vomiting, abdominal muscle tension, and diarrhea occurred in a child shortly after accidental inhalation of a **brass** powder (70% Cu and 30% Zn); the child later developed cyanosis and died from acute bronchopneumonia and pulmonary edema; at autopsy, necrosis of the renal tubules was noted [Browning].

Copper and its compounds have also caused poisoning of varying degree when ingested with food or water. For example, a group of people were poisoned after drinking water that contained copper in concentrations up to 44 mg/L; such poisoning may be due to the use of copper pipes for drinking-water supply, especially if the water has a high pH value (around 7.0) [Grushko, 1979]. In another instance, members of a family were complaining of recurring stomach ache after drinking tap water, and when samples of the latter were submitted to chemical analysis, they were found to contain copper in concentrations up to 7.8 mg/L; the family members also had greatly increased copper levels in their nails (up to 100 μg/g) and their hair (up to 1200 μg/g). Four outbreaks of acute intoxication as a result of drinking water with copper concentrations of 3 to 80 mg/L were reported [Craun].

Intractable vomiting occurred in 10 persons after drinking lemonade delivered by a vending machine with copper tubing inside and containing 29 ppm Cu [McDonald & Cook].

Poisoning may also be caused by food that has been cooked or stored in a container made of copper. For instance, symptoms of acute intoxication developed following the ingestion of a fruit jelly that contained 224 mg Cu/100 g. In occupational settings, copper-based insecticides were sometimes ingested during their use for spraying plants; the main symptoms in such cases were a metallic taste in the mouth, nausea, colicky abdominal pains, and blood-stained watery stools, although liver damage, hemolytic anemia, hematuria, and necrotic nephrosis were also noted if the poisoning was severe [Krasnyuk].

Cases of accidental or suicidal oral poisoning by **copper(II) sulfate** seem to have been the subject of more clinical study than any other cases of copper-related intoxication. Its ingestion was followed immediately by nausea, vomiting (often with blood), and abdominal pain; diarrhea, cyanosis, motor incoordination, and collapse then ensued. Hemoglobin appeared in the blood plasma and urine, and not uncommon findings were methemoglobinemia or sulfhemoglobinemia, bilirubinemia, uremia, and metabolic acidosis. Serum copper levels attained 510 µg%. Death occurred in the presence of acute renal failure. Post-mortem examination showed hemorrhages into the gastric and intestinal mucosas (which in some cases were covered with a bluish deposit) and focal or diffuse necrosis in the liver and kidneys. Cases of poisoning following irrigation of wounds with copper sulfate have also been reported.

Other examples of acute copper intoxication worth mentioning involve the use of an intrauterine contraceptive device made of copper whose prolonged sojourn in the uterus led to a large increase of copper in the woman's hair; lethal poisoning of a woman after the intrauterine insertion of blue vitriol for pregnancy interruption [Zhukov & Novoselov]; and poisoning of patients on hemodialysis by the copper released from the tubing of the hemodialysis system [Klein].

Chronic exposure

Animals

Long-term inhalation exposure of rabbits to dusts of **electrolytic copper** at 200–350 mg/m^3 for 5 h daily caused hypersalivation in the first few days of exposure; the animals still alive after a month were lethargic and thirsty. Rabbits and rats inhaling these dusts in much lower concentrations (10–20 mg/m^3, 5 h/day over 8 months) showed some inhibition of erythropoiesis, elevation of copper levels in the liver, bones, and blood, and increases of collagen in the lungs by 65% at 1 month and by 100% at 6 months. The main findings at autopsy were desquamative bronchitis, peribronchial and perivascular infiltration, areas of emphysema in the lungs, and cellular dystrophy in the liver and kidneys. **Copper** dust concentrations of 1 to 10 mg/m³ caused no detectable blood changes; a small increase in collagen was noted, but not until after 4 months of exposure [Brakhnova].

Exposure to a **copper-containing condensation aerosol** (70–80 mg/m³, 3 h/day, for 4–9 months) led to pulmonary alveolar proteinosis in rats [Arutyunov *et al.*]. In a 12-month study, aerosols generated in the welding of copper articles and consisting essentially of **copper oxides** caused partial mortality in rats exposed to their concentrations of 120–150 mg/m³ for 3 h per day. The surviving animals appeared exhausted, their leukocytes exhibited weakened phagocytic activity, and their blood was low in sulfhydryls; at autopsy, alveolar epithelial necrosis and diffuse pneumosclerosis were noted. The presence of silica in the aerosol generated during copper welding enhanced the fibrogenicity of the latter [Vlasova & Pryadilova].

Chronic inhalation exposure of mice and rats to a **copper ore** dust (1.1% Cu) or a **copper concentrate** dust (17.5% Cu) at 200 mg/m³ decreased sialic acids in the blood, reduced cholinesterase activity, and altered the content of free amino acids in the blood serum, urine, and lung tissue; the dusts also elicited fibrogenic responses, which were most strongly marked by months 12–15 of exposure. Inhalation of **copper chlorides** at 0.02–0.0075 mg/m³ over 3 months impaired gonadal function in male rats; the no-observed-effect level on this criterion was 0.0033 mg/m³ [T. Yermachenko & A. Yermachenko]. With 4-month inhalation exposure, the threshold copper chloride concentration for rats was 0.02 mg/m³ as defined from changes in chronaxy and in sulfhydryl, cholinesterase, alkaline phosphatase, and copper oxidase levels in the blood [Kamildzanov].

Copper concentrations comparable to those present in the atmospheric air at distances of 2.5 to 3.5 km from a mining and smelting complex (1.3 mg/m³ on average) caused substantial alterations in rats after 90 to 100 days of continuous exposure. Hematologic changes, including shifts in total protein and in the ratio of protein fractions and an increase in alkaline phosphatase activity, were induced in rats by a **copper(II) disintegration aerosol** in the concentration range of 0.1 to 1.0 mg/m³, while its concentrations as low as 0.01 mg/m³ caused testicular atrophy and impaired spermatogenesis. Female rats were less susceptible than males [Ginoyan].

Ruminant animals, in particular sheep, are very sensitive to copper. After dietary dosing of sheep with **copper** at 100 mg/kg feed for 9 days, the animals appeared strongly inhibited and weak, refused to eat, and responded poorly to external stimuli. Their visible mucous membranes were yellow and anemic and their urine was dark red. Muscle tremor, painfulness in the abdominal region on palpation, a weak and rapid pulse, and frequent, superficial, and labored breathing were all noted. Some of the animals died, their death being preceded by convulsions, hind leg paresis, a comatose state with extinction of reflexes and the complete disappearance of reactions to external stimuli. Erythropenia, leukocytosis, decreased hemoglobin concentrations, and increased bilirubin levels were recorded in the blood, as were elevations of lactate dehydrogenase, aspartate transaminase, and alanine transaminase activities along with inhibition of acetylcholinesterase activity. On pathologic examination of the dead animals, generalized yellow discoloration of serous membranes, fasciae, subcutaneous fatty tissue, and some other tissues was seen. The pleural and abdominal cavities were filled with fluid, and there were vascular congestion and areas of hemorrhage and necrosis in the stomach and intestine. The kidneys were dark and enlarged, their capsules were yellowish with evidence of hemorrhage beneath them, their vessels were dilated, and the boundary between the cortical and medullary layer was indistinct. In some parenchymal organs copper concentrations were as high as 700 mg/kg [Krol & Talanov]. According to these authors, the dietary concentration of copper for sheep should not exceed 20 mg/kg feed.

For sheep, copper levels in soil from 6×10^{-3} to $7 \times 10^{-2}\%$ have been reported as excessive and carrying a risk of anemia and hemolytic jaundice; concentrations of $(1.3-1.7) \times 10^{-3}\%$ are regarded as normal and those below $1 \times 10^{-3}\%$ as too low [Goncharuk; Kovalsky].

Sheep fed daily diets containing **copper sulfate** at 20 mg/kg (as Cu) and **copper iodide** at 750 mg/kg developed chronic intoxication manifested as hemolytic anemia and renal dysfunction; autopsy showed cellular necrosis in the renal tubular epithelium and dystrophic changes in the liver and spleen characteristic of hemolytic disease. The severity of clinical and morphologic manifestations correlated with copper levels in the blood [Sutter *et al.*]. Hemolysis occurred in sheep after 6 weeks of **copper(II) sulfate** feeding in a dose of 1 g daily [Ishmael *et al.*].

In young pigs, copper salts added to the diet increase gains in body-weight but may produce a toxic response if their level in the feed is large; thus, signs of toxicosis were evident in pigs with a copper salt dose of 250 g per tonne of dry feed, while the dose of 750 g per tonne led to severe poisoning [Suttle & Mills].

In rats, the dietary intake of **copper(II) sulfate** in a dose of 5 or 50 mg/kg reduced hemoglobin concentration and ceruloplasmin activity, caused disturbances in hepatic protein metabolism and function, and depressed antibody production; with the dose of 0.5 mg/kg, the only notable change was an increase in alkaline phosphatase activity. High embryonal mortality rates and significantly reduced fertility were recorded for female rats that had been mated with intact males after a 5-month exposure to **copper** through their drinking water in relatively low doses (10.1, 1.0, or 0.1 mg/L) [Nadeyenko *et al.*].

In rabbits, dietary dosing with **copper(II) acetate** in amounts equal to about 0.1% or 0.5% of the diet by weight, caused liver damage (including periportal fibrosis and mononuclear infiltration of Glisson's capsule) and degenerative changes in brain cells. Liver function was impaired (as judged by the rate of sulfobromophthalein excretion) in rats fed a 0.125% copper acetate solution for 1 month, although no morphologic changes were detected in that organ. A 0.2%

solution of the acetate fed to rats for 320 days resulted in significantly raised copper levels in the kidneys, brain, and liver. No alterations in the function or structure of the thyroid gland were seen in rats given copper acetate in doses of 0.01 or 0.05 mg for 1 to 6 months [Novokhatskaya]. The effect of copper on rabbit erythrocytic porphobilinogen synthesis has been described [Farant & Wigfield].

For feedstuffs, the following maximum permissible copper levels have been proposed (in mg/kg): 80 in mixed feeds for pigs and poultry; 30 in mixed feeds for large and small cattle; 30 in grain, grain forage, coarse and succulent fodders, and in root and tuber crops; 80 in feeds prepared through microbial synthesis; 500 in mineral additives [Antonov et al.].

Man

In chronic intoxication by copper or its salts, functional disturbances of the nervous system, kidney, and liver may occur, as well as ulceration and perforation of the nasal septum; affinity of copper for the sympathetic nervous system was reported [Browning].

Among workers manufacturing articles from copper or its alloys, a number of abnormalities have been recorded, including cerebral angioneurosis, reduced phagocytic activity of leukocytes, lowered lysozyme titers and bactericidal activity in blood serum, and gastrointestinal disorders along with 28% to 38% increases in the blood content of copper [Gorbunova; Nikogosian et al.; Nikultseva]. Autoimmune reactions, reduced lysozyme activity, and elevated blood levels of copper and ceruloplasmin were observed in workers inhaling aerosols of refined copper in concentrations of 0.45 to 0.89 mg/m^3; the prevalence of abnormal findings was highest among those with a long period of employment (10–15 years) [Lychagin et al.].

Health examinations of workers employed in electrolytic copper refining and exposed to a combination of copper and other adverse agents (e.g. sulfuric acid vapor) revealed a number of abnormalities, including central nervous system dysfunction, peripheral nervous system disorders of the lumbosacral radiculitis type, diminished immunobiologic reactivity, dental and oral mucosal lesions, gastritis, and gastric ulcers [Melnikova et al.]. No signs of pneomoconiosis or chronic intoxication were detected in copper cutters and welders exposed to copper aerosol concentrations of 1.6 to 45 mg/m^3 for periods of 1 to 3 years [Vorontsova & Pryadilova].

Among workers handling copper powders, rhinitis, pharyngitis, and gastritis occurred at higher frequencies and lasted longer than in the general population. Some of the workers complained of heartburn, stomach aches, and poor appetite; in those employed for more than 5 years, reduced vital capacity of the lungs, bilirubinemia, and liver dysfunction were often noted. Increases in hemoglobin, erythrocyte counts, and blood content of copper were described as characteristic findings in copper-mine workers; other abnormalities included accentuated bronchovascular markings, hilar thickening, and basal emphysema on X-ray examination, elevated blood levels of sialic acids, and gingival inflammation and fibrosis [Bagian; Lutsenko, 1970, 1976].

Histopathologic examinations of miners who died from various causes have indicated that two prominent features of the pneumoconiosis induced by copper-containing dusts are bronchial inflammation of early onset and a nodular or diffusely sclerotic silicosis, and that the inflammatory response mainly depends on how much copper the dust contains, whereas the fibrotic changes are dependent on its content of free silica [Zhakenova].

Many copper smelter workers exposed for years to the combined action of copper and silica aerosols and other adverse factors (e.g. unfavorable microclimate) complained of dull pain in the right hypochondrium and were found to have enlarged livers, positive tests for urobilin in urine, and bilirubinemia [Litkens]. Various gastric disorders, including gastritis with achylia in some cases, were detected in 93% of copper smelters thus exposed. Other findings were elevated serum copper levels (1–1.8 μg/ml), urinary excretion of coproporphyrin in large amounts, and depleted alkaline reserves in the blood [Zhislin et al.]. There is evidence that occupational cancer may develop [Shabad].

It has been shown that the length and intensity of occupational exposure to copper correlate well with its concentration in the workers' hair. Thus, hair samples from workers in various occupations exposed to copper for short periods (6–8 months or less) contained the metal in concentrations that did not virtually differ from the average value of 1.93 mg% in the control group, whereas those from workers with longer lengths of employment contained significantly higher concentrations, for example 4.19 mg% (foundry workers), 6.5 mg% (polishers and electric welders), and 13.5 mg% (copper platers). The hair of persons who complained of rapid fatiguability and of a metallic taste in the mouth by the end of the shift had copper levels ranging from 18.5 to 23.6 mg%. Increased copper levels were also found in the hair and urine of children living near copper smelters [Milham].

In urban areas, the major source of copper (and other metals) in the hair is thought to be street dust [Doi et al.]. Elevated copper levels (up to 94 μg/L) were detected in the sweat of individuals taking oral contraceptives [Stauber & Florence]. In Germany five children of different farmers, younger than one year old, suffered from severe liver cirrhosis — they had all been fed with Cu-containing (2.5–10 mg/L) acid water from similar private wells [Scheinberg].

Patients with Wilson's disease exhibit a life-long deficiency of the plasma copper protein ceruloplasmin, and an excess of hepatic copper which is the result of impaired lysosomal excretion of hepatic copper into bile and is associated with diminished hepatic synthesis of ceruloplasmin. This genetic defect causes the net retention of 1% of dietary intake of copper, which results in progressive and fatal copper toxicosis [Scheinberg].

Effects of the pesticide Cupronil and the fungicide Cuprozan in animals

Cupronil

This pesticidal preparation contains 35% copper(II) carbonate hydroxide and 15% zinc N,N-dimethyldithiocarbamate; it is readily wettable with water and dissolves well in saliva and gastric juice.

With oral administration by gavage, LD_{100} and LD_{50} values of 400 mg/kg and 150 mg/kg, respectively, were obtained for mice. In rats and rabbits exposed to Cupronil concentrations of 16 to 40 mg/m^3 by inhalation for 3 h/day over a month, the following manifestations of toxicity were observed: irritation of the upper respiratory tract mucosa, reduced hemoglobin levels and erythrocyte counts, reductions in the phagocytic activity and phagocytic index of leukocytes, eosinophilia, increased permeability of the histohematic and hematoencephalic barriers, and elevated copper levels in interior organs; the main findings on autopsy were catarrhal bronchitis, bronchiectasis, and areas of interstitial inflammation in the lungs. Exposure to 0.6–0.8 mg/m^3 for 3 h daily during 4 months was attended by progressive inhibition of phagocytic and cholinesterase activities in the blood [Medved].

Multiple dermal application of a 1.5% Cupronil suspension in water resulted in epidermal cornification in rabbits. Also in rabbits, a 0.5% suspension of this pesticide applied to the nasal mucosa produced a sharp increase in leukocyte migration and in the number of shed cells. Both of these suspensions caused mild conjunctivitis when dropped into rabbit eyes [Medved].

Cuprozan

This is a fungicide containing 37.5% copper(II) chloride hydroxide and 15% zineb (zinc ethylene-bis-dithiocarbamate). Its LD_{50} was 373 mg/kg for mice after oral administration and 30 mg/kg for rats after introduction into the trachea. The clinical picture of acute poisoning was dominated by symptoms attributable to inhibition of the central nervous system. After a single intratracheal application at 10 mg/kg (the lowest dose lethal to rats), decreased weight gains and reductions of ascorbic acid content and carbonic anhydrase activity in the blood were noted, as was lymphopenia in some animals. A single intragastric administration at 30 mg/kg caused a big rise in the urinary excretion of 17-ketosteroids and a substantial reduction in blood cholesterol [Smirnova].

Exposure to Cuprozan by inhalation at 1.5 to 3.5 mg/m^3 for 3 h/day caused partial mortality among rats after 1 month. In the surviving animals, a 40% rise in erythrocyte counts, lymphocytosis, a 5-fold increase in the chronaxy of extensor muscles of the legs, and a 2-fold reduction in the phagocytic activity of leukocytes with a 4-fold decrease in the phagocytic index were recorded. Inhalation exposure to 0.1–0.4 mg/m^3, 3 h/day for 2 months, resulted in slight erythrocytosis and strongly marked lymphocytosis. In both chronic and acute experiments, evidence for the ability of this fungicide to cross the histohematic and hematoencephalic barriers was obtained. Autopsy of acutely exposed animals showed necrotic tracheal mucosas, exudate-filled alveoli, and dystrophic changes in the liver, kidneys, and myocardium. Subcutaneous injection of 100 mg Cuprozan once a weak for 12 months induced the development of tumors in about 50% of the test (male) rats [Neiman].

In rabbits, Cuprozan produced no irritant effects on the skin; it caused mild conjunctival hyperemia when dropped into their eyes [Medved].

Dermal and ocular effects

Prolonged occupational exposure to copper compounds often leads to greenish-yellow or greenish-black discoloration of the facial skin, hair, and conjunctivae, and to the appearance of a dark-red or purple-red line on the gangivae. Copper and its salts and oxides irritate the skin. Contact of the skin with copper salts may result in an itching eczema of papulovesicular nature. Their dusts irritate the eyes and may cause conjunctivitis, edema of the eyelids, and ulceration and turbidity of the cornea.

Many cases of allergic dermatitis resulting from contact with copper, copper compounds, or brass have been described, for example, in workers exposed to **copper(II) nitrate**, those handling telephone wires, and those who carried bags with **copper(II) sulfate**.

In health surveys of coppersmiths and workers of copper electrolysis departments, skin tests for copper were positive in nearly a half of individuals with dermatitis or eczema at the time of the survey or in the past and also in some of those free from these diseases [Izrailet & Eglite; Kolpakov & Momot]. The marked allergic properties of copper have been attributed to its high capacity for penetrating the epidermal barrier to reach the dermis [Kolpakov & Mamot].

In workers spraying vineyards with Bordeaux mixture (which contains copper(II) sulfate and hydrated lime), a 4-fold increase in the number of leukocytes on the nasal mucosa was recorded after 2 to 3 h of work, indicating that the sulfate can produce mucosal lesions. Copper(II) sulfate may cause acute toxic dermatitis and even necrosis when in contact with a moist skin surface [Tsyrkunov].

Patients suffering from vitiligo were reported to have greatly decreased copper levels in the affected skin areas as compared to the skin of normal subjects (25 ± 5 *versus* 65 ± 8 µg/100 mg) while, on the contrary, having significantly elevated copper and ceruloplasmin concentrations in the blood serum (by 30% and 60%, respectively) [Genov *et al.*].

No penetration of copper through undamaged skin was observed [Kolpakov & Momot; Odintsova].

ABSORPTION, DISTRIBUTION, AND ELIMINATION

Under ordinary conditions, the human body receives, on average, 2 to 5 mg of copper per day, mostly in the diet. The inhalation route contributes only a small fraction of this amount — for example 0.02 mg at most in urban areas given that about 23 m^3 of air is breathed in daily. The adult human body normally requires 1.5–3 mg/day of copper; the demand for this metal may be increased to 4–5 mg by strenuous muscular activity [Rusin *et al.*; Venchikov]. Highest copper levels recorded in various human foods are given in Table 2.

Table 2. Highest copper levels found in various foods

Foods	Cu content (mg/kg)	Foods	Cu content (mg/kg)
Cereals:		Mushrooms	3
Corn	4	Berries:	
Barley groats	10	Raspberry	2
Buckwheat groats	11	Strawberry	7
Rye grain	12	Gooseberry	8
Rye flour	15	Black currant	11
Barley	18		
Oats	23	Fruits:	
Wheat grain	25	Apples	2
Wheat flour	10	Pears	2
Vegetables:		Milk	2
Radish	1	Curds	3
Cucumber	1	Meats:	
Carrots	2	Mutton	5
Tomatoes	2	Pork	6
Cabbage	3	Beef	7
Beets	4	Fish (river)	12
Dill	5	Eggs (hen's)	2
Horseradish	5		
Onions	6		
Potatoes	6		
Peas	13		
Kidney beans	13		

Source: Nozdryukhina.

Copper is a trace element essential to the nutrition of animals as well as plants. In higher animals it is a component of various proteins, including hematocuprein, ceruloplasmin, erythropoietin, turacin, cytochrome-*c*-oxidase, tyrosinase, etc. Its level in environmental media (e.g. soil) may be a limiting factor in the development of many organisms. Both deficiency and excess of copper are conducive to disease.

The average human being weighing 70 kg contains 50 to 120 mg of copper. Its common concentration ranges are 100–140 µg% in whole blood, 60–150 µg% in plasma, and 65–130 µg% in formed elements of the blood. Its content in the blood·progressively decreases with age. The urine contains from 0.016 to 0.033 mg Cu/L. Average levels in organs and tissues are as follows (in mg% wet weight) : liver, 0.3–1.3; brain, 0.1–0.6; heart, 0.1–0.6; kidney, 0.17–0.3; spleen, 0.26; lung, 0.25; bone, 0.1–0.6; muscle, 0.54 [Kovalsky]. Nails contain 11–53 µg/g; skin, 65 µg/100 mg; breast milk, 0.24–0.5 µg/ml. A sensitive and readily accessible indicator of the total body burden of copper is the hair, and its measurement there is therefore of particular interest to occupational physicians, toxicologists, and medical examiners; normal values in hair have been reported to be 14.0±2.5 µg/g for males and 16.4±3.2 µg/g for females [Ageyev et al.]. An important depot of copper is considered to be the cerebral cortex [Reitses].

The amount of copper absorbed in the body constitutes only some 30 to 50% of its daily intake in the diet; the rest is converted to the unavailable copper(II) sulfide and excreted. Absorption of ingested copper occurs in the stomach and the upper segments of the small intestine, and its plasma concentration peaks in about 2 h. Once in the blood, copper binds to albumin and transcuprein to be transported by them to the liver where it is incorporated into ceruloplasmin [Siegemund et al.]; in the latter, 90 to 96% of the plasma copper circulates. A copper-binding capacity is also shown by L-histidine [Levina].

About 80% of the absorbed copper is excreted by the liver via bile, 16% by the intestine, and 4% at most by the kidneys. The biological half-time of copper in the body is about 4 weeks [Bergqvist; Nozdryukhina].

HYGIENIC STANDARDS

Exposure limit values adopted for copper and its compounds in the USSR are given in Table 3. In the USA, the TLV-TWA has been set at 0.2 mg/m³ for copper fumes and at 1.0 mg/m³ for copper dusts and mists (as Cu). The highest permissible copper levels established in the USSR for human foods are listed in Table 4.

METHODS OF DETERMINATION

In air: polarographic determination (detection limit 0.1 µg/L in the assayed air volume) and spectral analysis (detection limit 0.2 mg/m³) [Muraviyeva]. In water: spectrographic determination after chloroform extraction of copper complexed with diethyldithiocarbamate and 8-hydroxyquinoline (detection

Table 3. Exposure limits for copper and its compounds

	Workroom air		Atmospheric air		Water sources		Soil	
	MAC_{wz} (mg/m³)	Aggregative state	MAC_{hm} (mg/m³)	MAC_{ad} (mg/m³)	MAC_w (mg/L)	MAC_{wf} (mg/L)	MAC_s (mg/kg)	Hazard class
Copper	1.0/0.5[a]	Aerosol	—	—	1.0[b]	0.001	3.0	2
Copper(I) chloride	0.5	Aerosol	—	0.002	—	0.001	—	2
Copper(I) or (II) oxide	—	—	—	0.002	—	0.001	—	2
Copper phosphide	0.5	Aerosol	—	—	—	—	—	2
Copper(II) sulfate	0.5	Aerosol	0.003	0.001	—	—	—	2
Copper(I) or (II) sulfide	—	—	0.003	0.001	—	—	—	2
Copper-nickel ore	4.0	Aerosol	—	—	—	—	—	2

[a] 1.0 mg/m³ is the concentration not to be exceeded at any time, while 0.5 mg/m³ is the average maximum allowable concentration per shift.
[b] Based on organoleptic criteria.

limits 4–20 μg) [Novikov et al.]. Inversion electrochemical methods have also been used, by which copper is determined together with zinc, cadmium, and lead [Kamenev et al.]. For copper measurement in dry residues of water, a technique of emission spectral analysis has been described [Petrova & Tabatchikova]. In soil: photometry [Makarova]. In biological media: colorimetry using dithizone, diphenylcarbazone, or sodium diethylthiocarbamate; luminescent and spectrographic methods are also available [Gadaskina et al.; Krylova]. (See also Sarkar.)

For copper determination in plant materials, a method of atomic absorption spectroscopy has been proposed (detection limit, 0.002–0.004 mg/kg; relative error ≤16%) [Sokolova]. For measuring copper in grapes, stewed fruits, juices, jams, and marinades derived from plants treated with the Bordeaux mixture, a method based on diethyldithiocarbamate formation has been used (detection limit 2 μg Cu/sample) [Klisenko]. Particle-induced X-ray emission analysis or synchrotron radiation have been used to study mobilization of copper in root tips [Scheinberg].

Other methods that have been used for the determination of copper in biological and environmental samples include neutron activation analysis and electrochemical methods such as pulse polarography, anodic stripping voltammetry, and high performance chromatography [Scheinberg].

Table 4. Maximum permissible levels of copper in foods (mg/kg)

Cereal grains, groats, flour, candies and pastries	10.0
Sugar	1.0
Bread	5.0
Milk and sour milk products	1.0
Condensed milk, canned	3.0
Butter	0.5
Margarine and cooking fats	1.0
Vegetable oils	0.5
Cheese, curds, casein	4.0
Vegetables, fresh or canned	5.0
Fruits, fresh or canned	5.0
Berries, fresh or canned	5.0
Spices	5.0
Tea	100.0
Meat and poultry, fresh or canned	5.0
Eggs	3.0
Fish, fresh or canned	10.0
Mollusks and crustaceans	30.0
Mineral waters	1.0
Beverages based on infusions or essences	3.0
Beer, wine, and other alcoholic beverages	5.0
Infant foods:	
milk-based	1.0
grain- and milk-based	5.0
vegetable- or fruit-based	5.0
meat- or poultry-based	5.0
fish-based	10.0

MEASURES TO CONTROL EXPOSURE

At the workplace. The basic engineering methods for preventing the exposure of workers to copper and its compounds may be considered to include the introduction of processes that will keep the emission of heat, gases, and dusts to a minimum, and mechanization and automation with remote handling of the operations involved. A primary consideration in securing a healthy occupational environment is the provision of adequate air exchange both in the breathing zone and in the working premises as a whole. Local exhaust ventilation should ensure the containment and removal of contaminants at the sources of their production. Thermal insulation of the heat-generating equipment is necessary for the maintenance of a normal microclimate. Soundproofing of the noise-generating machinery by means of appropriate covers is recommended. Silencers (e.g. those made of cermet material) may be used to attenuate the aerodynamic noise from pneumatic devices. In combination, the control measures must assure that the permissible levels of noxious agents are not exceeded. At least once every 3 months, the concentration of airborne harmful substances should be measured in the working area.

For *personal protection*, workers should be provided with appropriate devices to protect respiratory organs and particularly the skin. The dust from contaminated clothing must be removed. All work clothes should be laundered at the factory. For prophylactic purposes, the workers should undergo health examinations whose scope and periodicity will depend on the nature and extent of the risks involved. At the end of each shift, alkaline or oily inhalations should be administered to workers for preventing adverse effects of air contaminants on the upper air passages. For those exposed to high temperatures, oxygen cocktails, infusions prepared from grasses and berries, or other beverages such as 'kvas' (a national Russian beverage prepared from bread) may be recommended to slake thirst and improve the general condition.

Protection of the general environment. In plants where copper-containing raw materials are processed, efforts should be made to prevent or minimize the pollution of water, soil, food, and atmospheric air. The most difficult problems relate to the management of metal-containing solid wastes (slags, slurries, sludges, etc.). The dumping or burial of wastes containing toxic but industrially valuable materials should be regarded as a temporary measure to be adopted pending the development and introduction of technologies enabling solid wastes to be utilized to the maximal possible extent.

All waste liquids produced in the processing of copper-containing ores should be purified. In copper mines, copper can be removed from mine water in various ways, for example by cementation (on iron scrap or 'nickel sand'), neutralization, or sedimentation. Provision should be made for the construction of a storm-water sewage system and of facilities for mechanical and chemical (by neutralization) treatment of the rain and melt waters collected from the workplace.

The most efficient way of dealing with air pollution is to contain pollutants inside closed systems or enclosures, or to capture them by suction hoods and then remove them from the air, before releasing it to the outside, by means of adequate and well-operated and maintained air-cleaning devices such as dust separators (for particulate matter) and gas purifiers (for gaseous pollutants). An important contribution to the prevention of atmospheric pollution by copper can be made by technologies whereby valuable components are recovered from the captured particulate matter. Systematic monitoring of pollutant sources and levels is necessary, as is the observance of existing hygienic standards and of sound working practices at all stages of raw-material processing.

In regions with elevated environmental levels of copper, measures for the prevention of anemia among the population should be carried out. Where the soil has a high copper content, the application of gypsum or manure to farmlands is recommended to block the copper there and limit its entry into agricultural products.

The environmental burden of copper-containing pesticides and fertilizers can be appreciably reduced through the utilization of alternative methods for plant protection or soil treatment, such as the use of suitable weed-eating insects (herbiphages), the breeding of weed-resistant plants by selection, the introduction of improved soil-management practices, etc.

Measures to protect the environment from pollution by wastes generating in electroplating and containing metals, including copper, consist primarily in appropriate treatment of the spent wash waters. The discharge of these and other effluents into water streams should be systematically monitored through analyses of water samples in the laboratory. The effluents can be effectively purified by a method of two-phase separation that enables the solid phase of the sediment to be utilized for the manufacture of, for example, bricks or tiles. In this way, the copper burden on the soil can be decreased as well.

FIRST AID IN POISONING

For treatment of copper-fume fever, acetylsalicylic acid orally (in single doses of 1 g) and euphylline (synonym: aminophylline) intravenously should be administered, along with symptomatic therapy. In the event of poisoning by ingestion, the stomach should be washed out with copious quantities of water and diuretics be given. As antidotal therapy, unithiol (sodium 2,3-dimercapto-1-propane sulfonate) is indicated — 10 ml of a 5% solution subcutaneously in a single injection followed by 5 ml of this solution every 3 h for 2 or 3 days by the same route. Morphine and atropine may be injected subcutaneously to relieve severe abdominal pain [Artamonova; Mogosh].

REFERENCES

Ageyev, V. P. et al. (1983) Lab. Delo, No. 12, 28—31 (in Russian).
Alam, I. B. & Sadiq, M. (1989) Env. Poll., 57, 167—178.
Alloway, B. J. (1990) Heavy Metals in Soils. Halstead Press, New York.
Antonov, B. I. et al. (1989) Laboratornye issledovaniya v veterinarii: khimiko-toksikologicheskie metody. Spravochnik [Laboratory Studies in Veterinary: Chemical and Toxicologic Methods. A Reference Handbook]. Agropromizdat, Moscow (in Russian).
Artamonova, V. G. (1981) Neotlozhnaya pomoshch pri professionalnykh intoksikatsiyakh [Emergency Medical Care in Occupational Intoxications]. Meditsina, Leningrad (in Russian).
Arutyunov, V. D. et al. (1976) Gig. Truda Prof. Zabol., No. 9, 33—37 (in Russian).

Bagian, R. G. (1964) In: *Materialy 2-oi itogovoi nauchnoi konferentsii Instituta gigiyeny truda i professionalnykh zabolevaniy* [Proceedings of a conference organized by the Institute of Occupational Health]. Yerevan, pp. 42—44 (in Russian).

Bandura, Yu. D. (1969) *Byul. Eksp. Biol. Med.*, No. **10**, 105—108 (in Russian).

Berinia, D. Zh. *et al.* (1980) In: *Zagryaznenie prirodnoi sredy vybrosami avtotransporta* [Environmental Pollution by Automobile Exhausts]. Riga, pp. 16—27 (in Russian).

Berisha, A. *et al.* (1983) *Ontogenez*, No. **2**, 173—179 (in Russian).

Bergqvist, U. (1983) *New metals.* University of Stockholm (Institute of Physics).

Bergqvist, U. & Sundbom, M. (1978) *Copper: Health and Hazard.* University of Stockholm (Institute of Physics).

Bolshakov, V. A. *et al.* (1978) *Zagriaznenie pochv i rastitelnosti tyazhelymi metallami* [Contamination of Soils and Plants by Heavy Metals]. Moscow (in Russian).

Brakhnova, I. T. (1971) *Toksichnost poroshkov metallov i ikh soyedineniy* [Toxicity of Metals and their Compounds in Powder Form]. Naukova Dumka, Kiev (in Russian).

Bresler, V. M. (1959) *Vopr. Onkologii*, No. **12**, 663—668 (in Russian).

Brodsky, J. F. *et al.* (1930) In: *Gigiyena, bezopasnost i patologiya truda* [Occupational Hygiene, Safety, and Pathology]. No. 7, pp. 32—38 (in Russian).

Browning, E. (1969) *Toxicity of Industrial Metals.* Appleton-Century Crofts, London.

Chand, M. (1970) *Antiseptic*, **67**, 267—270.

Craun, G. (1981) *J. Amer. Water Works Assoc.*, **73**, 360—369.

Csasta, S. *et al.* (1968) *Z. Ges. Inn. Med.*, **23**, 345—348.

Davydova, V. I. *et al.* (1986) *Gig. San.*, No. **1**, 74 (in Russian).

Deodhar, L. & Deshpande, C. (1968) *J. Postgrad. Med.*, **14**, 38—41.

Dive, D. G. *et al* (1988) *Sci. Total Environ.*, **87/88**, 355—364.

Dixon, D. G. & Sprague, Y. B. (1981) *Can. J. Fish Aquat. Sci.*, **38**, 880—888.

Dmitriev, M. T. & Tarasova, L. N. (1985) *Gigiyena: Ekspress-informatsiya* [Hygiene: Current Awareness Information]. No. 5, p. 20 (in Russian).

Dobrovolskaya, E. V. (1983) *Izv. Akad. Nauk SSSR. Ser. Biol.*, No. **2**, 302—305 (in Russian).

Dobrovolsky, V. V. (1983) *Geografiya mikroelementov. Globalnoe rasseyanie* [Geography of Trace Elements. Global Dispersion]. Mysl, Moscow (in Russian).

Doi, R. *et al.* (1988) *Sci. Total Environ.*, **77**, 153—161.

Dzhangozina, D. M. & Slutsky, L. N. (1968) *Gig. Truda Prof. Zabol.*, No. **1**, 18—23 (in Russian).

Farant, J. P. & Wigfield, D. C. (1990) *J. Anal. Toxicol.*, **14**, No. 4, 222—226.

Gadaskina, I. D. *et al.* (1975) *Opredeleniye promyshlennykh neorganicheskikh yadov v organizme* [Determination of Inorganic Industrial Poisons in the Organism]. Meditsina, Leningrad (in Russian).

Genov, D. *et al.* (1972) *Clin. Acta*, **37**, 207—211.

Ginoyan, M. M. (1976) *Gig. San.*, No. **6**, 8—12 (in Russian).

Goncharuk, E. I. (1977) *Sanitarnaya okhrana pochvy ot zagryazneniya khimicheskimi veshchestvami* [Sanitary Measures to Protect Soil from Chemical Pollution]. Zdorovya, Kiev (in Russian).

Gorbunova, I. I. (1968) *Gig. Truda Prof. Zabol.*, No. **9**, 56—60 (in Russian).

Grushko, Ya. M. (1972) *Yadovitye metally i ikh neorganicheskiye soyedineniya v promyshlennykh stochnykh vodakh* [Poisonous Metals and their Inorganic Compounds in Industrial Waste Waters]. Meditsina, Moscow (in Russian).

Grushko, Ya. M. (1979) *Vrednye neorganicheskie soyedineniya v promyshlennykh stochnykh vodakh* [Harmful Inorganic Compounds in Industrial Waste Waters]. Khimiya, Leningrad (in Russian).

Helman, R. *et al.* (1983) *Toxicol. Appl. Pharmacol.*, **67**, 238—245.

Ilyaletdinov, A. N. (1984) *Mikrobiologicheskie prevrashcheniya metallov* [Microbiological Transformations of Metals]. Alma-Ata (in Russian).

Ishmael, I. *et al.* (1971) *Res. Veterin. Sci.*, **12**, 358—366.

Izmerov, N. F. *et al.* (1977) *Parametry toksikometrii promyshlennykh yadov pri odnokratnom vozdeistvii* [Toxicometric Parameters of Industrial Poisons Following Single Exposures]. Meditsina, Moscow (in Russian).

Izrailet, L. I. & Eglite, M. E. (1978) *Promyshlennost i immunologicheskoe sostoyanie organizma: Obzornaya informatsiya* [The Industry and the Immune Status: A review]. Moscow (in Russian).

Kabata-Pendias, A. & Pendias, H. (1986) *Trace Elements in Soils and Plants*. CRC Press, Boca Raton (Fl.).

Kamenev, P. I. *et al.* (1988) In: *Sovremennye fiziko-khimicheskie metody issledovaniya v gigiyene* [Present-day Physicochemical Methods of Investigation in the Field of Hygiene]. Moscow, pp. 50–54 (in Russian).

Kamildzhanov, A. Kh. (1985) *Gig. San.*, No. 10, 90–91 (in Russian).

Klein, W. *et al.* (1972) *Arch. Intern. Med.*, 129, 578–582.

Klisenko, M. A. (ed.) (1983) In: *Metody opredeleniya pestitsidov v produktakh pitaniya, kormakh i vneshnei srede* [Methods for Measuring Pesticide Levels in Foods, Feeds, and the Environment]. Moscow, pp. 232–234 (in Russian).

Kolpakov, F. I. & Momot, V. I. (1969) In: *Voprosy profpatologii* [Selected Aspects of Occupational Pathology]. Krasnoyarsk, pp. 62–65 (in Russian).

Kovalsky, V. V. (1974) *Geokhimicheskaya ekologiya* [Geochemical Ecology]. Nauka, Moscow (in Russian).

Krasnyuk, E. P. (1983) In: *Rukovodstvo po professionalnym zabolevaniyam* [A Handbook of Occupational Diseases]. Vol. 1, Meditsina, Moscow, pp. 273–289 (in Russian).

Krol, M. Yu. & Talanov, G. A. (1988) *Veterinariya*, No. 8, 52–54 (in Russian).

Krylova, A. N. (1975) *Issledovanie biologicheskogo materiala na metallicheskie yady drobnym metodom* [Testing of Biological Materials for Metallic Poisons by a Fractional Technique]. Meditsina, Moscow (in Russian).

Levina, E. I. (1972) *Obshchaya toksikologiya metallov* [General Toxicology of Metals]. Meditsina, Leningrad (in Russian).

Litkens, V. A. (1973) *Gig. Truda Prof. Zabol.*, No. 4, 99–100 (in Russian).

Lutsenko, L. A. (1970) *Gig. Truda Prof. Zabol.*, No. 10, 11–15 (in Russian).

Lutsenko, L. A. (1976) In: *Voprosy gigiyenicheskoi otsenki organicheskikh i mineralnykh pylei slozhnogo khimicheskogo sostava* [Hygienic Evaluation of Organic and Mineral Dusts having Complex Chemical Composition]. Moscow, pp. 24–30 (in Russian).

Lychagin, V. V. *et al.* (1984) *Gig. Truda Prof. Zabol.*, No. 10, 31–34 (in Russian).

Makarova, V. P. (1967) *Trudy Samarkandskogo Meditsinskogo Instituta* [Transactions of the Samarkand Medical Institute], 37, 308–313 (in Russian).

Manchuk, V. A. & Ryabov, N. A. (1983) In: *Metally: Gigiyenicheskie aspekty otsenki i ozdorovleniya okruzhayushchei sredy* [Metals: Hygienic Aspects of Evaluation and Environmental Health Promotion]. Moscow, pp. 194–199 (in Russian).

McDonald, J. & Cook, R. (1974) *Environ. Health*, 82, 153–154.

Medved, L.I. (ed.) (1965) *Gigiyena i toksikologiya pestitsidov i klinika otravleniy* [Pesticides: Hygiene, Toxicology, and Clinical Features of Poisoning]. Zdorovya, Kiev. (in Russian).

Mehra, A. (1989) *Availability and Uptake of Metals by Plants in Selected Areas of North Wales*. PhD thesis, University of London.

Melnikova, N. A. *et al.* (1972) *Trudy NII krayevoi patologii Kazakhskoi SSR* [Transactions of the Institute for Regional Pathology of the Kazakh SSR]. 23, 25–227 (in Russian).

Merenyuk, G. V. & Medzhibovskaya, E. E. (1970) *Gig. San.*, No. 1, 108–110 (in Russian).

Metelev, V. V. *et al.* (1971) *Vodnaya toksikologiya* [Aquatic Toxicology]. Kolos, Moscow (in Russian).

Milham, S. (1977) *Environ. Health Perspect.*, 19, 131–132.

Mogosh, G. (1984) *Ostrye otravleniya* [Acute Poisonings]. (Russian translation from Roumanian published in Bucharest).

Mullins, T. (1977) In: Bockris, J. O'M. (ed.) *Environmental Chemistry*. Plenum Press, New York, pp. 331–400.

Muraviyeva, S. I. (ed.) (1982) *Sanitarno-khimicheskiy kontrol vozdukha promyshlennykh predpriyatiy* [Sanitary and Chemical Control of Indoor Air in Industrial Establishments]. Meditsina, Moscow (in Russian).

Nadeyenko, V. G. *et al.* (1980) *Gig. San.*, No. **3**, 8—10 (in Russian).

Neiman, I. M. (1972) *Kantserogeny i pishchevye produkty* [Carcinogens and Foods]. Meditsina, Moscow (in Russian).

Nikogosian, Kh. A. *et al.* (1970) In: *Materialy nauchno-prakticheskoi konferentsii vrachei Kuibyshevskoi oblasti* [Proceedings of a workshop for physicians from the Kuibyshev region]. Kuibyshev, pp. 425—427 (in Russian).

Nikultseva, A. A. (1972) In: *Trudy Kuibyshevskogo NII Gigiyeny* [Transactions of the Kuibyshev Institute of Hygiene]. No. **7**, pp. 134—135 (in Russian).

Novikov, Yu. V. *et al.* (1981) *Metody opredeleniya vrednykh veshchestv v vode vodoyemov* [Methods for Measuring Hazardous Substances in Bodies of Water]. Meditsina, Moscow (in Russian).

Novokhatskaya, Z. V. (1960) *Vrach. Delo*, No. **2**, 155—158 (in Russian).

Nozdryukhina, L. R. (1977) *Biologicheskaya rol mikroelementov v organizme cheloveka i zhyvotnykh* [Biological Roles of Trace Elements in Humans and Animals]. Nauka, Moscow (in Russian).

Odintsova, N. A. (1981) *Vestn. Dermatol. Venerol.*, No. **2**, 18—22 (in Russian).

Patin, S. A. (1979) *Vliyanie zagryaznitelei na biologicheskie resursy i produktivnost mirovogo okeana* [The Impact of Pollutants on the Biological Resources and Productivity of the World's Ocean]. Moscow, (in Russian).

Petrova, N. N. & Tabatchikova, T. N. (1988) In: *Sovremennye fiziko-khimicheskie metody issledovaniya v gigiyene* [Present-day Physicochemical Methods of Investigation in the Field of Hygiene]. Moscow, pp. 67—72 (in Russian).

Raitses, V. S. (1981) *Neirofiziologicheskie osnovy deistviya mikroelementov* [Neurophysiological Principles Underlying the Action of Trace Elements]. Meditsina, Leningrad (in Russian).

Reich, D. *et al.* (1976) *Water Res.*, **10**, 299—302.

Ringenberg, O. *et al.* (1988) *South. Med. J.*, **81**, 1132—1139.

Roshchin, A. V. (1977) *Gig. Truda Prof. Zabol.*, No. **11**, 28—35 (in Russian).

Rusin, V. Ya. *et al.* (1980) *Vopr. Pitaniya*, No. **4**, 15—19 (in Russian).

Ryabinin, A. I. & Lazareva, E. A. (1982) In: *Monitoring fonovogo zagryazneniya prirodnykh sred* [Monitoring the Background Pollution of Natural Environmental Media]. No. 1, Leningrad, pp. 165—174 (in Russian).

Sarkar, B. (1988) In: Seiler, H. C. *et al.* (eds) *Handbook on Toxicity of Inorganic Compounds*. Marcel Dekker, New York, pp. 265—276.

Savitsky, A. V. & Golovan, V. I. (1970) In: *Gigiyena truda. Respublikanskiy mezhvedomstvennyi sbornik* [Occupational Hygiene. A collection of Republican interdepartmental papers]. No. 6, Zdoroviya, Kiev, pp. 42—49 (in Russian).

Scheinberg, I. H. (1991) In: Merian, E. (ed.) *Metals and their Compounds in the Environment—Occurrence, Analysis and Biological Relevance*. VCH, New York, pp. 893—908.

Shabad, L. M. (1978) In: *Itogi nauki i tekhniki, seriya Toksikologiya* [Recent Advances in Science and Technology, Toxicology series]. Vol. 9, VINITI publication, Moscow, pp. 7—58 (in Russian).

Shivanandappa, T. *et al.* (1983) *Poultry Sci.*, **62**, 405—408.

Sidorenko, G. I. (ed.) (1985) *Gigiyena okruzhayushchei sredy* [Environmental Hygiene]. Meditsina, Moscow (in Russian).

Siegemund, R. *et al.* (1988) *Zbl. Pharm. Pharmak. Laboratoriumsdiagn.*, **127**, 211—214.

Smirnova, V. M. (1968) *Farmakol. Toksikol.*, No. 1, 112—113 (in Russian).

Sokolova, G. N. (1989) *Gig. San.*, No. **5**, 52—54 (in Russian).

Stauber, Y. & Florence, T. (1988) *Sci. Total Environ.*, **74**, 235—247.

Sutter, M. *et al.* (1958) *Amer. J. Veter. Res.*, **19**, 890—892.

Suttle, N. & Mills, C. (1966) *Brit. J. Nutr.*, **20**, 135—148.

Tsoi, L. M. (1982) In: *Voprosy gigiyeny truda i professionalnykh zabolevaniy* [Selected Topics of Occupational Hygiene and Industrial Diseases]. Karaganda, pp. 227—228 (in Russian).

Tsyrkunov, L. P. (1982) *Vestn. Dermatol. Venerol.*, No. **10**, 42—46 (in Russian).

Vadkovskaya, I. K. & Lukashev, K. P. (1977) *Geokhimicheskie osnovy okhrany biosfery* [Protection of the Biosphere: Geochemical Principles]. Nauka i Tekhnika, Minsk (in Russian).

Venchikov, A. I. (1982) *Printsipy lechebnogo primeneniya mikroelementov v kachestve biotikov* [Principles for the Therapeutic Use of Trace Elements as Biotics]. Ylym, Ashkhabad (in Russian).

Vlasova, N. V. & Pryadilova, N. V. (1971) *Gig. Truda Prof. Zabol.*, No. 3, 20–24 (in Russian).

Vorontsova, E. I. & Pryadilova, N. V. (1977) *Vestn. Akad. Med. Nauk SSSR*, No. 2, 34–41 (in Russian).

Wang, P. K. & Chang, L. (1991) *Env. Poll.*, **72**, 127–139.

Yermachenko, T. P. & Yermachenko, A. B. (1987) In: *Gigiyena naselennykh mest* [Hygiene in Populated Areas]. Zdorovya, Kiev, No. 26, pp. 86–88 (in Russian).

Zakharova, L. N. & Udelnova, T. M. (1977) *Usp. Sovr. Biol.*, **83**, 274–276 (in Russian).

Zhakenova, R. K. (1966) *Zdravookhranenie Kazakhstana* [Health Care in Kazakhstan]. No. 8, 46–47 (in Russian).

Zhukov, V. F. & Novoselov, V. P. (1983) *Sudebno-Meditsinskaya Ekspertiza*, No. 3, 55–56 (in Russian).

Zhulidov, A. V. *et al.* (1980) *Dokl. Akad. Nauk SSSR*, **252**, 1018–1020 (in Russian).

Zislin, D. M. *et al.* (1972) *Kombinirovannoe deistvie khimicheskikh i fizicheskikh faktorov proizvodstvennoi sredy* [Joint Action of Chemical and Physical Factors in the Occupational Environment]. Sverdlovsk, pp. 124–128 (in Russian).

Silver and its compounds

Silver bromide (bromargyrite [min.]); **s. chloride** (cerargyrite or horn silver [min.]); **s. nitrate** (lunar caustic); **s. oxide; s. sulfide** (acanthite (α) [min.]), argentite (β) [min.])

IDENTITY AND PHYSICOCHEMICAL PROPERTIES
OF THE ELEMENT

Silver (Ag) is a metal in group I of the periodic table, with the atomic number 47. It is a mixture of two stable isotopes, ^{107}Ag and ^{109}Ag.

Silver is more malleable and ductile than any other metal except gold, and it has the highest thermal and electric conductivity of all metals. In compounds it is present in the oxidation state +1 and, much less commonly, +2 or +3. In air, oxygen adsorbs on its surface to form a layer of the oxide Ag_2O; molten silver takes up O_2 in considerable amounts. Halogens form protective films of halides on its surface at ordinary temperatures. Chlorine containing traces of water reacts well with silver at 80°C, while dry chlorine reacts at 300°C. Fluorine will only react at a red heat. Hydrogen halides do not react at ordinary temperatures in the absence of an oxidizing agent. On heating, silver reacts with sulfur, selenium, and tellurium. It does not react with hydrogen or nitrogen. Carbon and phosphorus interact with it at a red heat. Silver reacts readily with nitric acid, with hot concentrated sulfuric acid, and, in the presence of O_2, with dilute sulfuric acid as well. Its reaction with aqua regia is hindered by the silver chloride that is formed. It does not react with alkalies. It reacts with molten nitrates, sulfates, thiosulfates, chlorides, and cyanides of other metals. It forms many complex compounds. See also Appendix.

NATURAL OCCURRENCE AND ENVIRONMENTAL LEVELS

Silver is one of the least abundant elements. It occurs native and also combined in ores in the form of insoluble or difficultly soluble salts, e.g. lead, zinc, and copper ores in which it occurs as silver sulfide (argentite). More than 50 silver-containing minerals are known [Petering & McClain].

The clarkes of silver in the earth's crust and the granite layer of the continental crust are estimated at $0.07 \times 10^{-4}\%$ and $0.048 \times 10^{-4}\%$, respectively [Dobrovolsky]. Its concentrations range from 40 to 100 µg/kg in magmatic rocks, and from 50 to 250 µg/kg in sedimentary rocks, the higher levels being found in limestones and sandstones [Kabata-Pendias & Pendias].

River waters contain silver in concentrations of 0.01 to 3.5 µg/L, and its global run-off from rivers amounts to 11 000 tonnes annually. Its average concentration in seawater is about 0.04 µg/L. Coal and oil have average silver contents of 0.5 µg/kg and 0.0001 µg/kg, respectively [Dobrovolsky].

In soils, the common range of silver concentrations is 0.03–0.09 µg/g. Soils in mineralized areas are enriched in silver, but its content does not often exceed 1 µg/g, although levels up to 44 µg/g have been reported. The plow layer of mineral soils contained about 0.7 µg/g while soils rich in organic matter contained 2–5 µg/g [Kabata-Pendias & Pendias].

Reported silver concentrations in plants ranged from 0.03 to 0.5 µg/g dry weight, while in plant foodstuffs they ranged from 0.07 to 2.0 µg/g dry weight. The values depend heavily on the plant species and the time of plant sample collection (e.g. plants sampled in September contained much less silver than those sampled in May) [Kabata-Pendias & Pendias].

The amounts of silver emitted into the atmosphere from natural sources were reported to be as follows (tonnes/year): 150 from forest fires, 50 from wind erosion, 17 from volcanic eruptions, 9 from sea salt. In the biosphere, the metal is highly dispersed [Bowen].

PRODUCTION

Approximately 80% of the silver produced worldwide is extracted as a by-product from complex ores and from the ores of gold or copper. To recover silver from a silver or gold ore, the ore is treated with an alkaline solution of sodium cyanide followed by separation of the metal from solutions of the formed complex cyanides by reduction with zinc or aluminum. From copper ores, silver is smelted together with blister copper and is then separated from the anode slime formed during electrolytic refining of the copper. Silver nitrate is prepared by dissolving silver in nitric acid followed by refining.

The world output of silver, which amounted to 9430 tonnes in 1973, has been predicted to reach 12 000 tonnes in 2000 [Kabata-Pendias & Pendias]. It is estimated that 31% of all silver refined in 1985 originated from secondary waste sources [Petering & McClain].

USES

Silver is mainly used in the form of alloys, in particular for jewelry, tableware, and a variety of domestic articles. Silvering of radio components increases their electric conductivity and corrosion resistance. Silver solders are employed for soldering titanium and its alloys. Silver is used in printed circuits, electrical contacts, in silver-zinc and silver-cadmium storage batteries, as a catalyst, in the food industry, and in the manufacture of stained porcelain (see also Petering & McClain; Nordberg & Gerhardsson). Silver ions are used in low concentration to sterilize water. **Silver halides** and **silver nitrate** are widely used in photography. Silver compounds have found much use in medicine, while metallic silver is employed in surgery and in the making of various surgical instruments and is an important component of alloys used as filling materials in dentistry.

Experiments on the seeding of clouds with silver iodide to prevent precipitation have been reported [Bigg].

MAN-MADE SOURCES OF EMISSION INTO THE ENVIRONMENT

When coal is burned, about 50% of the silver contained in it is emitted into the atmosphere. The total amount of silver entering the atmosphere in this way was estimated to be approximately 800 tonnes in 1900, 2000 tonnes in 1970, and 4100 tonnes in 1980; for 2000, the release of 6500 tonnes has been predicted [Dobrovolsky]. Annually, some 800 tonnes of silver are emitted to the ambient air by thermal power stations and 1400 tonnes by industry and transport.

TOXICITY

Soil microbiota
Silver present in the soil as Ag^+ at a concentration of 5 μmol/g was toxic to nitrifying bacteria, as judged by inhibited production of inorganic nitrogen in the soil [Kabata-Pendias & Pendias].

Plants
The highest concentration of silver in top soil not toxic to plants was 2 μg/g dry weight of the soil. On a dry weight basis, its concentrations in plant leaves were

considered to be normal at levels around 0.5 µg/g and toxic at 5 to 10 µg/g.
Silver concentrations of 9.8 µg/L and 4.9 µg/L in the water used for irrigation
were destructive to corn and lupin plants, respectively. Silver nitrate was lethal
for bean plants at concentrations of 10^{-4} mol/L water; this salt was found to
inhibit the formation of ethylene in apples [Lis *et al.*]. The toxicity of silver to
plants is mainly attributed to the alteration it causes in the permeability of their
cell membranes. Aquatic plants tend to concentrate silver from their environments
several hundredfold [Petering & McClain].

Aquatic organisms

Data from various sources on the toxicity of different silver concentrations to fish
and some other hydrobionts are given in Table 1.

Table 1. Toxicity of silver for hydrobionts

Silver concentration (µg/L)	Organisms	Effect
3	Stickleback	Lethal (silver nitrate)
4	Various fish	Lethal for 50%
4.3	Guppy	Lethal for 50%
4.8	Stickleback	Lethal
5	Various fish	Lethal
5.1	Daphnia	Lethal
6	Guppy	Lethal for 50% after 24 h
10	Bacteria	Lethal
30	Daphnia	Lethal for 48 h
40	*Escherichia coli*	Harmful
50	Algae	Harmful
100	Stickleback	Lethal after 24 h
290	Carp	Toxic

Gastropod mollusks exposed to **silver nitrate** in water at concentrations of 1,
5, or 10 µg/L were found, particularly the males, to accumulate considerable
amounts of silver in all tissues. Females exposed to 10 µg/L over that period
gave significantly reduced larval yields [Calabrese *et al.*].

Animals and man

Acute exposure

Animals
Metallic silver dust administered to mice orally (by gavage) in doses of about 5 g/kg or intraperitoneally at 4 g/kg was lethal to some animals, but the LD_{50} could not be established. Exposure of mice to this dust by inhalation resulted in lung tissue changes very similar to those observed in response to a foreign body, indicating that the dust had low toxicity.

Man
A case of severe occupational poisoning with **silver fumes was** described [Forycki *et al.*]; the clinical picture resembled that seen in the adult respiratory distress syndrome (shock lung). Cases of death after intravenous injection of 50 mg collargol (colloidal silver) for therapeutic purposes have been described [Nordberg & Gerhardsson]; autopsy demonstrated pulmonary edema and areas of hemorrhage and necrosis in the bone marrow, liver, and kidneys. A woman died shortly after intrauterine administration of **silver nitrate** in a dose of about 25 g [Fowler & Nordberg]. The lethal dose of silver nitrate for humans is approximately 10 g. See also Petering & McClain.

Repeated and chronic exposures

Animals
Rabbits receiving **silver** in an aqueous solution at 0.25 or 0.025 mg/kg for prolonged periods showed reduced activity of the immune system, altered conditioned reflexes, and histopathologic changes in the vascular, nervous, and glial tissues of the brain and spinal cord. Doses of 0.0025 mg/kg or less were without effect. None of the doses mentioned above altered hemoglobin level, erythrocyte and differential blood counts, protein-forming function of the liver, or the blood content of sulfhydryl groups [Barkov & Elpiner].

Silver nitrate administered by oral gavage to rats daily in an amount of 1 ml (1% solution) for 2 to 3 weeks caused weight loss in all animals (by 20% on average); some of them had reduced erythrocyte counts and hemoglobin levels and some showed foci of fatty degeneration, small areas of necrotized tissue, and signs of toxic edema in the liver. Moderate nephrosis and a slight increase in thyroid function were also noted occasionally. Silver nitrate had virtually no effect on the blood level of sulfhydryl groups in rats, but reduced it in rabbits and guinea pigs which were similar to man in this respect.

Rabbits with experimentally induced silver deposition in tissues (argyria) showed weight loss. Turkeys that had been receiving **silver nitrate** in the diet, at 900 mg/kg body weight for 18 weeks, had enlarged hearts and ventricular hypertrophy; about 30% of them died in the course of the study. Hyperchromic anemia has been a common finding in animals chronically ingesting silver compounds in amounts of several hundred mg/kg in the diet daily. Rats exposed to 0.2% silver nitrate in their drinking water for 10 to 50 weeks were found to have deposits of silver granules in the renal glomeruli and tubules as well as pathologic changes in the latter [Fowler & Nordberg].

Man

Four cases of pulmonary emphysema discovered in silver covering workmen at autopsy were described [Barrie & Harding]; the authors, however, were hesitant in ascribing this condition to silver impregnation of the alveolar walls in view of the advanced age of the deceased persons.

Long-term occupational exposure to silver and its salts results in metallic silver being deposited in the connective tissue and capillary walls of various organs, including kidney, bone marrow, and spleen, as well as in reticuloendothelial elements, notably the stellate cells of the liver. Silver also accumulates in the skin and mucous membranes imparting to them a distinctive gray-green, bluish, or slaty-gray color, especially in areas exposed to light; occasionally the skin is so dark as to resemble that of a negro. The discoloration is enhanced by ultraviolet light.

This disease (argyria) develops slowly: its signs first appear after 2 to 4 years of exposure and do not usually become strongly marked until after dozens of years. Two forms of argyria — diffuse and localized (or limited) — have been described. The former was characterized by widespread discoloration of the skin and mucous membranes. The first to darken were the lips, temples, and conjunctiva, followed by the eyelids; the other hairless areas of the facial skin were less discolored. Similarly, the oral and gingival mucosas and the neck exhibited more discoloration than did the chest and back. In the upper extremities, the hands, particularly the nail beds, were discolored the most. The skin areas normally covered by clothing were usually affected to a lesser degree (if at all) than the uncovered areas. Localized argyria presented in the form of very small bluish-black spots on the skin or, sometimes, on the mucous membranes of the upper airways. The spots tended to remain unchanged for years; rarely, they decreased in size slightly with time.

The patient's general health may remain normal, although prolonged occupational exposure to silver salts is likely to impair it. For example, most of the 15 workmen with manifestations of argyria complained of pain in the right hypochondrial region, and the liver was enlarged and painful in 4 of the 5 workwomen with this condition. Cases of silver poisoning involving severe mental disturbances have been described [Breton *et al.*].

In some cases, gastrointestinal catarrhs and increased numbers of mononuclear cells have been observed. If the cornea is strongly discolored, visual acuity may decrease somewhat in dim light. In very long-standing cases, opacification in the anterior capsule of the crystalline lens, punctate inclusions in the lens, and changes in the ocular fundus may occur. There is evidence that patients with argyria do not suffer from infectious diseases, apparently because of the disinfecting properties of silver ions. Attempts to control argyria with 2,3-dimercaptopropanol (BAL), administered topically or intramuscularly, have proved unsuccessful and even detrimental. Further progression of the disease can be stopped by permanent removal of the patient from exposure to silver and its compounds.

Persons working with **silver bromide** or **silver sulfide** dust may complain of cough, a tickling sensation in the throat, sanguineous nasal discharge, and lacrimation; further exposure is likely to result in necrotization and desquamation of the nasal mucosa with accumulation of crusts and blood in the nose. **Silver nitrate** exerts caustic and astringent effects upon the skin and mucous membranes. Inflammatory diseases of the skin have also occurred among those handling this salt.

Persons occupationally exposed to **silver oxide** for about 2 years presented with ophthalmologic problems, such as altered color of the eyeballs, heavy eyelids, lacrimation, mucous or mucopurulent discharges from the eyes, and somewhat sticky eyelids in the morning [Larionov].

Studies of renal function in 27 workers (mean age 41 years) engaged for a mean period of 8.1 years in the preparation of powders containing the **chloride**, **nitrate** and **oxide** of silver, revealed elevated silver levels in the urine (in 26 cases; mean value 11.3 µg/L) and the blood (in 24 cases; mean value 1 µg/L), silver deposits in the conjunctiva (in 17) and in the cornea (in 6), and irritation of the nasopharyngeal and conjunctival mucosas (in 15); moreover, 8 of the workers complained of repeated nosebleeds and 9 complained of cough and wheezing. Renal function was reduced to varying degrees, but the cause of the reduction could not be ascertained [Rosenman *et al.*].

Examination of 36 workers who had been recovering silver from photographic and x-ray films for prolonged periods suggested the possibility of chronic poisoning with the hydrogen cyanide released during the processes involved in such work [Blanc *et al.*]. Chronic and antagonistic effects on animals and humans are covered in Petering & McClain.

ABSORPTION, DISTRIBUTION, AND ELIMINATION

The average amounts of silver ingested by human beings in food and water range from 60 to 80 µg per day. Parenterally administered silver is long retained at the site of entry, with only small amounts being absorbed into the blood, and is slowly excreted by the gastrointestinal tract (~90%) and kidneys. After oral administration, intestinal absorption in mice, rats, and monkeys amounted to 10%. Silver readily penetrates into erythrocytes and is bound by blood proteins to form a nondialyzable compound. Silver distribution in the body is fairly uniform. The metal does not pass through the dermal or placental barriers; some of it is excreted in milk [Fowler & Nordberg; Roshchin & Ordzhonikidze].

HYGIENIC STANDARDS

These are shown in Table 2.

Table 2. Exposure limits for silver and its compounds

	Workroom air		
	MAC_{wz} (mg/m^3)	Aggregative state	Hazard class
Silver, metallic	1.0 (0.1*)	Aerosol	2
Inorganic silver compounds	0.5	Aerosol	2
Soluble silver compounds	0.01*	—	—

* TLV-TWA in the USA.

METHODS OF SILVER DETERMINATION

Silver can be determined by spectrographic, colorimetric, and polarographic methods [Nordberg & Gerhardsson]. The reported detection limits are 0.01 mg/L per 20 ml sample for spectrographic and dithizone colorimetric methods and 5×10^{-6} mol/L for polarographic techniques [Fowler & Nordberg]. Atomic absorption spectrometry (detection limit 2 µg/L) and neutron activation analysis (detection limit 2 ng) have also been recommended [Suess]. A rapid assay for silver in drinking water derived from seawater has been described [Pilipenko *et al.*].

Silver measurement in human *saliva*, based on the ability of the metal to activate catalytically sulfanilic acid oxidation in the presence of $K_2S_2O_8$ has been proposed [Alexiev *et al.*]. Silver in *blood* and *urine* can be determined by atomic absorption spectrometry as described in Vince & Williams. Analytic procedures for silver in biologic materials are reviewed in *Toxic Metals and their Analysis* [Berman].

MEASURES TO CONTROL EXPOSURE

The primary engineering measures to be implemented for preventing exposure during the extraction or use of silver and its compounds are mechanization and automation of the processes involved. In areas where dusty materials are loaded or unloaded, appropriate facilities for suppression of the dust at source or its removal from the air must be provided if feasible. The removal of dust from dust collectors and its transportation should also be mechanized and prevent the dust from entering the environment. Enclosures and local air inlets should be integral parts of the production equipment and be so designed as to afford easy access to the latter for maintenance work. Equipment for the preparation of silver-containing pastes should be provided with local exhaust ventilation. Automatic machines or lines with built-in exhaust ventilation should be used for applying such pastes to articles.

For *personal protection*, respirators should be worn where effective dust control cannot be assured. Those who apply silver-containing pastes should wash their hands as they become dirty. Organic solvents must not be used for removing the pastes. The hands can be protected by an appropriate barrier cream. Personal protection against silver nitrate may require the wearing of protective clothing to avoid skin contact as well as chemical safety goggles for the protection of the eyes where spillage may occur.

Pollution of the *general environment* with emissions from plants producing or using silver and its compounds can be abated through effective treatment of the emissions and maximum possible utilization of the waste materials.

REFERENCES

Alexiev, A. A. *et al.* (1973) *Arch. Oral. Biol.*, **18**, 1461—1467.
Barkov, G. D. & Elpiner, L. I. (1968) *Gig. San.*, No. **6**, 16—21 (in Russian).
Barrie, H. J. & Harding, H. E. (1947) *Brit. J. Ind. Med.*, **4**, 225—228.
Berman, E. (1980) *Toxic Metals and their Analysis*. Heyden, London.
Bigg, E. K. (1985) *J. Weather Modif.*, **17**, 7—17.
Blanc, P. *et al.* (1985) *JAMA*, **252**, 367—371.
Bowen, H. J. M. (1985) In: *The Handbook of Environmental Chemistry*. Springer, Berlin, Vol. 3, Pt D, pp. 1—27.
Breton, J. *et al.* (1971) *Med. Legale Domm. Corp.*, **4**, 245—248.
Calabrese, A. *et al.* (1981) In: *Genetika i razmnozhenie morskikh zhivotnykh* [The Genetics and Reproduction of Marine Animals]. Vol. 2, Vladivostok, pp. 156—165 (in Russian).
Dobrovolsky, V. V. (1983) *Geografiya mikroelementov. Globalnoe rasseyanie* [Geography of Trace Elements. Global Dispersion]. Mysl, Moscow (in Russian).
Forycki, Z. *et al.* (1983) *Bull. Inst. Maritime and Tropical Medicine* (Gdynia), **83**, 199—203.
Fowler, B. A. & Nordberg, G. F. (1979) In: Friberg, L. *et al.* (eds) *Handbook on the toxicology of metals*. Elsevier, Amsterdam, pp. 579—586.
Kabata-Pendias, A. & Pendias, H. (1986) *Trace Elements in Soils and Plants*. CRC Press, Boca Raton (Fl.).
Larionov, L. N. (1984) *Vestn. Oftalmologii*, No. **1**, 54—56 (in Russian).
Lis *et al.* (1984) *Plant Sci. Lett.*, **33**, 1—6.
Nordberg, G. F. & Gerhardsson, L. (1988) *Silver* in: Seiler, H. G. *et al.* (eds) *Handbook on Toxicity of Inorganic Compounds*. Marcel Dekker, New York, pp. 619—624.

Petering, H. G. & McClain, C. J. (1991) Chapter II—26 *Silver* in: Merian, E. (ed.) *Metals and their Compounds in the Environment — Occurrence, Analysis and Biological Relevance.* VCH, New York, pp. 1191—1202.

Pilipenko, A. T. *et al.* (1986) *Gig. San.*, No. 1, 34—35 (in Russian).

Rosenman, K. D. *et al.* (1987) *Brit. J. Ind.* Med., **44**, 267—272.

Roshchin, A. V. & Ordzhonikidze, E.K. (1984) *Gig. Truda Prof. Zabol.*, No. **10**, 25—28 (in Russian).

Suess, M. J. (ed.) (1982) *Examination of Water for Pollution Control: A Reference Handbook.* Vol. 2, Pergamon, Oxford.

Vince, D. G. & Williams, D. F. (1987) *Analyst*, **112**, 1627—1629.

Gold and its compounds

Gold(I) cyanide; potassium dicyanoaurate(I)

IDENTITY AND PHYSICOCHEMICAL PROPERTIES OF THE ELEMENT

Gold (Au) is a metal in group I of the periodic table, with the atomic number 79. It has one stable isotope, ^{197}Au.

Gold is the most malleable and ductile metal, and can be forged or rolled into sheets as thin as 10^{-4}–10^{-5} mm. It is an excellent conductor of heat and electricity, being inferior only to silver and copper in this respect.

The chemical activity of gold is very low. In compounds it exists in the oxidation states of +1 and +3. It remains stable in air, does not react with halogens in the absence of moisture unless heated, and will not react with hydrogen, oxygen, nitrogen, or carbon even at high temperatures. With chlorine it forms the chloride $AuCl_3$ at 140–150°C and the chloride AuCl at 180–190°C. It does not dissolve in alkalies and sulfuric, hydrochloric, nitric, or hydrofluoric acid, nor in organic acids. It does, however, react with aqua regia and with mixtures of sulfuric and manganic or nitric acids and with selenic acid. It dissolves in cyanide solutions in the presence of oxygen. Gold compounds are unstable and readily undergo reduction to the metal. Gold easily forms complex compounds. It forms alloys with many metals. See also Appendix.

NATURAL OCCURRENCE AND ENVIRONMENTAL LEVELS

The clarke of gold is estimated at $0.04 \times 10^{-4}\%$ in the earth's crust and $0.0012 \times 10^{-4}\%$ in the granite layer of the continental crust [Dobrovolsky]. Its levels range from 0.5 to 5 µg/kg in volcanic rocks and from 2 to 7 µg/kg in sedimentary rocks. Hot waters in the earth's crust are the sources of hydrothermal gold deposits (e.g. gold-bearing quartz veins). In ores, gold is found chiefly in the free state (native gold). There are a few rare minerals in which it occurs together with selenium, tellurium, antimony, or bismuth; pyrite and other sulfides contain gold as an impurity. On average, 1 liter of sea or river water contained $\approx 4 \times 10^{-9}$ g of gold, while 1 liter of underground water near gold deposits contained 10^{-6} g.

In soils, reported levels of gold range from 0.8 to 8 µg/kg (higher levels within this range occur in chernozem and chestnut soils), the average values being 1–2 µg/kg. In mineralized areas, the gold content in forest soil reached up to 5 µg/kg. Gold migrates in soil and in water from which it is taken up by plants; some of these, such as corn and horsetail, are good concentrators of the metal and can serve as its bioindicators. Certain deciduous trees are able to accumulate gold to more than 10 mg/kg dry weight. In plant ash, gold concentrations were found to range between 0.0005 and 125 mg/kg. Anomalously high contents occur in plants from mineralized areas. Fruits and vegetables were reported to contain 0.01 to 0.4 µg Au/kg wet weight [Bowen; Kabata-Pendias & Pendias]. The geobiochemical cycle of gold is described in *The Handbook of Environmental Chemistry* [Bowen].

PRODUCTION

Gold from placer deposits is extracted by processes of amalgamation or cyanidation (less commonly by elutriation), while that from primary deposits is usually extracted by cyanidation; the extracted gold is refined electrolytically.

In 1974, the world production of gold was approximately 1467 tonnes; for the year 2000, 2000 tonnes has been forecast [Kabata-Pendias & Pendias].

USES

Gold is used for jewelry and decorative gilding; in the manufacture of chemically stable equipment; in electronics for electric contacts, printed circuits, resistors, and semiconductors; in spacecraft for coatings to reflect infrared and intensive solar radiation and to protect metals from corrosion; in dentistry and medicine (in particular, radioactive gold has found application in oncology and neurosurgery);

for coating reflectors and special glasses; for painting porcelain; etc. For most uses gold is alloyed with other metals.

Potassium dicyanoaurate is used for galvanic gilding in the instrument-making industry (see also Sutton & Di Martino).

TOXICITY

Plants

Some plant species are relatively resistant to high gold concentrations in their tissues. The toxicity of gold is manifested in wilting of leaves through loss of their turgidity, and in necrosis. The toxic effects are thought to be primarily due to a decrease in the permeability of cell membranes [Kabata-Pendias & Pendias].

Fish

The minimal lethal dose of gold for stickleback was 0.5 mg/L [Grushko].

Animals and man

Compounds of gold affect hematopoiesis: thrombocytopenia, leukopenia, agranulocytosis, and aplastic anemia may develop [Berman]. Neurotoxic effects have also been observed [Katz; Truhaut et al.], as has been inhibition of humoral immunity [Davis]. Toxic responses in occupationally exposed workers did not correlate with gold levels in their plasma [Berman]. Post-mortem examination of gold mine workers in South Africa indicated that exposure to the dusts present in the mines was a risk factor for chronic obstructive disease of the lungs [Becklake et al.]. Gold toxicity may result from chrysotherapy [Shaw].

Cutaneous effects

Animals

Dilute solutions of gold compounds may cause skin pigmentation in laboratory animals. Soluble salts produce caustic effects.

Man

Specific allergic dermatitis or eczema may occur in workers handling gold, its alloys, or its compounds for long periods of time. These conditions are manifested clinically in persistently recurring itchy papular eruptions on the hands, forearms, and face. Cases of nonoccupational contact dermatitis, resulting from the wearing of golden articles, are rare. After the cessation of contact with gold, full recovery ensues [Raben & Antonov], but the sensitivity to gold remains. In workers exposed to solutions of **gold cyanide**, the dermatitis was confined to open skin areas [Somov & Khaimovsky]. The occurrence of allergic dermatoses in employees of gold-extracting factories has been attributed to the use of cyanides in the work processes involved [Kolpakov & Prokhorenkov]. Allergic dermatitis and eczema can also be caused by **potassium dicyanoaurate** [Skripkin *et al.*].

ABSORPTION, DISTRIBUTION, AND ELIMINATION IN MAN

The ingested salts of gold are poorly absorbed [Sutton & Di Martino]. In the blood, gold usually remains for a long time in concentrations that depend on the compound used. For example, it was found in human blood 13 weeks after the last injection, and its traces were detectable in the liver, spleen, and kidneys much later. Following parenteral administration, the blood concentration of gold attained its maximum in 2 h and decreased by half in 7 days [Gottlieb]. Its soluble salts are eliminated mainly in the urine and the insoluble salts, in the feces. About 40% of the parenterally administered gold was present in the excreta, with 70% of this amount being contained in the urine and 30% in the feces. Gold is also excreted in breast milk and is able to cross the placental barrier; it is not deposited in the hair [Berman]. Normal levels of gold have been reported to be (g/ml) 4×10^{-10} in whole blood, 8×10^{-10} in erythrocytes, and 0.6×10^{-10} in plasma and serum [Bogdavadze *et al.*].

METHODS OF DETERMINATION

Determination of gold using colorimetry, fluorimetry, chromatography, emission spectroscopy, atomic absorption spectrometry, and other methods are reviewed in *Toxic Metals and their Analysis* [Berman]. The principle of atomic absorption spectrometry underlies a well-developed assay of blood and urine for gold described by Vrchlabský *et al.* Gold in biological materials can be determined without decomposing them by neutron activation analysis [Lobanov *et al.*] (see also Sutton & Di Martino).

MEASURES TO CONTROL EXPOSURE AT THE WORKPLACE

It is important that the use of metallic mercury for extracting gold be limited to the maximum possible extent. Preference should be given to sorption techniques of gold extraction without filtration, using a closed circuit system of water recycling. The use of gold-freed cyanide solutions for moistening ore in the ore-crushing department must be strictly forbidden. To filter the cyanide pulp, automatically operated vacuum filters should be employed. All equipment should be provided with mechanically vented enclosures.

In the working area, air should be delivered by forced ventilation systems to both the permanent work posts and the passageways. The flow rate of the air vented from the enclosures of tubs and apparatus for cyanidation as well as from the enclosures of tubs for acid treatment of the sediment should be at least 1.5 m/s, while the air flow rate in the openings of exhaust hoods for amalgam treatment and mercury packaging should be no less than 2 m/s. Light signalization to indicate how the ventilation equipment works should be fitted above the doors leading to the sorption, regeneration, electrolysis, and reagent rooms. Air ducts of the exhaust systems in the amalgam department should be covered with material that will not absorb mercury vapor.

Workers should receive ultraviolet radiation at appropriate intervals for prophylactic purposes. Properly conducted pre-placement and periodic medical examination of the workers are mandatory.

Work clothes should be laundered, repaired, and decontaminated on a centralized basis. Under no circumstances may they be allowed to be taken home for washing. Any dust on the clothes should be removed.

For *personal protection*, workers must by provided with appropriate footwear and personal protection devices, including respirators to be used to protection against dust; where vapors are present in the workroom air, gas masks must be worn.

REFERENCES

Becklake, M. R. *et al.* (1987) *Amer. Rev. Respir. Dis.*, **135**, 1234–1241.

Berman, E. (1980) *Toxic Metals and their Analysis.* Heyden, London.

Bogdavadze, N. V. *et al.* (1965) *Soobshcheniya AN Gruz. SSR* [Reports of the Academy of Sciences of the Georgian SSR]. **39**, 287–294 (in Russian).

Bowen, H. J. M. (1985) In: *The Handbook of Environmental Chemistry,* Vol. 3, Pt D. Springer, Berlin, pp. 1–27.

Davis, P. (1983) In: *Modern Aspects of Gold Therapy.* Karger, Basel, pp. 75–83.

Dobrovolsky, V. V. (1983) *Geografiya mikroelementov. Globalnoe rasseyanie* [Geography of Trace Elements. Global Dispersion]. Mysl, Moscow (in Russian).

Gottlieb, N. L. (1983) In: *Modern Aspects of Gold Therapy.* Karger, Basel, pp. 32–39.

Grushko, Ya. M. (1979) *Vrednye neorganicheskie soyedineniya v promyshlennykh stochnykh vodakh* [Harmful Inorganic Compounds in Industrial Waste Waters]. Khimiya, Leningrad (in Russian).

Kabata-Pendias, A. & Pendias, H. (1986) *Trace Elements in Soils and Plants.* CRC Press, Boca Raton (Fl.).

Katz, A. (1973) *Arch. Pathol.*, **96**, 133–136.

Kolpakov, F. I. & Prokhorenkov, V. I. (1978) *Vestn. Dermatol. Venerol.*, No. 7, 76–79 (in Russian).

Lobanov, E. M. *et al.* (1967) In: *Aktivatsionnyi analiz gornykh porod i drugikh obyektov* [Activation Analysis of Rocks and Other Objects]. Tashkent, pp. 147–157 (in Russian).

Raben, A. S. & Antonov, A. A. (1975) *Professionalnaya dermatologiya* [Occupational Dermatology]. Meditsina, Moscow (in Russian).

Shaw, C. F. (1991) Chapter II.12 Gold in: Merian, E. (ed.) *Metals and their Compounds in the Environment–Occurrence, Analysis and Biological Relevance.* VCH, New York, pp. 931–938.

Skripkin, Yu. K. *et al.* (1974) *Gig. Truda Prof. Zabol.*, No. 3, 44–46 (in Russian).

Somov, B. A. & Khaimovsky, G. D. (1964) *Vestn. Dermatol. Venerol.*, No. 10, 33–35 (in Russian).

Sutton, B. M. & Di Martino, M. J. (1988) *Gold* in: Seiler, H. G. *et al.* (eds) *Handbook on Toxicity of Inorganic Compounds*. Marcel Dekker, New York, pp. 307–314.

Truhaut, R. *et al.* (1975) In: *Rein et Toxique*. Paris, pp. 167–174.

Vrchlabský, M. *et al.* (1976) *Vnitrki Lék.*, **22**, 1206–1210.

Beryllium and its compounds

Beryllium bromide; b. carbide; b. carbonate tetrahydrate; b. chloride; b. fluoride; b. hydroxide (bechoite(β) [min.]); **b. iodide; b. nitrate tetrahydrate; b. orthophosphate; b. oxide** (bromellite(α) [min.]); **b. oxyfluoride** [b. fluoride–b. oxide (5/2)]; **b. sulfate**

IDENTITY AND PHYSICOCHEMICAL PROPERTIES OF THE ELEMENT

Beryllium (Be) is an alkaline-earth metal in group II of the periodic table, with the atomic number 4. It has one natural isotope, ^{9}Be.

Beryllium is strong and hard but fairly brittle, especially in the cold or when impure. It is the lightest of all chemically stable solid substances. It lends itself to extrusion and rolling at elevated temperatures. It is highly permeable to X-rays, reflects neutrons well, and emits neutrons itself when irradiated with α-particles. It remains chemically stable in air owing to the oxide (BeO) film coating its surface.

Beryllium occurs in two allotropic modifications (α and β). In compounds, it is present in the oxidation state +2. Most of its salts are soluble in water, and their solutions have an acid reaction. The hydroxide $Be(OH)_2$ is amphoteric. At high temperatures, the evaporating beryllium compounds form aerosols and decompose to give the oxide BeO. See also Appendix.

NATURAL OCCURRENCE AND ENVIRONMENTAL LEVELS

The commonest mineral of beryllium is beryl ($3BeO \cdot Al_2O_3 \cdot 6SiO_2$). Other minerals used for industrial production of metallic beryllium include phenakite Be_2SiO_4, chrysoberyl $BeAl_2O_4$, bertrandite $BeSi_2O_7(OH)_2$, beryllite $Be_2BaSi_2O_7$, and helvite $(Mn,Fe,Zn)_8(Be_6Si_4O_{24}S_2)$.

The clarke of beryllium is estimated at $(2.8-3.8) \times 10^{-4}\%$ in the earth's crust and at $2.5 \times 10^{-4}\%$ in the granite layer of the continental crust [Dobrovolsky]. From soils and rocks, beryllium and its compounds are released into the atmosphere by weathering and are also washed out by and migrate in ground and surface waters. Average beryllium concentrations in river and ocean waters are 0.006 µg/L and 0.0006 µg/L, respectively. In atmospheric air, its concentrations commonly lie in the range of 0.03 to 0.05 ng/m^3 in rural areas and 0.1 to 0.67 ng/m^3 in industrial cities. Marine organisms contain the metal at an average level of about 100 µg/kg wet weight.

Beryllium abundances in surface soils ranged from 1 to 15 mg/kg in the USA and from 1.2 to 13 mg/kg in the USSR and averaged 2.7 mg/kg in standard British soil samples. In plants, concentrations from 0.001 to 0.4 mg/kg fresh weight and from <2 to 100 mg/kg ash weight were recorded [Kabata-Pendias & Pendias].

PRODUCTION

The two most important methods of extracting beryllium from its ores are the sulfate process and the fluoride process. Metallic beryllium is then obtained by reduction of beryllium fluoride with magnesium or by electrolysis of beryllium chloride. Beryllium of very high purity is made by grinding the metal to powder and treating the powder with oxalic acid, and then pressing and sintering it at 1100–1200°C.

Beryllium chloride is manufactured by passing chlorine over a mixture of beryllium oxide and carbon, the **nitrate** is produced by the action of nitric acid on beryllium oxide, while the **oxide** is made by treatment of beryl ore or by heating beryllium hydroxide or nitrate.

USES

Beryllium is mainly used in x-ray tubes, neon tubes, and luminophores, and in nuclear engineering as a neutron reflector, in the jackets of fuel elements, and as a neutron source. It is also finding increasing application in electronics. Its alloys with various metals are employed in the rocket-building, airplane and instrument-making industries and in welding and soldering work. Its **oxide** is used as a refractory, in nuclear engineering, in turbine and rocket manufacture, as a catalyst in organic synthesis, and in the production of special glasses. It is also used as an electrical insulator and in power transistors. Among the oldest beryllium-containing materials are **beryllium bronzes** (~4% Be), which are used in nonspark tools, electrical contacts, watch springs, bushings for bearings, etc. [Keizer *et al.*]. Beryllium and its alloys with copper, nickel, and aluminum find widespread use for military applications. For other uses see WHO and Zorn *et al.*

MAN-MADE SOURCES OF EMISSION INTO THE ENVIRONMENT

Beryllium is released into the environment from the burning of fossil fuels (it is present in coal and oil), in automobile exhausts, and in emissions from industrial plants producing or using the metal and its compounds. The air, water, and soil are likely to be polluted in areas where beryllium ores are mined and concentrated or where the concentrates are stockpiled or subjected to further processing. About 5 kg of beryllium for every 1000 tonnes of processed beryllium were estimated to be released into the ambient atmosphere during beryllium production. According to the Clean Air Act adopted in West Germany, the sum of emissions of beryllium, benz(a)pyrene, and dibenzanthracene must not exceed 0.1 mg/m^3 in the flue gases from incinerators of hazardous wastes.

In the USA, coal combustion was estimated to be responsible for the emission of around 200 tonnes of beryllium in 1984 (790×10^6 tonnes of coal was burned, with a mean beryllium content of 1.4 mg/kg). Beryllium emissions into the atmosphere can be greatly reduced through the use of efficient dust-collecting equipment.

In 1980–84, the world output of beryllium minerals was around 10 000 tonnes per year. In the 1990s, the demand for beryllium is expected to increase at a rate of 4% annually. Beryllium–producing plants exist in only three countries — the USA, the former USSR, and Japan; other countries import this metal.

Cigarettes were found to contain from 0.47 to 0.74 µg of beryllium each, with up to 10% of this amount entering the lungs in the inhaled smoke [Zorn & Diem].

TOXICITY

Plants

Beryllium appears to be readily taken up by plants from soil, and it accumulates in their roots to a greater extent than in the stems or leaves. It impairs plant growth and development if its soil concentration is excessive. Its toxic concentrations in mature leaves ranged most often from 10 to 50 mg/kg but varied widely depending on the plant species and growth conditions [Kabata-Pendias & Pendias]. The metal is thought to produce toxic effects by inhibiting enzymes (phosphatases) and the uptake of essential ions. The translocation coefficient for beryllium was estimated to vary from 0.01 to 0.1 depending on the plant species and soil properties [Kloke *et al.*].

Aquatic organisms

LC_{50} values of beryllium for freshwater fish lie in the wide range of 0.15 to 32 mg/L depending on the species. The toxicity tends to rise with increasing hardness of the water.

A sharp decrease in biochemical oxygen demand was observed in bodies of water at beryllium concentrations of 0.5 to 1 mg/L. At 0.005 mg/L, bacterial growth and the biochemical oxidation of organic matter proceeded at rates close to control values. The concentration of 0.5 mg/L did not inhibit ammonification or nitrification processes. The level of 0.1 mg/L was considered as the threshold concentration required to affect the general sanitary condition of water bodies.

Beryllium did not alter the taste or transparency of water in concentrations of <1 mg/L.

Animals and man

General considerations

Beryllium is characterized by high biologic activity. The critical determinant of the toxicity exhibited by the metal and its compounds is the ion Be^{2+} which has been shown capable of exerting systemic, allergic, carcinogenic, and embryotoxic actions. Studies with cultures of somatic mammalian cells have indicated that beryllium can interact with DNA to cause mutations, chromosome aberrations, and sister chromatid exchanges; in contrast, no mutagenic effects were observed in bacterial test systems.

Soluble beryllium compounds, such as the chloride, fluoride, and sulfate, are generally more toxic than insoluble ones, and they can also act as irritants. When inhaled in the form of fume or dust, they can cause a productive interstitial process leading to the formation of specific granulomas in the lungs and known as berylliosis.

Beryllium and its compounds can alter the immune status of the body and the activity of many enzymes involved in energy metabolism (phosphoglucomutase, enolase, lactate dehydrogenase, alkaline phospatase, etc.) in the lung, liver, kidney, muscle tissue, and bone tissue. They have been found to cause very marked dysproteinemia (manifested in diminished albumin-globulin ratio) and severe imbalance of trace elements [Ivanova; Keizer *et al.*].

Metallic beryllium and various beryllium compounds administered to rabbits intravenously or intramedullarly induced the development of osteosarcomas or chondrosarcomas with metastases, most often to the lungs. Rats exposed to metallic beryllium or to soluble or insoluble beryllium compounds by inhalation or intratracheally developed malignant pulmonary neoplasms such as adenoma and adenocarcinoma. Particularly susceptible to pulmonary cancer after exposure to beryllium appear to be rats and monkeys; rabbits, hamsters, and guinea-pigs did not develop pulmonary tumors. Beryllium carcinogenicity has been reported to be associated with characteristic changes in antigens of the critical organ, with DNA and RNA damage, and with the appearance of certain low-molecular-weight proteins [Alekseyeva & Vasilyeva].

Beryllium interacts in a competitive and sometimes overtly antagonistic manner with ions of biologically important bivalent metals (Mg, Ca, Mn, etc.) that are natural enzyme activators [Bernshtein & Kichina; Nikitina]. Witschi & Aldridge have pointed out that beryllium is a potent necrotizing agent with oligodynamic action for which no consistent dose-response relationship could be established.

Komitowski distinguished three phases in the action of beryllium, referred to as initial, sclerotizing, and neoplastigenic. The hallmark of the first phase is inhibited alkaline phosphatase activity, often in the presence of elevated acid phosphatase activity, parficularly in the liver. A characteristic feature of the second phase is a high level of acid mucopolysaccharides in the region of connective-tissue formation. In the third phase, tumors of an osteogenic sarcoma type are seen to arise.

In the lungs, signs of general toxicity and allergic reactions are observed in the early phases of beryllium poisoning; in later phases, the lungs undergo hyaline degeneration of collagen fibrils and gross destructive changes in the parenchyma with impairment of air exchange; hypoxia and hypoxemia develop. The content of collagen proteins in the lungs increases during the early phases (as is evidenced by elevated hydroxyproline levels) and decreases later to rise sharply by months 9–13 of intoxication.

Acute/single exposure

Animals

LD_{50} values of several beryllium compounds are shown in Table 1.

In rats given a single intramuscular injection of **beryllium sulfate** at 4.5 mg/kg, oxidation processes were inhibited in the lungs and liver and vascular abnormalities were present. The toxic effect could be substantially reduced by preinjecting the animal with magnesium salts.

Intratracheal administration of **beryllium oxyfluoride** at 5 to 15 mg/kg was lethal to mice. Inhalation exposure to its fumes at concentrations of 0.9 to 25 mg/m^3 caused labored breathing and dyspnea in rabbits and dyspnea, salivation, cough, vomiting, and fever in dogs. The blood of poisoned animals showed increased erythrocyte counts, elevated hemoglobin levels, and leukocytosis with a shift to the left. Deaths occurred on days 3 to 9. Autopsy demonstrated pulmonary edema, peribronchitis, and detachment of the bronchial epithelial mucosa. Pulmonary edema and congestion were in evidence as early as 2 h post-administration; after 12 h, bronchopneumonia was noted.

In rats, typical beryllium granulomas with a necrotic center consisting of a beryllium-protein complex and surrounded by cellular elements developed in the lungs 2 to 3 days after a single intratracheal administration of **beryllium oxide** in an amount of 5 or 20 mg. Several days later the granulomas were made up essentially of partially hyalinized collagen fibers. The main clinical manifestations of poisoning were breathlessness, loss of appetite, and weight loss. The lungs presented a picture of pneumonitis with abundant exudation, proliferation, and desquamation of the alveolar epithelium. The granuloma eventually transformed into a sclerotic nodule. Later on, papillary outgrowths of the bronchial epithelium and its metaplastic transformation to stratified squamous epithelium were observed, as was metaplasia of the alveolar epithelium; eventually, malignant neoplasms were seen to have developed.

Table 1. Median lethal doses of some beryllium compounds for mice, rats, and guinea-pigs

Compound	Animal	Route of administration	LD_{50} (mg Be/kg)
Carbonate	Guinea-pig	Intraperitoneal	1.2
Chloride	Mouse	Intramuscular	1.3
		Intraperitoneal	0.15
	Rat	Oral	9.8
		Intraperitoneal	0.6
	Guinea-pig	Intraperitoneal	6.3
Fluoride	Mouse	Oral	19.1
		Intravenous	0.34
		Subcutaneous	3.8
Nitrate	Mouse	Intravenous	0.5
		Subcutaneous	10.8
	Guinea-pig	Intraperitoneal	3.48
Phosphate	Mouse	Intravenous	1.4
	Rat	Oral	6.5
		Intravenous	0.36
Sulfate	Mouse	Oral	6.95
		Intravenous	0.04
		Subcutaneous	0.13
	Rat	Oral	7.02
		Intravenous	0.62
		Intraperitoneal	1.54
		Subcutaneous	0.13

Source: WHO.

For rats, the lowest concentrations of soluble beryllium compounds and beryllium oxide required to induce autoallergy and alter lung structure were around 0.2 mg/m^3 with 1-hour exposure by inhalation and 0.003 mg/m^3 with chronic exposure by this route.

Generically determined susceptibility to beryllium was demonstrated using two guinea-pig strains [Barna *et al.*] : after an intravenous injection of beryllium oxide at 10 mg/kg, granulomatous lesions developed in only one strain, but not in the other.

Beryllium chloride injected subgerminally into chick embryos in a low dose $(0.03–0.3$ µg) caused a number of severe malformations; doses exceeding 0.3 µg were lethal to the embryos [Puzanova *et al.*]. The chloride also produced gene mutations in cultured Chinese hamster cells at concentrations of 2 and 3 mmol/L [Miyaki *et al.*] and marked chromosomal aberrations in cultured swine cells at concentrations of $0.5–10.0$ mmol/L [Talluri & Guiggiani].

Man

Cases of acute poisoning by inhalation have been described in workers exposed to soluble beryllium compounds (typically the **fluoride**) as well as in those engaged in the smelting, casting, or mechanical working of **metallic beryllium** [Griffiths & Skilleter; Zorn *et al.*].

Clinically, beryllium poisoning manifests itself as conjunctivitis, nasopharyngitis, tracheobronchitis, bronchiolitis, and pneumonitis. Two forms of the disease have been distinguished; in one, the clinical course resembles that of metal fume fever with chills, fever, and blood changes, while the other form runs the course of indolent pneumonitis.

The common complaints were cough, breathlessness, chest pain, and loss of appetite. The upper airway mucosa was inflamed; occasionally, moist rales and crepitation could be heard. Frequent x-ray findings were increased transparency of the lung fields and enhanced or indistinct lung markings; infiltrates and low mobility of the diaphragm were sometimes noted.

Moderate fatty degeneration and, occasionally, areas of necrosis were found in the liver. A case of acute interstitial myocarditis was described. A high incidence of acute berylliosis was recorded among those who had inhaled **beryllium sulfate** or **fluoride** in an amount of about 1 mg. Some 300 cases of acute disease occurred after inhalation exposure to **beryllium oxide, sulfate,** and/or **halides** in concentrations that averaged 0.67 mg/m^3 [Beliayeva].

Fatalities among heavily exposed individuals have occurred although complete recovery after several weeks or (sometimes) months has been the rule. Further exposure to beryllium or its compounds after recovery may bring on the disease again [Constantinidis].

Repeated exposure

Animals
Laboratory animals of various species inhaling the **fluoride, phosphate,** or **sulfate** of beryllium at concentrations of 1.1–8.3 mg/m^3 during 6-h daily for 8–15 days developed a clinical picture of intoxication, followed by partial mortality of the animals.

Repeated daily 6-h exposures to beryllium sulfate mist in a mean concentration of 47 mg BeSO$_4$ · 6H$_2$O/m^3 (or 2 mg Be/m^3) were lethal for goats (100% deaths), guinea-pigs (60%), monkeys (100%), rats (90%), dogs (80%), cats (80%), rabbits (10%), hamsters (50%), and mice (10%). Death occurred within the first week in the goats, guinea-pigs, and monkeys, in the second or third week in the rats, and after 1 or 2 months in the dogs and cats. Exposure to 0.95 mg BeSO$_4$ · 6H$_2$O/m^3 (0.04 mg Be/m^3) over 100 days did not lead to any deaths in the species tested [Stokinger et al.].

Chronic exposure

Animals
Lesions of lung tissue with focal interstitial infiltration and the development of granulomas and tumors (mostly adenocarcinomas) were observed in rats exposed by inhalation to **beryllium sulfate** for a period of 6 months. Rats proved to be more sensitive to the sulfate than other animal species used in the experiments (mice, dogs, cats, guinea-pigs, monkeys, goats, and pigs) [Reeves et al.]. Inhalation exposure to this compound at 10 mg/m^3 for 6 h/day over 96 days caused some mortality while the levels of 50 mg/m^3 and 100 mg/m^3 were lethal to most rats; no deaths occurred with exposure to 1 mg/m^3. Long-term oral administration of the sulfate was inhibitory to erythropoiesis and increased the leukoerythroblastic index in the bone marrow.

Inhalation of **beryllium fluoride** at 2.2 mg/m^3 for 23 weeks led to anemia in rabbits, while chronic interstitial pneumonia with pneumosclerosis but without typical granulomatous lesions was observed in rats exposed to the fluoride at 0.04 or 0.4 mg/m^3 or to **beryllium chloride** at 0.02 or 0.2 mg/m^3 (as Be) for 1 h daily during 4 months; after 18 months, adenomas or adenocarcinomas were found in the lungs of some animals [Litvinov et al.].

Long-term exposure (110 days) of rats to **beryllium oxide** dust at 2 or 20 mg/m^3 led to the appearance of round-cell infiltrates in the lungs and subsequent pulmonary sclerosis. Some rats also had tumors in their internal organs. Central lobar necrosis and pyknosis of many cell nuclei were seen in the liver. The kidneys showed foci of interstitial cellular infiltration and degenerative changes in the proximal convoluted tubules [Melnikov et al.].

Intravenous or intramedullary injections of beryllium or its compounds caused the development of bone tumors in rats, guinea-pigs, and rabbits [Zorn et al.]. In rabbits, the latent period before their appearance varied from 5.5 to 24 months after the last injection of beryllium [Groth].

Man

Cases of pulmonary granulomatosis have been described in workers occupationally exposed to beryllium compounds, in those who had handled the underwear of such workers, and also in those living near beryllium plants. The latent period preceding the onset of overt disease may extend from several weeks to 10 or more years after first exposure, and the disease may develop after the cessation of exposure [Beliayeva]. For example, the delay had been less than 1 year in about half of 310 patients with chronic berylliosis. The duration of occupational exposure among 100 patients ranged from several weeks to 17 years, but was 1 to 3 years in most. Factors known to precipitate or exacerbate berylliosis include concomitant disease, surgery, and pregnancy. Cases of berylliosis ranging from mild to rapidly progressing and life-threatening disease occurred among workers handling alloys containing 1.8% Be.

The common presenting complaints were breathlessness, cough, chest pain, weakness, rapid weight loss, migrating pains in joints, chills, and body temperature elevation to 38–39°C [Beliayeva]. Cyanosis, hemodynamic changes, reduced myocardial contractility, myocardial dystrophy, and diminished blood coagulability in the absence of hemorrhagic syndrome were reported [Shatskaya & Orlova; Mekhtiyeva; Orlova & Glotova; Rushkevich]. Auscultation findings often included bandbox sounds, restricted lung mobility, rales, and pleural friction sounds. Adhesive pleurisy was also common. X-ray examination often showed diffusely enhanced and reticular lung patterns.

The granulomas of berylliosis may be small (1–2 mm) or large (3–4 mm), the latter form of the disease being particularly severe; an interstitial form may also occur, in which no granulomas are seen on the roentgenogram and no organ-specific autoantibodies characteristic of the granulomatous process are detectable.

In severe beryllium disease involving bullous emphysema, extensive adhesions between pleura and diaphragm, and gross respiratory disturbances, cardiovascular abnormalities are also observed, such as cor pulmonale, edematous lower extremities, and ascitis. Progressive respiratory and cardiac insufficiency may be fatal.

In such cases, autopsy showed whitish-gray dense nodules, emphysema, and focal atelectasis in the lungs, while histologic examination revealed multiple granulomas composed of epithelial or sometimes multinuclear giant cells, and signs of chronic interstitial pulmonary inflammation. Granulomas were also found in regional lymph nodes, the liver, and the spleen. Epithelial hyperplasia and metaplasia were observed in the airways and lungs [Dvizhkov]. Although there is no conclusive evidence that malignant tumors are induced by beryllium in human beings, epidemiologic studies have shown significantly increased incidence rates of cancer among beryllium workers as compared to occupationally unexposed individuals [Mancuso; Wagner; Wagoner *et al.*]. The International Agency for Research on Cancer [IARC] has placed beryllium and its compounds in the group of probable human carcinogens.

Two important features of chronic beryllium intoxication are its systemic component and the absence of consistent exposure-effect relationships. This suggests that an immunologic mechanism is involved. Indeed, increased levels of immunoglobulins G, A, and M have been recorded in beryllium workers [Bencko *et al.*].

Effects on the skin and eyes

Diseases of the skin are caused most commonly by **fluoride-containing compounds** of beryllium or by its **sulfate**.

Direct contact with soluble beryllium compounds may cause contact dermatitis characterized by reddened, elevated, or fluid-filled lesions on exposed surfaces of the body. The symptoms develop after a period of 1–2 weeks, indicating a delayed allergic reaction. A concomitant conjunctivitis may occur. After cessation of exposure the lesions usually disappear, whereas on continued exposure bronchitis and pneumonitis may develop. Sensitized individuals react much more rapidly and to smaller amounts of beryllium. The beryllium hypersensitivity state usually persists for a long time (up to 20 years), even in the absence of any clinical signs of dermatosis.

The specificity of beryllium in allergic contact dermatitis was demonstrated by using a diagnostic patch test [Curtis]. Some investigators believe that the patch test is a reliable diagnostic test for chronic beryllium disease, whereas other doubt its value. There is considerable evidence that this test may be a hazardous procedure in individuals with a high degree of sensitivity to beryllium. Thus, the patch test has not been used much for diagnostic purposes.

Contact of damaged skin with beryllium may give rise to a beryllium ulcer. Intracutaneous introduction of beryllium may result in granuloma development with large areas of fibrinoid necrosis after a certain period of latency; the lesion will persist until the offending agent is removed.

Dusts and fumes of beryllium compounds can cause catarrhal conjunctivitis or chemical burns with strong edema and inflammation of the eyelid skin; the cornea is rarely involved.

ABSORPTION, DISTRIBUTION, AND ELIMINATION

In occupational exposure beryllium enters the body mainly by inhalation in the form of dusts or fumes; absorption through undamaged skin is of little practical importance. In the skin, beryllium ions were found to combine chiefly with epidermal alkaline phosphatase and nucleic acids [Belman]. For the general population, the principal mode of entry is by oral intake with food and water, which amounts to 0.07 μg Be/day on average.

Beryllium is poorly absorbed from the gastrointestinal tract. This is largely because its soluble compounds are converted into the relatively insoluble hydroxide in the alkaline medium of the intestine. In contrast, it is well absorbed from the lung to be widely distributed in the body, although the rate of uptake is greatly influenced by the chemical and physical characteristics of the inhaled material. Slightly soluble compounds such as the oxide may remain in the lung for a long time. After a single inhalation exposure of rats to beryllium oxide particles 0.17 to 1.1 µg in diameter, 30% and 70% of the material deposited in the lungs was cleared with half-times of 7 and 325 days, respectively. After inhalation of soluble compounds, most of the beryllium entering the blood from the lungs accumulates in the skeleton. In the blood, most of the beryllium circulates as colloidal phosphate, and only small amounts are transported as citrate or hydroxide [Reeves]. Large colloidal particles are retained in the reticuloendothelial system of the liver and spleen, from which they are gradually redistributed and deposited in the bones [Keizer et al.]. Beryllium may be detected in the lungs and bones many years after the cessation of occupational exposure.

Beryllium can cross the placenta and reach the fetus, and it may also be passed to infants in the breast milk from nursing mothers. Moreover, it is capable of penetrating the blood-brain barrier.

The elimination of absorbed beryllium occurs mainly in the urine and to a minor degree in the feces. Most of the beryllium taken up by the oral route passes through the gastrointestinal tract unabsorbed and is eliminated in the feces. A 'normal' level of beryllium in human urine is considered to be around 1 µg/L.

In a report of the International Commission on Radiological Protection [ICRP], the biological half-life of beryllium in human beings was calculated to be 180, 120, 270, 540, and 450 days for total body, kidneys, liver, spleen, and bone, respectively.

HYGIENIC STANDARDS

In the USSR, the following three exposure limits have been adopted for beryllium and its compounds (in terms of Be) in workplace air, atmospheric air of residential areas, and water sources, respectively: $MAC_{wz} = 0.001$ mg/m^3 (as aerosols, with the notation 'carcinogenic and allergenic activity'); $MAC_{ad} = 0.00001$ mg/m^3; and $MAC_w = 0.002$ mg/L (based on sanitary and toxicologic criteria). Beryllium and its compounds are included in hazard class 1. In occupational settings, the values of 0.01, 0.05, and 0.5 mg/m^2, respectively, are considered as the maximum permissible levels of contamination for the skin, underwear, and outer clothing of workers handling these substances [Keizer et al.].

In the USA, the TLV-TWA for beryllium has been set at 0.02 mg/m^3, this metal being classified as a suspected human carcinogen.

METHODS OF DETERMINATION

In *air*: atomic absorption spectrophotometry (detection limit 2.5 ng/m^3) [Zdrojewski *et al.*], graphite-furnace atomic absorption spectrophotometry (detection limit 0.05 ng/m^3) [Zdrojewski *et al.*] or 5 ng/sample [IARC]; emission spectroscopy using laser after collecting particles on a filter (detection limit 3.6 ng/filter) [Cremers & Radziemski]; gas chromatography with electron-capture detection (detection limit 0.04 pg/sample) [Ross & Sievers]; optical emission spectrometry (detection limit 5.3 µg/ml) [Scott *et al.*]; photometry (sensitivity 0.005 µg/sample) [Krivorutchko]. Beryllium can also be determined in air by measuring the fluorescence intensity of the stained solution that forms upon the interaction of the Be^{2+} ion with morin (sensitivity 0.05 µg/sample) [Keizer *et al.*].

In *biological materials*: atomic absorption spectrometry with nitrous oxide-acetylene flame (detection limit 2 µg/L of urine) [Bokovsky]; flameless atomic absorption spectrometry (detection limit 0.6 µm/L of blood, urine, or other liquids) [Stiefel *et al.*]; graphite-furnace atomic absorption spectrometry (detection limit 0.01 µg/L of urine or 1 µg/kg of hair, fingernail, or feces) [Hurlbut; Zorn *et al.*]. In *fresh water*: graphite-furnace atomic absorption spectrometry (detection limit 0.06 µg/L) and graphite-furnace atomic emission spectrometry (detection limit 2 µg/L) [Epstein *et al.*]. In *sea water*: gas chromatography with electron capture detection (detection limit 18 pg/kg) [WHO].

Physicochemical methods of beryllium measurement have been described by Gadaskina. An overview of methods for sampling, sample preparation, and the determination of beryllium in various media is presented in *Environmental Health Criteria 106: Beryllium* [WHO].

MEASURES TO CONTROL EXPOSURE

At the workplace

The main technical control measures to be implemented for preventing or keeping down exposure during the production or use of beryllium and its compounds and alloys include elimination or reduction of the air-pollution sources; complete segregation of the rooms where beryllium or alloys containing >20% Be are handled; mechanization and automation of the work processes (with preference given to continuous processes over batch techniques); and installation of ventilation systems appropriate to the nature of the operations involved. (A full listing of preventive measures specific for beryllium-related work is presented in the papers by Bobryshchev-Pushkin *et al.* and Mordberg & Filkova.) Beryllium-containing wastes should be processed to recover the metal.

Workers should receive a pre-placement medical examination followed by periodic examinations at one-year intervals throughout the period of employment.

For *personal protection*, the use of respirators in work areas containing airborne beryllium dust is mandatory, as is the wearing of gas masks in those where beryllium vapor, fume, or mist is present. Protective clothing should be provided for all beryllium workers. The essential hygiene measures include hand washing, taking food only in messrooms, canteens, or other specially designated places, showering at the end of the workday, and a complete change of work clothes followed by their decontamination. (Methods of decontamination and personal hygiene measures are described in detail in Keizer *et al.*)

Protection of the general environment

Hazardous wastes that can cause environmental pollution include gases and aerosols (ventilation exhausts), liquid wastes (industrial effluents), and solid wastes (tailings from hydrometallurgical processes, slags, metal scrap, etc.).

The air contaminated by beryllium or its compounds must be cleaned by methods appropriate for the particular industrial processes. Whenever a multistage system of air cleaning is employed, fine filters (made of electrostatic fabric) should be used as the last stage.

Liquid wastes from plants where beryllium-containing materials are produced or used must also be treated by suitable methods or be delivered to tailings dumps. Wastewaters from laundries, shower rooms and from the cleaning of working areas need to be purified as well.

Solid wastes that are not to be reprocessed may be disposed of in tailings dumps, in moisture-proof concreted trenches, or in undrainable gullies or depressions provided with a waterproof screen at the bottom and surrounded with an embankment whose ridges and outer slopes are protected from wind erosion by stonework, turfing, or grassing. The tailings dumps should be so designed as to be inaccessible to storm, melt, and flood waters.

It is recommended to wet dry beryllium-containing wastes during storage and before transportation; they should be transported in closed containers. If they are burned in incineration plants, the smoke from these must be purified before being emitted into the atmosphere.

Finished goods as well as any materials or equipment removed from beryllium factories must be cleaned and tested for beryllium.

Table 2. Emergency medical care in acute beryllium poisoning by inhalation

Clinical condition	Treatment given by the general practitioner (in a health center or polyclinic)	Specialized treatment in a polyclinic or hospital
Nasopharyngitis/tracheitis	Symptomatic treatment with antitussives; desensitizing agents. If laryngeal edema and spasm are present, the patient must be hospitalized without delay	If there is reflex edema of the larynx leading to mechanical asphyxia, desensitizing agents should be given, such as pipolphen (promethazine), dimedrol (diphenylhydramine), suprastin (chloropyramine), and hydrocortisone; if they are ineffective, immediate tracheostomy is required
Tracheobronchitis	Symptomatic treatment with antitussives; desensitizing agents. If there is bronchospasm, bronchodilators should be given. Admission to hospital if the patient does not improve	It is necessary to observe the patient in the hospital for at least 48–72 h since the tracheobronchitis may be the precursor of more severe respiratory disease. The patient should drink large quantities of liquid (tea or alkaline mineral water). Irritants (cupping glasses or mustard plasters) should be applied to the chest. Codeine to allay cough and a bronchodilator (ephedrine or aminophylline) to relieve the bronchospasm should be given. If indicated, antibiotics should also be administered.

Table 2. (cont'd)

Clinical condition	Treatment given by the general practitioner (in a health center or polyclinic)	Specialized treatment in a polyclinic or hospital
Acute bronchobronchiolitis/ pneumonitis	A 40% hypertonic glucose solution (20 ml) and a 10% calcium chloride solution (10 ml) should be given to raise ostomic pressure and prevent pulmonary edema. Prolonged oxygen therapy may be required. Cardiovascular drugs (camphor, cordiamine [nikethamide], and, if indicated, strophanthin), antitussives, and bronchodilators should be administered. The patient must be admitted to hospital at once	Strict bed rest is mandatory. Measures to prevent pulmonary edema should be taken. If infection has supervened, antibiotics must be administered. Symptomatic treatment with antitussives and other appropriate agents should be instituted. To eliminate alveolar hypoxia, artificial pulmonary ventilation via intubation tube or tracheostomy opening may be necessary. Treatment for pulmonary edema: (a) inhalation of a foam suppressor, (b) drugs reducing pressure in the pulmonary circulation (ganglionic blocking agents, diuretics, aminophylline), and (c) cardiac glycosides such as strophanthin

EMERGENCY TREATMENT IN POISONING

In case of upper respiratory tract irritation, warm inhalations using a 2% baking soda solution or a 1% menthol solution in oil are indicated. Persistent cough can be alleviated by codeine, dionine, and by mustard plasters applied to the chest. In conjunctivitis, zinc- and epinephrine-containing boric acid solutions and a 30% albucid (synonym: sodium sulfoacetamide) solution are beneficial. For dermatitis in the acute phase, cold lotions of lead water, Burow's solution, tannin, or resorcine should be applied for 1 to 3 days during daytime; an indifferent paste or ointment such as Lassar's paste or bismuth subgallate ointment may be applied for the night.

If beryllium particles have penetrated into the skin, the involved soft tissues must be excised. Microtraumas should be carefully treated with iodine tincture and a bactericidal plaster applied to the affected skin area.

In the event of eye contact with a beryllium compound, the eye(s) should be immediately irrigated with copious amounts of water.

In acute poisoning by inhalation (the main clinical forms are nasopharyngitis, tracheitis, tracheobronchitis, broncho-bronchiolitis [bronchiolitis], and pneumonitis), the victim must by promptly removed from the area of exposure and helped to take off the personal protective equipment and change the working clothes. He should then be dispatched to a health center where, as the initial step, warm inhalations using a 2–3% baking soda solution should be administered by mouth and a 2–3% ephedrine solution or a 2–3% novocain [synonym: procaine hydrochloride] solution together with a 0.1% epinephrine solution be dropped into the nose. Recommended medical measures to be taken subsequently are listed in Table 2.

After an acute beryllium intoxication (even if its symptoms have completely regressed), avoidance of any further exposure to beryllium and its soluble compounds is strongly recommended, for otherwise the disease may rapidly progress or recur because of the sensitization.

REFERENCES

Alekseyeva, O. G. & Vasilyeva, E. V. (1972) *Gig. Truda Prof. Zabol.*, **5**, 23–26 (in Russian).
Barna, B. P. *et al.* (1984) *Int. Arch. Allergy Appl. Immunol.*, **73**, 42–48.
Belman, S. (1969) *J. Occup. Med.*, **11**, 175–183.
Beliayeva, L. N. (1965) *Gig. Truda Prof. Zabol.*, No. 3, 28–32 (in Russian).
Bencko, V. *et al.* (1979) *J. Hyg. Epidemiol. Microbiol. Immunol.*, **23**, 361–367.
Bencko, V. *et al.* (1980) *Environ. Res.*, **22**, 439–449.
Bernshtein, F. Ya. & Kichina, M. M. (1968) In: *Mikroelementy v meditsine* [Trace Elements in Medicine]. No. 1, Zdorovya, Kiev, pp. 38–41 (in Russian).
Bobrishchev-Pushkin, D. M. *et al.* (1976) *Gig. Truda Prof. Zabol.*, No. 11, 6–9 (in Russian).
Bokovski, D. L. (1968) *Am. Ind. Hyg. Assoc. J.*, **29**, 474–481.
Constantinidis, K. (1978) *Br. J. Clin. Pract.*, **32**, 127–136.

Cremers, D. A. & Radziemski, L. J. (1985) *Appl. Spectrosc.*, **39**, 57—63.

Curtis, G. H. (1951) *Arch. Dermatol.*, **64**, 470—482.

Dobrovolsky, V. V. (1983) *Geografiya mikroelementov. Globalnoe rasseyanie* [Geography of Trace Elements. Global Dispersion]. Mysl, Moscow (in Russian).

Dvizhkov, P. P. (1967) *Arkh. Patol.*, No. **3**, 3—11 (in Russian).

Epstein, M. S. *et al.* (1978) *Anal. Chem.*, **50**, 874—880.

Gadashkina, I. D. *et al.* (1975) *Opredeleniye promyshlennykh neorganicheskikh yadov v organizme* [Determination of Inorganic Industrial Poisons in the Organism]. Meditsina, Leningrad (in Russian).

Griffiths, W. R. & Skilleter, D. N. (1991) Chapter II.4 *Beryllium* in: Merian, E. (ed.) *Metals and their Compounds in the Environment — Occurrence, Analysis and Biological Relevance*, VCH, New York, pp. 775—788.

Groth, D. H. (1980) *Environ. Res.*, **21**, 56—62.

Hurlbut, J. A. (1978) *At. Absorpt. Newsl.*, **17**, 121—124.

IARC (1986) In: O'Neill, I. K. *et al.* (eds) *Environmental Carcinogens: Selected Methods of Analysis*. International Agency for Research on Cancer. Lyon, pp. 283—288 (IARC Scientific Publications No. 71).

ICRP (1960) [International Commission on Radiological Protection] Report of ICRP Committee II on permissible dose for internal radiation, *Health Phys.*, **3**, 154—155.

Ivanova, A. S. (1971) *Gig. Truda Prof. Zabol.*, No. **6**, 32—35 (in Russian).

Kabata-Pendias, A. & Pendias, H. (1986) *Trace Elements in Soils and Plants*. CRC Press, Boca Raton (Fl.).

Keizer, S. A. *et al.* (1985) *Berilliy. Toksikologiya, gigiyena, profilaktika, diagnostika i lechenie berillievykh porazheniy. Spravochnik* [Beryllium: Toxicology, Hygiene, Preventation, Diagnosis, and Treatment of Beryllium-induced Diseases. A Handbook]. Energoatomizdat, Moscow (in Russian).

Kloke, A. *et al.* (1984) In: Nriagu, J. O. (ed.) *Changing Metal Cycles and Human Health. Report of the Dahlem Workshop. Berlin (FRG), 20—25 March, 1983*. Berlin & New York, Springer—Verlag, pp. 113—141.

Komitowski, D. (1971) *Vestn. Akad. Med. Nauk SSSR*, No. **8**, 10—11 (in Russian).

Krivorutchko, F. D. (1966) *Gig. San.*, No. **4**, 57—60.

Litvinov, N. N. *et al.* (1984) *Gig. Truda Prof. Zabol.*, No. **7**, 34—37 (in Russian).

Mancuso, T. F. (1980) *Environ. Res.*, **21**, 48—55.

Melnikov, V. V. *et al.* (1962) *Gig. San.*, No. **4**, 30—36 (in Russian).

Mekhtiyeva, D. M. (1974) *Gig. Truda Prof. Zabol.*, No. **11**, 24—26 (in Russian).

Miyaki, M. *et al.* (1979) *Mutat. Res.*, **68**, 259—263.

Mordberg, E. L. & Filkova, E. M. (1974) *Gig. San.*, No. **6**, 71—76 (in Russian).

Nikitina, A. S. (1970) *Gig. Truda Prof. Zabol.*, No. **5**, 50—52 (in Russian).

Orlova, A. A. & Glotova, K. V. (1969) *Gig. Truda Prof. Zabol.*, No. **2**, 32—36 (in Russian).

Puzanova, L. *et al.* (1978) *Folia Morphol.* (Prague), **26**, 228—231.

Reeves, A. L. (1986) In: Friberg, L. *et al.* (eds) *Handbook on the Toxicology of Metals*. 2nd ed. Elsevier, Amsterdam, Vol. II, pp. 95—116.

Reeves, A. L. *et al.* (1967) *Cancer Res.*, **27**, 439—445.

Ross, W. D. & Sievers, R. E. (1972) *Environ. Sci. Technol.*, **6**, 155—160.

Rushkevich, O. P. (1968) *Gig. Truda Prof. Zabol.*, No. **4**, 55—57 (in Russian).

Shatskaya, N. N. & Orlova, A. A. (1968) *Gig. Truda Prof. Zabol.*, No. **11**, 16—20 (in Russian).

Scott, D. R. *et al.* (1976) *Appl. Spectrosc.*, **30**, 392—405.

Stiefel, T. *et al.* (1976) *Anal. Chim. Acta*, **87**, 67—78.

Stokinger, H. E. (1981) In: Clayton, G. D. & Clayton, F. E. (eds) *Patty's Industrial Hygiene and Toxicology*. John Wiley & Sons, New York, Vol. 2A, pp. 1493—2060.

Talluri, M. V. & Guiggiani, V. (1967) *Caryologia*, **20**, 355—367.

Wagner, W. D. *et al.* (1969) *Toxicol. Appl. Pharmacol.*, **15**, 10—29.

Wagoner, J. K. *et al.* (1980) *Environ. Res.*, **21**, 15—34.

WHO (1990) *Environmental Health Criteria 106: Beryllium*. World Health Organization, Geneva.

Witschi, H. P. & Aldridge, W. N. (1967) *Biochem. Pharmacol.*, **16**, 163–278.
Zdrojewski, A. *et al.* (1976) *Sci. Total Environ.*, **6**, 165–173.
Zorn, H. & Diem, H. (1974) *Zentralbl. Arbeitsmed. Arbeitsschutz.*, **24**, 3–8.
Zorn, H. R. *et al.* (1988) *Beryllium* in: Seiler, H. G. *et al.* (eds) *Handbook on Toxicity of Inorganic Compounds*. Marcel Dekker, New York, pp. 105–114.

Magnesium and its compounds

Magnesium carbonate (magnesite [min.]; **m. chlorate hexahydrate; m. chloride** (chloromagnesite [min.]); **m. diboride; m. hydroxide** (brucite [min.]); **m. oxide** (magnesia, periclase [min.]); **m. polyboride; m. sulfate; m. sulfate heptahydrate** (Epsom salt, epsomite [min.])

Elektrons (magnesium alloys with aluminum [3–10%] and zinc [0.2–3%])

IDENTITY AND PHYSICOCHEMICAL PROPERTIES OF THE ELEMENT

Magnesium (Mg) is an alkaline-earth metal in group II of the periodic table, with the atomic number 12. Its natural isotopes are ^{24}Mg (78.6%), ^{25}Mg (10.11%), and ^{26}Mg (11.29%).

Magnesium is a paramagnetic metal. It possesses a relatively high malleability and ductility. Its mechanical properties strongly depend on how it has been worked.

In compounds magnesium is present in the oxidation state +2. It is a chemically active element. In air it is rapidly coated with a film of oxide that prevents its further oxidation, even on heating up to 350°C. When heated in air to 600–650°C, it ignites and burns to give off a white fume consisting of the oxide MgO and the nitride Mg_3N_2. The nitride is also produced when magnesium is heated in a nitrogen atmosphere to 500°C.

Magnesium forms the hydride MgH_2 on reacting with hydrogen at 400–500°C, and gives halides of the type $MgHal_2$ on reacting with chlorine in the cold and with other halogens on heating. At 500–600°C it forms the sulfide MgS with sulfur, sulfur dioxide, and hydrogen sulfide. It decomposes hydrocarbons to give the carbides MgC_2 and Mg_2C_3. It forms intermetallic compounds with most metals.

Magnesium is a powerful reducing agent and can displace many metals (including alkali metals, beryllium, and aluminum) from their oxides and halides. It also reduces boron, silicon, and carbon from their oxides and halides.

Cold water, if it contains no air, has little effect on magnesium, while hot water (>70°C) reacts with slow evolution of hydrogen and formation of the hydroxide $Mg(OH)_2$; water vapor reacts readily at 400°C. Dilute acids react well even in the cold.

Magnesium does not dissolve in hydrofluoric acid because of the formation of a strong surface film of the fluoride MgF_2, and it is almost insoluble in concentrated sulfuric acid and in mixtures of sulfuric acid and nitric acids. In the cold, aqueous solutions of alkalies have little or no effect on magnesium, while aqueous solutions of hydrogen carbonates and ammonium salts dissolve it. The lower explosive limit of magnesium dust in air is $10 \ g/m^3$; it ignites at 520°C. See also Appendix.

NATURAL OCCURRENCE AND ENVIRONMENTAL LEVELS

Magnesium is found in more than 100 minerals, including brucite $Mg(OH)_2$ (41.7% Mg), magnesite $MgCO_3$ (28.8%), and dolomite $MgCO_3 \cdot CaCO_3$ (18.2%). The clarke of magnesium in the earth's crust is 1.87%.

Magnesium is poorly retained in the biological cycle on the continents and most of it enters the oceans with river discharges. Its average content in seawater is 0.13%, which is less than that of sodium but more than that of any other metal. The global biomass contains an estimated 40 billion tonnes of magnesium; its average concentrations in organisms are (g/kg dry weight): marine algae, 5.2; terrestrial plants, 3.2; marine animals, 5.0; terrestrial animals, 1.0; bacteria, 7.0. During the evolution of life on the Earth, the magnesium content in living organisms has been declining, as judged from differences in dry weight contents between bacteria (0.525%), angiospermous plants (0.305%), and mammals (0.100%) [Kovalsky].

PRODUCTION

Magnesium is chiefly extracted from dolomites, but is also obtained from seawater, by electrolysis of fused carnallite or magnesium chloride, and by reduction of magnesium oxide with carbon or calcium carbide. Crude magnesium is purified by smelting, sublimation in vacuum, and electric or zone refining; these processes result in the generation of Cl_2, HCl, and SO_2.

Magnesium carbonate is obtained as a precipitate from a mixture of magnesium sulfate and sodium carbonate solutions. **Magnesium chlorate** is made by reaction of sodium chlorate and magnesium chloride. **Magnesium chloride** is extracted from seawater or the natural brines of certain lakes or from carnallite by salting out with hydrogen chloride. **Magnesium oxide** is made by roasting magnesite (at 700°C to obtain heavy magnesia for cements and building materials or at 1500–1800°C to obtain it for refractories) or by calcining magnesium carbonate or magnesium hydroxide carbonate (light magnesia). **Magnesium sulfate** is derived from natural sea brines and solid salt deposits, or by heating magnesite and dolomite in an ammonium sulfate solution.

USES

Magnesium is mainly used, chiefly in alloy form, for components of aircraft, rockets, ships, automobiles, and hand tools; as a deoxidizer in metallurgy; in organic synthesis; and in photography and pyrotechnics. Fine magnesium powders — all very reactive — are used in thin foils and wires.

Magnesium carbonate is used to produce the metal and its oxide; in paints and printing inks; as a filler in rubber and paper; in medicine and pharmacy. The **chlorate** finds use as a defoliant and insecticide. The **chloride** is used in producing magnesium metal and its oxide; as a fireproofing agent for wood; in the textile industry; in cements (in mixture with magnesium oxide and fillers); and in medicine. The **oxide** is used in magnesia cements and refractories; as a filler in rubbers; in medicine and pharmacy. The **sulfate** is employed in the tanning of skin; in the manufacture of refractory fabric and paper; as a component of some fertilizers; in medicine and veterinary applications.

MAN-MADE SOURCES OF EMISSION INTO THE ENVIRONMENT

The major anthropogenic sources of magnesium in the environment are emissions from magnesite plants. The level of magnesium compounds in such emissions exceeded 20% (as MgO), while magnesium concentrations in the vicinity of magnesium plants reached 78 mg/L in spring water (as compared to 36 mg/L in control areas) and were also elevated in the fruits and vegetables grown there [Groch *et al.*].

TOXICITY

The taste perception threshold of magnesium in water has been determined at 195 mg/L [Feofanov & Demidenko].

Plants

Magnesium levels above 2.5% are considered to be detrimental to most crops [Kovalsky].

Animals and man

Acute/single exposure

Animals

After a single intraperitoneal injection, the LD_{50} of **magnesium chloride** was 1040 mg/kg for mice and 760 mg/kg for rats; the LD_{50} of **magnesium sulfate** for mice was 150 mg/kg. The intravenous LD_{50} of the chloride for mice was 14.4 mg/kg (as Mg) [Liublina & Dvorkin]. In rabbits, the chloride injected subcutaneously at 1, 10, or 100 mg/kg reduced blood levels of sugar and of lactic and pyruvic acids.

Inhalation of **magnesium oxide** mist led to hyperthermia and neutrophilic leukocytosis; bronchitis and pneumonia also occurred in some animals.

After an intratracheal administration of **pure magnesium** powder in the amount of 50 mg to rats, elevated levels of hydroxyproline in the lungs and of sulfhydryls in internal organs were noted, as were signs of impaired protein metabolism [Brakhnova & Borodiuk]. At 3 to 6 months after an intratracheal insufflation of magnesite dust in the same amount, rats developed moderate fibrosis in the lungs with elevated lipid and hydroxyproline levels, while the introduction of calcined magnesite by the same route resulted in a 30% mortality among rats within several days, the cause of death being pulmonary edema [Zelenova et al.]. Magnesite dust is less fibrogenic than that of clay or chamotte. Appreciable fibrogenic properties are displayed by dusts of the mineral **brucite**, which also caused dystrophic changes in internal organs and the development of bronchiectasis [Kler & Karocharova].

Magnesium alloys with certain metals (notably $CuMg_2$, Zn_2Mg, and CdMg) display moderate fibrogenic activity as well as systemic toxicity, the latter being attributed to their blocking effects on active groups in protein molecules [Borodiuk].

Subcutaneously or intravenously administered magnesium and its compounds bring about a state of narcosis that is not preceded by excitation of the animals. In large doses, they produced paralysis of muscles and motor nerves and depressed cardiac activity to varying degrees, causing cardiac arrest in some cases [Veksler].

Man

Fumes of metallic magnesium may cause metal fume fever [Aikawa]. Individuals who had been inhaling magnesium oxide in a concentration of 4 or 6 mg/m^3 for 12 min developed signs and symptoms that resembled those seen in metal fume fever but were milder. Magnesium salts taken orally in low to medium doses act as 'osmotic' laxatives without producing any toxic side-effects. High doses may be hazardous.

Repeated and chronic exposures

Animals

Cats and guinea-pigs inhaling **magnesium fume** did not develop pathological changes in their lungs [Gardner]. Rats inhaling metallic **magnesium dust** in a concentration of 6.7 or 85 mg/m^3 for 1 to 6 months showed elevated levels of hydroxyproline in the lungs and of sulfhydryl groups in various internal organs as well as dysproteinemia and impaired nucleic acid metabolism; the concentration of 6.7 mg/m^3 was described as being close to the threshold level [Brakhnova & Borodiuk].

Inhalation exposure of rats to **magnesium oxide** at 1000 mg/m^3 (4 h daily for 50 days) resulted in reduced erythrocyte resistance and in increased magnesium levels in the liver and spleen; at autopsy, small dust-laden areas without signs of fibrosis, foci of atelectasis, and peribronchial infiltrates were seen in the lungs [Pazynin & Chinchevich]. Long-term exposure of rats to **magnesite** dust at 263 mg/m^3 (5 h/day for 12 months) or 375–400 mg/m^3 (2 h/day for 4 months) produced chronic bronchitis and proliferative responses in the lungs; some animals developed magnesite pneumoconiosis [Zeleneva et al.].

Magnesium sulfate administered to female rats orally in a daily dose of 150 mg/kg between days 17 and 21 of pregnancy caused gross changes in their hepatic mitochondria and led to a high embryonal mortality. In sheep, the sulfate added to the diet in large amounts impaired mineral metabolism, with reductions in calcium and inorganic phosphorus levels in the blood serum which was found to contain magnesium in an average concentration of 4.75 mg% [Loshchilova & Matantseva].

Man

Persons exposed to **metallic magnesium** for prolonged periods had a variety of abnormalities including chronic atrophic bronchitis, nasal bleeding, frequent colds, loss of hair, excessive sweating, cyanotic hands, tremor of the hands, tongue, and eyelids, red dermatographism, and exaggerated tendon reflexes, as well as elevations of magnesium concentrations in blood serum up to 5 mg% (normal range, 1.7–2.8 mg%) [Kakauridze et al.].

Workers occupationally exposed to **elektrons** (magnesium alloys with aluminum and zinc) experienced gastric troubles involving pain and sometimes nausea and vomiting. The recovery was slow and did not usually occur until a long time after the cessation of exposure. Highly toxic are **magnesium alloys** with mercury, thallium, or lithium [Brakhnova & Borodiuk; Roshchina].

Examination of workers exposed to **magnesium oxide** demonstrated no adverse effects except for slight irritation of the ocular and nasal mucosas. Workers employed in the production of **magnesite** powders and exposed to dusts of raw or calcined magnesite (at concentrations of 8–796 and 5–364 mg/m^3, respectively) complained of breathlessness on exertion, chest pain, and cough; roentgenologically, diffuse interstitial firbosis of the lungs combining with emphysema was noted [Zeleneva *et al.*]. The risk of developing coniosis by those exposed to magnesite dust was reported to be high [Ulrikh].

Pathologic changes in the upper respiratory tract mucosa and mild to moderate pulmonary fibrosis were observed in workers engaged in the production of **chloromagnesite** refractories [Smolnikov & Danilov].

Workers in magnesium plants may be exposed to a range of adverse agents. They often complained of headache, shortness of temper, pains in the legs and chest, fatiguability, sweating, and sleep disturbances. Frequent findings on examination of these workers were dysproteinemia, reticulocytosis, reduced immunologic resistance, dental caries, paradontosis, and manifestations of the 'asthenic-vegetative syndrome'. About one-third of the workers had roentgenologic signs of toxic pneumosclerosis that varied in degree depending on the duration of exposure and was running a benign course in most instances [Muzafarov; Rakhimova]. Increased incidence rates of respiratory and digestive disorders were recorded among those employed in the electrolysis departments of magnesium plants and exposed to **magnesium chloride** at a mean concentration of 1 mg/m^3 [Lyakh].

There is evidence that low magnesium levels in well water (around 6 mg/L) may be associated with a higher prevalence of malignant neoplasms in the local population than elevated levels (~12 mg/L). A relationship of impaired magnesium metabolism to elevated mortality from cardiovascular and gastrointestinal disorders has been reported [Shaumann & Bergmann].

Effects of magnesium chlorate

Acute toxicity studies in animals
Lethal doses of magnesium chlorate administered intratracheally as a 40%, 20%, or 15% aqueous solution were 500 mg/kg for mice and 700 mg/kg for rats and guinea-pigs; the LD$_{50}$ for mice was 380 mg/kg [Abayev; Demidenko]. The oral LD$_{50}$ by gavage was 5235 mg/kg for mice, 6348 mg/kg for rats, and 8660 mg/kg for rabbits; in the clinical picture of poisoning, symptoms attributable to the anticholinesterase activity of this compound predominated. Guinea-pigs injected with 1 ml of a 10% magnesium chlorate solution intraperitoneally died after 10 to 15 min with signs of complete atrioventricular block [Feofanov & Demidenko; Nazarova].

In pregnant rats, a single gavage administration of the chlorate at 127 mg/kg resulted in heightened pre- and post-implantation embryonal mortality rates and in increased numbers of resorbed fetuses; no teratogenic or mutagenic effects were observed [Feofanov & Demidenko; Nazarova]. The dose of 1 mg/kg given to female rabbits by this route on days 7–9 or pregnancy increased embryonal mortality, while the dose of 10 mg/kg was lethal to all embryos and caused degenerative changes in ovarian follicles [Rezhabek & Khalnazarov, 1968].

Multiple-dose and chronic toxicity studies in animals

Daily inhalation exposure of dogs with a pancreatic fistula to 100 ml of a 1% aqueous magnesium chlorate solution in aerosol form increased the amount of discharged pancreatic juice, caused a decline in the activities of its enzymes (trypsin, amylase, and lipase) followed by their elevation, and enhanced cholinesterase activity in the blood serum [Lapasov].

Chronic exposure of rats to a magnesium chlorate aerosol concentrations of 21, 11, 8, or 6 mg/L (4 h/day, up to 4 months) was lethal to a proportion of animals. Autopsy showed congestion and dystrophic changes in the liver, kidneys, and heart; the concentration of 0.56 mg/L was considered as being close to the threshold level [Demidenko].

Rats and rabbits inhaling a magnesium chlorate aerosol intermittently at 2.5 mg/L for 30 days exhibited regenerative anemia, changes in the bone marrow, and depressed immunoreactivity (these effects were more marked in the rats); no detectable hematologic or immunologic changes occurred in animals exposed to 0.25 mg/L. Continuous exposure of rats to 0.1 mg/L was accompanied by alterations in the chronaxie of muscles and cholinesterase activity in the blood. With the concentrations of 0.05 and 0.02 mg/L, the only abnormality was leukocytosis [Mirzayev].

Daily oral administration of magnesium chlorate to rats by gavage at 8 g/kg resulted, after 5–7 days, in deterioration of their general condition, adynamia, bloody nasal discharge, more frequent breathing, tachycardia, hypothermia, convulsions, and partial mortality. Rabbits and rats receiving the chlorate in their diets at 0.4 g/kg body-weight showed growth retardation, accelerated respiratory rate, tachycardia, increased vascular tissue permeability, hypochromic anemia, lowered cholinesterase activity in the blood serum and erythrocytes, and elevated blood protein levels. In rats, inflammatory and dystrophic changes in the ocular system and clear signs of regenerative anemia were also observed [Nazarova; Rezhabek & Khalnazarov, 1968, 1971].

Dietary dosing of female rats with magnesium chlorate at 3 g/kg body-weight for 6 months resulted in severe damage to organs and tissues of the embryos [Rezhabek & Khalnazarov, 1972]. Rats exposed to much lower doses (4 to 25 mg/kg) in their drinking water exhibited, after 10 months, altered blood catalase activity and vascular abnormalities in internal organs; no adverse effects on spermatogenesis were noted. The dose of 5 mg/kg was defined as the subthreshold level [Feofanov & Demidenko].

Man

In areas where magnesium chlorate was used as a defoliant to spray cotton plants, its concentrations in the working zone were in the range 0.03–0.04 mg/L, and anemia with pronounced reticulocytosis and reduced phagocytic activity of blood leukocytes was a common finding among the defoliators. Cracks on the hands were also observed, particularly in those preparing the working solution of magnesium chlorate [Demidenko].

Effects of local application

Animals

A 20% or 40% **magnesium chlorate** solution applied to the intact skin of rabbits made it grossly hyperemic, edematous, and ulcerated; when such a solution was dropped into their eyes, blepharospasm and profuse lacrimation were immediately observed. A 0.75% solution was not irritating to the skin. The 20% and 40% solutions applied repeatedly caused a loss of appetite and body-weight together with reductions of cholinesterase activity in the blood and erythrocytes, indicating that the chlorate can be absorbed through undamaged skin [Demidenko; Nazarova].

In mice, multiple painting of the oral mucosa with a 20% magnesium chlorate solution rendered it hyperemic and edematous on day 3 or 4; on day 70, a number of abnormalities were observed, including hyperkeratosis and acanthosis of the oral mucosal epithelium, swollen bone trabeculae, focally demineralized parodontal tissue, impaired circulation in the dental pulp, and altered dentin structure [Abayev].

Man

Painful intumescences or indolent purulent inflammatory processes have been observed in the skin of workers exposed to **magnesium** or **high-magnesium alloys**. Wounds contaminated with magnesium tend to heal very slowly. In workers directly involved in titanium and magnesium production, skin wounds were healing at rates 2.5 times lower than in other workers of the same plant. Skin disorders caused by **magnesium sulfate** have been described. Cutaneous reactions resulting from penetration of magnesium through the skin have been reported by Birch.

ABSORPTION, DISTRIBUTION, AND ELIMINATION IN MAN

The average human body contains between 20 and 25 g of magnesium, of which about half resides in the skeleton. The daily magnesium requirements of the adult human normally range from 250 to 300 mg but may increase to 500–700 mg under severe stress [Schaumann & Bergmann]. A dietary intake of 10 mg/kg body-weight is recommended [Kukes] and can be provided by a diet high in foods of plant origin, especially green vegetables. Internal absorption of ingested magnesium in humans is about 12–40%; the regulation of its body content depends on kidney activity [Birch].

The ingested magnesium is absorbed slowly; it is oxidized to MgO which is partially converted to $MgCl_2$ by the action of gastric juice.

Blood concentrations of magnesium are usually around 2 mg%, but have been shown to rise to 10–20 mg% when its dietary intake was high. Concentrations in the blood of workers exposed to magnesia dust did not exceed 5 mg%.

The principal routes of magnesium elimination from the body are the feces and urine [Levina]. The urine of children living near a magnesia plant was reported to contain 4.9 mmol/L, which is about twice the normal level [Groch et al.].

In recent years, the prevalence of magnesium deficiency has been rising in populations of industrialized European countries, which is attributed to the enrichment of soils and plants with calcium and the ability of alcohol to stimulate magnesium excretion from the body [Markiewicz].

HYGIENIC STANDARDS

Exposure limits adopted for magnesium and its compounds in the USSR are given in Table 1. In the USA, threshold limit values have been set for magnesite (TWA) and magnesium oxide fume (STEL). For magnesite, the TWA is 10 mg/m^3 for total dust containing no asbestos and <1% crystalline silica and 5 mg/m^3 for respirable dust; the STEL for magnesium oxide is 20 mg/m^3.

Table 1. Exposure limits for magnesium and its compounds

	Workroom air		Atmospheric air			Water sources		
	MAC_{wz} (mg/m^3)	Aggregative state	MAC_{hm} (mg/m^3)	MAC_{ad} (mg/m^3)	$TSEL_{aa}$ (mg/m^3)	MAC_w (mg/L)	MAC_{wf} (mg/L)	Hazard class
Magnesium	—	—	—	—	—	—	50.0	3
Magnesium chlorate	5.0	Aerosol	—	0.3	—	20.0[a]	—	3
Magnesium di-boride (as B)	1.0	Aerosol	—	—	—	—	—	3
Magnesium oxide	10.0	Aerosol	0.4	0.05	—	—	—	3
Magnesium polyboride	—	—	—	—	0.02	—	—	3

[a] Based on the general sanitary index of harmfulness that characterizes the impact of a compound on the self-purification capacity of the soil and on the soil microbiocenosis.

METHODS OF DETERMINATION

Colorimetric methods for the determination of magnesium in air [Pazynin & Chinchevich] and in blood and plasma [Gardner] have been described. A spectrographic method for its determination in blood has been proposed by Nifontova. Methods for measuring magnesium in water are described in Novikov *et al.* Tarasov and Mirzayev have developed a photometric procedure for determining magnesium chromate after its formation in an acid medium. Atomic absorption spectrometry can also be used [Birch].

MEASURES TO CONTROL EXPOSURE

At the workplace
Basic technical control measures to be implemented in factories where magnesium and its compounds are produced or used include the introduction of improved technologies and hermetic sealing of the equipment to minimize the emission of harmful substances and excessive heat into the working area. Where releases of gases, dusts, and heat cannot be prevented, efficient local exhaust ventilation of the process equipment should be provided to localize and remove noxious agents at source. Good general ventilation is also mandatory.

Increased heat insulation should be provided for the equipment, particularly electrolyzers, mixers, chlorinators, and smelting furnaces to ensure thermal comfort for the workers. If they are exposed to magnesium dust, contact lenses must be worn and eyewash facilities should be immediately available. Workers machining magnesium should wear overalls to which small fragments of the metal will not adhere.

Working clothes should be laundered, repaired, and decontaminated on a centralized basis and must not be allowed to be taken home for these purposes. Where the working clothes may be contaminated with dust, arrangements should be made for its removal in such as way as to prevent it from entering the environment or contaminating the inner surfaces of clothes and the skin of workers.

All workers should be made aware of the hazards associated with their specific jobs. Before employment and systematically during the period of employment, they should be fully instructed in, and encouraged to observe, the corresponding safety precautions and general and personal hygiene measures. Pre-employment and periodic in-employment medical examinations are also necessary.

For *personal protection*, workers engaged in the production or use of magnesium and its compounds should be supplied with appropriate clothes, footwear, and personal protective devices. Mirrors should be installed in the cloakrooms of factories where protective ointments before work or greasy creams after showering at the end of the workday need to be applied by workers to the face.

Protection of general environment

In factories producing or using magnesium and its compounds, provisions should be made for utilization and effective purification of the dust and gas emissions and of the liquid and solid wastes.

FIRST AID IN POISONING

In the event of eye contact with magnesium or its compound such as the chlorate, the eye should be irrigated with copious amounts of cooled water and a 30% albucid [synonym: sodium sulfacetamide] solution dropped into the conjunctival sac. In cases of skin contact, the affected skin area should be flushed with large quantities of water. Acute dermatitis can be alleviated by dressings impregnated with lead water or Burow's solution; for the night, an indifferent paste or ointment such as Lassar's paste or synthomycin ointment may be applied.

REFERENCES

Abayev, V. Yu. (1981) *Zdravookhranenie Turkmenistana* [Health Care in Turkmenistan], No. 11, 37—39 (in Russian).

Aikawa (1991) Chapter II.18 *Magnesium* in: Merian, E. (ed.) *Metals and their Compounds in the Environment — Occurrence, Analysis and Biological Relevance.* VCH, New York, pp. 1025—1034.

Birch, N. J. (1988) Chapter 33 *Magnesium* in: Seiler, H. G. *et al.* (eds) *Handbook on Toxicity of Inorganic Compounds.* Marcel Dekker, New York, pp. 397—403.

Borodiuk, T.M. (1978) *Gig. Truda Prof. Zabol.*, No. 9, 49—51 (in Russian).

Brakhnova, I. T. & Borodiuk, T. M. (1975) *Vrach. Delo*, No. 3, 131—135 (in Russian).

Demidenko, N. M. (1960) In: *Trudy Tashkentskogo meditsinskogo instituta* [Transactions of the Tashkent Medical Institute]. Vol. 16, pp. 189—196 (in Russian).

Feofanov, V. N. & Demidenko, N. M. (1983) *Gig. San.*, No. 4, 68—70 (in Russian).

Gardner, G. (1946) *Biochem. J.*, 40, 828—832.

Groch, J. *et al.* (1984) *Čs. Hyg.*, 29, 515—520.

Kakauridze, E. M. *et al.* (1956) *Gig. San.*, No. 11, 73—74 (in Russian).

Kler, O. V. & Karocharova, V. N. (1962) *Gig. Truda Prof. Zabol.*, No. 6, 51—54 (in Russian).

Kovalsky, V. V. (1974) *Geokhimicheskaya Ekologiya* [Geochemical Ecology]. Nauka, Moscow (in Russian).

Kukes, V. G. (1968) *Sov. Med.*, No. 4, 120 (in Russian).

Lapasov, Kh. (1974) *Vliyanie khlorat-gidrata i khlorat-khlorid kaltsiya na sekretornuyu funktsiyu podzheludochnoi zhelezy u sobak v eksperimente* [Effect of Calcium Chlorate Hydrate and Calcium Chlorate Chloride on Pancreatic Secretion in Experimental Dogs]. Samarkand (in Russian).

Levina, E. I. (1972) *Obshchaya toksikologiya metallov* [General Toxicology of Metals]. Meditsina, Leningrad (in Russian).

Liublina, E. I. & Dvorkin, E. A. (1983) In: *Gigiyenicheskaya toksikologiya metallov* [Hygienic Toxicology of Metals]. Meditsina, Moscow, p. 114 (in Russian).

Loshchilova, V. G. & Matantseva, S. L. (1963) In: *Materialy 3-ey Povolzhskoi konferentsii fiziologov, biokhimikov i farmakologov* [Proceedings of the Third Volga Region Conference of Physiologists, Biochemists, and Pharmacologists]. Gorky, pp. 265—267 (in Russian).

Lyakh, G. D. (1970) In: *Trudy NII kraevoi patologii Kazakhskoi SSR* [Transactions of the Institute for Regional Pathology of the Kazakh SSR]. Alma-Ata, pp. 58—62 (in Russian).

Markiewicz, J. (1985) *Folia Med.*, **26**, 5—8.

Mirzayev, Sh. Sh. (1983) *Gig. San.*, No. **7**, 69—71 (in Russian).

Muzafarov, A. I. (1977) In: *Trudy NII kraevoi patologii Kazakhskoi SSR* [Transactions of the Institute for Regional Pathology of the Kazakh SSR]. Alma-Ata, Vol. 29, pp. 59—62 (in Russian).

Nazarova, O. B. (1973) *Zdravookhranenie Turkmenistana* [Health Care in Turkmenistan], No. **5**, 3—5 (in Russian).

Nifontova, M. V. (1962) *Lab. Delo*, No. **2**, 27—30 (in Russian).

Novikov, Yu. V. *et al.* (1981) *Metody opredeleniya vrednykh veshchestv v vode vodoyemov* [Methods for Measuring Noxious Substances in Water Bodies]. Meditsina, Moscow (in Russian).

Pazynin, V. M. & Chinchevich, V. I. (1983) *Gig. San.*, No. **7**, 60—61 (in Russian).

Rakhimova, M. T. (1977) In: *Trudy NII kraevoi patologii Kazakhskoi SSR* [Transactions of the Institute for Regional Pathology of the Kazakh SSR]. Alma-Ata, Vol. 29, pp. 41—46 (in Russian).

Rezhabek, O. Ya. & Khalnazarov, K. A. (1968) *Zdravookhranenie Turkmenistana* [Health Care in Turkmenistan], No. **8**, 27—31 (in Russian).

Rezhabek, O. Ya. & Khalnazarov, K. A. (1971) *Zdravookhranenie Turkmenistana* [Health Care in Turkmenistan], No. **3**, 6—8 (in Russian).

Rezhabek, O. Ya. & Khalnazarov, K. A. (1972) *Zdravookhranenie Turkmenistana* [Health Care in Turkmenistan], No. **1**, 6—10 (in Russian).

Roshchina, T. A. (1980) *Gig. Truda Prof. Zabol.*, No. **10**, 9—13 (in Russian).

Schaumann, E. & Bergmann, W. (1984) *Z. Gesamte Hyg.*, **30**, 84—87.

Smolnikov, L. S. & Danilov, V. I. (1966) In: *Voprosy truda i profpatologii v khimicheskoi i mashinostroitelnoi promyshlennosti* [Occupational Health in the Chemical and Machine-Building Industries]. Kharkov, pp. 90—91 (in Russian).

Tarasov, V. V. & Mirzayev, Sh. Sh. (1982) *Gig. San.*, No. **1**, 39 (in Russian).

Ulrikh, L. (1963) *Gig. Truda Prof. Zabol.*, No. **4**, 22—27 (in Russian).

Veksler, I. L. (1938) *Vliyanie magniya na organizm v svete kliniki i eksperimenta* [Health Effects of Magnesium: Clinical and Experimental Findings]. Rostov-on-Don (in Russian).

Zeleneva, N. I. *et al.* (1970) *Gig. Truda Prof. Zabol.*, No. **2**, 21—24 (in Russian).

Calcium and its compounds

Calcium carbide; **c. carbonate** (chalk, marble, limestone, aragonite, calcite, calcspar, Iceland spar [minerals]); **cement; c. chlorate chloride; c. chloride** (hydrophillite [min.]); **c. hydroxide** (slaked lime, hydrated lime, white lime [mixture of $Ca(OH)_2$, sand, and water], limewater [aqueous solution of $Ca(OH)_2$], milk of lime [$Ca(OH)_2$ suspension in water], lime paste, portlandite [min.]); **c. hypochlorite trihydrate; c. metaarsenite** (calcium arsenite); **c. orthoarsenate trihydrate** (calcium arsenate trihydrate); **c. orthoarsenate orthophosphate; c. oxide** (quicklime, unslaked lime, air-slaked lime); **c. oxide orthophosphate** (basic slag, Thomas slag); **c. sulfate** (anhydrite(α) [min.], calcined gypsum); **c. sulfate dihydrate** (gypsum [min.], selenite [min.])

IDENTITY AND PHYSICOCHEMICAL PROPERTIES
OF THE ELEMENT

Calcium (Ca) is an alkaline-earth metal in group II of the periodic table, with the atomic number 20. Its natural isotopes are ^{40}Ca (96.94%), ^{42}Ca (0.647%), ^{43}Ca (0.135%), ^{44}Ca (2.09%), ^{46}Ca (0.003%), and ^{48}Ca (0.187%).

Calcium exists in two allotropic modifications. It is a ductile metal that is readily moldable and cuttable and can be rolled into sheets.

In compounds, calcium is present in the oxidation state +2. It is a highly reactive element. It is a strong reducing agent and can displace almost any metal from its oxide, sulfide, or halide. When heated in air or oxygen, it ignites and gives the oxide CaO. It decomposes carbon dioxide and, at 550–650°C, carbon monoxide. It reacts with cold water at a fast rate initially and much slower later because a layer of the hydroxide $Ca(OH)_2$ forms on its surface. It reacts vigorously with hot water and with acids (except concentrated nitric acid) with the evolution of hydrogen. It will interact with chlorine, bromine, and fluorine in the cold; at temperatures $\geqslant 400°C$, the chloride $CaCl_2$ and the bromide $CaBr_2$ are produced at a fast rate. It gives the sulfide CaS with sulfur on heating, the hydride CaH_2 with dry hydrogen at $300-400°C$, and the nitride Ca_3N_2 with nitrogen at $500°C$; the latter compound is also produced, together with the hydride, when calcium is heated in an ammonium medium. Heating calcium with phosphorus in the absence of air supply will yield the phosphide Ca_3P_2. On heating with graphite, calcium forms the carbide CaC_2. Its reaction with silicon gives rise to silicides. Calcium forms intermetallic compounds when alloyed with Al, Ag, Au, Cu, Li, Mg, Ni, Pb, Sb, Sn, Tl, or Zn. See also Appendix.

NATURAL OCCURRENCE AND ENVIRONMENTAL LEVELS

Calcium is the fifth most abundant element after oxygen, silicon, aluminum, and iron, its content in the earth's crust being estimated at 2.96% by mass. It migrates readily and accumulates in a variety of geochemical systems, forming a large number of minerals. Most of the calcium is contained in feldspars. In waters with a high CO_2 content, calcium occurs in solution while in those low in CO_2, the mineral calcite ($CaCO_3$) precipitates to form eventually deposits of limestone, chalk, and marble. Among other major calcium minerals are anhydrite $CaSO_4$, gypsum $CaSO_4 \cdot 2H_2O$, fluorite CaF_2, and apatite $3Ca(PO_4)_2 \cdot Ca(F,Cl)$.

Although rivers discharge large quantities of calcium to the oceans, it is not retained in the oceanic water but concentrates in the skeletons of organisms and, after their death, settles to the bottom predominantly as $CaCO_3$. An important role in calcium migration is played by underground waters (karst). Natural waters in the European part of the USSR contain calcium at an average concentration of 67.2 mg/L.

Calcium is the most abundant metallic element found in living matter. Organisms exist in which calcium makes up more than 10% of the body weight and whose skeletons largely consist of $CaCO_3$ (calcareous algae, mollusks, corals, echinoderms, etc.).

PRODUCTION

Metallic calcium is obtained by electrolysis of molten calcium chloride or fluoride or by metallothermic reduction of calcium oxide in vacuum.

C. carbide is made by heating a mixture of crushed calcium oxide (quicklime) and coke or anthracite at 2000°C. The principal sources of **calcium carbonate** are the minerals calcite and argonite. **C. chlorate** is commonly obtained in making potassium chlorate as a mixture with calcium chloride ($CaCl_2 \cdot 6H_2O$) that contains 22–23% calcium chlorate, 42% calcium chloride, and crystallization water. **C. chloride** is obtained by dissolving limestone in hydrochloric acid or as a by-product in sodium carbonate and potassium chlorate manufacture. **C. hydroxide** (slaked lime) is produced by hydration of calcium oxide, the slaking process being accompanied by the evolution of much heat. Technical-grade **c. hypochlorite** (chlorinated lime, bleaching powder), which also contains c. hydroxohypochlorite and c. hydroxochloride, is obtained by chlorination of dry calcium hydroxide. **C. oxide** (quicklime) is prepared by thermal decomposition of limestone ($CaCO_3$) at 900–1200°C; when phosphoric iron is melted in the presence of lime, Thomas slag containing 45–50% of CaO is produced. The principle source of **calcium sulfate** is gypsum.

USES

Calcium is mainly used as a reducing agent in thermal reduction processes, as a scavenger in metallurgy, as a dehydrant of organic solvents, and in alloys.

Calcium carbonate finds use as a building material; as a raw material in the manufacture of lime and Portland cement; as a flux in metallurgy; in the chemical industry; in agriculture; and as an antacid in medicine. From **C. carbide**, c. cyanamide and acetylene are prepared. **C. chlorate** is a defoliant used to remove leaves from cotton plants (usually as a 42% chlorate–chloride solution). **C. chloride** is used for the production of calcium and its alloys, for drying gases and liquids, as an accelerator of concrete curing, as a de-icing agent for airfields and railroads, and for various purposes as a drug in medicine. **C. hypochlorite** is employed as a bleaching and oxidizing agent in the textile industry; as an oxidizer in the production of chloroform, chloropicrin, and other chemicals; in the refining of acetylene and some petroleum products; and as a disinfectant and a degasifying agent. **C. oxide** and **C. hydroxide** are used in cements (Portland cement contains 62–76% CaO, Roman cement 34%, and slag cement 54–60%); for fettling the hearths of furnaces; as binders in building construction; as fluxes in metallurgy; in the leather and food industries; in glass manufacture; in agriculture; in wastewater treatment; and as fertilizers (in Thomas slag). **C. sulfate** is used in building construction, in sculpturing and in making casting molds, in the cement industry, in agriculture, and in medicine.

MAN-MADE SOURCES OF EMISSION INTO THE ENVIRONMENT

The main sources are the quarrying and mining of natural calcium compounds, cement manufacture, and effluents from various industries such as paper, chemical, glass, pharmaceutical, leather, paints and varnishes, and brewing industries; much calcium may be contained in effluents from laundries [Grushko]. The presence of calcium in water at high concentration and the associated increase in its alkalinity (up to pH 12.0) and hardness may be due to the use of cemented or concreted water storage reservoirs and pipes [Ilichkina]. Storage of drinking water in cemented containers also deteriorates its organoleptic properties.

TOXICITY

Aquatic organisms

LC_{50} values of **calcium** for lower aquatic organisms ranged from 3000 to 7000 mg/L on 48 hour exposure. Daphnids exhibited decreased reproductive capacity at 116 mg Ca/L [Grushko]. Calcium-associated reproductive problems of fish in acidified environments have been described [Munkittrick].

Animals and man

General remarks

Calcium is a major element and can only produce adverse health effects on entering the body in very large doses; hypercalcemia, calciuria, augmented calcification, and weakened regenerative processes are then observed [Balgabekov; Rezvaya *et al.*]. Dusts and aerosols of **calcium oxide** and **hydroxide** exert strong cauterizing effects on the skin and mucous membranes. The oxide is more dangerous; it saponifies fats, absorbs moisture from the skin, dissolves proteins, and irritates and cauterizes tissues. These compounds can also cause severe lesions in the lower airways and the lungs.

Calcium carbide

Man

Calcium carbide exerts pronounced irritant effects due to the formation of calcium hydroxide upon reaction with moist air or sweat. Dry carbide in contact with skin may cause dermatitis — not only at the application site but also in the surrounding skin. Contact with moist skin and mucous membranes leads to inflammatory and ulcerative lesions similar to those caused by lime. The carbide is especially hazardous to the eyes. Burns caused by hot calcium carbide are particularly frequent in the carbide industry. The tissues are generally damaged to depths of 1–5 cm, and the burns heal very slowly, are difficult to treat, and often require excision.

Calcium carbide workers, particularly those employed for a short time, complained of skin itching, and all workers had a very dry skin on their hands. Among carbide furnacemen, hypohidrosis, striation of the dental enamel, increased brittleness and fragility of the fingernails, and a thick brownish-yellow coating at the gingival margins were frequent findings, and a high prevalence of caries was recorded. Examination of 1020 calcium carbide workers showed that 763 of them had melanoderma of the skin on the hands, arms, face, and/or neck; in some of them, melanoderma with strong hyperpigmentation was noted. Other common abnormalities were swelling, hyperemia, and desquamation of the lips (on the varmilion border) and thickening of the nail walls and nail plates [Antoniev et al.].

Calcium carbonate (limestone)

Animals
In a 10-month chronic toxicity study, bronchial lesions, focal accumulations of dust-laden cells, mild fibrosis, emphysema, and increased hydroxyproline levels were observed in the lungs of rats exposed to limestone dust by inhalation at 84 mg/m^3 for 4 h/day; in addition, impaired calcium metabolism and lesions of the urinary bladder and liver were noted, indicating absorption of the carbonate through the lungs [Domnin]. Moderate and slowly progressing nodular pneumoconiosis occurred in rats administered limestone dust intratracheally on a single occasion as well as in those exposed to it by inhalation at 250–300 mg/m^3 for 2 h/day over 6 to 12 months [Rumyantsev & Kochetkova].

Repeated oral exposure of rats to limestone for 1 to 2 months caused the appearance of papillary outgrowths in the esophageal epithelium and desquamation of the epithelium together with considerable hyperplasia of the parietal glandular cells in the stomach [Domnin et al.].

Man
Workers employed in the mining or processing of limestone were often observed to have atrophic catarrhs of the upper airways and bronchitis in combination with pulmonary emphysema; in addition, gastritis or gastroduodenitis and liver dysfunction were detected in 21 out of 53 workers examined. Common complaints were epigastric pain, a bitter taste in the mouth, and poor appetite; some also complained of heartburn, nausea, and occasional vomiting unrelated to food intake. No cases of pneumoconiosis were reported [Domnin]. Beckenkamp & Hardeck, too, did not find any clinical or roentgenologic evidence of pneumoconiosis in limestone-exposed workers; only dilatation of the pulmonary hilus and aorta, lymph node calcification, and accentuated lung markings were observed. However, cases of pneumoconioses among limestone mine workers have been described [Smirnova], as well as among those employed in the production of precast ferroconcrete [Krapukhina et al.]. The occurrence of pneumoconiosis in limestone miners and grinders has been attributed to the presence of silica in the limestone dust [Domnin & Angelova; Smirnova].

Workers handling a fluxed agglomerate whose major component was limestone complained of irritation in the nose and throat and nosebleeds; on examination, mucosal hyperemia and chronic catarrh in the upper airways were seen, but no evidence of pneumosclerosis or pneumoconiosis could be obtained. Toxic and fibrogenic effects of airborne limestone dust were detected in the lungs of those working in underground limestone mines where the dust was present in concentrations as high as $500-1500$ mg/m^3 [Spiridonova *et al.*].

Cement

Cement is a fine powder usually obtained by grinding the clinker of a clay and limestone mixture calcined at high temperatures. Triethanolamine, anthracite, naphtha soap, or lignin may be added to improve the grinding process, fluorides of alkaline or alkaline-earth metals to accelerate the calcining, and substances such as sodium triphosphate to reduce the viscosity of the slurry.

Animals

In rats, inhalation of cement dust $(250-300$ mg/m^3, 2 h/day) for 6 to 12 months induced a moderate and slowly progressing nodular pneumoconiosis together with catarrhal or purulent bronchitis. In rabbits, inhalation exposure to a silica-containing cement $(24.7\%-31.6\%$ $SiO_2)$ for 6 h/day over 4 to 5 months resulted in a fibrotic process in the lungs; in addition, body-weight loss, reduced erythrocyte counts and hemoglobin levels, leukocytosis, increased erythrocyte sedimentation rate, and abnormal regeneration of osseous tissue were observed, indicating that the inhaled material had been absorbed through the lungs. Moreover, hair loss and pustules on the skin were seen [Ternopolskaya & Osetinsky].

The severity of pulmonary lesions is believed to depend on the levels of free and bound silica in the cement dust [Dvizhkov]. According to Kolev & Shumkov, the most typical manifestation of exposure to airborne cement dust is peribronchial fibrosis with perifocal and subpleural emphysema so that this condition may be appropriately called 'cement pneumopathy'.

Evidence for systemic toxicity of Portland cement dust after its intratracheal administration to mice was provided by increases in histamine levels and cholinesterase and diamine oxidase activities and decreases in total glutathione and phagocytic activity of the leukocytes [Kolev & Mandadzhiev].

Dusts obtained from marbles of various grades used in cement production and containing various proportions of free silica induced diffuse sclerotic fibrosis of the lungs in experimental animals after a month of exposure. Loss of wool in sheep and the formation of gastric phytobezoars in cows occurred as a result of their consuming vegetation covered with a thick crust of cement dust which had led to decreased carbon dioxide uptake by the plants; the soil was found to contain high levels of the fluorides added to the cement [Lindberg].

Inhalation exposure of male and female rats to cement dust for 3 months before mating followed by further exposure of the females during gestation resulted in increased rates of fetal resorptions and stillbirths.

Man

Employees of cement works, including those employed for short periods, complained of pain and oppression in the chest, breathlessness, cough, dryness in the mouth, hoarseness, reduced sense of smell, and nosebleeds [Dragomiretsky *et al.*]. Some marble crushers were reported to have diffuse sclerotic forms of pneumoconiosis [Sedov *et al.*]. Cases of pneumoconiosis, usually of a benign variety with few clinical symptoms, have been described repeatedly in cement workers. Depending on the chemical composition of cement dust, either typical silicosis or interstitial sclerosis may develop [Dvizhkov], as may pneumoconiosis of a mixed type, characterized roentgenologically by both irregular and round opacities, the former being due to the free silica and the latter, to the silicates contained in the cement; for this condition, the term 'pneumoconiosis of cement workers' has been proposed [Popovič]. Some cement workers had rhinoliths on the posterior pharyngeal wall, on the amygdalae, in the larynx, or in the nasal cavity, while some presented with fissures and even perforation of the nasal septum. Bronchitis, emphysema, pleural adhesions, chronic inflammations in the antrum of Highmore, and polyposis of the nasal mucosa have also been reported [Bariliak *et al.*], as well as increased prevalence of chronic nonspecific disease of the lungs [Kalacic]. Temporary deafness also occurred occasionally.

Cement production by a wet process was found to be associated with a higher prevalence of dust diseases than its production by dry processes because the former process (in which kilns were used) generated more airborne dust than the latter [Makulova *et al.*]. Pneumoconiosis was more prevalent among workers producing pozzolana or acid-resistant cement than among those producing Portland cement or Portland-slag cement [Sadkovskaya].

A morbidity survey of workwomen aged 19 to 70 years employed in cement and brick factories for an average period of 12 years, showed that 12% of them were apparently healthy, 33.7% had cervical erosions, 27.2% had vaginitis, 19.4% had inflamed uterine appendages, and 0.62% had uterine or vaginal cancer [Glowinski & Dudkiewicz].

Radiographs taken in cement workers, even young ones employed for only 1 to 4 years, often revealed appreciably narrowed dental canals. Gastric and duodenal ulcers were also not uncommon, apparently as a consequence of cement dust acting specifically on the gastrointestinal mucosa [Domnin *et al.*]. All cements are freely soluble in gastric juice (up to 72%). Increased urinary excretion of calcium is conducive to cystitis.

Cases of bronchial asthma and Quincke's edema, also observed among cement workers, have been explained by sensitizing properties of cement which have in turn been ascribed to the presence of chromium(VI) compounds in the cement. Sex differences in responses to allergenic cement components have been reported [Mikhalev & Pustotin].

The air, soil, water, and plants sampled at considerable distances from cement factories may be polluted by their emissions that contain not only cement dust but also various metals; thus, so-called 'technogenic' lead anomalies were identified at distances up to 30–40 km around a cement plant, and it has been claimed that the contribution of lead and manganese from the cement industry to air pollution is comparable to those of automotive transport and metallurgical industries [Nikiforov et al.].

Health surveys of children living 0.2 to 2 km away from a cement factory for 5 years or more showed that correlations exist between the intensity of ambient air pollution and the prevalence of respiratory, digestive, and ENT diseases. The surveys also showed reduced excitability of the autonomic nervous system and the olfactory apparatus, increased leukocyte migration, and epithelial desquamation in the nasal and ocular mucous membranes [Davydov; Straus et al.].

Dermal and ocular effects. Cement has been placed among the more aggressive chemicals that are responsible for most cases of dermatosis in the population [Bagnova]. Dermatoses caused by cement exposure have been widely reported in the literature and may account for 25% (or more) of all occupational skin diseases [Prodan].

The frequency and severity of skin disorders in cement workers depend on the amount of lime or calcium carbonate present in the cement. A factor predisposing to skin lesions is profuse perspiration. The most characteristic cutaneous manifestations are 'cement itch', ulcers, and 'bricklayer's eczema'. Cement itch is more common than the two other conditions and is marked by the emergence of small itchy nodules on open skin areas, especially in the interdigital folds, on the back of the hand, and on the face. Not infrequently, bleeding and slowly healing deep fissures appear on the hands. The lesions occasionally spread to involve nearly all skin of the body. Skin necrosis affecting the anterior aspect of the shins occurred in a worker who had been kneeling on moist cement [Rowe & Williams]. Dermatitis can be caused by the calcium hydroxide washed out from cement by water.

Cement dust contacting skin wounds complicates the wound process [Grigorian]. Contact with the eyes may cause severe conjunctivitis and even necrosis of some conjunctival areas with subsequent formation of adhesions; scars and opacities may appear on the cornea. Perforation of the eyeball is also possible.

Calcium chlorate chloride
The taste perception threshold for calcium chlorate chloride in drinking water has been determined at 5 mg/L.

Animals
The LD_{50} of calcium chlorate chloride for rats was 152 mg/kg with intratracheal administration and 1112 mg/kg with administration by the oral route

[Demidenko]. Rats given a 10% aqueous solution orally experienced hemorrhages and extensive necrosis of the gastric mucosa. Dogs that had been inhaling the aerosol of a 0.5% aqueous solution for only 1 min developed reduced blood coagulability, leukocytosis, and marked hemolysis accompanied by hemosiderosis in the lungs, spleen, and liver; 40 days later, their lungs contained areas of atelectasis and emphysema together with accumulations of a brown pigment in macrophages of the interstitial tissue [Rakhimova et al.].

Inhalation exposure of mice, rats, and rabbits to a calcium chlorate chloride aerosol at 1 mg/L for 2.5 months was not accompanied by any apparent clinical manifestations of poisoning; autopsy, however, showed congestion and hemorrhages in internal organs, hemosiderosis and areas of atelectasis and emphysema in the lung, and dystrophic changes in the liver. Animals exposed to 24 mg/L lost weight and developed eosinophilia and leukocytosis with lymphocytosis. In dogs with a pancreatic fistula, lowered cholinesterase activity and reduced pancreatic juice secretion in response to any kind of stimulus was noted after 20 days of inhalation exposure to the aerosol produced by spraying a 1% aqueous solution of the chlorate chloride; the juice contained elevated trypsin, amylase, and lipase activities.

In rats, long-term (for up to 6 months) ingestion of calcium chlorate chloride in the drinking water at 70 mg/kg produced marked alterations in the summated threshold index, catalase activity in the blood, prothrombin time, and in the rate of Bromsulphalein excretion; autopsy showed edema and dystrophic changes in internal organs and in the brain. The dose of 70 mg/kg led to much milder changes, while the dose of 7 mg/kg was considered to be the no-observed-effect level.

Man
Some airfield employees engaged in preparing calcium chlorate chloride solutions and filling aircraft tanks with them complained of eye irritation and had fissures of the skin on their hands [Avezbakieva & Demidenko].

Calcium chloride
The taste perception threshold for calcium chloride in drinking water ranged from 150 to 350 mg/L.

Aquatic organisms
Toxic concentrations were 500–1000 mg/L for fish, 650–920 mg/L for daphnids, and 3500 mg/L for plants [Grushko].

Animals
The oral LD_{50} was 4000 mg/kg for rats and 1384 mg/kg for rabbits. The intraperitoneal LD_{50} was 600 mg/kg for mice and 500 mg/kg for rats.

Man

Workers employed in the production of calcium chloride presented a number of complaints including burning and itching in the exposed parts of the body, dryness and scaling of the skin, pinching pain in the eyes, lacrimation, burning and pain in the nose, and nosebleeds. On examination, papular or pustular eruptions, scleral hyperemia, eyelid edema, purulent discharge from the eyes, and hyperemia of the nasal mucosa were observed. Skin biopsy demonstrated necrotic areas in the superficial dermal layers with deposits of calcium salts in the dermis. Perforation of the nasal septum was also seen in some cases [Barsky; Molchanov].

Calcium hypochlorite

The taste perception threshold for calcium hypochlorite in drinking water has been determined at 5 mg/L.

Man

Dust of calcium hypochlorite and the chlorine gas released from the latter can produce strong irritation of the skin, upper respiratory tract, and conjunctivae, and have provoked the development of bronchial asthma in some occupationally exposed workers. The hypochlorite also damages teeth. A person who had been stacking jars containing the hypochlorite developed toxic hepatitis in additon to conjunctival and upper respiratory tract irritation.

In workers using the hypochlorite to remove paints from surfaces, increased sweating of the palms and skin softening on the hands were noted; with further exposure, atrophic changes occurred in the skin which became smooth, white, and shiny. Most of the workers were young.

Calcium oxide and calcium hydroxide

Fish

Calcium hydroxide causes severe damage to the gills, as do other alkalies. The fish become restless, begin to breathe rapidly, and may develop convulsions; destruction of the respiratory epithelium results in asphyxia. Toxic concentrations of the hydroxide ranged from 20 to 120 mg/L. Threshold pH values (i.e. those at which toxic effects just began to occur) were found to be 9.2 for brook trout, rainbow trout, perch, and ruff; 10.4 for roach; 10.7 for pike; 10.8 for carp and tench. Deaths among fish placed in newly built ponds having concrete walls have been described, the cause of death being intoxication with the cement leached out from the walls [Metelev *et al.*].

Animals and man: acute poisoning

Animals. With intraperitoneal administration to mice, the LD_{50} of calcium oxide was 3059 mg/kg while its threshold dose, as determined by recording effects on body-weight and blood sulfhydryls, was 259 mg/kg [Levina].

Short-term inhalation exposure of rats to dust of an agglomerate containing 5–10% calcium oxide resulted, several days later, in the death of a proportion of animals from aspiration pneumonia. Pathologic examination of the surviving rats showed catarrhal or purulent bronchitis, emphysema, and nodular pneumoconiosis. Rapidly progressive tracheobronchitis that led to destruction of the mucous membranes occurred in rats following an intratracheal administration of lime dust; bronchiectasis and pulmonary emphysema were also present in some animals.

Man. Inhalation of Thomas slag dust may cause severe pneumonia. The pneumonia starts suddenly with chills, fever, and pain in the lateral parts of the trunk, but may be preceded by fatigue, headache, loss of appetite, bronchitis, and attacks of coughing. At the height of the disease, mental confusion, severe dyspnea, and blood-stained sputum were often observed; pulmonary hemorrhage also occurred in some cases. Case fatality rates have been reported to be 25 to 30%. Recovery was usually very slow [Molokanov *et al.*].

Clinical manifestations of acute poisoning with calcium hydroxide following accidental ingestion are similar to those caused by other alkalies and include burns to the mucous membranes of the mouth, esophagus, and stomach, sharp pains along the course of the upper digestive tract, nausea, racking vomiting, bloody stools, and anuria; the victim's skin may be of a grayish color and cold to touch. Circulatory failure and collapse tend to occur later if no proper treatment is given. Perforative peritonitis has been observed in some instances [Artamonova].

Animals and man: chronic exposure

Animals. Daily exposure of rats to calcium oxide dust by the oral route (gavage) led, after a month, to focal necrosis and ulceration in the esophagus and to epithelial desquamation, inflammatory edema, and muscle fiber breakdown in the stomach. After 2 months of exposure, cornification of the esophageal mucosa, complete desquamation of the gastric mucosal epithelium, and areas of deep necrosis in both the esophagus and stomach were observed. Increased thyroid function and impaired iodine metabolism were recorded in rats following prolonged consumption of water containing 340 or 1000 mg Ca/L [Rakhov].

Man. Health examinations of 230 construction workers handling lime showed the presence of atrophic change in the upper airway mucosa, chronic bronchitis, and pulmonary emphysema in approximately one-third of them, often (in ≈50% of cases) in combination with pneumosclerosis. The calcium oxide contained in cement dust has been implicated as the cause of gastritis, gastroduodenitis, and gastric ulcers frequently seen in the cement industry [Birch].

High prevalence rates of gastritis, gastric ulcers, enteritis, and colitis have been noted among people consuming water supplied from cemented reservoirs; such individuals often complained of epigastric pain, irregular bowel action, poor sleep, and rapid onset of fatigue [Voitenko].

Prolonged exposure to Thomas slag dust led to chronic bronchitis and pulmonary emphysema and aggravated the course of pulmonary tuberculosis to a much greater extent than did dusts of cement, limestone, gypsum, clay, iron, or bronze [Molokanov *et al.*].

Dermal and ocular effeçts in man

Thomas slag produces a pricking or burning sensation in the skin and may cause inflammation as well as eczema; skin burns have also occurred occasionally, as have conjunctivitis and keratitis. Unslaked lime (calcium oxide) causes severe skin burns and has been placed on the list of particularly aggressive chemicals that account for most cases of occupational dermatosis [Bagnova]. Prolonged exposure of the skin to slaked lime (calcium hydroxide) has often resulted in the appearance of small nodules covered with a blackish crust, followed by the development of sharply circumscribed, rather deep, and very painful ulcers with a smooth bottom; the surrounding skin appeared normal. Eczema has also occurred, especially in those with hyperhidrosis. In chronic conditions, the skin is dry and hard and undergoes exfoliation in the form of large scales or sheets. The nails become brittle and thin and come to have longitudinal fissures; the skin between the fingers is often grossly irritated and the digital joints are surrounded by crusts.

A variety of skin lesions, but particularly dermatitis of a purulent follicular type, have been described in plasterers and modelers. The site of predilection were the hands and the lower third of the forearms, while the most frequent presenting complaints were swelling of the hands and intense pains at night. Chemical skin lesions may be complicated by supervening infection. Entry of lime into an eye, even in small amount, results in gross hyperemia of the conjunctiva. Severe ocular burns can be produced by plaster [Zagora].

Calcium sulfate

Fish

The lowest calcium sulfate concentration toxic to fish has been reported to be 633 mg/L [Grushko].

Animal

After an intratracheal administration of gypsum ($CaSO_4 \cdot 2H_2O$) dust in amounts of 50 to 100 mg, rats developed catarrhal or purulent bronchitis, lymphoid tissue hyperplasia, and interstitial pneumonia, and extensive hemorrhagic areas were seen in their lungs; 6–9 months later, peribronchial and perivascular interstitial sclerosis was observed [Barsky & Kochetkova; Snegs *et al.*].

Prolonged (3–9 months) inhalation exposure of rats to gypsum dust at 300–320 mg/m³ produced catarrhal bronchitis, emphysema, and diffuse pneumoconiosis of a sclerotic type.

Man

Some of the workers engaged in underground gypsum mining had pathologic changes in the mucous membranes of the upper respiratory tract and intensified lung markings on their chest radiographs. In addition, dryness in the nose, hoarseness, reduced senses of smell and taste, conjunctivitis, and nonspecific respiratory diseases were common findings in workers exposed to high gypsum dust concentrations (up to 375 mg/m³); some of them showed radiologic evidence of reticular or nodular pulmonary fibrosis. Two cases of allergic reactions to gypsum were described; in one case, asthma developed in a worker who had been handling gypsum for 40 years; in the other, allergic dermatitis and stomatitis occurred in a patient following repeated exposure to gypsum during its use to make impressions for dental prostheses [Kozdoba *et al.*; Tara].

ABSORPTION, DISTRIBUTION, AND ELIMINATION

The requirements of an adult human body for calcium can be met by the intake of 600–700 mg of this element per day. Calcium is contained in all diets, and its metabolism in the body is closely related to that of magnesium and strontium. The body of an adult weighing 70 kg contains, on average, 1700 g of calcium, of which about 1200 g is contained in the skeleton. Studies with isotopes indicate that in an adult about 700 mg are resorbed from bone and replaced daily. The blood contains calcium in an average concentration of 9.47%. Elimination is via the kidneys (26%) and intestine (74%) [Levina; Williams].

HYGIENIC STANDARDS

Exposure limits adopted for calcium compounds in the USSR are shown in Table 1. In the USA, TLV-TWA values have been set for the following calcium compounds (mg/m³) : carbonate (marble) at 10 mg/m³ for total dust (containing <1% quartz) and 5 mg/m³ for respirable dust; chromate (as Cr) at 0.001 (classified as a possible human carcinogen); cyanamide at 0.5 mg/m³; hydroxide at 5.0 mg/m³; oxide at 2.0 mg/m³; silicate at 10 mg/m³ for total dust (containing

<1% quartz) and at 5 mg/m³ for respirable dust; and sulfate at 10 mg/m³ for total dust containing no asbestos and <1% crystalline silica. Also, a TLV-STEL value of 20 mg/m³ has been set for calcium carbonate.

METHODS OF DETERMINATION

In water, calcium can be determined by a method based on the formation of a complex between the Ca^{2+} ion and the EDTA anion, followed by titration (sensitivity 0.4–0.6 mg/L) [Novikov *et al.*]. For its measurement in blood, a colorimetric procedure using a murexide-calcium complex fixed with glycerol was proposed [Vishnevskaya & Lyashevskaya].

Table 1. Exposure limits for calcium compounds

Compounds	Workroom air		Atmospheric air		
	MAC_{wz} (mg/m³)	Aggregative state	MAC_{ad} (mg/m³)	$TSEL_{aa}$ (mg/m³)	Hazard class
Calcium alumochromo- phosphate (as CrO_2)	0.01	Aerosol	—	—	1
borate	—	—	0.02	—	3
carbide and oxide	2.0	Aerosol	—	0.3	4
cyanamide	1.0/0.5[a]	Aerosol	—	—	—
hydroxide	5.0	Aerosol	—	—	—
metaarsenite, orthoarsenate, and orthoarsenate– orthophosphate	0.04/0.001[a]	Aerosol	—	—	1

[a] The value in the numerator is the concentration not to be exceeded at any time, while that in the denominator is the average maximum allowable concentration per shift.

MEASURES TO CONTROL EXPOSURE

At the workplace

Health hazards during the production or use of calcium and its compounds mainly arise from exposure to dust. Effective dust control can be best achieved by mechanizing and automating work processes, fully enclosing dust-generating equipment, and fitting it with exhaust ventilation. The crushing and grinding equipment, which is the major source of dust, should be remotely controlled with complete exclusion of manual loading and unloading operations. Calcium and its compounds should be transported in such a way as to rule out their spillage and consequent contamination of the environment. Where cement is produced or used, safe working conditions can only be ensured by observing the hygiene regulations laid out for industries where powdery building materials or other dusty materials (such as ceramics, asbestos, or talc) are processed or handled.

The air of premises in which calcium carbide is processed must be monitored for the presence of calcium hydroxide, carbon monoxide, and, if the carbide contains impurities such as calcium phosphate or calcium arsenate, also for the presence of phosphine and arsine [Sadkovskaya & Ivanov].

Preplacement and subsequent periodic examinations of the workers are necessary, with due regard to the working conditions existing in the particular factory.

Workers exposed to dusts of calcium or its compounds should use respiratory and skin protection equipment (respirators, gloves, protective clothing). Adequate respiratory, eye, and especially skin protection is most important for calcium carbide workers who should be provided with face masks, respirators, goggles, clothing made of resistant material, waterproof gloves, and hydrophobic and greasy ointments. Strict observance of personal hygiene measures, including hand washing after work, is mandatory.

Where calcium workers are exposed to radiant heat and flying sparks or hot particles, they should wear aprons and foot and leg protection devices lined with insulating material; for eye and face protection, a helmet fitted with a metal shield with an inset of safety glass or fine wire mesh is used [Sadkovskaya & Ivanov].

Workers applying calcium chlorate chloride should use respirators, hermetically sealed goggles, gloves, and protective pastes. Those working with cement should use disposable respirators and wear gloves, overalls, and goggles.

Protection of the general environment

Prevention of dust emissions into the general atmosphere is a major consideration in the quarrying or underground mining of calcium and its compounds as well as during their transportation over considerable distances. It is particularly important to make arrangements for effective purification of emissions from cement works because these may pollute the soil, water, and plants not only with cement dust but also with various metals. The discharge of calcium arsenates and arsenites into the environment can be significantly reduced through effective utilization of solid wastes from copper and lead smelters and observance of all relevant precautions when using these compounds for pest control. A proper method for the disposal of calcium arsenate is its burial in concrete bunkers. Defoliation of cotton plants with calcium chlorate chloride should be carried out not later than 7 to 12 days before harvest.

EMERGENCY TREATMENT IN POISONING

When calcium hydroxide dust has been inhaled, inhalations of water vapors (after adding several crystals of citric acid to the water) should be carried out and mustard plasters applied to the chest. Cardiacs are given if indicated. In the event of eye contact, the eyes should be flushed with jets of water for 10–30 min while keeping the lids wide apart, followed by washing with a 5% ammonium chloride solution or a 0.01% solution of $CaNa_2$-EDTA (calcium disodium edetate); a 0.5% dicaine [synonym: tetracaine hydrochloride] solution should then be dropped into the conjunctival sac. In case of a skin burn, the adherent remnants of lime should be removed with mineral or vegetable oil, and lotions of citric, tartaric, acetic, or hydrochloric acid applied to the affected area. Following ingestion, the stomach should be washed with 8–10 L of water; morphine may be injected subcutaneously if necessary. Further treatment is symptomatic [Artamonova].

In cases of eye contact with calcium chlorate chloride, the eyes should be immediately washed with water. Zinc drops are not recommended.

REFERENCES

Antoniev, A. A. *et al.* (1974) *Gig. Truda Prof. Zabol.*, No. **6**, 15–18 (in Russian).

Artamonova, V. (1981) *Neotlozhnaya pomoshch pri professionalnykh intoksikatsiyakh* [Emergency Care in Occupational Poisoning]. Meditsina, Leningrad (in Russian).

Avezbakieva, I. I. & Demidenko, N. M. (1979) *Gig. San.*, No. **5**, 11–14 (in Russian).

Bagnova, M. D. (1984) *Professionalnye dermatozy* [Occupational Dermatoses]. Meditsina, Leningrad (in Russian).

Balgabekov, K. I. (1963) In: *Trudy Semipalatinskogo Meditsinskogo instituta* [Transactions of the Semipalatinsk Medical Institute]. Vol. 3, Semipalatinsk, pp. 276–291 (in Russian).

Bariliak, R. A. *et al.* (1967) In: *Aktualnye voprosy otorinolaringologii* [Current Problems in Otorhinolaryngology]. Kiev, pp. 52–53 (in Russian).

Barsky, I. P. (1960) *Gig. Truda Prof. Zabol.*, No. **8**, 48 (in Russian).

Barsky, V. D. & Kochetkova, T. A. (1971) *Gig. San.*, No. **2**, 103–105 (in Russian).

Beckenkamp, H. & Hardeck, W. (1967) *Int. Arch. Gewerbepathol. Gewerbehyg.*, **23**, 175–196.

Birch, N. J. (1988) Chapter 15 *Calcium* in: Seiler, H. G. *et al.* (eds) *Handbook on Toxicity of Inorganic Compounds*. Marcel Dekker, New York, pp. 175–179.

Davydov, S. A. (1965) *Gig. San.*, No. **10**, 7–11 (in Russian).

Demidenko, N. M. (1963) *Med. Zhurnal Uzbekistana* [The Medical Journal of Uzbekistan], No. **7**, 41–43 (in Russian).

Domnin, S. G. (1969) In: *Voprosy profpatologii v eksperimente i klinike* [Occupational Diseases: Experimental and Clinical Aspects]. Sverdlovsk, pp. 175–182 (in Russian).

Domnin, S. G. & Angelova, O. S. (1967) In: *Professionalnye bolezni pylevoi etiologii* [Occupational Disorders Caused by Dust]. Sverdlovsk, pp. 241–250 (in Russian).

Domnin, S. G. *et al.* (1969) In: *Voprosy profpatologii v eksperimente i klinike* [Occupational Diseases: Experimental and Clinical Aspects]. Sverdlovsk, pp. 171–174 (in Russian).

Dragomiretsky, V. D. *et al.* (1967) In: *Aktualnye voprosy otorinolaringologii* [Current Problems in Otorhinolaryngology]. Kiev, pp. 50–52 (in Russian).

Dvizhkov, P. P. (1965) *Pnevmokoniozy* [Pneumoconioses]. Meditsina, Moscow (in Russian).

Glowinski, M. & Dudkiewicz, J. (1967) *Med. Pr.*, **18**, 83—91.

Grigorian, A. G. (1972) *Zh. Eksper. Klin. Med.*, No. **5**, 24—29 (in Russian).

Grushko, Ya. M. (1979) *Vrednye neorganicheskie soyedineniya v promyshlennykh stochnykh vodakh* [Harmful Inorganic Compounds in Industrial Waste Waters]. Khimiya, Leningrad (in Russian).

Ilichkina, A. G. (1972) *Gig. San.*, No. **3**, 110 (in Russian).

Kalacic, J. (1973) *Arch. Environ. Health*, **26**, 84—85.

Kolev, K. & Mandadzhiev, I. (1970) *Khigiena i Zdraveopoznavane* [Hygiene and Public Health], No. **1**, 42 (in Bulgarian).

Kolev, K. & Shumkov, G. (1975) *Probl. Khigienata* [Problems of Hygiene], No. **1**, 111—118 (in Bulgarian).

Kozdoba, A. A. *et al.* (1981) *Stomatologiya*, No. **2**, 86 (in Russian).

Krapukhina, E. P. *et al.* (1966) In: *Voprosy gigieny truda i profzabolevaniy* [Occupational Health: Current Issues]. Moscow, pp. 254—260 (in Russian).

Levina, E. I. (1972) *Obshchaya toksikologiya metallov* [General Toxicology of Metals]. Meditsina, Leningrad (in Russian).

Lindberg, Z. Ya. (1971) In: *Sanitarnaya okhrana pochvy* [Sanitary Protection of Soils]. USSR Ministry of Health publication, Moscow, pp. 51—52 (in Russian).

Makulova, I. D. *et al.* (1981) *Gig. Truda Prof. Zabol.*, No. **2**, 36—37 (in Russian).

Metelev, V. V. *et al.* (1971) *Vodnaya toksikologiya* [Aquatic Toxicology of Metals]. Kolos, Moscow (in Russian).

Mikhalev, A. L. & Pustotin, N. I. (1988) in: *Aktualnye problemy teoreticheskoi i prikladnoi toksikologii* [Current Topics of Theoretical and Applied Toxicology]. Moscow, pp. 128—137 (in Russian).

Molchanov, I. A. (1967) *Gig. Truda Prof. Zabol.*, No. **8**, 58 (in Russian).

Molokanov, K. P. *et al.* (1973) In: *Professionalnye bolezni* [Occupational Diseases]. Meditsina, Moscow, pp. 378—451 (in Russian).

Munkittrick, K. B. (1991) *Environ. Toxicol. Chem.*, **10**, No. 8.

Nikiforov, B. *et al.* (1979) *Gig. San.*, No. **4**, 58—62 (in Russian).

Novikov, Yu. V. *et al.* (1981) *Metody opredeleniya vrednykh veshchestv v vode vodoyemov* [Methods for Measuring Noxious Substances in Water Bodies]. Meditsina, Moscow (in Russian).

Popovič, D. (1964) *Protect., Sécur., Hyg. Trav.*, **44**, 11—20.

Prodan, L. (1983) In: *Encyclopaedia of Occupational Health and Safety*. International Labour Office, Geneva, pp. 436—439.

Rakhimova, M. K. *et al.* (1967) In: *Trudy Samarkandskogo meditsinskogo instituta* [Transactions of the Samarkand Medical Institute]. Vol. 37, Samarkand, pp. 271—274 (in Russian).

Rakhov, G. M. (1964) *Gig. San.*, No. **7**, 12—13 (in Russian).

Rezvaya, E. A. *et al.* (1971) In: *Trudy GIDUV* [Transactions of the State Institute for Advanced Medical Studies], No. **98**, Leningrad, pp. 185—189 (in Russian).

Rowe, R. & Williams, G. (1963) *Arch. Environ. Health*, **7**, 709—710.

Rumyantsev, G. I. & Kochetkova, T. A. (1966) In: *Voprosy gigiyeny truda i profpatologii* [Occupational Health: Current Issues]. Moscow, pp. 268—274 (in Russian).

Sadkovskaya, N. I. (1959) *Gig. Truda Prof. Zabol.*, No. **1**, 39—44 (in Russian).

Sadkovskaya, N. I. & Ivanov, N. G. (1983) In: *Encyclopaedia of Occupational Health and Safety*. International Labour Office, Geneva, pp. 358—360.

Sedov, K. P. *et al.* (1973) *Gig. San.*, No. **12**, 41—44 (in Russian).

Smirnova, E. B. (1965) In: *Materialy 10-oi nauchno-prakticheskoi konferentsii gigiyenistov i sanitarnykh vrachei* [Proceedings of the Tenth Workshop of Hygienists and Sanitarians]. Moscow, pp. 138—139 (in Russian).

Snegs, R. N. *et al.* (1972) In: *Trudy GIDUV* [Transactions of the State Institute for Advanced Medical Studies], No. **115**, Leningrad, pp. 104—108 (in Russian).

Spiridonova, L. V. *et al.* (1982) *Gig. San.*, No. **12**, 50—52 (in Russian).

Straus, K. *et al.* (1960) In: *Trudy Leningradskogo sanitarno-gigiyenicheskogo meditsinskogo instituta* [Proceedings of the Leningrad Institute of Sanitation and Hygiene]. Vol. 58, Leningrad, pp. 102−113 (in Russian).

Tara, S. (1957) *Arch. Malad. Profess.*, 18, 278−280.

Tarnopolskaya, M. M. & Osetinsky, T. G. (1957) *Sov. Med.*, No. 8, 90−94 (in Russian).

Vishnevskaya, T. M. & Lyashevskaya, T.N. (1976) *Lab. Delo*, No. 7, 44−46 (in Russian).

Voitenko, A. M. (1966) *Gig. San.*, No. 3, 93−94 (in Russian).

Williams, D. R. (1971) *The Metals of Life*. Van Nostrand Reinhold Company, London.

Zagora, E. (1961) *Promyshlennaya oftalmologiya* [Industrial Ophthalmology]. Medgiz, Moscow (in Russian).

Strontium and its compounds

Strontium carbonate (strontianite (α) [min.]); **s. chloride**; **s. chromate**; **s. ferrite** (solid solution consisting of 10% SrO and 90% Fe_2O_3); **s. fluoride**; **s. hydroxide**; **s. metaphosphate**; **s. nitrate**; **s. oxide**; **s. sulfate** (celestite [min.])

IDENTITY AND PHYSICOCHEMICAL PROPERTIES OF THE ELEMENT

Strontium (Sr) is an alkaline-earth metal in group II of the periodic table, with the atomic number 38. Its natural isotopes are ^{84}Sr (0.56%), ^{86}Sr (9.96%), ^{87}Sr (7.02%), and ^{88}Sr (82.56%).

Strontium is a soft, highly malleable and ductile metal that can be easily cut with a knife. It exists in three allotropic modifications.

In compounds strontium is present in the oxidation state +2. It resembles calcium and barium in chemical properties. It rapidly oxidizes in air to form a surface film consisting of the oxide SrO, peroxide SrO_2, and nitride Sr_3N_2. It ignites on being heated in air and is pyrophoric in the powder form. It reacts vigorously with water to form the hydroxide $Sr(OH)_2$ with the evolution of hydrogen. It gives the hydride SrH_2 with hydrogen (at >200°C), the nitride Sr_3N_2 with nitrogen (at >400°C), the phosphide Sr_3P_2 with phosphorus, the sulfide SrS with sulfur, and halides of the type $SrHal_2$ with halogens. Strontium is a strong reducing agent. When melted it readily forms solutions with many metals (Ag, Al, Ba, Ca, Cd, Mg, Pb, Sb, Sn, Zn, and others). It gives many intermetallic compounds (e.g. SrAl, $SrAl_4$, $SrMg_2$, $SrMg_4$, $SrMg_9$, SrSn, $SrSn_3$, $SrSn_5$, $SrPb_3$, $SrZn_5$, $SrZn_{12}$). Water-soluble strontium salts include the halides (with the exception of the fluoride), nitrate, and chlorate. The solubilities of these salts are intermediate between those of the respective calcium and barium salts. See also Appendix.

NATURAL OCCURRENCE AND ENVIRONMENTAL LEVELS

Twelve strontium minerals are known, the most prevalent of which are celestite ($SrSO_4$) and strontianite ($SrCO_3$).

The clarke of strontium is estimated at $(340-375) \times 10^{-4}\%$ in the earth's crust and at $230 \times 10^{-4}\%$ in the granite layer of the continental crust. The oceans contain an estimated 11 097 000 million tonnes of strontium, at an average concentration of 8100 µg/L; ferromanganese oxide deposits of the Pacific take up an estimated 4.9 thousand tonnes of strontium annually. Its concentration in river water averages 80 µg/L, and the amount discharged by rivers to the oceans is estimated at 2600 thousand tonnes per annum [Dobrovolsky].

The global biomass contains about 2 billion tonnes of strontium. The continental vegetation contains an estimated 100 million tonnes, at average wet weight, dry weight, and ash weight concentrations $16.0 \times 10^{-4}\%$, $40.0 \times 10^{-4}\%$, and $800 \times 10^{-4}\%$, respectively; the amount of strontium taken up annually through the increment in phytomass amounts to 6900 thousand tonnes, or 46 mg/km^2 of land area (K_b = 3.48) [Dobrovolsky; Kovalsky].

In soils, strontium occurs at an average level of 0.035% (in the USSR, levels above 0.2% have been recorded in some areas). Normal strontium concentrations for plants are considered to be around 600 mg/kg soil; higher concentrations (600–1000 mg/kg) are excessive and may be associated with an increased risk of Kashin-Beck's disease (osteoarthritis deformans). Among plants, highest strontium levels are found in the families Umbelliferae (0.044%), Vitaceae (0.037%), and Elaeagnaceae (0.036%), lower levels in Rosaceae (0.024%) and Leguminosae (0.023%), and still lower in Piperaceae (0.012%), Gramineae (0.011%), and Solanaceae (0.009%).

Tens of millions of tonnes of strontium become involved in biological migration each year. Reported strontium concentrations in organisms are (mg/100 g dry weight); marine algae, 26–140; terrestrial plants, 2.6; marine animals, 2–50; terrestrial animals, 1.4; bacteria, 0.27–30 [Kovalsky].

PRODUCTION

Strontium is usually obtained by electrolysis of its fused chloride or by thermal reduction of its oxide with aluminum. **Strontium carbonate** is commonly made by passing carbon dioxide through a strontium hydroxide solution; **s. chloride**, by reaction of strontium oxide or carbonate with hydrochloric acid; **s. fluoride**, by reaction of strontium chloride with potassium fluoride or sodium fluoride; **s. hydroxide**, by reaction of strontium compounds with sodium hydroxide;

s. nitrate, by reaction of strontium hydroxide with dilute nitric acid; s. oxide, by roasting strontium nitrate or carbonate. The principal source of strontium sulfate is the mineral celestite.

USES

Strontium is mainly used for deoxidizing copper and bronze, in the production of electronic (vacuum) tubes to absorb residual gas, and in pyrotechnic compositions. Its alloys with lead and tin are used in the manufacture of storage batteries. The artificial radioactive isotopes ^{89}Sr and ^{90}Sr are employed, respectively, for detecting defects in cables and as a source of β-radiation.

Strontium carbonate is mainly used in weatherproof glazes; s. chloride, in refrigeration equipment, cosmetics, and medicine; s. chromate, as a corrosion-resistant agent and as a pigment for paints; s. hydroxide, in making greases and recovering sugar from molasses; s. oxide, as a source of metallic strontium; s. nitrate, in pyrotechnics and in making other strontium compounds; s. sulfate, as an electrolyte additive in accelerated chrome-plating. The minerals strontianite and celestite are used in heavy liquids for well drilling. (See also Wennig & Kirsch.)

MAN-MADE SOURCES OF EMISSION INTO THE ENVIRONMENT

The main anthropogenic sources of environmental strontium and its compounds are wastewaters from the metallurgical, electrical, glass, ceramics, and beet-sugar industries. In large industrial cities, raw sludges in the primary settling tanks of biological wastewater-treatment plants had strontium concentrations up to 226 mg/kg [Goncharuk; Naishtein; Pavletskaya et al.].

TOXICITY

For strontium chloride, the lowest concentration (threshold level) affecting the taste of water was determined at 12 mg/L, while those affecting its odor were found to be 20 to 30 times higher. In concentrations close to the threshold level for the taste of water, the chloride did not affect the gustatory quality of the fish [Shafirov]. Bespamyatnov & Krotov reported that subthreshold concentrations for effects on organoleptic properties of water were 13 mg/L for the chloride, 12 mg/L for the nitrate, and 11.5 mg/L for the sulfate of strontium. These authors also found that the subthreshold concentrations for effects on the sanitary condition of water bodies were >13 mg/L for the chloride, 26 mg/L for the nitrate, and >11.5 mg/L for the sulfate, and that 26 mg/L was the highest concentration of these three compounds as well as of **strontium carbonate** that did not influence biochemical processes in water bodies however long any of these compounds remained there.

Aquatic organisms

The toxicity of strontium compounds for fish is relatively low. To crucian carp, **strontium chloride** was lethal at 1538 mg/L and **strontium nitrate**, at 3200 mg/L after exposure for 8 days. To daphnids, the chloride was lethal at 114 mg/L after 64 h of exposure and the nitrate at 1200 mg/L [Bespamyatnov & Krotov].

Animals and man

An excess of strontium in the body will primarily affect the osseous tissue, liver, and hematopoietic organs [Krasovsky *et al.*]. The most characteristic manifestations of strontium poisoning are those of Kashin-Beck's disease (osteoarthritis deformans endemica) and include increased fracturability and malformation of bones [Kovalsky & Samarina]. Strontium is able to produce a disease referred to as 'strontium rickets', and it has been suggested that it induces this disease by blocking the biosynthesis of 1.25-dihydroxy-cholecalciferol, which is the metabolically active form of vitamin D in the intestine [Omdahl & DeLuca]. There is also evidence that strontium is goiterogenic and can act as a nerve and muscle poison, and that its chloride is able to stimulate the production of thromboxane B_2 by human platelets and act as a local anesthetic [Best *et al.*].

Acute/single exposure

Animals

LD_{50} values of strontium compounds are given in Table 1. The laboratory animals most sensitive to strontium and its compounds are mice; rats and rabbits and particularly guinea-pigs are more resistant [Shafirov].

For mice and rats, the highest tolerated oral dose of **strontium chromate** was 1000 mg/kg. **Strontium carbonate** was not toxic in oral doses up to 14 000 mg/kg (as Sr). The threshold intraperitoneal dose of **strontium oxide** for effects on body-weight and the blood content of sulfhydryl groups in mice was 50 mg/kg.

Animals acutely poisoned with the **chloride** or **nitrate** of strontium exhibited short-term excitation and tremor followed by listlessness, general inhibition, motor incoordination, and profuse urination and defecation; deaths were recorded during the interval of 0.5 h to 20 h post-administration. On histopathologic examination, no microscopic changes were detectable [Izmerov *et al.*; Levina; Shafirov; Shubochkin & Pokhodzei; Ziuziukin & Makolkina].

A single inhalation exposure of rats to a **strontium hydroxide** aerosol at 80 mg/m^3 resulted in increased values of the summated threshold index, lowered rectal temperature and muscle strength, impaired orienting reflexes, altered respiratory frequency, and exaggerated cellular responses in the upper airways and the lungs. The concentration of 44 mg/m^3 increased the summated threshold index and induced exaggerated cellular responses without producing any other effects; the concentration of 20 mg/m^3 was considered as the threshold level. A single inhalation exposure of rats to a **strontium nitrate** aerosol at 98 mg/m^3 increased the summated threshold index and lowered rectal temperature. The threshold concentration was determined at 74 mg/m^3 for the **nitrate** and 45.3 mg/m^3 for the **fluoride**. The threshold concentration of **strontium hydroxide** for acute irritant effects on inhalation exposure was 20 mg/m^3 [Ziuziukin & Milushkina].

Table 1. Median lethal doses of strontium compounds

Compound	Animal	Route of administration	LD$_{50}$ (mg/kg)
Chloride	Mouse	Oral	1036
		Intravenous	148
	Rat	Oral	1796
	Rabbit	Oral	1512
	Guinea-pig	Oral	2843
Chromate	Rat	Oral	3118
Nitrate	Mouse	Oral	1028
	Rat	Oral	1892
	Rabbit	Oral	1600
	Guinea-pig	Oral	3077
Oxide	Mouse	Intraperitoneal	667

A single inhalation exposure of rats to **strontium chromate** at 40 mg/m^3 led to appreciable alterations in body temperature, motor activity, and the summated threshold index; the concentration of 20 mg/m^3 induced only slight changes in body temperature and this index, while the concentration of 10 mg/m^3 was regarded as subthreshold [Shubochkin *et al.*].

Rats administered 50 mg of **strontium carbonate** or **sulfate** intratracheally, developed moderate sclerosis of the bronchial and vascular walls 9 to 12 months later; at that time, elevated hydroxyproline levels were detected in their lungs [Ziuziukin & Makolkina]. Highly toxic to rats by this route proved to be **strontium chromate**, which caused 100% mortality in doses of 50, 25, and even 10 mg; on autopsy, toxic pneumonia and pulmonary edema were seen. The dose of 0.5 mg was not lethal to any rats [Shubochkin & Pokhodzei]. A fibrotic process was elicited in the lungs of rats by **strontium ferrite** dust.

Chronic exposure

Animals

In a 30-day inhalation toxicity study on rats, daily 4-hour exposure to **strontium nitrate** at 44.6 mg/m^3 altered the summated threshold index, decreased cholinesterase activity in the blood, and increased the urinary output of calcium; autopsy demonstrated interstitial pneumonia and myocardial dystrophy. After 4 months of exposure to a lower concentration (14.7 mg/m^3, 4 h/day, 60 days/week, for 4 months), rats exhibited impaired liver function, blood changes (lowered cholinesterase activity, elevated alkaline phosphatase activity, and elevated levels of β-lipoproteins and calcium), and increased urinary excretion of calcium and chlorides; autopsy showed marked histologic changes in bones and in bronchial cartilages as well as peribronchial and perivascular sclerosis in the lungs. The concentration of 3.2 mg/m^3 was described as the threshold level [Ziuziukin].

Inhalation exposure of rats to a **strontium chromate** aerosol at 5 mg/m^3 over 2 months led to reductions in body-weight, rectal temperature, and erythrocyte and leukocyte counts in the blood; autopsy revealed moderate inflammation and signs of fibrosis in the lungs and granular degeneration in the liver and kidneys. Similar but much milder changes were seen in rats exposed to 0.5 mg/m^3 [Shubochkin & Pokhodzei].

When rats were exposed for 72 days to a **strontium hydroxide** aerosol at 20 mg/m^3 (which was the threshold level for acute effects on inhalation), gonadotropic activity of this compound was recorded in male rats in addition to its systemic toxicity. Thus, the percentage of motile spermatozoa, the osmotic resistance of these cells and the time they remained motile were all decreased and, moreover, as shown by electron microscopy, the epithelial linings of seminiferous tubules were loosened and swollen, the cytoplasm of Sertoli's cells was strongly vacuolated, and large numbers of free-lying spermatids and spermatocytes were seen in the lumens of seminiferous tubules. When the males thus exposed were mated to intact females, their progeny was less viable than that of control rats. The concentration of 1.18 mg/m^3 was interpreted as subthreshold [Silayev et al.].

Rats exposed repeatedly to **strontium chloride** at 180 mg/kg or **strontium nitrate** at 190 mg/kg by the oral route (gavage) for 4 months showed increases in reticulocute counts, in blood levels of sulfhydryl group and cholinesterase, and in prothrombin time. Similar changes plus elevations in alkaline phosphatase activity and calcium levels in the blood were observed in rats and rabbits given strontium chloride at 80 mg/kg by this route over 8 months. The dose of 0.65 did not produce any effects in rabbits but tended to raise alkaline phosphatase activity in rats. The dose of 0.13 mg/kg was referred to as subthreshold. No roentgenologic changes were detectable in the bones of any animals [Shafirov].

Rats receiving **strontium nitrate** by oral gavage in doses of 300 or 600 mg/kg for periods of 1 to 6 months were highly voracious; autopsy showed dystrophic changes in parenchymal organs and toxic endocarditis. Rats given much lower doses of the nitrate (5000, 50, or 10 μg/kg) for 1 month, exhibited 30% to 50% increases in catalase activity in the blood; the dose of 0.01 μg/kg raised carbonic anhydrase activity in rabbits. Dietary feeding of **strontium** resulted in decalcification and increased fracturability of bones.

Embryotoxic effects were observed in female rats given **strontium chloride** in their drinking water at 5, 10, or 20 mg/L during pregnancy [Sergeyev & Kuchma]. There is evidence that strontium may be teratogenic [Shepard].

Man

When chest radiographs taken in 40 apparatus operators of a strontium factory were examined, five of them (all employed for more than 5 years) were found to have moderate changes in the pulmonary interstitium, presumably caused by exposure to dusts of strontium compounds. An analysis of morbidity with temporary disability among strontium workers revealed rather high incidence rates of diseases affecting the respiratory and nervous systems; signs of vascular dystonia and significantly reduced cholinesterase and acetylcholinesterase activities in the sera and erythrocytes were also noted. Many workers complained of unpleasant sensations and periodic pains in the cardiac region. Clinical manifestations of cardiovascular abnormalities correlated with the length of employment and the blood levels of strontium [Pyatak; Ziuziukin].

Prolonged consumption of drinking water containing strontium in concentrations of around 10 mg/L or more has been reported to affect the growth of children, apparently through stimulation of calcium metabolism [Knizhnikov & Novikova]. In young children (aged 1 to 3 years) living in areas with such high strontium levels in drinking water, the time required for the anterior fontanelle to undergo ossification was longer and the number of teeth erupting by the age of 1 was smaller than in children from control areas [Sergeyev & Kuchma]. A weak positive correlation between frequency of malignant disease and strontium levels in the soil has been noted [Dubikovsky].

Dermal and ocular effects

The **chloride** and **nitrate** of strontium produced marked dermal irritation in rats, guinea-pigs, and rabbits when applied in lanolin to the shaved skin of their back for 4 h daily over 30 days. The **carbonate, chromate, sulfate,** and **hydroxide** were not irritating. Following placement into conjunctival sacs of rabbits in an amount of 50 mg, these three salts did not cause any visible ocular changes, whereas the hydroxide produced conjunctivitis and necrosis. Ocular application of the chloride and nitrate gave rise to lacrimation, blepharospasm, hyperemia of the eyelids, and dilatation of conjunctival vessels [Ziuziukin & Milushkina].

ABSORPTION, DISTRIBUTION, AND ELIMINATION

Strontium is a skeletal component in both higher and lower animals and occurs in all human organs and tissues. It influences bone formation and the activities of several enzymes, in particular catalase, carbonic anhydrase, and alkaline phosphatase. Strontium acts on isolated organs in the same way as does calcium and can completely replace the latter. In fact, Sr^{2+} ions are so similar to those of calcium in their characteristics as to be metabolized together with them; however, because strontium ions are metabolized at higher rates and differ from calcium ions considerably in size, they will gradually impair the normal skeletal calcification and lead to Kashin-Beck's disease [Yershov & Kononov].

Average strontium concentrations found for some food and feed plants were as follows (mg/kg): wheat, 3.0; rye, 7.8–70.2; barley, 43.2; buckwheat, 2.7–6.8; pea, 6.5–39.1; corn, 0.0008–0.02; beet, 0.02–0.14; oats, 0.46; various grasses, 0.5–7.8; clover, 1.3; fodder beet, 33.0; fodder soybean, 138.0; silage, 63.1. Much strontium (up to 4 mg%) is contained in aromatic greens such as dill and parsley.

An adult human receives, on average, 14–22 mg of strontium in food and drinking water daily. The metal is chiefly absorbed in the small intestine and is eliminated mainly through the feces (60%) and urine (32%). It is transported in the bloodstream both in ionic and bound forms. It accumulates principally in the bones; smaller amounts accumulate in the kidneys, liver, and brain. The strontium level of $3.77 \times 10^{-3}\%$ found in the distal epiphysis of femoral bones from cadavers of rural people has been proposed as the physiologic 'norm' for the present human generation [Dubrovina et al.].

HYGIENIC STANDARDS

Exposure limits adopted in the USSR for strontium and its compounds are shown in Table 2. In the USA, a TLV-TWA of 0.001 mg/m^3 has been set for strontium chromate (as Cr) which is classified as a possible human carcinogen.

METHODS OF DETERMINATION

For the determination of strontium in air, water, foods, and biological media, spectrographic methods have been employed [Khotomlyansky et al.; Wenning & Kirsch]. Strontium in air and water samples can also be determined by flame photometry [Novikov et al.]. For its determination in foodstuffs, a method of flame emission spectrophotometry has been described (detection limit 0.21 mg/kg for a 10-g sample) [Evans & Read].

Table 2. Exposure limits for strontium and its compounds

	Workroom air			Water sources	Hazard class
	MAC_{wz} (mg/m^3)	TSEL (mg/m^3)	Aggregative state	MAC_W (mg/L)	
Strontium	—	—	—	7.0[a]	2
Strontium carbonate	6.0	—	Aerosol	—	4
hydroxide	1.0	—	Aerosol	—	2
metaphosphate	—	8.0	Aerosol	—	3
nitrate	1.0	—	Aerosol	—	2
oxide	1.0	—	Aerosol	—	2
phosphate	6.0	—	Aerosol	—	4
sulfate	6.0	—	Aerosol	—	4

[a] Based on sanitary and toxicological criteria.

MEASURES TO CONTROL EXPOSURE

At the workplace
Basic technical control measures to be implemented where strontium or its compounds are processed should be directed at the elimination of sources polluting the ambient air and include mechanization and automation of the work processes and provision of the process equipment with adequate exhaust ventilation and thermal insulation. Strontium should be processed in enclosed equipment.

Radioactive strontium isotopes should be handled strictly in accordance with the adopted safety practices for the use of radioactive materials.

All workers should undergo preplacement and periodic medical examinations. Eating, drinking, and smoking in the workrooms must not be allowed, and there should also be a ban on entering the lunchrooms or canteen in working clothes as well as on storing any clothes or personal belongings in the workrooms.

For *personal protection*, respirators should be provided to workers at risk of being exposed to airborne dust of strontium or its compounds. For hand protection gloves should be worn. Materials available in the first-aid boxes at the workplace should be used for treating injuries to the skin of the hands.

Protection of the general environment. Effective removal of strontium from the industrial wastewaters is necessary, as is strict observance of the regulations concerning the stockpiling, transportations, decontamination, and burial of strontium-containing solid wastes.

In areas where hydrogeochemical conditions result in elevated strontium content in drinking water, measures aimed at preventing adverse health effects of strontium on the population should be taken and may include, for example, monitoring strontium levels in the drinking water, installation of a central drinking water-supply system using water from sources low in strontium, and provision of children, especially infants, with foods that favor the normal formation and growth of osseous tissue.

REFERENCES

Bespamyatnov, G. P. & Krotov, Yu. A. (1985) *Predelno dopustimye kontsentratsii khimicheskikh veshchestv v okruzhayushchei srede* [Maximum Allowable Concentrations of Chemicals in the Environment]. Khimiya, Leningrad (in Russian).

Best, L. *et al.* (1981) *Biochem. Pharmacol.*, **6**, 635–637.

Dobrovolsky, V. V. (1983) *Geografiya mikroelementov. Globalnoe rasseyanie* [Geography of Trace Elements. Global Dispersion]. Mysl, Moscow (in Russian).

Dubikovsky, G. P. (1982) *Khimiya v selskom khozyaistve* [Chemistry in Agriculture], No. **3**, 33–34 (in Russian).

Dubrovina, Z. V. *et al.* (1967) *Gig. San.*, No. **4**, 43–46 (in Russian).

Evans, W. H. & Read, J. L. (1985) *Analyst*, **110**, 619–623.

Goncharuk, E. I. (1977) *Sanitarnaya okhrana pochvy ot zagryazneniya khimicheskimi veshchestvami* [Sanitary Measures to Protect Soil from Chemical Pollution]. Zdorovya, Kiev (in Russian).

Izmerov, N. F. *et al.* (1982) *Toxicometric Parameters of Industrial Toxic Chemicals under Single Exposure.* Published by the USSR Commission for the United Nations Environmental Programme (UNEP), Moscow.

Khotomlyansky, G. Ya. (1976) *Gig. San.*, No. **11**, 100–101 (in Russian).

Knizhnikov, V. A. & Novikova N. Ya. (1964) *Gig. San.*, No. **8**, 93–95 (in Russian).

Kovalsky, V. V. (1974) *Geokhimicheskaya Ekologiya* [Geochemical Ecology]. Nauka, Moscow (in Russian).

Kovalsky, V. V. & Samarina, I. A. (1960) *Dokl. Akad. Nauk SSSR*, No. **6**, 1378–1381 (in Russian).

Krasovsky, G. N. *et al.* (1977) *Gig. San.*, No. **7**, 11–16 (in Russian).

Levina, E. I. (1972) *Obshchaya toksikologiya metallov* [General Toxicology of Metals]. Meditsina, Leningrad (in Russian).

Lukashev, K. I. *et al.* (1984) *Chelovek i priroda: Geokhimicheskiye i ekologicheskiye aspekty ratsionalnogo prirodopolzovania* [Man and Nature: Geochemical and Ecological Aspects of Environmental Management]. Nauka i Tekhnika, Minsk (in Russian).

Naishtein, S. Ya. (1975) *Aktualnye voprosy gigigeny pochvy* [Important Problems of Soil Hygiene]. Kishinev (in Russian).

Novikov, Yu. V. *et al.* (1981) *Metody opredeleniya vrednykh veshchestv v vode vodoyemov* [Methods for Measuring Noxious Substances in Water Bodies]. Meditsina, Moscow (in Russian).

Omdahl, J. & DeLuca, H. (1971) *Science*, No. **4012**, 949–951.

Pavletskaya, F. I. *et al.* (1965) *Gig. San.*, No. **11**, 54–63 (in Russian).

Pyatak, O. A. *et al.* (1978) *Vrach. Delo*, No. **6**, 29–30 (in Russian).

Sergeyev, E. P. & Kuchma, N. Yu. (1979) *Gig. San.*, No. **6**, 11–13 (in Russian).

Shepard, T. H. (1986) *Catalog of Teratogenic Agents.* Johns Hopkins University Press, Baltimore & London.

Shafirov, Yu. B. (1965) *Gig. San.*, No. **11**, 17 (in Russian).

Shubochkin, L. N. & Pokhodzei, Yu. I. (1980) *Gig. San.*, No. **10**, 76–77 (in Russian).

Silayev, A. A. *et al.* (1979) *Gig. Truda Prof. Zabol.*, No. **1**, 43–44 (in Russian).

Wennig, R. & Kirsch, N. (1988) Chapter 57 *Strontium* in: Seiler, H. G. *et al. Handbook on Toxicity of Inorganic Compounds.* Marcel Dekker, New York, pp. 631–638.

Yershov, Yu. A. & Kononov, A. M. (1981) In: *Aktualnye problemy farmatsii* [Topics of Current Interest in Pharmacy]. Moscow, pp. 60–62 (in Russian).

Ziuziukin, Yu. V. (1974) *Gig. San.*, No. 8, 99–100 (in Russian).

Ziuziukin, Yu. V. & Makolkina, E. P. (1979) *Gig. Truda Prof. Zabol.*, No. 11, 53–54 (in Russian).

Ziuziukin, Yu. V. & Milushkina, T. A. (1973) *Gig. San.*, No. 12, 100–101 (in Russian).

Barium and its compounds

Barium carbonate (triple carbonate [50% $BaCO_3$, 45% $SrCO_3$, and 5% $CaCO_3$], witherite (α) [min.]), **b. chloride; b. ferrite; b. fluoride; b. hydroxide** (baryta water [aqueous solution of $Ba(OH)_2$], caustic baryta); **b. nitrate** (nitrobaryte [min.]); **b. oxide; b. sulfate** (baryta white, blanc fixe, permanent white; barite (heavy spar) [min.]); **b. sulfide; b. tetratitanate b. titanate; b. titanate zirconate**
Barium-aluminum titanate; b.-calcium aluminate, b.-calcium aluminosilicate; b.-calcium titanate

IDENTITY AND PHYSICOCHEMICAL PROPERTIES OF THE ELEMENT

Barium (Ba) is an alkaline-earth metal in group II of the periodic table, with the atomic number 56. Its natural isotopes are ^{130}Ba (0.101%), ^{132}Ba (0.097%), ^{134}Ba (2.42%), ^{135}Ba (6.59%), ^{136}Ba (7.81%), ^{137}Ba (11.32%), and ^{138}Ba (71.66%).

Barium is a ductile metal that is harder than lead and softer than zinc. It is present in compounds in the oxidation state +2. Barium and many of its compounds possess chemical properties similar to those of calcium and particularly strontium and radium but are more reactive.

Barium is rapidly oxidized in air, forming a surface layer containing its oxide, peroxide, and nitride. On heating in air it readily takes fire and burns with a greenish flame. It decomposes water more vigorously than calcium; in the reaction, hydrogen is liberated and the hydroxide $Ba(OH)_2$ is produced. Barium gives the oxide BaO with oxygen, the hydride BaH_2 with hydrogen, the nitride Ba_3N_2 with nitrogen (at 260–600°C), the carbide BaC_2 with carbon, the cyanide $Ba(CN)_2$ with carbon and nitrogen, and halides with halogens. See also Appendix.

NATURAL OCCURRENCE AND ENVIRONMENTAL LEVELS

The chief barium ores are barite, or heavy spar ($BaSO_4$), and witherite ($BaCO_3$). The clarke of barium is estimated at $(425–650) \times 10^{-4}\%$ in the earth's crust and at $680 \times 10^{-4}\%$ in the granite layer of the continental crust. The oceans contain approximately 29 000 million tonnes of barium, at an average concentration of 21 µg/L; the amount of its dissolved species taken up annually by ferromanganese oxide deposits of the Pacific is about 11 000 tonnes. Its average concentration in river water in 20 µg/L, and its global run-off from rivers to the oceans amounts to 740 000 tonnes annually [Dobrovolsky]. Its average abundance in soils is 0.05%. The earth's biomass contains an estimated 3 billion tonnes of barium. Its average concentrations in organisms are (mg/100 g dry weight): marine algae, 3.0; terrestrial plants, 1.4; marine animals, 0.02–0.3; terrestrial animals, 0.075; bacteria, 18–90 [Kovalsky].

PRODUCTION

Barium is obtained by reduction of barium oxide with aluminum at 1100–1200°C in vacuum.

Barium carbonate is made by bubbling carbon dioxide through an aqueous barium sulfide solution at 30–40°C or by mixing solutions of sodium carbonate and b. sulfide or chloride at 70–80°C. **B. chloride** is made by reaction of b. sulfide with hydrogen chloride or by fusion of b. sulfate with calcium chloride and carbon at 770–1100°C. **B. ferrite** is made by annealing powders of b. oxide and iron (III) oxide at 1000–1400°C. **B. hydroxide** is obtained by calcining b. carbonate and slaking the resultant b. oxide with water or by reacting a b. chloride solution with sodium hydroxide. **B. meta-aluminate** results from fusion of b. oxide with aluminum oxide. **B. nitrate** is obtained as a product of the exchange reaction between b. chloride and sodium nitrate (or nitric acid) in their aqueous solutions or by dissolving b. carbonate in nitric acid. **B. oxide** is made by heating either b. nitrate with coal at 1000–1050°C (oxides of nitrogen are released) or b. carbonate with coal at 1200°C (carbon monoxide is released). **B. sulfate** is obtained by several methods, for example in refining barite; by precipitation with sulfuric acid or with sulfate solutions from solutions of barium salts; or as a by-product during the purification of brines by a sulfate process. **B. sulfide** is produces by fusion of b. sulfate and carbon at 1000–110°C (the exiting gases contain about 5% CO). **B. titanate** is made by fusion of b. carbonate with titanium (IV) oxide and **b. zirconate**, by fusion of b. oxide, b. hydroxide, or b. carbonate with zirconium (IV) oxide.

USES

Barium is mainly used as a 'getter' to remove last traces of gases from vacuum tubes; in the equipment for making sulfuric acid; and, alloyed with lead, in typographic printing.

Barium carbonate is used in ceramics, optical glass, in glazes, and as a rodenticide; as a constituent of triple carbonate, it forms the basis of solid solutions used in electronics. Similar uses are made of **b.-calcium aluminate** and **b.-calcium aluminosilicate**. **B. chloride** is used in pest control, in the ceramic and textile industries, in the manufacture of mineral paints, and to remove sulfates from boiler water and brines. From **b. ferrite,** materials for electronic radio equipment and computers are prepared. **B. fluoride** is used as an insecticide and as an antiseptic for wood. **B. hydroxide** is used in making lubricating greases and in refining sugar in laboratory practice. **B. metatitanate** finds application in the manufacture of hydroacoustic devices, small high-power capacitors, electronic circuits, ultrasonic equipment, and sound pickups. **B. nitrate** is employed in pyrotechnics, in incendiaries, in the vacuum-tube industry, and occasionally in the glass industry in place of b. carbonate. **B. oxide** is used in electromagnet cores and in making b. peroxide and b. hydroxide. **B. sulfate** is used as a weighting compound in clayey muds in deep oil drilling, in making mineral paints, in the paper, rubber, textile, and ceramic industries, and as a radiopaque substance in medicine. **b. sulfide** finds use in the leather industry and, in a highly purified form, in luminous paints.

MAN-MADE SOURCES OF EMISSION
INTO THE ENVIRONMENT

Major sources of barium and its compounds in the environment are effluents from the chemical/pharmaceutical, petrochemical, metallurgical, paint and varnish, soap-making, printing, and rubber industries. In large industrial cities, raw sludges in the primary settling tanks of biological wastewater-treatment plants contained barium in concentrations of up to 230 mg/kg (its concentrations in wastewaters discharged into municipal sewage systems should not exceed 1 mg/L) [Goncharuk].

Pesticides and fertilizers may be significant sources of barium in soils and, hence, in plants where barium levels up to 5 mg/kg were detected. The greatest hazard in terms of accumulation in plants is presented by the so-called mobile (ionic) forms of barium whereas its immobile forms (phosphates and sulfates) are not hazardous in this respect [Grushko; Kovalsky].

Barium emissions into the atmosphere from the mining, refining, and processing of barium ores can occur during loading and unloading operations, stockpiling, materials handling, and grinding and refining of the ore. For example, the mining and processing of barite ore was responsible for the release of an estimated 3200 tonnes of particulates into the air in 1976 in the USA. Fossil fuel combustion may also release barium into the atmosphere. It was found that over 90% of the barium additive in diesel fuels is emitted in vehicle exhaust (in the form of barium sulfate) [WHO].

Occupational exposure to soluble barium compounds occurred in workers exposed to welding fumes during arc welding using electrode wiring that contained 20–40% of such compounds; urine samples from the workers contained barium in concentrations of 31 to 234 µg/L after 3 h of exposure (against 1.8–4.7 µg/L in control samples) [Dare *et al.*]. Welders using such wire were found to be exposed to concentrations of soluble barium ranging from 2200 to 6200 µg/m^3 of air. In a mineral processing plant, barium exposure resulting from the grinding and mixing of several grades of barium-containing ore ranged from 0.8 to 1.92 mg/m^3 [WHO].

TOXICITY

The highest concentration of **barium chloride** that did not affect the taste or odor of water was 4 mg/L [Bespamyatnov & Krotov]; for the **carbonate** and **nitrate**, the corresponding (organoleptically defined) subthreshold concentration was 5 mg/L. For both the carbonate and chloride, the highest concentration that did not influence the sanitary condition of water bodies was 10 mg/L, and this was also their concentration found not to influence biochemical processes in a water body however long these compounds remained there [Bespamyatnov & Krotov].

Aquatic organisms

Mollusks that are the intermediate hosts of *Opisthorchis* trematodes died when exposed to barium salts in the water at a concentration of 1 : 5000 for 2–3 h or 1 : 50 000 for 24–48 h [Drozdov]. **Barium chloride** adversely affected the nervous system of young salmon at 50 mg/L after 72 h of exposure; exposure to 150 mg/L during this period was lethal to 92% of the fish. **Barium nitrate** was lethal for stickleback at 500 mg/L on 7-day exposure. For daphnids, the chloride was lethal at 17 mg/L with 48-hour exposure. For the aquatic plant *Elodea canadensis*, it was lethal at 10 mg/L [Grushko].

Animals ans man

General considerations

Water-soluble salts of barium such as the chloride, nitrate, and sulfide, are highly poisonous. In acute poisoning, the myocardium, nervous system, and vessels suffer most, while chronic poisoning mainly affects the bones, bone marrow, gonads, and liver. The action of barium on the myocardium is similar to that of digitalis. It blocks a group of enzyme systems and causes acute degenerative changes thereby impairing cardiac conductivity and rhythm and reducing myocardial contractility. It is sorbed, for the most part, by the cell membrane without penetrating into the cytoplasm. It can produce a cholinolytic effect and induce hypokalemia. Its polysulfide derivatives inhibit cellular respiration similarly to cyanides.

In poisoning by barium chloride, vascular permeability increases, and this may lead to hemorrhages and edemas. Nervous system disorders are manifested in encephalopathy, paresis, and paralysis [Mogosh]. At the basis of some of these manifestations of barium poisoning lie gross changes in vessels and cells of the brain (including the anterior cental gyrus) and of the anterior horns of the spinal cord [Mesko et al.; Mogosh; Silayev; Spoor].

Barium was present at high concentrations in organs of humans (0.004–0.03% on an ash weight basis) and animals (0.0083–0.208%) that died from spontaneous acute leukemia. High-level intakes of barium have been shown to result in a powerful leukemoid response of the bone marrow and to be conducive to true leukemia [Kapran; Pyatak et al.].

Barium is able to cross the placental and blood-brain barriers, and there is some evidence that it may be teratogenic and mutagenic.

Barium can displace phosphorus and calcium from bones, leading to osteoporosis. Its chemical and physiological properties enable it to compete with and replace calcium in processes mediated by the latter. Excess of barium in soil, water, and animal feeds, especially in combination with the presence of excess strontium in these media, may result in impaired calcium metabolism and a severe disease of the peripheral joints and spine known as Kashin-Beck's disease (osteoarthritis deformans endemica).

Acute/single exposure

Animals

LD$_{50}$s of some barium compounds for mice and rats are indicated in Table 1.

For mice, the threshold dose of **barium oxide**, as defined by recording effects on body-weight and the blood content of sulfhydryl groups after an intraperitoneal injection, was 25 mg/kg [Akinfieva & Gerasimova; Liublina & Dvorkin].

In small laboratory animals, the main early signs of poisoning by most barium compounds, observed at 20–30 min post-administration, were sluggishness, smooth muscle spasm, diarrhea, spermatorrhea, and convulsions; later, adynamia, paralysis of the hind paws, and rare deep breathing were added, and the animals assumed a lateral position. Dogs given a lethal dose (700–1000 mg/kg by oral gavage or 50–100 mg/kg intravenously) exhibited excitation, persistent vomiting, reduced pain sensitivity, unsteady gait, hypothermia and paralysis of the hind legs; autopsy showed diffuse hemorrhagic encephalitis [Dubeikovskaya & Nechayeva; Kharitonov].

Table 1. Median lethal doses of barium compounds

Compound	Animal	Route of administration	LD$_{50}$ (mg/kg)
Carbonate	Mouse	Oral	200
		Intraperitoneal	50
	Rat	Oral	418
		Intraperitoneal	50
Chloride	Mouse	Oral	150
		Intraperitoneal	56
	Rat	Oral	397
Hydroxide	Rat	Oral	308
		Intraperitoneal	255
Nitrate	Rat	Oral	390
		Intraperitoneal	293
Oxide	Mouse	Intraperitoneal	300

The insecticide Ba-Neopol, composed of sulfur-containing barium compounds (BaS, BaS$_x$, BaS$_2$O$_3$), was lethal to rats when administered orally by gavage at 320, 640, 1280, or 2560 mg/kg; it impaired blood coagulability, caused hepatic lesions, and inhibited cellular respiration [Jobba & Rengei].

In order of decreasing acute toxicity to mice on subcutaneous injection, readily soluble barium compounds have been arranged as follows: **chloride, nitrate, oxide, peroxide**; difficulty soluble salts such as **barium carbonate** are less toxic, although on multiple oral exposure the carbonate was as toxic as the chloride and nitrate [Dubeikovskaya & Nechayeva; Rumyantsev].

Barium chloride and **barium hydroxide** have strong irritant effects on the upper airways; their threshold concentrations for acute irritancy in rats were 4 mg/m^3 and 3 mg/m^3, respectively, as judged by the change they caused in the breathing rate [Akinfieva & Gerasimova]. **Barium carbonate** was lethal to rats at 33 mg/m^3 on 4-h inhalation exposure.

Rats administered 50 mg of **barium carbonate** dust intratracheally were found, 6 to 9 months later, to have developed moderate pulmonary sclerosis with mucosal necrosis in the large bronchi. Intratracheal administration of **barium sulfate** dust in the same amount produced a roentgenologic picture of fibrosis; autopsy revealed nodules around dust particles, delicate collagen fibers, and gross hypertrophy or hyperplasia of the peribronchial and perivascular tissues; the lesions were less strongly marked in rats administered dust of crude barite ore [Rumyantsev]. Histopathological examination of **barite**-poisoned rats showed no evidence of reticular fibrosis or adverse effects on macrophages; the only notable finding was a chronic inflammatory response [Zajusz]. The relative unaggressiveness of barite dust has also been reported by other authors [Kakauridze & Narsiya].

Each of the three kinds of dust generated in the smelting of **barium ferrosilicon** — dust of barium, that of slag, and that of sublimate — was lethal to rats following intratracheal administration in a dose of 50 mg. The lungs of rats given lower doses (25 mg of the barium or slag dust or 10 mg of the sublimate dust) had elevated levels of hydroxyproline and lipids 3 to 9 months later. The sublimate dust was the most fibrogenic and the barium dust the least; all dusts were found to contain readily soluble barium compounds [Sokolnikov].

When **barium chloride** was injected in an amount of 20 mg into the yolk sac of chick embryos on day 8 of development, developmental defects were observed in the chicks; no defects were noted when the injection was made on day 4 of development [Ridgeway & Karnofsky].

Man

Acute oral poisoning with barium may be accompanied by a variety of clinical signs and symptoms, including salivation, a burning sensation in the mouth, gastric pain, nausea, vomiting, diarrhea, colic, blood pressure elevation, hard irregular pulse, convulsions, profuse excretion of cold sweat, muscular weakness, uncertain gait, vision and speech disorders, dyspnea, vertigo, and tinnitus. The victims did not usually lose consciousness. Paralysis of the lower limbs, followed by that of the trunk and lower limbs occasionally developed. In severe cases, death occurred several to 24 h after the oral intake [Machata].

According to Kazakevich, there are the following three stages of acute barium poisoning: (1) initial manifestations of disordered brain activity, acute gastroenteritis, cardiovascular weakness, and blood changes (leukocytosis, lymphopenia, a left shift of the blood formula, elevated erythrocyte sedimentation rate); (2) markedly disordered brain activity, impaired function of the vestibular apparatus, loss of speech for short periods, weakened vision and hearing, increased vegetative and emotional lability, general weakness, reactive depression, and hysteric manifestations; (3) persistent neuropsychic changes and hysteric reactions in the presence of dystonia.

Barium chloride is toxic in doses of 0.2–0.5 g when ingested; its lethal doses by this route are 0.8–0.9 g. Cases of severe poisoning with **barium carbonate** have been described (toxic doses, 0.2–0.5 g; lethal doses, 2–4 g). The poisoning was characterized by acute gastroenteritis, loss of tendon reflexes, paresthesia, convulsions, paralysis of the extremities, and hypokalemia. This clinical picture resembles than seen in botulism. Cases of lethal poisoning resulting from the use of **barium sulfate** with an admixture of barium carbonate for radiographic examination have occurred. Accidental ingestion of **barium sulfide** was followed by extrasystoles and cardiac arrest in 5 h. Post-mortem examinations showed hemorrhagic and necrotic areas in the gastrointestinal mucosa, hemorrhages to the meninges, hemorrhagic pneumonia, and pulmonary edema. Barium levels in various organs were greatly increased [Govindian & Bhaskar; Savostin].

Chills, fever, a feeling of prostration, and allergic reactions were experienced by 96 out of 196 patients following bronchography using sulfoiodol and **barium sulfate** [Neimark & Trubnikov]. A patient who aspirated barium sulfate while vomiting after its diagnostic use died 2 months later; autopsy showed multiple granulomas containing the sulfate. 'Barium' peritonitis was reported to develop, with a 50% case fatality rate, following the entry of this contrast medium into the peritoneal cavity through the accidentally perforated gastric or intestinal wall; in such cases immediate surgery is required [Ludewig & Lohs].

A case of acute inhalation poisoning by an insecticide based on **barium sulfide** was described; the patient died after a month. Autopsy demonstrated an area of bronchopneumonia in the upper lobe of the right lung and lesions of the yellow atrophy type with diffuse necrosis in the liver. A serious hazard was presented by a fire-extinguishing mixture consisting of 20% NaCl, 28% KCl, and 52% $BaCl_2$ and releasing the latter compound in high concentration when used in fire fighting [Spoor].

Chronic exposure

Animals

Rats exposed to **barium chloride** concentrations of 20 to 250 mg/L through their drinking water for periods of 1 to 16 months developed hypertension and cardiovascular dysfunctions and accumulated barium in the liver, myocardium, muscles, and bones [Tardiff *et al.*].

In a 6-month oral toxicity study, rats administered **barium chloride** at 0.5 mg/kg exhibited abnormal conditioned-reflex activity, altered activity of several enzymes, and increased incidence of chromosomal aberrations; in males, the number of spermatozoa and the time they remained motile were decreased. Embryotoxic effects were also observed. The dose of 0.05 mg/kg was considered as the threshold level and that of 0.005 mg/kg as the subthreshold (no-observed-effect) level. Prolonged (up to 4 months) exposure of rats to this compound by the subcutaneous route at 10–20 mg/kg resulted in increased leukocyte counts and the appearance of myelosis in the liver and spleen [Bebeshko]. A typical picture of leukemia was produced in calves by feeding them with the chloride at 50 mg/kg daily over months for a total dose of 10–16 g.

In a 4-month inhalation toxicity study, exposure of rats to **barium carbonate** at 5 mg/m³ for 4 h/day led to weight loss, arterial blood pressure elevation, decreased blood levels of hemoglobin, platelets, and sugar, and to increased levels of phosphorus in the blood and of calcium in the urine; at the end of that period, the blood serum contained a lowered level of total protein, cholinesterase and alkaline phosphatase activities were inhibited, and the barrier function of the liver and cardiac conductivity were impaired. Autopsy showed dystrophic changes in the myocardium and liver and moderate perivascular and peribronchial sclerosis with focal thickening of interalveolar septa in the lung. The concentration of 1 mg/m^3 was taken as the threshold level [Tarasenko & Pronin].

Barium carbonate has also been shown to exert gonadotropic and embryotoxic effects. Male rats exposed to its concentration of 5.2 mg/m^3 by inhalation for 70 days exhibited reduced libido together with lowered spermatozoal activity and morphological changes in the testes [Popova; Popova & Peretolchina; Popova *et al.*]. In female rats inhaling the carbonate at 13.4 mg/m^3 for 4 months, estrous cycle abnormalities and morphological alterations in the ovaries were noted. For these females as well as for unexposed ones but mated to the exposed males, embryonal and postembryonal mortality rates 2.5 times higher than those in the control groups were recorded. The embryotoxicity of barium carbonate, which was due to its passage through the placental barrier, was manifested in the pups by abnormal activities of the cardiovascular, nervous, and hematopoietic systems and by alterations in calcium metabolism and in liver and kidney functions. The concentration of 1.15 mg/m^3 was considered as the threshold level [Silayev & Tarasenko]. The gonadotoxicity and embryotoxicity shown by barium compounds are believed to result from their general toxic actions, in particular on intrinsic vessels of organs, rather than being specific effects of these compounds.

Dusts of **barium sulfate** or **barite** administered to rats by inhalation at 250–300 mg/m^3 for 1 h/day for up to 6 months induced hypertrophy of the peribronchial tissue and lymph nodes, diffuse thickening of the interalveolar septa, and the formation of collagen fibers [Rokva & Kipiani]. The only findings in the lungs of rats exposed to the sulfate dust at 40 mg/m^3 for 2 months were proliferation of alveolar macrophages and specific changes in the bronchial epithelium; the barium content of lymph nodes remained unchanged, while that of osseous tissue rose to 1500 µg/g [Einbrodt *et al.*]. Experimental baritosis is held to be a variety of pneumoconiosis running a benign course without the formation of true fibrous nodules. Such a benign condition was observed even in rats and guinea-pigs after exposure to barite dust at concentrations as high as 500 mg/m^3 for 5 h/day over 14 months [Viderli].

In a 9-month study on the inhalation toxicity of a **lead-barite ore** (which contained 50% Ba, 25% SiO$_2$, and 5% lead compounds with admixtures of cadmium, calcium, and zinc), strongly marked inhibition of the central nervous system and impaired pulmonary and hepatic functions were observed in rats exposed to a dust of this ore at 10 mg/m^3 (4 h/day, 5 days/week); the animals also had elevated urinary levels of δ-aminolevulinic acid, which was interpreted as a manifestation of lead toxicity. Autopsy revealed dystrophy in parenchymal organs, nodular/interstitial pneumoconiosis, and increased contents of hydroxyproline and lipids in the lung tissue [Burkhanov & Bazaliuk].

Man

Workers handling **barium salts** and exposed to their disintegration aerosols at concentrations varying from 12 to 276 mg/m^3 complained of irritation in the upper airways, eyes, and skin, and of headache, dizziness, sleep disturbances, and aching cardiac pains; **barium chloride** workers also complained of liquid stools, insomnia, and lack of appetite. Frequent findings in these workers were vegetovascular dystonia (vasoneurosis), blood changes (anemia, leukopenia, lymphocytosis, eosinophilia, degenerative changes in neutrophils), persistent bradycardia, polyneuritis, and toxic encephalopathy [Pronin & Pashkovsky; Pyatak *et al.*; Sichko & Nechayeva; Tarasenko & Pronin]. Upper respiratory tract and cardiovascular diseases were present in all barium workers employed for periods of 10 to 20 years [Bakulina]. Female barium workers had increased incidence of ectopic pregnancy, while decreased sexual activity was noted in males [Popova *et al.*].

In a **barium sulfate** factory, workers exposed to this sulfate's dust at 17 to 313 mg/m^3 developed laryngitis, bronchitis, emphysema, and respiratory abnormalities; their sputums contained barite crystals. The question of whether or not the classic pneumoconiosis can result from inhalation of barium-containing dust remains debatable. Most authors regard baritosis (occupational pneumoconiosis) as a benign condition with mild clinical symptoms. The diagnosis is largely based on radiographic findings, but the dense and often irregular shadows seen on the radiographs may disappear after the cessation of barium exposure [Dvizhkov; Doig]. According to Zorn, no pneumoconiosis will arise unless the barite dust contains a substantial admixture of silica. As indicated by results of long-term health surveillance of workers employed in barite mining or processing, baritosis *is* a variety of pneumoconiosis. Its clinical manifestations have usually been scanty, the main complaints being chest pain, shortness of breath, and cough [Kakauridze & Narsiya; Rokva & Kipiani]. Recently, four cases of pneumoconiosis occurring among barium miners in Scotland have been described [Seaton *et al.*]; three of the miners developed progressive massive fibrosis, from which two died. The radiological and pathological changes observed in their lungs were those of silicosis, which was confirmed by the presence of much quartz in the lungs where no barium could be detected. This may suggest that a high proportion of the inhaled barium is not absorbed into the pulmonary tissue, but remains in alveolar macrophages to be eventually removed by the mucociliary escalator [WHO].

An epidemiological study indicated that in communities having high barium levels in the drinking water (2–10 mg/L), death rates from all cardiovascular diseases were higher than in those consuming water with low barium levels (<0.2 mg/L) [Brenniman *et al.*, 1979]. In a later study, these authors concluded that high levels of barium in drinking water did not significantly elevate blood pressure levels [Brenniman *et al.*, 1981].

An analysis of 204 leukemia cases indicated that the prevalence of leukemia in rural areas where barium chloride was used for pest control was 8 times higher than elsewhere. Increased morbidity has also been observed among building construction workers handling a barium-containing stucco [Dubikovsky].

Barium-calcium aluminate, barium-calcium aluminosilicate, and triple carbonate

Barium-calcium aluminate: $2.5 \, BaO \cdot 1.5 \, CaO \cdot Al_2O_3$
Barium-calcium aluminosilicate: $3 \, BaO \cdot 0.5 \, CaO \cdot Al_2O_3 \cdot SiO_2$
Triple carbonate: a complex compound containing 50% $BaCO_3$, 5% $CaCO_3$, and 45% $SrCO_3$

Acute exposure of rats
The oral LD_{50}s of barium-calcium aluminosilicate and triple carbonate were determined at 708 and 545 mg/kg, respectively. Intratracheal administration of triple carbonate in an amount of 50 mg produced a weak fibrotic effect. Four-hour inhalation exposure to dusts of barium-calcium aluminosilicate at 94 mg/m^3, barium-calcium aluminate at 46 mg/m^3, or triple carbonate at 135 mg/m^3 was accompanied and followed by increased values of the summated threshold index, deterioration of orienting reflexes, and alterations in muscle strength and breathing frequency. The triple carbonate concentration of 36 mg/m^3 was rated as subthreshold for these effects, while the 35 mg/m^3 and 14 mg/m^3 levels were considered as thresholds for the aluminosilicate and aluminate, respectively. The latter concentration was also found to be the threshold level for irritant effects of the aluminate on mucous membranes [Shubochkin & Gerasimova].

Chronic exposure of rats
On long-term inhalation exposure to triple carbonate at 10 mg/m^3, decreased body-weight gains and depressed cholinesterase and alkaline phosphatase activities in the blood were noted starting with the third month of exposure; autopsy showed edema of the lung stroma and dystrophy of the renal tubules. In animals exposed to 1 mg/m^3, the only finding was altered activity of these two enzymes [Shubochkin *et al.*].

Barium ferrite
In a 9-month inhalation toxicity study, exposure of rats to barium ferrite at concentrations of 200 to 400 mg/m^3 for 4 h/day, 6 days/week led to elevated blood levels of several enzymes (peroxidase, catalase, cholinesterase, and transaminases) and to increased collagen content in the lungs; autopsy revealed

signs of pneumoconiosis. In rats exposed to $10-30$ mg/m^3, only focal accumulations of dust-laden cells were noted at autopsy [Kosova & Gershovich].

Continuous inhalation exposure of rats to a barium ferrite aerosol at $0.41-0.76$ mg/m^3 led to increased leukocyte counts, elevated cholinesterase activity and lowered calcium levels in the blood as well as to diminished osmotic resistance of spermatozoa and increased numbers of spermatozoa with pathological changes in male rats; autopsy showed histopathological changes in their testes. The concentration of 0.095 mg/m^3 only induced changes in germ cells, while that of 0.038 mg/m^3 did not produce any effects [Yermachenko et al.].

Three months after a single intratracheal administration of barium ferrite dust to rats, their lungs showed a well-defined fibrotic reaction that did not increase during the next $3-6$ months. Continuous inhalation exposure of female rats to this dust throughout pregnancy resulted in increased post-implantation and embryonal death rates; the concentration of 0.0048 mg/m^3 was described as the highest level without effects on pregnant rats, their fetuses, and their first-generation progeny [Govorunova & Grin].

Barium fluoride

Barium fluoride is a protoplasmic poison that affects a number of enzymes and blocks sulfhydryl groups. On acute exposure, its primary targets are the nervous system and musculature. The toxic effects of this compound are due to the toxicity of both its cation and its anion. Because of the presence of fluorine in its molecule, the gonadotoxic and embryotoxic activities of barium fluoride are higher than those of other barium compounds.

Acute exposure of rats

The LD$_{50}$ of barium fluoride by oral gavage was determined at 250 mg/kg, and its threshold concentration for toxic effects on inhalation exposure, at 4.42 mg/m^3. Intratracheal administration of its dust in an amount of 50 mg gave rise to an inflammatory process resembling that seen in chronic interstitial pneumonia, and to increased hydroxyproline levels in the lungs. Irritant effects on the skin and mucous membranes were also noted.

Long-term and short-term exposure of rats

Animals exposed to barium fluoride by inhalation at 5.66 mg/m^3 (4 h/day, 6 days/week, up to 4 months) showed increased values of the summated threshold index, reduced body-weight gains, dental fluorosis, and several hematological changes including decreases in hemoglobin levels, erythrocyte counts, and serum cholinesterase activity along with increases in serum levels of urea; roentgenograms showed pathological fractures of femoral and tibial bones.

Inhalation exposure of female rats to a barium fluoride concentration of 0.61 mg/m^3 throughout pregnancy increased antenatal mortality of their offspring. Pups born alive to such rats showed abnormal electrocardiograms, low hemoglobin levels and elevated alkaline phosphatase activity in the blood; at autopsy, hemorrhages were seen in various organs. Long-term exposure of males to this concentration impaired functional properties of their spermatozoa and caused morphological changes in their testes.

Embryotoxic effects were observed in pregnant rats following repeated exposure to barium fluoride by oral gavage in doses equal to 1/10 or 1/20 of the LD$_{50}$; the dose of 1/30 LD$_{50}$ was defined as the sub-threshold level for such effects.

Barium titanates

The toxicity of composite materials based on barium titanates is due to the barium and is not significantly altered by the ions of other metals present there.

Acute and short-term exposures: animals

Intraperitoneal LD$_{50}$s of barium-aluminum titanate and barium-calcium titanate for rats were 3.2 and 1.7 g/kg, respectively. In doses equal to 0.2 or 0.13 of their LD$_{50}$s, these compounds induced behavioral changes. After an intramuscular injection of barium titanate zirconate at 40 mg/kg, signs of neuroparalysis were observed in guinea-pigs [Dubeikovskaya & Nechayeva].

After a single intratracheal administration to rats in the form of dust (50 mg), barium titanate, barium tetratitanate, barium titanate zirconate, barium-aluminum titanate, and barium-calcium titanate induced similar roentgenological and morphological changes in the lungs: 1 month post-administration, nodules of dust-laden cells, delicate networks of argentophilic fibers, and inflammatory changes in the bronchi were seen; these initial signs of fibrosis did not progress during the subsequent 6–9 months. When barium titanate or barium titanate zirconate were administered repeatedly by this route (50 mg at monthly intervals for 3 months), alterations in mineral metabolism, structural changes in long bones, reduced gonadal weight, and enhanced gonadotropic activity of the pituitary were observed in the rats [Nechayeva et al., 1982, 1983; Sichko & Nechayeva; Talalaenko & Degonsky].

Chronic exposure: man

Some of the workers engaged in the crushing, grinding, mixing, or synthesis of materials based on barium titanates and exposed to their dusts (mean concentration, 8.4 mg/m^3) had functional changes in the central nervous system and polyneuritis and/or neuralgia in the upper extremities.

Effects of local application

Animals
Barium carbonate dust had irritating effects on the skin and mucous membranes of rabbits and rats. **Barium chloride** was less irritating, while the **nitrate** did not irritate either the skin or mucous membranes. The latter two salts were not absorbed through the skin. A strongly marked cauterizing effect on the skin and cornea was exerted by **barium hydroxide** [Akinfieva & Gerasimova].

Man
Workers handling **barium carbonate** or **hydroxide** had a dry and itchy skin on the hands and arms as well as fissures in the skin of the hands.

ABSORPTION, DISTRIBUTION, AND ELIMINATION

Barium has not been shown to be an essential trace element, although it is found in small amounts in many tissues of the healthy human body. Barium ions, like those of strontium, are so much similar to calcium ions in their characteristics that they are involved in mineral metabolism together with the latter. However, because barium ions are metabolized at faster rates and differ from calcium ions in size, they can impair the normal skeletal calcification, resulting in damage to the epiphyseal and articular cartilages, which is a characteristic sign of Kashin-Beck's disease [Kovalsky; Shvaikova].

Barium enters the body mainly through inhalation and ingestion. The barium absorbed from the intestine is transported in the bloodstream being bound to proteins (but not to globulin); human blood has been reported to contain barium in concentrations of 0.006–0.009 mg% [Levina]; other authors [Pyatak *et al.*] could only detect barium in the blood of people occupationally exposed to its salts.

The 'standard man' of 70 kg contains about 22 mg of barium; the major part of it (up to 91%) is concentrated in the bones, the remainder being in soft tissues such as aorta, brain, heart, kidney, spleen, pancreas, and lung [WHO]. Relatively high levels were found in the gonads and uvea [Krasovsky *et al.*]. Normal concentration ranges are considered to be (mg%) 2.5–4.36 in the kidney, 0.66–8.7 in the gastric wall, 0.39–1.0 in the myocardium, and 2.1–6.0 in the cerebral cortex. The barium level of $1.65 \times 10^{-4}\%$, found in the distal epiphysis of femoral bones from cadavers of rural people, has been regarded as the physiological 'norm' for the present human generation [Dubrovina *et al.*].

The elimination of injected or ingested barium occurs principally in the feces; lesser amounts are excreted in the urine. The elimination occurs slowly, especially from bones. For example, dogs given a single dose of 100 mg/kg still had some barium in their bone tissue 5 to 7 years later. In cows, barium passed through the placenta and was excreted in milk and colostrum.

HYGIENIC STANDARDS

Exposure limits adopted for barium and its compounds in the USSR are shown in Table 2. In the USA, a TLV-TWA of 0.5 mg/m^3 has been set for soluble barium compounds (as Ba) and a TLV-TWA of 10 mg/m^3 for total dust of barium sulfate containing no asbestos and <1 % crystalline silica.

METHODS OF DETERMINATION

A readily available and widely used technique for determining several metals, including barium, in solution from various samples is atomic absorption spectrophotometry (AAS) [Machata]. Two AAS methods for barium — the direct aspiration method, and the furnace method — have been recommended. The optimal concentration range for determining barium by the direct aspiration method, using a wavelength of 553,6 nm, is 1–20 mg/L, with a sensitivity of 0.4 mg/L and a detection limit of 0.03 mg/L. An AAS direct aspiration procedure for measuring water-soluble barium-containing components in air has a sensitivity of 0.0004 mg/m^3 (the estimated working range of the method is 0.15–1.3 mg/m^3). For concentrations <0.2 mg Ba/L, the furnace technique is recommended; with it, the optimal concentration range for barium determination is 10–200 μg/L and the detection limit is 2 μg/L (a detection limit of 0.5 ng/ml using a 20-μl sample has also been reported). For measuring barium concentrations in plant tissue, emission spectrography has been employed; this method has also been used to measure barium in drinking water, with a detection limit of 10 μg/L. Emission spectrometry using an inductively coupled plasma source has also been used for measuring barium in water, with detection limits of ≤0.1 ng/ml. Analytical methods utilized for special applications include mass spectrometry, X-ray fluorescence spectrometry, and neutron activation analysis [WHO].

Barium carbonate in air can be determined gravimetrically [*Specifications for Methods of Determining Noxious Substances in Air*] and barium ferrite in air, by photometry involving dissolution of barium-containing dust in hydrochloric acid and photometric determination of the pink complex compound formed by iron with o-phenanthroline [Yermachenko *et al.*]. Photometry can also be used for barium determination in water — through the formation of barium chromate in neutral medium followed by photometry of the chromate (sensitivity 1.5 mg/L) [Novikov *et al.*]. For barium in biological materials, a method involving mineralization of the sampled material and subsequent application of a complexometric procedure using zinc chloride was proposed [Krylova].

MEASURES TO CONTROL EXPOSURE

In industries where barium and its compounds are produced or used, the main adverse factor is dust or the decomposition products formed in thermal processes. Equipment that excludes manual operations should be used to process raw materials for the production of ceramic masses and barium glass and to synthesize compositions based on barium or its compounds. The processes of loading and unloading the equipment should be enclosed to rule out spillage of dusty materials. The charging and discharging of furnaces for founding barium glass should also be completely mechanized, and balanced general ventilation should be provided in the furnace rooms.

Table 2. Exposure limits for barium and its compounds

	Workroom air		Atmospheric air		Water sources	
	MAC_{wz} (mg/m^3)	Aggregative state	MAC_{ad} (mg/m^3)	$TSEL_{aa}$ (mg/m^3)	MAC_w (mg/L)	Hazard class
Barium	—	—	—	0.004	0.1[a]	2
Barium aluminate	0.1	Aerosol	—	—	—	2
aluminosilicate	1.0/0.5[b]	Aerosol	—	—	—	2
carbonate	0.5[c]	Aerosol	0.004	—	—	2
chloride	0.3	Aerosol	—	—	—	2
fluoride	0.1	Aerosol	—	—	—	2
hydroxide	0.1	Aerosol	—	—	—	2
nitrate	0.5	Aerosol	—	—	—	2
oxide	—	—	—	0.004	—	2
phosphate, disubstituted	0.5	Aerosol	—	—	—	2
sulfate	6.0	Aerosol	—	0.1	—	3

[a] Based on sanitary and toxicological criteria.
[b] 1.0 mg/m^3 is the concentration not to be exceeded at any time, while 0.5 mg/m^3 is the average maximum allowable concentration per shift.

In factories producing ceramic articles or piezoelectric cells, preference should be given to wet methods of obtaining ceramic masses. Mechanical working of barium ceramic articles must be carried out in areas fitted with local exhaust ventilation.

Rest rooms with adequate relaxation facilities should be provided at the factory. Workers exposed to dusts of barium or its compounds should receive ultraviolet irradiations and appropriate inhalations for prophylactic purposes.

Before and systematically during employment, all workers should be instructed in technical safety and health precautions to be observed in work with barium and its compounds. Pre-employment and in-employment medical examinations are also required.

Adequate washing and other sanitary facilities should be provided for workers exposed to toxic soluble barium compounds.

There should be a ban on smoking, drinking, and storing and taking food in the work areas. Dust from work clothes should be removed in special facilities. Work clothes should not be allowed to be taken home for washing. The main personal protection equipment is respirators.

Measures to protect the *general environment* should be directed at developing effective methods for the treatment of wastewaters containing barium and its compounds. In areas where barium-containing pesticides or fertilizers are used, arrangements should be made to prevent their excessive accumulation in the soil and thus avoid the contamination of food, water, and atmospheric air.

EMERGENCY TREATMENT IN POISONING

For those who have ingested barium or its soluble compounds, the following is recommended. 1) Detoxicating and antidotal treatment: gastric lavage with a 1% sodium sulfate or magnesium sulfate solution or enemas consisting of a 10% sodium (or magnesium) sulfate solution; oral administration of a 30% magnesium sulfate solution (100 ml); forced diuresis and hemodialysis; administration of a 10% tetacin-calcium [synonym: sodium calcium edetate] solution (20 ml) in 500 ml of 5% glucose by intravenous drip. 2) Symptomatic treatment: intravenous injections of promedol [synonym: trimeperidine hydrochloride] (1 ml of 2% solution) and atropine (1 ml of 0.1% solution) in 5% glucose; calcium chloride by intravenous drip (2.5 g in 500 ml of 5% glucose) in cases of pronounced cardiac arrhythmia; cardiovascular agents (cardiac glycosides are contraindicated); and intramuscular injections of a 6% thyamine solution (10 ml) and a 5% pyridoxine solution (10 ml). Oxygen and antishock therapy may be required. Emetics may be given to remove the insoluble barium sulfate from the stomach. Milk and mucilaginous soups are indicated. A hot water bag or bottle may be applied to the feet.

Soluble barium compounds splashed on the skin or in the eyes should be flushed out with copious amounts of water.

REFERENCES

Akinfieva, T. A. & Gerasimova, I. L. (1984) *Gig. Truda Prof. Zabol.*, No. 6, 45—46 (in Russian).
Bakulina, A. V. (1962) In: *Trudy Permskogo meditsinskogo instituta* [Transactions of the Perm Medical Institute]. No. 37, Perm, pp. 300—309 (in Russian).

References 187

Bebeshko, V. G. (1965) In: *Trudy Kievskogo NII perelivaniya krovi* [Transactions of the Kiev Research Institute for Blood Transfusion]. Kiev, Vol. 1, pp. 222—227 (in Russian).

Bespamyatnov, G. P. & Krotov, Yu. A. (1985) *Predelno dopustimye kontsentratsii khimicheskikh veshchestv v okruzhayushchei srede* [Maximum Allowable Concentrations of Chemicals in the Environment]. Khimiya, Leningrad (in Russian).

Brenniman, G. R. *et al.* (1979) *Environ. Res.*, **20**, 318—324.

Brenniman, G. R. *et al.* (1981) *Arch. Environ. Health*, **36**, 28—32.

Burkhanov, A. I. & Bazeliuk, L. T. (1985) *Gig. Truda Prof. Zabol.*, No. **4**, 30—33 (in Russian).

Dare, P. R. *et al.* (1984) *Ann. Occup. Hyg.*, **28**, 445—448.

Doig, A. T. (1976) *Thorax*, **31**, 30—39.

Dobrovolsky, V. V. (1938) *Geografiya mikroelementov. Globalnoe rasseyanie* [Geography of Trace Elements. Global Dispersion]. Mysl, Moscow (in Russian).

Drozdov, V. N. (1966) In: *Trudy Omskogo meditsinskogo instituta* [Transactions of the Omsk Medical Institute]. No. 69, Omsk, pp. 72—73 (in Russian).

Dubeikovskaya, L. S. & Nechayeva, E. N. (1988) In: *Aktualnye problemy teoreticheskoi i prikladnoi toksikologii* [Current Topics in Theoretical and Applied Toxicology]. Mocow, pp. 94—101 (in Russian).

Dubikovsky, G. P. (1982) *Khimiya v selskom khozyaistve* [Chemistry in Agriculture], No. 3, 33—34 (in Russian).

Dubrovina, Z. V. *et at.* (1967) *Gig. San.*, No. **4**, 43—46 (in Russian).

Dvizhkov, P. P. (1965) *Pnevmokoniozy* [Pneumoconioses]. Meditsina, Moscow (in Russian).

Einbrodt, H. J. *et al.* (1972) *Intern. Arch. Arbeitsmed.*, **30**, 675—677.

Goncharuk, E. I. (1977) *Sanitarnaya okhrana pochvy ot zagryazneniya khimicheskimi veshchestvami* [Sanitary Measures to Protect Soil from Chemical Pollution]. Zdorovya, Kiev (in Russian).

Govindian, D. & Bhaskar, G. (1972) *Antiseptic*, No. 3, 237—244.

Govorunova, N. N. & Grin, N. V. (1984) *Gig. San.*, No. **4**, 72—74 (in Russian).

Grushko, Ya. M. (1972) *Yadovitye metally i ikh neorganicheskie soyedineniya v promyshlennykh stochnykh vodakh* [Poisonous Metals and their Inorganic Compounds in Industrial Waste Waters]. Medistina, Moscow (in Russian).

Jobba, . & Rengei, B. (1971) *Arch. Toxicol.*, No. 2, 106—110.

Kakauridze, E. M. & Narsiya, A. A. (1960) *Gig. Truda Prof. Zabol.*, No. **12**, 43—45 (in Russian).

Kapran R. G. (1978) *Klin. Med.*, No 7, 104—105 (in Russian).

Kazakevich M. A. (1948) *Klin. Med.*, No. **11**, 56—59 (in Russian).

Kharitonov, O. I. (1957) *Farmakol. Toksikol.*, No. 2, 68—74 (in Russian).

Kosova, L. V. & Gershovich, E. M. (1972) *Gig. Truda Prof. Zabol.*, No. **5**, 41—44 (in Russian).

Kovalsky, V. V. (1974) *Geokhimicheskaya ekologiya* [Geochemical Ecology]. Nauka, Moscow (in Russian).

Krasovsky, G. N. *et al.* (1977) *Gig. San.*, No. 7, 11—16 (in Russian).

Krylova, A. N. (1969) *Farmatsiya*, No. 4, 63—66 (in Russian).

Levina, E. I. (1972) *Obshchaya toksikologiya metallov* [General Toxicology of Metals]. Meditsina, Leningrad (in Russian).

Liublina, E. I. & Dvorkin, E. A. (1983) In: *Gigiyenicheskaya toksikologiya metallov* [Hygienic Toxicology of Metals]. Moscow, p. 114 (in Russian).

Ludewig, R. & Lohs, K. (1981) *Akute Vergietungen.* VEB Gustav Fischer, Jena.

Machata, G. (1988) Chapter *9 Barium* in: Seiler, H. G. *et al.* (eds) *Handbook on Toxicity of Inorganic Compounds.* Marcel Dekker, New York, pp. 97—101.

Mesko, K. *et al.* (1968) *Z. Gesam. Innere Med.*, No. 5, 151—154.

Mogosh, G. (1984) *Ostrye otravleniya* [Acute Poisonings]. (Russian translation from Roumanian published in Bucharest.)

Naimark, D. A. & Trubnikov, G. V. (1968) *Klin. Med.*, No. 12, 128—130 (in Russian).

Nechayeva, E. N. *et al.* (1982) *Gig. Truda Prof. Zabol.*, No. 9, 10—13 (in Russian).

Nechayeva, E. N. *et al.* (1983) In: *Gigiyenicheskaya toksicologiya metallov* [Hygienic Toxicology of Metals]. Moscow, pp. 30–38 (in Russian).

Novikov, Yu. V. *et al.* (1981) *Metody opredeleniya vrednykh veshchestv v vode vodoyemov* [Methods for Measuring Noxious Substances in Water Bodies]. Meditsina, Moscow (in Russian).

Popova, O. Ya. (1978) *Gig. Truda Prof. Zabol.*, No. **5**, 34–37 (in Russian).

Popova, O. Ya. & Peretolchina, N. M. (1976) *Gig. San.*, No. **2**, 109–111 (in Russian).

Popova, O. Ya. et. al. (1977) In: *Fiziologiya i gigiyena truda v nauchno-tekhnicheskom progresse* [Occupational Physiology and Hygiene in Relation to Scientific and Technological Progress]. Moscow, pp. 46–50 (in Russian).

Pronin, O. A. & Pashkovsky, V. G. (1973) *Gig. Truda Prof. Zabol.*, No. **5**, 36–37 (in Russian).

Pyatak, O. A. *et al.* (1980) *Gigiyena Truda* (Kiev), No. **6**, 89–92 (in Russian).

Ridgeway, L. P. & Karnofsky, D. A. (1952) *Ann. N. Y. Acad. Sci.*, **55**, 203–215.

Rokva, V. A. & Kipiani, S. P. (1970) In: *Borba s silikozom* [Silicosis Control]. Vol. 8, Meditsina, Moscow, pp. 213–215 (in Russian).

Rumyantsev, G. I. (1963) In: *Toksikologiya redkikh metallow* [Toxicology of Rare Metals]. Medgiz, Moscow, pp. 176–187 (in Russian).

Savostin, G. A. (1984) *Sudebno-Meditsinskaya Ekspertiza*, No. **2**. 52–53 (in Russian).

Seaton, A. *et al.* (1986) *Thorax*, **41**, 591–595.

Shubochkin, L. N. & Gerasimova, I. L. (1978) in: *Aktualnye problemy gigiyeny truda* [Current Issues in Occupational Hygiene]. Moscow, pp. 24–28 (in Russian).

Shubochkin, L. N. & *et al.* (1980) *Gig. Truda Prof. Zabol.*, No. **6**, 55 (in Russian)

Shvaikova, M. D. (1975) *Toksikologicheskaya khimiya* [Toxicologic Chemistry]. Moscow (in Russian).

Sichko, Zh. V. & Nechayeva, E. N. (1983) In: *Gigiyenicheskaya toksikologiya mettalov* [Hygienic Toxicology of Metals]. Moscow, pp. 38–43 in Russian).

Silayev, A. A. (1973) In: *Vliyanie soedineniy bariya na generativnuyu funktsiyu i ego gigiyenicheskoe znacheniye* [Impact of Barium Compounds on Reproductive Function and its Health Implications]. Moscow (In Russian).

Silayev, A. A. & Tarasenko, N. Yu. (1976) *Gig. Truda Prof. Zabol.*, No. **7**, 33–37 (in Russian)

Sokolnikov, E. A. (1973) In: *Gigiyena truda i profilaktika profpatologii v proizvodstve ferrosplavov* [Occupational Health in the Manufacture of Ferroalloys]. Sverdlovsk, pp. 41–43 (in Russian).

Specifications for Methods of Determining Noxious Substances in Air (1965) [Tekhnicheskiye usloviya na metody opredeleniya vrednych veshchestv v vozdukhe], No. 4. Publications of the USSR Ministry of Health, Moscow (in Russian).

Spoor, N. (1964) *Ann.Occup. Hyg.*, No. **4**, 369–372

Talalaenko, A. N. & Degonsky, A. N. (1974) *Gig. San.*, No. **6**, 102–103 (in Russian)

Tarasenko, A. N. & Pronin, O. A. (9174) *Gig. Truda Prof. Zabol.*, No. **1**, 27–31 (in Russian).

Tardiff, R. G. et al. (1980) *J. Environ. Pathol. Toxicol.*, **4**, 267–275

Viderli, M. M. (1954) *Eksperimentalnyi baritoz* [Experimental Baritosis]. Baku (In Russian).

Yermachenko, T. P. *et al.* (1983) *Gig.San.*, No. **7**, 68–69 (in Russian).

WHO (1990) *Environmental Health Criteria 107: Barium*. World Health Organization, Geneva

Zayusz, K. (1969) *Med. Pr.*, No. **5**, 464–483.

Zorn, O. (1966) *Ztbl. Arbeitsmed.*, No. **5**, 133–135.

Zinc and its compounds

Zinc carbonate (smithsonite [min.]); **z. chloride**; **z. hydroorthoarsenate** (acid zinc arsenate, disubstituted); **z. hydroorthophosphate** (acid zinc orthophosphate, disubstituted); **z. nitrate hexahydrate**; **z. phosphide** (diphosphide); **z. orthoarsenite**; **z. orthophosphate**; **z. oxide** (zinc white; zincite [min.]); **z. selenide** (schtilleite [min.]); **z. sulfate** (zincosite); **z. sulfate heptahydrate** (white vitriol); **z. sulfide** (sphalerite (α) or zinc blende [min.], wurtzite (β) [min.])

IDENTITY AND PHYSICOCHEMICAL PROPERTIES OF THE ELEMENT

Zinc (Zn) is a metallic element in group II of the periodic table, with the atomic number 30. Its natural isotopes are ^{64}Zn (48.89%), ^{66}Zn (27.81%), ^{67}Zn (4.11%), ^{68}Zn (18.57%), and ^{70}Zn (0.62%).

Zinc is ductile when pure and can be rolled into sheets and fine foil. In the commercial form (technical-grade zinc) it is brittle at ordinary temperatures, but becomes ductile on slight heating. It is a diamagnetic metal.

In compounds zinc is present in the oxidation state +2. It remains unchanged in dry air up to 225°C. In moist air containing carbon dioxide, it oxidizes even at ordinary temperatures; when heated in air it burns to the oxide ZnO. Zinc dust is pyrophoric. In the cold, zinc will not react with dry bromine, chlorine, or fluorine but in the presence of moisture may ignite to form the corresponding halide. Powdered zinc reacts explosively with sulfur to give the sulfide ZnS. In an acetylene current, it forms the carbide ZnC_2.

Zinc is a strong reducing agent. It can reduce the salts of Mn, Fe, Ni, Cu, and Cd; in reacting with nitric acid, it reduces the NO_3^- ion to NH_4^+. It reacts with other acids with formation of the corresponding salts. It dissolves in alkalies and ammonia. Most zinc salts are soluble in water. Zinc forms industrially important alloys with Al, Cu, and Mg. See also Appendix.

NATURAL OCCURRENCE AND ENVIRONMENTAL LEVELS

Of 64 zinc minerals, the most important ones are sphalerite (or zinc blende) ZnS, zincite ZnO, smithsonite $ZnCO_3$, wurtzite ZnS, calamine $Zn_4Si_2O_7[OH]_2 \cdot H_2O$, and goslarite $ZnSO_4 \cdot 7H_2O$.

The clarke of zinc is estimated at $(70-83) \times 10^{-4}\%$ in the earth's crust and at $51 \times 10^{-4}\%$ in the granite layer of the continental crust [Dobrovolsky]. Zinc is widely distributed in all geospheres. The bulk of it migrates via the hydrosphere. The oceans contain an estimated total of 6850 million tonnes of zinc, at an average concentration of 0.9 µg/L; ferromanganese oxide deposits of the Pacific take up an estimated 2.8 thousand tonnes annually. In river water, zinc is contained at an average concentration of 20 µg/L, and its global runoff from rivers to the oceans amounts to 740 thousand tonnes per annum (K_b = 3.27) [Dobrovolsky].

Zinc has come to be one of the most widespread toxic pollutants of the world's ocean, as is indicated by its high concentrations (up to 1020 µg/L) in the uppermost (60–100 µm) seawater layer, whereas the upper limit of ecologic tolerance for oceans and inland seas is considered to be 50 µg Zn/L.

In soils, the average abundance of zinc is $5 \times 10^{-3}\%$. Its ability to migrate in soil depends on the acid–base and oxidation–reduction characteristics of the soil medium. Zinc is more mobile than cadmium or lead and migrates at highest rates in eroded and light-textured soils. The zinc-fixing capacity of soils is the higher, the more organic matter and calcium they contain [Bolshakov et al.; Roshchin et al.].

A major factor of zinc migration and redistribution is the bios. This contains approximately 500 million tonnes of zinc, about a quarter of which is present in terrestrial plants which have been estimated to contain a total of 125 million tonnes of zinc, at average wet weight, dry weight, and ash weight concentrations of $20 \times 10^{-4}\%$, $50 \times 10^{-4}\%$, and $1000 \times 10^{-4}\%$, respectively. Some plants are capable of accumulating zinc to considerable levels; for example, caryophylaceous plants accumulated up to 4900 mg/kg and cruciferous plants, up to 13 630 mg/kg dry weight [Goncharuk]. The amount of zinc taken up annually through the increment in phytomass has been calculated to average 57.5 kg/km^2 of land area (K_b = 19.6) [Dobrovolsky].

Average dry-weight zinc concentrations in organisms were reported to be as follows (mg/100 g): marine algae, 15.0; terrestrial plants, 10.0; marine animals, 0.6–150.0; terrestrial animals, 16.0; bacteria, 0.1–28.0 [Kovalsky].

A powerful factor in zinc redistribution over the earth's surface has become human industrial activity. Of major concern here is the ability of zinc to be gradually concentrated in food chains. Large quantities of zinc are accumulated by aquatic plants, many of which show seasonal variation in this respect. In lakes of the Baltic region of the USSR, for instance, reed plants contained on average 1031 mg Zn/kg ash weight in May, 1012 in June and July, and 3527 in September; for horsetail, the respective figures were 724, 1259, and 1349 mg Zn/kg [Vadze *et al.*]. Good zinc concentrators are gastropods and particularly notonectid bugs which had concentrations up to 141 mg Zn/kg dry weight while living in water that contained only 5.1×10^{-3} mg Zn/L [Zhulidov *et al.*].

PRODUCTION

Zinc is produced chiefly by electrolyzing solutions of its salts. **Zinc chloride** is made by dissolving zinc wastes in hydrochloric acid. **Zinc orthoarsenite** and **hydroorthoarsenate** are waste products in nonferrous metal plants; **z. oxide** is obtained by roasting zinc carbonate or by burning metallic zinc; **z. phosphide**, by fusing red phosphorus with zinc dust; **z. selenide**, by passage of hydrogen selenide over zinc at a red heat, by reaction of hydrogen sulfide and zinc chloride vapor, or by reduction of zinc selenate with carbon or hydrogen. **Z. sulfate** is obtained by dissolving zinc wastes in sulfuric acid. The principal source of **z. sulfide** is the mineral sphalerite (zinc blende).

USES

Zinc is used in alloys with nonferrous metals (brasses, tombac, argentan); in galvanic cells and storage batteries; as an anticorrosive for steel and iron articles; and in the metallurgical and chemical industries.

Zinc chloride is used in the paper and pulp industry, as a flux in hot zinc plating, tin plating, soldering and welding, and in medicine. **z. oxide** is used as a filler for plastics and rubbers, in the manufacture of glass, ceramics, matches, and celluloids, in the production of photocopying paper, paints and chemicals, in cosmetic pharmaceutical preparations, and in powders and ointments for burns and skin infections. **Z. phosphide** is a constituent of rodenticides. **Z. selenide** and **z. sulfide** are employed in semiconductors, in photocells and photoresistors, and as luminous pigments. **Z. sulfate** finds use in the manufacture of viscose and in electroplating, in ore concentration, as a hardener in rayon manufacture, and as a supplement for humans, animals, and plants with zinc deficiency.

Zinc compounds are also used as pigments for paints (the chloride, oxide, sulfate, and sulfide), including anticorrosive pigments (the orthophosphate and hydroorthophosphate); as antiseptics for wood preservation (the chloride, hydroorthoarsenate, orthoarsenite, and sulfate), and as components for dental cements (the orthophosphate, hydroorthophosphate, and oxide). Zinc sulfide is also used as a phosphor in X-ray and TV screens and in luminous watch faces. See also Bertholf and Ohnesorge & Wilhelm.

MAN-MADE SOURCES OF EMISSION INTO THE ENVIRONMENT

The major sources of man-made emissions of zinc into the environment are its atmospheric emissions from high-temperature industrial processes. In the period 1963–75, approximately 500 thousand tonnes of zinc were calculated to have been dispersed worldwide from these sources together with less substantial contributions from transportation, beneficiation, and grading processes. As a result of coal burning alone, an estimated 137.5 thousand tonnes of zinc were released into the atmosphere in 1980 and nearly 220 thousand tonnes has been forecast for the year 2000 [Dobrovolsky]. According to one estimate, the total annual release of zinc into the environment from all anthropogenic sources is around 315 thousand tonnes [Roshchin *et al.*].

In the vicinity of a plant recovering and processing nonferrous metals from secondary sources (scrap), average atmospheric zinc concentrations were (mg/m^3 of air) 0.350 at a distance of 300 m from the plant, 0.285 at 500 m, 0.148 at 1000 m, and 0.52 at 2000 m.

In the northern hemisphere, atmospheric zinc concentrations were reported to range from 0.1 to 4.8 mg/m^3 in rural areas and from 1 to 4000 mg/m^3 in cities [Roshchin *et al.*].

Useful bioindicators of atmospheric pollution by zinc are mosses; mosses growing near nonferrous metals plants had an average zinc content of 0.86 mg/g [Nemenko *et al.*]. A peculiar indicator of air pollution may be zinc level in the hair of domestic dogs (which was found to contain up to 207 µg Zn/g in big cities).

Metallic zinc is oxidized by atmospheric oxygen and precipitates as the oxide ZnO. It has been calculated that annually 72 kg of zinc are deposited in precipitation per km^2 of the earth's surface, which is about 3 times more than the estimate for lead and 12 times more than that for copper [Dmitriev & Tarasova].

Emissions from metallurgical works and particularly zinc smelters can give rise to serious pollution problems in local areas. In particular, the zone of soil pollution may extend to 30–40 km. In the environs of one zinc smelter, for example, the stand of oak and maple trees had become very thin and there were only a few young trees as a result of heavy soil pollution with zinc (its levels amounted 80 mg/g of top soil within 0.8 km of the smelter); an area of about 470 hectares in the immediate vicinity of the smelter was virtually devoid of vegetation [Bolshakov *et al.*].

It was calculated in one study that some 9 mg of zinc per kg soil per year, i.e. 450 mg/kg over 50 years, would be received by the plow soil layer in the environs of a metallurgical works whose solid emissions contained zinc at levels of 3000–4800 mg/kg. At the time of the study, samples of that layer had an average zinc content of 255 mg/kg, which was about 5 times higher than in soil samples from unpolluted areas. As a result, high zinc concentrations were also found in vegetables grown in the contaminated soil. Thus, potato leaves contained, on average, 82.7 mg/kg dry weight (vs 42.8 in control samples); potato tubers, 20.4 mg/kg (vs 11.0), carrot tops, 60.7 mg/kg (vs 33.7); carrot roots, 27.0 mg/kg (vs 16.5); tomato leaves, 95.2 mg/kg (vs 28.6); tomato fruits, 28.3 mg/kg (vs 20.7) [Garmash]. (See also Hewitt & Candy and Ohnesorge Wilhelm.)

Considerable harm to nature and agriculture is done by electric power plants operating on brown coal; zinc emissions in fly ash from one such plant were reported to have reached 8.68 tonnes/km^2 per year [Doncheva et al.].

Much care needs to be exercised in the use of wastewater and sludge in agriculture for irrigation and fertilization [Adamu et al.], for they may well contain very high concentrations of heavy metals, including zinc. In one instance up to 5000 mg Zn/kg dry residue was recorded [Perelygin]. Such concentrations (and even much lower ones) interfere with the biologic processes of wastewater treatment to result in deteriorated quality of the effluent. Some authors consider zinc-containing wastewaters to be unsuitable for irrigating agricultural lands [Grushko, 1972, 1979]. The threshold zinc level in wastewater, defined as the concentration reducing the efficiency of its treatment by 5%, has been found to lie between 5 to 10 mg/L. The maximum permissible zinc levels in sludges and soils, established several years ago in the East European countries and the USSR, are 3000 mg/kg for sludge and 300 mg/kg for soil (recommended levels are 2500 mg/kg and 150 mg/kg, respectively) [Leschber]. In Japan, an outbreak of severe disease affecting the skeleto-muscular system occurred among people consuming rice grown in fields irrigated with water heavily polluted by zinc sulfide containing cadmium [Lag].

For natural bodies of waters, particularly the oceans, considerable hazards are created by effluents from chemical, wood-working, textile, paper, and cement industries, from mines, and from ore-concentrating, smelting, and various metallurgical works [Prahalad & Seenayya]. An important source of zinc in water supplies may be zinc-plated pipes [Alam & Sadiq]; from such pipes, 1.2 to 2.9 mg Zn were found to be washed out daily by hot water per square decimeter of surface area.

TOXICITY

Subthreshold concentrations of **zinc chloride** and **zinc oxide** in water were 5 mg/L and 30 mg/L, respectively, for effects on its organoleptic properties and

5 mg/L for effects on its sanitary condition. The highest concentration at which these two compounds did not affect biochemical processes in a water body however long they remained there was found to be 10 mg/L for the chloride and 5 mg/L for the oxide [Bespamyatnov & Krotov].

Soil microorganisms and plants

The growth of most microorganisms was significantly retarded at zinc levels of 100–200 µg/kg soil; fungi were more resistant. The adverse effects of zinc on soil microorganisms and microfauna can result in reduced fertility of the soil; in a temperate region, for example, the yield of cereal crops decreased by 20–30%, that of beet by 35%, that of bean by 40%, and that of potato by 47%. Zinc levels reducing the yield or height of plants by 5–10% are regarded as toxic and have been reported to range from 435 to 725 ppm for oats, 210 to 290 ppm for clover, and 240 to 275 ppm for beet [Bolshakov et al.].

It has been stated that zinc concentrations in soils below 30 mg/kg may be regarded as too low for most plants, those between 30 and 70 mg/kg as normal, and those above 70 mg/kg as excessive [Kovalsky].

Aquatic organisms

Zinc compounds can cause severe damage to the gills of fish. Lethally poisoned fish first become excited and begin to breathe rapidly and then die from asphyxia caused by destruction of the respiratory epithelium. They can recover on transfer to clean water at the narcosis stage of intoxication. The toxicity of zinc is enhanced by copper and nickel ions.

Eight-hour exposure to a **zinc** concentration of 15 mg/L was lethal within 8 h for all fish species that were tested. Roach did not tolerate concentrations in excess of 1 mg/L. For rainbow trout, the LC_{50} of zinc in hard water was determined at 4.76 mg/L, while in soft water zinc was obviously toxic in a concentration as low as 0.15 mg/L. **Zinc chloride** was toxic for snails and crustaceans at 0.2 mg/L [Metelev et al.].

For sharks, the 24-hour and 48-hour LC_{50}s of zinc in seawater were 175 mg/L and 80 mg/L, respectively. A high capacity for accumulating zinc is possessed by oysters; rats fed oysters with high zinc content showed clear signs of poisoning [Thrower & Olly].

Animals and man

General remarks

Zinc and many of its compounds are very toxic and can cause both acute and chronic poisoning with a wide variety of clinical manifestations. At the basis of many of these lie competitive relationships of zinc with other metals, notably calcium, lead, cadmium, and copper. For example, serum levels of calcium significantly below normal were recorded in zinc smelter workers and zinc oxide packers, while excessive zinc intakes by experimental animals led to lowered calcium levels not only in the blood but also in the bones, and to impaired phosphorus uptake, resulting in osteoporosis [Roshchin]. The antagonism between zinc and calcium is believed to occur in part at the cell membrane level. Even relatively low zinc doses have inhibitory effects on formed elements of the blood, presumably as a result of interference with the uptake and retention of copper by the cells. Lead poisoning has been shown to be accompanied by greatly increased urinary excretion of zinc [Burkhanov & Shchelkunova]. The toxicity of zinc oxide is thought to be largely determined by its catalytic activity [Beeckmans & Brown].

Zinc may present mutagenic and carcinogenic hazards [Roshchin et al.]. The mutagenic activity is probably associated with the ability of zinc ions to insert into the active sites of DNA and RNA polymerases—enzymic proteins that are essential for the normal replication and transcription of DNA. One indication of zinc carcinogenicity has been provided by the finding that many erythrocytes with altered membranes were present in both inbred and non-inbred mice following zinc administration [Martinson et al.]. Another indication is that the incidence of cancer in a human population has been shown to correlate with the content of total zinc in the soil [Dubikovsky]. Zinc also displayed gonadotoxic activity by reducing the motility of spermatozoa and their ability to penetrate into the ovum.

Acute/single exposure

Animals

For rats, the oral and intravenous LD_{50}s of zinc chloride were 750 mg/kg and 25 mg/kg, respectively; the intraperitoneal LD_{50} of zinc sulfate was 23 mg/kg. The sulfide and oxide of zinc were practically nontoxic because of their poor solubility. In mice, the oral threshold dose of zinc oxide for effects on body-weight and blood sulfhydryls was 7.6 mg/kg [Rabotnikova].

Zinc salts of strong mineral acids act as astringents and in higher concentrations as corrosives. Ingestion of larger doses results in serious damage to the upper alimentary tract and severe shock symptoms. Lower doses result in nausea, emesis, stomach cramps, diarrhea, and fever [Ohnesorge & Wilhelm].

A short single inhalation exposure of cats to zinc dust led to pulmonary edema and hemorrhage and to the accumulation of leukocytes and macrophages in the bronchioles and alveoli. Findings in cats inhaling zinc dust in subacute tests included nodules of epithelial cells in the lungs, increased zinc content and cirrhosis in the pancreas, presence of sugar in the urine, and proliferation (or, in some animals, degeneration) of beta cells in the islets of Langerhans. Rabbits with experimentally induced zinc fume fever developed anemia. Cats exposed to zinc oxide fumes at concentrations of 110 to 600 mg/m^3 for 15 min became lethargic and had lowered body temperature, whereas cats exposed for 45 min exhibited complete prostration, tremor, respiratory distress, hypothermia, and reduced erythrocyte counts. Cats sacrificed immediately after exposure did not have pronounced changes in the lungs; those sacrificed after 24 h had congested bronchi filled with exudate and areas of induration with large numbers of leukocytes in the alveoli, while those sacrificed after 4 days had developed pneumonia. Rats and rabbits were less sensitive than cats [Beeckmans & Brown]. Severe pulmonary edema was noted in guinea-pigs that had been inhaling zinc oxide at 25 mg/m³ for 3 h [Conner *et al.*].

Inhalation exposure of guinea-pigs to a **zinc sulfate** aerosol at 1.1 mg/m³ for 1 h produced irritation of the upper airways.

In rats administered 40 mg of **zinc** intratracheally, considerable changes in the bronchi, hyperplasia of lymphoid elements, intensive connective tissue growth, and pulmonary emphysema were observed after 8 months. Observations made in rats 18 to 24 months after a single administration of 50 mg **zinc oxide** by this route demonstrated bronchial deformation, hyperplasia and sclerosis of lymphatic follicles, and peribronchial pneumonia [Mogilevskaya].

When rats were administered a finely divided **zinc dust** into the trachea in a dose of 25 or 50 mg on a single occasion (or in doses of 2 or 5 mg repeatedly), sarcomas in the lungs and/or testicular tumors developed in 15% of the animals 18 to 24 months later [Dvizhkov & Potapova]. These authors also found that 1 mg of zinc dust added to 25 mg of silica dust enhanced fibrogenicity of the latter.

Man

In industrial settings, considerable hazards are presented by fumes containing finely divided particles of **zinc oxide** which, when inhaled in high concentration, cause an acute illness of short duration known as zinc fume fever (or zinc chills) — a metal fume fever that tends to recur on repeated exposure and is more common than those caused by fumes of other metals. Already at work, a sweetish metallic taste may appear in the mouth of an exposed person, followed a few hours later by loss of appetite and the onset of thirst, fatigue, tightness and constrictive pains in the chest, drowsiness, and dry cough. This period, which usually lasts 1 to 5 h depending on the severity of poisoning, is succeeded by severe chills 1 to 1.5 h in duration followed by a body temperature elevation to 37–38°C (sometimes to 40°C or more) persisting for several hours and accompanied by pupil dilatation and hyperemia of the conjunctivae, pharynx, and face; sugar and, often, hematoporphyrin and urobilin appear in the urine (which may also contain increased zinc and copper levels), and the blood content of sugar rises; liver enlargement has also been noted occasionally.

The condition usually clears within 24 h but may last 2–3 days or longer. Its manifestations depend on the concentration of zinc oxide fume and individual susceptibility. Zinc fever occurred in a photographer who had been using a zinc oxide-containing paint to color portraits [Pazdarowa].

Recurrent episodes of zinc fume fever weaken the organism and have been shown to activate the tuberculous process and increase the susceptibility to other respiratory diseases.

Metal fume fever is thought to result from the entry into the blood circulation of submicroscopic tetragonal zinc crystals whose size and shape make it easy for them to penetrate the cell and come in contact with cellular proteins to cause their denaturation.

In cases of lethal poisoning, edema of the interstitial pulmonary tissue and destruction and metaplasia of the alveolar epithelium were found at autopsy [Knaan et al.].

A survey of workers employed in zinc dust production showed that most of them had experienced zinc fume fever in the past. Shortly after an attack of fever, many workers had painful and edematous joints and hemorrhagic eruptions on the feet [Mogilevskaya]. Acute intoxications with typical manifestations of zinc fume fever have frequently resulted from such operations as electric welding and gas cutting of zinc-containing metallic structures. Workers exposed to welding fumes containing fine zinc particles in amounts of 18 to 58 mg/m^3 (depending on the thickness of the zinc coating in the structures being welded) had greatly elevated zinc and copper levels in their urine and often suffered from dysuria [Vorontsova et al.]. In electric welders a number of diseases have been detected, including chronic catarrhal inflammations of the upper airways and the digestive tract, conjunctivitis, dermatitis, anemia, bilirubinemia, and gastritis due to hypochlorhydria.

Inhalation of **zinc chloride** fume for 5 to 30 min caused paroxysmal coughing, nausea and occasionally vomiting; 1 to 24 h after the cessation of exposure, breathlessness and fever appeared. Inflammation and edema in the lungs are possible. Complications are most likely to arise in the next 5–12 days. This syndrome has been designated 'acute chemical pneumopathy' [Mogosh]. Postmortem examination of those who died on days 6 to 11 post-exposure showed necrotizing tracheitis, bronchitis, confluent bronchopneumonia with thrombosis of small vessels, and obliterating bronchiolitis.

Ingestion of **zinc sulfate** resulted in nausea, vomiting, and diarrhea (in several cases with blood), and abdominal cramps. The emetic dose of the sulfate was 1–2 g (225–450 mg of zinc). In cases of lethal poisoning, severe lesions of the gastric mucous membrane, including necrosis in some instances, and evidence of disrupted cerebral circulation were observed at autopsy. An outbreak of mass poisoning with zinc sulfate was described, resulting from the consumption of food that had been prepared and stored in galvanized containers where the sulfate formed under the action of acids contained in the food [Brown et al.]. Numerous instances of poisoning with foods stored or prepared in zinc-plated containers have been cited by Pollak et al., including poisoning with kvas [a national Russian beverage] (zinc concentration 187.6 mg%), milk (31.3 mg%), tomato juice (89 mg%), and porridge (650 mg%).

Zinc chloride can cause severe damage to the perioral skin and buccal and gastroenteric mucous membranes. Severe oral poisoning was marked by mucosal burns, gastric colic, vomiting (with an admixture of blood), bloody diarrhea, and strong excitation; these symptoms were followed later by jaundice, pain in the extremities, anuria, and elevation of residual nitrogen (up to 280 mg%). Esophageal stenosis also occurred in some cases. In fatal cases, autopsy showed evidence of damage to the liver, kidney, and myocardium. A case of death from tracheal hemorrhage that arose suddenly one month after poisoning was described [Yudov].

Chronic and subchronic exposures

Animals

In rats exposed to a **zinc oxide** concentration of 15 mg/m^3 through inhalation for 1.5 or 8 h/day up to 84 days, alterations correlating with the magnitude and duration of exposure were observed in several parameters of external respiration and in the histologic appearance of the lungs. Young rats exposed to a zinc oxide condensation aerosol at 5 mg/m^3 for 5 h/day over 4 months showed slowed weight gains, anemia, abnormalities in carbohydrate metabolism and cardiac activity, and alterations in blood serum cholinesterase and aldolase activities; exposure to 0.5 mg/m^3 did not result in significant pathologic changes after 4 months. Some rats died from purulent bronchitis or pneumonia when the exposure period was prolonged up to 12 months [Vorontsova et al.].

Weanling pigs fed diets containing 0.1% to 0.8% of **zinc carbonate** showed depressed growth rates and food intakes and developed arthritis and extensive hemorrhages in the axillary region; some animals had died before termination of the 42-day study period and were found at autopsy to have gastritis, enteritis, and hemorrhages to lymph nodes, internal organs, and the brain ventricles; the dose of about 0.05% did not produce toxic effects [Brink et al.]. In ducks maintained on diets containing zinc carbonate in doses of 3000, 6000, or 9000 ppm, retardation of body-weight gain proportional to the dose, diarrhea, paralysis of the limbs, reduced hemoglobin levels in the blood, and a high mortality rate were observed, along with increases in zinc levels by factors 15–20 in the pancreas, 6–9 in the liver, 2 in the pectoral muscles, and 2–6 in the genitalia. An increased incidence of miscarriages and marked copper deficiency occurred in pregnant sheep fed zinc in doses of 30, 150, or 750 mg/kg of the diet [Campbell & Mills].

Other effects reported from studies on mice, rats, and rabbits administered zinc or its compounds by various routes for periods of 1 to 5 months included the following:

– altered activity of many enzymes, in particular cytochrome oxidase, alkaline phosphatase, succinate dehydrogenase, lactate dehydrogenase, glucose-6-phosphate dehydrogenase, and duodenal enterokinase [Roshchin *et al.*; Yamaguchi *et al.*];
– effects on carbohydrate and mineral metabolism: decreases of calcium and potassium in bones and of calcium in blood, elevation of alkaline phosphatase activity in blood and of zinc content in bone, hypoglycemia, decrease of glycogen in liver, inhibited insulin secretion, and reduced glucose tolerance;
– gonadotoxic effects: depressed spermatogenesis, decreased osmotic resistance of spermatozoa and shortened time of their motility, altered contents of RNA and sulfhydryl groups in seminiferous tubules, inhibited ovarian function, and morphologic changes in female genitalia [Bonashevskaya *et al.*];
– mutagenic and carcinogenic effects manifested, respectively, in increased incidence of chromosomal aberrations in bone marrow cells of mice and in the appearance of carcinomas in both inbred and non-inbred mice consuming drinking water with zinc levels of 10–20 g/L [Grushko, 1979]. Evidence for teratogenicity of **zinc chloride** in mice has also been obtained [Shepard].

Farm animals are regarded as quite zinc tolerant, but horses, sheep and cattle have been intoxicated by grazing in the vicinity of a zinc smelter. In ruminants, typical clinical manifestations are loss of condition, diarrhea, subcutaneous edema, profound weakness, and jaundice [Ohnesorge & Wilhelm].

Ohnesorge & Wilhelm also showed that rats could tolerate 0.55–1.0 g Zn per kg diet without showing any visible symptoms, but above 4 g Zn per kg diet there was a reduction in litter size and in the number of offspring born alive. At levels of 1000 mg Zn per kg diet pigs showed a reduction in body weight and above 2000 mg Zn per kg diet they developed hemorrhages in the intestine and brain, had swollen joints, and some of them even died. Dogs and cats could tolerate 175–1000 mg Zn per day in their diet without any visible symptoms.

Man

Workers exposed to **zinc** dust frequently complained of irritability, insomnia, memory deterioration, sweating at night, reduced hearing, noises in the ears, and gastrointestinal troubles. Many were found to have developed hypochromic anemia and subatrophic catarrhs of the upper airways after 2–3 years of occupational exposure; roentgenographic findings included accentuated lung markings, emphysema, and initial signs of pneumoconiosis [Mogilevskaya].

Zinc has been shown to produce cumulative toxic effects even when its levels in the air were fairly low [Jeszke *et al.*].

Hair samples from workers of a galvanizing department contained **zinc** in concentrations severalfold higher than those from control subjects, mean values being 25.5 mg% in solderers, 22.9 mg% in painters, and 30.04 mg% in galvanizers as compared to 7.76 mg% in the controls; hair samples from those who complained of weakness and poor sleep had even higher zinc concentrations (mean, 57.5 mg%). Increased mortality from lung cancer was observed among Swedish zinc miners [Axelson & Sundell].

Frequent findings in workers employed in **zinc oxide** production were hyperglycemia, hypocholesterolemia, elevated urobilin and porphyrins in the urine, impaired pancretic and hepatic functions, and pulmonary fibrosis [Dzukayev & Kochetkova]. After work for 1 year or more in an atmosphere polluted with zinc oxide dust, even those who had used respirators had altered levels of polysaccharides, peroxidases, and acid phosphatases in their blood cells; after 10 years or so, anemia was observed. Persons chronically exposed to zinc oxide complained of indigestion [Hamdi].

Individuals handling **zinc-containing fertilizers** complained of general weakness, dryness in the nose, cough, and noises in the ears, and on examination were often found to have chronically inflamed mucosas in the upper airways [Lienko]. Women engaged in the production of **zinc white** (ZnO) and exposed to zinc concentrations of 2.4 to 7.1 mg/m^3 for 5 years, had lowered hemoglobin and iron levels in the blood and serum, respectively, together with elevated transferrin and erythropoietin levels. Occupational exposure to zinc chloride may produce mucosal lesions in the upper respiratory tract; in some workers, perforation of the nasal septum and gastrointestinal disorders were also noted (after 1 year of work), as well as gastric or duodenal ulcers (after 5–20 years).

In a zinc factory, a range of adverse factors may act on employees apart from zinc compounds themselves. For example, workers in the leaching department were exposed to such substances as sulfur trioxide, sulfur dioxide, hydrogen fluoride, hydrogen arsenide, and chlorine, while those in the electrolysis department were exposed, in addition, to a fairly strong magnetic field (100–200 oersteds) and to explosive claps of a powerful aperiodic noise (125–155 decibels). Workers employed in either of these departments for a long time complained of headache, numbness in the fingers, sweating, noise in the ears, deterioration of hearing and memory, hoarseness, and nosebleeds. Medical examinations revealed a number of abnormalities, the more common of which were chronic catarrhs of the upper respiratory tract, erosions and ulcerations in the nasal mucosa, cochlear neuritis, impaired innervation of the brain, hyposmia or anosmia, and signs of the asthenic–vegetative syndrome. Nearly a half of the workers exhibited narrowing of the visual fields (which was of cortical origin) and decreased immunity. The teeth of many workers were seriously damaged by caries and erosions and had a dull enamel, often with a dark-brown pigmentation. Roentgenographic findings included intensified lung markings,

Zinc orthoarsenite and hydroorthoarsenate

Animals

For rats, the oral LD_{50} of the orthoarsenite was 1503 mg/kg and that of the hydroorthoarsenate, 1020 mg/kg; for mice, the oral LD_{50} of the latter compound was 601 mg/kg. The main manifestations of poisoning were hypodynamia, dyspnea, diarrhea, and elevated pyruvic acid levels and depressed concentrations of sulfhydryl groups in the blood; autopsy showed hemorrhagic areas along the digestive tract. The threshold oral (intragastric) dose for acute irritant effects in rats was 14 mg/kg with the orthoarsenite and 54 mg/kg with the hydroorthoarsenate. On repeated oral administration at 27 mg/kg and 102 mg/kg, respectively, these compounds caused vascular disorders and impaired central nervous system functions, thermoregulation, and porphyrin metabolism; autopsy findings included ulcers on the gastrointestinal mucosa, hepatitis, and a relatively high content of arsenic in the liver [Davydova *et al.*].

Zinc phosphates (orthophosphate and hydroorthophosphate)

Animals

In rats administered either of these phosphates intratracheally in an amount of 50 mg, inflammation and moderate reticular sclerosis were present in the lungs 3 months later, but disappeared within the next 12 months. Their oral doses up to 10 g/kg were not lethal to any rats. The intraperitoneal LD_{50} was determined at 55 mg/kg for the orthophosphate and 600 mg/kg for the hydroorthophosphate [Spiridonova & Shabalina].

Zinc phosphide

The high oral toxicity of zinc phosphide (Zn_3P_2) is due to the phosphine PH_3 that forms in the stomach as a result of the phosphide reacting with the hydrochloric acid of the gastric juice. Phosphine shows a high degree of neurotoxicity. Some of the phosphide is oxidized in the blood, converting to phosphoric acid, and some is eliminated unchanged through the lungs; it is not usually found in the blood and organs of lethally poisoned animals or humans. Zinc phosphide is toxic on entering the body by any route but especially when ingested [Pollak *et al.*].

Animals

Lethal oral doses of zinc phosphide were 50–100 mg/kg for mice, around 40 mg/kg for guinea-pigs, and 20–25 mg/kg for rabbits and dogs. Lethally poisoned rabbits and dogs had a rapid and weak pulse, breathed frequently, were thirsty, and exhibited general inhibition followed by excitation before dying; the death of many animals was preceded by convulsions. Protein, sugar, and acetone were present in the urine. On autopsy, the brain, lungs, and tracheal and small intestinal mucosas were hyperemic, and venous congestion was seen in the kidneys, liver, and spleen [Nesterov].

Man

Reported manifestations of severe oral poisoning by zinc phosphide included thirst, nausea, gastric pains, vomiting and diarrhea, tightness in the chest, dyspnea, pain in the occipital region, a feeling of fear, pupil dilatation, convulsions, and coma. On examination of the victims, impaired cardiac activity, profound acidosis, tetanus due to hypocalcemia, anuria, and signs of renal and hepatic insufficiency were observed. Death usually occurred 7 to 60 h post-ingestion in the presence of asphyxia and gross respiratory and circulatory abnormalities. Post-mortem examination revealed hyperemia and edema in the brain and lungs, large hemorrhagic areas in the lungs and pancreas, and degenerative changes in all parenchymal organs [Stephenson]. The lethal dose for an adult human is approximately 25 mg [Pollak *et al.*]. A case of poisoning with the phosphide applied to a wound on the tip of a knife has been described [Blisnakov & Iskrov]. The presence of any zinc phosphide residues in human foods is not permitted.

Zinc selenide and zinc sulfide

Animals
The threshold concentration of zinc selenide for acute effects on body-weight gains and rectal temperature in rats with inhalation exposure was determined at 44.5 mg/m^3. The toxic effects of these compounds on intratracheal administration to rats were confined to the lungs. The selenide was not lethal to any rats in oral doses up to 8 g/kg, nor was it absorbed through their skin.

Man
Workers occupationally exposed to zinc selenide or sulfide complained of headache, rapid onset of fatigue, dizziness, dryness in the mouth, liquid stools, pain in the liver and joints, and excessive hair loss. In some work areas, the formation of hydrogen selenide and hydrogen sulfide was found to occur [Shabalina & Spiridonova; Spiridonova *et al.*].

Effects of local application
Zinc chloride produces severe irritative and cauterizing effects on human skin and mucous membranes. Skin contact results in the appearance of small round ulcers with an area of mummified tissue in the center. It has weak sensitizing properties although cases of allergic dermatitis involving extensive body areas have been described. Ulcerations, mainly on the backs of the hands, were caused by dry **zinc sulfate** and its concentrated solutions.

There is evidence that zinc salts, particularly those insoluble in water, can be absorbed through undamaged skin [Burkhanov & Mirzoyan]. The orthoarsenite and hydroorthoarsenate were found to be absorbed through undamaged skin without irritating it [Davydova *et al.*].

Zinc orthophosphate and **hydroorthophosphate** did not irritate the skin of rats, but did penetrate through the intact skin of their tails when applied for 4 h daily over 10 days in the form of a 70% aqueous suspension; in these rats, reductions in the summated threshold index and in erythrocyte and leukocyte counts were recorded [Spiridonova & Shabalina].

ABSORPTION, DISTRIBUTION, AND ELIMINATION IN MAN

The daily requirement for zinc is estimated at 3 mg in infants up to 4 months in age, 5 mg in those aged 5–12 months, 10 mg in children aged 1 to 10 years, and 15 mg in adults (20–25 mg in pregnant and lactating women). On average, about 0.1 mg of zinc enters the body in inspired air, 0.5 mg in the drinking water, and 12 mg in the diet. The major sources of dietary zinc are of animal origin. Among vegetables, the main sources are likely to be buckwheat, rye, wheat, and corn [Skurikhin].

Zinc is an essential trace element that exerts its physiological activities primarily by virtue of being bound to many enzymic proteins. More than 80 zinc-proteins have been identified, most of which are metalloenzymes. Zinc is a constituent of dehydrogenases, aldolases, DNA and RNA polymerases, peptidases, phosphatases, and isomerases, which is an indication of its active participation in the metabolism of nucleic acids, proteins, fats, and carbohydrates. It plays an important physiological role in bone calcification, as is evidenced by its high levels at calcification sites during the development of osteons. Zinc deficiency results in impaired ossification, while its excess is conducive to osteoporosis [Khakimov & Tatarskaya; Yamaguchi *et al.*]. Other dangerous manifistations of zinc deficiency are atrophy of lymphoid organs, reduced secretion of thymic hormone, and impaired function of helper T cells.

The adult human body contains 1 to 2.5 g of zinc, of which about 30% is deposited in bones and 60% in muscles. The ingested zinc is absorbed from the duodenum and the upper part of the small intestine. In the liver, part of the zinc is deposited and part enters metal–protein complexes, including metalloenzymes. In the bloodstream, zinc is transported in protein complexes, with only a small proportion being present in the ionic form. In whole blood, its concentrations commonly range from 700 to 800 µg%, of which 75–85% occurs in erythrocytes, 12–22% in plasma, and about 3% in leukocytes. Its blood concentration was found to decrease by an average of 33% from 8 o'clock a.m. to 7 p.m. [McMillan & Rowe].

Zinc was reported to be distributed in the body as follows (µg/g) : adrenals, 6; bones, 66; brain, 13; gastrointestinal tract, 21; heart, 27; kidneys, 37; liver, 38;

lymph nodes, 14; muscles, 48; ovaries, 12; prostate, 87; skin, 6; sperm, 125; hair, up to 175 [Roshchin et al.]. The body burden of zinc increases with age.

Zinc is mainly excreted via the intestine ($\sim$10 mg/day); it is also excreted in urine (0.3–0.6 mg), sweat (2–3 mg in hot weather), and milk; its concentration in human milk averages 1.63 mg/L (3–5 mg/L in cow's milk).

Table l. Exposure limits for zinc compounds

	Workroom air			Atmospheric air			Water sources		Soil	
	MAC_{wz} (mg/m^3)	TSEL $(mg/m^3$	Aggreg. state	MAC_{ad} (mg/m^3)	$TSEL_{aa}$ (mg/m^3)	MAC_w (mg/L)	MAC_{wf} (mg/L)	MAC_s (mg/kg)	Hazard class
Zinc	—	—	—	—	—	1.0^a	0.01	23.0^b	—
carbonate	—	2.0	Aerosol	—	—	—	—	—	3
chloride	2.0	—	Aerosol	—	0.005^c	—	—	—	—
nitrate	—	0.5	Aerosol	—	0.005^c	—	—	—	—
oxide	0.5	—	Aerosol	0.05	—	—	—	—	2
phosphates	—	0.5	Aerosol	—	—	—	—	—	—
phosphide	0.1	—	Aerosol	—	—	—	—	—	2
selenide	—	2.0	Aerosol	—	—	—	—	—	3
sulfate	5.0	—	Aerosol	—	0.005^c	—	—	—	3

[a] Based on the general sanitary index of harmfulness that characterizes the impact of a compound on the self-purification capacity of the soil and on the soil microbiocenosis.
[b] Based on the translocation index of harmfulness that characterizes the transfer of a compound from the arable soil layer to the green mass and fruits of the plants through the root system.
[c] As zinc.

HYGIENIC STANDARDS

Exposure limits adopted for zinc compounds in the USSR are shown in Table 1. In the USA, TLV-TWA values have been set for the following zinc compounds (mg/m^3): chromate at 0.01 (as Cr), classified as a confirmed human carcinogen; dichloride at 1.0 (fume); oxide at 5.0 (fume) and 1.0 (total dust containing no asbestos and <1% crystalline silica); and stearate at 20.0. TLV-STEL values of 2.0, 10.0, and 20.0 have been set for zinc dichloride (fume), zinc oxide (fume), and zinc stearate, respectively.

The highest permissible levels established in the USSR for zinc in human foods are indicated in Table 2.

Table 2. MACs for zinc in foods (mg/kg)

Cereal grains, groats, and flour	50.0
Starch	30.0
Bran	130.0
Candies, pastries, and pectin	50.0
Sugar	3.0
Salt	10.0
Milk, sour milk products and butter	5.0
Condensed milk, canned	15.0
Margarines and cooking fats	10.0
Animal fats and vegetable oils	5.0
Cheese, curds, and casein	50.0
Vegetables, fresh or canned	10.0
Fruits, fresh or canned	10.0
Berries, fresh or canned	10.0
Meat and poultry, fresh or canned	100.0
Fish, fresh or canned	40.0
Eggs	50.0
Mineral waters	5.0
Beverages based on infusions or essences	10.0
Beer, wine, and other alcoholic beverages	10.0
Infant foods:	
milk-based	5.0
grain- and milk-based	10.0
vegetable- or fruit-based	10.0
meat- or poultry-based	50.0
fish-based	30.0

METHODS OF DETERMINATION

One method for determining zinc and its compounds in *air* is based on the formation of a complex upon interaction of the Zn^{2+} ion with the hydrochloride of diantipyrylmethylmethane in the presence of ammonium thiocyanate (sensitivity 1 µg in the volume analyzed) [*Specifications for Methods of Determining Noxious Substances in Air*]. Another method is based on the reduction of zinc ions on a mercury-dropping electrode in the presence of acetic acid and ammonium acetate solutions followed by the determination of zinc with alternating-current polarography (detection limit 0.08 µg/L) [Nekrasova & Tashi]. A photometric method for determining zinc in *water* is based on the formation of its red-colored compounds with dithizone and extraction of the zinc dithizonate into a carbon tetrachloride layer at pH 4.5–4.8 followed by photometry (sensitivity 0.005 mg/L) [Novikov *et al.*]. In *foods*, zinc can be determined by a chromatographic procedure based on the formation of a complex between zinc cations (at pH 4.5–5.0) and sodium diethyldithiocarbamate (sensitivity 0.005 mg/L) [Katayeva]. For determination in *plants*, an x-ray fluorescence technique may be used [Sorokin & Bolshakov]. Zinc in *biological materials* can be measured by colorimetry using dithizone or other reagents forming complexes with zinc. In *seminal fluid*, zinc can be determined colorimetrically by its reaction with (4-pyridylazo)-resorcinol (sensitivity 2 mg/L) [Lampugnali & Maccheroni]. A fluorimetric method for zinc determination with 8-hydroxyquinoline has been described [Gadaskina *et al.*]. Other analytical techniques in use are emission spectrography, neutron activation, and particularly atomic absorption spectrometry which has become the most widely used method for the determination of zinc in recent years (see also Bertholf and Ohnesorge & Wilhelm).

MEASURES TO CONTROL EXPOSURE

At the workplace

Technical measures to protect workers from harmful exposure to zinc and its compounds should be mainly directed at minimizing air pollution by their aerosols, which can be best achieved through mechanization and enclosure of the processes generating these. In zinc production, leaching of the zinc calcine and thickening of the zinc pulp should be remotely controlled, and there must be a ban on manual performance of such operations as charging zinc dust into the agitators, disassembling and reassembling the filter presses, stripping zinc from the cathodes, removing slag from the lead anodes and electrolysis baths, and discharging sludge from the apparatuses and removing it from their surfaces.

Adequate general ventilation of the work premises and local exhaust ventilation of the equipment and permanent work posts is essential. All workers must be instructed in technical safety and personal hygiene measures. Consumption and storage of food in the working areas should be prohibited. Work clothes should be laundered on a centralized basis and must not be allowed to be taken home. Any dust on the clothes should be removed using appropriate devices.

Rest rooms with adequate relaxation facilities should be provided at the plant. All zinc workers are subject to preplacement and periodic medical examinations.

For *personal protection*, respirators or gas masks should be worn where effective air pollution control cannot be assured. For hand protection against zinc and its compounds, a barrier cream or ointment should be applied before starting work, followed by the application of a skin-conditioning cream after washing at the end of the work shift.

Protection of the general environment

The essential measure for preventing the atmospheric air from pollution with zinc and its compounds is effective cleaning of gaseous and particulate emissions from thermal electric power stations, zinc smelters, and other installations where high-temperature industrial processes involving zinc are used. At each industrial establishment, the operation of gas-purifying and dust-collecting equipment should be constantly monitored by local maintenance personnel as well as by workers of special services.

In areas where zinc-containing industrial wastewaters are used for irrigation purposes, monitoring of the soil is necessary to see that the prescribed permissible levels of zinc and other ingredients of the wastewaters are not exceeded.

FIRST AID IN POISONING

Persons who have developed zinc fume fever or have been acutely poisoned by inhaling fume or dust of zinc, its oxide or its chloride should be removed from further exposure into fresh air (oxygen may be required) and then given alkaline inhalations and an intravenous injection of 5% glucose, dextran, and physiologic saline to normalize venous pressure. In severe cases, penicillamine and then antibiotics and corticosteroids should be given [Ludewig & Lohs; Mogosh].

After the ingestion of a soluble zinc salt, the stomach should be washed out promptly with a 0.5% tannin solution followed by the intake of milk with beaten-up egg whites and activated charcoal. As the antidote, unithiol (sodium 2,3-dimercapto-1-propane sulfonate) injected subcutaneously is effective. Antishock therapy is carried out if indicated [Artamonova; Ludewig & Lohs].

REFERENCES

Adamu, C. A. et al. (1989) Env. Poll., 56, 113–126.
Alam, J. B. & Sadiq, M. (1989) Env. Poll., 57, 167–178.
Artamonova, V. G. (1981) Neotlozhnaya pomoshch pri professionalnykh intoksikatsiyakh [Emergency Care in Occupational Intoxications]. Meditsina, Leningrad (in Russian).
Axelson, G. & Sundell, L. (1978) Scand. J. Work. Environ. Health, No. 1, 46–52.
Beeckmans, J. & Brown, J. (1963) Arch. Environ. Health, 7, 346–350.
Bertholf, R. L. (1988) Chapter 71 Zinc in: Seiler, H. G. et al. (eds) Handbook on Toxicity of Inorganic Compounds. Marcel Dekker, New York, pp. 787–800.
Bespamyatnov, G. P. & Krotov, Yu. A. (1985) Predelnodopustimye kontsentratsii khimicheskikh veshchestv v okruzhayushchei srede [Maximum Allowable Concentrations of Chemicals in the Environment]. Khimiya, Leningrad (in Russian).
Blisnakov, C. & Iskrov, G. (1961) Folia Med., 2, 73–77.
Bolshakov, V. A. et al. (1978) Zagryaznenie pochv i rastitelnosti tyazhelymi metallami [Soil and Plant Pollution by Heavy Metals]. Moscow (in Russian).
Bonashevskaya, T. I. et al. (1978) In: Materialy 1-oi respublikanskoi konferenstii gigienistov i sanitarnykh vrachei [Proceedings of the 1st Republican Conference of Hygienists and Sanitarians]. Ashkhabad, pp. 138–141 (in Russian).
Brink, M. et al. (1959) J. Anim. Sci., 18, 836–842.
Brown, A. et al. (1964) Arch. Environ. Health, 8, 657–660.

Burkhanov, A. I. & Mirzoyan, I. M. (1982) *Vestn. Dermatol.*, No. **5**, 26—29 (in Russian).

Burkhanov, A. I. & Shchelkunova, O. V. (1983) *Gig. San.*, No. **7**, 81—82 (in Russian).

Campbell, J. & Mills, C. (1979) *Environ. Res.*, No. **1**, 1—13.

Conner, M. *et al.* (1982) *Toxicol. Appl. Pharmacol.*, **20**, 434—442.

Davydova, V. I. *et al.* (1968) *Khimiya v selskom khozyaistve* [Chemistry in Agriculture], No. **1**, 81 (in Russian).

Dmitriev, M. T. & Tarasova, L. N. (1985) *Gigiyena: Ekspress-informatsiya* [Hygiene: Current Awareness Information], No. **5**, 20 (in Russian).

Dobrovolsky, V. V. (1983) *Geografiya mikroelementov. Globalnoye rasseyaniye* [Geography of Trace Elements. Global Dispersion]. Mysl, Moscow (in Russian).

Doncheva, A. V. *et al.* (1982) *Khimiya v selskom khozyaistve* [Chemistry in Agriculture], No. **3**, 8—10 (in Russian).

Dubikovsky, G. P. (1982) *Khimiya v selskom khozyaistve* [Chemistry in Agriculture], No. **3**, 33—34 (in Russian).

Dvizhkov, P. P. & Potapova, I. N. (1968) *Gig. Truda Prof. Zabol.*, No. **1**, 23—27 (in Russian).

Dzukayev, Z. E. & Kochetkova, T. L. (1970) *Gig. Truda Prof. Zabol.*, No. **10**, 38—41 (in Russian).

Gadaskina, I. D. *et al.* (1975) *Opredeleniye promyshlennykh neorganicheskikh yadov v organizme* [Determination of Inorganic Industrial Poisons in the Organism]. Meditsina, Leningrad (in Russian).

Garmash, G. A. (1983). *Khimiya v selskom khozyaistve* [Chemistry in Agriculture], No. **10**, 45—48 (in Russian).

Gemke, G. R. & Brevdo, V. A. (1969) *Gig. Truda Prof. Zabol.*, No. **4**, 44—45 (in Russian).

Goncharuk, E. I. (1977) *Sanitarnaya okhrana pochvy ot zagryazneniya khimicheskimi veshchestvami* [Sanitary Measures to Protect Soil from Chemical Pollution]. Zdorovya, Kiev (in Russian).

Grushko, Ya. M. (1972) *Yadovitye metally i ikh neorganicheskiye soyedineniya v promyshlennykh stochnykh vodakh* [Poisonous Metals and their Inorganic Compounds in Industrial Wastewaters]. Khimiya, Leningrad (in Russian).

Grushko, Ya. M. (1979) *Vrednye neorganicheskiye soyedineniya v promyshlennykh stochnykh vodakh* [Harmful Inorganic Compounds in Industrial Wastewaters]. Khimiya, Leningrad (in Russian).

Hamdi, E. (1969) *Brit. J. Ind. Med.*, **26**, 126—134.

Hewitt, C. N. & Cardy, G. B. B. (1990) *Env. Poll.*, **63**, 129—136.

Jeszke, M. *et al.* (1984) *Med. Pr.*, **35**, 289—292.

Katayeva, S. E. (1982) *Gig. San.*, No. **5**, 60—61 (in Russian).

Khakimov, Kh. Kh. & Tatarskaya, A. Z. (1985) *Periodicheskaya sistema i biologicheskaya rol elementov* [The Periodic System and the Biological Role of Elements]. Meditsina UzSSR, Tashkent (in Russian).

Knaan, G. *et al.* (1970) *Harefuah*, No. **11**, 536.

Kovalsky, V. V. (1974) *Geokhimicheskaya ekologiya* [Geochemical Ecology]. Nauka, Moscow (in Russian).

Lag, J. (1984) *Ambio*, **13**, 287—289.

Lampugnali, L. & Maccheroni, M. (1984) *Clin. Chem.*, **30**, 1366—1368.

Leschber, . (1983) *Zbl. Bact. Hyg.*, No. **178**, 186—193.

Lienko, I. D. (1968) *Gig. San.*, No. **1**, 51—54 (in Russian).

Ludewig, R. & Lohs, K. (1981) *Akute Vergietungen*. VEB Gustav Fischer, Jena.

Martinson, T. G. *et al.* (1983) In: *Gigiyenicheskaya toksikologiya metallov* [Hygienic Toxicology of Metals]. Moscow, pp. 67—77 (in Russian).

McMillan, E. & Rowe, D. (1982) *Clin. Exp. Dermatol.*, No. **6**, 629—632.

Metelev, V. V. *et al.* (1971) *Vodnaya toksikologiya* [Aquatic Toxicology]. Kolos, Moscow (in Russian).

Mogilevskaya, O. Ya. (1963) In: *Toksikologiya redkikh metallov* [Toxicology of Rare Metals]. Medgiz, Moscow, pp. 187—195 (in Russian).

Mogosh, G. (1984) *Ostrye otravleniya* [Acute Intoxications]. (Russian translation from Roumanian published in Bucharest).

Nekrasova, S. V. & Tashi, P. F. (1988) *Gig. San.*, No. 5, 48—49 (in Russian).

Nemenko, B. A. (1981) *Gig. San.*, No. 1, 74—75 (in Russian).

Nerovnaya, R. P. (1972) In: *Trudy NII krayevoi patologii Kazakhskoi SSR* [Transactions of the Research Institute for Regional Pathology of the Kazakh SSR], Vol. 23, pp. 177—179 (in Russian).

Nesterov, G. F. (1961) In: *Trudy Instituta Eksperimentalnoi Veterinarii* [Transactions of the Institute of Experimental Veterinary], Vol. 24, pp. 190—193 (in Russian).

Novikov, Yu. V. *et al.* (1981) *Metody opredeleniya vrednykh veshchestv v vode vodoyemov* [Methods for Measuring Noxious Substances in Water Bodies]. Meditsina, Moscow (in Russian).

Ohnesorge, F. R. & Wilhelm, M. (1991) Chapter II. 36 Zinc in: Merian, E. (ed.) *Metals and their Compounds in the Environment — Occurrence, Analysis and Biological Relevance.* VCH, New York, pp. 1309—1342.

Pazdarowa, J. (1970) *Prac. Lek.*, No. 8, 281—284.

Perelygin, V. M. (1984) *Gig. San.*, No. 8, 12—15 (in Russian).

Pollak, G. F. *et al.* (1960) In: *Trudy Kazakhskogo instituta epidemiologii* [Transactions of the Kazakh Institute of Epidemiology], Vol. 4, pp. 590—596 (in Russian).

Prahalad, A. K. & Seenayya, G. (1989) *Env. Poll.*, **58**, 139—154.

Rabotnikova, L. V. (1966) *Materialy k uskorennym metodam opredeleniya toksichnosti okislov metallov* [Rapid Methods for Determining the Toxicity of Metal Oxides]. Dissertation, Leningrad (in Russian).

Rakhimova, M. T. (1971) *Zdravookhraneniye Kazakhstana* [Health Care in Kazakhstan], No. 1, 47—48 (in Russian).

Ringenberg, O. *et al.* (1988) *South. Med. J.*, No. 9, 1132—1139.

Roshchin, A. V. (1977) *Gig. Truda Prof. Zabol.*, No. 11, 28—35 (in Russian).

Roshchin, A. V. *et al.* (1982) *Gigiyena: Obzornaya informatsiya VNIIMI* [Hygiene: Reviews of the All-Union Institute for Medical Information]. Moscow, No. 3 (in Russian).

Shepard, T. H. (1986) *Catalog of Teratogenic Agents.* Johns Hopkins University Press, Baltimore & London.

Shabalina, L. P. & Spiridonova, V. S. (1978) In: *Aktualnye problemy gigiyeny truda* [Current Issues in Occupational Hygiene]. Moscow, pp. 40—44 (in Russian).

Skurikhin, I. M. (1981) *Vopr. Pitaniya*, No. 2, 10—16 (in Russian).

Sorokin, S. E. & Bolshakov, V. A. (1982) *Khimiya v selskom khozyaistve* [Chemistry in Agriculture], No. 3, 38—39 (in Russian).

Specifications for methods of determining noxious substances in air (1968) [Tekhnicheskiye usloviya na metody opredeleniya vrednykh veshchestv v vozdukhe], No. 5. Publication of the USSR Ministry of Health, Moscow (in Russian).

Spiridonova, V. S. & Shabalina, L. P. (1982) *Gig. San.*, No. 8, 23—25 (in Russian).

Spiridonova, V. S. *et al.* (1985) *Gig. San.*, No. 2, 74—75 (in Russian).

Stephenson, J. (1967) *Arch. Environ. Health*, No. 1, 83—88.

Thrower, S. & Olly, J. (1982) *J. Appl. Toxicol.*, 2, 110—115.

Vadze, D. R. *et al.* (1972) *Uchenye zapiski Latviyskogo gosudarstvennogo universiteta* [Annals of the Latvian State University], Riga, **176**, 92—104 (in Russian).

Vorontsova, E. I. *et al.* (1976) *Gig. San.*, No. 7, 49—51 (in Russian).

Yamaguchi, M. *et al.* (1984) *Toxicol. Lett.*, **22**, 175—180.

Yudov, N. N. (1959) *Vestn. Otorinolaringol.*, No. 4, 92—93 (in Russian).

Zhulidov, A. V. *et al.* (1980) *Dokl. Akad. Nauk SSSR*, No. 4, 1018—1020 (in Russian).

Zlateva, M. & Gelebova, V. (1974) *Khigiena i zdraveopoznavane* [Hygiene and Public Health], No. 1, 17—20 (in Bulgarian).

Cadmium and its compounds

Cadmium chloride; c. oxide; c. nitrate tetrahydrate; c. sulfate; c. sulfide
(greenockite [min.]); **c. sulfide selenide** (c. sulfoselenide)

IDENTITY OF THE ELEMENT

Cadmium (Cd) is a metallic element in group II of the periodic table, with the
atomic number 48. Its natural isotopes are ^{106}Cd (1.215%), ^{108}Cd (0.875%),
^{110}Cd (12.39%), ^{111}Cd (12.75%), ^{112}Cd (24.07%), ^{113}Cd (radioactive;
12.26%), ^{114}Cd (28.86%), and ^{116}Cd (7.58%).

PHYSICAL AND CHEMICAL PROPERTIES

Cadmium is malleable and ductile. It has no allotropic modifications. In air it
becomes covered with an oxide (CdO) film that protects it from further oxidation.
It exhibits the oxidation state +2 in most compounds. When strongly heated, it
burns to the oxide CdO. Solutions of cadmium salts have an acid reaction; when
an alkali is added to them, the hydroxide $Cd(OH)_2$ precipitates. Cadmium halides
are readily soluble in water. Cadmium forms a large number of complex
compounds. See also Appendix.

NATURAL OCCURRENCE AND ENVIRONMENTAL LEVELS

Cadmium belongs to the group of rare and highly dispersed elements. It is
contained as an isomorphous impurity in many minerals and is found in all zinc
ores. No ore of commercial importance is mined solely for cadmium. Very rare

cadmium minerals are greenockite (cadmium blende) CdS, otavite $CdCO_3$, montemponite CdO, and cadmoselite CdSe.

Cadmium abundances in the earth's crust, soils, and natural waters range from $n \times 10^{-5}$ to $n \times 10^{-6}\%$; in plants, its content is $n \times 10^{-4}\%$ on a dry weight basis. Inside the earth's crust, cadmium is transported in hot underground waters together with other chalcophile elements. Its migration in the environment depends on the compound in which it is present and on the pH of the medium. It is most mobile in acidic soils within the pH range of 4.5 to 5.5; its mobility in alkaline soils is relatively low. The average cadmium content of soils is between 0.07 and 1.1 mg/kg dry weight.

Cadmium eventually accumulates in soils either in the ionic form (in acid soil solutions) or as the water-insoluble hydroxide $Cd(OH)_2$ and carbonate $CdCO_3$. It may also occur in soils in the form of complex compounds such as cyanides and tartrates.

Up to 70% of the cadmium entering the soil combines with its chemical complexes that are available for assimilation by plants. Although cadmium is considered to be a nonessential element for plants, it is effectively absorbed by both the root and leaf systems [Stoeppler]. Cadmium is especially absorbed by plants in the vital rooting zones [Lamersdorf].

Most of the dissolved cadmium entering freshwater streams and seas settles and accumulates in sediments, while some is taken up and concentrated by aquatic plants and animals. Sediments of unpolluted rivers and lakes have cadmium contents ranging from 0.04 to 0.8 mg/kg, whereas in polluted rivers its levels range from 30 to 400 mg/kg [Breder]. Cadmium levels in mollusks and crustaceans usually range from $n \times 10^{-5}$ to $n \times 10^{-3}\%$ on a dry weight basis.

Cadmium is discharged into the atmosphere as a result of volcanic eruptions and by being released from plants [Mikhaleva]. Cadmium mobility in the environment has been examined by Xu *et al.*

PRODUCTION

Metallic cadmium is obtained as a by-product in the refining of zinc and other metals, particularly copper and lead. Pyrometallurgical and hydrometallurgical methods for the recovery of cadmium exist, as well as various combinations of these methods. The hydrometallurgical technique that is used most commonly involves leaching out of the pulp, its separation into a cadmium solution (containing 8–10 g Cd/L) and solid residues of the satellite metals (copper and others), precipitating the cadmium electrolytically or by cementation with zinc dust from a cadmium-enriched solution, and subsequent remelting and refining of the metal. If a metal of very high purity is to be obtained, methods such as zone refining, vacuum distillation, rectification, or ionic exchange are used.

USES

Cadmium is used in the manufacture of control, shim, and safety rods for nuclear reactors; for electroplating of other metals (mainly steel and iron) to provide corrosion-resistant or decorative coatings; as one of the electrodes in nickel-cadmium batteries; in semiconductors; and in various alloys, including low-melting alloys, precious alloys (with gold or silver), and alloys used as solders or in the manufacture of bearings, typographic clichés, and welding electrodes.

Cadmium compounds are constituents of numerous pigments, catalysts, pyrotechnic compositions stabilizers, laser materials, etc. Sulfur-containing cadmium compounds are components of many luminophors. (For other uses see Dunnick & Fowler; Fergusson.)

MAN-MADE SOURCES OF EMISSION INTO THE ENVIRONMENT

Emissions of cadmium from man-made sources have been estimated to exceed about threefold those from natural sources [Mikhaleva]. Anthropogenic sources may be classified into two groups: local sources such as industrial plants producing or using cadmium, and miscellaneous sources of varying capacities diffusely scattered throughout the Earth and ranging from thermal power plants and internal combustion engines through mineral fertilizers to tobacco smoke.

Substantial water and soil pollution by cadmium is often observed in the vicinity of lead and zinc mines and particularly smelters. Their wastewaters treated by conventional methods and then discharged into water streams have been shown to increase cadmium levels there by dozens of times. Cadmium in a water body inhibits its process of self-purification.

Although current methods of waste treatment may remove about 90% of cadmium [Yost], wastewaters from metallurgical works and from factories manufacturing dyes, cadmium-nickel storage batteries, or mineral fertilizers sometimes contain cadmium in substantial amounts.

In European countries, mean cadmium levels in urban air ranged from 0.01 to 0.05 $\mu g/m^3$. Much higher values (weekly means of 0.2–0.6 $\mu g/m^3$) were recorded around certain cadmium-emitting industrial plants [Friberg *et al.*]. In the vicinity of some metallurgical works, cadmium deposited from the atmospheric air was present in the top (2.5 cm) soil layer at levels 20 to 50 times higher than in control areas. Once in the soils, cadmium may be retained there for prolonged periods. In the United Kingdom, for example, soil cadmium concentrations elevated by factors of hundreds were found at smelter sites that had been inactive since the Middle Ages [Yost].

Soil may also be contaminated through the use of mineral fertilizers (e.g. superphosphate was found to contain 721 µg Cd/100 g, potassium phosphate 471 µg, and saltpeter [potassium nitrate] up to 66 µg [Berinia *et al.*]), as well as of sewage sludge for soil fertilization (sewage sludge may contain 100 mg Cd/kg dry weight [Friberg *et al.*]).

Air, soil, and plants may be seriously polluted by cadmium contained in exhaust gases from cars and tractors. For instance, plants growing at distances up to 30 m from highways with dense traffic carried 2 to 3 times more cadmium on their leaves than did those in control areas [Berinia *et al.*].

Because cadmium is readily available to plants from both air and soil, its concentration rapidly increases in plants grown in areas polluted by this metal [Alloway *et al.*]. Thus, plants grown in areas with heavy soil pollution contained 20 to 30 times more cadmium in their above-ground parts than did those in unpolluted areas. Considerable quantities of cadmium can be taken up from the soil by wheat and rice plants. Rice grown in fields polluted by cadmium-containing wastewater from an ore-dressing factory contained 10–12 times more cadmium than that harvested in unpolluted areas. Cadmium toxicity results in a decrease in productiveness, rates of photosynthesis, transpiration and enzymatic activities in plants [Stoeppler] and also produces changes in vitamin levels. A high cadmium-accumulating capacity is shown by tobacco leaves, and tobacco smoke may therefore be an important route of cadmium exposure for the general population [Adamu *et al.*]. It has been estimated that smoking one cigarette results in the inhalation of about 0.1–0.2 µg of the metal, and that smoking two packs per day may increase its body burden by 15 mg over a 20-year period [Lewis *et al.*]. By age 50–60, heavy smokers accumulated twice as much cadmium as nonsmokers [Bernard & Lauwerys].

A health risk for the general population may also be presented by drinking water contaminated with cadmium impurities contained in zinc-plated pipes or in fittings, water heaters, water coolers, and taps. Leaking of cadmium to ground water from dumped cadmium oxide sludge has also been reported [Friberg *et al.*].

In industry, significant cadmium exposure may occur in lead, zinc, and cadmium smelters, in cadmium–nickel battery factories, in the electroplating industry, in the welding of cadmium-plated materials, and in welding work using cadmium-containing electrodes [Friberg *et al.*].

In factories where cadmium was produced, workers handling cadmium solutions and pulp were exposed to airborne cadmium concentrations of 0.04 to 0.017 mg/m^3, while those involved in further stages of the production process were exposed to increasingly higher concentrations that reached an average value of 0.16 mg/m^3 per 8-hour shift. In a factory manufacturing cadmium-containing luminophors, workroom concentrations reached 3.7 mg Cd/m^3 and the metal was detected in nearly all air samples; in contrast, its concentrations did not exceed 0.1 mg/m^3 and it was present in only 30% of air samples taken in a factory manufacturing kinescopes [Peregudova; Shcherbakov].

TOXICITY TO ANIMALS AND MAN

General aspects

Cadmium and its compounds may be very toxic by ingestion and inhalation. Ingestion of highly contaminated food results in acute gastrointestinal effects, while high-level exposure via inhalation may cause acute or chronic lung disease and chronic renal disease [Fergusson]. The latter may also develop as a result of prolonged exposure to cadmium via food. For example, high cadmium levels in deer livers in New Jersey have been reported to present a potential threat to human health through ingestion [Stansley *et al.*]. Repeated exposure leads to cumulative effects. Cadmium can impair mineral metabolism, producing mineral depletion in bones and osteomalacia [Dunnick & Fowler]. It may also cause damage to the central and peripheral nervous systems and to the heart, liver, and skeletal muscle. In animal experiments, gonadotoxic and teratogenic effects have been observed [Dunnick & Fowler]. There is evidence suggesting that occupational exposure may increase the risk of prostatic and pulmonary cancer.

Toxic effects of cadmium appear to be largely due to its ability to inhibit various enzyme systems by blocking the carboxyl, amine, and especially sulfhydryl groups in protein molecules. Cadmium reduces the activity of the digestive enzymes trypsin and (to a lesser extent) pepsin. It stimulates catalase activity in the blood and liver tissue in low doses and inhibits it in high doses. Complexes of cadmium with enzymes such as alkaline phosphatase, carboxypeptidase, and cytochrome oxidase, are known. Cadmium activates urease and arginase (replacing the natural metallic component in the enzyme), affects carbohydrate metabolism (and can cause hyperglycemia), and inhibits glycogen synthesis in the liver [Yeremenko].

Metals such as zinc and selenium have been shown to modify cadmium toxicity, apparently by competing with it for binding sites in certain biological substrates [Sharma *et al.*]. It has been suggested that the competition with zinc determines the type of many of the effects produced by cadmium in the body as well as the protective action of zinc observed in cadmium intoxication [Pettering *et al.*].

Animal experiments have indicated that cadmium can exert inhibitory actions on some of the body defenses mediated by the immune system [Greenspan; Dunnick & Fowler; Kawamura; Müller *et al.*].

Acute poisoning

Animals

In acute toxicity studies where mice and rats were administered metallic cadmium, its oxide, and a number of its salts by the oral route (gavage), the **nitrate** and **sulfate** proved particularly toxic (LD_{50} = 47 mg/kg for mice), followed by the **chloride** and **oxide** (LD_{50} = 67 mg/kg for mice). The LD_{50} of metallic cadmium was 890 mg/kg. The lowest toxicity was displayed by its **sulfide** and **sulfide selenide** [Vorobiyeva].

Rats, guinea-pigs, and rabbits administered cadmium oxide by inhalation, into the trachea, or subcutaneously developed central nervous system abnormalities, hypotension, and interstitial pneumonia with subsequent manifestations of fibrosis. For rats, the LC_{50} was 45 mg/m^3 with 1-hour inhalation exposure. Dogs and monkeys were more resistant to the oxide than the three species mentioned above.

Parenteral administration of soluble cadmium salts in doses of 1–3 mg/kg caused severe damage (including necrosis) to the testes in several animal species.

Man

Inhalation of cadmium oxide fume or dust at concentrations above 1 mg Cd/m^3 for 8 h (during the entire work shift) or at higher concentrations for shorter periods led to chemical pneumonia and later to pulmonary edema. Their symptoms generally appeared within 1 to 8 h after exposure, but occasionally the period of latency extended to 24 h. Such acute poisoning most frequently resulted from the inhalation of fumes generated by welding, soldering, or smelting cadmium-containing materials. In cases of lethal poisoning, death usually ensued on days 3 to 5.

Balmer & Rothwell described 15 cases of acute poisoning with cadmium oxide fumes inhaled at concentrations of 8–10 mg/m^3; two of the victims died on days 4 and 8. The first sign of poisoning was laryngeal irritation, which was followed after a latent period of several hours to days by severe cough and dyspnea, fever, chest pain, nausea, vomiting, cyanosis, and pulmonary edema.

Poisoning via ingestion of drinks contaminated with **cadmium** at concentrations exceeding 15 mg/L gave rise to nausea, vomiting, abdominal pains, and sometimes diarrhea. Sources of food contamination may be pots and pans with cadmium-containing glazing and cadmium soldering used in vending machines. Oral intoxication has been observed after cadmium contamination of tap water and beverages from soldering points in water pipes and faucets, as well as from cadmium-plated heating and cooling tubes and kitchen utensils [Stoeppler].

Cases of lethal poisoning by ingestion have been described. For example, three men died after drinking half a glass of a liquid containing 17.5% of **cadmium chloride**; death occurred in 5–6 h in two cases and in 56 h in the third. In all victims, the face, auricular conchae, and tongue became edematous 3 to 5 min post-ingestion, followed by loss of consciousness. Autopsy revealed congested internal organs, edemas and hemorrhages in the upper respiratory tract and gastric mucosas, dystrophic changes in the myocardium and liver, and brain edema. In addition, necrotic hepatitis, nephrosis, dystrophic changes in cells of the cortex, subcortex and stem of the brain, hemorrhagic pneumonia, pulmonary edema, and hemorrhages under the epicardium and endocardium and in the myocardium were seen in the individual who died after 56 h.

A woman died after accidentally drinking 20 ml of a 0.9% **cadmium sulfate** solution. Its intake was immediately followed by a burning sensation in the mouth and the epigastrium and then throughout the abdominal region; vomiting appeared 15 to 20 min later and was followed by diarrhea. The electrocardiogram showed signs of myocardial ischemia. Death occurred at 33 h in the presence of progressive cardiovascular insufficiency. Autopsy showed congested vessels in internal organs, edema and hemorrhages in the lungs, necrotic changes in the liver, and necrotic areas in gastric and intestinal mucous membranes [Zelengurov]. It has been reported that lethal oral doses of cadmium range from 350 to 8900 mg, and that the dose of about 3 mg did not cause any signs of acute poisoning [Lauwerys *et al.*].

Chronic poisoning

Animals

In rats, long-term ingestion of drinking water containing **cadmium chloride** at 5 µg/ml lowered blood levels of sulfhydryl groups and depressed cytochrome oxidase and succinate dehydrogenase activities in the liver, kidneys, and skeletal muscles, while stimulating their activities in the adrenals which were also found to contain elevated levels of ascorbic acid and reduced glutathione [Soloviyev *et al.*]. Rats injected with the chloride intravenously in doses of 1 to 3 mg/kg developed connective tissue tumors.

Cadmium salts added to the diet and drinking water of rats retarded bone growth and led to osteoporosis; in such rats, increased urinary and fecal excretion of calcium and depressed alkaline phosphatase activity were recorded. Similar effects were observed in parous female rats on chronic exposure to **cadmium sulfate** by inhalation [Tsvetkova].

In chronic inhalation toxicity tests of **cadmium oxide** on rats concentration of 0.5 mg/m^3 was determined as the threshold level for gonadotoxic effects and that of 0.23 mg/m^3, as the no-observed-effect level for these effects [Parizek].

Ultrastructural changes were detected in the adenohypophysis, adrenals, and thyroid of rats shortly after the start of their chronic inhalation exposure to **cadmium** at concentrations around 0.1 mg/m^3 [Polyakova]. The lungs of rats chronically exposed to a cadmium aerosol by inhalation showed signs of cytotoxicity whose severity depended on the amount of cadmium in the aerosol, but not on the duration of exposure [Glaser & Otto]. Among early signs of experimental cadmium intoxication in rabbits, elevated serum levels of cholesterol and free fatty acids were noted [Yoshikawa]. Cadmium stimulated both the glomerular and fascicular zones in the adrenals and increased the urinary output of 17-ketosteroids [Kollmer]. Animal experiments with young rats have shown that if it the femurs of such rats contain more than 5–10 ppm cadmium, their mechanical strength is reduced [Stoeppler].

Several recent studies on mice, rats, and hamsters have provided evidence for carcinogenicity of cadmium compounds administered by injection or inhalation [Dunnick & Fowler; Fergusson; Glaser *et al.*; Heinrich; Oldiges *et al.*; Oberdoerster]. Dunnick & Fowler also discussed the impact of cadmium on the reproductive performance of rodents.

Man

Long-term exposure to cadmium may affect various body organs and systems. With prolonged low-level exposure the critical organ (the site of earliest functional disturbances) is the kidney. Classically, the functional disturbances involve the proximal tubule, giving rise to a tubular-type proteinuria with increased excretion of low-molecular-weight proteins such as β_2-microglobulin. As the renal dysfunction progresses, amino acids, glucose, and minerals such as calcium and phosphorus are also lost into the urine [Herber *et al.*].

The relative severity of renal *versus* lung damage depends on the level of exposure (and probably also on individual susceptibility) so that with higher-level exposures the lungs are relatively more affected, whereas low-level exposures results predominantly in renal damage. Various types of lung damage (emphysema, obstructive pulmonary disease, diffuse interstitial fibrosis) have occurred in workers exposed for long periods to cadmium dust and cadmium oxide fumes. Late manifestations of severe chronic intoxication may be bone lesions (osteomalacia, osteoporosis, spontaneous fracture).

Frequent symptoms and signs of chronic occupational intoxication with cadmium are anosmia, yellow coloration of the dental neck, hypochromic anemia, impaired calcium metabolism, and diminished blood content of sulfhydryl groups [Tarasenko & Vorobiyeva]. Slightly decreased hemoglobin levels have also been a common finding in cadmium-exposed workers, probably as a consequence of inhibited iron absorption from the diet and the resultant reduction of iron supply to the bone marrow [Schaefer *et al.*].

Long-term inhalation of **cadmium fumes** or dusts resulted in chronic rhinitis and pharyngitis accompanied by dryness and irritation in the larynx, burning in the nose and ulceration of its cartilaginous part, and nasal bleeding. Most of the workers exposed to cadmium dust in concentrations of 0.1 to 1.0 mg/m^3 for a long time (1 year at least) developed proteinuria, with the amount of protein in the urine varying from 70 to 2600 mg/day. Autopsy findings in exposed individuals demonstrated severe kidney damage. Proteinuria may be associated with other manifestations of impaired renal tubular function, such as glucosuria, aminoaciduria, decreased urine-concentrating capacity of the kidney, increased urinary excretion of calcium and phosphorus, and elevated plasma creatinine. Calciuria may lead to nephrocalcinosis. Examination of 285 cadmium workers showed upper respiratory tract diseases in 70% of them, yellow coloration of dental necks in 43%, impaired sense of smell in 66%, and liver disease in 10%; also, structural changes in osseous tissue were observed in 18 out of the 32 workers who complained of pains in joints and bones [Tarasenko & Vorobiyeva].

Epidemiological studies have indicated an increased risk of prostatic and respiratory tract cancer in cadmium workers [Kazantzis; Oberdoerster; Piscator]. Significantly increased mortality rates from prostatic and pulmonary cancer were observed in the USA among 292 smelter workers with at least 2 years of exposure to cadmium oxide dust and fumes [Lemen *et al.*], and a similar tendency was noted in Sweden with regard to prostatic cancer and nasopharyngeal cancer among 228 cadmium–nickel battery workers with exposure periods of more than 5 years [Kjellström *et al.*]. There is evidence that cadmium may produce teratogenic effects [Shepard].

Long-term cadmium poisoning by ingestion is only known for Japan and has mainly occurred in the form of itai-itai disease — a combination of severe renal tubular damage and osteomalacia – in people (mostly elderly women) consuming rice grown on soil contaminated by cadmium-polluted irrigation water. The disease was marked by skeletal deformities, severe pains in the back and legs, a 'goose' gait, and a tendency to bone fractures in various parts of the skeleton even on moderate exertion (e.g. rib fracture on coughing); in addition, pancreatic dysfunction and hypochromic anemia developed, and the blood contained reduced iron, calcium, and phosphorus levels along with sharply elevated alkaline phosphatase activity [Shiroishi *et al.*]. Renal tubular dysfunction was of the same type as that seen in chronic occupational intoxication with calcium.

ABSORPTION, DISTRIBUTION, AND ELIMINATION

Cadmium enters the human body by inhalation (mainly in occupational settings) and ingestion (in food). Food chains via which it is ingested form in the regions of increased soil and water pollution with cadmium. In industry, cadmium may enter foodstuffs from contaminated hands. The daily human diet was estimated to contain 200 μg of cadmium on average [Schroeder *et al.*].

Gastrointestinal absorption of ingested cadmium is about 2% to 6% under normal conditions, but may amount to 20% in individuals with low body stores of iron. Pulmonary absorption of inhaled respirable cadmium dust is estimated at 20% to 50% depending on particle size. The highest airborne cadmium concentrations in urban areas and in the vicinity of intense point sources of this metal (e.g. zinc and lead smelters) rarely exceeded 100 ng/m^3; assuming that an adult breathes 20 m^3 air per day, the resulting cadmium intake by inhalation is only 2 μg/day at most [Yost]. The chemical species of cadmium absorbed from the air is unknown. Dermal absorption is slight.

Whether or not cadmium plays a specific biological role as a trace element has not been established.

Cadmium distributes throughout the body, but its main depots are the kidneys and liver where about 50% of the absorbed cadmium concentrates. (Cadmium distribution models have been considered by Herber *et al.*) It has been detected in sweat, milk, hair, and nails. Its transfer across the placenta is relatively small: neonatal blood had concentrations only half those found in maternal blood. Its levels in the blood do not usually exceed 10 µg/L in people exposed to natural background concentrations of this element but may be higher by an order of magnitude under conditions of occupational exposure or other special circumstances. In the blood, more than 70% of cadmium occurs in erythrocytes. The total body burden of cadmium increased with age; individuals aged about 50 were found to contain 20–30 mg [Bogomazov & Veranian].

Elimination of cadmium is very slow, which may explain its accumulation in the body with age and time of exposure. It is mainly excreted via the kidneys, at an average rate of 2 µg/day. The urinary cadmium concentration rises substantially in individuals with severe dysfunction of renal glomeruli. Considerable excretion in feces is only observed after intensive exposure.

HYGIENIC STANDARDS

Exposure limits adopted for cadmium and its compounds in the USSR are shown in Table 1. In the USA, a TLV-TWA value of 0.05 mg/m^3 has been set for cadmium dust and salts (as Cd) and a TLV-C of 0.05 mg/m^3 for cadmium oxide fume. A TLV-STEL of 0.2 mg/m^3 has been set for cadmium dust and salts.

The highest permissible levels established in the USSR for cadmium in human foods are indicated in Table 2.

Table 1. Exposure limits for cadmium and its compounds

	Workroom air		Atmos-pheric air	Water sources		
	MAC_{wz} (mg/m^3)	Aggregative state	MAC_{ad} (mg/m^3)	MAC_w (mg/L)	MAC_{wf} (mg/L)	Hazard class
Cadmium	—	—	—	0.001[a]	0.005	2
Cadmium and its inorganic compounds	0.05/0.01[b]	Aerosol	—	—	—	1
Cadmium oxide (as Cd)	—	—	0.0003	—	0.005	1
Cadmium stearate	0.1	Aerosol	—	—	—	1

[a] Based on sanitary and toxicological criteria.

[b] 0.05 mg/m^3 is the concentration not to be exceeded at any time, while 0.01 mg/m^3 is the average maximum allowable concentration per shift.

The highest permissible levels established in the USSR for cadmium in human foods are indicated in Table 2.

Table 2. MACs for cadmium in foods (mg/kg)

Grain, groats, flour, bran, candies, and pastries	0.1
Sugar	0.05
Bread	0.05
Milk, sour milk products, and butter	0.03
Condensed milk, canned	0.1
Margarines, cooking fats, and vegetable oils	0.05
Cheese, curds, and casein	0.2
Vegetables, fresh or canned	0.03
Fruits, fresh or canned	0.03
Berries, fresh or canned	0.03
Spices	0.2
Tea	1.0
Meat and poultry, fresh or canned, and sausages	0.05
Eggs	0.03
Fish, fresh or canned	0.2
Mollusks and crustaceans	2.0
Mineral waters	0.01
Beverages based on infusions or essences; beer, wine, and other alcoholic beverages	0.03
Infant foods:	
milk-based	0.02
grain- and milk-based	0.02
vegetable- or fruit-based	0.03
meat- or poultry-based	0.03
fish-based	0.1

METHODS OF DETERMINATION

Cadmium in air can be determined by a nephelometric method, based on the ability of a complex (iodide) cadmium anion to give difficultly soluble compounds with triphenyltetrazolium chloride (sensitivity 0.5 µg/5 ml sample), or a colorimetric method based on a color reaction of cadmium ion with bromobenzthiazo (sensitivity 1 µg in the volume analyzed) [Peregud *et al.*]. A method for measuring cadmium in dusts from cadmium-zinc luminophors relies on the production of a monosubstituted cadmium dithizonate during the interaction of a dithizone solution with cadmium in carbon tetrachloride in an alkaline medium (sensitivity 0.5 mg Cd in the volume analyzed) [Petrov & Vernigorov]. For cadmium determination in environmental and biological media, a polarographic technique (sensitivity 0.01 µg/ml) [Goldberg] and a thin-layer chromatographic procedure (sensitivity 0.01 µg/ml) [Katayeva] have been used. With atomic absorption spectroscopy, cadmium could be determined in blood serum and urine with a sensitivity of 0.012 µg% and 0.06 µg%, respectively [Bec & Sychra; Klaus]. Versieck *et al.* have described determination of cadmium in serum and packed blood cells using neutron activation analysis. Analytical techniques for its determination in environmental and biological samples have been reviewed by Fergusson and Kafritsas *et al.*

MEASURES TO CONTROL EXPOSURE

At the workplace
Cadmium presents a permanent hazard wherever it is used or produced. The cardinal measure for preventing the adverse health effects of cadmium and its compounds in both the working and general environments is to limit their use and replace them with less dangerous substances.

The process equipment should be carefully sealed; where practicable, furnaces, reactors, and other installations should work under vacuum.

Work processes and operations that may release cadmium fumes or dust into the workroom air should be designed to keep concentration levels to a minimum and, if feasible, fitted with exhaust ventilation. Good general ventilation is also important.

If adequate ventilation is impossible to maintain (e.g. during welding and cutting), respirators or gas masks must be worn.

Cadmium-containing material should be manufactured in the form of granules, liquids, or pastes as this will diminish or rule out the formation of aerosols.

Industrial processes in which cadmium and its compounds are used should be preferably centralized, for this would make it easier to implement the set of measures required to prevent their harmful effects on workers as well as on the general population.

Cadmium-containing wastes should be transported in specially equipped vehicles and be disposed of in burial grounds designed to receive toxic industrial wastes.

Adequate sanitary facilities should be provided and workers encouraged to wash hands thoroughly before meals and to shower and change work clothes at the end of the work shift. Smoking, eating, and drinking in work areas should be forbidden. Food may be taken only in canteens or separate rooms designated for this purpose.

For *personal protection* workers should be supplied with appropriate protective equipment including respirators and/or gas masks. For hand protection, barrier ointments or pastes are recommended. Special devices should be available for decontaminating working clothes and footwear. Separate ventilated lockers should be provided for work clothing.

All workers at risk of exposure to cadmium or its compounds should receive a medical examination at the time of recruitment and at regular intervals (at least once every 2 years) during the period of employment. Each examination should include at least a full-size chest X-ray, lung function tests, and urinalysis. Medical contraindications to work associated with exposure to cadmium and its compounds are chronic diseases of the upper respiratory tract (including ozena and deviation of the nasal septum if they interfere with respiratory function), chronic bronchitis, bronchial asthma, chronic pneumonia, and renal disorders. Persons with these conditions and also those with signs of osteoporosis or osteomalacia of any origin should be excluded from such work.

Protection of the general environment

Since cadmium has become an important pollutant of the biosphere, effective measures to prevent its entry into bodies of water, soils, the atmosphere, and foods should be implemented in industrial undertakings that are sources of environmental pollution with this element. Such measures mainly include, apart from the use of air-cleaning devices, the wide introduction of recirculated water supply systems and of technologies that produce little or no wastes.

FIRST AID IN POISONING

A person acutely poisoned with cadmium-containing vapor or fume should be removed from the polluted atmosphere into fresh air and put to complete rest (for at least 24 hours even if the poisoning seems to be mild) and kept warm; the intake of liquids should be restricted. The victim should be transported in the recumbent position only. If signs of pulmonary edema appear, osmotherapy and oxygen therapy are needed. Special care should be taken to maintain the airways patent. Further treatment is symptomatic. Antibiotics are indicated.

Those who have ingested a cadmium solution should be made to vomit and then given a mucilaginous decoction, isotonic sodium sulfate solution, and activated charcoal. The use of Strizhizhevsky's antidote (up to 500 ml) followed by gastric lavage with an activated charcoal suspension and 0.2% tannin solution is indicated.

There is no specific drug treatment for cadmium poisoning. Although chelating agents may form nephrotoxic complexes with cadmium, some of them, such as sodium calcium edetate and unithiol (sodium 2,3-dimercapto-1-propane sulfonate), may be used with caution; BAL is not recommended [Ludewig & Lohs].

Recovery is slow. If signs of osteoporosis have appeared, vitamin D and calcium preparations are recommended [Mizyukova].

REFERENCES

Adamu, C. A. *et al.* (1989) *Environ. Poll.*, **56**, 113–126.
Alloway, B. J. *et al.* (1990) *Sci. Total Environ.*, **91**, 223–236.
Balmer, F. M. R. & Rothwell, H. E. (1938) *Can. Public Health J.*, **29**, 19–26.
Bec, F. & Sychra, V. (1971) *Chem. Listy*, **65**, 1233–1255.
Berinia, D. Zh. *et al.* (1978) In: *Opyt i metody ekologicheskogo monitoringa* [Environmental Monitoring: Experience and Methods]. Gidrometeoizdat, Pushchino, pp. 171–173 (in Russian).
Bernard, A. & Lauwerys, R. (1986) *Experientia*, **50**, 114–123.
Bogomazov, M. Ya. & Veranian, O. A. (1988) *Vopr. Pitaniya*, No. 3, 38–40 (in Russian).
Breder, R. (1988) In: Hutzinger, O. & Safe, S. H. (eds) *Environmental Toxins, Vol. 2. Cadmium*, Stoeppler, M. & Piscator, M. (vol. eds) Springer, Berlin, pp. 159–169.
Dunnick, J. K. & Fowler, B. A. (1988) Chapter 14 *Cadmium* in: Seiler, H. G. *et al.* (eds) *Handbook on Toxicity of Inorganic Compounds*. Marcel Dekker, New York, pp. 155–174.
Fergusson, J. E. (1990). *The Heavy Elements: Chemistry, Environmental Impact and Health Effects*. Pergamon Press, Oxford.
Friberg, L. *et al.* (1979) In: Friberg, L. *et al.* (eds) *Handbook on the Toxicology of Metals*. Elsevier, Amsterdam, pp. 355–381.
Glaser, U. & Otto, F. (1986) *Proceedings of the Second International Aerosol Conference. Berlin, 1986*. Oxford, pp. 260–262.
Glaser, U. *et al.* (1989) *Environ. Chem.*, **27**, 153–162.
Greenspan, B. J. (1984) *Appl. Toxicol.*, **50**, 48–56.
Heinrich, U. (1988) In: Hutzinger, O. & Safe, S. H. (eds) *Environmental Toxins. Vol. 2., Cadmium*, Stoeppler, M. & Piscator, M. (vol. eds). Springer, Berlin, pp. 13–25.
Herber, R. F. M. *et al.* (1988) *Ibid*, pp. 115–133.
Kafritsas, I. A. *et al.* (1989) *Toxicol. Environ. Chem.*, **20–21**, 169–173.
Katayeva, S. E. (1976) *Gig. San.*, No. 10, 60–61 (in Russian).
Kawamura, R. (1983) *Arch. Toxicol.*, **54**, 289–296.
Kazantzis, G. (1988) In: Merian, E. *et al.* (eds) *Carcinogenic & Mutagenic Metal Compounds. 3: Interrelation Between Chemistry & Biology*. Gordon & Breach Science Publishers, New York, pp. 499–506.
Kjellström, T. *et al.* (1979) *Environ. Health Persp.*, **28**, 199–204.
Klaus, L. (1966) *Ztschr. Klin. Chem.*, **4**, 299–302.
Kollmer, W. E. (1985) *Nutr. Res.*, No. 1, 637.
Kryukova, T. A. *et al.* (1959) *Polyarograficheskiy analiz* [Polarographic Analysis]. Goskhimizdat, Moscow (in Russian).
Lamersdorf, N. P. (1989) *Toxicol. Environ. Chem.*, **18**, 239–247.
Lauwerys, R. *et al.* (1984) *Toxicol. Lett.*, **23**, 287–289.
Lemen, R. A. *et al.* (1976) *Ann. N. Y. Acad. Sci.*, **271**, 273–279.
Lewis, G. P. *et al.* (1972) *Lancet*, **1**, 291–292.
Ludewig, R. & Lohs, K. (1981) *Akute Vergietungen*. VEB Gustav Fischer, Jena.
Mikhaleva, L. M. (1980) *Arkh. Patol.*, **50**, 81–85 (in Russian).
Mizyukova, I. G. (1982) In: M. L. Tarakhovsky (ed.) *Lechenie ostrykh otravleniy* [Treatment of Acute Intoxications]. Zdorovie, Kiev, pp. 166–167 (in Russian).
Müller, S. *et al.* (1979) *Experientia*, **35**, 909–910.
Oberdoerster, G. (1989) *Toxicol. Environ. Chem.*, **23**, 41–51.
Oldiges, H. *et al.* (1988) In: Hutzinger, O. & Safe, S. H. (eds) *Environmental Toxins, Vol. 2, Cadmium*, Stoeppler, M. & Piscator, M. (vol. eds). Springer, Berlin, pp. 33–38.
Parizek, J. (1983) *Reproductive and Developmental Toxicity of Metals*. Plenum Press, New York, pp. 301–313.

Peregud, E. A. *et al.* (1970) *Bystrye metody opredeleniya vrednykh veshchestv v vozdukhe* [Rapid Methods for Determining Harmful Substances in the Air]. Khimiya, Moscow (in Russian).

Peregudova, T. I. (1983) In: *Metally. Gigiyenicheskie aspekty otsenki i ozdorovleniya okruzhayushchei sredy* [Metals. Hygienic Aspects of Evaluation and Environmental Health Promotion]. Moscow, pp. 242—245 (in Russian).

Petrov, V. E. (1982) In: *Sudebno-meditsinskaya ekspertiza otravleniy. Sbornik nauchnykh trudov kafedry sudebnoi meditsiny I LMI* [Forensic—medical Evaluation of Intoxications. A collection of papers by workers of the Department of Forensic Medicine, First Leningrad Medical Institute]. Leningrad, pp. 70—73 (in Russian).

Petrov, A. M. & Vernigorov, N. F. (1977) In: *Gigiyena truda v proizvodstve luminoforov* [Occupational Hygiene in the Manufacture of Luminophors]. Stavropol, 67—72 (in Russian).

Pettering, H. G. *et al.* (1971) *Arch. Environ. Health*, **23**, 93—101.

Piscator, M. (1988) In: Hutzinger, O. & Safe, S.H. (eds) *Environmental Toxins, Vol. 2, Cadmuim*, Stoeppler, M. & Piscator (vol. eds). Springer, Berlin, pp. 3—12.

Polyakova, N. A. (1983) In: *II Vsesoyuznaya konferentsiya 'Endokrinnaya sistema organizma i vrednye faktory vneshnei sredy'* [Second All-Union Conference on the Endocrine System and Adverse Environmental Agents]. Leningrad, pp. 162—163 (in Russian).

Schaefer, S. G. *et al.* (1988) In: Hutzinger, O. & Safe, S. H. (eds) *Environmental Toxins, Vol. 2, Cadmium*, Stoeppler, M. & Piscator, M. (vol. eds). Springer, Berlin, pp. 27—31.

Schroeder, H. A. *et al.* (1967) *J. Chron. Dis.*, **20**, 179—210.

Sharma, A. *et al.* (1985) *Curr. Sci.*, **54**, 539—549.

Shcherbakov, G. G. (1979) *Gig. Truda Prof. Zabol.*, No. 3, 15—18 (in Russian).

Shepard, T. H. (1986) *Catalog of Teratogenic Agents*. Johns Hopkins University Press, Baltimore & London.

Shiroishi, K. *et al.* (1977) *Environ. Res.*, **13**, 407—424.

Soloviyev, A. V. *et al.* (1979) In: *I Vsesoyuznaya konferentsiya Endokrinnaya sistema organizma i toksicheskie faktory vneshnei sredy* [First All-Union Conference on the Endocrine System and Adverse Environmental Agents]. Leningrad, pp. 187—189 (in Russian).

Stansley, W. *et al.* (1991) *Sci. Total. Environ*, **107**, 71—82.

Stoeppler, M. (1991) *Cadmium* in: *Metals and their Compounds in the Environment*. Merian, E. (ed.). VCH, New York, pp. 803—851.

Tarasenko, N. Yu. & Vorobiyeva, R. S. (1973) *Vestn. Akad. Med. Nauk SSSR*, No. 10, 37—42 (in Russian).

Tsvetkova, R. P. (1970) *Gig. Truda Prof. Zabol.*, No. 3, 31—35 (in Russian).

Versieck, J. *et al.* (1988) In: Hutzinger, O. & Safe, S. (eds) *Environmental Toxins, Vol. 2, Cadmium*, Stoeppler, M. & Piscator, M. (vol. eds). Springer, Berlin, pp. 195—212.

Vorobiyeva, R. S. (1979) *Gigiyena i toksikologiya kadmiya* [Cadmium: Hygienic and Toxicological Aspects]. Publication of the All-Union Institute for Medical Information, Moscow (in Russian).

Xu, H. *et al.* (1989) *Sci. Total Environ.*, **81/82**, 653—660.

Yeremenko, L. I. (1977) In: *Gigiyena truda v proizvodstve luminoforov* [Occupational Hygiene in the Manufacture of Luminophors]. Stavropol, pp. 53—57 (in Russian).

Yoshikawa, H. (1973) *Industr. Health*, **11**, 113—167.

Yost, K. J. (1984) *Experientia*, **40**, 157—164.

Zelengurov, V. M. (1969) In: *Voprosy sudebnoi travmatologii* [Selected Topics of Forensic Traumatology]. Zdorovya, Kiev, pp. 95—97 (in Russian).

Mercury and its compounds

Mercury (1) (mercurous) compounds: **acetate**; **chloride** (calomel) [min.]; **nitrate dihydrate**; **sulfate**
Mercury(II) (mercuric) compounds: **acetate**; **amidochloride** (ammoniated mercury; infusible white precipitate); **bromide**; **chloride** (corrosive sublimate); **fulminate** (fulminating mercury); **iodide**; **nitrate semihydrate**; **oxide**; **sulfate**; **sulfide** (cinnabar [min.]); **thiocyanate**

IDENTITY AND PHYSICOCHEMICAL PROPERTIES
OF THE ELEMENT

Mercury (Hg) is a metallic element in group II of the periodic table, with the atomic number 80. Its natural isotopes are ^{196}Hg (0.15%), ^{198}Hg (10.12%); ^{199}Hg (17.04%), ^{200}Hg (23.13%), ^{201}Hg (13.18%), ^{203}Hg (29.8%), and ^{204}Hg (6.72%).

Mercury is the only metal that is liquid throughout the usual temperature range. At temperatures below -39°C, it solidifies to become white and ductile.

Metallic mercury is volatile. A saturated atmosphere of mercury vapor contains approximately 18 mg Hg/m^3 at 24°C. In compounds mercury is present in the oxidation state +1 (mercury(I), or mercurous, compounds) or +2 (mercury(II), or mercuric, compounds). It is chemically stable at ordinary temperatures. At elevated temperatures, its oxidation by dry air gives rise to mercury(II) oxide (HgO); the oxidation proceeds much faster in the presence of moisture and traces of metals (e. g. zinc or lead). In the oxidation process, a gray film of oxides forms on its surface. At ordinary temperatures, mercury readily combines with halogens to give mercury(I) and mercury(II) halides, and it gives mercury(II) sulfide (HgS) when triturated with sulfur in the cold. It forms alloys (amalgams) with all metals except iron, nickel, cadmium, aluminum, cobalt, and platinum. See also Appendix.

NATURAL OCCURRENCE AND ENVIRONMENTAL LEVELS

Mercury is found in nature in the minerals cinnabar (HgS) and calomel (Hg_2Cl_2) and, rarely, as the metal (e. g. in the form of geodes of liquid mercury).

The major natural sources of mercury are the degassing of the earth's crust, emissions from volcanoes, and evaporation from natural bodies of water. According to recent estimates, natural atmospheric emissions are of the order of 2700 to 6000 tones per year [WHO, 1991].

The clarke of mercury is estimated to be $0.08 \times 10^{-4}\%$ in the earth's crust and $0.033 \times 10^{-4}\%$ in the granite layer of the continental crust. The oceans contain an estimated 206 million tons of dissolved mercury and receive from rivers up to 2.6 thousand tons of soluble mercury species annually [Dobrovolsky].

Estimated average abundances of mercury are $0.1 \times 10^{-4}\%$ in soils, $0.4 \times 10^{-4}\%$ in sediments, and $0.2 \times 10^{-4}\%$ (dry weight) in coals. Continental vegetation contains an estimated 0.03 million tones, at average fresh weight, dry weight, and ash weight concentrations of $0.005 \times 10^{-4}\%$, $0.012 \times 10^{-4}\%$, and $0.25 \times 10^{-4}\%$, respectively; the amount of mercury taken up annually through the increment in phytomass is estimated to be 2.16 thousand tones, or 0.014 kg/km^2 of land area [Dobrovolsky].

Mercury is characterized by high water absorption and biological absorption coefficients ($K_w = 17.59$; $K_b = 7.58$ [Dobrovolsky]. From the aquatic medium, soluble mercury species enter sediments where they concentrate in nonbiogenic clay muds.

In soil, mercury is firmly bound to its particles, forming slightly mobile complexes with humic acids; its half-life in soil is estimated to be 250 years. The quantity of mercury entering the biologic cycle per year is estimated to be about 40 thousand tonnes, of which some 10 thousand tonnes enter the aquatic ecosystems [Izrael].

Recent measurements of mercury in aquatic systems have given the following concentration ranges, which may be considered as representative for dissolved mercury, although local variations are considerable: open ocean, 0.5–3 ng/L; coastal seawater, 2–15 ng/L; rivers and lakes, 1–3 ng/L [WHO, 1989].

In the lower troposphere, the background level of mercury is about 2 ng/m^3 in the northern hemisphere and about 1 ng/m^3 in the southern hemisphere, at least over oceanic areas. In European areas remote from industrial sources, concentrations in the air are most often in the range from 2 to 3 ng/m^3 in summer and from 3 to 4 ng/m^3 in winter; in urban air, the concentrations may be higher [WHO, 1989].

Mercury is present in the atmosphere as vapor and aerosol in approximately equal amounts, and an equilibrium between these two forms is reached in about 5 days [Izrael].

A major factor in removing mercury from the atmosphere is precipitation. The lowest concentrations of mercury in rainwater, around 1 ng/L, have been reported from a coastal site in Japan and from the islands of Samoa; most other reported values lie in the range between 5 and 100 ng/L [WHO, 1989].

PRODUCTION

Mercury metal is usually obtained industrially by roasting its ore cinnabar (HgS) in air and condensing the resulting mercury vapor.

The industrial production of inorganic mercury compounds is based on the synthesis of a water-soluble mercury(II) nitrate which serves as the key intermediate product for the preparation of all other mercurials, both water-soluble and insoluble. In the laboratory, mercury salts are obtained by reacting the appropriate acid (e. g. sulfuric, nitric, or hydrochloric) with the yellow mercury(II) oxide.

USES

A major use of mercury is as a cathode in the electrolysis of sodium chloride to produce caustic soda and chlorine gas. It is also widely used in the synthesis of mercury-containing inorganic and organic compounds; in the electrical industry (in electrical bulbs, arc rectifiers, and mercury battery cells); in measurement and control instrument (switches, thermostats, barometers, thermometers, blood pressure meters, etc.); in the extraction of noble metals (particularly gold) from their ores; as an alloy additive, heat-transfer agent, and catalyst in the chemical industry; for tooth amalgam fillings in dentistry; in the manufacture of antifouling paints and of agents for preventing the rotting of wood; in laboratory and medical practice.

MAN-MADE SOURCES OF EMISSION
INTO THE ENVIRONMENT

The estimated total global release of mercury to the atmosphere due to human activities has been reported to be of the order of 2000–3000 tonnes per year [WHO, 1991], i. e. is comparable to its emissions from natural sources. The main anthropogenic sources of environmental pollution are the mining of mercury and its separation from the ores, the combustion of organic fuels (coal, main anthropogenic sources of environmental pollution are the mining of mercury and its separation from the ores, the combustion of organic fuels (coal, oil, peat, wood), and a number of miscellaneous activities not directly related to mercury, such as smelting of metals from their sulfide ores, gold refining, cement manufacture, industrial applications of metals, refuse incineration, and disposal of sewage [WHO, 1976, 1991]. The quantity of mercury emitted into the environment annually from coal burning has been estimated to have increased from about 100 tonnes in 1900 to 600 tonnes in 1980 (which roughly corresponds to a sixfold increase in the mercury burden per km^2 of land area), and is expected to reach 900 tonnes in the year 2000 [Dobrovolsky]. The amount of mercury discharged into the atmosphere daily from stack emissions of a coal-fired 700-megawatt electric power plant averaged 2.5 kg.

It has been calculated that about 70% of the total mercury produced worldwide per annum (10 000–15 000 tonnes) is irrevocably lost, with only 15% being reused [Harris & Hohenebser]. Some 450 g of mercury was found to be emitted per tonne of caustic soda produced in chloralkali plants [WHO, 1991]. One chlorine plant had been losing mercury at a rate of 15 kg/day for at least 20 years, i. e. the cumulative loss over that period must been more than 100 tonnes. River sediments near the outfall of effluent from that plant contained up to 0.17% mercury [Brooks]. In a copper smelter, the production of 1 tonne of copper resulted in the release of 2.1 tonne of dust containing 4% mercury.

The mercury dispersed from anthropogenic sources as vapor, in water-soluble salts, or in organic compounds generally displays greater geochemical mobility than naturally occurring mercury compounds (which are for the most part poorly soluble sulfides of low volatility) and is therefore more hazardous from an ecological point of view.

The mercury vapors emitted into the atmosphere are absorbed by particulate matter and then washed down by precipitation onto soil and water where the mercury undergoes various processes such as ionization, conversion into salts, and methylation, and is taken up by plants and animals. In the course of its migration in the air, water, soil, and food, the metallic mercury Hg^0 is converted to the Hg^{2+} species which is soluble. The residence time of Hg^0 (mercury vapor) in the atmosphere is estimated to be between 0.4 and 3 years and, as a consequence, mercury vapor is globally distributed (enters the global mercury cycle). The soluble species Hg^{2+} is thought to have the residence time of only several days or weeks, and therefore the distance over which it may be transported is limited. In industrial emissions, the average ratio of mercury vapor to the particulate fraction of mercury is 12:1.

The most common occupational exposure to mercury is that to its vapors. It occurs in a variety of industries including mercury mining, chloralkali factories, and instrument manufacturing, as well as in laboratories of physics and medicine. Air levels of mercury up to 5 mg/m^3 have been reported to have occurred during mining operations. Exposure has not infrequently resulted from spillage of mercury compounds on the work clothes from which the entrapped metallic mercury evaporates and is inhaled. In chloralkali factories, the presence of chlorine in combination with mercury gives rise to the formation of mercury(II) chloride aerosols. Mercury pollution in the vicinity of a chloralkali plant has been reviewed by Panda *et al.* and Wild *et al.* Exposure to aerosols of inorganic mercury may also occur in chemical factories where mercury(II) salts are handled [Berlin].

The general population is primarily exposed to mercury through the diet. Recent studies have shown that a substantial source of mercury exposure may be the mercury released in the mouth as vapor from dental amalgam restorations [WHO, 1991].

TOXICITY

Mercury is highly toxic to all forms of life.

Plants

Mercury vapor exhibits phytotoxicity, which is manifested in stunted growth of shoots and roots and accelerated aging of the plants [Besel; Trakhtenberg *et al.*], although terrestrial plants are generally considered to be rather insensitive to inorganic mercury compound [WHO, 1991].

Aquatic organisms

Although mercury is an important pollutant of the atmosphere and lithosphere, it is in the hydrosphere that its ecological effects are most significant. Mercury inhibits life processes in unicellular marine algae (at concentrations $\geqslant 0.1$ mg/L), reduces photosynthesis, interferes with the assimilation of nitrates, phosphates, and ammonia by aquatic plants, and brings about alterations in the structure and functioning of natural communities (at concentrations ≈ 1.0 µg/L). Within that range lie its threshold concentrations for many aquatic invertebrates (their susceptibility to mercury decreases in the order: crustaceans, mollusks, worms, bryozoans). Mercury concentrations of 5–10 µg/L or higher are toxic to fish in the early phases of development, reduce their growth rates and impair their olfactory function, cellular respiration in the gills, and enzyme systems of the liver; reproduction is also adversely affected by mercury [WHO, 1991].

The 96-hour LC_{50} values vary between 33 and 40 µg Hg/L for freshwater fish and are higher for seawater fish. Toxicity is affected by temperature, salinity, dissolved oxygen, and water hardness. A wide variety of physiological and biochemical abnormalities have been observed after exposure of fish to sublethal concentrations of mercury [WHO, 1989].

Animals and man

General aspects

Clinical manifestations of mercury toxicity in warm-blooded animals are many and varied but depend on the form in which mercury enters the body (whether as vapor or in inorganic or organic compounds), the route of entry, and the dose. At the basis of the mechanism by which it acts lies blockade of biologically active groups in the protein molecule (sulfhydryl, amine, carboxyl, and other groups) and of low-molecular-weight compounds, with the formation of reversible complexes that possess nucleophilic ligands.

Mercury(II) is capable of being incorporated into the transfer RNA which plays a central role in protein biosynthesis. Exposure to mercury has been shown to cause structural alterations in membranes of the endoplasmatic reticulum [Ivanov; Trakhtenberg & Ivanova].

Early in the course of low-level inhalation exposure to mercury, considerable amounts of hormones are released by the adrenals, with a consequent activation of hormonal synthesis there. Phasic changes in the content of catecholamines have been observed in these glands, as have been increases in the monoamine oxidase activity of hepatic mitochondria.

Mercury also produces a number of other biochemical effects — for example it impairs oxidative phosphorylation in hepatic and renal mitochondria, interferes with transamination processes, and upsets the balance between the activities of enzymes catalyzing glycogen resynthesis and breakdown. Mercury has also been shown to cause hypocoagulability involving hyperheparinemia and lowered fibrinogen levels.

Mercury in the vapor form displays neurotoxicity, particularly toward the higher divisions of the nervous system. Under exposure to mercury the initial increase in the excitability of the cerebral cortex is succeeded later by inertness of cortical processes. Subsequently, supraliminal (protective) inhibition is seen to develop.

Inorganic mercury compounds are highly nephrotoxic. Gonadotoxic, embryotoxic, teratogenic, and genotoxic effects of mercury compounds have also been reported [Gale; Ignatiyev; Shepard; WHO, 1991].

Mercury vapor

Acute/single and repeated exposures

Animals. Guinea-pigs exposed to mercury vapor at concentrations of 10–16 mg/m³ for 2–4 h per day did not survive beyond three exposures, while those continuously exposed to 3.5–5 mg/m³ survived for 4–6 days. In contrast, intermittent exposure to these and even higher concentrations (up to 8 mg/m³) was not lethal to most guinea-pigs for 35 to 70 days. With 8-hour daily exposure, dogs died in 1 to 3 days at $15-20$ mg/m^3 and 6 to 16 days at 12.5 mg/m^3; dogs exposed to $3-6$ mg/m^3 developed a typical picture of mercury poisoning, whereas those exposed to 1.9 mg/m^3 did not show visible signs of intoxication over the 40-day study period. Reversible changes in conditioned reflexes and in a number of functional, biochemical, and immunological parameters occurred in rats, rabbits, and cats exposed to $0.01-0.14$ mg/m^3. Mice, rats, and rabbits are more sensitive to mercury vapor than guinea-pigs; the least sensitive animals are dogs.

Man. Acute human poisoning has usually occurred as a result of industrial accidents (e. g. fires in mercury mines) or of a gross violation of occupational safety rules. Typically, clinical manifestations of poisoning appear 8 to 14 h after exposure and include general weakness, headache, painful swallowing, fever, and catarrhal lesions in the airways (rhinitis pharyngitis, less commonly bronchitis). A hemorrhagic syndrome then develops; later, a number of pathological changes may be added, including soreness of the gingivae, severe inflammatory lesions in the oral cavity (mercurial stomatitis with ulceration of the gingival mucosa), abdominal pain, gastric disturbances, signs of renal damage, and, less often, pneumonitis.

Cases of acute poisoning with such manifestations have occurred, for example, among electric welders working with mercury-contaminated equipment and in laboratory workers after a container with elemental mercury had been inadvertently left standing on an electric hot plate. An outbreak of poisoning has been described among workers engaged in the flame-cutting of a tubular heat exchanger delivered from an acetaldehyde-producing factory. The first signs of poisoning such as malaise, headache, and tremor of the hands appeared as early as $25-30$ min after the start of work; somewhat later, the subjects experienced nausea, multiple vomiting, watery blood-stained stools, and colicky epigastric pains. Impaired hearing, constriction of the visual fields, ataxia, and peripheral neuropathy in the absence of renal involvement have also been reported [Kagawa *et al.*].

Fitters working inside a melting pot that had previously contained mercury experienced nausea, multiple vomiting, chest pain, and fever on the same day; one of them developed severe tremor. Their urine contained mercury in concentrations of $0.23-0.27$ mg/L, while the air they had breathed contained 5 mg/m^3. In two other workers, symptoms of poisoning appeared 3 h after the start of riveting operations in a similar melting pot. Their urine contained 0.19 mg Hg/L. The rust covering the pot was found to have large amounts of mercury which was also present in the clothes of the victims (0.06 mg/100 g).

Milne *et al.* reported four cases of acute poisoning by mercury vapor in high concentration; the symptoms (excitement, chills, cough, tightness in the chest) appeared a few hours after the commencement of work. Acute poisoning with mercurial pneumonitis was observed in four men who had been exposed to mercury vapor while attempting gold ore purification at home using a gold-mercury amalgam and sulfuric acid [Levin *et al.*]. Fatal cases of acute poisoning by mercury vapor have also occurred.

Chronic exposure

Animals. Manifestations of chronic intoxication in laboratory animals developed slowly. The first to appear were behavioral changes (excitation succeeded by inhibition and weakness); later, hematological changes, nervous disorders, and signs of kidney and liver damage were observed. Dogs in the early phases of intoxication exhibited anxiety and aggressiveness accompanied by leukocytosis and a sharply increased erythrocyte sedimentation rate. Continued exposure led to impairment of living and then kidney functions. The dogs refused to eat meat, became emaciated, and developed tremor. Still later, changes in the oral mucosa were seen. After 2 months of exposure to low mercury vapor levels (0.005–0.02 mg/m^3), muscular weakness was recorded.

Autopsy of lethally poisoned animals revealed ulcerations in the lower parts of the colon, fatty degeneration of the kidney and liver, necrotic areas in the liver, and alterations in the cells bodies of spinal ganglia.

Morphological changes in the nervous system of animals during long-term exposure to elemental mercury vapor were first reversible and confined to interneuronal axonal and dendritic connections and to cell receptors, predominantly in the upper layers of the cerebral cortex (within the system of association fibers). With continued exposure, substantial and sometimes irreversible changes also occurred in cells lying in the lower portions of the brain, including the subcortical ganglia and the thalamus-hypothalamus region, and also in cells of the spinal cord. Still later, peripheral nerves were affected.

The prominent morphological changes in the circulatory system were vascular dilatation, greatly increased permeability of vessel walls, and endothelial swelling and vacuolation.

The morphological changes mentioned above were paralleled or followed in some animals by degenerative changes in a number of internal organs, manifested at the cellular level by cytoplasmic granularity and cloudiness in the liver, bronchial tree, pancreas, and kidney, which indicated that the protein metabolism suffered most. The lungs developed inflammatory processes in the early period of chronic exposure.

In animals exposed to low mercury vapor concentrations, different parts of the brain exhibited similar structural changes in the vessels, neurons, and neuroglia. In the neurons, changes of a reactive nature predominated early during the exposure period; with further exposure, degenerative changes could be seen.

In rats, strong inhibition of synthetic processes in the adrenal medulla, reduced monoamine oxidase and histaminase activities in various organs, and depressed serotonin levels in the small intestine were observed at the end of a 6-month exposure period. These changes were all reversible.

Morphological changes were also noted in endocrine glands such as the hypophysis, adrenals, and thyroid. Animals exposed to low mercury vapor concentrations exhibited signs of hypoxia and trophic disturbances as a result of vascular and hemodynamic changes.

In general, these and other observations have indicated that prolonged low-level exposure to mercury vapor leads not only to functional alterations, but, often, also to moderate and as a rule reversible morphological changes in the brain, endocrine glands, and various internal organs. In most cases the latter changes were observed in the absence of overt clinical manifestations of micromercurialism (see below).

Man. Clinical features of mercury poisoning vary in magnitude and in the rapidity of onset and development depending on the intensity and duration of mercury exposure and individual susceptibility. Mild chronic intoxications occurred in workers occupationally exposed to mercury vapor concentrations of $0.2-1.3$ mg/m^3 for 6 months or of 0.035 mg/m^3 for 3 years; some symptoms of intoxication were present in 2 out of 15 of those exposed to mercury vapor levels as low as $0.01-0.06$ mg/m^3. Psychophysiological examination of 20 workers following low-level occupational exposure to mercury vapor showed deteriorated motor coordination, early onset of fatigue, and impaired short-term memory.

In general, chronic intoxications tend to develop in an insidious way, without overt signs of the disease being present for a long time (compensatory phase). With further progression of mercury intoxication, two stages have been identified—an initial stage marked by signs and symptoms similar to those seen in neurasthenia and vegetovascular dystonia (angioneurosis), and an advanced stage characterized by more pronounced manifestations of poisoning that resemble those occurring in the asthenic–vegetative (psychoneurotic) syndrome [Milkov & Dumkin].

The stage of transition from the compensatory phase to the initial stage of mercurialism has been referred to as micromercurialism (see below). A demarcation line between these stages of the disease is, however, difficult to draw since the transition from one stage to another takes place gradually, often in an inperceptible manner. Helpful guides in this regard may be (i) the change from a fine and asymmetric tremor of fingers in the initial stage to a coarse hand tremor characteristic of the advanced stage of mercurialism and (ii) the presence or predominance in the latter stage of affective disturbances (e.g. emotional instability and explosiveness), hypothalamic dysfunctions, vagotonic reactions, and visceroneurotic manifestations such as cardiac pain, palpitation, and intestinal and vesicular dyskinesia. Although mercurial encephalopathy, not uncommon in the past, is rarely encountered at present, some of its signs may be observed at that stage [Piikivi].

A sensitive indicator of long-term mercury exposure in occupational settings has proved to be elevation of lysosomal enzymes in the blood, and tests for these enzymes may therefore be used in the diagnosis of chronic mercury poisoning.

Mercurialism. Characteristic early symptoms of this condition are fatiguability, weakness, somnolence, apathy, emotional instability, headache, and dizziness ('mercurial neurasthenia'). Their onset is followed by the development of trembling ('mercurial tremor') which affects the hands, tongue, eyelids, and in severe cases the legs and then the entire body. Parallel to the development of tremor, a state of excessive excitability (mercurial erethism) sets in, expressed in the form of shyness, timidity, fearfulness, and general depression. With progression of the disease, the patient becomes highly irascible, gloomy, and tends to weep frequently.

The actions of mercury on the autonomic nervous system are manifested in hypersalivation, impaired secretory function of the stomach, cyanosis, profuse sweating, rapid or slowed heart rate, and frequent urination; early in the course of the disease, signs attributable to increased excitability of the sympathetic portion of the autonomic nervous system predominate and combine with hyperfunction of the thyroid gland. Disorders of the peripheral nervous system usually manifest themselves as multiple neuralgia; polyneuritis has also been observed. Neurotic manifestations of intoxication usually take the form of pains in the extremities and in the trigeminal nerve region; sensitivity in the distal parts of the body may be slightly impaired. A degree of facial asymmetry has been seen in some cases.

An important sign of chronic mercurialism is diminished strength of the extensors in the most used hand. This sign may appear during the first few months of exposure independently of the other signs of poisoning.

Changes in the digestive system and kidneys are usually slight or absent. Body temperature is sometimes elevated.

There is evidence that prolonged exposure to mercury is associated with an increased incidence of tuberculosis and a high mortality from this disease.

Not uncommon findings in female workers with chronic mercurialism are disorders of the generative function and menstrual cycle (e.g. painful and profuse menstruations). The incidence of menstrual disturbances rises with increasing length of occupational exposure, and so do the rates of spontaneous abortion, premature delivery, and mastopathy. Chronic maternal exposure to mercury tends to result in water retention and occult edemas developing in the neonate. Infants born to chronically exposed mothers often have malformations and underdeveloped protective and adaptive mechanisms.

Micromercurialism. This term has been proposed to designate the sum of signs and symptoms arising during protracted human exposure to low mercury concentrations (usually of the order of several hundredths of a milligram per m^3). Micromercurialism of first degree is manifested predominantly in reduced working capacity, rapid onset of fatigue, and heightened excitability; mild swelling of the nasal mucous membrane is seen occasionally. In second-degree micromercurialism, progressive memory deterioration, anxiety, lack of self-confidence, irritability, and headache are added to the above-mentioned manifestations which become more strongly marked. Catarrhal phenomena in the upper air passages, inflammatory changes in the oral mucous membrane, and sore gums that bleed easily may also be observed. Occasionally, the patients complain of unpleasant sensation in the cardiac region, slight trembling, frequent urination and a tendency to diarrhea. The signs and symptoms of third-degree micromercurialism resemble those outlined above for chronic mercurialism but are less pronounced.

Micromercurialism may be difficult to recognize and is often misdiagnosed as a respiratory or nervous system disease depending on the prevailling symptoms; many patients present with a diagnosis of neurasthenia or hysteria. Individuals occupationally exposed to mercury for prolonged periods were found to have thyroid hyperfunction, elevated plasma levels of 11-hydroxycorticosteroids, increased androsterone levels in the urine, and considerably reduced urinary excretion of bound 17-hydroxycorticosteroids along with increased excretion of these steroids in the free form.

In workers exposed to mercury for relatively short periods, increased norepinephrine and decreased dopamine levels were noted in the urine, as well as an elevation of blood serotonin followed by augmented urinary excretion of its metabolite 5-hydroxyindoleacetic acid and a rise in serum monoamine oxidase activity. The glucocorticoid and androgen-producing functions of the adrenal cortex were often inhibited, and this was accompanied by lethargy, adynamia, and hyperglycemia. Increases in the activity of hepatic aminotransferases (GOT and GPT) and cholinesterase were also observed.

Micromercurialism has not infrequently been detected in workers at research laboratories exposed to low mercury concentration (at MAC or somewhat higher levels) for prolonged period — 8–10 years in most cases and for 3 years or so in some. The disease manifested itself essentially as a neurasthenia syndrome involving vegetative dysfunction ans psychic disturbances of varying severity such as irritability, emotional instability, fatiguability, reduced capacity for mental work, decreased ability to concentrate, memory deterioration, etc. Nearly all subjects exhibited a characteristic fine tremor of fingers on outstretching the arm; tremor of the eyelids and tongue was also common. In addition, some subjects had an enlarged thyroid gland, readily bleeding gums, hypersalivation, and/or gingivitis. Hematological findings included reduced hemoglobin levels and erythrocyte counts, leukopenia, and, less often, leukocytosis, altered differential blood counts, and elevated or lowered levels of sulfhydryl groups.

Inorganic mercury compounds

Acute exposure

Animals. Acute oral poisoning with mercury compounds was first manifested in general weakness and adynamia followed shortly by anorexia, strong thirst, hypersalivation, and vomiting, to which bloody diarrhea was added later. The ocular mucous membranes showed catarrhal inflammation. Ataxia and increased general excitability were also noted, as were tremor, paresis, and occasionally convulsions.

Lethal oral poisoning with **mercury(II) chloride** resulted in a sudden death (within 1 h) of some animals from damage to the bundle of His. In animals that died later (1 to 3 h), lesions were observed in the gastrointestinal tract, while those surviving for longer periods (up to day 5) developed lesions in the kidneys which are the main pathways for mercury elimination from the body. Characteristic features of poisoning were inflammations and hemorrhagic ulcerations in the stomach and intestine, degenerative and inflammatory changes in the kidney, and cellular degeneration in the liver.

Rabbits administered **mercury(I) chloride** orally in doses 20–50 mg/kg for 2 months exhibited body-weight loss, tremor, and diarrhea [Tolgskaya]. Microscopic examination of the nervous system revealed profound hemodynamic and degenerative (dystrophic) changes in the brain, subcortical ganglia, spinal cord horns, and peripheral nerve trunks. The histopathologic picture of intoxication was dominated by degenerative processes in a variety of structures including the thalamus and hypothalamus, dorsal nuclei of the vagus nerve, cells of the anterior and lateral horns of the spinal cord, and peripheral nerves and conducting pathways. In the gastrointestinal tract, desquamative catarrh of the mucous membranes and signs of ulcerative colitis were observed. The kidneys showed degenerative and sometimes necrotic changes in the epithelium of convoluted tubules. In the liver, fatty degeneration of hepatocytes was noted. The observations that the thalamus/hypothalamus region and the anterior and lateral horns and conducting pathways of the spinal cord were affected most, whereas other parts of the brain (including the cerebral cortex) and spinal cord were only moderately damaged, are similar to those made in human mercury poisoning. Despite the marked changes in the peripheral nerves and anterior spinal horns, no clinical manifestations of paresis were seen.

Inhalation exposure of mice, guinea pigs, and rabbits to **mercury(II) chloride** vapor caused general inhibition, adynamia, increased reflex-mediated excitability, rapid and then slow and superficial breathing, motor incoordination, paresis, and paralysis.

Man. Cases of acute poisoning have resulted from accidental or suicidal ingestion of readily soluble colorless mercury(II) salts such as the **chloride** or **nitrate**.

Poisoning by mercury salts is characterized by a diversity of signs and symptoms including headache, salivation, reddening, swelling, and ready bleeding of the gums (a dark line may appear on their border), stomatitis, swelling of lymphatic and salivary glands, dyspepsia, colitis, and tenesmus. Fever was common. In milder cases, the functional symptoms cleared in 2 to 3 weeks, whereas in severe cases gross renal lesions developed, including necrotizing necrosis which proved fatal in 5–6 days. Digestive disorders ranged from loss of appetite and slight nausea to vomiting (sometimes with blood) and mucous diarrhea, most often bloody. Multiple ulcerations appeared on the gastric and duodenal mucous membranes. In some cases atrophic changes developed in the liver. Urinary output was increased in the early stage of intoxication followed by oliguria or even anuria later.

In nephroses caused by **mercury(II) chloride** (corrosive sublimate), edemas were characteristically absent, and the blood content of total protein was increased. The kidneys exhibited total necrosis of the convoluted tubular epithelium which underwent breakdown and desquamation, not infrequently with calcification of the dead tissue elements. Lime was retained in the body because of the damage caused by the sublimate to the large intestine.

Necropsy of those who had been lethally poisoned by mercury salts revealed grossly congested internal organs and numerous hemorrhages in the pleura, pericardium, peritoneum, pia mater, and occasionally also in brain and spinal cord tissues — predominantly in the cerebral cortex and spinal gray matter in the form of *purpurae cerebri*. The brain was edematous; its ventricles were dilated and filled with a sanguineous fluid. Microscopic studies showed acute swelling of nerve cells in various parts of the nervous system. Severe lesions in cases of sublimate poisoning were also found in the gastrointestinal tract, but the kidneys were more affected than any other organ [Tolgskaya].

Chronic exposure

Man. Chronic intoxications have resulted from prolonged exposure to mercury compounds through inhalation. The overall clinical picture resembled that outlined above, but there were certain specific features, particularly in poisoning with **mercury fulminate**. Its dust irritates the conjunctivae, producing a burning sensation, and may cause conjunctivitis as well as catarrh of the upper air passages and pharyngeal irritation. Poisoning with the fulminate is usually milder than that with mercury vapor, and mercurial erethism is seldom encountered.

Exposure to mercury fulminate dust may cause dermal sensitization and lead to dermatitis. The skin then becomes reddened, edematous, and itchy; the edema tends to be most conspicuous around the eyes, in the occipital region, and on the forearms. Papules and nodules may appear in the affected skin areas; pustular folliculitis and small ulcers also developed in some cases. During the recovery period, desquamation of the skin occurs for 1 to 2 weeks. The disease may recur.

ABSORPTION, DISTRIBUTION, AND ELIMINATION

Mercury is a trace element that is continuously present in the human body. On post-mortem examination of unexposed individuals, 67% and 33% of them had mercury concentrations of about 1 mg% in the kidney and liver, respectively.

The main route for the entry of metallic mercury into the body is the inhalation of its vapor or dust, particularly in occupational settings. About 80% of inhaled vapor is retained in the respiratory tract (the retention occurs almost entirely in the alveoli); if the mercury vapor concentration in the air does not exceed 0.25 mg/m^3, virtually all vapor is retained in the airways.

Ingested liquid metallic mercury is absorbed from the digestive tract in very small or even negligible amounts, while inorganic mercury compounds are absorbed from foods to a level of less than 10% [WHO, 1991]. The average daily intake of mercury with food is in the range of a few micrograms. Absorption of mercury or some of its compounds through the skin is possible, at a low rate. Studies on human volunteers have indicated that the absorption of metallic mercury vapor via the skin is only about 1% of its uptake by inhalation [Hursh et al.]. However, the use of skin-lightening creams containing inorganic mercury salts results in substantial absorption and accumulation in the body [WHO, 1991].

Absorbed mercury accumulates in the thyroid, pituitary, brain, kidney, liver, pancreas, testes, ovaries, prostate, and other organs, but its distribution is not uniform. The main mercury depository after the entry of elemental mercury vapor or mercury salts is usually the kidney. In cases of lethal poisoning with mercury vapor, the highest concentrations occurred in the brain, kidney, liver, and heart. Following poisoning with a single dose of mercury(II) chloride, the greatest amount was found in the kidneys, followed by the liver and spleen. Mercury can pass through the placental barrier.

The uptake of metallic mercury vapor from inhaled air into the blood depends on the dissolution of mercury vapor as it passes through the pulmonary circulation. About 90% of mercury in the blood occurs in erythrocytes.

Mercury is eliminated from the body by all glands of the gastrointestinal tract, by the kidneys, by salivary, sweat and mammary glands, and by the lungs. The urinary route dominates if the exposure is high. Elimination begins several hours after poisoning and continues for a long time, even after a single exposure. On average, 40% was found to be excreted by the kidneys, 30–35% by glands of the large intestine, and 20–25% by salivary glands. The elimination is nonuniform, both during the period of intoxication and subsequently.

In cells, the major site of mercury accumulation is the nucleus, followed by the microsomes, cytoplasm, and mitochondria in that order. Mercury has damaging effects on all subcellular structures [Dieter *et al.*]. After a single administration of inorganic mercury compounds to rats, 54% and 30% of the mercury found in the kidney was present in the soluble and nuclear fractions of this organ, respectively, while 11% and 6%, respectively, were in the mitochondrial and microsomal fractions. Elimination of mercury from all subcellular structures is relatively rapid and uniform, but it is retained somewhat longer in the mitochondria and microsomes than in the other structures [Levina].

Mercury levels in blood and urine can be used as indicators of exposure. Following exposure, mercury concentrations rarely exceed 5 µg% in the blood and 2.5 µg% in the urine, but elevated concentrations in biological media do not necessarily correlate with poisoning and are usually of diagnostic importance only when considered in conjunction with clinical manifestations of intoxication. If clinical symptoms are present, urinary concentration of ≥1 µg% may be regarded as significant [Gadaskina *et. al.*].

Table 1. Exposure limits for mercury and its compounds

	Workroom air		Atmospheric air	Water sources		Soil	
	MAC_{wz} (mg/m^3)	Aggregative state	MAC_{ad} (mg/m^3)	MAC_w (mg/L)	MAC_{wf} (mg/L)	MAC_s (mg/kg)	Hazard class
Mercury, metallic	0.01/0.005[a]	vapor	0.0003	0.0005	0.0001	2.1[b]	1
Inorganic mercury compounds (as Hg):							
(I) & (II) acetate	0.2/0.05	Aerosol	0.0003	—	—	—	1
(II) amidochloride	0.2/0.05	Aerosol	0.0003	—	—	—	1
(II) bromide	0.2/0.05	Aerosol	—	—	0.001	—	1
(I) & (II) chloride	0.2/0.05	Aerosol	0.0003	—	—	—	1
(II) iodide	0.2/0.05	Aerosol	0.0003	—	0.001	—	1
(I) & (II) nitrate	0.2/0.05	Aerosol	0.0003	—	0.001	—	1
(II) oxide	0.2/0.05	Aerosol	0.0003	—	—	—	1
(I) & (II) sulfate	0.2/0.05	Aerosol	—	—	0.001	—	1
(II) thiocyanate	0.2/0.05	Aerosol	—	—	0.001	—	1

[a] The value in the numerator is the concentration not to be exceeded at any time, while that in the denominator is the average maximum allowable concentration per shift.
[b] Based on the translocation index of harmfulness that characterizes the transfer of a compound from the arable soil layer to the green mass and fruits of the plants through the root system.

HYGIENIC STANDARDS

Exposure limits adopted for mercury and its inorganic compounds in the USSR are shown in Table 1. In the USA, a TLV-C of 0.05 mg/m^3 has been set for mercury vapor (with the "skin" notation to indicate the possibility of absorption by the cutaneous route, including the mucous membranes and eyes) and a TLV-TWA of 0.1% for inorganic mercury compounds (as Hg).

The highest permissible levels established in the USSR for mercury in human foods are listed in Table 2.

For oceanic waters, the highest permissible concentration for soluble forms of mercury and those that can pass through a filter with 0.5-μg pores is considered to be 0.001 mg/L on the basis of toxicological and biogeochemical criteria [Patin & Morozov].

METHODS OF DETERMINATION

Mercury vapor in air can be determined using automatic instruments operating, for example, on the basis of a reaction between mercury and selenium sulfide. Various colorimetric procedures are available, such as those based on the absorption of mercury vapor with an aqueous solution of iodine and potassium iodide followed by determination of the HgI_4^{2-} anion according to the intensity of the yellow−pink color of the sedimented complex salt ($Cu_2[HgI_4]$); on the interaction of mercury with iodine, copper chloride, and sodium sulfite (sensitivity of this method is 0.02 μg in the assayed solution volume; iron interferes with the assay). [Peregud & Gorelik]; or on the use of antipyrine or solid sorbents [Peregud]. Concentrations of mercury chloride vapors and aerosols can be estimated by a colorimetric technique based on the determination of mercury ions by measuring the pink color produced by a complex salt (sensitivity 0.5 μg in the volume analyzed; interference may come from organic mercury compounds [*Specifications for Methods of Determining Noxious Substances in Air*].

The determination of mercury by colorimetric measurement of a mercury dithizonate complex was the basis of most of the methods used in the 1950s and 1960s. These methods all make use of wet oxidation of the sample followed by extraction of mercury into an organic solvent as a dithizonate complex and finally the colorimetric determination of the complex itself. Copper is the metal most likely to interfere with mercury measurement by the dithizone procedure which has an absolute sensitivity of about 0.5 μg of mercury [WHO, 1976].

More recently, wide use for the determination of mercury in air as well as in other media has been made of atomic absorption procedures. With them, mercury (inorganic or total) has been determined in foods [Fergusson], blood, urine, hair [Pineau et al.], and other biological tissues with detection limits between 2.0 and 4.0 ng/g. For routine analyses, various forms of flameless atomic absorption spectrophotometry have been employed most frequently. Procedures for neutron activation analysis of total mercury have also been developed and used [WHO, 1991]. Detailed information concerning analytical methods for mercury can be found in *Environmental Health Criteria 1: Mercury* [WHO, 1976]. Procedures for mercury determination in environmental media and biological materials have also been reviewed by Angener & Schaller; Dmitriyev & Tarasova; Dubinskaya; Gadaskina *et al.*; and Magos.

Table 2. MACs for mercury in foods (mg/kg)

Cereal grains, groats, and bran	0.03
Flour, starch, and pastries	0.02
Sugar and candies	0.01
Bread	0.01
Salt	0.01
Milk and sour milk products	0.05
Condensed milk, canned	0.15
Butter and vegetable oils	0.03
Margarines and cooking fats	0.05
Cheese and curds	0.02
Vegetables, fresh or canned	0.02
Fruits, fresh or canned	0.02
Berries, fresh or canned	0.02
Mushrooms, fresh	0.05
Nuts	0.03
Spices	0.02
Tea	0.1
Meat and poultry, fresh or canned	0.03
Sausages	0.03
Fish, fresh or canned	0.3–0.6
Eggs	0.02
Mineral waters	0.005
Beverages based on infusions or essences	0.005
Infant foods:	
milk-based	0.005
grain- and milk-based	0.01
vegetable- or fruit-based	0.01
meat- or poultry-based	0.02
fish-based	0.15

MEASURES TO CONTROL EXPOSURE

At the workplace

Efforts should be made to replace mercury by less hazardous substances wherever possible. During the manufacture and use of mercury-containing instruments, measures should be taken to prevent contact with this metal. Special attention should be given to ensuring the appropriate design, equipment, and operation of mercury-producing plants. It is essential to prevent the escape of mercury from containers, the dispersion of its droplets in the air, and its infiltration into crevices and gaps in the floors and working tables. It is highly desirable that mercury be handled in hermetically sealed systems. Where there is a danger of mercury volatilization, an exhaust ventilation system should be installed if feasible.

For *personal protection* against the mercury vapors (as well as those of organic mercury compounds) present in concentrations that do not exceed the permissible level more than 15-fold, filter-type gas masks must be worn; those exposed to condensation or disintegration aerosols of inorganic mercury compounds should wear particulate-filter respirators or valve respirators with an anti-mercury cartridge.

Strict personal hygiene should be enforced at the workplace. Eating and smoking in premises where vapors of mercury or its compounds may be emitted into the air must be prohibited. Before a meal, workers exposed to mercury or its compounds should take off their overalls, wash their hands, and gargle a 0.25% potassium permanganate solution. The working clothes should be made of light-colored cotton fabric and decontaminated on a centralized basis. Other personal hygiene measures include changing the working clothes, showering, gargling, and cleaning the teeth at the end of the work shift. All workers who may be subject to mercury exposure should undergo preplacement and periodic medical examination.

Protection of the general environment

The basic measures include removal of mercury and its compounds from ventilation exhausts and waste waters and utilization of solid wastes; regulations should be developed to specify how municipal and industrial sewage and slags be treated and used in agriculture [Stepanov].

Based on a consideration of the global mercury balance and the patterns of mercury migration in the environmental media (with the aquatic medium being given priority in assessing the ecologic cosequences of the earth's pollution by mercury from anthropogenic sources), the maximal permissible global mercury emission into the atmosphere from these sources has been determined at $(4-7) \times 10^4$ tonnes per year [Izrael]. This is much more than the estimated total global release of mercury into the atmosphere due to human activities at the present time (of the order of 2000–3000 tonnes/year) [WHO, 1991]; however, situations where mercury exposure presents real hazards to human health can arise locally, particularly if atmospheric pollution combines with the direct entry of mercury compounds (especially organic) into bodies of water.

CASE MANAGEMENT IN POISONING

Specific treatment consists in the use of substances that combine with mercury to form relatively innocuous water-soluble complex compounds which are excreted by the kidneys. As such, dithiols are used including BAL (dimercaprol; chemical name 2,3-dimercaptopropanol) and, especially in the former USSR, unithiol (sodium 2,3-dimercapto-1-propane sulfonate) and succimer (meso-2,3-dimercaptosuccinic acid). Unithiol (5% solution) is administered subcutaneously or intravenously at 5 mg per kg of patient's body weight—three or four injections at 6–8-hour intervals on the first day, two or three at 8–12-hour intervals on the second, and one or two injections subsequently (at least for 4 days); the number of injections depends on the patient's condition.

Succimer is available in 0.5-g tablets for oral use and in powder form for intramuscular injections (0.3 g of the powder is dissolved ex tempore in 6 ml of 5% sodium hydrocarbonate solution). It is administered intramuscularly in single doses of 0.3 g twice daily for up to 5–7 days in mild cases and four times on the first day, three times on the second, and once or twice subsequently (5 days) in severe cases (total dose ≈5 g). In all patients with mercurialism whatever its severity, succimer therapy markedly enhances urinary excretion of mercury, improves the general condition of the patient, reduces tremor, greatly diminishes vegetative vascular disturbances, and improves biochemical parameters. The drug is well tolerated.

Other antidotes against mercury are D-penicillamine and oxathiol. D-penicillamine, the molecule of which contains a carboxyl, an amine, and a thiol group, forms water-soluble complexes with mercury. It is marketed in gelatin capsules (0.15 g daily). Oxathiol combines high efficiency with low toxicity. The efficacy of various antidotal therapies in mercury poisoning has been reviewed by Borisov et al.

A patient with severe acute intoxication caused by mercury ingestion should have his stomach washed out immediately with water containing 20–30 g of activated charcoal and then given milk and egg whites beaten up with water, followed by a cathartic.

In acute poisoning, particularly by inhalation, with mercury vapor, complete rest is necessary for the victim after his removal from the area of exposure. In addition to unithiol, 10 ml of a 10% calcium chloride solution, 20–40 ml of a 40% glucose solution, and 10 ml of a 20% sodium thiosulfate solution should be given intravenously; symptomatic treatment is then carried out.

In conjunction with the specific antidotal treatment, general supportive therapy and cardiac tonics are widely used, especially in chronic mercury poisoning, including dibazol [synonym: bendazole hydrochloride], small doses of bromide in combination with caffeine, and intravenous infusions of calcium chloride solution alternating with those of glucose solution with ascorbic acid and thiamine. Good results have been observed after intravenous infusions of a 20% sodium thiosulfate in amounts of 10–15 ml on alternate days. Physical therapy is also recommended, in particular, the taking of hydrogen sulfide baths and ultraviolet irradiation in combination with warm coniferous baths. Spa treatment (e.g. in Matsesta or Pyatigorsk) is beneficial.

Workers with mild or initial chronic mercury intoxication should be kept away from mercury exposure for 3 to 4 weeks and those with moderate intoxication for 6 to 9 weeks; people with severe or recurrent intoxication must not be further employed in occupations involving mercury exposure.

REFERENCES

Angener, J. & Schaller, K. H. (1988) *Analyses of Hazardous Substances in Biological Materials* Vol. 2. VCH, New York, pp. 195–211.

Bezel, V. S. (1982) *Ekologiya*, No. 5, 65–71 (in Russian).

Berlin, M. (1979) In: Friberg, L. *et al.* (eds) *Handbook on the Toxicology of Metals.* Elsevier, Amsterdam, pp. 503–530.

Borisov, V. P. *et al.* (1981) *Gig. San.*, No. 4, 88–91 (in Russian).

Brooks, R. R. (1979) In: Bockris, J. O'M. (ed.) *Environmental Chemistry.* Plenum Press, New York, pp. 429–476.

Dieter, M. P. *et al.* (1983) *Toxicol. Appl. Pharmacol.*, **68**, 218–228.

Dmitriyev, M. T. & Tarasova, L. N. (1985) *Zagryaznenie okruzhayushchei sredy metallami i metody ikh opredeleniya: Obzor literatury* [Environmental Pollution by Metals and Methods of their Determination. A review], No. 9. Moscow (in Russian).

Dobrovolsky, V. V. (1983) *Geografiya mikroelementov. Globalnoe rasseyanie* [Geography of Trace Elements. Global Dispersion]. Mysl, Moscow (in Russian).

Dubinskaya, N. A. (1980) *Izv. Akad. Nauk Latv. SSR* (Ser. Fiz.–Tekhn., Khim. i Geol. Nauk), No. 2, 16–22 (in Russian).

Fergusson, J. E. (1990) *The heavy Elements: Chemistry, Environmental Impact and Health Effects.* Pergamon Press, New York.

Fitzgerald, W. F. & Watras, C. J. (1988) *Sci. Total. Environ.*, **87/88**, 223–232.

Gadaskina, I. D. *et al.* (1975) *Opredeleniye promyshlennykh neorganicheskikh yadov v organizme* [Determination of Inorganic Industrial Poisons in the Organism]. Meditsina, Leningrad (in Russian).

Gale, T. F. (1981) *Environ. Res.*, **24**, 152–161

Harris, R. & Hohenebser, C. (1978) *Environment*, **20**, 25–36.

Hursh, J. B. *et al.* (1989) *Arch. Environ. Health.*, **44**, (2), 120–127.

Ignatiyev, V. M. (1980) *Gig. San.*, No. 3, 72–73 (in Russian).

Ivanova, L. A. (1982) *Gig. Truda Prof. Zabol.*, No. 2, 27–30 (in Russian).

Izrael, Yu. A. (1984) *Ekologiya i kontrol sostoyaniya prirodnoi sredy* [Ecology and Environmental Monitoring]. Gidrometeoizdat, Moscow (in Russian).

Kagawa, J. *et al.* (1974) *Jap. J. Ind. Health*, **16**, 11–22.

Levin, M. *et al.* (1988) *Chest*, **94**, 554–556.

Levina, E. I. (1972) *Obshchaya toksikologiya metallov* [General Toxicology of Metals]. Meditsina, Leningrad (in Russian).

Lundbergh, K. & Paulsson, K. (1988) *Sci. Total Environ.*, **87/88**, 495–507.

Magos, L. (1988) Chapter 35 *Mercury* in: Seiler, H. G. *et al.* (eds) *Handbook on Toxicity of Inorganic Compounds.* Marcel Dekker, New York, pp. 419–436.

Milkov, L. E. & Dumkin, V. L. (1983) *Rukovodstvo po professionalnym zabolevaniyam* [Occupational Diseases (Manual)]. Meditsina, Moscow (in Russian).

Milne, J. *et al.* (1970) *Brit. J. Ind. Med.,* **27,** 334–338.

Panda, K. K. *et al.* (1990) *Sci. Total Environ,* **96,** 281–296.

Patin, S. A. & Morozov, N. P. (1981) *Mikroelementy v morskikh organizmakh i ekosistemakh* [Trace Elements in Marine Organisms and Ecosystems]. Legkaya Promyshlennost, Moscow (in Russian).

Peregud, E. A. (1976) *Khimicheskiy analiz vozdukha (novye i usovershenstvovannye metody* [Chemical Analysis of Air (New and Improved Methods)]. Khimiya, Leningrad (in Russian).

Peregud, E. A. & Gorelik, D. O. (1981) *Instrumentalnye metody kontrolya zagryazneniya atmosfery* [Instrumental Methods of Atmospheric Pollution Monitoring]. Khimiya, Leningrad (in Russian).

Piikivi, L. (1989) *Acta Universitatis Ouluensis,* Series D. Medica 183, Oulu.

Pineau, A. *et al.* (1990) *J. Anal. Toxicol.,* **1/4,** 235–238.

Sager, P. & Matheson, D. (1988) *Toxicology,* **49,** 479–492.

Shepard, T. H. (1986) *Catalog of Teratogenic Agents.* Johns Hopkins University Press, Baltimore & London.

Shrestha, K. P. & Ruiz de Quilarque, X. (1989) *Sci. Total Environ.,* **79,** 233–240.

Stepanov, A. S. (1980) *Ochistka gazovykh vybrosov ot rtuti i eye soyedineniy: Obzornaya informatsiya* [Purification of Gaseous Emissions from Mercury and its Compounds. A Review]. Moscow (in Russian).

Specifications for Methods of Determining Noxious Substances in Air (1971) [Tekhnicheskie usloviya na metody opredeleniya vrednykh veshchestv v vozdukhe], No. 6. Publication of the USSR Ministry of Health, Moscow, pp. 9–11 (in Russian).

Tolgskaya, M. S. (1967) *Morfologicheskie izmeneniya v nervnoi sisteme pri professionalnykh neirointoksikatsiyakh* [Morphological Changes in the Nervous System in Occupational Neurological Intoxications]. Meditsina, Leningrad (in Russian).

Trakhtenberg, I. M. & Ivanova, L. A. (1984) *Gig. San.,* No. **5,** 59–63 (in Russian).

Trakhtenberg, I. M. *et al.* (1985) *Gig. San.,* No. **2,** 60–63 (in Russian).

WHO (1976) *Environmental Health Criteria 1: Mercury.* World Health Organization, Geneva

WHO (1989) *Environmental Health Criteria 86: Mercury—Environmental Aspects.* World Health Organization, Geneva.

WHO (1991) *Environmental Health Criteria 118: Inorganic Mercury.* World Health Organization, Geneva.

Wild, S. R. *et al.* (1992) *Env. Poll.,* **76,** 33–42.

Boron and its compounds

Boron fluoride (boron trifluoride) **b. nitride** (borazon [diamond-like modification], white graphite [hexagonal modification]); **b. oxide** (boric anhydride)
Barium metaborate; calcium hexaboride, calcium metaborate; chromium diboride; decaborane(14); diborane(6); dimolybdenum pentaboride; lead metaborate hydrate; magnesium diboride, magnesium dodecaboride (polyboride); **molybdenum boride; orthoboric acid** (boric acid); **pentaborane(9); potassium pentaborate tetrahydrate; sodium metaborate, sodium perborate, sodium tetraborate decahydrate** (borax [min.], tincal [min.]); **tetraboron carbide** (boron carbide); **titanium-chromium diboride**

IDENTITY AND PHYSICOCHEMICAL PROPERTIES
OF THE ELEMENT

Boron (B) is an element in group III of the periodic table, with the atomic number 5. Its natural isotopes are ^{10}B (19%) and ^{11}B (81%).

Crystalline boron is extremely hard, being inferior only to diamond in this respect. It can also exist as an amorphous powder. It is diamagnetic and a poor conductor of electricity.

Boron is fairly inert chemically, particularly in the crystalline form. Its oxidation state in most compounds is +3. In air it burns only at 700°C, to give the oxide B_2O_3. It reacts slowly with water vapor at a red heat, converting to the orthoboric acid H_3BO_3. It does not react with hydrogen even at high temperatures, but will combine with fluorine at room temperature and with the other halogens on heating, to form volatile halides of the type $BHal_3$. At 900°C it reacts with nitrogen to form nitride. When heated with sulfur or selenium it gives the sulfide B_2S_3 or the selenide B_2Se_3. On heating with coal in an electric furnace it gives the carbide B_4C. It does not react with phosphorus, arsenic, or antimony. It gives borides with metals. It reduces many oxides, halides, and sulfides.

Nonoxidizing acids do not react with boron. Concentrated nitric acid and aqua regia oxidize it to orthoboric acid. Boron reduces concentrated sulfuric acid to sulfur(IV) oxide at 250°C and orthophosphoric acid to phosphorus at 800°C. In the presence of boron, hydrogen fluoride and hydrogen chloride release their hydrogen at a red heat. Boron does not react with aqueous solutions of alkalies but will give borates on fusion with alkalies. It reacts vigorously with molten sodium peroxide and with a mixture of potassium nitrate and sodium carbonate. Boron(III) has a tendency to complex formation; its characteristic coordination number is 4. See also Appendix.

NATURAL OCCURRENCE AND ENVIRONMENTAL LEVELS

The clarke of boron is estimated at $(10-12) \times 10^{-4}\%$ in the earth's crust and at $10 \times 10^{-4}\%$ in the granite layer of the continental crust [Dobrovolsky]. Its estimated abundances in sedimentary and igneous rocks are $56 \times 10^{-4}\%$ and $13 \times 10^{-4}\%$, respectively [Brooks]. The major boron minerals are borax $N_2B_4O_7 \cdot 10H_2O$, kernite $N_2B_4O_7 \cdot 4H_2O$, sassoline H_3BO_3, asharite $2MgO \cdot B_2O_3 \cdot 3H_2O$, and datolite $2CaO \cdot B_2O_3 \cdot 2SiO_2 \cdot H_2O$.

The oceans contain an estimated 6 096 500 million tonnes of dissolved boron, at an average concentration of 4450 µg/L, and receive up to 370 000 tonnes in river discharges annually. In river water, boron concentrations average 10 µg/L [Dobrovolsky].

In the USSR, elevated levels of boron were found in surface and ground waters in Northwest Kazakhstan (1.8–15.7 mg/L), some regions of Western Siberia (up to 9.4 mg/L), and several Transcarpathian and Transcaucasian areas (up to 40 mg/L and 203.7 mg/L, respectively).

The average abundance of boron in soils is $1 \times 10^{-3}\%$, of which about 10% is usually water-soluble boron, except in areas with high soil levels of boron salts where the soluble fraction may make up as much as 80% of the total boron [Vinogradov]. Boron levels in the soil and water of a biogeochemical province were found to correlate well with those in all plants (including cereal crops) and in farm animals of that province [Kovalsky & Shakhova]. A weak positive correlation between the incidence of cancer and the overall soil content of boron has been reported [Dubikovsky].

Continental vegetation contains an estimated 62.5 million tonnes of boron at average fresh weight, dry weight, and ash weight concentrations of $10.0 \times 10^{-4}\%$, 25.0×10^{-4} and $500 \times 10^{-4}\%$ respectively; the amount of boron taken up annually worldwide through the increment in phytomass is estimated to be 4312 thousand tonnes, or 28.8 kg/km^2 of land area (K_b=50) [Dobrovolsky]. Reported average boron concentrations in organisms are (mg/100 g dry weight) : marine algae, 12; terrestrial plants, 5; marine animals, 2–5; terrestrial animals, 0.05; bacteria, 0.55 [Kovalsky].

PRODUCTION

Boron is obtained by decomposition of borates with sulfuric acid at 80–95°C, dehydration of the resulting orthoboric acid to boron oxide and reduction of the latter; by thermal decomposition of boron halides; or by electrolysis of borates.

Borides of metals are obtained by reacting oxides of the corresponding metals with boron carbide; by electrolysis of molten mixtures of alkaline or alkaline-earth metal borates and oxides of high-melting metals; or by thermometallurgical reduction of mixtures of metal oxides and boron. **Boron fluoride** is made by reaction of boron halides with fluorine, of boron oxide with coal in a fluorine atmosphere, or of sodium or potassium tetrafluoroborate with boron oxide in the presence of sulfuric acid. **Boron nitride** is obtained by heating boron and boron oxide in a flow of ammonia. **Diborane(6)** is obtained by reaction of sodium boron hydride or lithium aluminum hydride with boron fluoride, or from boron trialkyls and hydrogen at temperatures of 140–200°C and pressures of 19.6–25.5 MPa. **Pentaborane(9)** and **decaborane(14)** are produced chiefly by pyrolysis of diborane(6). **Sodium tetraborate** is extracted from tincal, kernite, and some other minerals by their recrystallization and from the water of salt lakes by fractional recrystallization; it is also produced by reaction of orthoboric acid with sodium carbonate. **Tetraboron carbide** is made by heating boron or boron oxide with coal.

USES

Boron is added in small amounts to steels and certain alloys of nonferrous metals to increase their resistance to heat and corrosion. It is also used in the nuclear power industry (e.g. in control rods and as a constituent of neutron shielding materials for nuclear reactors), in uranium and solar batteries, and as a semiconductor material.

Boranes are used in welding and boronizing metals, as rocket fuels, in the pharmaceutical and perfume industries, and as reducing agents in organic synthesis. **Decaborane(14)** finds use as a vulcanizing agent in the rubber industry and as an initiator in styrene polymerization. **Borides** are used as alloys with certain transition metals to make critical components for machines; for coating steel and other metals to increase their hardness and resistance to wear and corrosion; as catalysts and semiconductors. **Boron fluoride** is employed as a highly active catalyst in organic synthesis, as a flux to magnesium, as a fumigant, and in nuclear engineering. **Boron nitride** in the α modification is used to make fittings for furnaces and sockets for electron tubes, and in rocket and nuclear engineering; in the β modification, it is utilized as an abrasive. **Boron oxide** is used as a soldering flux, as a dehydrating agent, in the manufacture of glass, enamels, and glazes and to obtain tetraboron carbide. The **metaborates** of calcium, lead, and barium and the **pentaborates** of potassium and sodium are used in the manufacture of glass fibers, glazes, enamels, rubbers, and plastics, in the textile industry. Boron and peborates are used in the production of bleaching agents and as additives to washing powders. Borates are also added to various materials to reduce their combustibility. **Orthoboric acid** (boric acid) and **sodium tetraborate** (borax) are used in the glass and ceramic industries, as preservatives for wood and food, and in medicine. In addition, orthoboric acid serves as a raw material for the production of other boron compounds, while sodium tetraborate finds use in the soldering and welding of metals and in the textile, soap, leather, and rubber industries. **Tetraboron carbide** is mainly used as an abrasive.

MAN-MADE SOURCES OF EMISSION INTO THE ENVIRONMENT

Important sources of emission are wastewaters from the metallurgical, machine-building, textile, glass, ceramic, and leather goods industries as well as household sewage saturated with dissolved washing powders [Grushko; Larsen]. Boron levels in wastewaters may be as high as 15 mg/kg of dry sediment as compared to the highest permissible level of 0.75 mg per liter of water used for irrigating agricultural soils [Perelygin]. Local soil pollution may result from the mining and processing of boron ores and from the application of boron-containing fertilizers. Hazards for humans and animals are presented by natural waters high in boron or its compounds.

The main sources of air contamination by boron and its compounds in work environments are ore processing, the manufacture and application of fertilizers, boric acid production, and the processing and use of boron-containing materials.

TOXICITY

Soil microbiota
Some microorganisms isolated from soil or activated sludge were found to have a high boron-concentrating capacity [Kovalsky].

Plants
Boron concentrations above 700 mg/kg dry tissue weight are toxic to most crops while those above 100 mg/kg are toxic for the more sensitive ones including

legumes, cabbage, and some fruit frees [Goncharuk]. Boron concentrations of 1 mg/L in irrigation water were harmful for many fruit (especially citrus) plants while those of 4 mg/L caused injury to most crops [Grushko]. The toxicity of boron is usually more common in arid and semi-arid regions with high levels of this element. Especially toxic to crops grown in an arid area may be elevated boron levels in irrigation water. Toxic effects are also likely to arise in plants through excessive use of boron-containing fertilizers.

Good boron accumulators are beet and cotton plants, which accumulated 1000–3300 mg B/kg   and   500–1600 mg B/kg   dry   weight,   respectively [Goncharuk]. Biological effects of boron on plants have been reviewed by Larsen.

Fish

The toxicity of boron and its compounds to fish is low. For three-spined stickleback, toxic concentrations of **sodium tetraborate** were as high as 3000–3500 mg/L in soft water and 7000–7500 mg/L in hard water. For gambusias, the tetraborate appeared harmless at 1800 mg/L but was lethal at 3200 mg/L within 4 days. Skin irritation, profuse slime production, and intestinal inflammation were observed in various fish species exposed to **orthoboric acid** at a concentration of 2500 mg/L for 24–48 h; at 6250 mg/L, roach died in 48 h and rudd died in 18 h [Metelev *et al.*].

Animals and man

General remarks

Boron entering the body in excessive amounts has been shown to lower the activities of various oxidative and digestive enzymes, to inhibit the microsomal respiratory chain and nucleic acid and carbohydrate metabolism, and to alter the metabolism of B-group vitamins. Boron oxide and orthoboric acid are highly toxic substances with multiple sites of action; in particular, they display marked hepatotoxicity and gonadotropic activity and can also exert considerable embryotoxic effects by virtue of being able to pass the placental barrier with ease [Kasparov, 1967]. There is evidence that orthoboric acid may be carcinogenic [Shepard]. Orthoboric acid and the borates of metals are potent cholinolytic agents, which may explain the enhanced toxicity of boron following its entry into the body together with alcohol, hexobarbital, or other barbiturates [Schramel & Magour].

Among experimental animals, rats proved to be about twice as resistant to boron and its compounds as cats, dogs, rabbits, or guinea-pigs [Rakhmanov].

Acute/single exposure

Animals

LD$_{50}$ values of some boron compounds for mice and rats are shown in Table 1 (see also the sections on boranes and borides).

Table 1. Median lethal doses of boron compounds

Compound	Animal species	Route of administration	LD$_{50}$ (mg/kg)
Barium metaborate	Mouse	Oral	640*
	Rat	Oral	3800
Boron oxide	Mouse	Oral	3163
		Intraperitoneal	1868
	Rat	Oral	3150
Lead metaborate	Rat	Oral	1000
Orthoboric acid	Mouse	Oral	3450
	Rat	Oral	2660 — 3450
Potassium pentaborate	Rat	Oral	1700
Sodium metaborate	Rat	Oral	3700
Sodium perborate	Mouse	Oral	1060
	Rat	Oral	1200
Sodium tetraborate	Mouse	Intraperitoneal	2817
	Rat	Oral	4500

* Lowest lethal dose = 100 — 200 mg/kg; LD$_{100}$ = 1200 mg/kg.
Sources: Kasparov, 1966, 1967; Silayev; Zvezdai et al.

For rats, a suspension of powdered **boron** in peach oil was not lethal in an oral dose as high as 7000 mg/kg; **calcium metaborate** was virtually nontoxic by the oral route.

In mice given **barium metaborate** by oral gavage in doses of 100 mg/kg or higher, an initial increase in breathing frequency was followed by its sharp decrease, convulsions, and paresis of the hind paws; autopsy revealed congested vessels, myocardial edema, and dystrophic changes in the liver and kidneys. In rats, the metaborate administered by this route at 730 mg/kg caused anxiety, a fall in muscle tone, and a comatose state.

In acute inhalation toxicity studies on rats, partial mortality was recorded on 2-hour exposure to **boron oxide** at 150 mg/m^3 and on 4-hour exposure to **orthoboric acid** at 28 mg/m^3 [Izmerov et al.]. A short-term inhalation exposure of rats to **sodium perborate** dust at 3.7 mg/m^3 or 11.3 mg/m^3 resulted in decreased neuromuscular excitability; the concentration of 39.2 mg/m^3 was considered as the threshold level for irritant effects on respiratory organs [Silayev].

Observations made in rats 6 months after a single intratracheal administration of amorphous or crystalline **boron** (50 mg) in the form of suspension showed a substantial increase of hydroxyproline in the lungs, epithelial desquamation and mucosal ulceration in the bronchi, destructive bronchiolitis, marked inhibition of free respiration in hepatic mitochondria and microsomes, and clear evidence of neurotoxicity [Matkhanov]. Rats administered 50 mg of **barium metaborate** dust into the trachea died from pulmonary edema, while those exposed to 25 mg by this route exhibited focal pneumonia, purulent bronchitis, and destructive changes in the bronchial walls when examined 3 to 9 months later; in addition, signs of vascular lesions in the myocardium and of moderate albuminous degeneration in the liver and kidneys were noted at autopsy.

Man

The lethal oral dose of **orthoboric acid** for an adult human being varies from 15 to 20 g. The early symptoms of acute poisoning were nausea, severe vomiting (sometimes with blood), gastric pains, and diarrhea; later, erythematous eruptions and desquamation of the skin were evident. In fatal cases, autopsy demonstrated severe gastroenteritis (with necrotic areas in the gastric mucosa in some instances), hepatitis, and congestion and edema in the brain and myocardium [Kasparov, 1965, 1967; Larsen].

Cases of severe and even fatal poisoning following lavage of the urinary bladder with 3% orthoboric acid have been described. Its main manifestations, which appeared on the second day after this procedure, were erythroderma, headache, tachycardia, a blood pressure fall, oliguria, and azotemia; at autopsy, fatty degeneration of the kidneys, liver, and myocardium as well as edema or swelling of the brain were observed. In one fatal case, boron concentrations were 21 mg% in the brain 14.5 mg% in the kidney, 2.7 mg% in the liver, and 7 mg% in the urine [Schmidt *et al.*].

Repeated exposure

Animals

Rabbits receiving **orthoboric acid** daily by oral gavage at 2000 mg/kg died on days 3–5, while those receiving it at 1000 mg/kg and 500 mg/kg died on days 15–20 and 30–36, respectively.

Growth retardation was observed in rats exposed to **boron** or **boron oxide** dust at 150–230 mg/m^3 for 2 h/day over 15 days; in addition, rats exposed to the oxide dust exhibited a sanious discharge from the nose during most of the exposure period. At autopsy, signs of moderate inflammation were seen in the lungs [Kasparov, 1965].

In fistulized dogs, repeated intragastric introduction of water containing more than 2 or 3 mg of **boron** per liter depressed the secretory and acid-forming functions of the stomach. As a result, gastric juice production was reduced and enterokinase activity decreased 1.5-fold in the gastric juice and 5- to 6-fold in the feces, indicating that the boron had adversely affected the digestive process [Khachatrian, 1970, 1971; Nikolayeva *et al.*].

Reduced body weight, substantially decreased weights of the testes and seminal vesicles, and azoospermia were observed in male rats given **sodium tetraborate** for several weeks in the drinking water at 150 or 300 mg/L or in the diet at 54 μg/g; the content of boron in their testes amounted to 6–8 μg/g [Bruce & Weeth].

Testicular atrophy with evidence of necrobiosis in the seminiferous epithelium occurred in rats exposed to **orthoboric acid** by the oral route repeatedly at 800 mg/kg or once in a dose of 3000–4000 mg/kg; the surviving male pups thus exposed remained sterile throughout life [Bouissou & Castagnol].

Man
Repeated ingestion of **orthoboric acid** small doses led to reduced appetite, nausea, and in some cases to vomiting and the appearance of erythematous eruption [Kasparov, 1967].

Chronic exposure

Animals
Rats exposed to amorphous or crystalline **boron** dust by inhalation at 150–178 mg/m^3 for 4 h/day over 4 months exhibited growth retardation and reduced antitoxic function of the liver without marked changes in the hemogram or serum proteins. Autopsy findings included moderate perivascular and peribronchial sclerosis, emphysema, and dystrophic changes in the renal tubular epithelium. Increased diuresis, urinary excretion of creatine and boron, and reduced pH of the urine were noted in rats after 24 weeks of inhalation exposure to a **boron oxide** aerosol (highest concentration, 470 mg/m^3) and in dogs after 23 weeks of such exposure (highest concentration, 57 mg/m^3); the chemical and morphologic composition of the blood or the strength of the bones were not affected, nor were any roentgenologic changes seen in the animals [Wilding *et al.*].

Rats exposed to an **orthoboric acid** aerosol at 120–150 mg/m^3 for 2 h daily over 10 weeks showed retarded growth and raised levels of nitrogen-containing substances in the urine; autopsy showed chronic bronchitis and tracheitis. In rats exposed to lower concentrations of this aerosol (10–15 mg/m^3) but for 4 h daily and over a much longer period (4 months), the manifestations of poisoning were more numerous and included growth retardation, altered cholinesterase activity in the blood, lowered urine pH, atrophic changes in the testes and epididymides, reductions in the total number and motility of spermatozoa, and the presence of many abnormal spermatozoa; autopsy revealed epithelial desquamation in the trachea, perivascular and peribronchial infiltrates in the lungs, and swelling of the convoluted renal tubules together with atrophy of the seminiferous epithelium and the presence or necrotic cells in many seminiferous tubules. These findings were interpreted as an indication that the gonadal toxicity of orthoboric acid is higher than its systemic toxicity [Tarasenko *et al.*].

Emaciation, inhibition of external respiration, and progressive pneumoconiosis were observed in rats on long-term inhalation of a dust from datolite ore (which contains a basic calcium borosilicate) [Kasparov & Yakubovsky].

Chronic exposure of rabbits to **borates** by the oral route (44 mg/kg for 4 months) led to moderate leukocytosis and, as found at autopsy, to edema of the brain and myocardium; all borates, except that of calcium, also caused hypoglycemia. In rats, no increase in mortality occurred after 12 months of oral exposure to **sodium tetraborate** at 60 mg/kg. Rabbits exposed to the tetraborate at 270 mg/kg or to **orthoboric acid** at 175 mg/kg for 3 months had decreased erythrocyte and leukocyte counts and elevated hemoglobin and inorganic phosphorus levels in the blood.

The addition of **boron** to the drinking water of female rats in doses of 6.5 or 0.225 mg/kg throughout the period of pregnancy resulted in partial embryonal mortality and decreased body weight and body size of the embryos; the dose of 0.05 mg/kg was considered as the highest level at which boron did not produce detectable embryotoxic effects.

Increased incidence and severity of dental caries were observed among rats consuming, for 4 months, fluorine-containing water (0.1 mg F/L) to which **orthoboric acid** had been added in amounts of 1.5 to 3 mg per liter; increased solubility of the dental enamel in lactic acid and diminished fluorine assimilation by the teeth and bones were also noted in the test rats [Gabovich & Stepanenko].

A common disease among sheep in areas where boron levels in the soil exceed 30 mg/kg is boric enteritis, the underlying cause of which in inhibition of several digestive enzymes. Diseased sheep were found to accumulate 3.7 times as much boron in their body as did healthy sheep at the same level of daily boron intake [Kovalsky].

Man

After prolonged occupational exposure to **boron**, signs of reduced sexual activity along with pathologic changes in the sperm were detected in males, as were menstrual disturbances without reproductive failure (the ability to conceive) in females. Neurasthenia, vegetative dystonia, neuralgia, subtrophic changes in the upper respiratory tract mucosa, arthralgia, and arthropathy were frequently observed in workers long exposed to **orthoboric acid** dusts or aerosols at concentrations of 20 to 50 mg/m^3 (which are 2 to 5 times higher than the MAC); many of the female workers thus exposed had inflammatory disease of the genital organs. Common findings among male workers of reproductive age exposed to orthoboric acid aerosols at 10 to 35 mg/m^3 for 10 years or more were reduced ejaculate volumes, low sperm counts, and decreased proportions of motile spermatozoa [Kasparov, 1967, 1971; Tarasenko *et al.*]. In one case, described by Tan, nervous irritability, headaches, facial flushing, appetite depression, and body-weight loss occurred in a male worker after 6 months of occupational exposure to **sodium tetraborate** (at which time boron was detected in his urine); 3 months later, he became completely bald.

Complaints of weight loss, nausea, and poor appetite were often heard from workers employed in the manufacture of fertilizer from a datalite concentrate and exposed to its dust at concentrations of 5 to 58 mg/m^3 [Kasparov, 1971].

In workers extracting or producing **borates**, the two main clinical forms of dust-induced disease were dust bronchitis and pneumoconiosis. Frequent presenting complaints were cough (either dry or productive of sputum in small amount) and (in those with longer periods of exposure) breathlessness and chest pain; asthmatiform phenomena were more prevalent among subjects with bronchitis. Dilatation, induration, and infiltration of the lung hili were noted, as well as a high prevalence of adhesive lesions in the diaphragm [Beloskurskaya *et al.*]. For **sodium perborate**, the threshold concentration producing irritant effects on the human respiratory tract has been determined at 21 mg/m^3 [Silayev].

For the general population, risk of chronic peroral intoxication with **orthoboric acid** or **sodium tetraborate** may arise because of their use as food preservatives or of their consumption with drinking water from water sources rich in boron. It has been shown for both children and adults that the consumption of drinking water containing 4–6 mg B/L reduces gastric juice acidity as well as enterokinase and alkaline phosphatase activities in the feces [Nikolayeva *et al.*; Verbitskaya].

In men, long-term consumption of water with a boron content of 3–6 mg/L (or 1 mg/L for 5 years or more) was consistently associated with reduced sexual potency; in particular, there was a high percentage of those with inadequate erections and diminished libido [Borisov]. Evidence of reduced sexual function was also obtained in questionnaire surveys conducted among men living in areas with boron levels of 0.3 mg/L in drinking water supplies, and it has therefore been recommended to decrease the MAC for boron from 0.5 to 0.3 mg/L [Krasovsky *et al.*].

Cases of severe chronic poisoning resulting from repetitive therapeutic use of boron-containing drugs (tranquilizers or antiepileptics) have been reported; in one instance, cachexia, eczematous dermatitis, hypoplastic anemia, gastric ulcer, changes in the nails and teeth, and alopecia were observed [Herren & Wyss].

Effects of local application

Animals
In rabbits, **boron oxide** produced irritation on application to the skin of the back (1 g per 25 cm^2) and caused immediate inflammation when dropped into the eyes

[Kasparov, 1967]. **Orthoboric acid** is not irritating to the skin but can penetrate through it to cause systemic poisoning.

Sodium perborate was not absorbed through undamaged skin of rabbits but produced acute conjunctivitis when dropped into their eyes [Silayev]. **Barium metaborate** has been reported to be both irritating to, and absorbed through, the skin [Zvezdai *et al.*].

Man

Cases of intoxication with **orthoboric acid** or **sodium tetraborate** absorbed through undamaged or damaged skin have been described. In their review Skipworth *et al.* cited 109 cases of poisoning, with a fatality rate of 70%, occurring in infants treated with powders, emulsions, or solutions containing these substances.

Boranes

Boranes (boron hydrides, hydrogen borides) are many times more toxic than other boron compounds and even more toxic than phosgene and hydrogen cyanide. The toxicity of pentaborane(9) is higher than that of diborane(6) and decaborane(14). Like phosgene, diborane(6) produces chemical pulmonary edema. Pentaborane(9) and decaborane(14) cause severe damage to the nervous system, kidneys, and liver [Bogdanov; Rashevskaya *et al.*]. Repeated exposure results in cumulative effects. Boranes are highly volatile and present considerable fire and explostion hazards in addition to health hazards.

Animals. The oral LD_{50} of **decaborane(14)** for rats was 64 mg/kg, while its intraperitoneal LD_{50} was 23 mg/kg for rats, 33 mg/kg for mice, and 28 mg/kg for rabbits. In rabbits, signs of intoxication were nystagmus, motor incoordination, and convulsions; injection of a high dose was soon followed by prostration, coma, and death. For monkeys, decaborane(14) was lethal in an intraperitoneal dose of 6 mg/kg.

In dogs and rabbits acutely poisoned with a decaborane fuel in a dose of 15 mg/kg, convulsions appeared and boranes were detected in the blood at levels of 4.0 to 4.8 µg/L.

The intravenous LD_{50} of a boron hydride rocket fuel was determined at 13 mg/kg for rats and 4–6 mg/kg for rabbits.

On inhalation exposure to **diborane(6)**, the major signs of poisoning were excitation, rapid breathing, and irritation of the lungs; for rats, the 4-hour LC_{50} was between 0.06 and 0.09 mg/L. The main manifestations of inhalation poisoning with **pentaborane(9)** and **decaborane(14)** were weakness, loss of appetite, motor incoordination, tremor, and coma; autopsy showed pulmonary edema, intra-alveolar hemorrhages, and lesions in the liver, myocardium, and kidneys [Stumpe]. For rats, the 4-hour LC_{50}s of pentaborane(9) and decaborane(14) were 0.018 mg/L and 0.11–0.23 mg/L, respectively; for mice, the LC_{50} of decaborane(14) was 0.03 mg/L.

Rats inhaling **pentaborane(9)** vapor at the LD_{50} level (0.018 mg/L) for 30 min were found to have a drastically reduced serotonin level in the brain [Weir & Mayers]. In dogs, inhalation of the pentaborane at 0.067 mg/L for 5 min, 0.03 mg/L for 15 min, or 0.0078 mg/L for 60 min produced no visible signs of intoxication; convulsions, conjunctival hyperemia, and strongly marked constriction of the pupils were, however, observed when any on these exposures was repeated five times [Weir *et al.*].

Rabbits exposed to **decaborane(14)** vapor daily at 0.022 mg/L for 5.5–6 h/day died after the first or second exposure; mice and rats were more resistant when so exposed, surviving for 10–100 days and up to 135 days, respectively (in rats, changes in nucleic acid metabolism were noted).

Hamsters, rabbits, rats, and dogs died within 4 weeks on repeated exposure to **pentaborane(9)** vapor at 0.0026 mg/L. Chronic exposure of these animals and also of monkeys to a lower concentration of 0.0005 mg/L led to weight loss, decreased activity of the animals, and (less frequently) irritation of the ocular and nasal mucous membranes; in rats, raised erythrocyte counts were recorded. At autopsy, pathological changes in the liver, kidney, and various parts of the central nervous system were seen.

Man.   Ingestion of **pentaborane(9)** or **decaborane(14)** led to vomiting, paresthesia, meningism, tremor, and, in more severe cases, to generalized convulsions and coma; in some cases, death occurred within 24 to 48 h post-ingestion in the presence of central and peripheral circulatory disorders [Ludewig & Lohs].

Examination of 14 individuals 4 to 12 weeks after their brief exposure to pentaborane gas accidentally released during the detoxification of old canisters, provided evidence by which seven of them were diagnosed as having post-traumatic stress disorder and seven as having mild brain dysfunction, although the results of physical and routine laboratory evaluations were normal. In particular, reduced ability to concentrate, impairment of short-term memory, electroencephalographic changes, and elevated levels of neurotransmitter metabolites in the cerebrospinal fluid were recorded [Silverman *et al.*]. As stated by these authors, the results of their study contradict previous reports in the literature that most symptoms of intoxication with pentaborane resolve within the first week after exposure.

The minimal effective concentrations were 0.002–0.004 mg/L of air for **diborane(6)**, 0.0025 for **pentaborane(9)**, and 0.0003 for **decaborane(14)**, while the mean concentrations of these three boranes for odor perception were (mg/L of air) 0.001, 0.007, and 0.002, respectively.

The main symptoms of mild intoxication with pentaborane(9) by inhalation were drowsiness, mental confusion, a feeling of tightness in the chest, headache, and tremor and twitching of muscles. Severe intoxication manifested itself by a variety of abnormalities, including motor incoordination, double vision, pendulous jowls, Romberg's sign, slurred speech, salivation, profuse sweating, nystagmus, attacks of severe convulsions lasting 0.5 to 2 min, leukocytosis and erythrocytosis, tachycardia, fever, blood pressure elevation, and considerable electroencephalograhic and electrocardiographic changes; neuritis and partial loss of hearing were observed in some cases, as was a laryngeal spasm that required surgical intervention [Mindrum]. In the acute phase of poisoning, boron levels in the blood rose up to 0.24 μg/ml (normal: 0.04–0.05 μg/ml).

In one study, most of the 83 individuals occupationally exposed to boranes over a 3-year period, complained of headache, lassitude, cough, nausea and chills. When some of them were hospitalized, various signs of damage to the respiratory organs were seen and pathologic changes were found in the liver and kidneys; on the day of admission, boron levels in the urine ranged to 2.9 mg/L [Lowe & Freeman].

Signs of brain dysfunction were still present in 14 persons 2 months after moderate poisoning with **pentaborane(9)** [Hart et al.]. Characteristic effects produced by **decaborane(14)** were reported to be those on the nervous and cardiovascular systems and on serum transaminases [Naeger & Leibman]. Autopsy of persons lethally poisoned with boranes showed necrosis and fatty degeneration in the central hepatic lobules and moderate damage to the renal tubules [Bogdanov].

Effects of local application
Pentaborane(9) and **decaborane(14)** are very irritating to the skin and mucous membranes [Fried]. In rats, necrosis developed in the skin at the site of decaborane(14) application. In rabbits, introduction of 0.2 ml of either of these compounds into the conjunctival sac resulted in severe keratoconjuntivitis, followed by corneal ulceration and opacification, even though the eye had been washed at 5 min post-application. Pentaborane(9) and especially decaborane(14) may cause systemic poisoning by being absorbed through undamaged skin. With 4-hour dermal application, the LD_{50}s of decaborane(14) were (mg/kg) 317–502 for mice, 158–251 for guinea-pigs, 740 for rats, 57–105 for rabbits, and 126 for cats. Pentaborane vapors were not absorbed through the intact skin of dogs.

Borides of metals

Acute/single exposure: animals
LD_{50} values of several borides for mice and rats are given in Table 2. The main symptoms of **calcium hexaboride** poisoning in mice were general inhibition, motor incoordination, and (on day 3 or 4) paralysis of the hind limbs, with deaths following shortly after its onset. Autopsy of animals given **molybdenum boride** revealed gross dystrophy of the liver. Animals poisoned with the other borides

appearing in Table 2 became weak, apathic, and lethargic; deaths occurred on days 3–5, and autopsy demonstrated lesions in the liver, kidneys, adrenals, and brain.

Table 2. Median lethal doses of borides

Compound	Animal species	Route of administration	LD$_{50}$ (mg/kg)
Calcium hexaboride	Mouse	Intraperitoneal	1045[a]
Magnesium diboride	Mouse	Oral	567
		Intraperitoneal	444
	Rat	Oral	826
		Intraperitoneal	423
Magnesium polyboride	Mouse	Intraperitoneal	6000
	Rat	Oral	>10000
		Intraperitoneal	6000
Molybdenum boride	Mouse	Intraperitoneal	1377[b]
Titanium-chromium diboride	Mouse	Intraperitoneal	2722
	Rat	Oral	>10000
		Intraperitoneal	3494
Metal-ceramic alloy based	Mouse	Intraperitoneal	3998
on this compound	Rat	Oral	>10000
		Intraperitoneal	2000

[a] Lowest lethal dose = 10 mg/kg.
[b] Highest tolerated dose = 1000 mg/kg.

With intratracheal administration, borides caused damage to the liver, kidneys, and respiratory organs. Symptoms of poisoning were motor incoordination, muscle twitching, and paralysis of the hind limbs. Of the borides tested, the highest toxicity was shown by those of calcium and molybdenum. In rats, a **molybdenum boride** suspension administered intratracheally caused 100% mortality in the dose of 50 mg and 50% mortality in the dose of 25 mg. With most of the borides introduced by this route at 50 mg, thickening of the interalveolar septa, moderate fibrosis, emphysema, collagenosis, and detachment of bronchial epithelium were observed in the lungs of animals 3 to 9 months later; **titanium diboride** produced purulent destructive bronchiolitis and bronchial wall ulceration, while **molybdenum boride** led to purulent bronchitis and to hyperplasia and proliferation of the bronchial mucosa. The **borides of molybdenum, titanium, zirconium,** and **chromium** caused, in addition, dystrophic changes in the liver, kidney, and (in the case of chromium diboride) the myocardium [Brakhnova].

After a single intratracheal administration, metal **hexaborides** also decreased blood coagulability and the content of sulfhydryl groups in blood, serum, and internal organs, while **magnesium polyboride** dust was moderately fibrogenic [Olefir]. Borides also reduced the natural immunity of animals, for periods of up to 1 year.

Chronic/multiple exposure

Animals. In a chronic oral toxicity study, rabbits given **calcium hexaboride** at 100 mg/kg daily by gavage for 4 months exhibited leukocytosis, lymphocytosis, and reductions in serum albumins and prothrombin time; autopsy showed dystrophic changes in the liver, thickening of the interalveolar septa, and peribronchial pneumonia [Mogilevskaya *et al.*]. Chronic gastritis, hepatitis, and nephrosonephritis, but no deaths, were observed among rats after oral exposure to 30 high doses of **magnesium polyboride** (5000 mg/kg by gavage).

Long-term inhalation exposure of rats to **molybdenum boride** dust (60–70 mg/m^3, 2 h daily, 5 months) was accompanied by decreases in the color index and increases in the nitrogen of amino acids in the blood serum. In rats exposed to **chromium diboride** dust (300–350 mg/m^3, 2 h on alternate days, for 3 months), the most conspicuous change in the blood was reduced hemoglobin. Autopsy revealed thickened interalveolar septa and vessels in the lungs, swollen convoluted tubules in the kidney, and granular degeneration in the liver [Mogilevskaya *et al.*]. The threshold concentration of **magnesium polyboride** dust for rats, as determined by recording both systemic toxic effects and specific gonadotoxicity, was in excess of 100 mg/m^3 [Konstantinova *et al.*].
Man. Workers exposed to aerosols of metal borides had reduced blood levels of albumins [Bezvershenko].

Cutaneous and ocular effects

In rabbits, **magnesium diboride** produced only slight irritation when applied to the skin but severe irritation and inflammation (even keratitis in some animals) following introduction into the conjunctival sac; **magnesium polyboride** did not irritate the skin. Neither of these compounds was absorbed through the intact skin.

Boron carbide (tetraboron carbide) and boron nitride

Animals

When rats were administered 50 mg of a boron carbide and boron nitride suspension intratracheally, no substantial changes in their general condition or body weight were observed 6 to 12 months later as compared to nonexposed controls, although the content of collagen proteins in the lungs had risen by 20–50%; autopsy findings in the lungs included detachment of the bronchial mucosa, peribronchitis, focal emphysema, and a torpid interstitial inflammatory process with signs of moderate peribronchial and interlobular sclerosis [Brakhnova].

Long-term inhalation exposure of rats to dusts of boron carbide at 300–350 mg/m^3 (4 h/day for 12 months) or of boron nitride at 100–200 mg/m^3 (4 h/day for 6 months) led to catarrhal desquamative bronchitis, emphysema, and moderate diffuse sclerosis in the lungs; in addition, the lungs of rats exposed to the boron nitride dust had significantly elevated (by 130%) levels of collagen proteins. In rats exposed to concentrations of 1 to 10 mg/m^3, no appreciable morphologic changes were seen [Brakhnova & Samsonov; Brunshtein & Danilova].

Man
Acute and chronic inflammatory diseases of the upper respiratory tract were detected in workers employed in the boron carbide-producing department of a factory for 12 to 14 years; some of the workers had also developed pneumoconiosis or pneumosclerosis [Brunshtein & Salikhodzhayev].

Boron fluoride

Animals
Inhalation exposure to boron fluoride at 2.1 mg/L for 5.5 h caused partial mortality in mice and rats and 100% mortality in guinea-pigs. Some animals did not survive the exposure even to a much lower boron fluoride concentration (0.35 or 0.021 mg/L). The concentration of 0.008 mg/L produced irritation of respiratory organs [Torkelson et al.].

The inhalation LC$_{50}$ of boron fluoride was determined at 3.46 mg/L for mice (2-hour exposure), 1.18 mg/L for rats (4-hour exposure), and 0.11 mg/L for guinea-pigs (4-hour exposure) [Izmerov et al.]. The main manifestations of intoxication were severe irritation of the mucous membranes in the upper air passages and eyes and respiratory disorders; autopsy demonstrated congestion and dystrophy in internal organs and edema and interstitial inflammation in the lungs [Kiriy].

Exposure to boron fluoride concentrations of 0.03–0.04 mg/L for 15 to 30 days led to partial mortality among mice and guinea-pigs, whereas rats, rabbits, cats, and dogs all survived. Several days after the start of exposure, inhibition of cholesterase activity in the blood was observed. Multiple exposures to 0.28 mg/L produced fluorosis in all rats. At autopsy, the connective tissue of internal organs appeared edematous and swollen and the blood vessels showed increased permeability. Exposure of rats, guinea-pigs, and rabbits to 0.01 mg/L for 4 h/day over 4 months or to 0.03 mg/L for 7 h/day over 2 months led to marked irritation of the upper air passages, leukocytosis, impaired protein-forming function of the liver, and elevated fluorine levels in the teeth; the most conspicuous changes seen at autopsy were chronic tracheobronchitis and increased weights and edematous appearance of internal organs. In animals exposed to 0.003 mg/L, the changes were similar but slight and disappeared within a month [Kasparov & Kiriy].

Boron fluoride has been reported to display marked nephrotoxicity [Rusch *et al.*] and to produce embryotoxic effects [Popova & Peretomchina].

Man

In factories producing boron fluoride or polyisobutylene (using the fluoride as a catalyst), workers exposed to boron fluoride fumes (at 0.035–0.61 mg/L) and boron oxide dust (≈ 320 mg/m^3) complained of irritation, dryness, and bleeding of the nasal mucosa, itching in the open areas of the skin, easy fracturability of the teeth, and joint pains. Examinations revealed functional disturbances of the nervous system, increased blood vessel permeability, and high rates of upper respiratory tract and skin diseases [Kiriy].

ABSORPTION, DISTRIBUTION, AND ELIMINATION

Boron and its compounds can enter the body by inhalation, ingestion, and by absorption through mucous membranes or skin burns. Absorption of boron and soluble borates from the gastrointestinal tract and from skin burns is almost complete. Boron oxide appears to be absorbed only partially.

In the blood, boron distributes between the erythrocytes and plasma in similar proportions, but is rapidly transferred to tissues; its blood concentration was found to have decreased by half in 26 h [Popov & Angelieva]. About 10% of the ingested dose is concentrated in soft tissues (predominantly the brain, liver, and fatty tissue).

Boron is a component of all tissues but its role in the body is not clear. Its main repository is the skeleton (where, e.g., >86% of the administered orthoboric acid was found). Normal levels are considered to be around 0.8 mg/L in the blood and 0.715 mg/L in the urine. In cases of poisoning, its blood concentrations rose to 40 mg/L or more. Boron is highly cumulative on repeated administration. In cats and dogs, boranes cleared from the blood in 6–7 days after an oral dose of 7 mg/kg and in 1–2 days after 1–3 mg/kg. In cases of accidental human poisoning with decaborane fuel, blood serum concentrations of boranes were 0.5–0.6 µg/ml after 4 h and virtually zero after 48 h. After acute poisoning with pentaborane(9), serum borane levels of 0.7–1.3 µg/ml were recorded [Miller *et al.*]. Boron, boron oxide, and orthoboric acid can easily cross the placental barrier.

Healthy sheep had nearly the same boron levels in their lungs, thyroid glands, and wool as did diseased sheep (those living in a boron province) but significantly lower levels in most other tissues (e.g. 0.042 *vs* 0.258 mg% on a wet weight basis in the kidney and 0.052 *vs* 0.464 mg% in the brain [Kovalsky].

Nearly all boron found in tissues may be bound to carbohydrates; for example, the carbohydrate fraction of the liver contained 97%, while the remaining boron was present tn the fatty tissue (protein fractions did not contain any boron at all) [Shakhova].

Elimination of most boron compounds is predominantly by way of the gastrointestinal tract. Orthoboric acid is mainly (up to 85%) excreted by the kidneys.

HYGIENIC STANDARDS

Exposure limits adopted for boron and its compounds in the USSR are given in Table 3. In the USA, threshold limit values have been set for boron oxide at 10.0 mg/m^3 (TWA), boron tribromide at 10.0 mg/m^3 (ceiling limit), boron trifluoride at 2.8 mg/m^3 (ceiling limit), and sodium tetraborates (TWA) at 1.0 mg/m^3 for the anhydrous form, 5.0 mg/m^3 for the decahydrate, and 1.0 mg/m^3 for the pentahydrate.

METHODS OF DETERMINATION

One method of determining boron in *air* is based on interaction with quinalizarin in an acid medium followed by spectrophotometry (detection limit, 2.5 mg/m^3 of air) [Muraviyeva]. Another method relies on the reaction of orthoboric acid with 1,1-dianthrimide in concentrated sulfuric acid with the formation of a blue compound followed by its photometry (sensitivity 1 µg in the volume analyzed) [Trusova]. A method for the determination of boron fluoride and of the boron hydrolysis products is based on the production of a colored complex with methylene blue followed by extraction of this complex and its colorimetry (sensitivity 2 µg/sample) [Peregud]. Boron nitride and boron carbide nitride can be determined gravimetrically [*Specifications for Methods of Determining Noxious Substances in Air*].

For measuring boron in *water*, a photometric method has been described, based on the ability of orthoboric acid to alter, in concentrated sulfuric acid, the color of carmine from red to violet-blue though the formation of an orthoboric acid ester (sensitivity 0.1 mg/L) [Novikov *et al.*].

In *urine* and *blood*, boron can be determined by colorimetry after mineralizing the sample with sulfuric and nitric acids and reacting the boron with quinalizarin [Rozenberg & Byalko]. Procedures for measuring boron in *soil* are described in *Methods for the Determination of Trace Elements in Environmental Media*. Analytical methods of determination of boron are described by Larsen.

Table 3. Exposure limits for boron and its compounds

	Workroom air		Atmospheric air		Water sources		
	MAC_{wz} (mg/m^3)	Aggregative state	MAC_{ad} (mg/m^3)	$TSEL_{aa}$ (mg/m^3)	MAC_w (mg/L)	Hazard class	Note
Boron, amorphous	—	—	—	0.01	—	—	—
crystalline	—	—	—	—	0.5^a	2	—
Boron carbide	6.0	Aerosol	—	—	—	4	F^b
Boron fluoride	1.0	Vapor	—	—	—	2	—
Boron nitride	6.0	Aerosol	—	0.02	—	4	F^b
Boron tribromide	30.0	—	—	—	—	—	—
Calcium metaborate	—	—	0.02	—	—	4	—
Orthoboric acid	10.0	Aerosol	—	—	—	4	—
Titanium-chromium diboride (as B)	1.0	Aerosol	—	—	—	3	—

[a] Based on sanitary/toxicologic criteria.
[b] Predominantly fibrogenic activity.

MEASURES TO CONTROL EXPOSURE

At the workplace

It is important to prevent the air of work premises from becoming polluted with vapors and aerosols of boron and its compounds, including borides of metals, by tightly enclosing the equipment and providing it with local exhaust ventilation. Effective dust control can be best achieved by installing automatic lines with remote control.

Particular care needs to be taken to prevent toxic exposure and fire and explosion hazards in areas where boron hydrides (boranes) are stored, handled, or processed. It is advisable to locate storage tanks with these compounds outdoors in areas clearly marked as danger zones. Pentaborane should be kept under a dry nitrogen atmosphere and diborane should be refrigerated to prevent its decomposition. Any containers situated indoors must be adequately ventilated and away from any source of ignition. All process equipment should be fully enclosed and purged before being opened. Automatic leak detectors with visual and acoustic warning signals should be fitted. Where possible, production processes should be operated by remote control. The workroom air should be sampled regularly to determine borane concentrations [Kasparov & Balynina].

Local exhaust ventilation is required in areas where boron compounds are used in soldering or welding operations, and effective general ventilation should be provided in those where any operations involving boron or its compounds are carried out.

It is necessary to avoid contamination of the hands and work clothes, and any dust on these should be removed without delay. All work clothes should be washed in a centralized manner and not allowed to be taken home. Adequate sanitary facilities should be provided and the workers encouraged to wash before meals and at the end of the shift. Smoking, eating, and drinking in work areas should be prohibited.

For *personal protection*, respirators must be used where the air contains dusts or aerosols, and goggles, gloves, aprons, and protective footwear must be worn if there is a risk of thermal or chemical burns.

In view of the extremely hazardous nature of boron hydrides, the use of personal protective equipment may be necessary even where adequate technical safety measures are applied; the workers should be supplied with appropriate filter-type respirators, overalls with long sleeves, and eye, hand, and foot and leg protection [Kasparov & Balynina].

All workers exposed to boron and its compounds are subject to a preplacement and subsequent periodic medical examinations.

Protection of the general environment

The main measures to reduce the environmental burden of boron and its compounds include the introduction of more efficient methods of treating effluents from the industries where they are used or produced; enforcement of the existing regulations for wastewater use to irrigate crops; strict control over of the application of boron-containing fertilizers to soils; and monitoring of boron levels in drinking-water supplies.

FIRST AID IN POISONING

Persons poisoned with orthoboric acid should be given dextran, 5% glucose, or physiologic saline by intravenous infusion (until the venous blood pressure returns to normal) in conjunction with vasodilators, cardiacs, and corticosteroids [Mogosh]. Gastric lavage or induction of vomiting may be required, and an exchange transfusion, intravenous injection of an electrolyte solution, and cortisone application against dermatitis may be indicated [Kasparov & Balynina]. In the event of splashes into the eyes, they should be immediately washed with clean water.

In case of skin exposure to boranes, these should be promptly flushed off with a dilute ammonium hydroxide solution and the affected area wiped with a sponge impregnated with this solution or with 1% triethanolamine solution.

REFERENCES

Beloskurskaya, G. I. *et al.* (1979) In: *Trudy NII krayevoi patologii Kaz. SSR* [Transactions of the Research Institute for Regional Pathology of the Kazakh SSR]. Alma-Ata, Vol. 37, pp. 133–147 (in Russian).
Bezvershenko, A. S. (1972) *Vrach. Delo*, No. 10, 128–131 (in Russian).
Bogdanov, N. A. (1961) *Voprosy toksikologii raketnogo topliva* [Toxicologic Aspects of Rocket Fuels]. Leningrad (in Russian).
Borisov, A. I. (1976) *Gig. San.*, No. 1, 11–16 (in Russian).
Bouissou, H. & Castagnol, R. (1965) *Arch. Malad. Profess.*, 26, 293–306.
Brakhnova, I. T. (1971) *Toksichnost poroshkov metallov i ikh soyedineniy* [Toxicity of Metals and Their Compounds in Powder Form]. Naukova Dumka, Kiev (in Russian).
Brakhnova, I. T. & Samsonov, G. V. (1970) *Gig. Truda Prof. Zabol.*, No. 12, 48–50 (in Russian).
Brooks, R. R. (1979) In: Bockris, J. O'M. (ed.) *Environmental chemistry*. Plenum Press, New York, pp. 429–476.
Bruce, S. & Weeth, H. (1980) *Bull. Environ. Contam. Toxicol.*, 25 782–789.
Brunshtein, I. P. & Danilova, R. I. (1969) *Med. Zh. Uzbekistana* [Medical Journal of Uzbekistan], No. 12, 40–42 (in Russian).
Brunshtein, I. P. & Salikhodzhayev, S. S. (1967) *Gig. Truda Prof. Zabol.*, No. 2, 50–52 (in Russian).
Dobrovolsky, V. V. (1983) *Geografiya mikroelementov. Globalnoe rasseyanie* [Geography of Trace Elements. Global Dispersion]. Mysl, Moscow (in Russian).
Dubikovsky, G. P. (1982) *Khimiya v selskom khozyaistve* [Chemistry in Agriculture]. No. 3, pp. 33–34 (in Russian).
Fried, R. (1959) *Arch. Ind. Health*, 20 448–455.
Gabovich, R. D. & Stepanenko, T. A. (1973) *Gig. San.*, No. 12, 20–22 (in Russian).
Goncharuk, E. I. (1977) *Sanitarnaya okhrana pochvy ot zagryazneniya khimicheskimi veshchestvami* [Sanitary Measures to Protect Soil from Chemical Pollution]. Zdorovya, Kiev (in Russian).
Grushko, Ya. M. (1979) *Vrednye neorganicheskie soyedineniya v promyshlennykh vodakh* [Harmful Inorganic Compounds in Industrial Waste Waters]. Khimiya, Leningrad (in Russian).
Hart, R. *et al.* (1984) *Amer. J. Ind. Med.*, 6, 37–44.
Herren, Ch. & Wyss, F. (1964) *Schweiz. Med. Wochenschr.*, 94, 1815–1818.
Izmerov, N. F. *et al.* (1982) *Toxicometric Parameters of Industrial Toxic Chemicals under Single Exposure*. Published by the USSR Commission for the United Nations Environment Programme (UNEP), Moscow.
Kasparov, A. A. (1965) *Gig. Truda Prof. Zabol.*, No. 8, 50–51 (in Russian).
Kasparov, A. A. (1966) In: *Gigiyenicheskaya otsenka khimicheskikh faktorov vneshnei sredy* [Hygienic Evaluation of Environmental Chemical Agents]. Moscow, pp. 108–110 (in Russian).
Kasparov, A. A. (1967) In: Z. I. Izraelson (ed.) *Novye dannye po toksikologii redkikh metallov i ikh soyedineniy* [Recent Data on the Toxicology of Rare Metals and Their Compounds]. Meditsina, Moscow, p.p. 126–142 (in Russian).
Kasparov, A. A. (1971) *Gig. Truda Prof. Zabol.*, No. 8, 11–15 (in Russian).
Kasparov, A. A. & Balynina, E. S. (1983) In: *Encyclopaedia of Occupational Health and Safety*. International Labour Office, Geneva, pp. 317–322.
Kasparov, A. A. & Kiriy, V. G. (1972) *Farmakol. Toksikol.*, No. 3, 369–372 (in Russian).
Kasparov, A. A. & Yakubovsky, A. K. (1970) In: *Materialy konferentsii gigiyenicheskikh kafedr Moskovskogo meditsinskogo instituta* [Proceedings of a Conference of the Departments of Hygiene of a Moscow Medical Institute]. Moscow, pp. 115–116 (in Russian).
Khachatrian, T. S. (1970) *Gig. San.*, No. 7, 93–94 (in Russian).
Khachatrian, T. S. (1971) *Gig. San.*, No. 1, 11–15 (in Russian).

Kiriy, V. G. (1968) *Ftoristyi bor kak promyshlennyi yad* [Boron Fluoride as an Industrial Poison]. Dissertation, Moscow (in Russian).

Kolotilov, N. N. (ed.) (1973) *Okhrana truda v aviatsionnoi promyshlennosti* [Occupational Safely in the Aircraft Industry]. Moscow (in Russian).

Konstantinova, L. I. *et al.* (1986) *Gig. San.*, No. 1, 81–82 (in Russian).

Kovalsky, V. V. (1974) *Geokhimicheskaya ecologiya* [Geochemical Ecology]. Nauka, Moscow (in Russian).

Kovalsky, V. V. & Shakhova, I. K. (1962) *Dokl. Akad. Nauk SSSR*, No. 4, 967–970 (in Russian).

Krasovsky, G. N. *et al.* (1974) In: *Materialy 1-go Sovetsko-Amerikanskogo simpoziuma po probleme "Gigiyena okruzhayushchei sredy"* [Proceedings of the 1st Soviet-American Symposium on Environmental Hygiene]. Riga, pp. 88–96 (in Russian).

Larsen, L. A. (1988) Boron In: Seiler, H. G. *et al.* (eds) *Handbook on Toxicity of Inorganic Compounds*. Marcel Dekker, New York, pp. 129–143.

Lowe, H. & Freeman, G. (1957) *Arch. Ind. Health*, **16**, 523–533.

Ludewig, R. & Lohs, K. (1981) *Akute Vergietungen*. VEB Gustav Fischer, Jena.

Matkhanov, E. I. (1970) *Gigiyenicheskaya otsenka pyli amorfnogo i kristallicheskogo bora* [Dusts of Amorphous and Crystalline Boron: Hygienic Evaluation]. Moscow (in Russian).

Metelev, V. V. *et al.* (1971) *Vodnaya toksikologiya* [Aquatic Toxicology]. Kolos, Moscow (in Russian).

Methods for the Determination of Trace Elements in Environmental Media (1976) [Metody opredeleniya mikroelementov v prirodnykh obyektakh]. Nauka, Moscow in Russian).

Miller, D. *et al.* (1960) *Toxicol. Appl. Pharmacol.*, **2**, 430.

Mindrum, G. (1964) *Arch. Intern. Med.*, **114**, 364–374.

Mogilevskaya, O. Ya. *et al.* (1967) in: Z. I. Izraelson (ed.) *Novye dannye po toksikologii redkikh metallov i ikh soyedineniy* [Recent Data on the Toxicology of Rare Metals and their Compounds]. Meditsina, Moscow, pp. 142–152 (in Russian).

Mogosh, G. (1984) *Ostrye otravleniya* [Acute Poisonings]. Russian translation from Roumanian published in Bucharest.

Muraviyeva, S. I. (ed.) (1982) *Sanitarno-khimicheskiy kontrol vozdukha promyshlennykh predpriyatiy* [Physical and Chemical Monitoring of Air Quality in Industrial Workplaces]. Meditsina, Moscow (in Russian).

Neizvestnova, E. M. et al. (1983) In: *Gigiyena truda i professionalnaya patologiya v tsvetnoi i chernoi metallurgii* [Occupational Health in Ferrous and Nonferrous Metallurgical Industries]. Moscow, pp. 79–84 (in Russian).

Nikolayeva, T. A. *et al.* (1970) *Gig. San.*, No. 11, 11–14 (in Russian).

Naeger, L. & Leibman, K. (1972) *Toxicol. Appl. Pharmacol.*, **22**, 517–527.

Novikov, Yu. V. *et al.* (1981) *Metody opredeleniya vrednykh veshchestv v vode vodoyemov* [Methods for Measuring Noxious Substances in Water Bodies]. Meditsina, Moscow (in Russian).

Olefir, A. I. (1967) *Vrach. Delo*, No. 11, 95–97 (in Russian).

Peregud, E. A. & Gorelik, D. O. (1981) *Instrumentalnye metody kontrolya zagryazneniya atmosfery* [Instrumental Methods of Environmental Pollution Monitoring]. Khimiya, Leningrad (in Russian).

Perelygin, V. M. (1984) *Gig. San.*, No. 8, 12–15 (in Russian).

Popov, T. & Angelieva, R. (1969) *Gig. San.* No. 1, 78–80 (in Russian).

Popova, O. Ya & Peretomchina, N. M. (1976) *Gig. San.* No. 2, 109–111 (in Russian).

Rakhmanov, Yu. A. (1969) In: *Materialy III syezda gigienistov i sanitarnykh vrachei Gruzii* [Proceedings of the 3rd Congress of Georgian Hygienists and Sanitary Physicians]. Tbilisi, pp. 92–94 (in Russian).

Rashevskaya, A. M. *et al.* (1973) In: *Professionalnye bolezni* [Occupational Diseases]. Meditsina, Moscow, pp. 185–258 (in Russian).

Rozenberg, P. L. & Byalko, N. K. (1969) *Khimicheskiye metody issledovaniya biologicheskikh substratov v profpatologii* [Chemical Methods for the Study of Biologic Substrates in the Field of Occupational Health]. Meditsina, Moscow (in Russian).

Rusch, G. *et al.* (1986) *Toxicol. Appl. Pharmacol.*, **83**, 69–78.

Schmidt, F. *et al.* (1972) *Schweiz. Med. Wochenschr.*, **102**, 83—88.

Schramel, P. & Magour, S. (1981) *J. Clin. Chem. Clin. Biochem.*, **49**, 830—831.

Shakhova, I. K. (1961) *Biokhimiya*, **26**, 315—318 (in Russian).

Shepard, T. H. (1986) *Catalog of Teratogenic Agents.* Johns Hopkins University Press, Baltimore.

Silayev, A. A. (1984) *Gig. Truda Prof. Zabol.*, No. **6**, 44 (in Russian).

Silverman, J. *et al.* (1985) *JAMA*, **254**, 2603—2608.

Skipworth, G. *et al.* (1967) *Arch. Dermatol.*, **95**, 83—86

Specifications for Methods of Determining Noxious Substances in Air (1965) [Tekhnicheskiye usloviya na metody opredeleniya vrednykh veshchestv v vozdukhe], No. 4. Publication of the USSR Ministry of Health, Moscow (in Russian).

Stumpe, A. (1960) *Arch. Ind. Health*, **21**, 519—524.

Tan, T. (1970) *Acta Dermato-Venerol.*, **50**, 55—58.

Tarasenko, N. Yu. (1972) *Gig. Truda Prof. Zabol.*, No. **11**, 13—16 (in Russian).

Torkelson, T. *et al.* (1961) *Amer. Ind. Hyg. Assoc. J.*, **22**, 263—270.

Trusova, L. N. (1982) *Gig. San.*, No. **7**, 44—45 (in Russian).

Verbitskaya, G. V. (1973) In: *Aktualnye voprosy gigiyeny naselennykh mest* [Hygiene in Populated Areas: Current Topics]. Moscow, pp. 26—29 (in Russian).

Verbitskaya, G, V. (1975) *Gig. San.*, No. **7**, 49—53 (in Russian).

Vinogradov, A. P. (1957) *Geokhimiya redkikh i rasseyanykh khimicheskikh elementov v pochvakh* [Geochemistry of Rare and Dispersed Chemical Elements in Soils]. Moscow, publication of the USSR Academy of Sciences (in Russian).

Wilding, J. *et al.* (1959) *Amer. Ind. Hyg. Assoc. J.*, **20**, 284—289.

Weir, F. & Mayers, F. (1966) *Ind. J. Med. Surg.*, **35**, 696— 701.

Weir, F. *et al.* (1964) *Toxicol. Appl. Pharmacol.*, **6**, 121— 131.

Zvezdai, G. I. *et al.* (1983) *Gig. San.*, No. **6**, 91 (in Russian).

Aluminum and its compounds

Aluminum chloride; a. hydroxide (gibbsite [min.]); a. nitrate nonahydrate;
a. nitride; a. oxide (alundum [molten Al_2O_3], alumogel, alumina, corundum
[min.], emery, monocorundum, electrocorundum [synthetic corundum]);
a. sulfate (sulfated alumina, aluminum coagulant, kaolin coagulant); a. sulfate
octadecahydrate
Aluminum-potassium sulfate dodecahydrate [aluminum sulfate-potassium sulfate-
water (1/1/24)] (alum, potassium alum, potash alum); alunite (alumstone);
bauxite; duralumins (aluminum alloys with copper [3.8–4.9%], magnesium
[0.4–1.8%], and manganese [0.3–0.9%]); kaolin [min.]; lanthanum
orthoaluminate-calcium metatitanate; mullite [min]; silumins (aluminum alloys
with silicon [8–13%])

IDENTITY AND PHYSICOCHEMICAL PROPERTIES
OF THE ELEMENT

Aluminum (Al) is a metallic element in group III of the periodic table, with the
atomic number 13. Its natural isotope is ^{27}Al (100%).

Aluminum is a very malleable and ductile metal that readily lends itself to
plastic working, rolling, forging, stamping, and drawing, and is a very good
conductor of heat and electricity. It exists in one allotropic modification.

In compounds aluminum is present in the oxidation state +3. In air it becomes
covered with a strong layer of the oxide Al_2O_3 that protects it from further
oxidation. Oxidation proceeds faster in a flow of oxygen or when the metal is
melted. Fluorine reacts with aluminum weakly at ordinary temperatures and
vigorously at red heat, chlorine and bromine react at both ordinary and elevated

temperatures while iodine reacts on heating. At a red heat, aluminum combines with sulfur (in a hydrogen flow) to form the sulfide Al_2S_3. It does not react with hydrogen. It gives the nitride AlN with nitrogen at 800°C and the carbide Al_4C_3 with carbon on heating. At a red heat it reacts with CO and CO_2 to form aluminum carbide, aluminum oxide, and carbon. It is resistant to both dilute and concentrated nitric acid, is sparingly soluble in sulfuric acid, reacts relatively well with hydrochloric acid, and does not react with phosphoric acid. It readily dissolves in alkalies, forming aluminates. See also Appendix. The chemistry of aluminum in the environment has been described by Driscoll & Schecher.

NATURAL OCCURRENCE AND ENVIRONMENTAL LEVELS

Aluminum is the most abundant metal and the third most abundant element (after oxygen and silicon) in the earth's crust, comprising 8.8% of its mass [Bertholf *et al.*]. In nature it is found in combination with other elements (mostly in the form of various silicates) and never in the metallic state. Its principal minerals are bauxite (a mixture of diaspore $HAlO_2$, bellite AlOOH, hydrargyllite $Al(OH)_3$, and oxides of other metals), alunite $(Na,K)_2SO_4 \cdot Al_2(SO_4)_3 \cdot 4Al(ON)_3$, nepheline $(Na,K)_2O \cdot Al_2O_3 \cdot 2SiO_2$, kaolinite $Al_2O_3 \cdot 2SiO_2 \cdot 2H_2O$, and some other aluminosilicates that occur in clays.

Aluminum concentrations reported for bodies of water range widely from 2.5 μg/L to about 1 mg /L, and are generally much higher (by a factor of 10 on average) in rivers and lakes then in oceans. Levels of 150–600 mg/kg have been reported for soil, up to about 10 μg/m^3 for urban air, and usually below 0.5 μg/m^3 for nonurban air [Sorensen *et al.*].

The living matter of the Earth has been estimated to contain 5 billion tonnes of aluminum; its reported average levels are (mg/100 g dry weight) : marine algae, 6; terrestrial plants, 0.5–400; marine animals, 1–5; terrestrial animals, 4–10; bacteria, 21 [Kovalsky]. See also *Aluminum and its role in biology* [Sigel].

Food plants were found to contain the following aluminum concentrations (mean values in mg/kg dry weight) : barley (grains), 10 or 135; oats (grains), 82; cabbage (leaves), 8.8; spinach (leaves), 104; lettuce (leaves), 73; carrot (roots), 7.8; onion (bulbs), 63; potato, 13 or 76; tomato (fruits), 20; apple (fruits), 7.2; orange (fruits), 15 [Kabata-Pendias & Pendias]. Voinar reported aluminum levels (mg/kg) of 42 in wheat, 36 in peas, 16 in corn, 1.6–20 in meats and meat products, and 850–1400 in tea leaves. (The daily intake of 2 liters of tea containing up to 200 mg aluminum is believed to be harmless for people with healthy kidneys [Schlagmann].).

PRODUCTION

Aluminum metal is usually produced from bauxite, the production process comprising two main stages: (1) digestion of bauxite at high temperature and pressure in a hot sodium hydroxide solution, followed by crystallization and calcination of the resulting hydrate to obtain alumina (aluminum oxide); and (2) reduction of the alumina to aluminum by electrolysis in a bath of molten cryolite using carbon electrodes; small amounts of magnesium fluoride and sodium chloride are added to the bath in addition to the fluoride-containing cryolite. The indoor air of aluminum-producing plants is likely to contain fluorine-bearing dust, CO_2, CO, and HF (purified waste gases contained fluorine at about 3 mg/m^3).

Aluminum chloride is obtained from a mixture of alumina, kaolin, bauxite, or clays with coke and tar by treatment with chlorine gas at 650–850°C; pure a. chloride can be obtained by reacting aluminum with chlorine. **A. nitrate** is made by dissolving aluminum hydroxide in nitric acid followed by evaporation and crystallization of the nitrate, and **a. nitride**, by reduction of aluminum oxide with coal in a nitrogen atmosphere at 1600–1800°C or by nitridation of powdered aluminum at 800–1200°C. **A. oxide** can be obtained from bauxite by treatment with a molten alkali, leaching out the alkali with water, decomposing the material with carbon dioxide, and calcining the resultant precipitate of aluminum hydroxide. **Synthetic corundum** is prepared by melting bauxite with coke and iron filings. **A. sulfate** is made by dissolving aluminum hydroxide in hot concentrated sulfuric acid.

USES

Aluminum is used widely in numerous industries and in larger quantities than any other nonferrous metal. It is alloyed with a variety of elements including copper, zinc, silicon, magnesium, manganese, nickel, etc. Aluminum and its alloys are very extensively used in the building industry (e.g. for door and window frames), shipbuilding (e.g. for internal fittings and superstructures), the aircraft industry (e.g. for airframes and aircraft skin), the automobile industry (e.g. for bodywork and pistons), the electrical equipment industry (e.g. for wires and cables), in nucleonics, electronics, and metallurgy, and in light industry to make office equipment and household utensils and appliances.

Aluminum chloride is used as a catalyst in wool dyeing and in the making of lubricating oils; **a. hydroxide** as a source of aluminum salts; **a. nitrate** in the textile and petroleum industries and in the production of catalysts; **a. nitride** as a lining material for baths, elecrolyzers, and reservoirs, in chemical engineering, as an additive to aluminum and its alloys, in electric and thermal insulations. **A. oxide** finds use as a source of metallic aluminum, alums, and other salts, as an abrasive and refractory, in electric insulations, in ceramic articles for radio engineering, and in the manufacture of vacuum tubes; electrocorundum [synthetic corundum] is used in the production of abrasive materials, artificial gems, and step bearings for precision instruments. A. sulfate is used in water purification, in papermaking, in the textile and leather-goods industries, and for wood preservation.

Alunite is used as a source of alumina, alums, and potassium salts. Bauxite is the principal source of aluminum and is also used as a raw material for the production of alumina, refractories, cements, and abrasives. Lanthanum orthoaluminate-calcium metatitanate is a ferroelectric. Mullite and its alloy with alumina serve as refractory linings for glass furnaces.

Bentonite clays are used in drilling muds and as mineral pigments (ocher, umber); low-iron kaolinite clays, in the production of fireclay refractories, porcelain, glazed pottery, and acid-resistant articles, in the paper, rubber, and perfumery industries and (kaolin) in casting alloys; red ferruginous-montmorillonite clays, in making bricks, cements, and pottery) and to purify and decolorize petroleum products and cooking fats.

Aluminum and its compounds are also used in medicine.

MAN-MADE SOURCES OF EMISSION INTO THE ENVIRONMENT

The main anthropogenic sources of aluminum in the environment are mining activities; solid and liquid emissions from the aluminum and some other industries; and the manufacture and use of aluminum components, structures, containers, and utensils. In the 1960s, aluminum was used in the production of approximately 4000 different articles, and their number has probably increased since then. Wastewaters highest in this metal were reported to be those from aluminum mining, chemical/pharmaceutical, paint and varnish, paper, textile, and synthetic rubber industries [Grushko, 1979]. In large industrial cities, raw sludges in the primary settling tanks of biologic wastewater-treatment plants had aluminum concentrations up to 2715 mg/kg [Goncharuk]. Two specific sources of aluminum and its compounds in the human body arise from their growing use in the food industry (in the processing, packaging, and preservation of food and as food additives) and in pharmacology [Sorensen *et al.*]; the aluminum content of some foods was found to double if cooked in an aluminum container.

In the vicinity of aluminum-producing works, fluorine was present in concentrations of $0.10-0.14$ mg/m^3 in the atmospheric air and at 135 mg/100 g, on average, in the soil and vegetation, and it also occurred at elevated levels in the blood, urine, hair, teeth, and nails of animals and people living there. Moreover, depressed phagocytic activity of leukocytes, appreciable alterations in some biochemical parameters, and increased pediatric morbidity were recorded in members of the local population [Kvartovkina *et al.*].

TOXICITY

Subthreshold concentrations of **aluminum chloride** and **aluminum nitrate** for effects on organoleptic properties of water were 4 mg/L and 0.1 mg/L, respectively, while subthreshold concentrations for effects on the sanitary condition of water bodies were 50 mg/L for aluminum chloride, 2 mg/L for aluminum nitrate, and 2.8 mg/L for **aluminum sulfate**. The highest concentrations at which the nitrate and sulfate did not disturb biochemical processes in a water body however long they remained there were found to be 2 mg/L and 5 mg/L, respectively [Bespamyatnov & Krotov].

Plants

In several countries, limits have been set or recommended for levels of aluminum and its compounds in wastewaters used for crop irrigation. In the USSR, the maximum permissible concentration of aluminum in such wastewaters is 5 mg/L; in the USA, 1 mg/L is the recommended limit value for long-term irrigation of any soil and 20 mg/L, for short-term irrigation of soils with low sensitivity to contamination [Grushko, 1979].

Wheat, oats, beans, and some other plants were adversely affected by aluminum at concentrations of ~1 mg/L in aqueous solutions applied to the soil [Grushko, 1979]. For corn, aluminum sulfate was toxic at 10 mg/L and aluminum chloride at 132 mg/L [Goncharuk]. In concentrations of 5.4 to 40.5 mg/L, aluminum sulfate inhibited the growth and development of spruce trees grown in an artificial nutrient medium.

Aquatic organisms

Rainbow trout exposed to an **aluminum nitrate** concentration of 5 mg/L exhibited uncoordinated movements and then tipped onto their side or back. Three-spined stickleback exposed to the nitrate in soft water died in 5 h at 10 mg/L, in 12 h at 0.5 mg/L, in 3 days at 0.12 mg/L, and in 6 days at 0.1 mg/L. The toxicity of **aluminum sulfate** was similar to that of the nitrate. The **chloride** was toxic at concentrations of ⩾0.5 mg/L. Other aluminum compounds were less toxic for fish than those mentioned above. For daphnids, the most toxic compound was $Al(NO_3)_3$ (first signs of toxicity appeared at 5 mg/L), followed by $AlCl_3$; the toxicity of other aluminum compounds was low. No deaths occurred among fish in a water stream on 1-week exposure to wastewaters discharged into the stream if they contained aluminum at concentrations of ⩽14 mg/L [Grushko, 1972, 1979; Metelev et al.].

Animals and man

General considerations

Aluminum is a metallic element of relatively low toxicity, although it may induce marked alterations in animals and humans on long-term exposure. It can adversely affect metabolic processes (particularly mineral metabolism) and nervous system functions, and is able to act on cells directly, interfering with their reproduction and growth. Prolonged exposure to dusts of aluminum and some of its compounds is conducive to pulmonary fibrosis [Bertholf *et al.*]. Aluminum produces many of its toxic effects by direct action on the nuclear chromatin or in an indirect way — by replacing other elements or modifying the activity of enzyme systems. It readily forms complexes with enzymes and can reduce the activity of lactate dehydrogenase, alkaline phosphatase, carbonic anhydrase, and catalase, as well as the enzymatic activity of ceruloplasmin, and has been shown to block active sites of enzymes involved in hematopoiesis [Khakimov & Tatarskaya].

An important role in the pathogenesis of intoxications caused by aluminum is played by its competitive relationships with calcium and phosphorus. When a large excess of aluminum salts is present in the body, calcium retention and phosphorus absorption are both reduced with the results that the blood level of adenosine triphosphate (ATP) falls and the phosphorylation processes are depressed; in such cases, aluminum has been found to accumulate to high levels (exceeding normal values 10- to 20-fold) in the parathyroid glands, leading to osteomalacia [Kerr] and reduced blood concentrations of the parathyroid hormone. There is evidence that this hormone plays an important part in the aluminum-induced impairment of phosphorus and calcium metabolism by facilitating aluminum absorption from the gastrointestinal tract [Mayor *et al.*]. Increased intake of aluminum has been shown to result in reduced iron absorption, indicating that aluminum competes with iron for binding sites on cells [Lione]

The major manifestations of aluminum neurotoxicity are considered to be impaired motor activity, convulsions, memory loss (partial or complete), and psychotic reactions manifested as myoclonic encephalopathy. One variety of this syndrome is dialysis encephalopathy — a degenerative disease of the brain accompanied by dementia and associated with long-term use of a dialysis fluid high in aluminum; in such patients, aluminum levels ranged from 0.180 to 0.500 mg/L in the blood and were very high in the brain, particularly in the gray matter.

It has been suggested that high aluminum levels in the human environment and especially in the drinking water may be associated with the occurrence of Alzheimer's disease (presenile dementia). Moreover, the excess mortality from this disease has been shown to be directly related to increases in the amount of acid precipitation, which usually resulted in elevated aluminum levels in the drinking water; it has also been found for some geographic areas that during periods when acid rainfall was high, an increased number of patients with Alzheimer's disease were admitted to mental hospitals [Zatta *et al.*].

The neurotoxicity of aluminum may be due to its action on the DNA-dependent RNA synthesis, the synthesis of polypeptide chains in protein molecules, axonal transport, the synthesis and inactivation of neurotransmitters, or/and dendrite structure. However, neurotoxic effects of aluminum are seen only in conditions involving impairment of the blood-brain barrier — such as Down's disease, parathyroid gland hyperactivity, hypervitaminosis D, and alcohol-induced encephalopathy [Boegman & Bates; Lai & Blass]. Mechanisms of aluminum toxicity have been reviewed by Ohmer. The mechanism of aluminum neurotoxicity with respect to human disease is discussed by Knick & McLachlan.

Dusts of some aluminum compounds are fibrogenic, whereas those of others are not, and its had been suggested that the fibrogenicity depends on the number of free OH groups at the surface of dust particles [Domnin *et al.*]. The fibrogenicity of aluminum dusts correlates with their hemolytic activity [Lagocz & Paradowski].

Evidence for mutagenic activity of aluminum has been provided by detecting significantly increased numbers of mammalian cells with chromosome aberrations following exposure to aluminum concentrations of 1 mg/L or higher in cell culture [Bryk]. Aluminum may also present carcinogenic hazards, as is indicated by the findings that specific alterations occurred in the erythrocyte membranes of mice administered its salts [Martinson & Stroikov] and that subcutaneous implantation of mice with disks made of aluminum foil resulted in malignant degeneration of the tissues involved [Leonard & Leonard; Williams]. In rats, teratogenic, but not embryotoxic, effects of aluminum have been demonstrated [Patermaen *et al.*].

Acute/single exposure

Animals

LD_{50} values for some aluminum compounds are shown in Table 1. The main symptoms of acute poisoning were excitation, respiratory distress, profuse salivation, rigidity of the tail and hind limbs, and tonic convulsions; at autopsy, necrosis of the gastric mucosa and fatty degeneration of the liver were noted.

After a single oral dose of **aluminum chloride** rats developed signs of disturbance in carbohydrate metabolism (depressed glycogen levels in liver and elevated aldolase activity in blood serum) and of damage to phosphorylation mechanisms (lowered ATP levels and raised ADP and AMP levels in blood). In mice, the threshold intraperitoneal dose of **aluminum oxide** for effects on body-weight and blood sulfhydryls was determined at 300 mg/kg. The lowest effective oral dose of **aluminum** was 17 mg/kg for rats and guinea-pigs and 9 mg/kg for rabbits [Krasovsky *et al.*].

Table 1. Median lethal doses of aluminum compounds

Compound	Animal	Route of administration	LD_{50} (mg/kg)
Chloride	Mouse	Oral	1333
		Intraperitoneal	96
	Rat	Intraperitoneal	333
Nitrate	Mouse	Oral	370
	Rat	Oral	204 or 280
Oxide	Mouse	Intraperitoneal	>3600
Sulfate octadecahydrate	Mouse	Oral	980
	Rat	Oral	370

Sources: Korolev & Krasovsky; Liublina & Dvorkin.

Signs of renal damage, including electrolyte disbalance in the urine, polyuria or oligouria, and elevated blood serum levels of urea, appeared in all rats given, at ages of 5 to 105 days, a single intraperitoneal injection of aluminum at 1 to 2 mg/kg in a solution that also contained tartaric acid and sodium hydroxide [Bräunlich et al.].

A subcutaneous injection of aluminum into pregnant rabbits in a dose of 400 or 800 µmol/kg, resulted in partial mortality of their offspring by day 10 after birth; the surviving pups showed low weight gains. The milk of their mothers contained up to 6 µg Al/g as compared to 1.3 µg Al/g in the milk of control females [Yokel].

Inflammatory changes in the tracheal and bronchial mucous membranes as well as interstitial, peribronchial, and perivascular sclerosis in the lungs were seen in rats 3 to 10 months after a single intratracheal administration of **alumina** or **corundum** dust in an amount of 50 mg [Bikmullina; Brunshtein & Danilova]. In a study on rats with intratracheal administration of **alumina** and **aluminum hydroxide** dusts, their biologic activity was found to decrease in the order: γ-alumina > aluminum hydroxide > technical-grade alumina > α-alumina [Domnin et al.].

A single intratracheal administration of **lanthanum orthoaluminate-calcium metatitanate** led in rats to a proliferative cellular response in the lungs with only moderate development of reticulin fibers but a pronounced chronic inflammation in the bronchi.

When dusts of a **bauxite** ore, a limestone-bauxite melting stock, or a limestone-soda-bauxite clinker were administered intratracheally to three groups of rats (50 mg per animal), the highest toxicity was displayed by the clinker dust—one-third of the rats died from pulmonary edema several hours to days post-administration; in the surviving animals of this and the two other groups, a slowly evolving pulmonary sclerosis was observed [Oshchepkov et al.].

Continuous inhalation of an **alunite** dust (which contained $\approx 1\%$ SiO_2) by rats for 18 h led to a moderate perivascular and peribronchial fibrosis. Observations made in rats 9 months after a single intratracheal administration of this dust showed nodular fibrosis and elevated levels of collagenous proteins in their lungs. Nodules resembling those developing in response to metallic dusts were seen in the lungs of rats 21 months after their single exposure to mullite (a silicate of aluminum) by this route [Dvizhkov].

The dust collected in the foundry department of a factory producing aluminum alloys proved to be very toxic: rats administered 50 mg of this dust intratracheally showed marked growth retardation and greatly increased rates of hippuric acid synthesis during the subsequent 7 months and, as found at autopsy, developed dystrophic changes in parenchymal organs and moderate diffuse sclerosis in the lungs [Nikogosian et al.]. Nodular sclerosis occurred in the pulmonary tissue of rats following their intratracheal exposure to **silumins** (aluminum alloys with silicon) [Izraelson].

Increased levels of collagenous proteins, focal thickening of the interalveolar septa, areas of emphysema, and bronchitis were observed in the lungs of rats 1 month after an intratracheal administration of **aluminum nitride** in amounts of 50 to 100 mg; after 6 months, mild pneumosclerosis and dystrophic changes in liver cells were seen [Brakhnova].

Highly aggressive behavior was displayed by aerosols generated by welding operations using aluminum-magnesium alloys [Leonicheva].

Repeated and chronic exposures

Animals

In dogs injected with **aluminum** subcutaneously at 1 mg/kg once daily for 5 days, the blood plasma level of aluminum rose from 7.5 to 16.6 µg/L immediately after the first injection (plasma clearance was 4.4 ml/min and renal clearance, 1.9 ml/min) and ranged from 351 to 5148 µg/L after the fifth [Di Domenico et al.]. In rats, impaired ribosomal function in brain cells and altered transferase activity were noted after two intraperitoneal injections of **aluminum chloride** [Magour & Mäser].

Abnormal behavioral reactions were observed in young rabbits (aged 7 to 11 weeks) born to mothers that had been receiving repeated subcutaneous injections of **aluminum** at 100 µmol/kg between days 2 and 27 of pregnancy [Yokel].

A pulmonary fibrosis distinct from silicosis developed in guinea-pigs exposed by inhalation to an aluminum dust (18 000 particles per cm^3) for 44 h per week during 6 weeks. The lungs of rabbits that had been inhaling **aluminum dust** or **aluminum bronze dust** for 1 to 2 h daily over 20 to 40 days showed interstitial inflammation, thickened interalveolar septa, and accumulations of cells filled with aluminum dust particles; later, hyaline degeneration of the interstitial tissue was seen.

In long-term inhalation studies on rats and rabbits exposed to aluminum dust for 0.5 to 8 h daily, intensive dust uptake by alveolar epithelial cells was observed during the first few weeks of exposure. Only a minor proportion of the inhaled dust entered the lymph nodes: most of the dust was retained in the pulmonary tissue and accumulated in the alveolar and bronchial lumens which became obliterated by the desquamated or proliferating epithelium. Areas of atelectasis and emphysema were gradually forming in the lungs followed later, in the absence of any inflammatory response, by a relatively rapid development of fibrotic induration accompanied in many animals by hyaline degeneration of the interalveolar septa, vessel walls, bronchi, and pleura. A not uncommon finding was pneumonia. Pathologic changes progressed after the cessation of exposure.

Typical nodular forms of pneumoconiosis were reported to have occurred in rats that had been inhaling aluminum dust at 300–400 mg/m^3 for 5 h daily over 6 to 10 months [Viderli]. Rabbits exposed to aluminum dust at 80 mg/m^3 for 20 min daily over 4 to 6 months developed nodular or diffuse pneumosclerosis and obliterating endovasculitis and sclerosis of the pulmonary blood vessels as well as dystrophic changes in the liver; in addition, a number of hematologic changes were recorded, including elevated hemoglobin levels and erythrocyte counts, poikilocytosis, and anisocytosis.

Rats exposed to a condensation aerosol generated during the processing of aluminum scrap (205 mg/m^3, 5 h/day, 5 days/week, up to 6 months) exhibited signs of interstitial sclerosis in their lungs which were found to have elevated lipid and hydroxyproline levels; their blood contained less sugar and more cholesterol compared with that of unexposed controls [Domnin et al.].

Thickened alveolar septa, small focal accumulations of dust-laden cells, and growth of collagen fibers were seen in the lungs of rats that had been inhaling an aerosol resulting from aluminum welding (120–140 mg/m^3 for 3 h/day up to 12 months).

In a 12-month inhalation study, daily 1-min exposures of rats to fumes generated by burning electrodes and containing 28×10^9 aluminum particles per m^3, gave rise to catarrhal alveolitis and lipid pneumonia [Christie et al.]. In another study, most of the rats and rabbits exposed to **aluminum oxide** fumes at a concentration of 0.2 mg/L of air for 5 h daily, failed to survive the 7-month exposure period, the causes of death being pulmonary and bronchial inflammations and pulmonary edema. At autopsy, their lungs presented a picture similar to that described above for animals inhaling aluminum dust, but the fibrotic changes were less pronounced. The aluminum content of the lungs ranged from 0.04% to 1% on a dry weight basis [Miller].

Prolonged inhalation of **aluminum hydroxide** dust (at 10 mg/m^3, calculated as Al) elicited a stronger proliferative response from the alveolar epithelium of rats than did aluminum dust. The lungs showed intensive connective tissue growth with the formation of sclerotic nodules whose center, composed of macrophages, had undergone hyalinization. After 3 months of inhalation exposure of rats to a dust of **synthetic corundum** at 200 mg/m^3 for 2 h daily, loosening and desquamation of the upper respiratory tract epithelium and atypical outgrowths of its basal layer were seen; after 6 months, desquamation of the nasal epithelium and dystrophic changes in the tracheal cartilages were added, while after 8 months there were erosions in the nasal and tracheal mucous membranes, necrosis of the tracheal cartilages, and emphysema and diffuse or focal sclerosis in the lungs [Frangulian].

No fibrotic changes were seen in the lungs of guinea-pigs after 12 months of their intratracheal exposure, at weekly intervals, to 0.5 ml of a 5% suspension of dust particles that had sedimented at **bauxite**-smelting furnaces (the particles ranged in size from ultramicroscopic to 40 µg and the dust contained 6.8% SiO$_2$, 50% Al$_2$O$_3$ [corundum], 9% Fe$_2$O$_3$, 4% TiO$_2$, and minor proportions of other metal oxides). In contrast, accumulations of dust cells and fibrosis occurred in the lungs of guinea-pigs after 2 to 4 months of similar exposure to dust particles present in the fumes collected above these furnaces (the particles ranged from 0.05 to 0.5 µm in diameter and contained 32.3% SiO$_2$ and 56% Al$_2$O$_3$); autopsy performed at the end of the 12-month exposure period demonstrated pleuropulmonary adhesions, scars, and severe diffuse fibrosis. Prolonged inhalation exposure of rats to dusts of a limestone-soda-bauxite clinker at 0.9–91.7 mg/m^3 for 5 h/day over 10 months resulted in purulent inflammation of the nasal mucosa, metaplasia of the tracheal epithelium with its transformation into a multirowed squamous epithelium, and slow developing pulmonary sclerosis at sites of dust deposition [Oshchepkov *et al.*]. In a similar study, purulent inflammations of the nasal, laryngeal, and tracheal mucosas and, less frequently, inflammatory changes in the lungs developed in rats inhaling dusts of a clinker obtained by calcining a mixture of bauxites and soda at 1200°C

Chronic inhalation exposure of rats to **alunite** dust (400–500 mg/m^3, 5 h/day, for 6 months) was reported to have produced a roentgenologic picture seen in second-stage pneumoconiosis [Viderli].

Rats inhaling **lanthanum orthoaluminate-calcium metatitanate** dust exhibited reductions in erythrocyte counts, hemoglobin levels, blood cholinesterase activity, and oxygen consumption along with increases in the peroxidase index; 6 to 9 months after the start of exposure, significantly elevated collagen levels were present in their lungs. In these chronic tests, the concentration of 60 mg/m^3 was considered as the threshold level for systemic toxic effects [Zhilina]. In male rats, prolonged exposure to this dust at 60 mg/m^3 for 4 h/day, 6 days per week, produced irreversible gonadotoxic effects manifested as reduced sperm motility, increased numbers of degenerate spermatozoa, diminished numbers of spermatogonia, and increased numbers of seminiferous tubules with desquamated epithelium; in rats similarly exposed to 6 mg/m^3, the changes in spermatogenesis were much less marked and reversible, and this concentration was taken to be the threshold level for gonadotoxic effects on chronic exposure [Shcherbakov].

Repeated administration of **aluminum chloride** to mice at 19 mg/kg and rats at 200 mg/kg by oral gavage over an extended period of time resulted in impaired carbohydrate metabolism, retention of aluminum in the bones, liver and testes, and a negative phosphorus balance, as well as in decreased body-weight gains in the progeny; no untoward effects on the reproductive performance of animals were detected, however [Dixon et al.].

Rats and rabbits repeatedly exposed to **aluminum hydroxide** at 250–320 mg/kg by the same route as above, showed reduced capacity for antibody production without developing any histologically detectable abnormalities in their internal organs [Nesterenko & Ostrovskaya]. After 6 months of repeated oral exposure of rats and rabbits to **aluminum nitrate** in much lower doses (10 mg/kg), both the secretory activity of their gastric glands and the digestive capacity of the gastric juice were strongly depressed while their blood contained significantly less erythrocytes and hemoglobin than before the exposure.

Rats repeatedly exposed to **aluminum-potassium sulfate** at 2.5 mg/kg (as Al) by oral gavage over a 6 month period developed severe morphologic changes in their testes where abnormal seminiferous epithelia, depressed activities of several oxidative enzymes, and sharply diminished RNA levels in the parenchyma were noted; the dose of 0.25 mg/kg was only slightly gonadotoxic [Vasyukovich et al.].

Roentgenologic examination of piglets that had been receiving daily intravenous injections of **aluminum chloride** at 1.5 mg/kg from the age of 5 weeks, demonstrated substantial aluminum deposits in bones and sclerotic areas in the femoral and tibial bones [Fernandez et al.].

Cases of lethal poisoning were observed among cattle consuming water containing **aluminum hydroxide** at a concentration of 42.9 mg/kg (as Al) [Goncharuk].

Man

The disease caused by the inhalation of aluminum dust or fume and now known as aluminosis (or aluminosis pulmonum) was previously ascribed to the admixture of silica present in such dusts. Severe aluminosis developed in workers employed in the spraying of aluminum paints or in the production of pyrotechnic aluminum powders [Swensson et al.]. After a year of employment, such workers complained of weight loss, rapid onset of fatigue, breathlessness, and cough; dry and moist rales were heard in their lungs, and the chest radiographs showed substantial pulmonary opacities in some cases. The disease tended to progress even after the cessation of exposure, and aluminum was detectable in the sputums for a long time [Swensson et al.]. Subsequently, workers in these occupations complained of poor appetite, indigestion, nausea, pains in the stomach and/or in the whole body, and dry or wet cough. Common clinical findings were lymphocytosis, eosinophilia, blood levels of aluminum exceeding normal values severalfold, lowered acid phosphatase and ATP levels in the blood serum, and lowered alkaline phosphatase levels in the intestine; lung changes were present in 25–50% of the subjects [Waldron-Edward et al.].

The production of pyrotechnic powders was infrequently associated with the occurrence of spontaneous pneumothorax and even fatal disease. The increased prevalence and severity of intoxications among workers in this occupation has been attributed to the fact that the airborne dust was very fine and that the solubility of its particles might have been reduced by the stearic acid added to the powder in small amount.

Cases of pulmonary fibrosis accompanied with neurologic disturbances have been described in aluminum workers [Dvizhkov; Molokanov et al.], as has been a case of hypersensitivity with attacks of bronchial asthma in a worker engaged in the drilling of aluminum sheets. A worker who developed encephalitis and epileptiform seizures after 13 years of exposure to aluminum dust and died from terminal pneumonia, was found to contain a 20-fold excess of aluminum in his brain and a 122-fold excess in his liver [McLaughlin et al.].

Workers employed in the production of **secondary aluminum** may be exposed to dusts of complex chemical composition. Of special concern in the aluminum recovery industry have been exposures to elevated concentrations of benzpyrene and to the nickel and chromium contained in such dusts; a retrospective study of cancer mortality has indicated the existence of an increased carcinogenic risk in this industry [Roslyi].

The production of **alumina** by the wet alkaline method (Bayer process) has been associated with a 5- to 10-fold increase in the prevalence of catarrhs affecting the upper respiratory tract—even in workers employed for less than 1 year; the most strongly marked mucosal changes were present in workers in the calcining department. Many of those employed for a long time had developed aluminosis; in some cases, small nodules were seen on the chest radiographs against a background of diffuse fibrosis. Upper respiratory tract catarrhs were also common among workers engaged in alumina production by a dry process; cases of pneumosclerosis, perforation of the nasal septum, and inflammation of the acoustic nerve have been described [Domnin et al.; Oshchepkov et al.]. Rather frequent complaints of workers engaged in alumina loading operations were bleeding gums, hoarseness, dryness in the mouth and nose, lacrimation, smarting eyes, loss of appetite, and belching; blood samples taken before and after workshift gave mean aluminum concentrations of 42.8 µg% and 206 µg%, respectively. A fairly common condition in alumina production workers was chronic bronchitis, often with an asthmatic component. The most characteristic bronchoscopic finding was mucosal atrophy in the bronchial tree, while bronchography not infrequently revealed bronchial atony and bronchiectasis [Semennikova et al.]. The occurrence of bronchial asthma, which in some cases combined with allergic dermatitis and nasal septal perforation, was reported among workers producing alumina from a red bauxite containing small amounts (0.03–0.1%) of trivalent chromium that was converted into the more toxic hexavalent chromium in the baking process [Budanova].

Cases of severe disease occurred among workers attending furnaces in which **bauxites** were smelted (together with iron, quartz, and coal) to obtain artificial abrasive materials. The dusts and fumes generated in this process contained 41–62% Al_2O_3 and 30–44% SiO_2 as well as small quantities of other metal oxides; the concentrations of total dust in the workroom air ranged from 10 to 41 mg/m^3 (with silicon levels of 15–25%), and particles in the dusts and fumes did not usually exceed 1 µg in diameter. In workers employed for 3 to 8 years at furnace ports or as crane operators, the disease was marked by a rapid evolution of diffuse pulmonary fibrosis, severe marginal emphysema, and the formation of spontaneous pneumothorax, not infrequently fatal. The predominating complaints were cough (dry or moist) and rapidly progressing shortness of breath. In some instances, coarse or fine rales were heard in the lungs. The clinical picture of the disease did not generally correspond to its roentgenologic appearance and was similar to that seen in workers engaged in the manufacture of pyrotechnic aluminum powders, but in some advanced cases the dyspnea was more severe and the respiratory movements were grossly restricted. In a group of workers employed for 2 to 10 years, the prevailing complaints were pleural or retroperitoneal pains and especially a morning cough productive of dark sputum [Dvizhkov].

Histopathologic examination of the lungs in those who died from severe disease revealed thickening of the pleura, vessels, and bronchial tubes. In some instances there were areas of acute pneumonitis or of fibrous tissue completely replacing the lung tissue; thickened alveolar septa (in sites where lung tissue remained); slit-like spaces surrounded by infiltrates composed of giant cells; and either fibrohyaline strands passing through the lung or areas of continuous fibrohyaline tissue. The ash of the lungs usually had a high content of aluminum (up to 45%) and silicon (up to 30%).

A high prevalence (20%) of pulmonary fibrosis (silicosis of grade 1 or 2) was recorded among workers employed for 10 to 15 years in the manufacture of grinding wheels from synthetic abrasive materials (mainly **corundum**). The roentgenologic appearance of the disease in its early phases resembled that seen in some forms of pulmonary tuberculosis. Those exposed to fine corundum dusts (which contained small amounts of dichromates) experienced irritation of the upper respiratory tract mucosa and nasal bleeding shortly after the onset of exposure [Frangulian; Sturm & Paukert].

Examination of 43 workers employed in the crusting of pure corundum showed the presence of dust-induced pulmonary fibrosis (silicosis of grade 1 or 2) in eight individuals, all of whom had been on the job for 8 years or more.

Frequent complaints of corundum workers were general weakness, insomnia, sweating, paresthesia, and impotence [Oganesian].

Pneumoconioses seen in **bauxite** miners tended to be of late onset (after 11–15 years of employment), were usually accompanied by scanty clinical manifestations, and were rarely complicated by tuberculosis; autopsy of those who died revealed connective tissue proliferation in the perivascular and peribronchial spaces, in the alveolar septa, and in the pleura [Ragolskaya]. Moderate pulmonary fibrosis and basal emphysema have also been detected among bauxite miners with 5 to 10 years of employment, and the fibrosis caused by bauxite dust was said to be distinct from that induced by dusts of bauxite soot or metallic aluminum [Elyashev]. In a **mullite** factory, a 2–3-fold increase in the prevalence of chronic rhinitis and pharyngitis was discovered, and several cases of first-stage pneumoconiosis of a distinctive diffuse sclerotic type were identified among workers employed there for 7 to 20 years [Chubarian]. The term 'mullite lung' has been proposed to designate this condition [Dvizhkov].

Alunite dust is destructive to the enamel and dentine of teeth and may cause parodontosis and desquamative dystrophic or atrophic lesions in the gingival mucosa [Pleshcheyev et al.]. Chronic inflammations of the upper air passages, atrophic gingivitis, dystrophic forms of parodontosis, and carieses were all fairly common among workers engaged in the production of **silumin** by an electrothermal method and exposed to a composite dust containing coal, alumina, kaolin, silica, aluminum fluoride, and fluorine [Barannik; Golomidov]. Considerable respiratory and cardiodynamic abnormalities were exhibited by workers in **aluminum chloride** manufacture [Varnakova].

Of obvious interest to occupational physicians and medical examiners are data on aluminum levels in the hair of aluminum-exposed workers in various occupations. As an example, here are average values found in one study (mg%): founders, 15.85%; foundry cleaners, 18.8; sand mixers and molders, 7.8; control (unexposed) workers, 2.93.

A special occupational group are workers employed in the electrolysis departments of aluminum works, for they are exposed to a range of adverse factors including alumina dust, coal-tar pitch, carbon monoxide, high air temperature, magnetic fields, and (the most important factor) fluorides. The characteristic abnormalities reported for this group are the following:

— inflammatory changes in the upper respiratory tract mucosa, bronchial asthma or asthmatic bronchitis, and pneumoconioses of the aluminosis type [Miller; Molokanov et al.; Vermel];

—neurocirculatory disturbances and alterations in the pyramidal system, electroencephalographic abnormalities, and tremor of the fingers [Volkova et al.];

—narrowing of the visual fields (especially toward the end of the shift), vascular changes in the ocular fundus, and opacity of the crystalline lens;

—changes in the locomotor apparatus, such as spondylosis, decalcification of the intervertebral ligaments, exostosis, and focal osteosclerosis [Bogatin & Makarova];

—cutaneous manifestations in the form of erythematous patches or telangiectasias (on the arms, neck, and/or anterior thoracic wall), and toxic melanoderma [Aleksperov *et al.*];

—impaired reproductive function, manifested in reduced potency in men and in a variety of ways in women, including miscarriages, premature delivery, weak uterine action, late gestational toxicosis, and nonspecific colpitis;

—cytogenetic changes in lymphocytes [Sedova].

In addition, workers in this occupation may be subject to carcinogenic hazards created by sublimates of the coal-tar pitch used as a binding material in the production of electrode masses [Konstantinov].

The human body may receive rather large amounts of aluminum in foods and beverages that have been prepared or stored in containers made of this metal. For example, as much as 200–300 mg aluminum was found to be ingested daily by most brewery workers (up to 400 mg by some) with the beer stored in aluminum flasks [Evenshtein]. Yet no deviations from normal were detectable in general health, roentgenographic appearances of bones, or phosphorus and calcium metabolism in any of the workers. Aluminum in the colloidal form, unlike its soluble compounds, appears to be non-toxic.

According to Osmolovskaya & Kovalev, aluminum containers are unsuitable for storing or preparing sauerkraut, clabbered milk, and other acid foods.

Since the body burden of aluminum and its toxicity increase with age, elderly people are advised to curtail the use of aluminum-containing therapeutic agents such as antacids, anti-diarrheals, and rectal suppositories.

Intoxications with aluminum have occurred as a result of its entering the body in excessive amounts during hemodialysis [Piscator]. Aluminum accumulation/toxicity in dialysis patients has been studied more recently by D'Haese *et al.* Aluminum toxicity in relation to chronic renal failure has been discussed by Wills & Savory.

Toxic effects of clays

Animals

In animals experiments, **kaolin** dust produced diffuse interstitial and nodular lesions with less extensive development of fibrous connective tissue than in silicosis. This condition (kaolinosis) differed from the fibrosis caused by **bentonite** clays in that only nodules without formation of collagen fibers were seen in the latter disease. Fired Cambrian clay (claydite) was more toxic than the unfired, probably because it contains more free silica [Oksova *et al.*]. Rats administered 50 mg of claydite intratracheally developed bronchial lesions and pneumonia, followed later by the appearance of fibrous nodules of a silicotic type; a diffuse sclerotic form of pneumoconiosis also occurred in rats after the inhalation of claydite dust at a concentration of 50 mg/m^3 [Selivanov]. In rats administered 50–100 mg of gumbrin (bleaching clay) intratracheally, a moderate peribronchial and perivascular fibrosis was seen to have developed in the lungs within 6 to 12 months, superimposed on emphysema, catarrhal desquamative pneumonia, and bronchiectatic abscesses [Kipiani]. The lungs of rats administered 50 mg of **chamotte** (fireclay) dust by this route were found to contain increased macrophage numbers after 6 months and elevated cholesterol levels after 9–12 months; chronic inhalation exposure of rats to this dust (at 266 mg/m^3, 5 h/day, 5 days/week, for 12 months) resulted in chronic bronchitis and pulmonary fibrosis that were more pronounced than those caused by magnesite dust at the same concentration [Belobragina & Yelnichenykh].

Man

Occupational exposure to **bentonite** dusts at high concentrations for several years led to dysproteinemia, depressed phagocytic activity of leukocytes, and bronchospasm; some workers also developed a diffuse interstitial pneumoconiosis [Kipiani].

In workers manufacturing ceramics and abrasives, a relatively benign kaolinosis characterized by scanty clinical manifestations, a slow evolution of diffuse interstitial reticular fibrosis, and mild pulmonary emphysema was observed (not infrequently in combination with pulmonary tuberculosis) after 10 or more years of exposure to dusts of various clays, including chamotte [Karmazin et al.]. Cases of moderate pneumoconiosis in pottery workers [Warraki & Herant] and of massive pulmonary fibrosis resulting from inhalation of kaolin dust [Sheers] have been reported.

Kaolinosis was identified in female workers employed in the sifting and packaging of pure powdered **kaolin** in a pharmaceutical factory. After 16–20 years of work, they became dyspneic and showed pathologic changes in the chest roentgenogram. In many cases of long-lasting kaolinosis there was a diffuse sclerotic form of fibrosis characterized by pronounced interstitial sclerosis, the presence of only a few typical sclerotic nodules in the lungs, and their invariable presence in the hilar lymph nodes [Dvizhkov]. Cases of pneumoconiosis also occurred among workers producing a cosmetic powder that contained kaolin, talc, and smaller amounts of zinc oxide, chalk, and starch. Already in the early phases of the disease, emphysema developed in the basal lung regions and the fibrotic process involved the pleura. A number of workers also developed pneumoconiosis after 10 to 15 years of employment in the manufacture of ceramic tiles [Pigolev; Shatalov].

Dermal and ocular effects

Animals

Repeated skin application of a 10% **aluminum chloride** or **aluminum nitrate** solution resulted in epidermal hyperplasia and in the appearance of microabscesses and ulcerations in the skin of mice and rabbits; aluminum deposits were found in the horny layer. **Aluminum hydroxide** was not irritating to the skin.

Man

After the entry of **aluminum** particles into the eye, focal necrosis of the cornea, alterations in corneal pigmentation and in the lenticular capsule, and opacification of the vitreous body have been observed. **Aluminum** and **duralumin** dusts irritate mucous membranes of the eyes, nose, mouth, and genitals; occasionally, ulcerative lesions were seen in the nose at sites of dust deposition. Moreover, acne and acute or chronic eczema and dermatitis may develop on the skin exposed to these dusts. The eczema and dermatitis were accompanied by itching and burning of the skin and sometimes recurred upon exposure to other irritants. Under the action of aluminum and duralumin dusts, even minute skin cuts and wounds take a long time to heal. In areas of skin damage, phlegmons have often developed (in some cases after a slight trauma such as that caused by scratching), as have inflammation of lymphatic pathways and glands and even septicopyemia.

No irritation of the skin or eyes was observed in workers exposed to aluminum dust concentrations around 13 mg/m^3 in the workroom air [Elyashev].

Lanthanum orthoaluminate-calcium titanate exerted a slight allergizing effect on the skin of people with hypersensitivity [Zhilina]. Cases of eczema and dermatitis have been reported among workers of alumina departments [Antonieva et al.]. Workers exposed to corundum dusts complained of skin itching on the feet, legs, hands, and arms, which was sometimes accompanied by perspiration and impaired skin sensitivity; the clinical picture was said to resemble that seen in 'nickel itch' [Oganesian]. The skin exposed to a solution of potash alum becomes hard and acquires a tanned appearance. Izrailet & Eglite have classified aluminum among contact dermal sensitizers.

ABSORPTION, DISTRIBUTION, AND ELIMINATION IN MAN

Aluminum is an essential trace element. It influences the activity of several enzymes, the reproductive system, and both embryonal and postembryonal development [Sorensen et al.]; participates in the formation of epithelial and connective tissues and of nerve cell membranes in the brain; and plays important roles in various biochemical processes either by complexing with enzymic proteins or by acting as a cofactor in enzyme-substrate interactions [R. Bourdon & S. Bourdon; Day].

The adult human body requires from 35 to 50 mg aluminum daily. The main routes of its entry into the body are normally via ingestion (in food and water) and inhalation (in air), although significant amounts may sometimes enter through other media, e.g. drugs, deodorants, or kitchenware. The average daily intake has been estimated to be between 20 and 60 mg with food and around 1 mg with water; considerable contributions to the dietary intake may be made by food dyes and food preservatives. The range of acceptable aluminum intakes is considered to be 2–100 mg [Schlagmann]. A significant source of aluminum may be the atmosphere; it has been estimated that this source alone may account for a 10-fold increase in the aluminum content of pulmonary epithelium over 70 years of human life.

Aluminum is well absorbed in acid media such as the stomach and proximal duodenum; in the intestine it is precipitated by phosphates, which may result in phosphorus deficiency if the aluminum intake is high [Khakimov & Tatarskaya].

The total body content of aluminum in an adult human varies from 50 to 140 mg. Its common concentration range in the blood is 0.024 to 0.070 mg% (most of the aluminum being found in serum); its average levels in other tissues or organs were (mg/g): 0.059 in lung, 0.5 in long bone, 0.056–0.210 in heart, 0.087 in intestine, 0.04–0.250 in brain, 0.015 in muscle.

Elimination of aluminum is mainly through the kidneys. Urinary excretion rates normally range from 10 to 15 µg/day, but were found to increase 20- to 40-fold when this metal was entering the body in high amounts [Alfrey].

HYGIENIC STANDARDS

Exposure limits adopted in the USSR for aluminum and its alloys and compounds are shown in Table 2. In the USA, the following TLV-TWA values have been set (mg/m^3): 10.0 for aluminum metal and oxide (for the oxide, the value is for total dust containing no asbestos and <1% crystalline silica), 5.0 for pyro powders, 2.0 for soluble salts, 5.0 for welding fumes, and 2.0 for aluminum compounds not otherwise classified.

The highest permissible levels established in the USSR for aluminum in human foods are given in Table 3.

METHODS OF DETERMINATION

In air. A method for the determination of aluminum oxide in air is based on colorimetry of the colored solution that forms in the reaction of aluminum ion with alizarin or arsenazo I in a weakly acid medium (sensitivity 0.5 µg in the

Table 2. Exposure limits for aluminum and its alloys and compounds

	Workroom air		Atmospheric air	Water sources		
	MAC_{wz} (mg/m^3)	Aggregative state	$TSEL_{aa}$ (mg/m^3)	MAC_w (mg/L)	Hazard class	Note
Aluminum and its alloys (as Al)	2.0	Aerosol	—	0.5^a	3.2	F^b
Aluminum hydroxide	6.0	Aerosol	0.04	—	4	F
nitride	6.0	Aerosol	—	—	4	F
oxide containing $\leqslant$20% Cr_2O_3 (as Cr)	1.0	Aerosol	—	—	3	—
oxide as disintegration aerosol (alumina, electrocorundum, monocorundum)	6.0	Aerosol	—	—	4	F
oxide containing $\leqslant$20% free SiO_2 and $\leqslant$10% iron oxide (condensation aerosol)	6.0	Aerosol	—	—	4	F
oxide containing SiO_2 (condensation aerosol)	2.0	Aerosol	—	—	3	F
oxide in mixture with $\leqslant$15% nickel alloy (electrocorundum)	4.0	Aerosol	—	—	3	F

^{a} Based on sanitary and toxicologic criteria.
^{b} F=fibrogenic.

Table 3. MACs for aluminum in foods (mg/kg)

Cereal grains and bread	20.1
Milk and milk products	1.0
Meat and meat products	10.0
Fish and fish products	30.0
Vegetables and fruits	30.0
Juices	10.0

volume analyzed) [*Specifications for Methods of Determining Noxious Substances in Air*, 1968]. Aluminum nitride and lanthanum orthoaluminate-calcium metatitanate can be determined by a gravimetric method [*Specifications for Methods of Determining Noxious Substances in Air*, 1965]. Krivda & Makeyeva have described a photometric method of aluminum nitride determination using chromazurol S (sensitivity 0.05 µg/sample). A spectral method (a detection limit 0.2 mg/m^3 of air) [Muraviyeva] and a gas-liquid chromatographic procedure [Yavoskaya & Grinberg] have also been used for aluminum determination in air.

In water: a spectrographic technique has been proposed for the determination of aluminum metal after its extraction with chloroform as a complex with diethyldithiocarbamate and 8-hydroxyquinoline (sensitivity limits 0.6–80 µg) [Novikov *et al.*]. Analytical techniques in studies of aluminum species in aqueous solutions have been reviewed by Salbu.

For measuring aluminum in *biologic materials*, a colorimetric method may be used, based on the ability of the ammonium salt of aurintricarboxylic acid (aluminon) to form a bright red lacquer with aluminum (iron should be preliminarily removed as it gives the same color). Various procedures for atomic absorption spectroscopy are also available and have been extensively used for routine determination of aluminum in biologic materials; with flameless atomic absorption, detection limits of ~0.03 ng were obtained [Fuchs *et al.*]. Additional analytical information can be found in Baxter *et al.*; Bertholf *et al.*; Ramsey *et al.*; Savory & Wills; Way *et al.*

MEASURES TO CONTROL EXPOSURE

Hazardous exposures at *workplaces* are usually confined to the production of aluminum and the making of aluminum abrasives.

The measures to be implemented for securing a more healthy environment in the electrolysis departments of aluminum-producing plants are mainly of a technical nature. It is particularly important to ensure the following: automatic feeding of alumina to the electrolyzers; complete mechanization of operations involved in servicing the electrolyzers and in the removal of coal foam from the electrolyte; minimization of anode effects; use of prebaked anodes; and substitution of granulated for powdered fluorides. The electrolyzers should be provided with enclosures and efficient local exhaust ventilation. Good general ventilation of the workrooms is also important.

In the production of secondary aluminum, mechanized equipment should be used for unloading aluminum filings and large-sized scrap from trucks and vans; the sorting of scrap should also be mechanized.

If safe working conditions cannot be ensured by technical control measures, personal protection devices should be used, with special attention being given to respiratory and skin protection. Workers should be encouraged to observe a strict personal hygiene routine.

Those involved in the production of aluminum metal should be made aware of the hazards associated with handling caustic soda. In all sites of risk, eyewash bottles and basins with running water should be provided. Personal protective equipment (e.g. goggles, gloves, aprons, and boots) should be supplied. Showers and double locker accommodations (one locker for work clothing, the other for personal clothing) should be provided and all employees encouraged to wash thoroughly at the end of the shift. All furnacemen and carbon electrode workers should be supplied with visors, respirators, gauntlets, aprons, armlets, and spats to protect them against burns, dusts, and fumes. Adequate protection against fluorine fumes can be given by respirators with charcoal filters. Efficient dust masks are necessary for protection against carbon dust [Dinman].

In premises where the working clothes may be contaminated with dust, facilities should be available for its removal in such a way as to prevent it from entering the environment or contacting the inner surfaces of the clothes and the skin of workers. Working clothes should be laundered and repaired on a centralized basis and must not be allowed to be taken home.

It is considered important that a regular and sufficient supply of food and drink be available to aluminum workers during the working hours. Their diets should be such as to promote the detoxification and elimination of the fluorides that have entered the body, increase the body defenses, and improve the functioning of organs and systems [*Guidelines on the Organization of Catering for Aluminum Workers Exposed to Inorganic Fluorine Compounds*].

Aluminum workers are subject to a pre-employment medical examination, followed by regular medical examinations during the period of employment.

Environmental protection measures should be primarily directed at reducing aluminum concentrations in bodies of water. The content of residual aluminum in purified drinking water can be considerably lowered by replacing the conventionally used aluminum sulfate with basic sulfate as a coagulating agent in waterworks. In regions where crops are irrigated with industrial wastewaters, the recommended permissible levels of aluminum and its compounds in such wastewaters should be taken as guides in monitoring soil pollution to prevent excessive accumulation of aluminum in the soil.

EMERGENCY TREATMENT OF SKIN LESIONS

Small skin lesions should be treated with alcohol and covered with an aseptic dressing, while larger lesions should be excised and the wound edges sutured.

REFERENCES

Aleksperov, I. I. *et al.* (1966) *Azerbaidzhan. Med. Zh.* [Azerbaijanian Medical Journal], No. 10, 69—73 (in Russian).
Alfrey, A. (1984) Bull. *New York Acad. Med.*, No. 2, 210—212.
Antonieva, A. A. *et al.* (1970) *Klin. Med.*, No. 3, 92—95 (in Russian).
Barannik, N. G. (1978) *Zabolevaniya tverdykh tkanei i parodonta u rabochikh siluminovogo proizvodstva* [Diseases of the Hard Tissues and Parodontium in Workers Manufacturing Silumin]. Kiev (in Russian).

Baxter, M. J. *et al.* (1989) In: Massey, R. C. & Taylor, D. (eds) *Aluminum in Food and the Environment.* Royal Society of Chemistry, Cambridge.

Belobragina, T. V. & Yelnichnykh, L. N. (1973) *Gig. Truda Prof. Zabol.*, No. 11, 48–51 (in Russian).

Bertholf, R. L. *et al.* (1988) Chapter 6 *Aluminum* in: Seiler, H. G. *et al.* (eds) *Handbook on Toxicity of Inorganic Compounds.* Marcel Dekker, New York, pp. 55–64.

Bespamyatnov, G. P. & Krotov, Yu. A. (1985) *Predelno dopustimye kontsentratsii khimicheskikh veshchestv v okruzhayushchei srede* [Maximum Allowable Concentrations of Chemicals in the Environment]. Khimiya, Leningrad (in Russian).

Boegman, R. & Bates, L. (1984) *Can. J. Physiol. Pharmacol.*, 8, 1010–1014.

Bikmullina, S. K. *et al.* (1966) In: *Materialy 24-i godichnoi nauchnoi sessii Sverdlovskogo meditsinskogo instituta* [Proceedings of the 24th Annual Meeting of Sverdlovsk Medical Institute]. Sverdlovsk, pp. 145–146 (in Russian).

Bogatin, D. Ya. & Makarova, R. M. In: *Trudy Novokuznetskogo instituta usovershenstvovaniya vrachei* [Transactions of the Novokuznetsk Institute for Continuing Education of Physicians], Vol. 28. Novokuznetsk, pp. 257–262 (in Russian).

Bourdon, R. & Bourdon, S. (1985) *Toxicology*, No. 11, 85.

Brakhnova, I.T. (1971) *Toksichnost poroshkov metallov i ikh soedineniy* [Toxicity of Metals and their Compounds in Powder Form]. Naukova Dumka, Kiev (in Russian).

Bräunlich, H. *et al.* (1988) *Pharmazie*, No. 9, 634–637.

Brunshtein, I. P. & Danilova, R. I. (1969) *Med. Zh. Uzbekistana* [Medical Journal of Uzbekistan], No. 12, 40–42 (in Russian).

Bryk, M. V. (1977) In: *Gigiyenicheskiye aspekty okhrany okruzhayushchei sredy* [Hygienic Aspects of Environmental Protection] Moscow, No. 5, pp. 60–62 (in Russian).

Budanova, L. F. (1976) *Terap. Arkhiv*, No. 2, 30–33 (in Russian).

Christie, H. *et al.* (1963) *Amer. Ind. Hyg. Assoc. J.*, No. 1, 47–56.

Chubarian, A. L. (1961) *Gig. Truda Prof. Zabol.*, No. 2, 16–20 (in Russian).

Day, J. P. (1990) *Env. Geochem. & Health*, 12, Nos. 1/2, 75–76.

D'Haese, P. C. *et al.* (1991) In: Merian, E. *et al.* (eds) *Carcinogenic & Mutagenic Metal Compounds. 3: Interrelation Between Chemistry & Biology.* Gordon & Breach Science Publishers, New York, pp. 585–594.

Di Domenico, N. *et al.* (1982) *Kidney Intern.*, No. 1, 27.

Dinman, B. D. (1983) In: *Encyclopaedia of Occupational Health and Safety.* International Labour Office, Geneva, pp. 132–135.

Dixon, R. L. *et al.* (1977) In: *Materialy itogovogo Sovetsko-Amerikanskogo simpoziuma po probleme "Gigiyena okruzhayushchei sredy"* [Proceedings of the Second Soviet-American Symposium on Environmental Hygiene]. Maryland, pp. 41–47 (in Russian).

Domnin, S. G. *et al.* (1976) In: *Professionalnye bolezni pylevoi etiologii* [Occupational Diseases Caused by Dust], No. 3, Moscow, pp. 146–155 (in Russian).

Domnin, S. G. *et al.* (1983) In: *Professionalnye bolezni pylevoi etiologii* [Occupational Diseases Caused by Dust]. Moscow, pp. 22–26 (in Russian).

Driscoll, C. T. & Schecher, W. D. (1990) *Env. Geochem.& Health*, 12, 28–50.

Dvizhkov, P. P. (1965) *Pnevmokoniozy* [Pneumoconioses]. Meditsina, Moscow (in Russian).

Elyashev, L. I. (1960) *Gig. Truda Prof. Zabol.*, No. 4, 28–32 (in Russian).

Evenshtein, Z. M. (1975) *Vopr. Pitaniya*, No. 3, 68–71 (in Russian).

Fernandez, R. *et al.* (1987) *Skelet. Radiol.*, No. 3, 209–215.

Frangulian, F. A. (1971) *Zh. Ushn. Nos. Gorl. Bolezn.*, No. 4, 6–9 (in Russian).

Fuchs, C. *et al.* (1974) *Clin. Chem. Acta*, 52, 71–80.

Golomidov, N. F. (1975) *Sostoyaniye slizistoi obolochki verkhnikh dykhatelnykh putei u rabochikh tsekha elektrotermicheskogo silumina* [State of the Mucous Membrane of the Upper Airways in Workers Employed in Electrothermal Silumin Production]. Denpropetrovsk (in Russian).

Goncharuk, E. I. (1977) *Sanitarnaya okhrana pochvy ot zagryazneniya khimicheskimi veshchestvami* [Sanitary Measures to Protect Soil from Chemical Pollution]. Zdorovya, Kiev (in Russian).

Grushko, Ya. M. (1972) *Yadovitye metally i ikh neorganicheskiye soyedineniya v promyshlennykh stochnykh vodakh* [Poisonous Metals and their Inorganic Compounds in Industrial Waste Waters]. Meditsina, Moscow (in Russian).

Grushko, Ya. M. (1979) *Vrednye neorganicheskiye soyedineniya v promyshlennykh stochnykh vodakh* [Harmful Inorganic Compounds in Industrial Waste Waters]. Khimiya, Leningrad (in Russian).

Guidelines on the Organization of Catering for Aluminum Workers Exposed to Inorganic Fluorine Compounds (1988) [Organizatsiya pitaniya rabotayushchikh v aluminiyevoi promyshlennosti, kontaktiruyushchikh s neorganicheskimi soyedineniyami ftora]. Leningrad (in Russian).

Izraelson, Z. I. (ed.) (1967) *Novye dannye po toksikologii redkikh metallov* [Recent Advances in the Toxicology of Rare Metals]. Meditsina, Moscow (in Russian).

Izrailet, L. I. & Eglite, M. E. (1978) *Promyshlennost i immunologicheskoe sostoyanie organizma. Chast 1: Professionalnaya allergiya* (Obzornaya informatsiya — Gigiyena) [Industry and the Immunological Status of the Body. Part 1: Occupational allergy (Reviews: Hygiene section)], Moscow (in Russian).

Kabata-Pendias, A. & Pendias, H. (1986) *Trace Elements in Soils and Plants*. CRC Press, Boca Raton (F1.).

Karamzin, V. P. *et al.* (1968) *Gig. Truda Prof. Zabol.*, No. **12**, 8—11 (in Russian).

Karr, D. N. S. (1988). In: Sigel, H. & Sigel, A. (eds) *Metal Ions in Biological Systems. Vol. 24: Aluminum and its Role in Biology.* Marcel Dekker, New York, pp. 217—258.

Khakimov, Kh. Kh. & Tatarskaya, A. E. (1985) *Periodicheskaya sistema i biologicheskaya rol elementov* [The Periodic System and the Biological Role of Elements]. Tashkent (in Russian).

Kipiani, S. P. (1968) *Gig. Truda Prof. Zabol.*, No. **11**, 50—53 (in Russian).

Knick, T. P. A., & McLachlan, D. R. (1988). In: Sigel, H. & Sigel, A. (eds) *Metal Ions in Biological Systems. Vol. 24: Aluminum and its Role in Biology.* Marcel Dekker, New York, pp. 285—314.

Konstantinov, V. G. (1979) *Gig. Truda Prof. Zabol.*, No. **12**, 26—29 (in Russian).

Korolev, A. A. & Krasovsky, G. N. (1978) *Gig. San.*, No. **4**, 12—14 (in Russian).

Kovalsky, V. V. (1974) *Geokhimicheskaya ekologiya* [Geochemical Ecology]. Nauka, Moscow (in Russian).

Krasovsky, G. N. *et al.* (1977) In: *Materialy itogovogo Sovetsko-Amerikanskogo simpoziuma po probleme* "Gigiyena okruzhayushchei sredy" [Proceedings of the Second Soviet-American Symposium on Environmental Hygiene]. Maryland, pp. 34—40 (in Russian).

Krivda, G. I. & Makeyeva, E. P. (1987) *Gig. Truda Prof. Zabol.*, No. **2**, 46—48 (in Russian).

Kvartovkina, L. K. *et al.* (1968) *Gig. San.*, No. **4**, 94—96 (in Russian).

Lai, J. & Blass, J. (1984) *J. Neurochem.*, No. **2**, 438—446.

Leonard, A. & Leonard, E. D. (1991). In: Merian, E. *et al.* (eds) *Carcinogenic & Mutagenic Metal Compounds. 3: Interrelation Between Chemistry & Biology*. Gordon & Breach Science Publishers, New York, pp. 595—600.

Leonicheva, V. D. (1965) *Gig. Truda Prof. Zabol.*, No. **10**, 11—16 (in Russian).

Ligacz, J. & Paradowski, Z. (1972) *Med. Pracy*, No. **1**, 41—45.

Lione, A. (1983) *Food Chem. Toxicol.*, No. **1**, 103—109.

Liublina, E. I. & Dvorkin, E. A. (1983) In: *Gigiyenicheskaya toksikologiya metallov* [Hygienic Toxicology of Metals]. Meditsina, Moscow, pp. 25—29 (in Russian).

McLaughlin, A. *et al.* (1962) *Brit. J. Ind. Med.*, No. **4**, 253—263.

Magour, S. & Mäser, H. (1981) *Biochem. Soc. Trans.*, No. **1**, 100—101.

Martinson, T. G. & Stroikov, Yu. N. (1983) *Ibid*, pp. 67—77 (in Russian).

Mayor, G. *et al.* (1977) *Science*, No. **4309**, 1187—1198.

Metelev, V. V. *et al.* (1971) *Vodnaya toksikologiya* [Aquatic Toxicology]. Kolos, Moscow (in Russian).

Miller, S. V. (1961) In: *Rukovodstvo po gigiyene truda* [A manual of Occupational Hygiene], Vol. 3. Moscow, pp. 208—238 (in Russian).

Molokanov, K. P. *et al.* (1973) In: *Professionalnye bolezni* [Occupational Diseases]. Meditsina, Moscow, pp. 378—450 (in Russian).

Muraviyeva, S. I. (ed.) (1982) *Sanitarno-khimicheskiy kontrol vozdukha promyshlennykh predpriyatiy* [Physical and Chemical Monitoring of Air Quality in Industrial Workplaces]. Meditsina, Moscow (in Russian).

Nesterenko, K. A. & Ostrovskaya, L. P. (1967) In: *Trudy Irkutskogo meditsinskogo instituta* [Transactions of Irkutsk Medical Institute]. Irkutsk, No. 85, pp. 248—252 (in Russian).

Nikogosian, Kh. A. *et al.* (1971) In: *Trudy Kuibyshevskogo NII gigieny* [Transactions of the Kuibyshev Research Institute of Hygiene]. Kuibyshev, No. 6, pp. 83—85 (in Russian).

Novikov, Yu. V. *et al.* (1981) *Metody opredeleniya vrednykh veshchestv v vode vodoyemov* [Methods for Measuring Noxious Substances in Water Bodies]. Meditsina, Moscow (in Russian).

Oganesian, E. N. (1966) In: *Materialy nauchnoi sessii po voprosam gigigeny truda i profpatologii v khimicheskoi i gornorudnoi promyshlennosti* [Proceedings of a Workshop on Occupational Health in the Chemical and Mining Industries]. Yerevan, pp. 107—113 (in Russian).

O'Gara, R. & Brown, J. (1967) *J. Nat. Cancer Inst.*, No. 6, 947—952.

Ohmer, C. (1988) *DDR-Med.-Rept.* 17 (No. 9), 515—517.

Oksova, V. M. *et al.* (1972) In: *Trudy Leningradskogo instituta usovershenstvovaniya vrachei* [Transactions of the Leningrad Institute for Advanced Medical Studies]. Leningrad, No. 115, pp. 32—38 (in Russian).

Oshchepkov, V. I. *et al.* (1969) In: *Voprosy profpatologii v eksperimente i klinike* [Problems of Occupational Diseases (Experimental and Clinical Aspects)], Sverdlovsk, pp. 159—163 (in Russian).

Osmolovskaya, M. S. & Kovalyev, N. I. (1954) *Vopr. Pitaniya*, No. 5, 48—49 (in Russian).

Patermaen, J. L. *et al.* (1988) *Teratology*, 38, 253—257.

Pigolev, S. A. (1961) *Gig. Truda Prof. Zabol.*, No. 1, 50—51 (in Russian).

Piscator, M. (1985) *Nutr. Res. Suppl.*, No. 1, 375—379.

Pleshcheyev, A. D. *et al.* (1971) In: *Trudy Pervogo syezda patologoanatomov UkrSSR* [Proceedings of the First Congress of Ukrainian Pathological Anatomists]. Kiev, pp. 185—189 (in Russian).

Ragolskaya, F. S. (1969) In: *Voprosy profpatologii v eksperimente i klinike* [Problems of Occupational Diseases (Experimental and Clinical Aspects)]. Sverdlovsk, pp. 183—187 (in Russian).

Ramsey, M. H. *et al.* (1991) *Env. Geochem. & Health*, 13, 114—118.

Roslyi, O. F. (1983) In: *Gigiyena truda i profpatologiya v tsvetnoi i chernoi metallurgii* [Occupational Health in Ferrous and Nonferrous Metallurgy]. Moscow, pp. 56—57 (in Russian).

Salbu, B. (1990) *Env. Geochem. & Health*, 12, 3—6.

Savory, J. & Wills, M. (1988) In: Sigel, H. & Sigel, A. (eds) *Metal Ions in Biological Systems. Vol. 24: Aluminum and its Role in Biology.* Marcel Dekker, New York, pp. 347—371.

Schlagmann, C. (1987) *Z. Allgemeinmed.*, 33, 13—15.

Sedova, K. S. (1988) *Gig. Truda Prof. Zabol.*, No. 10, 50—51 (in Russian).

Selivanov, A. I. (1971) *Gig. Truda Prof. Zabol.*, No. 8, 40—42 (in Russian).

Semennikova, T. K. *et al.* (1980) In: *Professionalnye bolezni pylevoi etiologii* [Occupational Diseases Caused by Dust]. Moscow, No. 6, pp. 21—25 (in Russian).

Shatalov, N. N. (1964) *Trudy Pervogo Moskovskogo meditsinskogo instituta* [Transactions of the First Moscow Medical Institute, Vol. 28, pp. 100—106 (in Russian).

Sheers, G. (1964) *Brit. J. Ind. Med.*, No. 3, 218—225.

Shcherbakov, G. G. (1978) *Gig. San.*, No. 6, 107—108 (in Russian).

Sigel, H. (ed.) (1988) *Aluminum and its Role in Biology.* Marcel Dekker, New York.

Sorensen, J. *et al.* (1974) *Environ. Health Perspect.*, No. 8, 3—95.

Specifications for Methods of Determining Noxious Substances in Air (1965) [Tekhnicheskiye usloviya na metody opredeleniya vrednykh veshchestv v vozdukhe], No. 4. Publication of the USSR Ministry of Health, Moscow (in Russian).

Specifications for Methods of Determining Noxious Substances in Air (1968) [Tekhnicheskiye usloviya na metody opredeleniya vrednykh veshchestv v vozdukhe], No. 5. Publication of the USSR Ministry of Health, Moscow (in Russian).

Sviridov, N. K. (1966) *Lab. Delo*, No. 12, 699—702 (in Russian).

Sturm, W. & Peukert, W. (1966) *Z. Ges. Hyg.*, No. 8, 664—674.

Swensson, A. *et al.* (1962) *Intern. Arch. Gewerbepathol.*, No. 2, 131—148.

Vadkovskaya, I. K. & Lukashev, K. P. (1977) *Geokhimicheskie osnovy okhrany biosfery* [Protection of the biosphere: Geochemical Principles]. Nauka i Tekhnika, Minsk (in Russian).

Varnakova, S. V. (1976) *Azerbaidzhan. Med. Zh.* [Azerbaijanian Medical Journal], No. 9, 55—59 (in Russian).

Vasyukovich, L. Ya. *et al.* (1978) *Gig. San.*, No. 5, 101—102 (in Russian).

Vermel, A. E. (1986) *Professionalnaya bronkhialnaya astma* [Occupational Bronchial Asthma], Meditsina, Moscow (in Russian).

Viderli, M. M. (1967) *Azerbaidzhan. Med. Zh.* [Azerbaijanian Medical Journal], No. 7, 27—31 (in Russian).

Viderli, M. M. (1968) *Azerbaidzhan. Med. Zh.* [Azerbaijanian Medical Journal], No. 12, 7—11 (in Russian).

Viderli, M. M. (1968) *Vestn. Rentgenol.*, No. 2, 51—56 (in Russian).

Voinar, A. O. (1962) *Mikroelementy v zhivoi prirode* [Trace Elements in the Living Nature]. Vysshaya Shkola, Moscow (in Russian).

Volkova, V. M. *et al.* (1973) *Gigiyena truda* [Occupational Hygiene] (Kiev), No. 9, 162—165 (in Russian).

Waldron-Edward, D. *et al.* (1971) *Canad. Med. Assoc. J.*, 105, 1297—1299.

Wang, S. T. *et al.* (1991) *J. Anal. Toxicol.*, 15, 66—70.

Warraki, S. & Herant, Y. (1963) *Brit. J. Ind. Med.*, No. 3, 226—230.

Williams, D. R. (1971) *The Metals of Life*. Van Nostrand Reinhold Company, London.

Yavorskaya, S. F. & Grinberg, K. M. (1974) *Gig. San.*, No. 11, 54—57 (in Russian).

Yelnichenykh, L. N. (1973) *Gig. Truda Prof. Zabol.*, No. 1, 48—51 (in Russian).

Yokel, R. (1985) *Toxicol. Appl. Pharmacol.*, No. 1, 35—43 and 121—123.

Zatta, P. *et al.* (1988) *Med. Hypotheses*, No. 2, 139—142.

Zhilina, L. V. (1975) *Gig. Truda Prof. Zabol.*, No. 3, 42—43 (in Russian).

Gallium and its compounds

Gallium (III) (gallic) compounds: **antimonide; arsenide; chloride** (trichloride); **hydroxide; nitrate; nitride; oxide; phosphide; sulfate**

IDENTITY AND PHYSICOCHEMICAL PROPERTIES
OF THE ELEMENT

Gallium (Ga) is a metallic element in group III of the periodic table, with the atomic number 31. Its natural isotopes are ^{69}Ga (60.5%) and ^{71}Ga (39.5%).

Gallium is a rare metal that is hard and brittle at low temperatures but melts just above room temperature and expands on freezing.

Gallium resembles aluminum in chemical properties. It is not oxidized in air in the absence of moisture and reacts with oxygen at $\geqslant$ 260°C; the resulting oxide (Ga_2O_3) film prevents it from further oxidation. It exhibits the oxidation state of +3 in most compounds and +1 in some. With elements of the arsenic subgroup, it forms intermetallic compounds (GaAs and GaSb) that possess semiconductor properties. Gallium is easily fusible with many metals. When it is reacted with halogens (except iodine) in the cold or with mineral acids on heating, salts containing the ion Ga^{3+} are formed. In a gallium-containing sodium hydroxide solution, sodium metagallate is slowly produced. See also Appendix.

NATURAL OCCURRENCE AND ENVIRONMENTAL LEVELS

The clarke of gallium is estimated at $(15-19) \times 10^{-4}\%$ in the earth's crust and at $19 \times 10^{-4}\%$ in the granite layer of the continental crust [Dobrovolsky].

The oceans contain an estimated 41.4 million tonnes of dissolved gallium, at average concentrations of 0.03 µg/L in water and $0.00086 \times 10^{-4}\%$ in total salts, and receive annually up to 3.3 thousand tonnes of this metal in river discharges; the amount of dissolved gallium species taken up by ferromanganese oxide deposits of the Pacific is estimated at 0.06 thousand tonnes per year [Dobrovolsky]. In seawater, the principal species of gallium is $Ga(OH)_4^-$ and its residence time is 1×10^4 years [Henderson]. In river water, gallium concentrations average 0.09 µg/L. Coals have an average gallium content of $4.5 \times 10^{-4}\%$ [range $(0.6-18) \times 10^{-4}\%$] on a dry weight basis; the ash of coal and oil contained $64 \times 10^{-4}\%$ and $(3-30) \times 10^{-4}\%$, respectively [Dobrovolsky].

Gallium is distributed rather uniformly in the major types of rocks. Its values in magmatic and sedimentary rocks commonly range from 5 to 25 mg/kg and are usually around 3 mg/kg in ultrabasic and calcareous rocks. It is widely distributed in small or trace amounts in aluminum clays, feldspars, coal, and in the ores or zinc, iron, manganese, and chromium. The gallium mineral gallite $(CuGaS_2)$ is extremely rare [Kabata-Pendias & Pendias; Vadkovskaya & Lukashev].

In weathering, gallium behaves like aluminum and is usually closely associated with aluminum minerals such as bauxites. The ionic form of gallium in natural environmental media is Ga^{3+}. In soils, its migration is limited and it displays a tendency to accumulate in the organic matter of soil. Its levels in soil correlate positively with those of the clay fraction, and its distribution also shows a relation to iron and manganese oxides. Its concentrations in soils of various countries range from 1 to 70 mg/kg, the mean value being 28 mg/kg. Its concentrations were reported to range from 16 to 48 mg/kg for soils derived from basalts and andesites in New Zealand and from 6 to 17 and 11 to 30 mg/kg for various soils in the USSR and USA, respectively. British standard soil samples contained on average 21 mg Ga/kg, while calcareous soils in China contained up to 50 mg Ga/kg, the average value being 21 mg Ga/kg [Kabata-Pendias & Pendias].

Continental vegetation contains an estimated 0.13 million tonnes of gallium, at average fresh weight, dry weight, and ash weight concentrations of $0.02 \times 10^{-4}\%$, $0.05 \times 10^{-4}\%$, and $1 \times 10^{-4}\%$, respectively; the amount of gallium taken up annually worldwide through the increment in phytomass was estimated to be 8.63 thousand tonnes, or 0.057 kg/m^2 of land area ($K_b = 0.05$) [Dobrovolsky].

Gallium is commonly found in plants. Its concentrations ranged from 0.02 to 5.5 mg/kg dry weight in the native herbage of the USSR and from 3 to 30 mg/kg ash weight in a variety of native plant species in the USA. Its highest concentrations among plants were found for lichens (2.2–60.0 mg/kg dry weight) and bryophytes (mosses) (2.7–30 mg/kg dry weight) [Kabata-Pendias & Pendias].

PRODUCTION

Metallic gallium is obtained as a by-product in the extraction of aluminum (usually from bauxite) or, less commonly, of zinc from zinc ores. In the Bayer process, bauxite is dissolved in sodium hydroxide and aluminum oxide trihydrate is precipitated. After this process is repeated often enough, gallium builds up in the recycled liquor. Following removal of the alumina, gallium is precipitated by carbon dioxide treatment. The metal is recovered from the precipitate by electrolysis [Roscina].

Gallium(III) oxide is made by dehydration of gallium(III) hydroxide. **Gallium(III) sulfate** crystallizes from sulfate solutions as the hydrate which is then dehydrated by heating.

USES

Gallium and its compounds, particularly those with semiconductor properties, are mainly used in high-temperature rectifiers, in infrared technology, in ultraviolet lamps, in the manufacture of optical mirrors and high-temperature thermometers, and in nuclear power reactors. Gallium compounds are also used in the making of luminescent materials, in microcircuits for electronic instruments, as catalysts in chemical industries, and in medicine. In the early 1980s, the world production of gallium was estimated to be in the order of 20 tonnes annually [Hayes].

MAN-MADE SOURCES OF EMISSION INTO THE ENVIRONMENT

Gallium is released into the atmosphere with industrial emissions. Dust emissions from metallurgical works and power plants, for example, contained this metal at levels of $0.5-1.0 \times 10^{-2}\%$ and $2-5 \times 10^{-2}\%$, respectively [Vasilenko *et al.*].

As a result of coal burning, a total of 5.9 thousand tonnes of gallium, or 39 g/km^2 of land area, was estimated to have been emitted worldwide in 1970 and twice as much in 1980 (12.4 thousand tonnes, or 83 g/km^2); for the year 2000, 19.6 thousand tonnes, or 131 g/km^2, has been predicted [Dobrovolsky].

In industrial premises where intermetallic compounds were synthesized, considerable quantities (tens of mg/m^3) of gallium-containing airborne dust were present. Workers engaged in the production of gallium concentrate and metallic gallium by various methods may come in contact with solutions and suspensions of gallium compounds. These compounds can be easily ingested via contaminated hands, especially when eating, drinking, or smoking at the workplace. The skin and mucous membranes of workers may also be acted upon by gallium and its salts directly as a result of splashing or spilling of their solutions on the work clothes. Less commonly, workers are exposed to gallium by inhalation — when much dust is generated in such processes as pyrometallurgical extraction of gallium in the form of oxide, the production of powdered gallium salts (sulfate, nitrate, chloride), and the production or treatment of monocrystals of semiconductor gallium compounds (arsenide, phosphide, antimonide). Gallium fumes are rarely released into the air of working premises because the vapor pressure of gallium is low and because the industrial processes in which it is used often proceed under vacuum [Tarasenko & Fadeyev].

Gallium can also enter the environment in wastewaters. For instance, industrial and domestic wastewaters from the city of Kiev (Ukraine) used to irrigate soils where vegetables were grown, contained 0.525 mg Ga/kg [Leshchenko et al.].

TOXICITY

Acute/single exposure

Animals

LD_{50} values of several gallium compounds for mice are shown in Table 1. Acutely poisoned animals exhibited short-term excitation succeeded by general inhibition with motor incoordination, adynamia, loss of reflexes, and slow and irregular breathing; in the presence of these clinical manifestations, paralysis of the hind paws ensued followed by coma and death. The threshold for acute toxicity of **gallium(III) arsenide** to mice inhaling its aerosol was 152.5 mg/m^3 as determined on the basis of integral indicators (rectal temperature, orienting reflex, and summated threshold index). With oral administration of this compound by gavage, the acute toxicity threshold on these criteria was 7000 mg/kg. No deaths occurred among mice or rats during the 3-week observation period with oral gallium arsenide doses of up to 15 000 mg/kg. The cumulation coefficient of the arsenide was determined to be 5.2.

In rats administered **gallium(III) oxide** intratracheally, inflammatory changes were observed in the lungs with subsequent development of a sclerotic process at the sites of inflammation. After a single administration of the **arsenide** by the same route, collagen and lipid levels remained elevated in the lung tissue for 3 to 6 months; subsequently, only collagen levels were elevated, for periods up to 12 months. In general, the sclerotic process was more severe and widespread after inhalation exposure to the dust of gallium arsenide or phosphide (intermetallic compounds) than after that to gallium oxide dust [Fadeyev; Tarasenko & Fadeyev]. Although **gallium(III) oxide** and **gallium(III) nitride** have very low water solubilities, intratracheal administration of either of them in an amount of 50 mg led to the death of many rats within a short time and to significant weight losses in the surviving animals; the nitride was more toxic than the oxide, with a mortality rate of 75% after a month. Deaths were attributed to toxic pneumonia. Cloudy swelling (granular degeneration) of the convoluted tubular epithelium was observed in the kidneys.

Table 1. Median lethal doses of gallium compounds for mice

Compound	Route of administration	LD_{50} (mg/kg)
Gallium arsenide	Intraperitoneal	4700
Gallium nitrate	Oral	4400
	Subcutaneous	600
Gallium oxide	Oral	10000
Gallium phosphide	Oral	8000

Source: Tarasenko & Fadeyev.

Man

Inhalation exposure to a **gallium-containing aerosol** in a concentration of 50 mg/m^3 caused renal damage, as did intravenous administration of gallium salts at 10–25 mg/kg. Proteinuria, azotemia, and impaired clearance of urea were noted [Bergqvist].

Repeated exposure

Animals

During a monthly period of oral exposure of guinea-pigs to **gallium(III) nitrate** (intragastric administration on 10 occasions for a total dose of 4400 mg/kg body weight per animal), weight loss with increases in the relative weight of the spleen was recorded, as were hematologic changes such as reductions in acetylcholinesterase activity and in sulfhydryl groups and albumins in the serum along with elevations in serum α_1-globulin. Histopathologic examination of the exposed animals demonstrated epithelial desquamation in the stomach and small intestine and albuminous degeneration of the liver and of the convoluted tubular epithelium in the kidneys.

Man

During a course of treatment with gallium preparations given intravenously in doses of 10–25 mg/kg, renal dysfunction developed in 11 out of the 38 treated patients, for which reason the treatment had to be discontinued or the dose reduced in 7 patients; subsequently 4 patients died from renal disease. Manifestations of poisoning included nausea, vomiting, liquid stools, and changes in red cells [Bergqvist; Goering & Fowler].

In industry, the greatest potential health hazard comes from ingestion of soluble gallium compounds or from their contact with the skin and mucous membranes; insoluble compounds may produce toxic effects, especially on lung tissue, when their dusts or powders are inhaled.

Chronic exposure

Animals

In a 4.5-month oral toxicity study, rabbits given **gallium(III) nitrate** daily by gavage in doses of 220 mg/kg displayed hematologic signs of chronic intoxication including decreases in albumins, increases in α_2- and γ-globulins and in acetylcholinesterase activity, and inhibition of alkaline phosphatase activity. Histopathologic examination revealed degenerative changes in the liver parenchyma and in the epithelium of convoluted renal tubules and desquamative catarrh of the pulmonary tissue. Signs of chronic intoxication (decreased weight gains, abnormal protein metabolism, dystrophic changes in parenchymal organs) developed in rabbits on prolonged intragastric administration of **gallium(III) chloride**.

In a 4-month inhalation toxicity study, rats and guinea-pigs exposed to a **gallium arsenide** aerosol in a concentration of 12.0 mg/m^3, showed altered rectal temperature, slowed body-weight gains, and altered values of the summated threshold index as compared to unexposed controls. Erythrocyte counts were elevated after 2 months of exposure and began to decrease progressively after 3 months, as did the total blood protein as a result of falls in the albumin fraction. Blood serum levels of uric acid and urea nitrogen were markedly increased after 1 and 4 months of exposure and remained rather high during the recovery period; creatinine was also elevated at the end of the exposure period. In guinea-pigs, reduced erythrocyte numbers and elevated potassium and calcium levels occurred in the peripheral blood. Histopathologic findings included impregnation of vessel walls by plasma protein and their homogenization, fibrosis of interalveolar septa in the lungs, vacuolar and granular degeneration of the convoluted tubular epithelium in the kidneys, the presence of fat droplets in hepatocytes, and epithelial proliferation in the liver; these changes all persisted after the recovery period. The concentration of 4.2 mg/m^3 was considered as the threshold or near-threshold level for chronic inhalation exposure to gallium arsenide [Fadeyev]. Mutagenic and teratogenic activities of gallium have been described by Goering & Fowler.

Dermal and ocular effects

Gallium(III) nitrate (a soluble compound) applied to the conjunctiva of rabbits in powder form produced cauterizing effects: the conjunctiva and cornea became grossly edematous and underwent necrotic changes, followed by the appearance of permanent scars in a week, with the deformation of palpebral margins and the formation of leukoma. Its application to depilated rabbit skin resulted in dermatitis. No cauterizing effects were observed after dermal or conjunctival application of powdered **metallic gallium** or insoluble compounds such as the **nitride** or **oxide**.

Powdered **gallium arsenide** introduced into the conjunctival sac of rabbits in an amount of 50 mg, produced blepharospasm and episcleral injection which both cleared within 48 h.

ABSORPTION AND DISTRIBUTION IN THE BODY

Ingested gallium is poorly absorbed from the gastrointestinal tract. In a chronic experiment on rats, gallium accumulated to a concentration of 2.2 mg% in the bones, 0.5 mg% in the liver, and 0.7 mg% in the kidneys. In the media of the digestive tract, its soluble salts form poorly absorbed insoluble compounds such as hydroxide, gallates, or protein precipitates.

HYGIENIC STANDARDS

In the USSR, occupational exposure limits (MAC_{wz}) have been set for gallium arsenide at 2.0 mg/m^3 (aerosol) and gallium oxide at 3.0 mg/m^3 (aerosol); these compounds have been placed in hazard class 3.

METHODS OF DETERMINATION

In *air*: a photometric method based on the reaction of gallium with sulfochlorophenol to form a blue complex compound (measurable concentration range, 2–10 μg/25-ml sample) [Peregud]. In *plant materials*: graphite furnace atomic absorption spectrometry in the presence of nickel [Xiao-quan *et al.*]. Traces in *biological materials*: a fluorometric method based on the reaction of gallium with salicylaldehyde thiocarbohydrazone (detection limit 2 ng/ml, measurable concentration range 3–30 ng/ml, error ±1.7%) [Ureña *et al.*].

MEASURES TO CONTROL EXPOSURE AT THE WORKPLACE

In the production of gallium, health hazards can arise from the pollution of air not only by gallium itself, but also by chemical reagents (acids, alkalies) used in the production process. The physical properties of metallic gallium are such that its fumes are unlikely to be released. Evolution of gallium-containing dust is possible during the manufacture of semiconductors and transistors, and it is therefore necessary to segregate dust sources. The work posts and dust-generating equipment should be provided with efficient local exhaust ventilation.

Where gallium concentrates are produced, measures should be taken to minimize direct contact of the workers with solutions and suspensions that can contaminate their skin and clothing; exhaust ventilation is required for processes that are sources of airborne particulate matter.

Workers should wash hands and take off overalls before meals. Food may only be taken in the canteen or a specially designated room.

Medical surveillance of workers engaged in the production of gallium or its compounds is necessary, with special attention to the condition of the respiratory system, liver, kidneys, and skin.

Respirators must be worn where the ambient air contains aerosols of gallium and its compounds in high concentration. Gloves must be provided to workers whose hands are likely to be contaminated. For eye protection during electrolysis processes, goggles should be used.

REFERENCES

Bergqvist, U. (1983) *New metals*. University of Stockholm, Institute of Physics.
Dobrovolsky, V. V. (1983) *Geografiya mikroelementov. Globalnoe rasseyanie* [Geography of Trace Elements. Global Dispersion]. Mysl, Moscow (in Russian).
Fadeyev, A. I. (1980) *Gig. Truda Prof. Zabol.*, No. **3**, 45 (in Russian).
Goering, P. L. & Fowler, B. A. (1991) Chapter II.10 *Gallium* in: Merian, E. (ed.) *Metals and their Compounds in the Environment—Occurrence, Analysis and Biological Relevance*, pp. 909–920.
Hayes, R. L. (1988) *Gallium* in: Seiler, H. G. *et al.* (eds) *Handbook on Toxicity of Inorganic Compounds*. Marcel Dekker, New York, pp. 297–300.
Henderson, P. (1982) *Inorganic Geochemistry*. Pergamon Press, Oxford.
Kabata-Pendias, A. & Pendias, C. (1986) *Trace Elements in Soils and Plants*. CRC Press, Boca Raton (Fl.).
Leshchenko, P. D. *et al.* (1972) *Vopr. Pitaniya*, No. **1**, 81–85 (in Russian).
Peregud, E. A. (1976) *Khimicheskiy analiz vozdukha (novye i usovershenstvovannye metody)* [Chemical Analysis of the Air (new and improved methods)]. Khimiya, Leningrad (in Russian).
Roscina, T. A. (1983) In: *Encyclopaedia of Occupational Health and Safety*, Vol. 2. International Labour Office, Geneva, pp. 938–939.
Tarasenko, N. Yu. & Fadeyev, A. I. (1980) *Gig. San.*, No. **10**, 13–16 (in Russian).
Ureña, E. *et al.* (1985) *Anal. Chem.*, **57**, 2309–2311.
Vadkovskaya, I. K. & Lukashev, K. P. (1977) *Geokhimicheskie osnovy okhrany biosfery* [Protection of the Biosphere: Geochemical Principles]. Nauka i Tekhnika, Minsk (in Russian).
Vasilenko, V. N. *et al.* (1985) *Monitoring zagryazneniya snezhnogo pokrova* [Monitoring the Pollution of the Snow Cover]. Gidrometeoizdat, Leningrad (in Russian).
Xiao-quan, S. *et al.* (1985) *Anal. Chem.*, **57**, 857–861.

Indium and its compounds

Indium antimonide; i. arsenide; i. arsenide phosphide (i. arsenophosphide);
i.(III) chloride (i. trichloride); i. cyanide; i. hydroxide; i. nitrate; i. nitrate
trihydrate; i. nitride; i.(III) oxide; i. sulfate

IDENTITY AND PHYSICOCHEMICAL PROPERTIES
OF THE ELEMENT

Indium (In) is a metallic element in group III of the periodic table, with the
atomic number 49. Its natural isotopes are ^{113}In (4.33%) and ^{115}In (95.67%),
the latter being radioactive.

Indium is a soft, malleable and readily fusible metal. It remains essentially
unchanged in air but corrodes in water in the presence of air. It is a chemical
analogue of gallium. In compounds, it is present in the oxidation state +1, +2, or
(usually) +3. The latter compounds are more stable than the others. See also
Appendix.

NATURAL OCCURRENCE AND ENVIRONMENTAL LEVELS

The clarke of indium is estimated at $(0.1-0.25) \times 10^{-4}\%$ in the earth's crust and
at $0.25 \times 10^{-4}\%$ in the granite layer of the continental crust [Dobrovolsky].
Indium occurs most frequently together with zinc minerals, particularly
sphalerite (0.1–1% In), which is its chief commercial source. It is also found in
the ores of tin, manganese, tungsten, copper, iron, lead, cobalt, and bismuth, but
generally in small amounts (<0,1%). The two indium ores, roquesite ($CuInS_2$)
and indite (In_2S_4), are extremely rare, as is native indium.

In seawater, indium is present at an average concentration of 1×10^{-7} mg/L, its principal species being $In(OH)_2^+$ [Henderson]. In soil, its concentrations ranged from 0.2 to 0.5 mg/kg in the USA and averaged 0.01 mg/kg in a number of other countries. In the atmosphere, concentrations were (pg In/m^3): 0.05 over the South Pole, 20 over the Shetland Islands, 30–360 over West Germany, 1200 over Japan, and 20–140 over North America [Kabata-Pendias & Pendias]. Distribution of indium in environmental media is further considered by Fergusson and Fowler & Goering, and its distribution in foods and living organisms is covered in Fowler & Goering.

PRODUCTION

Indium is obtained commercially from wastes or intermediates in the production of lead, tin, cadmium, and particularly zinc with which it is often closely associated in nature. Of the many methods available for recovering indium from zinc residues and smelter slags, the most common one involves dissolution of indium in acid, from which a crude preparation of the metal is made by sponging with zinc or by forming insoluble hydroxides, sulfides, or orthophosphates. The crude sponge or cake is then repurified in solution and the metal is usually electroplated on rods; in this way, a purity of >99.999% can be achieved [Thomas].

In 1976, the world production of indium was reported to be about 48 tonnes [Fowler].

USES

Indium is mainly used in industry for surface coatings and in alloys. A thin coat of indium increases considerably the resistance of metals to corrosion and wear. Owing to its corrosion resistance, indium is extensively used in making motion picture screens, cathode ray oscillographs, mirrors, and reflectors. Indium-coated mirrors and reflectors do not tarnish with time and their coefficient of reflection remains constant. It is also much used in the metallurgical and instrument-making industries, in electronics (for diodes, transistors, lasers, electricity conducting coatings), in nucleonics, and in the aircraft and automobile industries (mainly in bearings to prolong the life of moving parts).

The more common industrial compounds of indium are the **chloride** $InCl_3$ used in electroplating, the **oxide** In_2O_3 used in glass manufacture, and the **antimonide** InSb which is a semiconductor.

MAN-MADE SOURCES OF EMISSION
INTO THE ENVIRONMENT

A major anthropogenic source of indium in the environment are coal-burning power stations; dust emissions from these contained indium at levels of $1 \times 10^{-3} - 1 \times 10^{-2}\%$ [Vasilenko *et al.*]. The atmosphere may also be polluted with indium-containing dusts or aerosols generated by industrial processes of indium production. In plants where metallic indium was produced, for example, the content of total dust in the ambient air ranged from 2.67 to 3.95 mg/m^3 during sublimation, recovery, and transportation processes, and the dust contained indium at an average level of 0.068%.

In the top soil near lead and zinc works, elevated indium concentrations (up to 4.2 mg/kg) were recorded. Plants from polluted industrial regions contained up to 2100 µg In/kg fresh weight while those from unpolluted sites contained up to 710 µg/kg. One source of indium in soil may be sewage sludge. Beet plants grown in a soil treated with such sludge contained 80 to 300 µg In/kg dry weight [Kabata-Pendias & Pendias].

TOXICITY

Microorganisms and plants

Indium inhibited the activity of nitrate-forming bacteria in soil at concentrations of 5 to 9 mg/kg. Its concentrations of 1–2 mg/kg in culture solutions produced signs of toxicity in roots of various plants grown there [Kabata-Pendias & Pendias].

Animals

Acute/single and repeated exposures

Toxic doses of indium and several of its compounds for laboratory animals are shown in Table 1.

After an oral (intragastric) administration of **indium(III) oxide** to rats at 10 000 mg/kg or its intraperitoneal injection into mice in the same dosage, no deaths occurred among the animals during a 2-week observation period. In rats, the threshold concentration of this compound for effects on rectal temperature, the orienting reflex, and the summated threshold index was determined at 193 mg/m^3 with 4-h inhalation exposure and at 3000 mg/kg with oral administration. A single intratracheal administration of the oxide led to inflammatory, sclerotic, and fibrotic lesions in the lungs and to dystrophic changes in several internal organs, over a 12-month observation period. Similar changes were caused by the **arsenide, arsenide phosphide,** and **antimonide** of indium introduced by this route.

Table 1. Toxic doses of indium and its compounds

Compound	Animal	Route of administration	Dose (mg/kg)
Indium ion[*]	Mouse	Intravenous	LD_0=7.5
	Mouse	Intravenous	LD_{50}=12.5
Indium antimonide	Mouse	Intraperitoneal	LD_{50}=3700
arsenide	Mouse	Intraperitoneal	LD_{40}=10 000
	Rabbit	Oral	LD_0=10 000
arsenide phosphide	Rabbit	Oral	LD_0=10 000
hydroxide	Mouse	Oral	LD_{100}=2000
colloidal	Mouse	Intravenous	LD_0=0.103
	Mouse	Intravenous	LD_{50}=0.323
nitrate	Mouse	Oral	LD_{50}=3350
	Mouse	Intraperitoneal	LD_{50}=100
sulfate	Rat	Oral	LD_{100}=2000
	Rabbit	Oral	LD_{100}=1300

[*] A soluble indium salt was used.

The minimal lethal dose of both **indium sulfate** and **indium(III) chloride** with subcutaneous injection was 10 mg/kg for rats and 2 mg/kg for rabbits; with the intravenous route, deaths among rabbits were recorded starting with 0.64 mg/kg.

In rats, the threshold concentration of **indium arsenide** for acute inhalation toxicity, as assessed on the basis of integral indicators after a single 4-h exposure, was 138 mg/m^3.

In mice, acute oral poisoning by **indium nitrate** was marked by adynamia, anorexia, and paralysis of the hind paws.

A single intravenous injection of **indium(III) chloride** into rats and mice at 3.6 mg/kg and 16.5 mg/kg, respectively, resulted in extensive necrosis of the convoluted renal tubules. Colloidal indium hydroxide given by this route produced necrosis of cells in the liver and spleen [Fowler]. In rabbits, orally administered **indium sulfate** and **nitrate** lowered blood sugar and raised glycogen levels in the liver and muscles.

In guinea-pigs given **indium nitrate** intragastrically on multiple occasions over a month in a total dose of 3.3 g, body-weight loss, leukocytosis, lowered serum levels of globulins and sulfhydryls, and increased relative weights of the heart and kidneys were noted at the end of this exposure period. Histopathologic findings included degeneration and desquamation of the surface epithelium in the gastric and small intestinal mucosa, cloudy swelling of the convoluted renal tubules, and fatty degeneration of hepatocytes with the presence of necrotic areas in the liver.

Teratogenic and embryotoxic effects of indium have been reported [Ferm & Carpenter]. Indium arsenide has a moderate capacity for cumulation (cumulation coefficient = 4.1).

Teratogenic and embryotoxic effects of indium nitrate in hamsters have been reported [Ferm & Carpenter]. Indium arsenide has a moderate capacity for cumulation (cumulation coefficient = 4.1).

On the basis of acute toxicity studies in animals, indium antimonide has been placed in hazard class 3 and indium arsenide and arsenide phosphide in hazard class 4 [Tarasenko & Fadeyev].

Chronic exposure

In rats exposed to **indium(III) oxide** 5 times/week for 4 h daily in a 4-month inhalation toxicity study, reductions in the summated threshold index were recorded after 1 and 4 months of exposure, lowered rectal temperature and elevated erythrocyte counts in the peripheral blood after 4 months, and raised hemoglobin levels and leukocyte counts after 2 weeks. Histopathologic findings in the lungs during the exposure and post-exposure periods included inflammatory lesions in the bronchi, epithelial degeneration and metaplasia, and several abnormalities in the alveolar parenchyma such as the accumulation of a golden pigment (in the form of fine granules), macrophages, and lymphoid cells and the presence of loose fibrous connective tissue; in other organs, granular degeneration of their functional elements was observed. The indium(III) oxide concentration of 11.5 mg/m^3 was considered as the threshold level for effects on the parameters measured [Pokhodzei, 1985]. In rabbits, daily oral exposure to **indium nitrate** by gavage at 0.16 g/kg for 4.5 months lowered α_2- and γ-globulin levels and produced degenerative changes in the liver and kidneys as well as congestion in a number of internal organs.

Dermal and ocular effects

No signs of skin irritation, absorption, or sensitization were in evidence in rats and guinea-pigs following cutaneous application of **indium(III) oxide** in the form of a 50% lanolin ointment or 95% aqueous suspension. Nor were any changes detectable in the ocular mucous membrane or eyeball when 50 mg of the oxide were placed in the conjunctival sac of rabbits. In contrast, the **chloride, cyanide,** and **nitride** of indium caused irritation, necrotic changes, and scar formation following local application to the skin or mucous membranes. **Metallic indium** and its insoluble compounds produced no such effects [Pokhodzei, 1985].

The highest concentration of **indium(III) chloride** that was not irritating to the shaved skin of guinea-pigs on repeated daily application was 5%. This compound has been shown to induce allergy on epidermal and subdermal application [Roshchin et al.].

Man

Workers engaged in indium production complained of pains in joints and bones, decay of teeth, nervous and gastrointestinal upsets, cardiac pain, and general weakness.

In occupational settings, indium may enter the body during the preparation of concentrated indium solutions, the cementation, smelting, refining, and electrolysis of indium, and welding and soldering operations in the manufacture of electronic components.

Solutions of indium salts (e.g. sulfate or chloride) may contaminate the clothing, skin, and mucous membranes, for example during the cementation of indium sponge from such solutions, the extraction of cathodes from the electrolyte, or the cleaning of cathodes or anodes. These substances may then be ingested (e.g. during smoking or eating in the workroom). Significant exposure by inhalation is relatively uncommon and mainly occurs during the production or treatment of indium salts (chloride, sulfate, nitrate) or of semiconducting indium alloys (exposure to indium antimonide, arsenide, or phosphide). The hazard from inhalation exposure to indium present in small amounts in mixed dusts generated during indium extraction by pyrometallurgical processes is low; of much greater importance in such cases is exposure to the major components of these dusts (zinc, lead, cadmium). The risk of inhalation exposure to the fumes of melted metals is not high either since the indium vapor pressure remains low even at temperatures above 1000°C (smelting is carried out at lower temperatures under a layer of flux). Because metallic indium is very soft and adhesive, exposure of the skin is much more likely but does not seem to present a serious hazard [Pokhodzei, 1984, 1985].

The greatest potential toxicity hazard in occupational environments is likely to arise from the use of indium together with antimony and germanium in the electronics industry, primarily as a result of exposure to fumes given off during welding and soldering operations [Thomas]. Effects of acute and chronic exposure to indium have been discussed by Fergusson.

ABSORPTION, DISTRIBUTION, AND ELIMINATION

In rats, only about 0.5% of the orally administered dose was absorbed from the intestine [Smith *et al.*]. In the blood, indium ions bind to and are transported by transferrin; in mice ionic indium was cleared from the blood within 3 days [Fowler].

Distribution of indium in the body largely depends on its chemical form. Ionic indium is accumulated chiefly by the kidneys, whereas colloidal indium is accumulated by the liver, spleen, and other organs of the reticuloendothelial system.

In rabbits that had been chronically exposed to indium nitrate *per os* for 4.5 months, $4 \times 10^{-4}\%$ and $4 \times 10^{-5}\%$ of the administered dose was detected in the liver and kidneys, respectively. Teeth from healthy persons were found to contain 0.001–0.01 mg In/g ash weight [Pokhodzei, 1985].

Ionic indium is excreted mainly by the kidneys (in mice, 52% of an administered dose of ionic indium was removed in the urine) and colloidal indium, by the intestine (53% in mice) [Fowler].

Studies of the biological half-time of indium have shown that its elimination occurs in two phases, fast and slow. In mice given [114]indium chloride intravenously, the fast component (representing ~50% of the body burden) showed a half-time of 1.9 days, while the slow component had a half-time of 69 days; for [114]indium hydroxide, the fast and slow phases had half-times of 2 days and 73.8 days. The whole-body biological half-time in mice was determined at 14–15 days [Fowler]. In rats, approximately 60% of intratracheally administered radioactive indium disappeared from the lungs within 16 days [Smith et al.]. Uptake, absorption, transport and distribution, metabolism, and elimination of indium in animals and humans have been reviewed by Fowler & Goering.

HYGIENIC STANDARDS

Exposure limits adopted for indium compounds in the USSR are shown in Table 2. In the USA, a TLV-TWA value of 0.1 mg/m^3 has been set for indium and its compounds (as In).

Table 2. Exposure limits for indium compounds

	Workroom air		Atmospheric air	
	MAC_{wz} (mg/m^3)	Aggregative state	$TSEL_{aa}$ (mg/m^3)	Hazard class
Indium nitrate (as In)	—	—	0.05	—
Indium(III) oxide	4.0	Aerosol	—	3

METHODS OF DETERMINATION

Neutron activation analysis has been used for determining indium in seawater (detection limit 0.006 ng/L; coefficient of variation ~5%) and rocks. Indium in water was analyzed with a detection limit of 1 µg/L by polarography. A spectrochemical method for determining indium in biological material has a detection limit of 0.002 µg indium (coefficient of variation ~10–15%) [Fowler]. Various methods of indium determination have been described by Busev, Fedorov et al., Fergusson, Fowler & Goering, and Smirnov et al.

MEASURES TO CONTROL EXPOSURE AT THE WORKPLACE

Processes involving sublimation, recovery, and transportation of metallic indium should be mechanized to the greatest possible extent and exclude the possibility of air pollution with dust. Where concentrated indium solutions are prepared and the metal is obtained by cementation, smelted and refined, efforts should be primarily directed at preventing the release of indium salt fumes. Similarly, a prime consideration in soidering and welding operations is to prevent the formation of aerosols containing indium and other toxic metals. The equipment used to obtain indium or its compounds should be vented and located in rooms provided with general ventilation.

Workers employed in the production or use of indium and its compounds should be issued with personal protective equipment that meets the regulations adopted for the particular work processes, and be instructed in the proper use and care of this equipment. They should receive periodic medical examinations. A preplacement medical checkup is also required.

REFERENCES

Busev, A. I. (1958) *Analiticheskaya khimiya indiya* [Analytical Chemistry of Indium] Khimiya, Moscow (in Russian).
Dobrovolsky, V. V. (1983) *Geografiya mikroelementov. Globalnoe rasseyanie* [Geography of Trace Elements. Global Dispersion]. Mysl, Moscow (in Russian).
Fedorov, P. I. *et al.* (1977) *Khimiya galliya, indiya, talliya* [The Chemistry of Gallium, Indium, and Thallium]. Novosibirsk (in Russian).
Fergusson, J. E. (1990) *The Heavy Elements: Chemistry, Environmental Impact and Health Effects.* Pergamon Press, Oxford.
Ferm, V. H. & Carpenter, S. J. (1970) *Toxicol. Appl. Phamacol.*, **16**, 166—170.
Fowler, B. A. (1979) In: Friberg, L. *et al.* (eds) *Handbook on the Toxicology of Metals. Elsevier*, Amsterdam, pp. 429—434.
Fowler, B. A. & Goering, P. L. (1991) Chapter II.13 *Indium* in: Merian, E. (ed) *Metals and their Compounds in the Environment — Occurrence, Analysis and Biological Relevance.* VCH, New York, pp. 939—944.
Hayes, R. L. (1988) Chapter 27 *Indium* in: Seiler, H. G. *et al.* (eds) *Handbook on Toxicity of Inorganic Compounds.* Marcel Dekker, New York, pp. 323—336.
Henderson, P. (1982) *Inorganic Geochemistry.* Pergamon Press, Oxford.
Kabata-Pendias, A. & Pendias, H. (1986) *Trace Elements in Soils and Plants.* CRC Press, Boca Raton (Fl.).
Pokhodzei, Yu. I. (1984) *Gig. Truda Prof. Zabol.*, No. 4, 43—45 (in Russian).
Pokhodzei, Yu. I. (1985) *Gig. Truda Prof. Zabol.*, No. 12, 38—39 (in Russian).
Roshchin, A. V. *et al.* (1982) *Gig. Truda Prof. Zabol.*, No. 2, 5—8 (in Russian).
Smirnov, V. A. *et al.* (1986) *Khimiya odnovalentnogo indiya* [The Chemistry of Univalent Indium]. Khimiya, Moscow (in Russian).
Smith, G. A. *et al.* (1960) *Health Phys.*, **4**, 101—108.
Tarasenko, N. Yu. & Fadeyev, A. I. (1980) *Gig. San.*, No. **10**, 13—16 (in Russian).
Thomas, R. G. (1983) In: *Encyclopaedia of Occupational Health and Safety*, Vol. 1. International Labour Office, Geneva, p. 1103.
Vasilenko, V. N. *et al.* (1985) *Monitoring zagryazneniya snezhnogo pokrova* [Monitoring the Pollution of the Snow Cover]. Gidrometeoizdat, Leningrad (in Russian).

Thallium and its compounds

Thallium(I) (thallous) compounds: **bromide; carbonate; chloride; iodide; nitrate; oxide; sulfate**
Thallium(III) oxide (thallic oxide)

IDENTITY AND PHYSICOCHEMICAL PROPERTIES
OF THE ELEMENT

Thallium (Tl) is a metallic element in group III of the periodic table, with the atomic number 81. Its stable natural isotopes are ^{203}Tl (29.50%) and ^{205}Tl (70.50%); its radioactive natural isotopes are ^{206}Tl (half-life = 4.19 min), ^{207}Tl (4.79 min), ^{208}Tl (3.1 min), and ^{210}Tl (1.32 min). There are also artificial radioactive isotopes, the most important of which is ^{204}Tl (half-life = 3.56 years).

Thallium is a malleable metal that resembles lead in physical properties but slightly softer. It exists in two modifications, α and β (transition temperature = 234°C).

In compounds thallium is present in the oxidation states +1 (thallium(I), or thallous, compounds) and +3 (thallium(III), or thallic, compounds), the latter being less stable than the former. On exposure to air, thallium forms a coating of oxide. It readily reacts with mineral acids except hydrochloric acid, in which it becomes covered with a poorly soluble film of thallium(I) chloride. It does not react with alkalies, water (in the absence of O_2), nitrogen, ammonia, or dry carbon dioxide. It gives thallium(I) and (III) oxides with O_2 and thallium(I) hydroxide in O_2-containing water. Like alkali metals, thallium in the oxidation state +1 forms water-soluble hydroxide, nitrate, carbonate, and other compounds. Like the univalent Ag, Cu, Au, and Hg and the bivalent Pb and Hg, thallium(I) forms halides, sulfides, and chromates that are difficultly soluble in water. Thallium(III) is readily reducible to thallium(I) and gives complex compounds that are more stable than those formed by other trivalent elements such as Bi, Sb, Fe, Cr, and Al. Thallium alloys with most metals. See also Appendix.

NATURAL OCCURRENCE AND ENVIRONMENTAL LEVELS

The clarke of thallium is estimated at $(1.0-0.43) \times 10^{-4}\%$ in the earth's crust and at $1.8 \times 10^{-4}\%$ in the granite layer of the continental crust [Dobrovolsky].

Thallium is widely distributed in the earth's crust, though in very low concentrations. It is also found as an accompanying substance of other heavy metals in pyrites and blendes and in the ferromanganese oxide deposits on the ocean floor. It occurs combined, more commonly in the oxidation state +1, in rare dispersed minerals such as crookesite ($\leqslant 19\%$ Tl), lorandite ($\leqslant 60\%$), markasite and ardaite ($\leqslant 32\%$), and avicennite (80%). Its levels in crystalline and copper-zinc ores are in the ranges $4 \times 10^{-4}-5.5 \times 10^{-3}$ and $5 \times 10^{-4}-6 \times 10^{-4}$, respectively; in some areas of manganese deposits, manganese ores contain it at much higher levels (up to 0.01%). Thallium often occurs as a satellite of lithium, potassium, rubidium, or cesium, and has also been found in arsenic deposits.

The distribution of thallium in the earth's crust indicates that its concentrations rise with increasing acidity of magmatic rocks and rising clay content of sedimentary rocks. Common thallium contents range from 0.05 to 0.4 mg/kg in basic rocks and from 0.5 to 2.3 mg/kg in acid rocks, while calcareous sedimentary rocks contain only 0.01 to 0.4 mg/kg. In the process of weathering, thallium is readily mobilized and is transported together with alkali metals, but is often found fixed *in situ* by clays and gels of manganese and iron oxides. It is readily sorbed by organic matter, particularly under reducing conditions [Kabata-Pendias & Pendias].

Seawater contains thallium at concentrations of around 0.01 µg/L, while its levels in fresh water ranged from 0.01 to 14.0 µg/L [Kazantzis]. Underground waters in areas of thallium deposits contained up to 2.7 mg/L. Levels of up to 430 ppb have been reported for some aquatic organisms [Whanger].

In soil, thallium concentrations average $10^{-5}\%$. In the USA, surface soils had concentrations from 0.02 to 2.8 mg/kg (soils over sphalerite veins contained up to 5 mg/kg). Soils within zones of mercury mineralization contained between 0.03 and 1.1 mg/kg. In garden soil samples, values in the range of 0.17 to 0.22 mg/kg were recorded [Kabata-Pendias & Pendias].

Thallium is present in cabbage, spinach, lettuce, and leek. Its content in plants depends on that in the soil. Herbaceous and woody plants usually contain more thallium that other plant species. Concentrations ranged from 0.02 to 1.0 mg/kg dry weight in herbage and from 2 to 100 mg/kg ash weight in pine trees, being higher in needles that in stems. On a dry weight basis, edible plants contained 0.02–0.125 mg/kg, clover 0.008–0.01 mg/kg, meadow hay

0.02–0.025 mg/kg. Thallium is accumulated to high levels by herbaceous plants growing in zones of its mineralization; thus, flowers of *Galium* sp. from the Rubiaceae family had values up to 17 000 mg/kg ash weight in flowers, while leaves and stalks of other plants contained about 100 mg/kg. Elevated thallium levels (up to 2.8 mg/kg dry weight) occurred in plants growing in the vicinity of potash fertilizer works and factories processing bituminous coal [Kabata-Pendias & Pendias]. Thallium ions are easily taken up by plants through roots [Ewers].

In the atmosphere, thallium levels of $(0.04–0.48) \times 10^{-9}$ g/m^3 were recorded over Nebraska, USA [Kazantzis], and $(18–83) \times 10^{-15}$ g/m^3 over the South Pole [Zoller *et al.*]. Normally ambient air contains less then 1 ng Tl per m^3 [Ewers].

PRODUCTION

Thallium is rarely extracted from the ores or concentrates that contain it, but is usually obtained as a by-product from sublimates or flue dusts generated by pyrite burners or by lead and zinc smelters and refiners. The production process involves decomposition of the starting material, transfer of the thallium to solution and its precipitation from the solution as thallium chloride, iodide, sulfate, chromate, bichromate, or hydroxide. The resulting concentrate is freed from metal impurities by extraction and ionic exchange and sequential precipitation of the difficulty soluble compounds. Thallium is then separated from the purified solution by various methods (cementation on zinc, anodic oxidation, etc.); the spongy metal that results is washed, briquetted, and melted. Metallic thallium of high purity meeting the requirements of semiconductor technology is made by a combination of chemical, electrochemical, and crystallization methods of purification and by amalgam refining. The purified thallium contains lead, copper, cadmium, and nickel at levels of $(1–4) \times 10^{-3}\%$.

Thallium oxide can be prepared by heating thallium in air at about 350°C or by heating thallium hydroxide to 100°C. Thallium sulfate is produced by dissolving thallium in hot concentrated sulfuric acid or by neutralizing thallium hydroxide with dilute sulfuric acid followed by crystallization.

USES

Thallium and its compounds find a variety of applications in the electrical-equipment, electronic and semiconductor industries. They are used as alloying additives to cadmium compounds, germanium, and silicon to impart acceptor properties; in the manufacture of photoresistors, highly sensitive photocells, and infrared detectors; in various types of scintillation counters; for the activation of luminescent alkali-halide crystals and for stabilizing the luminescence process. In the optical industry, monocrystals of solid thallium halide solutions are utilized in the making of various lenses and optical instruments. The artificial radioactive isotope [204]Tl is used in the instrument-making industry as a source of continuous β-radiation in devices designed for the study and monitoring of production processes. Thallium salts find application as antiknock agents in internal-combustion engines. In the chemical industry, thallium and its compounds are used as catalysts in the oxidation of hydrocarbons and olefins, in polymerization and oxide-coating processes, and in the manufacture of luminous paints and of artificial gems, pearls, and diamonds. Thallium is a component of depilating creams used for cosmetic purposes, in animal husbandry, and in medicine.

Thallium has been rather widely employed in some countries as a rodenticide or insecticide (reviewed by Saddique & Peterson).

MAN-MADE SOURCES OF EMISSION INTO THE ENVIRONMENT

The major anthropogenic sources of thallium in the environment are found in industry and arise from the burning of organic carbonaceous fuels (oil, coal, mazut, etc.) and the smelting of copper, lead, and zinc. Thallium and its compounds are emitted into the atmosphere as fumes, dusts, and aerosols from the air of work premises in various industries, and they are also released into bodies of water with industrial effluents. Depending on the source, thallium can enter the environment in the form of sulfate and hydroxide (TlOH) (liquid and solid wastes from ore mining and processing activities), oxide (emissions from coal-burning plants), or sulfide and oxide (various thermal processes). The monovalent form (Tl^+) is the most stable, and predominates, in environmental media. Tl^{3+} is found in seawater and fresh waters; it is much less stable than, and occurs in equilibrium with, Tl^+ ions [Manzo et al.].

According to one projection, in 1990 the release of thallium from all coal-fired power plants in member states of the European Community should have totalled 240 tonnes, including 7.0 tonnes discharged into the atmosphere from stack emissions; calculations by which these anticipated atmospheric emissions were arranged according to particle size indicated that 2.6 tonnes would be contributed by particles <1.5 µg in diameter, 2 tonnes by 1.5–3 µg particles, 2.2 by 3–7 µg particles, and 0.2 tonne and 0.07 tonne by those sized 5–15 µg and 15–40 µg, respectively [Sabbioni et al., 1984]. The greatest capacity for penetrating the bloodstream from the alveolar region is possessed by particles in the 0.3–0.6 µg range.

In the production of metallic thallium, the indoor air contained the metal and its oxides in concentrations up to 0.18 mg/m^3 in the smelting department, while in the casting department, the air concentrations of the oxides in aerosol form ranged from 13 to 17.3 mg/m^3. During the production and packaging of thallium salts, indoor air concentrations of metallic dust reaching 0.136 and 0.354 mg/m^3, respectively, were recorded. The production of thallium monocrystals and various crystal systems was associated with indoor air contamination at levels between 0.004 and 0.007 mg Tl/m^3. Washings from the walls of work premises and the surfaces of equipment contained up to 12.5 mg Tl/m^2, while those from the palms of workers contained 300–350 mg. In some coal-burning industries, employees were found to receive 150–180 ng Tl/kg body weight by the oral route daily [Sabbioni et al., 1980].

Thallium contamination of drinking water is likely in the vicinity of copper, zinc, and cadmium mining, smelting, and refining activities. For example, concentrations of 0.7–88.0 µg/L occurred in the water of rivers draining a metal mining area, and algae and mosses from these rivers contained thallium at levels between 9.5 and 162.0 µg/kg dry weight [Kazantzis].

TOXICITY TO ANIMALS AND MAN

Thallium is a systemic poison whose main targets are the central and peripheral nervous systems, the gastrointestinal tract, and the kidneys.

Acute exposure

Animals

Lethal doses of four thallium(I) compounds for mice and rats are shown in Table 1.

Table 1. Lethal doses of thallium(I) compounds

Compound	Animal	Route of administration	LD$_{100}$ (mg/kg)	LD$_{50}$ (mg/kg)
Bromide	Rat	Oral	35.0	—
Carbonate	Mouse	Subcutaneous	32.0	27.0
	Rat	Subcutaneous	25.0	18.0
	Mouse	Oral	25.0	21.0
	Rat	Oral	—	15.0
		Skin application	—	117.3
Chloride	Mouse	Oral	30.0	23.7
	Rat	Oral	55.0	—
Sulfate	Mouse	Oral	35.0	23.5
		Skin application	—	57.7

Clinically, acute poisoning by **thallium salts** was characterized by mucous discharges from the nose (starting with days 3–4), body-weight losses, bloody and mucous diarrhea, increasing adynamia, somnolence, and weakened breathing. Death usually occurred on day 5 or 6. Mice and rats were somewhat more sensitive to the salts than guinea-pigs and rabbits. Cats and dogs that had fed on thallium-poisoned rodents, developed hepatic and renal derangements and hemorrhagic enteritis, often with a fatal outcome; in animals that survived, extensive inflammatory lesions in the skin with infiltrations and hair loss were observed.

Histopathologic studies demonstrated tubular necroses and glomerular hyalinization in the kidneys, local chromatolysis and edema in the brain, degenerative changes in peripheral nerves, necrotic areas in skeletal and cardiac muscle fibers, pulmonary edema and bronchopneumonia, hemorrhages and ulcerations in the gastrointestinal tract, and centrolobular necrosis in the liver.

Thallium and its salts are readily absorbed through the skin and can produce systemic toxic effects. When extensive areas of mouse skin were exposed to a 20% solution of thallium(I) carbonate, 50% of the animals died after 1 h of contact and all were dead after 3 h.

Man

Thallium can enter the human body by ingestion, inhalation, or absorption through the skin. Acute thallium poisoning in industry is rare, but the acute toxic action of thallium and its salts has been well documented by reports of nonoccupational (not uncommonly fatal) poisoning by ingestion.

The symptoms and signs of acute thallitoxicosis are many and varied, and the diagnosis is often difficult to make. The first symptoms to appear were usually digestive disorders including anorexia, gastroenteritis with diarrhea (sometimes bloody), nausea, vomiting, and paroxysmal abdominal pain; oliguria was also observed. Later, 8 to 40 h after the onset of these early symptoms, nervous system disorders were added, manifested typically as insomnia, extreme weakness, anxiety, paresthesias, abnormal gait (ataxia), and tremor; muscle pain and convulsions also occurred in some cases. Mental disturbances, including delirium, developed occasionally. Fever was common. Progressive respiratory and circulatory involvement led to death, usually on days 7–10 after thallium ingestion.

Those surviving beyond 1 week or so experienced pronounced neurologic disturbances, with headache, ataxia, tremor, paresthesias, and muscular atrophy predominating. Hypertension and acute myocardial infarction developed in some cases. The victims refused to eat, remained wakeful at night, and felt sleepy during the day. Dysfunction of the cranial nerves, ptosis, retrobulbar neuritis, facial nerve paralysis, and strabismus were observed, as was body-weight loss.

Subsequent manifestations of poisoning include hypertension, tachycardia, swollen and painful joints (predominantly in the upper extremities), various psychic disturbances, and cutaneous lesions such as desquamation of the skin, cracks at the corners of the mouth, brown pigmentation, and hyperkeratosis of the palms and soles. Alopecia develops, sparing pubic and axillary hair and the medial third of the eyebrows, and alopecia is regarded as the most characteristic symptom of thallium intoxication, which often enables the definitive diagnosis to be established. Another notable feature is the emergence on fingernails of white transverse lines ('Mees lines') that advance as the nails grow. The impairment of carbohydrate metabolism may be manifested by so-called latent diabetes and by glucosuria. Signs of renal damage include hematuria, urobilinuria, and the appearance of cylinders, acetone bodies, and sometimes porphyrin in the urine. Hemoconcentration with altered hematocrit values and anemia have also been observed, as well as visual disturbances.

Recovery may be complete or neurologic defects may persist with mental abnormalities, ataxia, and tremor.

The severity of poisoning can be judged by the clinical picture and the amount of thallium in the urine: the poisoning is taken to be severe if clinical symptoms are accompanied by the excretion of more than 10 mg thallium daily. Urinary excretion of thallium may continue for 3 to 5 months and even longer.

Histopathological examinations of those who died from acute thallitoxicosis showed mucosal inflammation in the intestine, ecchymoses and edema in the myocardium, atrophic lesions in the skin and subcutaneous fatty tissue, dystrophic and degenerative changes in parenchymal organs, and degeneration of motor and sensory peripheral nerve fibers. In the brain, edema, multiple diapedetic hemorrhages, focal proliferation of the glia, neuronal degeneration, and chromatolysis of neurons in the motor cortex and some subcortical centers were observed. Metaplasia of the bronchial epithelium and productive interstitial myocarditis were also seen. Microscopic examination of the lost hair showed fusiform swellings at the hair roots with abundant deposits of a brownish-black pigment. Lethal thallium doses for adults vary from 0.5 g to 3.0 g, depending mainly on individual susceptibility [Davis *et al.*; Manzo *et al.*; Moeschlin; Thompson] (See also Fergussion; Kemper & Bertram; Manzo & Sabbioni.)

Repeated exposure

Animals

In rats administered **thallium sulfate** by oral gavage at 1.25 mg/kg for 15 days, the major findings were marked general inhibition, progressive weight loss, reduced catalase activity, impaired liver function, and disbalance of protein fractions in the blood serum.

A 3-week inhalation toxicity study in which rats were daily exposed to a mixed **thallium iodide** and **bromide** dust at concentrations of 0.0005–0.0007 mg/m^3 clearly demonstrated that thallium is a cumulative poison.

Chronic exposure

Animals

In a 6-month toxicity study of **thallium sulfate,** rabbits given this compound daily at 0.35 mg/kg by oral gavage began to display aggressiveness along with signs of general inhibition after 4 months of exposure; some animals developed paralysis of the hind paws, dysproteinemia (lowered albumins and raised globulins), and reduced levels of alkaline phosphatase activity and sulfhydryl groups in the blood serum. Histopathologic examination showed degenerative changes in the liver with the presence of multiple magnicellular infiltrates along the course of hepatic ducts; grossly engorged glomeruli and cloudy swelling in the kidneys; and lymphoid infiltration of the gastric mucosa. Similar findings were made in rabbits administered thallium sulfate subcutaneously at the same dose rate over the same period.

Rats receiving **thallium** orally by gavage at 5×10^{-4} mg/kg for periods up to 8 months exhibited alterations in conditioned-reflex activity as well as reductions in the levels of sulfhydryl groups, alkaline phosphatase, lactate dehydrogenase, glucose-6-phosphate dehydrogenase, and delta-aminolevulinic acid in the blood and of DNA and RNA in the spleen; the dose of 5×10^{-5} mg/kg produced similar, but less marked, changes while the dose of 5×10^{-6} mg/kg was without detectable effects. Moreover, mutagenic and gonadotropic effects were observed in rats administered thallium at the 5×10^{-4} and 5×10^{-5} mg/kg dose levels over 8 months: increased percentages of chromosomal aberrations and aberrant bone marrow cells and functional abnormalities in spermatozoa together with morphological changes in the testes. Furthermore, the embryos of pregnant females dosed with thallium at 5×10^{-4} mg/kg had reduced weights, and the offspring of these females (as well as of those dosed at 5×10^{-5} mg/kg) showed functional disorders and decreased survival rates, indicating that the metal possesses embryotoxic and fetotoxic activities. Mortality among thallium-poisoned animals could be increased by 60% to 90% through their loading with alcohol, which was attributable to the formation of the soluble compound thallium alcoholate [Eitingon; Spiridonova & Shabalina].

Man

Cases of chronic occupational intoxication have mainly resulted from inhalation of thallium-containing aerosols and dusts or absorption of the metal through the skin in the form of solution or powder. Long-term ingestion of thallium in food or drinking water may also occur, in both occupational and nonoccupational settings.

Cases of chronic poisoning have been described among patients who had received thallium for the treatment of syphilis, gonorrhea, or dysentery; among individuals using thallium-containing depilatories for the treatment of ringworm; and among workers exposed to thallium during the production of rodenticides, optic glass, or paints [Manzo *et al.*]. The initial symptoms of intoxication were excitation and insomnia. Those exposed for weeks or months experienced joint pains, weakness, and early manifestations of polyneuritis [Fergusson]. Other symptoms and signs included nausea, loss of appetite and body weight, cardiac abnormalities, and albuminuria; in some cases, mental disturbances in the form of delirium were also noted, with occasional episodes of hysterical laughter. The above authors have described the existing concepts regarding the mechanism of action of thallium and the nature of neurologic changes in thallitoxicosis.

A variety of complaints were presented by workers of a plant where the ambient air contained thallium at concentrations of 0.0039–0.066 mg/m^3 in the form of condensation and disintegration aerosols. The more common complaints were increased irritability and fatiguability, periodic headache, bad sleep, excessive perspiration, vague pains in the limbs, pains in the cardiac and epigastric regions and along the large intestine, irregular bowel action, hair loss, and brittleness of nails. On examination, many of the workers showed tremor of the fingers on holding out the hands, pronounced hyperhidrosis, diffuse red dermatographism, exaggerated tendon reflexes, tachycardia, and a tendency to hypotension. A quarter of those examined had thallium levels of 0.003 to 0.56 mg/L in their urine. The severity of neurologic disturbances correlated directly with the length of occupational exposure. It has also been shown that chronic exposure to thallium-containing aerosols may cause catarrhal conjunctivitis as well as alterations in the activity of enzymes regulating phosphorus metabolism [Eitingon].

Elevated thallium levels in the urine and hair of individuals chronically exposed to this metal were found by Brockhaus *et al.* (1980). These authors examined 1254 males and females of various ages (from 1 to 85 years) who lived in the vicinity of a plant producing special grades of cement using thallium-containing ore materials. Thallium concentrations in the urine ranged from 10 to 80 µg/L in 12% of the subjects and from 0 to 10 µg/L in 88% (as compared to 0.1–1.2 µg/L in the controls). The hair of 40% of the subjects had thallium levels 2 to 60 times higher than in the controls. It was also noted that the group of subjects eating fruits and vegetables harvested near the plant in areas where both the soil and air were contaminated with thallium, had urinary levels 2 to 5 times as high as the group that did not consume such products. The urinary content of thallium, which was found to decrease with increasing distance of those areas from the plant, showed a direct correlation with the frequency of neurological and neurovegetative disorders (sleep disturbances, tiredness, weakness, nervousness). No correlation of this kind was observed for other manifestations of poisoning such as gastrointestinal troubles, hair loss, or alopecia.

Exposures to thallium of those employed in cement works and the resulting effects, including elevated urinary levels of this metal, have also been reported by other authors [Apostoli *et al.*; Kemper & Bertram; Schaller *et al.*].

Effects of local application

Thallium and its salts can be readily absorbed through the skin to produce systemic effects, and they also act as strong local irritants of the skin and mucous membranes. In rabbits, **thallium carbonate** produced hyperemia, purulent conjunctivitis, and corneal opacity on ocular application and epidermal necrosis and dermal edema and infiltration when applied to the intact skin. In rats, its skin application resulted in hyperemia and punctate hemorrhages followed by crust formation.

ABSORPTION, DISTRIBUTION, AND ELIMINATION

Normally, the amount of thallium entering the adult human body with food and water ranges from 1.6 to 2.0 µg per day [Ewers]; in addition, 0.05 µg enters, on average, in the inspired air. The total amount of thallium in a healthy human being weighing 75 kg was calculated to be about 0.1 mg [Kazantzis]. In health, its concentration ranged 4.8–15.8 ng/g in hair, 0.72–4.93 ng/g in nails, 0.56–5.4 ng/g in the large intestinal wall, 0.12×10^{-3}–29.5×10^{-3} µg/g in parenchymal organs, 0.13–1.69 µg/L in urine. The biological half-time of thallium in the human body is 3.3 to 4.0 days [Eitingon; Spiridonova & Shabalina].

According to some authorities, thallium is not a normal ingredient of blood, tissues, and other biological structures, and its presence in them is a result of exposure to the metal contained in the environment [Berman]. Inhabitants of Chicago had the following thallium concentrations (µg/g): brain, 0–0.02; kidney, 0–0.002; liver, 0–0.19; spleen, 0.002; stomach, 0–0.04 [Berman]. Frequent absence of detectable thallium in the bodies of intact animals (and plants) has been pointed out by Whanger.

Thallium and its salts are readily absorbed through the gastric and intestinal mucous membranes, the skin, and the epithelium of the upper respiratory tract [Kemper & Bertram; Manzo & Sabbioni]. In terms of decreasing affinity for ^{204}Tl, animal organs and tissues were arranged as follows: kidney > testes > liver > prostate > brain > hair. Absorbed thallium rapidly disappears from the blood to be distributed in various organs but tends to concentrate in the kidneys and salivary glands. In rats, thallium appeared in the urine within 1 h of oral administration of thallium(I) sulfate [Kazantzis]. With the course of time it undergoes redistribution and is deposited in the bones and hair. In intercellular fluids it combines with amino acids and in bones is deposited in the form of phosphate.

In experiments on rats where ^{204}Tl was administered as a thallium(I) nitrate solution by various routes, its concentration in the kidney was found to be 5.5 times higher than in salivary glands, which had the second highest concentration. These were followed by testis, muscle, bone, lymph nodes, gastrointestinal tract, heart, spleen, and liver, all of which showed relatively small differences in concentration among themselves. On successive days during the first week, the relative thallium concentrations in these tissues, but not in hair, remained essentially the same. In hair, thallium concentrations were rising with time so that after 3 weeks this tissue contained up to 60% of the body burden of thallium [Lie et al.]. In man, too, thallium is widely distributed in the tissues, with the highest concentration in the kidneys, but the proportionate distribution may show considerable variation. In one fatal case, the second highest concentration was in the heart; in the brain, which ranked low in terms of thallium content, the concentration in gray matter was three times that in white matter [Kazantzis]. Brain areas densely populated with neurons have been found to accumulate thallium more than other areas [Manzo & Sabbioni]. Thallium is also accumulated in testicles, which leads to reduced sperm motility [Manzo & Sabbioni].

After an intravenous injection, 61% of the thallium present in the blood of rats was in erythrocytes and 39% in plasma. Muscles had the highest concentration 4 to 8 h post-injection and the brain, in approximately 24 h [Eitingon]. After injection into pregnant rats, [201]thallium sulfate soon appeared in the fetal brain where it rapidly reached a high level in parallel with accumulation in the maternal brain [Edel Rade et al.].

After an intraperitoneal injection of ^{201}Tl into rats, its distribution in their organs at different times post-injection was as shown in Table 2 [Sabbioni et al., 1980]. The time course of thallium sulfate concentrations in golden hamsters after oral administration is shown in Table 3.

Table 2. Thallium levels in rat organs as percentages of the intraperitoneally injected dose (2.0 µg/rat) per g tissue weight (calculated as Tl ion)

Organ	Time after injection (hours)		
	2	40	192
Kidney	2.57	4.43	2.50
Heart	0.63	0.37	0.96
Salivary glands	0.42	0.39	0.30
Brain	0.02	0.16	0.15
Testis	0.20	0.69	0.68
Stomach	0.68	0.43	0.27
Small intestine	1.46	0.95	0.35

Source: Sabbioni et al. (1980).

The intracellular distribution of thallium in the liver and testes of rats 16 h after intraperitoneal injection in a dose of 2.0 µg per rat was as follows (% of the dose in the hepatic and testicular cells, respectively): nucleus, 37.4 and

29.5; mitochondria, 7.9 and 11.8; lysosomes, 8.5 and 6.7; microsomes, 5.8 and 3.5; cytosol, 30.4 and 48.5 [Byczkowski & Sorenson].

Table 3. Thallium sulfate concentrations (μg/g wet weight) in organs or tissues of goiden hamsters following oral administration at 12.35 mg/kg body weight

Organ or tissue	1	12	24	72	163
		\multicolumn{4}{c}{Time after administration (hours)}		
Brain	0	3.9±0.3	3.7±0.3	2.2±0.3	1.3±0.1
Heart	11.4	9.9±1.0	7.2±0.7	4.1±0.4	3.0±0.4
Liver	14.3±7.6	7.8±2.4	6.6±2.1	3.3±0.8	3.0±0.4
Kidney	58.5±8.4	70.5±12.0	51.3±14.0	25.7±4.7	16.4±2.2
Testis	0	12.9±1.8	14.7± 0.7	13.1±1.7	7.6±0.9
Muscle	0	9.0±0.1	9.1± 2.4	8.2±1.0	4.2±0.6
Blood	1.5±0.6	1.2±0.1	1.0± 0.1	0.5±0.1	—

Source: Aoyama

Experiments with rats have shown that thallitoxicosis is accompanied by reductions in the number and acidic resistance of erythrocytes and by their increased breakdown along with increases in the level of 2,3-diphosphoglyceric acid in these cells, which results in lowered hemoglobin affinity for oxygen [Reut & Ardavichene]. It has been shown that thallium can replace potassium and competes with it for binding sites in biological membranes, and that infused potassium increases the renal clearance of thallium and its elimination from tissues. The LD_{50} of thallium for rats became higher when their diet contained elevated potassium levels [Reut & Ardavichene].

Excretion of thallium in both animals and man is mainly by way of the kidneys and intestine [Kemper & Bertram]; small amounts are also eliminated via hair and in milk [Kazantzis]. Elimination from the human body has been found to continue for up to 1 year [Eitingon].

HYGIENIC STANDARDS

Exposure limits adopted in the USSR for thallium and its compounds are given in Table 4. In the USA, a TLV-TWA of 0.1 mg/m^3 has been set for soluble thallium compounds (as Tl), with the 'skin' notation to indicate the possibility of absorption by the cutaneous route, including the mucous membranes and eyes.

Table 4. Exposure limits for thallium and its compounds

	Workroom air		Atmospheric air	Water sources	
	MAC_{wz} (mg/m^3)	Aggregative state	$TSEL_{aa}$ (mg/m^3)	MAC_w (mg/L)	Hazard class
Thallium	—	—	—	0.0001*	1
Thallium(I) bromide and iodide (as Tl)	0.01	Aerosol	—	—	1
Thallium carbonate (as Tl)	—	—	0.004	—	—

* Based on sanitary and toxicologic criteria.

METHODS OF DETERMINATION

For the determination of thallium in air, colorimetric, photometric, and spectrographic methods have been used. In biological materials, thallium has been measured by colorimetric and polarographic methods or by atomic absorption spectrophotometry [Brockhaus et al., 1981; Eitingon; Kazantzis; Muraviyeva; Reis; Kemper & Bertram, 1991; Fergusson, 1990; Manzo & Sabbioni, 1988].

MEASURES TO CONTROL EXPOSURE AT THE WORKPLACE

All processes that can pollute the air of industrial premises with thallium or its compounds should be mechanized and automated to the greatest possible extent, and the process equipment should be hermetically sealed and provided with suction devices to capture toxic dust. Preference should be given to continuous processes to ensure a closed production cycle without any manual operations. The packaging of liquids and dry salts must be carried out in airtight sleeve boxes equipped with suction hoods. For operations involving mechanical working of thallium materials (e.g. polishing crystals), machines provided with local exhaust ventilation should be used.

In places where thallium and its compounds are produced or used, special attention should be given to the decontamination of waste materials taking into account their toxicity (thallium and its compounds such as the bromide and iodide are placed in hazard class 1, i.e. classified as highly dangerous).

Showering at the end of the shift is mandatory. Those who have been carrying out remedial work in an emergency should undergo decontamination in the shower room. Hands should be washed without delay as they become soiled. Working clothes must be renewed immediately after the termination of repair or emergency operations and at least once a week under ordinary circumstances. Workers must not be allowed to leave the workroom in overalls; there should be a ban on smoking, drinking, eating, and keeping food in the workrooms, and these should be maintained scrupulously clean.

All workers are subject to preplacement and periodic medical examinations. It is desirable to test their urine for thallium once a month. For prophylactic as well as therapeutic purposes, diets rich in protein, methionine, choline, cystine, calcium, magnesium, iron, and vitamins are recommended.

For *personal protection*, dust-proof overalls, aprons, polyethylene sheaths for footwear, and other protective clothing as necessary should be used to prevent skin contact. For eye protection, goggles should be worn (a single piece of equipment to protect both the eyes and respiratory organs may be used). Acid-resistant rubber gloves are recommended for hand protection.

EMERGENCY TREATMENT IN POISONING

In acute thallium poisoning (thallitoxicosis), the following emergency therapeutic measures have been recommended: a 1% potassium iodide solution and Berlin blue orally; gastric lavage with water containing activated charcoal or with a 0.3% sodium thiosulfate solution together with administration of a saline laxative; histamine and 10 ml of a 20% sodium thiosulfate solution intravenously; activated charcoal and potassium chloride orally for 4 to 24 days. Forced diuresis and prolonged hemodialysis may be required. Lobeline and oxygen should be given to correct impaired respiration; caffeine, cordiamine [synonym: nikethamide], and corglycone (a preparation containing several glycosides from lily of the valley leaves) are recommended to eliminate circulatory disturbances; morphine to relieve colic; dibazol [synonym: bendazole hydrochloride], proserine [synonym: neostigmine methylsulfate], and vitamins B_1, B_2, and B_{12} to combat nervous system disorders [Eitingon].

REFERENCES

Aoyama, H. (1989) *Bull. Environ. Contam. Toxicol.*, 42, 456—463.
Apostoli, P. *et al.* (1988) *Sci. Total Environ.*, 71, 513—518.
Berman, E. (1980) *Toxic Metals and their Analysis.* Heyden, London, pp. 201—208.
Brockhaus, A. *et al.* (1980) In: Holmstedt, B. *et al.* (eds) *Mechanisms of Toxicity and Hazard Evaluation.* Elsevier, Amsterdam, pp. 565—568.
Brockhaus, A. *et al.* (1981) *Int. Arch. Occup. Environ. Health*, 48, 375—389.
Byczkowski, J. Z. & Sorenson, J. K. J. (1984) *Sci. Total Environ.*, 37, 133—162.
Davis, L. E. *et al.* (1981) *Ann. Neurol.*, 10, 38—42.
Dobrovolsky, V. V. (1983) *Geografiya mikroelementov. Globalnoe rasseyanie* [Geography of Trace Elements. Global Dispersion]. Mysl, Moscow (in Russian).
Edel Rade, J. *et al.* (1982) *Toxicol. Lett.*, 11, 275—280.
Eitingon, A. I. (1983) *Thalliy* [Thallium], published by the Center of International Projects at the State Committee for Science and Technology (in Russian).
Ewers, U. (1988) *Sci. Total Environ.*, 71, 285—292.

Fergusson, J. E. (1991) *The Heavy Elements: Chemistry, Environmental Impact and Health Effects*. Pergamon Press, Oxford.

Kabata-Pendias, A. & Pendias, H. (1986) *Trace Elements in Soils and Plants*. CRC Press, Boca Raton (Fl.).

Kazantzis, G. (1979) In: Friberg, L. *et al.* (ed.) *Handbook on the Toxicology of Metals*. Elsevier, Amsterdam, pp. 599–612.

Kemper, F. H. & Bertram, H.P. (1991) Chapter II.29 *Thallium* in: Merian, E. (ed.) *Metals and their Compounds in the Environment — Occurrence, Analysis and Biological Relevance*. VCH, New York, pp.1227–1241.

Lie, R. *et al.* (1960) *Health Phys.*, 2, 334–340.

Manzo, L. & Sabbioni, E. (1988) Chapter 62 *Thallium* in: Seiler, H. G. *et al.* (eds) *Handbook on Toxicology of Inorganic Compounds*. Marcel Dekker, New York, pp. 677–688.

Manzo, L. *et al.* (1985) In: Blum, K. & Manzo, L. (eds) *Neurotoxicology*. Marcel Dekker, New York, pp. 385–391.

Moeschlin, S. (1980) *Clin. Toxicol.*, 17, 133–146.

Muraviyeva, S. I. (1960) *Gig. Truda Prof. Zabol.*, No. 8, 53–54 (in Russian).

Reis, N. V. (1957) *Lab. Delo*, No. 6, 12–16 (in Russian).

Reut, A. A. & Ardavichene, T. A. (1978) In: *Aktualnye voprosy ekologicheskoi toksikologii* [Current Topics in Ecological Toxicology]. Ivanovo, pp. 93–94 (in Russian).

Sabbioni, E. *et al.* (1980) In: Holmstedt, B. *et al.* (eds.), *Mechanisms of Toxicity and Hazard Evaluation*. Elsevier, Amsterdam, pp. 559–564.

Sabbioni, E. *et al.* (1984) *Sci. Total Environ.*, 40, 141–154.

Saddique, A. & Peterson, C. D. (1983) *Vet. Hum. Toxicol.*, 25, 16–22.

Schaller, K. H. *et al.* (1980) *Int. Arch. Occup. Environ. Health*, 47, 223–231.

Spiridonova, V. S. & Shabalina, L. P. (1978) *Gig. Truda Prof. Zabol.*, No. 7, 41–43 (in Russian).

Thompson, D. F. (1981) *Clin. Toxicol.*, 18, 979–990.

Whanger, P. D. (1982) In: Hatchcock, J. N. (ed.) *Nutritional Toxicology*, Vol. 1. Academic Press, New York, pp. 163–208.

Zoller, W. H. *et al.* (1974) *Science*, 183, 198–200.

Rare-earth elements and their compounds

Cerium(III) chloride; c.(III) fluoride; c. hexaboride; c.(III) oxide; c.(IV) oxide
(cerium dioxide, cerianite [min.]); **c.(III) sulfide**
Dieuropium dioxide sulfide (europium oxysulfide)
Diyttrium dioxide sulfide (yttrium oxysulfide)
Dysprosium chloride; d. oxide; d. sulfide
Erbium oxide
Gadolinium chloride; g. oxide
Holmium oxide
Lanthanum chloride; l. hexaboride; l. nitrate hexahydrate; l. oxide; l. sulfate
Lutetium chloride; l. oxide
Neodymium chloride; n. oxide
Praseodymium(III) chloride; p.(III) oxide
Samarium(III) oxide
Scandium nitrate; s. oxide
Thulium oxide
Ytterbium oxide
Yttrium chloride; y. fluoride; y. oxide; y. sulfide

IDENTITY AND PHYSICOCHEMICAL PROPERTIES OF THE ELEMENTS

The rare earth elements (rare earths) are metals in group III of the periodic table and include scandium, yttrium and the 15 elements of the lanthanum series (lanthanides). Their atomic numbers, mass numbers, and isotopic compositions are given in Table 1.

Scandium (Sc), yttrium (Y), lanthanum (La), cerium (Ce), praseodymium (Pr), neodymium (Nd), samarium (Sm), gadolinium (Gd), terbium (Tb), and ytterbium (Yb) can exist in various allotropic modifications. The most characteristic modifications at ordinary temperatures are α-Sc, α-Y, α-La, γ-Ce, α-Pr, α-Nd, α-Sm, α-Tb, and α-Yb.

Table 1. Atomic numbers and isotopic composition of the rare-earth elements

Atomic number	Element	Symbol	Mass numbers and (in parentheses) % contents of natural isotopes
21	Scandium	Sc	45 (100)
39	Yttrium	Y	89 (100)
57	Lanthanum	La	138 (0.89), 139 (99.911)
58	Cerium	Ce	136 (0.195), 138 (0.265), 140 (88.45), 142 (11.10)
59	Praseodymium	Pr	141 (100)
60	Neodymium	Nd	142 (23.17), 143 (12.2), 144 (23.87), 145 (8.29), 146 (17.18), 148 (5.72), 150 (5.60)
61	Promethium	Pm	No stable isotopes are known; the longest-lived radioactive isotope is ^{145}Pm
62	Samarium	Sm	144 (2.87), 147 (14.94), 148 (11.24), 149 (13.85), 150 (7.36), 152 (26.90), 154 (22.84)
63	Europium	Eu	151 (37.77), 153 (52.23)
64	Gadolinium	Gd	152 (0.20), 154 (2.15), 155 (14.78), 156 (20.59), 157 (15.7), 158 (24.78), 160 (21.79)
65	Terbium	Tb	159 (100)
66	Dysprosium	Dy	156 (0.052), 158 (0.0902), 160 (2.29), 161 (18.88), 162 (25.52), 163 (24.97), 164 (28.18)
67	Holmium	Ho	165 (100)
68	Erbium	Er	162 (0.136), 164 (1.56), 166 (33.41), 167 (22.94), 168 (27.07), 170 (14.88)
69	Thulium	Tm	169 (100)
70	Ytterbium	Yb	168 (0.14), 170 (2.03), 171 (14.34), 172 (21.88), 173 (16.18), 174 (31.77), 176 (12.65)
71	Lutetium	Lu	175 (97.5), 176 (2.5)

Rare-earth metals of high purity are malleable and ductile and thus readily lend themselves to forging and rolling. Their mechanical properties depend on the levels of impurities, particularly O_2, S, N_2, and C. According to several characteristics and properties, the rare earths are conventionally divided into two subgroups, the composition of which differs slightly in various works of reference, but commonly consists of the following: (1) the cerium subgroup including La, Ce, Pr, Nd, Pm, and Sm, and (2) the yttrium subgroup including Eu, Gd, Tb, Dy, Ho, Er, Tm, Yb, Lu, Sc, and Y. Metals of the cerium subgroup have much lower melting points than those of the yttrium.

Rare earths are highly active chemically. In air they are rapidly covered with a film of oxides of the R_2O_3 type that protects them from further oxidation. At temperatures above 180–200°C they are vigorously oxidized. On roasting, the R_2O_3 oxides form high-melting white powders. Cerium, praseodymium, and terbium may give oxides of the RO_2 type. The oxides of cerium, terbium, and erbium are pink, those of samarium, dysprosium, and holmium are yellow, those of praseodymium and thulium are green, and those of neodymium are blue. In reacting with water, the rare-earth oxides form hydroxides of the $R(OH)_3$ type that are insoluble in water and soluble in acids; such hydroxides are also produced by rare-earth salts in reacting with aqueous solutions of alkalis and ammonia. On heating, rare earths react with H_2, C, N_2, P, CO, CO_2, hydrocarbons, water, and acids. Their chlorides and nitrates are soluble in water, their carbonates and phosphates are difficultly soluble, while their fluorides are insoluble even in concentrated acids. On fusion with metals, rare earths readily form intermetallic compounds. The most characteristic oxidation state of all rare earths in compounds is +3. Cerium, praseodymium, and terbium may be present in the oxidation state +4 and samarium, europium, and ytterbium, in the +2 state. See also Appendix.

NATURAL OCCURRENCE AND ENVIRONMENTAL LEVELS

Rare earths of the cerium subgroup are found in such minerals as monazite, loparite, cerite, and bastnaesite; those of the yttrium subgroup occur in gadolinite, fergusonite, xenotime, and other minerals. Some naturally occurring isotopes of the rare earths are radioactive (e.g. ^{144}Nd, ^{145}Pm, ^{147}Sm, ^{176}Lu, and ^{136}La).

Clarke values of 16 rare earth elements in the earth's crust and its granite layer are given in Table 2. Contents of rare earths in surface soils, terrestrial plants, and atmospheric air are shown in Table 3, 4, and 5, respectively.

The total content of the lanthanides in the earth's crust is estimated at $1.78 \times 10^{-2}\%$ by weight. Cerium, praseodymium, and samarium are contained in the lithosphere at levels of $7 \times 10^{-3}\%$, $9 \times 10^{-4}\%$, and $8 \times 10^{-4}\%$, respectively. In seawater, individual rare-earth elements occur at average concentrations of 1×10^{-8} to 3×10^{-6} mg/L, mainly in the form of $R(OH)_3$ [Henderson]. Surface waters in the Upper Volga River basin contained the following concentrations of rare earths (µg/L): lanthanum, 0.6–0.7; cerium, 0.8–1.1; samarium, 0.025–0.03; europium, 0.003–0.006; ytterbium, 0.026–0.03; lutetium, 0.001–0.006 [Vadkovskaya & Lukashev; Venchikov].

The content of lanthanides in plants depends on that in the soil and tends to decrease with increasing atomic numbers of the elements (Table 4). The greatest capacity to take up lanthanides appears to be possessed by woody plants; among these, hickory trees (*Carya* sp., Juglandacea family) have been most often reported as lanthanide-accumulating plants [Kabata-Pendias & Pendias].

Table 2. Clarke values of rare-earth elements in the earth's crust as a whole and in the granite layer of the continental crust

Element	Earth's crust: $1 \times 10^{-4}\%$	Continental crust: $1 \times 10^{-4}\%$	Element	Earth's crust: $1 \times 10^{-4}\%$	Continental crust: $1 \times 10^{-4}\%$
Scandium	10–22	11	Gadolinium	5.4–8.0	9.0
Yttrium	29–33	36	Terbium	0.9–4.3	1.4
Lanthanum	29–30	46	Dysprosium	3.0–5.0	6.5
Cerium	70	83	Holmium	1.2–1.7	1.8
Praseodymium	8.2–9.0	7.8	Erbium	2.8–3.3	3.6
Neodymium	28–37	33	Thulium	0.27–0.48	0.3
Samarium	6.0–8.0	9	Ytterbium	0.3	3.6
Europium	1.2–1.3	1.4	Lutetium	0.5–0.8	1.1

Source: Dobrovolsky.

Table 3. Mean concentrations of rare-earth elements (lanthanides) in surface soils (mg/kg dry weight)

Soils	Elements							
	La	Ce	Nd	Sm	Eu	Tb	Yb	Lu
Podzols	40.2	63.0	21.4	6.32	1.34	1.01	2.28	0.48
Chernozems	30.2	57.3	24.3	5.18	1.17	0.74	2.35	0.31
Forest soils	39.1	61.5	20.2	6.22	1.1	0.92	2.52	0.40
Various soils	35-36	48.1	19-36	6-7	1.4	0.85	2.5	0.4

Source: Kabata–Pendias & Pendias.

Samples of atmospheric precipitation collected in 1972–73 in the European part of the USSR contained scandium, cerium, and samarium in concentrations of 0.15, 2.6, and 0.03 µg/L, respectively; scandium and cerium were also detected, in concentrations of 1.0 and 3.4 ng/m^3, respectively, in the particulate matter of the air near the land surface [Petrukhin; Vadkovskaya & Lukashev].

Scandium

Two scandium minerals — thortveitite and sterrettite — are known, but both are extremely rare. Scandium is also found in gadolinite and euxenite.

The lithosphere contains this element at an average concentration of $1 \times 10^{-3}\%$. Its average contents in coals and oils are $1.8 \times 10^{-4}\%$ (dry weight) and $4 \times 10^{-4}\%$ (ash weight), respectively.

Table 4. Levels of rare-earth elements (lanthanides) in terrestrial plants

Element	Approx. detection rate (%)	Lichens & bryophytes (ppb DW)*	Vegetables (ppb DW)*	Woody plants (ppm AW)*	Horsetail (ppm AW)*	Miscella- neous plants (ppb DW)*
La	100	400−3000	0.4−2000	30−300	1−30	3−15000
Ce	100	600−5600	2−50	−	1−90	250−16000
Pr	90	80−620	1−2	700	0.5−6	60−300
Nd	90	240−3000	10	−	3−50	300
Sm	90	60−800	0.2−100	200−700	2−40	100−800
Eu	80	20−170	0.04−70	−	1−2	30−130
Gd	80	60−560	<2	<100−300	3−8	2−500
Tb	70	6−70	0.1−0.1	−	1−2	1−120
Dy	70	40−360	−	50−300	2−9	50−600
Ho	70	4−70	0.06−0.1	150	1−2	30−110
Er	70	10−190	0.5−2	<100−300	2−7	80−380
Tm	50	1−26	0.2−4	−	1	1
Yb	50	10−900	0.08−20	300	1−2	20−600
Lu	40	1−20	0.01−60	−	−	30

* DW = dry weight; AW = ash weight.
Source: Kabata-Pendias & Pendias.

The oceans contain an estimated 1.4 million tonnes of scandium, at average concentrations of 0.001 μg/L in seawater and $0.000028 \times 10^{-4}\%$ in total salts; its residence time in seawater is 4×10^4 years. The amount of scandium taken up annually by ferromanganese oxide deposits in the Pacific is estimated at 600 tonnes. In rivers, its average concentration is 0.04 μg/L in water and $0.3 \times 10^{-4}\%$ in total salts; its global run-off from rivers amounts to 1.5 thousand tonnes per year ($K_w = 0.03$) [Dobrovolsky; Henderson].

Table 5. Concentrations of some rare-earth elements in air over regions or countries (ng/m^3)

							Region or country		
	South	Green-	Shetland	Norway	Germany			America	
Element	Pole	land	Islands		(West)	Japan	North	Central	South
Sc	0.06-0.21	30-40	15	5	30-700	5-1300	80-300	150-220	60-3000
La	0.2-1.4	50-110	200	30	610-3420	53-3000	490-9100	440	290-3400
Ce	0.8-4.9	—	100	60	360-14000	100-18000	20-13000	—	—
Sm	0.03-0.09	10-12	10	3	240-420	9.8-320	70-1000	30-80	30-630
Eu	0.004-0.02	—	4	—	5-80	7.3-27	10-1700	—	—
Tb	—	1-5	—	—	10	—	19-34	8-27	20-120

Source: Kabata-Pendias & Pendias.

Soils contain scandium at an average concentration of $7 \times 10^{-4}\%$. Its mean contents in surface soils ranged from 1.5 mg/kg dry weight (podzols in Poland) to 16.6 mg/kg dry weight (various soils in Bulgaria). In the USA, surface soils had mean values (mg/kg dry weight) ranging from 5 (sandy soils and lithosols on sandstones) to 16 (soils over volcanic rocks). For plants, the following scandium levels were reported (mg/kg dry weight): 0.002–0.1 in some edible plants growing in tropical forests, 0.005 in vegetables, 0.07 in grasses, 0.014–0.26 in flax plants, 0.012–0.07 in lettuce leaves, 0.3–0.7 in lichens and mosses, and up to 0.3 in fungi [Kabata-Pendias & Pendias]. See also Tables 2 and 5.

Yttrium

This element is incorporated (in the oxidation state +3) into several minerals, among which silicate, phosphate, and oxide forms predominate. Its average content in the lithosphere is $2.9 \times 10^{-3}\%$. Coals have average dry weight and ash weight concentrations of $4 \times 10^{-4}\%$ and $59 \times 10^{-4}\%$, respectively.

The oceans contain an estimated 17.8 million tonnes of yttrium; at average concentrations of 0.013 µg/L in seawater and $0.00037 \times 10^{-4}\%$ in total salts. The amount of yttrium taken up annually by ferromanganese oxide deposits in the Pacific is estimated at 9800 tonnes. In rivers, its average concentration is 0.7 µg/L in water and $5.8 \times 10^{-4}\%$ in total salts; its global run-off from rivers amounts to 26 thousand tonnes per year ($K_w = 0.16$) [Dobrovolsky].

In soils, yttrium concentrations average $5 \times 10^{-3}\%$. Its contents in soils of the USA ranged from <10 to 150 mg/kg dry weight; the highest values were found in forest soils, over granites and gneisses, and in clay, clay loamy, and light desert soils. Its average concentrations in uncultivated and cultivated soils of the

USA were estimated at 23 and 15 mg/kg dry weight, respectively [Kabata-Pendias & Pendias].

In edible plants, yttrium occurred at concentrations of 20 to 100 mg/kg ash weight, the highest values being recorded for cabbage. Food plants from a tropical forest region contained from 0.01 to 3.5 mg/kg dry weight, woody seed plants up to 700 mg/kg ash weight, and mosses and lichens between 2 and 200 mg/kg dry weight [Kabata-Pendias & Pendias]. Continental vegetation contains an estimated 1.9 million tonnes of yttrium, at average fresh weight, dry weight, and ash weight concentrations of $0.3 \times 10^{-4}\%$, $0.8 \times 10^{-4}\%$, and $15.0 \times 10^{-4}\%$, respectively; the amount of yttrium taken up annually worldwide through the increment in phytomass is estimated to be 138 thousand tonnes, or 0.92 kg/km^2 of land area ($K_b = 0.44$) [Dobrovolsky]. See also Table 2.

Distribution of yttrium in the environment and living organisms has also been discussed by Deuber & Heim and Bulman.

Lanthanum
The average content of lanthanum in the lithosphere is $2.9 \times 10^{-3}\%$. Its dry weight and ash weight concentrations in coals are $8 \times 10^{-4}\%$ and $27 \times 10^{-4}\%$, respectively.

In seawater, the average concentration of lanthanum is 3×10^{-6} mg/L and its residence time is 6×10^2 years [Henderson].

Soils have an average lanthanum content of $4 \times 10^{-3}\%$. In the USA, its mean concentrations in soils ranged from 30 mg/kg dry weight (silty prairie soils) to 55 mg/kg (soils over volcanic rocks, granites, and gneisses, and light desert soils) [Kabata-Pendias & Pendias]. Continental vegetation contains an estimated 1.9 million tonnes of lanthanum, at average fresh weight, dry weight, and ash weight concentrations of $0.3 \times 10^{-4}\%$, $0.8 \times 10^{-4}\%$, and $15 \times 10^{-4}\%$, respectively; the amount of lanthanum taken up annually worldwide through the increment in phytomass is estimated to be 138 thousand tonnes, or 0.92 kg/km^2 of land area ($K_b = 0.33$) [Dobrovolsky]. See also Tables 2–5.

PRODUCTION

Rare-earth metals are obtained from their ores. In a commonly used procedure, the ore is concentrated and the concentrate, in the form of fused rare-earth chlorides, is lixiviated, freed from impurities, treated with acids, filtered, calcined, and then dissolved in hydrochloric acid, after which the various compounds are separated out from the chloride mixture by ion-exchange or extraction processes. Individual metals in the free state are finally obtained by electrosmelting or metallothermic reduction.

Oxides of rare earth metals are obtained by calcining their oxygen-containing compounds (nitrates, oxalates, sulfates, hydroxides) or by reacting the metals with air or oxygen at 180–200°C. Cerium(IV) oxide readily forms upon further oxidation of cerium(III) oxide.

Rare-earth halides and sulfides are produced, respectively, by reacting the metals with the corresponding halogens and by heating the metals with sulfur vapors.

Cerium(III) fluoride is made by reacting cerium(III) oxide, hydroxide, or carbonate or cerium(IV) oxide with hydrofluoric acid. Scandium nitrate can be obtained by reacting anhydrous scandium chloride with nitrogen(V) oxide. Annual world production of lanthanide oxides is in the order of 25 000 tonnes [Bulman].

USES

Rare-earth elements are used in metallurgical industries as deoxidizers, degasifiers, and desulfurizers. The incorporation of misch metal in small amounts (fractions of a percent) into steels of various grades makes them more heat- and corrosion-resistant. Rare earths are added to cast irons to improve their mechanical properties. They are also added to iron-chromium-nickel alloys used in heating elements for industrial electric furnaces. Rare earths are used in the neutron-absorbing control rods of nuclear reactors (gadolinium, samarium, and europium have anomalously large cross sections for neutron capture). They are also used for polishing glasses (e.g. in the form of Polirit—an abrasive made of Ce, La, Nd, and Pr oxides), in the silicate industry for staining and decolorizing glasses, in the production of special glasses (e.g., high-strength and optical glasses, those resistant to heat, chemical agents, or X-rays, and those with high electric conductivity), and for coloring porcelain and ceramic ware. They also find application in electronics, radio engineering, illumination engineering, in the textile and leather industries, in X-ray technology, in the pharmaceutical industry, in computers and in laser systems.

Scandium alloys of low densities and high melting points are used as structural materials in the rocket-building and aircraft industries. Cerium alloys with iron, magnesium, and aluminum have low values of the expansion factor and are utilized in the manufacture of components for piston engines. Scandium compounds are used as constituents of luminophors, as catalysts in the chemical industry, in the chemical technology of nuclear fuels, in the petroleum-refining industry for catalytic cracking of petroleum, in the manufacture of synthetic fibers and plastics, in the synthesis of liquid hydrocarbons, and in nonferrous metallurgy.

Uses of lanthanide compounds are discussed by Goering *et al.* and those of yttrium compounds by Deuber & Heim.

MAN-MADE SOURCES OF EMISSION INTO THE ENVIRONMENT

Rare earths are released into the air of industrial premises during various stages of their production, processing, and use. For example, analyses of over 350 air samples collected at various sites in a plant where rare earths were extracted from their minerals and processed, showed the concentrations of these elements to be between 20 and 90 mg/m^3 in most samples and to exceed 100 mg/m^3 in those from certain sites; 50–70% of the airborne dust was found to consist of particles up to 2 μm in diameter. The following sources of air pollution were identified: processes in which rare-earth concentrates were charged into or discharged from reactors, filters and furnaces; evaporators, extractors, electrolyzers, and other types of inadequately enclosed or unenclosed equipment; untight connections of pipelines to containers; and in-plant transport, packaging, and storage of finished products. The workers most exposed to the fumes, gases, and aerosols of rare earths were apparatus operators [Spassky]. This author also found that in a factory where Polirit (an abrasive made of Ce, La, Nd, and Pr oxides) was used, its concentrations in the air were at times of the order of tens of milligrams per m^3, but noted that if Polirit was loaded into bunkers from bags immediately after opening them, its levels in the air were only 0.5–1.2 mg/m^3 15 to 20 min after the loading operation. Oxides of Ce, La, Pr, and Nd were also present at low levels (0.18–0.24 mg/m^3) in the aerosol that occurred in the air of the working zone during the calcining of a catalyst used in the cracking or hydrocracking of petroleum products; 0.2–0.4% solutions of these elements did not cause appreciable damage to the skin of workers [Spassky & Lashnev].

Tarasenko *et al.* found cerium(IV) oxide concentrations to range from 20 mg/m^3 upwards in the work areas where this compound was produced or used.

Rare earths appeared in small amounts (up to 0.2 mg/m^3) in the airborne particulate matter of work premises where they were used for the modification of cast iron. Yttrium in concentrations close to 80 mg/m^3 occurred in workroom air as a result of processes by which this metal was recovered from drilling fluids [Zamchalov *et al.*].

A source of rare earths in the general atmosphere may be the combustion of various hydrocarbon fuels in industrial plants. Different types or fractions of coal dust, for example, contained (in μg/g) 1.1–6.3 of scandium, 20.0–43.0 of cerium, 0.2–0.4 of europium, 0–3.0 of yttrium, and 0.9–2.1 of lutetium [Manchuk & Ryabov].

Data on some agricultural sources of soil contamination by rare earths are given in Table 6.

Table 6. Content of cerium and scandium in some materials added to agricultural soils (mg/kg dry weight)

Source	Cerium	Scandium
Sewage sludges	20	0.5–7.0
Phosphate fertilizers	20	7–36
Organic fertilizers	–	5

Source: Kabata-Pendias & Pendias.

TOXICITY

Microorganisms
Lanthanides (La, Pr, Nd, Eu, Tb) were found to inhibit, specifically and competitively, the accumulation of calcium in the mitochondria of microbial cells [Kabata-Pendias & Pendias].

Animals and man

Single and short-term exposures

Animals
LD_{50} values of rare earths for laboratory animals are shown in Table 7.
Compounds of the yttrium subgroup are more toxic than those of the cerium subgroup, and the oxides of rare earths are generally more toxic than their salts [Deuber & Heim]. Animal studies indicate that after oral or intraperitoneal administration, the toxicity of compounds that have been tested is moderate to low. With 4-hour exposure by inhalation, the threshold concentrations of **cerium(IV) oxide** and **yttrium oxide** for acute effects were 95 mg/m^3 and 92 mg/m^3, respectively, in rats and 32±10 mg/m^3 in mice; these compounds were therefore classified as highly dangerous on short-term inhalation exposure. Rats administered 50 mg of yttrium oxide intratracheally were found, 9 months later, to have developed emphysema and chronic interstitial pneumonia; their lungs had increased relative weights and contained elevated hydroxyproline levels [Spassky].

Mice and guinea-pigs given high doses of rare earths intraperitoneally developed diffuse peritonitis. Lethal doses of **yttrium chloride** for these animals ranged from 300 to 500 mg/kg, and two stages in the clinical manifestations of poisoning were discerned. The first stage was characterized by dyspnea and pulmonary edema and the second by edema of the liver, congestion in the portal vein, subpleural hemorrhages, and hyperemia of the lungs. Progression of local inflammatory lesions led to peritonitis with the formation of serous or hemorrhagic exudates [Andreyeva et al.].

Table 7. Median lethal doses of compounds of rare-earth elements

Compound	Animal	Route of administration	LD_{50} (mg/kg)
Cerium(III) chloride	Mouse	Subcutaneous	4 000[*]
	Rabbit	Intravenous	<200[*]
Cerium(IV) oxide	Rat	Oral	10 700−300
	Mouse	Intraperitoneal	465−60
Cerium sulfide	Mouse	Oral	8 600−1 350
		Intraperitoneal	2 340−960
Cerium subgroup:			
mixture of nitrates	Rat	Oral	1 125−110
		Intraperitoneal	590−75
misture of oxides	Mouse	Oral	14 400−2 800
		Intraperitoneal	3 080−400
Dieuropium and diyttrium			
dioxide−sulfide	Mouse	Intraperitoneal	1 200−380
Dysprosium chloride	Mouse	Intraperitoneal	342.8−26.4
Dysprosium sulfide	Mouse	Intraperitoneal	1 620−520
Gadolinium chloride	Mouse	Intraperitoneal	378.8−29.9
Lanthanum chloride	Mouse	Intraperitoneal	185−40
		Subcutaneous	4000
	Rat	Oral	2 370[*]
	Rabbit	Intravenous	<200
Lanthanum nitrate hexahydrate	Mouse	Oral	850[*]
	Rat	Oral	1 400[*]
Lanthanum oxide	Mouse	Oral	2 450[*]
		Intraperitoneal	530±[*]65
	Rat	Oral	>8 500[*]
Lanthanum sulfate	Mouse	Oral	1 450
	Rat	Oral	2 450
Lutetium chloride	Mouse	Oral	7 100
		Intraperitoneal	315
Neodymium chloride	Mouse	Subcutaneous	4 000
	Rabbit	Intravenous	<200
Polirit (abrasive):			
carbonate form	Mouse	Oral	21 050±3 450
		Intraperitoneal	15 550±3 500
oxide form	Mouse	Intraperitoneal	33 350±12 800
Praseodymium chloride	Mouse	Subcutaneous	2±500
Yttrium oxide	Mouse	Oral	9 480±330
		Intraperitoneal	425±65
	Rat	Oral	15 340±910
		Intraperitoneal	230±45

Table 7. (cont'd)

Compound	Animal	Route of administration	LD_{50} (mg/kg)
Yttrium sulfide	Mouse	Oral	9 100±270
		Intraperitoneal	1 410±630
Yttrium subgroup: mixture of oxides	Mouse	Oral	11 400±5 950
		Intraperitoneal	2 420±650

* Calculated as metal ion.

Sources: Andreyeva *et al.*; Grushko; Haley *et al.*; Spassky.

In cats, a single intravenous injection of **lanthanum sulfate** in the 20–40 mg/kg dose range resulted in an increased volume rate of coronary blood flow and a slight reduction in arterial blood pressure; the dose of 50 mg/kg was lethal. In rabbits, a single intravenous injection of this compound in the indicated dose range prevented the cardiovascular effects of Pituitrin, such as prolongation of the cardiac cycle, severe bradycardia, and electrocardiographic changes. This action of lanthanum sulfate was ascribed to its ability to increase the volume rate of coronary blood flow [Abdullayev & Guseinov].

The **hexaborides** of **lanthanum** and **cerium** reduced about tenfold the blood serum titer of lysozyme in rats, while their **sulfides** reduced it three- to fourfold. These compounds also markedly decreased complement titers which began to rise after a month; complete restitution occurred after 3 months in animals given the sulfides and after 6 months in those given the hexaborides. Also in rats, the **oxides** of **lanthanum** and **cerium** reduced the blood serum titer of α-and β-lysins; the β-lysin titer began to rise slowly after 15 days and returned to the baseline level within 6 months, whereas that of β-lysin failed to return over that period.

Yttrium oxide, **cerium oxide**, **cerium chloride**, **lanthanum chloride**, **Polirit**, or a combination of several elements from the cerium or yttrium subgroup administered to rats for 2 h daily by inhalation at concentrations of 100, 200, or 300 mg/m^3 over a month, produced similar effects that included lowered rates of oxidation-reduction processes, leukopenia, altered electrocardiographic parameters, shifts in potassium and calcium metabolism, and general inhibition of enzyme activities; on histopathological examination, pleuritis and abscesses were observed in the lungs.

Acute affects of the lanthanides on animals are discussed by Goering *et al.* and those of yttrium by Deuber & Heim.

Long-term exposure

Animals
Rats exposed by inhalation to an **yttrium oxide** aerosol at 10 mg/m^3 for 6 h/day, 5 days/week over 4 months, showed erythropenia, reduced hemoglobin and color

oxide was observed. Evidence for impairment of liver function was provided by reduced acetylcholinesterase and elevated asparagine transaminase activities. Both residual nitrogen and urea nitrogen were increased in the blood. On histopathological examination, the lungs were found to be affected most, showing desquamation of the alveolar epithelium and clear sings of pneumosclerosis with the presence of granulomatous nodules. Some degenerative changes also occurred in the liver, kidneys, and spleen. In rats exposed to the yttrium oxide aerosol concentration of 4 mg/m^3 using the same schedule as above, only slightly marked changes in biochemical parameters and in lung morphology could be seen, and this concentration was taken to be the threshold level for chronic inhalation toxicity [Spassky].

Rats exposed by inhalation to a **cerium(IV) oxide** aerosol at 50 mg/m^3 in a similar experiment (6 h/day, 5 days/week for 4 months), exhibited decreased body-weight gains, increased potassium and decreased calcium levels in the blood serum, reduced asparagine transaminase and alanine transaminase activities, increased albumins and decreased globulins in the blood (the total protein level remained unchanged), and depressed aldolase activity; autopsy demonstrated chronic inflammatory changes in the trachea and large foci of desquamative interstitial pneumonia as well as signs of moderate pneumoconiosis in the lungs. In rats similarly exposed to the cerium(IV) oxide aerosol at 10 mg/m^3, pathological changes were milder, appeared later, and largely regressed during the recovery period. The concentration of 10 mg/m^3 was regarded as the threshold level for chronic inhalation toxicity to this compound [Spassky].

In long-term studies on rats, Spassky examined the fibrogenic potential of **yttrium oxide** and a combination of rare earths from the yttrium subgroup. After 3, 6, or 9 months of intratracheal exposure, small nodules of a granulomatous nature that had appeared in the lung tissue were seen to be coalescing into larger lesions, but hydroxyproline, collagen, and elastin levels in the lungs of exposed animals did not differ from those in the control group, and no signs of pneumosclerosis were present. These findings were taken as evidence that the fibrogenic potential of rare earths is low. The studies also demonstrated that rare earths can cause damage to the liver, as was indicated by reduced excretory and antitoxic functions of that organ, disturbances of protein, lipid, and carbohydrate metabolism, and histopathological changes. Chronic exposure of rats to **scandium nitrate** resulted in a sharply decreased rate of tissue respiration and in depressed transketolase activity, the major enzyme of the pentose cycle; the rate of glycolysis was not increased.

Rare-earth elements have been placed in hazard class 2 on the basis of chronic toxicity data, whereas acute toxicity data warrant placing them in hazard class 3 or 4. Elements of the yttrium subgroup are generally more toxic than those of the cerium subgroup [Spassky].

For further discussion of chronic effects of yttrium and lanthanides on animals, see Deuber & Heim and Goering *et al.*, respectively.

Man

Human exposure to rare earths usually occurs in occupational environments through inhalation of individual airborne elements or their compounds. No cases of severe occupational poisoning have been reported. Lithographic workers exposed to **lanthanum-containing dusts** complained of nausea and headache, inflammations of the upper airways (chronic pharyngitis and rhinitis) occurred in workers extracting and processing rare-earth ores or minerals, and cases of dermatitis were observed among those working with **Polirit**. Persons chronically exposed to **cerium-containing dusts** had roentgenologic changes in the lungs that resembled those seen in nodular pneumoconiosis of degree I or II and that could be attributed to the specific action of that rare earth [Andreyeva *et al.*]. Other chronic effects of the lanthanides on humans are considered by Goering *et al.*

The lung tissues of 66 individuals who died from various causes after 5 to 10 of years retirement following employment as copper smelters, for a mean of 30 years contained, on average, 11 ppb of lanthanum as compared to 5.5 ppb in the lung tissues of control subjects; lanthanum levels in liver and kidney tissues of the exposed individuals were about 5.5 ppb *versus* 0.2 ppb in the control group [Gerhardsson *et al.*].

Among workers who had been long exposed to mixtures of rare earths from the yttrium subgroup, a number of abnormalities were observed, including reduced hemoglobin levels, erythropenia and thrombopenia, pathological changes in the upper respiratory tract, dryness and scaling of the skin on the hands, skin pigmentation, and hair loss. Morbidity among workers exposed to rare earths is largely determined by the state of body's defenses. It has been shown that the greatest capacity to depress natural immunity is displayed by the oxides of rare earths, with their hexaborides and sulfides ranking second and third in this respect [Brakhnova].

Effects of local application

Animals

Instillation of 0.1 ml of a 50% **lutetium chloride** solution in the rabbit eye resulted in conjunctivitis, chemosis, and lacrimation which appeared in an hour, were at their height in 24 h, and regressed within 2 weeks. When 0.5 g of this compound was applied to undamaged rabbit skin, no local effects were seen, whereas intracutaneous application of its aqueous solution (1:10, 1:100, or 1:1000 dilution) led to necrosis within an hour, and no complete healing of the lesion occurred during the 5-week observation period. Daily application of rare-earth **chlorides** or **nitrates** to the depilated skin of rabbits and guinea-pigs for

30 days produced hyperemia in the affected areas, followed by ulceration and the formation of granulations.

Man
Skin responses to contact with rare-earth **oxides** are slight; stronger local reactions are elicited by **nitrates** and **chlorides**.

ABSORPTION, DISTRIBUTION, AND ELIMINATION

Rare-earth elements enter the human body mainly by inhalation. The lighter elements are predominantly deposited in the liver and the heavier ones in the osseous tissue [Goering *et al.*]. Rare earths retained in the lungs can cause inflammatory responses of the desquamative pneumonia type and lead to the formation of granulomatous nodules in all lobes and subsequently also in the perivascular and peribronchial areas of the hilar regions. The unretained earths rapidly pass through the alveolar membranes into the blood where their ions bind to plasma α- and β-globulins, circulate as colloidal particles or protein complexes, are taken up by cells of the reticuloendothelial system and distributed by them throughout the body, and are then deposited in the liver, bones, and kidneys.

Elements from lanthanum to samarium accumulate mainly in the liver ($\approx 50\%$) and skeleton ($\approx 25\%$) and are excreted predominantly in the bile. Compounds of europium and gadolinium were found in the liver and skeleton in amounts equal to 30% and 40% of the administered dose; they are eliminated by the kidneys and gastrointestinal tract to equal extents. Compounds of the rare earths from terbium to lutetium concentrate in the skeleton (50–60%) and to a lesser extent in the liver; they are chiefly eliminated in the urine. The lanthanides are virtually not absorbed from the gastrointestinal tract into the blood [Goering *et al.*]. After intratracheal administration of promethium, most of this element was retained in the lungs, while the rest was absorbed and found in the liver and skeleton. When cerium chloride was administered by inhalation, its highest level in the liver was recorded at a later time than after an intravenous or intraperitoneal injection. In general, rare earths are absorbed into blood more rapidly from the lungs than from the site of a subcutaneous or intramuscular injection.

Studies in which the subcellular distribution of rare earths in the liver was examined after their intravenous injection have shown that they first concentrate in the cytoplasm before penetrating into cellular organelles; the lowest levels occurred in the nuclear fraction and the highest (in the case of cerium and ytterbium), in the microsomal fraction and in mitochondria.

The distribution of rare earths and their accumulation in tissues depend, in large measure, on the blood flow rate and the volume of the interstitial space in the tissue concerned. For example, the amount of ^{91}Y per gram tissue increased with increasing volume of this space and the concentrations of this element were higher in organs with high blood flow rates such as the kidney; they were also high in osseous tissue because of its high affinity for this metal.

The elimination of rare earths from the osseous tissue lasts for a long time (2 years or more). Their toxic effects are primarily due to their location in hepatocytes and the direct damage they cause to these cells thereby impairing the major functions of the hepatic tissue.

Rare-earth elements possess pharmacological properties that make them suitable for use as drugs. Thus, they have been shown to reduce blood coagulability by exerting antithromboplastin and antiprothrombin effects, to have antiinflammatory properties, and to act as calcium antagonists [Lakin *et al.*; Spassky].

HYGIENIC STANDARDS

Exposure limits adopted in the USSR for rare-earth compounds are shown in Table 8. In the USA, a TLV-TWA of 1.0 mg/m^3 and a TLV-STEL of 3.0 mg/m^3 have been set for yttrium.

Table 8. Exposure limits for compounds of rare-earth elements

	Workroom air			
	MAC$_{wz}$ (mg/m^3)	Aggregative state	TPC[a]	Hazard class
Cerium fluoride	2.5/0.5[b]	aerosol	—	3
Cerium(IV) oxide	5.0	aerosol	—	3
Erbium oxide	—	—	4.0	
Dysprosium oxide	—	—	4.0	
Gadolinium oxide	—	—	4.0	
Holmium oxide	—	—	4.0	
Lanthanum oxide	—	—	6.0	
Lutetium oxide	—	—	4.0	
Neodymium oxide	—	—	6.0	
Praseodymium oxide	—	—	6.0	
Samarium oxide	—	—	6.0	
Scandium oxide	—	—	4.0	
Thulium oxide	—	—	4.0	
Ytterbium oxide	—	—	4.0	
Yttrium oxide	2.0	aerosol	—	3

[a] Temporarily permissible concentration in the working zone.
[b] 2.5 mg/m^3 is the concentration not to be exceeded at any time, while 0.5 mg/m^3 is the average maximum allowable concentration per shift.

METHODS OF DETERMINATION

A commonly used spectrographic method for determining the oxides of individual rare-earth elements in *air* involves evaporation of the sampled material from the channel of a graphite electrode followed by spectrographic measurement of the oxides on a differential spectrograph (detection limit 100 µg in the analytical solution volume). Scandium oxide can be determined by complexometric titration of a scandium solution with Trilon B in the presence of an indicator dye such as complex orange (detection limit 1 mg/m^3; measurable concentration range, 1–20 mg/m^3; measurement error, ±10%) [*Guidelines for the Determination of Noxious Substances in Air*, 1980]. Yttrium oxide can be determinated by a flame photometric method (sensivitivity, 1 µg per ml of the analytical volume [*Guidelines for the Determination of Noxious Substances in Air*, 1979]. A photometric method for cerium(IV) oxide relies on the formation of a cerium(IV) complex with sodium citrate, resulting in yellow coloration of the solution (detection limit, 0.5 mg/m^3; measurable concentration range, 0.5–10 mg/m^3. [*Guidelines for the Determination of Noxious Substances in Air*, 1980].

In *biological materials*, the total rare earths of the yttrium subgroup can be determined by a photometric method proposed by Maltseva and Pavlovskaya (detection limit 50 µg; measurement error, 13.7%). For the determination of yttrium and ytterbium in *soils* and of lanthanum in *plant materials*, emission spectral analyses have been described [Bulman; Loseva *et al.*].

MEASURES TO CONTROL EXPOSURE AT THE WORKPLACE

Technical measures to be implemented in factories where rare-earth ores or powdery materials are processed should be primarily directed at preventing exposure to dusts generated in such operations as grinding, sifting, blending, and loading and unloading. Effective dust control can be achieved by mechanizing or automating the work processes and ensuring their continuity, providing airtight enclosures for the equipment, and installing efficient local ventilation at the work posts. Special protective measures need to be taken against possible radiation exposure during ore processing.

Adequate ventilation is required for sources of rare-earth degradation products, in particular fluorine-containing compounds when the fluorides of cerium, yttrium, and other rare earths are processed; direct contact of the workers with such substances should be avoided or kept to a minimum.

Preplacement and subsequent regular medical examinations are necessary for those working with rare earths to prevent occupational diseases.

Good personal hygiene is important and protective clothing should be provided, especially for workers in dusty operations. Where airborne dust is present, respirators must be worn. For hand and arm protection, barrier creams should be used.

REFERENCES

Abdullayev, R. A. & Guseinov, D. Ya. (1977) *Farmakol. Toksikol.*, No. **2**, 173–174 (in Russian).

Andreyeva, O. S. *et al.* (1975) *Redkozemelnye elementy. Radiatsionno-gigienicheskiye aspekty* [Rare Earth Elements. Aspects of Radiation Hygiene]. Atomizdat, Moscow (in Russian).

Brakhnova, I. T. (1971) *Toksichnost poroshkov metallov i ikh soyedineniy* [Toxicity of Metals and their Compounds in Powder Form]. Naukova Dumka, Kiev (in Russian).

Bulman, R. A. (1988) *Yttrium and the Lanthanides* in: Seiler, H. G. *et al.* (eds) *Handbook on Toxicity of Inorganic Compounds.* Marcel Dekker, New York, pp. 769–785.

Deuber, R. & Heim, T. (1991) Chapter II. 35 *Yttrium.* in: Merian, E. (ed.) *Metals and their Compounds in the Environment — Occurrence, Analysis and Biological Relevance.* VCH, New York, pp. 1299–1308.

Dobrovolsky, V. V. (1983) *Geografiya mikroelementov. Globalnoe rasseyanie* [Geography of Trace Elements. Global Dispersion]. Mysl, Moscow (in Russian).

Gerhardsson, L. *et al.* (1984) *Sci. Total Environ.*, **37**, 233–246.

Goering, P. L. *et al.* (1991) Chapter II. 15 *The Lanthanides* in: Merian, E. (ed.) *Metals and their Compounds in the Environment—Occurrence, Analysis and Biological Relevance.* VCH, New York, pp. 959–970.

Grushko, Ya. M. (1979) *Vrednye neorganicheskiye soyedineniya v promyshlennykh stochnykh vodakh* [Harmful Inorganic Compounds in Industrial Waste Waters]. Khimiya, Leningrad (in Russian).

Guidelines for the Determination of Noxious Substances in Air (1979) [Metodicheskiye ukazaniya po metodam opredeleniya vrednykh veshchestv v vozdukhe], No. 15. Publication of the USSR Ministry of Health, Moscow (in Russian).

Guidelines for the Determination of Noxious Substances in Air (1980) [Metodicheskiye ukazaniya po metodam opredeleniya vrednykh veshchestv v vozdukhe], No. 16. Publication of the USSR Ministry of Health, Moscow (in Russian).

Haley, T. J. *et al.* (1964) *Pharmaceut. Sci.*, **53**, 1186–1188.

Henderson, P. (1982) *Inorganic Geochemistry.* Pergamon Press, Oxford.

Kabata-Pendias, A. & Pendias, H. (1986) *Trace Elements in Soils and Plants.* CRC Press, Boca Raton (Fl.).

Lakin, K. M. *et al.* (1982) *Farmakol. Toksikol.*, No. 6, 88–100 (in Russian).

Loseva, A. F. *et al.* (1984) *Spektralnye metody opredeleniya mikroelementov v obyektakh biosfery* [Spectral Methods of Determining Trace Elements in the Biosphere]. Rostov-on-Don (in Russian).

Maltseva, M. M. & Pavlovskaya, N. A. (1976) *Gig. San.*, No. **5**, 87–90 (in Russian).

Manchuk, V. A. & Ryabov, N. A. (1983). In: *Metally. Gigiyenicheskiye aspekty otsenki i ozdorovleniya okruzhayushchei sredy* [Metals. Hygienic Aspects of Evaluation and Environmental Health Promotion] Moscow, pp. 194–199 (in Russian).

Petrukhin, V. A. (1982). In: *Monitoring fonovogo zagryazneniya prirodnykh sred* [Monitoring the Background Pollution of Environmental Media], No. 1. Gidrometizdat, Leningrad, pp. 147–165 (in Russian).

Spassky, S. S. (1978) *Gig. Truda Prof. Zabol.*, No. 7, 55–56 (in Russian).

Spassky, S. S. & Lashnev, M. P. (1978). In: *Aktualnye problemy gigiyeny truda* [Current Problems of Occupational Hygiene]. Moscow, pp. 88–91 (in Russian).

Tarasenko, N. Yu. *et al.* (1974) *Gig. Truda Prof. Zabol.*, No. **10**, 29–32 (in Russian).

Vadkovskaya, I. K. & Lukashev, K. P. (1977) *Geokhimicheskiye osnovy okhrany biosfery* [Protection of the Biosphere: Geochemical Principles]. Nauka i Tekhnika, Minsk (in Russian).

Venchikov, A. I. (1982) *Printsipy lechebnogo primeneniya mikroelementov v kachestve biotikov* [Principles for the Therapeutic Use of Trace Elements as Biotics]. Ylym, Ashkhabad (in Russian).

Zamchalov, A. I. *et al.* (1976) *Gig. Truda Prof. Zabol.*, No. **12**, 17–19 (in Russian).

Thorium and its compounds

Thorium chloride; thorium fluoride; thorium nitrate; thorium oxide (thorium dioxide, thoria, thorotrast, thorianite [min.])

IDENTITY AND PHYSICOCHEMICAL PROPERTIES OF THE ELEMENT

Thorium (Th) is a radioactive metallic element in the seventh period of the periodic table and the first element in the actinide family, with the atomic number 90. Its naturally occurring radioactive isotopes are ^{232}Th ($\approx 100\%$; half-life $= 1.39 \times 10^{10}$ years), ^{228}Th ($1.37 \times 10^{-8}\%$; 1.91 years), ^{230}Th (8.0×10^4 years), ^{234}Th (24.1 days), ^{227}Th (18.17 days), and ^{231}Th (25.64 hours). There are nine artificial radioactive thorium isotopes. The longest-lived daughter product of thorium is radium-228, with a half-life of 6.7 years; other daughter products (actinium-228, radium-224, radon-220, polonium-216, lead-212, bismuth-212, thulium-208) are all short-lived, with half-lives of several hours to seconds.

Thorium exists in two allotropic modifications. It is a malleable metal that is readily deformable in the cold. Its mechanical properties heavily depend on its purity.

Chemical properties of thorium are similar to those of rare-earth elements (lanthanides) and elements of the second subgroup in group IV of the periodic table (titanium, zirconium, and hafnium). In compounds it exhibits the oxidation state +4 almost exclusively. In air it oxidizes only slightly at $\leqslant 20°C$. In distilled water it forms a surface layer of the oxide ThO_2 that prevents it from further corrosion. The oxide is also produced when thorium burns in air. Thorium reacts with water vapor at 200–600°C to give ThO_2 and H_2 and with fluorine even at 20°C to give the fluoride ThF_4. It reacts vigorously with chlorine, bromine, and iodine at 450°C. The corresponding halide also forms when thorium is dissolved in heated hydrochloric, hydrobromic, or hydroiodic acid. It does not react with hydrofluoric acid because a strong film of thorium fluoride is formed on its surface. On heating thorium with sulfur or hydrogen sulfide, the sulfides ThS, Th_2S_3, Th_4S_7, and ThS_2 are produced. It reacts with H_2 at 400–600°C to give the hydride ThH_2, with nitrogen at 800°C to give the nitride ThN, and with phosphorus on heating to give phosphides of the compositions $ThP_{0.75}$ and Th_3P_4. When heated in an induction or arc furnace, it forms the carbides ThC and ThC_2 with carbon and, at 1000–1700°C, the silicides $ThSi_2$, Th_3Si_2, and $ThSi$ with silicon. In reacting with oxygen-containing acids, it yields salts, many of which are soluble in water. It alloys with many metals and forms intermetallic compounds with Cr, Mn, Fe, Co, Ni, and Cu. See also Appendix.

NATURAL OCCURRENCE AND ENVIRONMENTAL LEVELS

The clarke of thorium is estimated at $(9.6-13) \times 10^{-4}\%$ in the earth's crust and at $14 \times 10^{-4}\%$ in the granite layer of the continental crust [Dobrovolsky]. Thorium oxide is a constituent of many minerals, and those of industrial importance are thorite, thorianite, and particularly monazite which may contain 2.5 to 28% of thorium oxide mixed with rare-earth elements, zirconium, titanium, cadmium, niobium, and uranium. The isotopes ^{232}Th and ^{228}Th occur in thorianite (ThO_2) and thorite ($ThSiO_4$); ^{234}Th and ^{230}Th are present in naturally occurring uranium, while ^{231}Th and ^{227}Th occur in uranium minerals as members of the uranium-235 disintegration chain. The remaining isotopes are formed upon neutron bombardment of those just mentioned.

The content of thorium in various rocks (sandstones, limestones, shales, etc.) ranges from 1.1×10^{-4} to 4.4×10^{-4} Bq/g [Marei et al.]. Its concentrations in seas and oceans are between 0.2×10^{-8} and 50×10^{-8} mg/L [Shvedov & Patin]. The bulk of thorium in sea and ocean waters is believed to be derived from continental and volcanic sources. The mean contents of thorium in soils of different countries range from 3.4 to 10.5 mg/kg dry weight [Kabata-Pendias & Pendias]. In soils of the USSR, concentrations of ^{228}Th ranged from 0.12 to 0.87 pCi/g and from 0.28 to 2.6 pCi/g depending on the total concentration of naturally occurring radionuclides in the soil-forming rocks [Nikolayev et al.].

The wood, leaves, and coniferous needles of trees contained ^{232}Th at levels up to 5 µg/g $(2 \times 10^{-2}$ Bq/g) and the bodies of birds, up to 2.4×10^{-7} g/g $(1 \times 10^{-3}$ Bq/g). Lichens growing in northern regions were found to accumulate ^{228}Th to 0.23×10^{-9} Ci/kg (1 Ci = 3.7×10^{10} Bq) [Litver et al.].

PRODUCTION

Thorium is produced by thermal reduction of its oxide or halides with metals, by electolysis of its melted salts, or by thermal decomposition of the halides.

Thorium chloride is obtained by chlorination of thorium oxide in admixture with coke or coal. The **fluoride** is made by reacting fluorine or hydrogen fluoride with the oxide, hydroxide, carbide, or hydride of thorium. The oxide can be obtained by burning metallic thorium in air or by calcining its oxygen-containing compounds such as the nitrate, oxalate, or hydroxide.

USES

The most important compounds of thorium are the hydrides (ThH_2, ThH_3), oxide [dioxide] (ThO_2), fluoride (ThF_4), sulfate ($Th(SO_4)_2$), nitrate ($Th(NO_3)_4$), and oxalate ($Th(C_2O_4)_2$). **Thorium** is mainly used in the production of refractory materials, in alloys utilized for components of jet engines, in electronics, and as a catalyst in the chemical industry. ^{232}Th is used as a nuclear fuel in nuclear reactors. The **fluoride** is used to obtain metallic thorium by electrolysis. The **oxide** finds application in electrodes for arc welding, in the manufacture of ceramics, and as a catalyst in the production of liquid fuel.

MAN-MADE SOURCES OF EMISSION
INTO THE ENVIRONMENT

Considerable amounts of thorium appear to be introduced into the biosphere from fossil fuel-burning power plants and also from phosphate fertilizer factories [Kabata-Pendias & Pendias]. Pollution of soils from the application of certain fertilizers has been reported [Szabo]. Elevated thorium levels are recorded in areas where this element or uranium are obtained or processed [Métivier].

TOXICITY

The biological effects of thorium are due both to the ionizing radiation from this element and to its chemical toxicity. The latter is generally low although some thorium salts have strong irritating properties.

Acute/single exposure

Animals
The acute toxicity of thorium salts by all routes examined is fairly low [Burkart; Métivier]. The intraperitoneal LD_{50} of **thorium nitrate** was calculated at 1220 mg/kg for mature female rats and at >2000 mg/kg for weanling rats

[Stokinger]. Oral $LD_{50}s$ of the nitrate for rats ranged from 1.8 to 3.4 g/kg, the causes of death being intestinal dystonia and peritonitis [Zhuravlev]. Twelve months after an intratracheal administration of ^{228}Th chloride to rats at 64, 6.4, or 1.0 mCi (1 Ci = 3.7×10^{10} Bq), the doses absorbed from their lungs were 210, 23, and 3 rads (1 rad = 1.0×10^{-2} Gy), respectively [Zhuravlev]. Short-term inhalation exposure of rats to **thorium oxide** at 5 mg/m^3 produces no detectable effects.

The minimal blastomogenic dose of thorium in rats was 1 nCi per animal [Burykina et al.]. In small laboratory animals, osteosarcomas developed after 9 to 12 months whatever the thorium dose used; in rabbits and dogs, the latent period lasted 21 to 36 months [Zhuravlev].

Man

The colloidal preparation thorotrast (**thorium oxide**) was formerly used as a radiopaque agent in medicine. Its injection in a dose of 2.0 to 15.0 g caused rises in body temperature, nausea, and injury to tissues at the injection site, followed by anemia, leukopenia, and impairment of the reticuloendothelial system. After intravenous administration, thorotrast particles are taken up by reticuloendothelial cells of the liver and spleen. Thorotrast is virtually not eliminated from the body. The rate of occurrence of osteosarcoma in those who had received this preparation was dose dependent [Pavlovskaya & Zeltser]. Between 1947 and 1961, 33 cases of cancer of the liver, larynx, and bronchi and sarcoma of the kidneys developing 6 to 24 years after thorotrast administration were described in the literature [Tarasenko, 1983].

Repeated and chronic exposures

Animals

Short-term inhalation exposures (for 2–10 weeks) of dogs to the **fluoride, nitrate,** or **oxide** of thorium at concentrations of 11, 76, and 51 mg/m^3, respectively (mean particle diameters ~1 μm), showed abnormal leukocytes as the only evidence of toxicity [Stokinger].

Rats exposed to **thorium** by inhalation at concentrations of 10 to 80 mg/m^3 for 60 to 270 days developed changes in the blood and bone marrow as well as mild lesions in the pulmonary parenchyma; none of these effects were observed in rats exposed to 5 mg/m^3 [Zhuravlev].

Man

Radiation hazards from **thorium** are much more serious than the chemical dangers. Personnel working with thorium and its disintegration products may be exposed both to internal radiation as a result of their penetration into the body in the form of aerosols through the respiratory or digestive system (sometimes also through the skin) and to external radiation by β- or γ-rays. Thorium and its disintegration products may cause changes in the hematopoietic, nervous, and reticuloendothelial systems and functional and morphologic damage to lung and bone tissues. On prolonged exposure, symptoms resembling those seen in chronic radiation sickness appear, and lung cancer and bone tumors may develop [Tarasenko, 1963, 1983].

Occupational exposure to **thorium fluoride** may result in dermatitis, lesions in the upper airways, hyposmia, perforation of the nasal septum, hepatitis, disordered hematopoiesis (reticulocytosis, leukopenia), fluorosis, and osteosclerosis.

ABSORPTION, DISTRIBUTION, AND ELIMINATION

Thorium and its disintegration products can enter the body by inhalation, ingestion, and by absorption through undamaged skin. In man, its daily intake in food and water is, on average, about 3 µg. Thorium is retained longest when it has entered the body in the form of insoluble compounds. These are mainly retained in the bones, lungs, lymph nodes, and parenchymatous tissues. Following intratracheal administration, insoluble compounds were deposited in the lungs during the first month and in the bones subsequently [Albert; Pavlovskaya & Zeltser; Tarasenko, 1963, 1983; Zhuravlev].

In animal experiments, not more then $10^{-4}\%$ of thorium was absorbed from the gastrointestinal tract; after an intravenous injection of thorium nitrate in a dose of 2×10^{-9} mg/kg, 60–93% of the dose was detected in the skeleton and 3–7% in the liver [Albert].

The distribution and deposition patterns of thorium disintegration products are similar to those of the elements whose isotopes they are.

After intramuscular injection, thorium oxide is retained for a long time at the injection site. Circulating soluble thorium compounds accumulate mainly in osseous tissue, insoluble compounds tend to concentrate in tissues of the lungs and gastrointestinal tract, while colloidal solutions are concentrated by cells of the reticuloendothelial system and accumulate in the liver, hematopoietic organs, and kidneys.

The elimination of thorium is mainly by way of the feces.

HYGIENIC STANDARDS

In the USSR, an occupational exposure limit (MAC_{wz}) for thorium has been set at 0.05 mg/m^3 (as aerosol, hazard class 1). See also 'Limits for intake of radionuclides by workers' *(Annals of the ICRP* [Oxford]. International Commission on Radiological Protection Publication 30. Supplement to part 1, 1979, 3/1–4) and 'Normativy radiatsionnoi bezopasnosti' [Radiation Safety Standards] (Moscow, 1979) [in Russian]).

METHODS OF DETERMINATION

The content of thorium in biologic samples from the body and in environmental samples is estimated by measuring the radioactivity of these samples, with identification of the radiation energy. A method for determining ^{228}Th in osseous tissue is available, involving extraction of the iron ions that interfere with the assay, followed by extraction of the thorium with dibutyl phosphate and determination of the ^{228}Th concentration by recording gamma radiation with a gamma spectrometer. Neutron activation analysis has also been utilized for thorium determination. The method particularly suitable for routine analysis is the determination of ^{232}Th in bone samples by colorimetry of the complex formed by thorium with arzenaso III [Pavlovskaya]; a modification of this method described by Ponikarova et al. is also well suited for such analysis. See also Burkart and Métivier.

MEASURES TO CONTROL EXPOSURE AT THE WORKPLACE

Technical control measures should be mainly directed at preventing the contamination of the ambient air and the clothing and skin of workers. Where thorium and its compounds are produced or used on a large scale, all operations should be mechanized and the equipment enclosed. Efforts should be made to automate work processes as much as possible, with provision of remote control.

Monitoring the levels of thorium and its disintegration products in the air of work premises is essential, as is γ-radiation monitoring. The extent to which the skin and clothing of workers are contaminated with radioactive substances should also be monitored. γ-Radiation levels and the concentrations of thorium aerosols, thoron gas, and disintegration products must not exceed the exposure limits set for them.

Efficient general ventilation and local exhaust ventilation should be installed, especially in premises where thorium or waste products are stored, to avoid the accumulation of thoron and its daughter products. The contaminated and clean areas should be clearly separated from each other [Tarasenko, 1983].

Those working with thorium and its compounds should undergo preplacement and subsequent periodic medical examinations.

As regards *personal protection* equipment, special attention should be paid to respiratory and skin protection. Rules of personal hygiene must be strictly observed, including regular hand washing and showering. (More detailed descriptions of safety and health measures are contained in *Thorium and its Compounds* by Evans and in *Occupational Hygiene for Thorium Workers* by Tarasenko.)

FIRST AID IN POISONING

The hands or face contaminated with thorium or its compounds should be washed with water and soap, and water should also be used to wash the nasopharynx and oral cavity contaminated with thorium-containing particulate matter. In the event of inhalation exposure, 50 g of an antidote against heavy metals (*antidotum metallorum*) or activated charcoal should be given orally; emetics, gastric lavage with water, saline cathartics, and cleansing enemas are indicated, as are diuretics and slow intravenous injections of pentacin [synonym: calcium trisodium pentetate] (10 ml of a 5% solution at a time).

REFERENCES

Albert, R. E. (1971) *Thorium. Its Industrial Hygiene Aspects.* Academic Press, New York.

Burkart, W. (1988) Chapter 73 *Radiotoxicity* in: Seiler, H. G. *et al.* (eds) *Handbook on Toxicity of Inorganic Compounds.* Marcel Dekker, New York, pp. 805–827.

Burkart, W. (1991) Chapter II. 33 *Uranium, Thorium and Decay Products* in: Merian, E. (ed.) *Metals and their Compounds in the Environment — Occurrence, Analysis and Biological Relevance.* VCH, New York, pp. 1275–1287.

Burykina, L. N. *et al.* (1983) In: *Biologicheskie effekty malykh doz radiatsii* [Biological Effects of Low Radiation Doses]. Meditsina, Moscow, pp. 87–91 (in Russian).

Dobrovolsky, V. V. (1983) *Geografia mikroelementov.* Globalnoe rasseyanie [Geography of Trace Elements. Global Dispersion]. Mysl, Moscow (in Russian).

Evans, E. A. (1966) *Thorium and its Compounds.* Butterworths, London.

Kabata-Pendias, A. & Pendias, H. (1986) *Trace Elements in Soils and Plants.* CRC Press, Boca Raton (Fl.).

Litver, B. Ya. *et al.* (1976) *Svinets-210, poloniy-210, radiy-226, toriy-228 v biosfere Krainego Severa SSSR* [Lead-210, Polonium-210, Radium-226, and Thorium-228 in the Biosphere in the Northernmost Regions of the USSR]. Moscow (in Russian).

Marei, A. N. *et al.* (1984) *Radiatsionnaya kommunalnaya gigiyena* [Radiation Hygiene in the Community]. Energoatomizdat, Moscow (in Russian).

Métivier, H. J. (1988) Chapter 63 *Thorium.* in: Seiler, H. G. *et al.* (eds) *Handbook on Toxicity of Inorganic Compounds.* Marcel Dekker, New York, pp. 689–694.

Nikolayev, L. V. *et al.* (1981). In: *Toksikologicheskiy i radiologicheskiy kontrol sostoyaniya pochv i rasteniy v protsesse khimizatsii selskogo khozyaistve* [Toxicological and Radiological Monitoring of Soils and Plants in Connection with the Growing Use of Chemicals in Agriculture]. Moscow, pp. 131–133 (in Russian).

Pavlovskaya, N. A. (1966) *Gig. San.* No. 5, 54–57 (in Russian).

Pavlovskaya, N. A. & Zeltser, M. R. (1981) *Toriy i producty ego raspada* [Thorium and its Decay Products]. Energoatomizdat, Moscow (in Russian).

Ponikarova, T. M. *et al.* (1985) *Gig. San.*, No. 9, 67–68 (in Russian).

Shvedov, V. P. & Patin, S. A. (1968) *Radioaktivnost okeanov i morei* [Radioactivity of Oceans and Seas]. Atomizdat, Moscow (in Russian).

Stokinger, H. E. (1963). In: Patty, F. A. (ed.) *Industrial Hygiene and Toxicology. Vol II: Toxicology.* Interscience Publishers, New York, pp. 1143–1148.

Szabo, E. (1981) *Igiene* (Roumania), **30** (No. 2), 131–135 (in Roumanian).

Tarasenko, N. Yu. (1963) *Gigiyena truda pri rabote s toriyem* [Occupational Hygiene for Thorium Workers]. Moscow (in Russian).

Tarasenko, N. Yu. (1983). In: *Encyclopaedia of Occupational Health and Safety*, Vol. 2. International Labour Office, Geneva, pp. 2173–2175.

Zhuravlev, V. F. (1982) *Toksikologiya radioaktivnykh veshchestv* [Toxicology of Radioactive Substances]. Energoatomizdat, Moscow (in Russian).

Uranium and its compounds

Uranium carbide; u.(IV) chloride (uranous chloride, uranium tetrachloride);
u.(V) chloride (uranium pentachloride); **u. dicarbide; u.(IV) fluoride** (uranous
fluoride, uranium tetrafluoride); **u.(VI) fluoride** (uranic fluoride, uranium
hexafluoride); **u.(IV) oxide** (uranous oxide, uranium dioxide, uraninite [min.]);
u.(VI) oxide (uranic oxide, uranium trioxide); **u. peroxide dihydrate**

Uranyl acetate dihydrate (uranium(VI) oxyacetate dihydrate); **u. carbonate**
(uranium(VI) oxycarbonate); **u. fluoride** (uranium(VI) oxyfluoride); **u. nitrate**
dihydrate (uranium(VI) oxynitrate dihydrate); **u. sulfate trihydrate**
(uranium(VI) oxysulfate trihydrate)

Ammonium diuranate (ammonium heptaoxodiuranate(VI)); **sodium diuranate**
(sodium pyrouranate); **triuranium octoxide** (uranium(V) oxide-uranium(VI)
oxide(1/1))

IDENTITY AND PHYSICOCHEMICAL PROPERTIES
OF THE ELEMENT

Uranium (U) is a radioactive metallic element of the actinide family in group III
of the periodic table, with the atomic number 92. It is the heaviest element found
in the earth's crust. In the natural state, it is a mixture of three isotopes: ^{238}U
(99.2739%), which is an α-emitter with a half-life of 4.51×10^9 years; ^{235}U
(0.7024%), an α-emitter with a half-life of 7.13×10^8 years; and ^{234}U
(0.0057%), an α-emitter with a half-life of 2.48×10^5 years. Of the 11 existing
artificial radioactive uranium isotopes (with mass numbers 227 to 240), those
with the longest half-life are ^{233}U and ^{228}U (1.3 and 9.3 min, respectively).

Uranium is a ductile and malleable metal resembling steel in appearance; it can be melted and extruded at high temperatures and takes a high polish after cold working. Its radioactivity is low; for example, 2800 kg of natural uranium have a radioactivity equivalent to that of approximately 1 g of radium-226. It exists in three crystalline modifications.

Uranium is highly reactive. In compounds it may be present in the oxidation states +3, +4, +5, or +6. Its most stable oxidation state in solution is +6, the stability being due to the formation of the uranyl ion UO_2^{2+}, which is also present in solid compounds. In air uranium slowly oxidizes with the formation of a surface oxide (UO_2) film. This, however, does not prevent the metal from further oxidation; at room temperature, the yellow oxide film gradually darkens so that the metal appears black in 3 to 4 days.

Uranium is a strong reducing agent. In the powder form, it is pyrophoric and burns with a bright flame. With oxygen, it forms the oxides UO, UO_2, and UO_3 and the peroxide $UO_4 \cdot 2H_2O$ as well as a large number of intermediate oxides, the most important of which is the octoxide U_3O_8. It gives the fluorides UF_4 and UF_6 with fluorine at 500–600°C and many compounds with sulfur; the most important of these is the sulfide of bivalent uranium, US (nuclear fuel). Uranium gives the hydride UH_3 with hydrogen at 220°C, the nitride U_4N_7 with nitrogen at 450–700°C, and the carbides UC, UC_2, and U_2C_3 with carbon at 750–800°C. It forms alloys of various types with metals. It reacts slowly with boiling water and with water vapor in the temperature range of 150–250°C to give UO_2 and H_2. It dissolves in nitric acid at a moderate rate when pure and compact but violently in a finely divided form; the resultant nitrogen oxides and nitric acid vapor react exposively with uranium. It dissolves slowly in hot concentrated sulfuric acid with formation of the sulfate $U(SO_4)_2$. The uranyl ion UO_2^{2+} has an extraordinary capacity for complexing with both organic and inorganic substances in aqueous solutions.

Uranium is chemically similar to elements in group VI of the periodic table such as chromium, molybdenum, and tungsten, but, unlike these, exhibits predominantly basic rather than acidic properties. See also Appendix.

NATURAL OCCURRENCE AND ENVIRONMENTAL LEVELS

The clarke of uranium is estimated at $(2.5–2.7) \times 10^{-4}\%$ in the earth's crust and at 2.6×10^{-4} in the granite layer of the continental crust [Dobrovolsky]. Well over 100 uranium minerals are known, but only 12 of them are of industrial importance, including the uranium minerals proper and uranium–molybdenum, uranium–vanadium, and uranium–nickel–cobalt–bismuth–silver deposits. Some uranium minerals contain uranium at levels of 50–60%. These are carnotite $K_2O \cdot 2UO_3 \cdot V_2O_5 \cdot (1–3)H_2O$, tyuyamunite $CaO \cdot 2UO_3 \cdot V_2O_5 \cdot (5–8)H_2O$, autunite $Ca(UO_2) \cdot (PO_4)_2 \cdot (10-12)H_2O$,   torbernite $Cu(UO_2)_2$ $\cdot (PO_4)_2 \cdot (8-12)H_2O$, and coffinite $U(SiO_4)_{1-x} \cdot (OH)_x$. The mineral richest in uranium is pitchblende, or uraninite $(xUO_2 \cdot yUO_3 \cdot ZnPbO)$, which contains 66 to 85% uranium oxides.

No less than 98% of the uranium dissolved in the earth's hydrosphere (about 2 billion tonnes) is contained in the oceans, where it is present at an average concentration of 3.2×10^{-3} mg/L, mainly in the form of $UO_2(CO_3)_3^{4-}$ and has an estimated residence time of 3×10^6 years (Henderson). In the oceans, accumulations of nonbiogenic suspended particulate matter (clay muds) contain uranium at an average level of 3×10^{-4}%, while biogenic sediments (calcareous muds) contain it at levels of $n \times 10^{-4}$% [Dobrovolsky].

Uranium concentrations in underground waters range from 5×10^{-8} to 2.7×10^{-3} g/L. Its average concentration in river water is 0.5 µg/L, and its global run-off from rivers to the oceans is estimated to be 19 000 tonnes per year $(K_w = 0.96)$ [Dobrovolsky].

Natural waters are capable of dissolving, transporting, and redepositing uranium. Uranium migrates vigorously in cold and hot as well as neutral and alkaline waters in the form of simple or complex ions. It may be transported in water in the form of sulfate (as the readily soluble uranyl sulfate UO_2SO_4), in a colloidal form (as negatively charged sols of the hydroxide $[UO_2(OH)_2]_n$), and in the form of carbonate (as readily soluble complex carbonates of the type $Na[UO_2(CO_3)_3]$). Gases dissolved in the water may enhance the migration of uranium or precipitate it from solution. Oxygen and carbon dioxide enhance the migration, whereas methane has the opposite effect by promoting, in the presence of SO_4^{2-} ion, the formation of hydrogen sulfide which precipitates uranium.

Uranium compounds are readily soluble in acid waters and only sparingly soluble in alkaline waters (e.g. in hydrogen sulfide solutions). Tetravalent uranium forms compounds with low solubilities, and its migration is therefore limited.

Considerable amounts of uranium are taken up from water by algae; the best concentrators are freshwater green algae from the Characeae family which may accumulate uranium to levels 1000 times higher (on a dry weight basis) than those found in the water where they live. When algae die, the uranium they have concentrated accumulates in the muds. Benthic organisms concentrate uranium by factors of 10 to 100 on a dry weight basis, the best concentrators being mollusks. In aquatic food chains, the gradient of uranium accumulation in living organisms decreases from algae to benthic organisms to fish.

In the upper layers of the earth's crust, uranium-bearing minerals undergo degradation, oxidation, and dissolution. In the processes of weathering and translocation, they enter into continental deposits such as sands and clays. During weathering, some of the uranium forms difficultly soluble hydroxides and some goes into solution as readily soluble uranyl complexes. Soluble uranium compounds may give rise to secondary uranium minerals (phosphates, vanadates, etc.) or enrich soils by being adsorbed on gels of the hydroxides of iron, aluminum, or other metals. In natural environments, hexavalent uranium is readily hydrolyzed, forming salts containg the uranyl ion. In this latter form uranium readily migrates to, and accumulates in, soils. From ground waters uranium is adsorbed by clay and humus soil particles in the form of a carbonate complex.

The worldwide uranium content of soils was reported to range from 3.4 to 10.5 mg/kg dry weight. Mean levels in surface soils were calculated to be (mg/kg dry weight) 0.79 in Poland, 1.22 in Canada, 2.60 in Great Britain, 11.0 in India, and 3.70 in the USA [Kabata-Pendias & Pendias].

Continental vegetation contains an estimated 0.05 million tonnes of uranium, at average fresh weight, dry weight, and ash weight concentrations of $0.008 \times 10^{-4}\%$, $0.02 \times 10^{-4}\%$, and $0.4 \times 10^{-4}\%$, respectively; the amount of uranium taken up annually worldwide through the increment in phytomass is estimated to be 5.18 thousand tonnes, or 35 g/km^2 of land area ($K_b = 0.15$) [Dobrovolsky]. Trees in an area of uranium mineralization had concentrations up to 2.2 mg U/kg ash weight, while sagebrush plants grown near a phosphoric fertilizer factory contained up to 8 mg U/kg [Kabata-Pendias & Pendias].

In the atmospheric air, average uranium concentrations were $3-4 \times 10^{-9}$ ng/m^3 over the Atlantic Ocean, 0.02 ng/m^3 over the United Kindom, and 0.4 ng/m^3 in the New York City area [Berlin & Rudell]. The atmospheric air over Continental Europe contained uranium at an average concentration of 20 pg/m^3. Uranium concentrations in rainfall varied from 2×10^{-6} to 1.3×10^{-8} g/L depending on the geographic area and metereological conditions and on the season [Novikov, 1974]. In the Upper Volga River region of the USSR, the amount of uranium deposited with atmospheric precipitation was estimated at 0.383 kg/m^2 per year, while its mean concentrations in the precipitation were 0.55 μg/L in solution and 0.13 μg/L in suspension [Petrukhin].

PRODUCTION

Uranium is extracted from its ores by varius methods depending on their composition. In a commonly used process, the ore is leached with a sulfuric acid or (sometimes) soda solution so that the uranium goes into the acid solution in the form of UO_2SO_4 of the complex anion $[UO_2(SO_4)_3]^{4-}$ or into the soda solution as $[UO_2(CO_3)_3]^{4-}$. To extract and concentrate the uranium from these solutions and to remove impurities, ion-exchange resins or extraction with organic solvents are used. Next, an alkali is added and ammonium or sodium uranate or uranium hydroxide precipitates from the solutions. To obtain uranium compounds of high purity, the technical products are dissolved in nitric acid and purified by affinity methods, the end products of which are UO_3 or U_3O_8. These are then reduced to UO_2 and treated with gaseous hydrogen fluoride to obtain uranium(IV) fluoride (UF_4). Uranium is recovered from the latter by electrolysis or by thermic reduction with calcium or magnesium. This yields uranium ingots which are then refined in vacuum furnaces. An important process is the enrichment of uranium with its isotope ^{235}U which is the principal nuclear fuel. The enrichment is usually carried out by thermodiffusion or centrifugation of gaseous UF_6 and is based on the mass difference between ^{235}U and ^{238}U.

USES

Uranium is the major raw material in nuclear power engineering, being used as fuel in nuclear energy plants. It is also utilized as the starting material for the synthesis of transuranium elements and as a source of the light isotope ^{235}U.

^{235}Uranium is a source of nuclear energy in nuclear weapons. The thermal energy generated by fission of 1 g of ^{235}U is equivalent to that released by burning 2200 L of crude oil or 2.7 tonnes of coal. ^{238}U is a source of secondary nuclear fuel (plutonium).

Ammonium diuranate is used to make uranium(IV) oxide, triuranium octoxide, and uranium(VI) oxide. **Uranium(IV) fluoride** serves as nuclear fuel in reactors operating on fused uranium fluorides; **uranium(VI) fluoride**, which is the only highly volatile uranium compound, is used in the gaseous state for separating the isotopes ^{235}U and ^{238}U by diffusion or centrifugation, and it can also serve as a nuclear fuel. **Uranium peroxide** finds use in the chemical refining of uranium. **Uranyl nitrate** and **uranyl sulfate** are utilized in the processing of uranium ores.

MAN-MADE SOURCES OF EMISSION INTO THE ENVIRONMENT

Dusts containing highly aggressive isotopes of the uranium series are released into the ambient air when uranium ores are mined and precessed. Multicomponent dusts and gases containing uranium, radon, radium, polonium, silicon oxide, and other substances may be emitted, for example, from mines, ore storage bins, mine dumps, and airway mouths as well as during the crushing, loading, unloading, and transportation of uranium ores. Dusts vary in mineralogic composition and particle size depending on the process involved. For example, as found for several uranium mills, airborne particles 6–10 μm in diameter predominated in the ore-crushing section and 2–5 μm particles in the ore-sieving section, while as much as 72–88% of all dust particles were less than 2 μm in the areas where uranium salts were calcined. Concentrations of airborne dust decrease with increasing moisture content of the ore (e.g. they were highest at moisture levels of 3.3–5.2% and averaged 1.4–3.7 mg/m³ when the moisture content was 10.4–12%) [Andreyeva, 1971; Bykhovsky *et al.*]. In the finished-products storage area of uranium mills, where operations such as unloading and packaging were automated and the air flow velocity through the openings in enclosures was maintained at 1.5–2 m/s, the uranium concentration in the air of the working zone did not usually exceed 0.2 mg/m^3 [Bulygina *et al.*].

The main types of uranium-containing liquid waste are mine waters and effluents from hydrometallurgical (uranium ore-processing) works. Such wastes were produced in quantities of 2–2.5 m^3 per tonne of processed ore and contained uranium in concentrations of the order of 1–3 mg/L [Kornilov].

According to one projection, in 1990 the release of uranium into the environment from all coal-fired power plants in the member states of the European Community should have totalled 440 tonnes, including 5.1 tonnes released into the atmosphere from stack emissions; calculations by which these anticipated atmospheric emissions were arranged according to particle size indicated that 2.5 tonnes would be contributed by particles in the 0–1.5 μg size range, 0.9 tonne by 1.5–3.0 μm particles, 1.7 tonnes by 3–7 μm particles, and 0.05 tonne by 5–15 μm particles [Sabbioni *et al.*].

Another source of environmental pollution by uranium are phosphoric fertilizers. They have been reported to increase the uranium content of top soil to between 30 and 300 mg/kg dry weight [Harmsten de Haan].

The world reserves of uranium (excluding the USSR, China, and Japan) are estimated at 2.8×10^6 tonnes (as U_3O_8), and the annual per capita consumption, at 1.4 g (as U_3O_8) [Mann].

TOXICITY

Land plants

Effects of uranium in soil on some plants are discussed in the papers by Boileau *et al.* and Sheppard *et al.*

Aquatic organisms

The growth of saprophytic aquatic microflora was depressed at a uranium concentration of 0.5 mg/L and ceased completely within 8–10 days at 100 mg/L. Algae showed slowed growth at 10 mg U/L and failed to grow altogether at 100 mg U/L.

For daphnids, uranium was toxic at concentrations of 5 and 10 mg/L and lethal within 24 h at 100 mg/L. Guppies exposed to uranium at 50 mg/L all died within 84 days; with exposures to 5, 10, and 26 mg U/L, stillbirths were frequently observed [Grushko, 1979].

The biochemical oxygen demand in bodies of water began to decrease within 24 h when uranium was added at 5 mg/L and after 24 h when it was added at 1 mg/L. The lowest uranium concentration than altered the sanitary condition of water bodies was 1 mg/L.

Animals and man

General considerations

Uranium is a general cellular poison and can potentially affect any organs and tissues. There are two main hazards associated with uranium exposure: the damage due to the chemical toxicity of uranium and its compounds and the injury caused by the ionizing radiation resulting from the disintegration of uranium isotopes. Which of these two hazards will be the limiting factor for exposure to uranium compounds depends on the solubility of the compound concerned, its route of entry into the body, and its isotopic composition.

The chemical toxicity of uranium compounds is directly dependent on their solubility in biological media. Upon contact with these, soluble uranium compounds go into solution at rates that are proportional to their solubilities.

According to their transportability, uranium compounds can be classified as highly, moderately, and slightly transportable. The highly transportable compounds are water-soluble and include uranium fluorides and the carbonate, nitrate, and sulfate of uranyl. Moderately transportable compounds (e.g. uranium nitrates) are less soluble but may be transformed into hexavalent soluble compounds in the body. To the slightly transportable compounds belong the oxides, hydrides, and carbides of uranium [Berlin & Rudell].

The critical organ for the chemical toxicity of uranium and its soluble compounds is the kidney. A majority of uranium compounds are excreted through and accumulate in this organ, causing structural damage to its tubules and glomeruli and thus impairing its function. Uranium and its compounds can also produce pathologic changes in, and impair the function of, many other organs and systems, including among others the gastrointestinal tract, liver, lungs, thyroid, adrenals, the cardiovascular system, and the central and peripheral nervous systems; they also cause disorders of water, lipid, and carbohydrate metabolism.

Uranium is an emitter of α-particles of low specific activity, and for this reason its radiobiological effects usually become manifest — in the form of chronic radiation sickness — much later than its chemical toxicity.

Occupational exposure to uranium usually involves inhalation of aerosols carrying particles varying in size and density and containing a mixture of uranium compounds with different solubilities. Insoluble particles are deposited in the lungs, retained there for a long time, and can cause radiation damage of varying degree. Similarly, the deposition of such particles in bones results in radiation injury to medullary hematopoiesis; bone retains uranium for longer periods than any other tissue.

Among laboratory animals rabbits and cats are generally more sensitive to uranium than dogs and mice, and these two species are more sensitive than rats and guinea-pigs. Bodily resistance declines with age, and old animals are much more sensitive than young ones. There are also sex differences in sensitivity.

Acute/single exposure

Animals

Acute manifestations of uranium poisoning are preceded by a latent period of several hours to several days depending on the dose. In this period, the animals became lethargic and were losing appetite. Later, most often on days 3–4, they exhibited thirst and vomited; still later, with the onset of enteritis, they refused to eat and showed signs of general inhibition. Subsequently, particularly just before death, tremor and paresis of the hind limbs were observed in a proportion of animals. Body-weight losses of 25–30% were recorded.

The development of intoxication was accompanied by falls in body temperature, progressive tachycardia, and increasing general inhibition. Hematological findings included decreased erythrocyte resistance, increased erythrocyte sedimentation rate, and marked leukocytosis succeeded by leukopenia and increases in the number of rod neutrophils; lymphopenia, eosinophilia, and monocytosis were frequently noted.

The initial (during the first 24 h) increase in urinary output was followed by oliguria (accompanied by proteinuria, cylindruria, glucosuria, and increased urinary excretion of catalase, phosphatase, and amino acids), which progressed to anuria. The blood levels of residual nitrogen and urea rose sharply. Transamination processes were also disrupted, with a fall in transaminase activities. Carbohydrate metabolism was impaired as a result of inhibited activities of hexokinase, glucose-6-phosphatase, and other enzymes responsible for its regulation. Moreover, the bioelectric activity of the brain became weakened, and diminished reactivity of the cerebral cortex and partial disassociation of the cortical and subcortical centers were observed. Structural changes and functional impairment were also exhibited by the thyroid gland.

The intraperitoneal dose of **uranyl fluoride** after which 50% of the animals died on days 14 to 21 was 2.5 mg/kg for male rats and 1 mg/kg for female rats. The corresponding 14–21-day LD_{50}s of **uranyl nitrate** were 2.0 mg/kg for male rats, 1.0 mg/kg for female rats, and 6–8 mg/kg for mice of both sexes. With the intravenous route, lethal doses of uranyl nitrate were (mg/kg) 0.1 for rabbits, 0.3 for guinea-pigs, 1.0 for rats, and 10–20 for mice [Galibin & Novikov]. The 24-hour LD_{50}s of these two compounds and of **uranium(IV) chloride** are given in Table 1.

The main manifestations of acute poisoning with uranyl nitrate were diarrhea, a bloody discharge from the nose, a violet-colored swollen tongue, hemorrhagic exudates in the abdominal cavity, and roundish hemorrhagic areas 1 to 6 mm in diameter in the spleen and in the walls of the stomach and small gut.

Animals acutely poisoned with uranyl fluoride exhibited diarrhea and lethargy, followed by restlessness which was succeeded by a comatose state. After a large dose, deaths occurred within 0.5 to 4 h. Animals that survived for

3 days regained appetite and had a relatively normal appearance 2–3 weeks later. Autopsy findings were similar to those in uranyl nitrate poisoning.

Table 1. The 24-hour median lethal doses of uranium compounds for rats with intraperitoneal administration

Sex	Body Weight (g)	LD_{50} (mg/kg)
	Uranium(IV) chloride	
Male	100–200	434 (411–458)[a]
	300–400	298 (260–341)[a]
Female	100–200	368 (327–415)[a]
	200–300	350 (330–371)[a]
	200–300	190 (176–204)[b]
Male and female	50–100	335 (314–358)[b]
	150–200	615 (598–633)[b]
	Uranyl fluoride	
Male and female	50–100	78 (65–94)
	150–200	87 (79–85)
Male	300–400	40 (36–44)
	Uranyl nitrate	
Male	50–100	305 (287–324)
	200–300	204 (167–250)
	300–400	128 (120–136)
Female	200–300	132 (120–152)

[a] In a 5% sodium acetate solution.
[b] In a 10% aqueous solution.
Source: Haven & Hodge.

In acute poisoning with uranium(IV) chloride, death was preceeded by diarrhea, the appearance of red crusts around the nose, and severe convulsions; at autopsy, large amounts of hemorrhagic exudate were present in the abdominal cavity while the liver and spleen appeared gray and as if boiled. The surviving animals looked normal in 3 days.

In dogs, a subcutaneous injection of uranyl nitrate at 2–5 mg/kg caused gastroenteritis, hepatitis, albuminuria, glucosuria, and hemorrhages into various organs. Death occurred within several days [Galibin et al.].

Acute poisoning with **ammonium diuranate** manifested itself clinically in body-weight loss, anorexia, thirst, general inhibition, and oliguria progressing to anuria. Deaths occurred in the presence of generalized paralysis. LD_{50} values for this compound are shown in Table 2. Its LC_{50} for mice by the inhalation route was 110 mg/m^3 with 60-minute exposure.

Table 2. Median lethal doses of ammonium diuranate after single administration

Animal	Administration route	$LD_{50}/30$ days (mg/kg)
Mouse	Oral	2.2 (1.88–2.52)[*]
	Subcutaneous	28.0 (24.1–31.9)
	Intravenous	9.2 (7.2–10.7)
Rat	Oral	6.5 (5.2–7.8)
	Subcutaneous	34.0 (26.8–39.6)
	Intravenous	16.0 (13.0–19.0)
Rabbit	Subcutaneous	16.0 (13.9–18.0)
	Intravenous	6.5 (5.2–7.8)

[*] Figures in parentheses indicate range.
Source: Galibin & Novikov.

Histopathological studies showed that the greatest damage was caused by the diuranate after intravenous administration: dystrophic and necrobiotic changes in the tubular epithelium and a picture of diffuse glomerulonephritis were seen in the kidneys; engorged and grossly dilated vessels, hemorrhages, and edema in the lung; and structural impairment of tissues with clear evidence of hemodynamic disturbances in the liver and digestive tract.

Man
Cases of particularly severe acute poisoning were caused by accidental exposure to **uranium(VI) fluoride**. In one accident involving 18 individuals, sudden rupture of a tank containing gaseous UF_6 resulted in the death of two persons in the path of the UF_6 cloud; one who was exposed for 5 min died 10 min later, while the other, after a rapid escape, died 70 min later. Some persons had second or third-degree burns. Of three other persons seriously injured, two were in the vicinity of the accident and the third outside nearby, and all required hospitalization for recovery; the 13 other exposed individuals required only dispensary treatment [Howland].

The early symptoms of intoxication were marked weakness or prostration, labored breathing, and retrosternal pain. On examination, some of the survivors were dyspneic and cyanotic, and numerous moist rales were heard in their lungs. They experienced a rapidly progressive irritating cough (productive of a gray–green blood-stained sputum), nausea, and vomiting; one victim developed pulmonary edema and shock. Radiographs showed renal lesions and diffuse inflammatory changes in the lungs. The gastrointestinal tract was also affected. There was high fever (up to 40°C) during the first 12–72 h after the poisoning. Conjunctivitis and deep keratosis (although without changes in visual acuity) were noted, as were azotemia and albuminemia. The period of weakness or prostration was succeded in some cases by one of nervous tension with transient anxiety and irritability. In one case, sporadic manifestations of dementia with loss of tactile sensitivity were observed. It was presumed that the renal damage was caused by absorbed uranium, while the skin, the eyes, and the respiratory mucous membranes were damaged by the direct action of fluorine.

For an adult human, the lowest intravenous dose of uranium(IV) producing albuminemia was 0.1 mg/kg, and extrapolation of animal data indicated that its lethal dose by this route would be 1.0 mg/kg [Bennellick].

Repeated exposure

Animals

In a study described by Galibin & Novikov, male rats were exposed by inhalation to an **ammonium diuranate** aerosol at 36 mg/m^3 for 4 h daily over 1 to 1.5 months (specific activity, 3×10^4 Bq/mg; uranium content, 73%; particle size, 0.5–4.0 μm). Shortly after the start of exposure, the animals became lethargic, moved very slowly and reluctantly, and refused to eat. Thirst, periodic sanious discharges from the nose, and marked mucosal hyperemia were added in the next few days. The animals had lost body weight and failed to regain it in the recovery period. During the exposure period and for a long time thereafter, the daily urinary output remained low and the urine had increased density. Albuminuria and elevated sugar and lowered creatinine and hippuric acid levels in the urine were observed, as was slowed clearance of the dye bromophenol blue from the bloodstream. In the peripheral blood, hemoglobin levels and erythrocyte counts decreased and there were slight eosinophilia and a pronounced leukocytosis with increased numbers of segmented neutrophils and decreased leukocyte numbers. By the end of exposure and during the subsequent 5 months, reduced bactericidal activity of the skin and increased content of total microflora in the oral cavity were recorded. Histopathologic studies disclosed an interstitial pneumonia with grossly thickened interalveolar septa and moderate proliferation of histiocytic elements which had enveloped vessels and bronchi; evidence of severe hemodynamic disturbances, particularly in the kidneys and liver; and perivascular hemorrhages in a number of areas.

Ammonium diuranate, like other uranium compounds, is not only a renal poison but one with hepatotropic activity, and, in addition, its chemical toxicity is subsequently manifested in adverse effects on the blood and in immunobiological effects [Galibin & Novikov].

Repeated subcutaneous injections of **uranyl nitrate** into rats at 1 mg/kg for 10 days led to substantial reductions in alkaline phosphatase activity in the liver, mainly in its central portion. After 30 days of subcutaneous exposure to the nitrate at the indicated dose level, the liver was depleted of alkaline phosphatase almost completely, and histologic examination of the thyroid glands from these rats revealed abnormally shaped follicles and reduced heights of follicular epithelial cells which contained granular uranium inclusions in their protoplasm. The thyroid gland showed reduced rates of ^{131}I uptake. In young rats, the development of the thyroid gland was severely inhibited, and gross morphologic abnormalities, including the presence of continuous epithelial strands and the lack of distinct follicular structure, were seen in its parenchyma. Rabbits given repeated subcutaneous injections of uranyl nitrate at 0.025 mg/kg for 5 days, developed hypothyroidism with inhibited iodine metabolism in all its phases. The thyroid glands from these rabbits contained adenomas of parenchymatous structure and presented a picture of colloid goiter; autoantibodies to these tumors were identified. In pregnant rabbits, 3 or 4 subcutaneous injections of the nitrate at 0.5 mg/kg produced pronounced symptoms of poisoning 15–17 days later and resulted in the death of most fetuses which were found to be in various stages of decomposition by the time of autopsy [Galibin & Novikov; Novikov, 1974].

Dogs given three subcutaneous injections of **uranium(IV) fluoride** at 6- to 36-day intervals exhibited functional changes relatable to the cerebral cortex, such as attenuation of positive conditioned reflexes, impaired differential inhibition of trace conditioned responses, and signs of irritative weakness (a psychopathological syndrome involving afferent lability and irritability combining with decreased capacity for physical activity, reduced ability to concentrate, and increased fatiguability); 30 to 60 days later, the conditioned reflexes improved and stabilized, but remained reduced compared to the pre-exposure level. Subsequently, dogs with a weak type of higher nervous activity displayed symptoms (motor and vegetative disturbances) relatable to subcortical centers; animals with a strong, balanced nervous system were more resistant [Guskova; Novikov, 1974].

Chronic exposure

Animals
The toxic effects from chronic uranium exposure are by and large similar to those seen in chronic poisoning, but differ in magnitude and, moreover, radiation damage may be caused by slightly soluble uranium compounds after their prolonged deposition in the body.

In Novikov's experiments, rabbits and guinea-pigs were exposed to **uranyl nitrate** by oral gavage at 0.02, 0.2, 1.0, or 2.0 mg/kg (as U) for 12 months, while rats and other guinea-pigs were exposed to **uranium(VI)** in their drinking water at 0.05, 0.6, 6.0, or 60 mg/L, also for 12 months. Animals receiving the nitrate orally at 0.02 or 0.2 mg U/kg as well as those ingesting uranium(VI) at 0.6 or 6.0 mg/L did not accumulate appreciable amounts of the metal either in the kidneys or in the skeleton. Exposure to 2.0 mg/kg or 60 mg/L, however, led to its substantial accumulation in the skeleton and also to decreased urinary levels of amino acids and chlorides, indicating impairment of the tubular reabsorption mechanisms. This finding correlated with the results of histological studies in which morphological changes were found in the renal tubules.

In addition, rats exposed to uranium(VI) at 6 or 60 mg/L experienced slowed sexual maturation and abnormalities in the sex cycle. After 10 months of exposure, raised alkaline phosphatase activity was detected in the blood sera of these rats and raised acid phosphatase activity in splenic homogenates from rabbits dosed by gavage at 1.0 mg U/kg.

In animals dosed with the nitrate at 1.0 or 2.0 mg U/kg, nucleic acids declined in the renal, hepatic, and splenic tissues. Toward the end of the exposure period, the kidneys of rats in the 60 mg U/L group had a mean uranium content of 0.004 mg, and the irradiation dose received by the renal tissues did not exceed 7×10^{-5} Sv per week. No alterations were noted in the other parameters measured (body weight; hemoglobin, peripheral blood erythrocytes and leukocytes; blood urea, creatinine, and chloride; serum sialic acids, urinary coproporphyrin; serum proteins, protein fractions, and sulfhydryls; blood catalase and aldolase; mineral composition of osseous tissue). It was concluded that the harmful effects of uranium are primarily due to its chemical toxicity, and that its radioactivity is an important consideration at fairly high doses or concentrations [Novikov, 1974].

In an inhalation toxicity study, described by Novikov [1972, 1974], male rats were exposed to an **ammonium diuranate** aerosol at 0.73, 1.0, 6.0, or 8.0 mg/m³ for 4 h daily over 4 months (aerosol particles ranged from 0.1 to 5.0 μg, with 80% being in the 0.5–3.0 μm range). Rats of the two latter groups exhibited pronounced proteinuria, glucosuria, and azotemia; reduced urinary excretion of hippuric acid both during exposure and for a long time after its termination (indicating impaired synthetic and detoxicating functions of the liver); reduced bromophenol blue clearance from the blood (an indicator of impaired excretory function of the liver); changes in the peripheral blood (lowered hemoglobin, relative neutrophilic leukocytosis and lymphopenia, marked and persistent eosinophilia); and considerable blood pressure elevations of an undulatory character. The respiratory rate was markedly increased both during the exposure period and for a long time thereafter. Moreover, the latency of unconditioned defense reactions was altered, attesting to disordered nervous system functioning.

On histopathological examination, lesions characteristic of uranium poisoning were seen in the kidneys, and substantial changes were also found in the liver and lungs of animals in the high-exposure groups. In the liver of rats, the main findings were congestion initially and swelling and nuclear polymorphism of Kupffer cells subsequently, including 2 months of the recovery period. In the lungs, vascular congestion, exudation, and desquamated septal cells were evident as early as after 1 week of exposure to acquire a well-defined focal pattern 3 weeks later; by that time, vessel walls appeared structureless and a striking perivascular edema was seen, as were hemorrhagic areas in the presence of dilated capillaries and, in the peribronchial lymphoid tissue, clear signs of erythrocyte disintegration. By the end of the 4-month exposure period, the lungs contained emphysematous and atelectatic areas and granuloma-like structures in the parenchyma. The concentration of 0.73 mg/m^3 was described as the minimal effective concentration of ammonium diuranate with 4-hour daily exposure for 4 months [Novikov, 1972, 1974].

In another 4-month inhalation toxicity study [Galibin & Novikov], rats exposed to **uranium(IV) fluoride** at 12.0 mg/m^3 for 4 h/day became lethargic, refused to eat, and began consuming increased quantities of water after several days of exposure. To these early signs of poisoning, sanious discharges from the nose and pronounced mucosal hyperemia were soon added in most animals. By the end of the first month of exposure and later, weight losses were recorded, and the pre-exposure body weight was not regained in the recovery period. The notable findings during and after the exposure period included proteinuria, glucosuria, elevated blood levels of residual nitrogen, impaired synthetic and detoxifying functions of the liver, and reduced hemoglobin levels, neutrophilic leukocytosis, and eosinophilia in the peripheral blood. Exposure at the 1.2-mg/m^3 level did not appear to produce abnormalities in the functioning of organs or systems under the experimental conditions indicated above, but proteinuria, oliguria, and high urine density were detected when loading tests (using caffeine or physiologic saline) were performed on days 90 and 120 of exposure.

In a similar 4-month study, Galibin [1975] exposed rats to **triuranium octoxide** at 10.0 or 1.0 mg/m^3 (concentrations equivalent to 8.5 and 0.85 mg U/m^3, respectively) for 4 h daily and found its toxic effects at the 10.0 mg/m^3 level to be much like those described above for two other slightly soluble uranium compounds (ammonium diuranate and uranium(IV) fluoride). The concentration of 1 mg/m^3 was judged to be the threshold level for chronic inhalation exposure under the conditions used (the absorbed dose in the lungs was about 0.3 Sv per year at this level).

In dogs and monkeys chronically exposed to **uranium(IV) oxide** dust at 5.8 mg/m^3, histologic changes were only seen in the lungs and tracheobronchial lymph nodes, and irradiation doses received by these structures over 5 years of exposure were 9 and 100 Gy, respectively. After 2–3 years of exposure, accumulations of pigment suggesting induration were found in the lungs of both dogs and monkeys, and pulmonary fibrosis (which progressed after the termination of exposure) was diagnosed in the latter animals. The absence of detectable renal damage in the test animals led the authors to conclude that the observed histologic changes were due to the radiation effects of the uranium oxide [Galibin & Novikov].

Man

The hazards associated with exposure to natural uranium and its compounds are twofold — those due to their chemical toxicity and those arising from the disintegration of uranium isotopes which can cause radiation injury. Pure uranium does not constitute an external radiation hazard since it emits mainly α-radiation of low energy. It does, however, present an internal radiation hazard if it enters the body by inhalation or ingestion. The concentration of 1 mg U/g biological tissue corresponds to an absorbed dose of 0.006 Sv per year. Uranium compounds carry chemical hazards as liver and kidney poisons, and uranium(VI) fluoride can cause injury because of the irritant action of fluorine. Radiation exposure may occur in the mining of uranium ores, although the silicosis hazard to which the uranium miners are exposed is greater than the radiation hazard. Radiation hazard may also be present during the processing of uranium ores and when the metal and its compounds are used. The chemical toxicity of uranium compounds is high. Particularly toxic are compounds soluble in water and body fluids [Tarasenko].

Examination of 31 workers after their year-long inhalation exposure to dusts of **uranium(VI) oxide**, **uranium peroxide**, and **uranium chlorides** (at concentrations that at times reached 155 mg/m^3 in terms of uranium), did not reveal any symptoms or signs of chronic poisoning (urinary catalase, nitrogen, chlorides and total protein, urea, creatinine, and several other parameters including uranium were measured) [Howland].

When 130 persons chronically exposed to soluble uranium compounds were examined, seven of them were found to have nephrosis; their urine contained uranium at levels up to 3.46 mg/L. Examination of 237 uranium mine workers revealed anemia in 31% of them, marked leukopenia in 23%, and lymphocytosis in 14%. In another, similar group of uranium workers employed for about 10 years, reduced body weight and pathological changes in the lungs, kidneys, and blood were observed and shown to have resulted from radiation exposure. On the other hand, none of the 100 individuals who had been exposed for 5 years to slightly soluble uranium compounds (at concentrations of 0.5–2.5 mg/m^3), exhibited any signs of renal damage nor any pulmonary or blood changes. Employees of a factory producing chemical reagents from uranium salts such as uranyl nitrate, uranyl acetate, and uranium(VI) oxide, were diagnosed as having 'vegetative dystonia and unstable parameters of peripheral blood with a tendency toward leuko-, lympho-, and thrombopenia' [Andreyeva, 1960, 1971].

Results of extensive studies have demonstrated that the consumption of drinking water containing uranium at elevated levels of 0.04 to 0.05 mg/L is not detrimental to human health [Novikov, 1974].

Effects of dermal and ocular application

Animals

Soluble uranium compounds such as uranium(V) chloride, uranium(VI) fluoride, and the fluoride and nitrate of uranyl are capable of being absorbed through the intact skin of animals in sufficient amounts to cause systemic poisoning which may be fatal. LD$_{50}$ values for uranyl nitrate are shown in Table 3.

Table 3. Median lethal doses of uranyl nitrate
on dermal application (g/kg)

Animals	Uranyl nitrate	As uranium
Rabbits	0.125 (0.095–0.165)*	0.059 (0.045–0.078)
Rats	1.04 (0.63–1.73)	0.49 (0.30–0.82)
Guinea-pigs	4.44 (3.88–5.09)	2.11 (1.84–2.41)

* Figures in parentheses indicate range.

In rabbits, cutaneous application of uranium compounds resulted in erythema, coagulation necrosis, and ulceration at the application site. **Uranyl nitrate** produced superficial coagulation necrosis accompanied by a rapid development of hyperemia in the surrounding tissue and by scab formation. At nonlethal levels, complete recovery ensued after a week. Similar, but more slowly evolving effects were produced in rabbits by a hygroscopic powder of **uranium(V) chloride**. When applied to their skin in lanolin ointment, **uranium(IV) chloride** gave rise to a moderate local erythema that disappeared 24 to 48 h later.

At lethal dose levels, first signs of acute poisoning appeared in rabbits at 48 to 72 h post-applications and included decreased consumption of food and water, body-weight loss, anxiety, and hyperreflexia. Two to three days before death, the animals became exhausted and ataxic and did not respond to stimulation. Paraplegia progressing to tetraplegia, a fall in body temperature, respiratory distress, and convulsions were then observed, followed by death. In guinea-pigs, the clinical picture of acute lethal poisoning was marked by anorexia and progressive cachexia. Uranium was detected in the urine of all animals exposed to uranyl nitrate by the dermal route. Post-mortem examination showed renal lesions typical for uranium poisoning. Chronic intoxication caused by skin applications of uranyl nitrate was manifested in body-weight losses and proteinuria. The toxicity of soluble uranium compound decreases with their decreasing solubilities. The insoluble compounds uranium(IV) oxide, uranium peroxide, and triuranium octoxide did not cause systemic poisoning upon dermal application [Orcutt].

Some uranium compounds can also cause systemic poisoning and death when placed into the conjunctival sac of experimental animals (Orcutt). Thus, two out of four rabbits died from systemic intoxication caused by **uranium(V) chloride** absorbed from the conjunctival sac into which it had been placed in an amount of 1 mg. Local changes seen in these animals included necrosis and detachment of the conjunctiva, eyelids, and parabulbar tissue, followed by ulceration and perforation of the cornea. Systemic intoxication as well as conjunctivitis and keratitis that persisted for 48 h occurred after conjunctival application of **uranium(IV) chloride**. Similar, though less marked and less persistent, effects were produced by **uranyl fluoride, uranyl nitrate, sodium diuranate,** and **uranium(IV) fluoride.** **Ammonium diuranate** produced conjunctivitis that lasted for 6 days, and signs of systemic intoxication. No signs of local reaction were observed after conjunctival application of **uranium(VI) oxide** to four rabbits which, however, all died as the result of systemic poisoning.

Man

Individuals engaged for prolonged periods in the packaging and weighing of uranium reagents developed uranium dermatosis on the hands and exposed parts of the face and neck; protein traces were detected in their urine [Andreyeva, 1971].

ABSORPTION, DISTRIBUTION, AND ELIMINATION

Uranium can enter the human body by the oral route (in food and water), by inhalation, and through the skin and mucous membranes.

The daily dietary intake of uranium by inhabitants of three urban areas in the USA was estimated to be $(1.3-1.4) \times 10^{-6}$ g. For the United Kingdom, an average daily intake of about 1×10^{-6} g was recorded, while intakes of up to 4.55×10^{-6} g/day were reported for areas near uranium mines in Japan. Normally, some 70% of dietary uranium is contained in meat, fresh fish, and bakery products. Drinking water containing uranium in concentrations of $(0.025-0.04) \times 10^{-6}$ g/L contributes only 2–3% of the daily intake. In contrast, drinking water was the major source of uranium intake in areas where it contained high concentrations of this element $(2 \times 10^{-4}$ g/L), and in such areas the contribution of food was small (1–5%). A large contribution to the uranium content of prepared foods can be made by table salt for which a concentration of 40 µg U/kg was recorded. For milk and curds, values of 4.3 and 35 µg U/kg, respectively, were reported [Berlin & Rudell; Novikov, 1974]. Milk has the lowest uranium-concentrating capacity among staple foods of animal origin; milk products such as butter and curds concentrate 7 to 8 times as much, while bakery products concentrate up to 20 times as much.

Warm-blooded animals, including man, contain uranium in all organs and tissues that have been examined, and uranium has also been detected in embryonic tissues. Average uranium levels found for human organs and tissues are (g/g): liver, 6×10^{-9}; lung, $(6-9) \times 10^{-9}$; spleen, 4.7×10^{-7}; blood, 4×10^{-10}; kidney, 5.3×10^{-9} (cortex) and 1.3×10^{-8} (medulla); osseous tissue, 1×10^{-9}; bone marrow, 1×10^{-8}; hair, 1.3×10^{-7}.

The absorption of soluble uranium compounds through the gastrointestinal tract of rats amounted to only 1–4% of the administered dose [Svyatkina & Novikov]. Insoluble or slightly soluble compounds are absorbed to a much lesser extent. Thus, 0.5% of ammonium diuranate, 0.2% of uranium(IV) fluoride, and 0.1% of triuranium octoxide were absorbed [Novikov, 1972]; in some studies, far lower absorption levels (hundredths to thousandths of one percent) have been recorded for insoluble uranium compounds.

Uranium compounds, both soluble and insoluble, are absorbed much more readily from the lungs. In rats, 28% of the inhaled ammonium diuranate (insoluble compound) was taken up for 15 min after a 4-hour exposure [Galibin, 1967]. Many uranium compounds were also detected in the blood of experimental animals within minutes following cutaneous application, and some were found capable of entering the circulation from the conjunctival sac. Lethal uranium poisoning is possible with any of the exposure routes mentioned above.

Following inhalation exposure of intact rats to uranyl nitrate or acetate, only three out of the 30 tested tissues contained uranium in amounts exceeding 0.1 µg/g tissue and none contained more than 0.45 µg/g (the mean uranium level in the tissues was of the order of 0.02 µg/g) [Neuman].

In the blood of exposed animals, uranium occurs in two forms: as a nondiffusible complex with plasma proteins and as a diffusible hydrogen carbonate (bicarbonate) complex, and the two forms are in equilibrium with each other. The hydrogen carbonate complex is contained in the extracellular fluid washing the soft tissues and either does not penetrate into the cells or is not fixed there and so re-enters the bloodstream to be filtered by the glomerular apparatus of the kidneys. The filtration rate and the amount of eliminated or reabsorbed uranium depend on the alkali reserve: in alkalosis less uranium is deposited in the kidneys and more is eliminated in the urine, whereas in acidosis the reverse is true. About 47% of the uranium in blood is found complexed with plasma hydrogen carbonate, 32% is bound to plasma proteins and 20% to erythrocytes [Berlin & Rudell].

Autoradiographic studies show that the main sites of uranium deposition are the renal cortex and the liver. In bones most of the uranium is found in the epiphyseal part. Uranium deposition in soft tissues is almost negligible. Hexavalent uranium is deposited mostly in the kidneys and eliminated with the urine, whereas tetravalent uranium is preferentially deposited in the liver and eliminated with the feces. The greatest capacity for accumulation in bone is possessed by hexavalent uranium.

Uranium can pass the placental barrier and has been detected in the milk of animals. Animals exposed to triuranium octoxide by inhalation accumulated this compound in the lungs, skeleton, and particularly in the mediastinal and bronchial lymph nodes. The lungs of a person who had inhaled triuranium octoxide still contained 3% of the inhaled amount 1.5 years later. According to Bennellick, the longest biological half-time of uranium is 800 days. Young animals accumulate more uranium in their bones than old animals because their skeleton is better supplied with blood. In a 12-month oral study on rats, uranium reached a steady state after 3 months of dosing in the kidneys and after 6 to 12 months in the skeleton. When rats were chronically exposed to uranium by inhalation, a steady state in the lungs, kidneys, skeleton, and liver was reached after approximately 3 months of exposure.

The elimination of uranium absorbed into the blood occurs via the kidneys in urine, and most of it is cleared within 4 to 24 h (for man, dog, rat, and mouse, values of 84%, 70%, 60%, and 50%, respectively, were recorded); during that period, uranium levels in the feces did not exceed 1% of the dose. In rats after a single intravenous injection, 43% and 57% of the uranium were cleared from the kidneys with biological half-times of 1.8 days and 15.5 days, respectively; from the bones, 76% was eliminated with a half-time of 84 days and 24% with a half-time of 400 days. From the rat body as a whole, about 85% of the uranium found 24 h post-injection, was cleared with a half-time of 20 days and 15% with a half-time of 400 days. In bones, the insoluble compound ammonium diuranate had biological half-times of 200 to 300 days. For man, the biological half-time of soluble uranium compounds has been estimated to be about 15 days [Galibin & Novikov; Novikov, 1972].

Uptake, absorption, transport and distribution, metabolism, and elimination of uranium in plants, animals, and humans has been reviewed by Burkart.

HYGIENIC STANDARDS

In the USSR, occupational exposure limits (MAC_{wz}) have been set at 0.015 mg/m^3 for soluble uranium compounds (aerosols, hazard class 1) and at 0.075 mg/m^3 for insoluble uranium compounds (also aerosols, hazard class 1). In the USA, TLV-TWA and TLV-STEL values of 0.2 mg/m^3 and 0.6 mg/m^3, respectively, have been set for both soluble and insoluble compounds of natural uranium (as U).

METHODS OF DETERMINATION

For the determination of uranium in air, water, soil, and biological materials, various methods of radiochemical analysis and radiometry have been used [Berlin & Rudell; Guskova]. Uranium can also be determined fluorimetrically and by various techniques of neutron activation analysis. A method of thermal ionization mass spectrometry for determining picogram quantities of uranium in biologic tissues has been described [Kelly & Fassett]. See also Burkart, Byrne & Benedik, and Fisher.

MEASURES TO CONTROL EXPOSURE

At the workplace. The major adverse factor during the mining and processing of uranium and uranium-containing minerals is airborne dust. Technical control measures should therefore be primarily directed at dust control and include mechanization of work operations to the greatest possible extent, provision of airtight enclosures for the equipment, and use of wet processes wherever possible. Local exhaust ventilation is necessary for any operations associated with health hazards from radon-containing dust. Systematic monitoring of the working conditions by the appropriate authorities is essential.

All those who are to be employed on jobs involving the mining, processing, or concentration of uranium ores should undergo a preplacement medical examination. During the whole period of employment periodic medical examinations should be carried out at such intervals as the work situation requires.

For *personal protection,* personal protective devices (respirators, protective clothing, footwear, etc.) should be used by workers in uranium mines and mills with elevated levels of radon contamination. Where the ambient air contains radioactive aerosols at levels that do not exceed the maximum permissible concentration more than 200-fold, air-purifying respirators may be used. If the levels are higher, air-supplied respirators (hose masks) must be worn; for increasing work efficiency, combined use of a respirator and a hose mask is recommended in such situations, with the hose mask being worn only when working in a fixed place and the respirator being used when moving from place to place within the contaminated area. The workers must be made familiar with, and strictly observe, the rules of personal hygiene.

All health-protection measures and recommendations should be specific, with full regard to the actual conditions prevailing at the particular industrial enterprise.

Protection of the general environment. Measures should be taken to prevent uncontrolled emissions of uranium-containing aerosols from the industrial premises. A universally applicable and fairly effective method of reducing uncontrolled emissions into the atmosphere consists in planting the industrial area with shrubs and trees. Great importance needs to be attached to the collection, processing, and disposal of wastes. A set of environmental protection measures against radioactive contamination in the mining and processing of uranium ores has been described in detail by Andreyeva *et al.*

FIRST AID IN POISONING

Intoxication by uranium, uranium salts, or uranyl salts
The affected skin and mucous membranes should be promptly decontaminated with water and soap or a sodium hydrogen carbonate solution, and a sodium hydrogen phosphate solution (10 g per 200 ml) should be given orally, as well as mucilaginous decoctions, milk, and/or egg whites. Gastric lavage should then be carried out, followed by repeated oral administration of sodium hydrogen phosphate in the same amount as above. A 5% pentacin [synonym: calcium

trisodium pentetate] solution given intravenously is indicated (in the event of inhalation poisoning, this solution may also be administered by inhalation in the form of an aerosol). Within the first few hours after poisoning, fonurit [synonym: acetazolamide] is recommended as an agent preventing renal injury. Sodium hydrogen carbonate (1 g 3 times per day) and a 10% sodium citrate (1 tablespoon 3 times per day) are also indicated. Administration of sodium hydrogen carbonate by intravenous drip (50–100 ml) is recommended. Belladonna or atropine may be given to alleviate pain. For the control of persistent vomiting, vitmin B_1 with glucose and 5 ml of a 0.5% aminazine [synonym: chlorpromazine] solution may be administered intramuscularly. Corazol [synonym: leptazol] (10 ml of 10% solution), caffeine (1.0 ml of 10% solution), and camphor (1 ml of 10% solution) are indicated as central stimulants, as are sodium sulfate and magnesium sulfate to stimulate bowel action [Ilyin].

Intoxication with uranium(VI) fluoride
The affected skin and mucous membranes should be immediately washed with copious amounts of water or, preferably, a 2% sodium hydrogen carbonate solution. Sodium hydrogen carbonate inhalations, lotions, and baths are indicated. In cases of ingestion, magnesia, calcium gluconate, and slimy decoctions are helpful. Therapy with oxygen or carbogen may be required. Belladonna may be given to relieve severe pain and 0.5 ml of a 0.1% atropine solution to eliminate laryngospasm. Persistent vomiting can be controlled by injecting 5 ml of a 0.5% aminazine [synonym: chlorpromazine] solution intramuscularly. Calcium gluconate or calcium chloride (20 ml of 10% solution with 20 ml of 40% glucose) may be beneficial. As cardiac/central stimulants, cordiamine [synonym: nikethamide] (1–2 ml of 10% solution), caffeine (1–2 ml of 10% solution), and camphor (2 ml of 10% solution) are indicated. Dionine [synonym: ethylmorphine hydrochloride] is effective as an antitussive agent. Cleansing enemas and diuretics such as fonurit [synonym: acetazolamide] and hypothiazide [synonym: hydrochlorothiazide] are recommended [Ilyin].

Ludewig & Lohs warn against the use of potassium preparations and emphasize the need for much caution in the use of opiates.

REFERENCES

Andreyeva, O. S. (1960) *Gigyena truda pri rabote s uranom i ego soyedineniyami* [Occupational Hygiene for Workers Handling Uranium and its Compounds). Moscow (in Russian).
Andreyeva, O. S. (1971) In: *Voprosy gigiyeny truda na uranovykh rudnikakh i obogatitelnykh predpriyatiyakh* [Occupational Hygiene in Uranium Mines and Mills]. Moscow, pp. 18–23 (in Russian).

Andreyeva, O. S. *et al.* (1979) *Prirodnyi i obogashchennyi uran: radiatsionno-gigiyenicheskiye aspekty* [Natural and Enriched Uranium: Radiation Hygiene]. Atomizdat, Moscow (in Russian).

Bennellick, E. J. (1966) *A Review of the Toxicology and Potential Hazards of Natural, Depleted and Enriched Uranium.* United Kingdom Atomic Energy Authority, Health and Safety Branch (Report 58), Harwell.

Berlin, M. & Rudell, B. (1979) In: Friberg, L. *et al.* (eds) *Handbook on the Toxicology of Metals.* Elsevier, Amsterdam, pp. 647–658.

Boileau, L. J. R. *et al.* (1985) *Canad. Botany*, **63**, 384–397.

Bulygina, A. M. *et al.* (1971) In: *Voprosy gigiyeny truda na uranovykh rudnikakh i obogatitelnykh predpriyatiyakh* [Occupational Hygiene in Uranium Mines and Mills]. Moscow, pp. 23–27 (in Russian).

Burkart, W. (1991) Chapter II. 33 *Uranium, Thorium and Decay Products* in: Merian, E. (ed.) *Metals and Their Compounds in the Environment — Occurrence, Analysis and Biological Relevance.* VCH, New York, pp. 1275–1287.

Bykhovsky, A. V. *et al.* (1971) In: *Voprosy gigiyeny na uranovykh rudnikakh i obogatitelnykh predpriyatiyakh* [Occupational Hygiene in Uranium Mines and Mills]. Moscow, pp. 5–17 (in Russian).

Byrne, A. R. & Benedik, L. (1991) *Sci. Total Environ.*, **107**, 143–157.

Dobrovolsky, V. V. (1983) *Geografiya mikroelementov. Globalnoe rasseyanie* [Geography of Trace Elements. Global Dispersion]. Mysl, Moscow (in Russian).

Fisher, D. R. (1988) Chapter 68 *Uranium* in: Seiler, H. G. *et al.* (eds) *Handbook on Toxicity of Inorganic Compounds.* Marcel Dekker, New York, pp. 739–748.

Galibin, G. P. (1967) *Gig. San.*, No. **12**, 40–43 (in Russian).

Galibin, G. P. (1975) *Gig. San.*, No. **2**, 41–46 (in Russian).

Galibin, G. P. & Novikov, Yu. V. (1976) *Toksikologiya promyshlennykh soyedineniy urana* [Toxicology of Industrial Uranium Compounds]. Atomizdat, Moscow (in Russian).

Galibin, G. P. *et al.* (1966) *Gig. San.*, No. **12**, 22–26 (in Russian).

Guskova, V. N. (1972) *Uran. Radiatsionno-gigiyenicheskaya kharakteristika* [Uranium: Radiation Hygiene]. Moscow (in Russian).

Harmsten, K. & de Haan, F. A. M. (1980) *Neth. J. Agric. Sci.*, **28**, 40.

Haven, F. L. & Hodge, H. C. (1949) In: Voegtlin, C. & Hodge, H. C. (eds.) *Pharmacology and Toxicology of Uranium Compounds.* McGraw-Hill, New York, pp. 281–309.

Henderson, P. (1982) *Inorganic Geochemistry.* Pergamon Press, Oxford.

Howland, J. W. (1949) In: Voegtlin, C. & Hodge, H. C. (eds) *Pharmacology and Toxicology of Uranium Compounds.* McGraw-Hill, New York, pp. 993–1017.

Ilyin, L. A. (eds.) (1976) *Neotlozhnaya pomoshch pri ostrykh radiatsionnykh vozdeistviyakh* [Emergency Care Following Acute Radiation Exposure]. Meditsina, Moscow (in Russian).

Kabata-Pendias, A. & Pendias, H. (1986) *Trace Elements in Soils and Plants.* CRC Press, Boca Raton (Fl.).

Kelly, W. R. & Fassett, J. D. (1983) *Analyt. Chem.*, **55**, 1040–1041.

Kornilov, A. N. (1971) In: *Voprosy gigiyeny truda na uranovykh rudnikakh i obogatitelnykh predpriyatiyakh* [Occupational Hygiene in Uranium Mines and Mills]. Moscow, pp. 71–85 (in Russian).

Ludewig, R. & Lohs, K. (1981) *Akute Vergietungen.* VEB Gustav Fischer, Jena.

Mann, A. W. (1979) In: Bockris, J. O'M (eds.) *Environmental Chemistry*, Plenum Pres, New York, pp. 121–178.

Neuman, W. F. (1949) In: Voegtlin, C. & Hodge, H. C. (eds) *Pharmacology and Toxicology of Uranium Compounds.* McGraw-Hill, New York, pp. 701–718.

Novikov, Yu. V. (1972) *Gig. San.*, No. **9**, 13–17 (in Russian).

Novikov, Yu. V. (1974) *Gigiyenicheskiye voprosy izucheniya soderzhaniya urana vo vneshnei srede i ego vliyaniya na organism* [Environmental Levels of Uranium and its Health Effects]. Moscow (in Russian).

Orcutt, J. A. (1949) In: Voegtlin, C. & Hodge, H. C. (eds) *Pharmacology and Toxicology of Uranium Compounds.* McGraw-Hill, New York, pp. 377–422.

Petrukhin, V. A. (1982) In: *Monitoring fonovogo zagryazneniya prirodnykh sred* [Monitoring the Background Pollution of Environmental Media], No. 1. Moscow, pp. 147–165 (in Russian).
Sabbioni, E. *et al.* (1984) *Sci. Total Environ.*, **40**, 141–154.
Sheppard, M. J. *et al.* (1985) *Water, Air, Soil Pollut.*, **26**, 85–94.
Svyatkina, N. S. & Novikov, Yu. V. (1975) *Gig. San.*, No. 1. 43–45 (in Russian).
Tarasenko, N. Yu. (1983) In: *Encyclopaedia of Occupational Health and Safety.* International Labour Office, Geneva, pp. 2237–2239.

Carbon

IDENTITY AND PHYSICOCHEMICAL PROPERTIES

Carbon (C) is a nonmetallic element in group IV of the periodic table, with the atomic number 6. It has two stable natural isotopes, ^{12}C (98.89%) and ^{13}C (1.108%); in the upper atmosphere, small amounts (about 2.1^{-10}% by mass) of ^{14}C with a half-life of 5.6×10^3 years are constantly present.

Carbon exists in two crystalline modifications, as diamond and graphite; in addition, two other crystalline modifications — carbyne and lonsdeilyte — have been produced artificially and, moreover, lonsdeilyte has been detected in meteorites. Carbon also occurs in various states of relatively disordered structure that are designated by the collective term amorphous carbon; included under this term are coke, bituminous coal, anthracite, brown coal (lignite), charcoal, carbon black, activated carbon, and some other substances.

In an inert atmosphere or in vacuum at temperatures above 1000°C, diamond converts to graphite. Amorphous carbon converts to graphite at 1500–1600°C. Carbon is inert at ordinary temperatures, but combines with many elements and compounds at elevated temperatures, showing reducing properties. Its oxidation state in compounds is chiefly +4. The reactivity of carbon depends on its crystalline structure. Amorphous carbon is the most reactive and diamond is the least. Carbon forms carbides with elements that have a lower or the same electronegativity as itself.

Amorphous carbon and graphite react with fluorine to form mainly carbon(IV) fluoride (carbon tetrafluoride). Diamond does not react with fluoride. Sulfur reacts at 900–1000°C with diamond and at 700–800°C with graphite and amorphous carbon, to form carbon sulfide (CS_2). Diamond does not react with hydrogen, while graphite and amorphous carbon react very slowly and, depending on the temperature, yield various hydrocarbons (methane, acetylene, benzene, and others). Carbon burns in air to carbon monoxide (CO) and carbon dioxide (CO_2). It is not attacked by concentrated acids and alkalies. A chrome mixture (5% potassium bichromate solution in concentrated sulfuric acid) oxidizes it to CO at 180–200°C (in the case of diamond) or at lower temperatures (in the case of graphite or amorphous carbon). On heating, a mixture of concentrated nitric acid and potassium chlorate oxidizes graphite to mellitic (benzenehexacarboxylic) acid.

Graphite is greasy to touch and very soft. It is insoluble in water, acids, or alkalies. It reacts with hydrogen at 600°C in the presence of catalyst and with sulfur at > 700°C. It sublimes at ~3700°C under atmospheric pressure and converts into liquid carbon at higher temperatures and pressures >1.05×10^5 Pa. See also appendix.

Amorphous carbons

Coke occurs in the form of elongated hard gray pieces of porous material having a density of 1.80 to 1.95 g/cm^3 and containing >96% carbon and 0.4–1.9% sulfur; its ash content is 9–10.5% (pitch and petroleum cokes have a lower ash content, about 0.8%).

Bituminous coals are heterogeneous; they contain several main petrographic types of macroconstituents such as vitrain, clarain, durain, and fusain, and also water and mineral substances. **Brown coals** are sedimentary rocks of a humus nature. Mined coals contain various amounts of sodium, potassium, and calcium compounds (0.1–5.2, 0.1–0.8, and 0.1–1.1 mg/100 g, respectively), chlorides (0.2–6.9 mg/100 g), sulfates (0.2–1.1 mg/100 g) as well as quartz, dolomite, pyrite, feldspar, siderite, and kaolinite. Elemental compositions of fossil coals are shown in Table 1.

Table 1. Elemental composition of some fossil fuels (%)

Fossil	C	H	N	S	O
Peat	53–62	5–6.5	1–4.0	0.1–0.5	29–40
Slate	56–78	6–10.0	0.2–1.0	0.5–7.0	12–36
Brown coal	63–75	4.5–6.5	0.5–1.5	0.3–12.0	12–32
Bituminous coal	76–95	3.5–6.3	1.2–2.5	0.3–12.0	1.5–17
Anthracite	90–96	1.3–3.0	up to 1.3	–	1–2

Charcoal is a hard, highly porous product, having a density of between 110 and 185 kg/m^3 and an average ash content of 3%. It adsorbs oxygen at ordinary temperature, which explains its tendency to spontaneous ignition. **Activated carbon** (active carbon, activated charcoal) is a form of carbon having a porous internal structure and characterized by high adsorptive properties, particularly with respect to hydrocarbons and many of their derivatives. Lower alcohols are adsorbed by it less well, while water is adsorbed very slightly.

Carbon black is a general term used to designate any finely divided substance produced industrially as soot and consisting principally of carbon (88–99.5%) and variable amounts of other substances depending on the manufacturing process (0.3–11% of oxygen, 0.1–1% of hydrogen, some tarry matter and trace amounts of sulfur; other inorganic or organic materials may also be present). Carbon blacks (as channel black, furnace black, and thermal black) are to be distinguished from soots that are formed as waste products in chimney flues or are contained in diesel exhausts. Carbon black particles may be spherical or sickle-shaped and vary in size from 5 to 500 nm. For example, channel black used for making lacquers and paints had particles sized 10–15 nm, that used for rubber production had 30–35 nm particles, while that used other purposes had particles in the 23–350 nm range. In various carbon blacks and soots, 3,4-benzpyrene has often been detected as an impurity, in concentrations ranging up to 1% by weight. In furnace black samples, for instance, its average level was 0.3% [Shabad & Dikun]. Highest levels were found in carbon blacks from coal and lowest, in those from coke.

NATURAL OCCURRENCE AND ENVIRONMENTAL LEVELS

Carbon is contained in the earth's crust at an average level of about $2.3 \times 10^{-2}\%$ by mass, mainly in bound states as natural carbonates (limestones and dolomites). It is also present in 112 minerals, including all combustible minerals. Its average contents are 96% in anthracite, 85% in petroleum, 80% in bituminous coal, and 50% in wood; lower contents are found in peats, bitumens, bituminous shales, and combustible natural gases. It is also contained in the atmosphere, hydrosphere, plants as well as in all living organisms (at average levels of about 18%).

The circulation of carbon in nature comprises two cycles, geological and biological. The geological cycle involves carbon in the form of carbon dioxide emitted into the atmosphere during the burning of fossil fuels, from volcanic gases and hot mineral springs, from the surface layers or oceanic waters, and by the weathering of rocks, and also the carbon as a constituent of sedimented calcium and magnesium carbonates. This cycle is very long-lasting. The biological cycle, in which carbon assimilated by plants as CO_2 from the troposphere is returned from the biosphere to the geosphere, is short. It consists of the uptake and fixing of CO_2 by green plants through photosynthesis, its subsequent consumption in carbohydrate, protein, and fat by animals, and its return in the inorganic state to the atmosphere and soil through respiratory processes and the decay of excreta and animal and plant bodies. In soils, the amount of carbon in organic compounds varies widely depending on the soil and attains 8000 g/m^3 in equatorial jungles. Microorganisms and fungi inhabiting the upper 15 cm soil layer contain, on average, 10–60 g of carbon per m^2 of soil. The mean residence time of carbon compounds in terrestrial ecosystems is about 15 years. Some 88% of the carbon dissolved in oceanic waters returns to the biosphere within 26 days, while the rest sediments as carbonates. The oceans absorb from the atmosphere about 30% of the CO_2, and they also release CO_2 to the atmosphere, particularly in the regions of warm waters. The waters of rivers, seas, and oceans contain approximately 60 times as much CO_2 as the atmosphere [Vadkovskaya & Lukashev].

Data on the global budget and residence times of carbon in its reservoirs are summarized in Table 2.

Table 2. Global budget and residence times of carbon in its reservoirs

Reservoirs	Amount of carbon (billion tonnes)	Residence time (years)
ATMOSPHERE		
Carbon dioxide	648	4
Methane	6.24	3.6
Carbon monoxide	0.23	0.1
Other gases	3.4	—*
	Total: 657.87	
BIOSPHERE		
Plants	827	16
Other organisms	1200	40
(extant and extinct)	Total: 2027	
OCEANS		
Plants	17.4	0.07
Particles of insoluble organic carbon compounds	30	—*
Soluble organic carbon compounds	1000	—*
Carbon dioxide/bicarbonates	38 400	385
	Total: 39 447.4	
LITHOSPHERE		
Carbonates	60.9×10^6	
Organic compounds	12.48×10^6	342×10^6

* No data.

PRODUCTION

Diamonds are obtained from diamond-bearing rocks and also artificially, for example by subjecting graphite to very high temperatures and pressures. Natural **graphites** are of three types — lump, amorphous, and flake. Lump graphite is found in veins that cross various types of igneous and metamorphic rocks; amorphous graphite is usually associated with sandstone, slate, shale, limestone, and adjunct minerals of quartz and iron sulfides; flake graphite is commonly associated with metamorphosed sedimentary rocks such as gneisses, schists, and marbles. Artificial graphite is chiefly prepared from high-grade coal by heating at high temperatures (up to 3000°C). **Brown coals, bituminous coals,** and **anthracite** are obtained by open-pit, strip, or underground mining. **Charcoal** is made usually by charring various woods without access of air. **Carbon blacks** are usually made by thermal decomposition (controlled combustion) of natural gas or oil or their mixture. Soot is commonly the substance deposited in chimneys and flues of domestic and industrial fires fed usually by coal or, less often, by wood, peat, or oil, and it is also contained in exhaust gases of diesel engines. **Coke** (an artificial solid fuel) is obtained by heating a suitable natural fuel (such as coking coal) to 950–1050°C without access of air. Pitch coke is made from a high-melting coal pitch, while petroleum coke is obtained from liquid petroleum residues and also in the cracking and pyrolysis of petroleum distillation products. **Activated carbon** (or activated charcoal) is obtained by removing tarry substances from raw coal or by carbonization of carbonaceous materials such as wood.

USES

Diamonds are used for the drilling, cutting, faceting, and grinding of particularly hard materials such as metals and glass, in filters, wire-drawing dies, and abrasive powders, and in jewelry. The principal applications of **graphite** include its use in refractories, in electrical equipment (e.g. to make electrodes), as a construction material in chemical equipment, as a component of lubricating and antifriction materials, in paints, as the lead in pencils, as an agent preventing scale formation on boiler walls. Artificial graphite is used in the nozzles and combustion chambers of jet engines and in the nose cones and certain other components of rockets; blocks of particularly pure artificial graphite are used as neutron moderators in nuclear reactors. **Coals** are used as fuels, in ferrous and nonferrous metallurgy (e.g. in the production of aluminum and in refining copper), in the manufacture of carbon sulfide and activated carbon, and in the fabrication of coke and liquid coal products and (by underground gasification processes) of gaseous fuels. **Carbon blacks** find applications in the rubber industry (mainly to increase the resistance of rubber to abrasion and wear); as black pigments in carbon paper, paints, and in printing and other inks; in the manufacture of plastics; in the electric equipment industry for stabilizing polymers (e.g. polyethylene); in the manufacture of linoleum, oil cloth, and kersey; as components of certain polishing compositions; as heat-insulating materials; as additives to concrete in road construction; and in metallurgy. **Coal coke** is mainly employed in metallurgy for the production of metals from their ores (e.g. of pig iron in blast furnaces); it is also used in the chemical industry (e.g. in the direct synthesis of calcium carbide) and to produce generator and water gases. Pitch coke and petroleum coke are used to make graphite and electrodes. **Activated carbon** is used for adsorbing gases, as a decolorizing and deodorizing agent, as a catalytic additive, and for a number of other purposes.

MAN-MADE SOURCES OF EMISSION INTO THE ENVIRONMENT

Carbon is a widely distributed constituent of atmospheric aerosols where it occurs in various physical and chemical forms. Distinction has been made between total carbon, which encompasses all carbon-containing materials; elemental carbon, represented by carbon particles entering the atmosphere as soot produced as a result of incomplete fuel combustion in industrial installations, domestic appliances, and diesel engines; and secondary carbon, by which are meant organic carbon compounds formed in atmospheric reactions [Novakov].

The residence time of elemental carbon particles in the atmosphere depends on their size and concentration and on meteorological conditions; for example, it may range from less than 40 h in rainy regions to a week or more in arid areas. The size and concentration of particles depend in turn on the efficiency with which industrial emissions are purified. When the residence time is long or concentration is high, the particles aggregate, which means that the proportion of larger particles in the atmosphere increases.

Depending on their source, elemental carbon particles may be covered with a coat, which determines their hydrophilicity. On entering a cloud, such particles act as condensation nuclei, i. e. become particles on which the condensation of atmospheric water vapor takes place. Chemical reactions occurring in the coat may give rise to nonvolatile products from gaseous substances (e.g. to sulfates from sulfur dioxide) [Orgen & Charlson].

The adsorption of carbon particles by aerosols makes the atmosphere less transparent, which decreases the number of sunny days and affects the regional climate. Carbon particles absorb solar radiation, and this may result in a thermal effect; thus, a 3% decrease in the albedo of clouds may raise the temperature of the earth's surface by 1 K [Chylek *et al.*; Zehnder].

Exhaust gases from petrol (gasoline) and diesel engines were found to contain soot at levels of $0.01-1.0$ g/m^3 [Berinia & Lapinia]. Diesel exhausts (notably from heavy trucks), which are largely composed of carbon particles, account for approximately a half of all carbon particles entering the atmosphere of large cities [Trijonis]. In urban soot, particles had mean diameters of less than 0.15 μm and their concentration averaged 2.4 μg/m³ of air but reached up to 20 μg/m³ in winter, particularly in smog [Heintzenberg & Winkler]. The distribution of carbon particles varies according to their size and the level at which they are sampled. In streets with tall buildings, their concentrations were highest at a level of 1.5 m and lowest at 48 m, being 20 μg/m³ for 0.8 μm particles and 4 μg/m³ for 3.2 μm particles [Puxbaum & Baumann]. In the Arctic, concentrations were higher at a height of 1 km (1.75 μg/m³) than at the ground level (0.5 μg/m³).

Carbon particles suspended in aerosols can travel over very large distances from the industrial source. Their concentrations were 100–400 μg/m³ over combustion sources directly, 2.4–15.9 μg/m³ over cities, 0.5–0.8 μg/m³ over rural areas, around 0.2 μg/m³ over remote continental areas, 0.1–0.68 μg/m³ over the ocean near continents, and 0.01–0.16 μg/m³ over the ocean far away from these; the zone adjacent to oceanic water had carbon particles concentrations of 4 to 30 ng/m³, the upper troposphere had 4 to 15 ng/m³, and the stratosphere, 0 to 10 ng/m³ [Andreas et al.; Clarke et al.].

Processes such as open-pit coal mining, underground coal gasification, manufacture of coal concentrate, and coal combustion in thermal power plants emit into the atmosphere a number of other substances, including sulfurous compounds, carbon monoxide and dioxide, chlorides and bromides, radioactive compounds, polycyclic aromatic hydrocarbons, and also cadmium, nickel, lead, zinc, and selenium within the volatile fractions of soot. It has been estimated that from 6×10^9 to 15×10^9 Bq enter the atmosphere daily as $^{14}CO_2$ in gaseous and particulate emissions from nuclear power plants [Rublevsky].

TOXICITY

General considerations

Populations living in areas with high levels of carbon particles in atmospheric aerosols have increased prevalence of diseases, particularly those affecting the upper respiratory tract and lungs. The main occupational diseases associated with carbon exposure are anthracosis and dust bronchitis; the morbidity due to anthracosis and its progression are influenced by the silicon dioxide content in the coal dust [Hurley et al.]. It has also been shown that of major importance for the development of dust diseases in coal miners is the total mass of respirable dust to which they have been exposed; the bulk of particles in such dust are more than 10 μg in diameter and the proportion of fine particles varies from 5 to 13%. The most aggressive coal dust particles are sized under 5–7 μm, for they can penetrate deep into lung tissue to be retained there in greater numbers than larger particles. A very important factor is also the duration of dust exposure;

longer exposures to lower dust concentrations are generally more damaging than shorter exposures to higher ones.

Another major factor is the ash content in coal; according to their ash content, all coal dusts have been divided into two categories: low-ash dusts ($\leqslant 20\%$ ash) and high-ash dusts; the most aggressive dust in the latter category is that of anthracite [Borisenkova et al.; Bykhovskaya et al.].

The hazard of anthracosis is associated with the fibrogenic action of the main petrographic coal components and with the formation of phenolic and carboxylic groups, free radicals, and peroxides during coal oxidation in the body. A factor contributing to the biological aggressiveness of coal is the presence of nitrogen compounds on its surface [Kukharenko & Dinkelis].

The harmfulness of coal dust is greater the higher the stage of metamorphism of the coal concerned, and, generally speaking, decreases in going from anthracite to hard coal to soft (bituminous) coal, being low for noncoking coals such as lignite [Balyshev]. The fibrogenicity of anthracite has been shown to be roughly the same as that of bituminous coal dust containing 30% free silica [Szymczykiwicz].

The disease process in anthracosis is triggered by the oxidation of the organic coal mass by body fluids with the subsequent formation of soluble products. The prevalence of coalworkers' pneumoconiosis increases with the degree of coalification [Bennet et al.; Dinkelis et al.].

Coal dusts impair protective mechanisms in the airways and lungs, in particular by interfering with mucociliary clearance and goblet-cell secretion. Their harmful effects depend both on the dust burden and on the cytotoxicity of the dust itself; as a result, the bronchial epithelium may be damaged and undergo desquamation [Lavreneva et al.]. This and also the obstruction of the bronchi, especially by large dust particles, lead to chronic bronchitis and to emphysema. Small dust particles are cleared by both the bronchial and lymphatic systems, while large particles are eliminated via bronchi exclusively.

One cause of pulmonary fibrosis in anthracosis is the disturbance of normal relationships between the bioantioxidant system and the concentration of free radicals. A substantial role in its pathogenesis is played by disordered lipid metabolism manifested in elevated free phospholipids [Belobragina & Yelnichnykh]. Seal et al. have associated the development of massive pulmonary fibrosis with the re-entry of coal dust from lymph nodes into the lung tissue, bronchioles, and blood vessels.

Effects in animals

Prolonged exposure of rats to dusts of coal, activated charcoal, or bone coal by inhalation or intratracheal administration caused accumulation of dust cells in their lungs and led to the development of bronchitis, peribronchitis, and then of diffuse or nodular fibrosis; the kidneys, liver, and myocardium were also affected, and increased susceptibility to bacterial infection was observed [Filippov *et al.*; Mumford & Lewitas; Paustovskaya *et al.*].

In a 4-month study, inhalation by rats of a coal dust containing 3% free and 47% bound silica (800–1000 mg dust/m^3, 4 h/day) resulted in accumulation of biogenic amines, as reflected in elevated histamine and serotonin levels in the lungs and blood and elevated serotonin in the liver [Talakin *et al.*]. In chronic inhalation studies, the fibrogenic effect of coking coal was more marked than that of coals used for power generation [Gvozdeva]. Most of the inhaled coal dust ($\approx 65\%$) is deposited in the right lung of animals. Intraperitoneal injection of coal dust suspension produced inhibition of immune responses in mice.

Diffuse or nodular pulmonary fibrosis was also induced in rats by intratracheally introduced graphite dust. Dusts of natural and synthetic diamonds were only slightly fibrogenic and led to mild pneumoconiosis, principally of a diffuse sclerotic type with a late formation of small dust foci which were more pronounced in the case of synthetic diamonds. Mild pneumoconiotic changes were also produced in rats through intratracheal administration of soot. In mice inhaling soot dust for prolonged periods, dental and oral mucosal lesions were observed (mainly hyperkeratosis and atrophy). Inhalation exposure to soot formed in the combustion of shale mazut or solid shale fuel resulted in malignant tumors, and the carcinogenic effect increased with decreasing size of the soot particles. Malignant tumors were also noted in mice following dermal applications of soot produced in electrocracking natural gas.

Effects in man

Risk of contracting occupational disease is high in a number of industries or operations, including underground coal and graphite mining (where drill operators are at special risk), open-pit mining, coal loading, coal-concentrating processes, production of graphite articles and coke, sugar refining (during the use of coarse bone black and activated carbon), and manufacture and application of carbon black.

Dust bronchitis is found in most coal workers employed in coal mines for a long time, and may lead to pulmonary fibrosis; its development is favored by work in cool or hot environments and by exposure to gases generated by blasting (shot-firing) and to sulfur compounds and other elements present in coal [Vorontsova *et al.*]. Anthracosis, on the other hand, is seen in only 20-30% of coal miners and usually after 10–20 years of work. However, early progressive anthracosis developing after 4–6 years of work has also been observed, as has been anthracosis of late onset — 18 to 20 years after leaving the mining industry. Among the factors contributing to early onset of anthracosis and its unfavorable course are steeply dipping coal seams, high degree of coalification, and high temperature and humidity in the mine. Increased morbidity has been registered among miners handling coal with high levels of copper, iron, nickel, lead, and/or zinc. Anthracosis tends to occur more frequently and to be more severe in those employed at breakage faces of coal mines or who started to work in a mine before age 20 or after age 40 [Kaganovich & Kuznetsova; Liubomudrov, 1965, 1969]. The development of anthracosis is accelerated by smoking and alcohol consumption.

Distinction is made between three types of anthracoses: interstitial, fine-nodular, and nodular. Interstitial anthracosis, which has been mainly encountered in workers of breakage faces after 16 years of employment, is not infrequently accompanied by bronchitis and an asthma-like syndrome involving impaired respiratory function, emphysema, and left ventricular hypertrophy. The second type occurs in breakage-face workers and drifters and is frequently complicated by bronchitis and tuberculosis. Nodular anthracosis resembles silicosis and usually developed after 10–14 years of employment, predominantly in drifters [Vorontsova et al.].

Early symptoms of anthracosis include rapid onset of fatigue, cough, sweating, nervous irritability, headache, and breathlessness; subsequently, cyanosis of the lips, chronic bronchitis, emphysema, bronchial obstruction, and functional respiratory disorders may develop. The incidence of emphysema increases with age (notably in smokers) and with progression of the pneumoconiotic process.

Chronic dust bronchitis and pneumoconiosis may be viewed as two forms of a single coniotic disease. Already after 5–10 years of continuous work, blood circulation in the lungs is usually abnormal. In workers with anthracosis after less than 13 years of employment, the dust content was more or less the same in all regions of both lungs, and the latter often appeared uniformly black; those employed for longer periods had the greatest quantities of dust in the upper and middle lobes of the right lung. Among workers with equal lengths of employment, more dust was present in the lungs of older individuals [Einbrodt]. Characteristic findings in the early stages of anthracosis were coarsening of the argyrophilic lung structures, intensive phagocytic reaction, and impaired vascular elasticity; the later stages were marked by the appearance of diffuse fibrosis with the presence of small and larger nodules and a massive (primarily centrilobular) emphysema with dust accumulations on the walls of dilated bronchioles ('black honeycombed lung') [Mülleri & Werth]. In severe cases the bronchioli undergo breakdown and coal-filled cavities are formed (a condition described as black phthisis of coal miners); spontaneous pneumothorax may develop. If exposure is discontinued at an early stage, the dust foci in the lungs usually regress.

The most common complication of anthracosis is tuberculosis. Anthracosis also predisposes to the development of malignant neoplasms in the lungs and other organs [Dechoux & Pivateau]. Tumor-like enlargements of mediastinal lymph nodes (anthracomas, anthracosilicomas) are encountered even in cases of relatively mild pneumoconiosis and may be misdiagnosed as lung cancer.

Other complications of the coniotic pulmonary disease are myocardial lesions, hypotension, disordered peripheral circulation, and increased blood coagulability.

There are also reports of abnormal liver, kidney, and stomach functions, impaired androgenic and glucocorticoid functions of the adrenals, disordered metabolism of biogenic amines, and reduced immunobiologic activity. Bronchial asthma, rheumatoid polyarthritis, and affections of the ear, nose, and throat were observed in some cases [Gridneva *et al.*; Heise *et al.*; Liubomudrov, 1965; Vinarik & Novikova]. Increased blood coagulability and disturbances of protein and lipid metabolism have occurred not only in patients with anthracosis but also among miners suspected of having one as well as in apparently healthy miners working underground. It has been suggested that these latter changes may be regarded as predictors of anthracosis [Pichkhadze]. Similarly, substantial concurrent increases of urokinase and mitochondrial aspartate aminotransferase activities in blood serum are said to be an early indicator of liver disease in anthracosilicosis [Altynbekov *et al.*]. The appearance of frank anthracosis in some coal miners after 10 or more years of employment was sometimes preceded by chronic rhinosinusitis [Chernov].

Workers engaged in the mining or processing of natural **graphite**, or in the production of graphite articles, not infrequently complain of headache, coughing, depression, low appetite, and black sputum. After 10 or more years of work, impaired ventilatory function of the lungs was observed even among apparently healthy individuals, most often smokers. After prolonged exposure, chronic bronchitis is common, with an asthmatic syndrome in some cases. A graphite form of anthracosis (graphitosis) has been described, marked by diffuse or nodular pulmonary fibrosis, emphysema, and the presence of necrotic areas and of cavities with a graphite-containing fluid. The disease usually developed after more than 15 years of work, but has also been encountered after as little as 4 or 5 years. Increased morbidity, general and particularly gynecologic, has been recorded among women engaged in the manufacture of graphite electrodes [Gladkova & Abdyrakhmanova]. Defective functioning of the T-cell system of immunity and secondary immunodeficiency have also occurred in graphite workers.

Biochemical evidence of destructive processes in the lungs of graphite workers with pneumoconiosis and bronchitis has been obtained by noting alterations in a number of parameters, including β- and γ-globulins, sialic acids, β-lipoproteins, cholesterol, sugar, and total and direct bilirubin [Arkhipova *et al.*; Bezzub *et al.*; Gladkova *et al.*]. Workers in plants manufacturing graphite articles are at increased risk of developing preneoplastic conditions of the upper air passages (hyperplastic laryngitis, laryngeal angiofibroma) and malignant neoplasms by being exposed to benzpyrene-containing polycyclic aromatic hydrocarbons adsorbed onto graphite dust [Dymova *et al.*; Gladkova *et al.*]. Combined exposure to graphite dust, slightly fibrogenic coke dust, and polycyclic aromatic hydrocarbons resulted in impaired lipid peroxidation with a consequent accumulation of lipids in the body — both in workers with bronchitis and those who were apparently healthy [Bezrukavnikova & Gladkova].

Pneumoconiosis has been found among workers engaged in the production or use of **carbon black**   and related products, presenting radiologically as small round opacities and increased bronchial and vascular markings. Some cases of massive fibrosis have also occurred, probably as a result of combined exposure to carbon black and silica [Risquez-Iribarren]. Carbon black workers complained of easy fatiguability, cough, shortness of breath, and chest pain, and were found to have increased prevalence of respiratory, digestive, and cardiovascular diseases [Oleru *et al.*]. Particular hazards are associated with the production of channel black. The mean duration of employment before onset of pneumoconiosis was 16 years, but in some cases the disease was diagnosed after 5 years   of exposure. A case of 'black phthisis' with cavity formation developing in a worker after 12 years of employment in carbon black sieving was described [Gerasimov & Shatalova].

Neoplasms developed in people exposed to soots produced in the combustion of slate mazut or solid slate fuel or to the carbon black obtained from coal pitch or in the electrocracking of natural gas [Shabad *et al.*]. Some (slight) blastomogenic activity was possessed by extracts of soot-containing resins. Channel black is believed to be less carcinogenic than ordinary soot because it contains less polycyclic aromatic hydrocarbons and because, in addition, these are very slowly freed from their strong bonds with the carbon particles [Boyland].

Carcinogenic properties of carbon blacks (soots) are due to the presence of impurities, particularly 3,4-benzpyrene. In some studies this carcinogen was detected in all types of carbon black and in all industrial products to which it had been added as a filler [Barknecht; Pylev]. The observed mutagenic activity of soot has been attributed to the presence of 1-nitropyrene.

Pulmonary emphysema caused by soot occurred after prolonged (for decades) and consistent occupational exposure, and it has also been observed in town dwellers living in atmospheres heavily laden with coal smoke. Pneumoconiosis due to soot has been described in furnace operators [Lawther].

Chronic inflammatory diseases of the upper respiratory tract were observed among workers of **coke** plants after 2–6 years of employment [Kandus]. The incidence of cancer of the bronchi was found to be significantly increased in coke-plant workers on certain jobs, apparently as a result of exposure to high atmospheric concentrations of 3,4-benzpyrene (4–200 mg/L of air). Pneumoconiosis from coal or coke dust is not a significant health hazard in coke plants except for those whose bronchial epithelium has already been damaged by disease or smoking [Zorn].

Coal pitch produces irritant lesions in the skin and mucous membranes, and it can also induce cutaneous neoplasms, both benign and malignant. Occupational cancer developed, for instance, in a worker engaged in the briquetting of coal fines using coal pitch.

Local actions of coal, graphite, and soot dusts

These dusts can give rise to tattooing of the conjunctiva, conjunctivitis, and corneal irritation. Their skin effects include dryness, pustulous lesions of the skin and subcutaneous tissue, dermatitis, epidermophytosis, allergic dermatoses, mycoses, and eczema. Psoriasis has also been encountered among exposed workers (more often in males), as have been pigmented spots on the skin, which are sometimes thought to be precancerous lesions. Coal-derived pitch can act as a photosensitizer — cases of acute photodermatitis have been described [Gladkova et al.]. Oral mucosal and nasopharyngeal lesions such as gingivitis, keratosis, and leukoplakia have also occurred, mainly in males under 40 (after 3–5 years of exposure) [Sorinson]. Increased prevalence of caries and paradontosis has been recorded among chimney sweeps and coke-plant workers [Karaboicheva & Velikov].

DISTRIBUTION AND ELIMINATION IN MAN

Chemical analysis of lungs affected with anthracosis shows a high percentage content of coal dust and a low percentage content of silica. Coal dust has also been found in small amount in the spleen, liver, bone marrow, and kidneys of anthracotic patients. Coal dust is eliminated from the lungs over many years after the cessation of occupational exposure.

HYGIENIC STANDARDS

Exposure limits adopted in the USSR for carbon dusts and carbonaceous fibrous materials are shown in Table 3. In the USA, a TLV-TWA value of 3.5 mg/m^3 has been set for carbon black.

MEASURES TO CONTROL EXPOSURE IN COAL MINING AND CARBON PRODUCTION

Health problems of particular concern in coal mining are those associated with adverse microclimatic conditions in mines and with airborne dust. Climatic conditions are generally improved through optimal distribution of ventilation air and avoidance of excessive humidity in the pits, roadways, and coal faces. Dust control, especially in underground mines, still involves considerable efforts despite the progress that has been made in recent decades. Detailed descriptions of dust control and other health measures can be found in the health and safety regulations adopted in various countries.

Table 3. Exposure limits for carbon dusts and
carbonaceous fibrous materials

	Workroom air		Atmospheric air			
	MAC_{wz} (mg/m^3)	Aggregative state	MAC_{hm} (mg/m^3)	MAC_{ad} (mg/m^3)	Hazard class	Note
Carbon dusts:						
Cokes (coal, pitch, petroleum, & slate cokes	6.0	Aerosol	—	—	4	F[a]
Other fossil coals containing free silicon dioxide:						
at <5%	10.0	Aerosol	—	—	4	F
at 5–10%	4.0	Aerosol	—	—	3	F
Diamonds:						
natural & synthetic	8.0	Aerosol	—	—	4	F
metallized	4.0	Aerosol	—	—	3	F
Carbon blacks (industrial), with benzpyrene levels of <35 mg/kg	4.0	Aerosol	0.15	0.05	3	F, C[b]
Carbonaceous fibrous materials based on hydrated cellulose fibers	4.0/2.0[c]	Aerosol	—	—	4	
on polyacrylnitrile fibers	4.0/2.0[c]	Aerosol	—	—	4	

[a] Predominantly fibrinogenic.
[b] Fibrogenic and carcinogenic.
[c] The MAC value in the numerator is the concentration not to be exceeded at any time, while that in the denominator is the maximum allowable average concentration per shift.

In hole-drilling operations such as the drilling by means of perforators for shotfiring, airborne dust formation can be effectively reduced by watering the drill holes, particularly if surface-active agents (surfactants) are added to the water. This method of dust suppression in combination with adequate ventilation will keep dust concentrations in the mine air within permissible limits.

An efficient means of suppressing dust even before it is released from the coal seam is water infusion into the latter either perpendicular or parallel to the front of its face. To control dust at its source, i.e. when the particles are formed, sufficient amounts of water should be supplied to the coal-winning tools before the particles become airborne (wet drilling). Where water cannot be used for some reason (e.g. low temperature of the air and rock, leaching out of valuable minerals, or scarcity of water), dry methods of dust collection should be used. In the USSR, a technique of dry dust collection was devised by which dust and drilling fines are suctioned with a vacuum pump from several drill holes at a time and the dust–air mixture is then transported by pipeline to an abandoned blind working (dust collector). This hermetically sealed collector also serves as a dust-settling chamber in which large and small particles are precipitated; the finest unsedimented particles are trapped by a special filter. The air cleaned in the collector re-enters the circulating air.

In blasting operations (shotfiring), adequate dust control has been achieved through the use of hydraulic stemming — even in northern and northeastern regions of Russia where low temperatures prevail. The cloud of gas and dust raised by the blast is suppressed by spraying.

A simple and cost-effective method of dust control during loading/unloading and transport operations in mines is spraying of all dust sources from stationary or portable spraying devices. These should be so designed as to provide sufficient quantities of water at a relatively high pressure. The wetting process can be accelerated and dust binding improved by adding surfactants to the water. The addition of surfactants should also be considered in cases where there is not enough water to ensure effective spraying.

In opencut (opencast) coal mining, the main measures against airborne dust formation in mines include dry dust collection, wet drilling, and spraying, and these measures are of great importance also for people living in the vicinity of opencut mines; above all, action should be taken to minimize dust formation on the routes to the mines — by sprinkling, greening, use of concrete or ferroconcrete slabs for paving the routes, and treatment of these with granulated calcium chloride, water–asphalt emulsion, magnesium sulfate, or other suitable agents. The principal method of dealing with the dust that has become airborne is ventilation, natural or artificial.

The cabs of mining machines and mechanisms as well as of transport vehicles must be reliably protected from dust and be fitted with ventilation devices in which provision is made for air cleaning.

In plants where carbon is obtained, technical control measures should be mainly directly at dust control and excluding the mutual influences on the workers' health of noxious substances present in the various departments and working areas. Preference should be given to wet methods of preparing pellets. Air ducts, the channels of gas exhausters, flues, and air purifiers should be fitted with devices for their periodic mechanized cleaning with removal of the dust and sludge. Packing machines should be provided with local exhaust ventilation with automatically controlled exhaust rates.

All workers should be made aware of the hazards associated with their specific jobs. Before employment and systematically during the period of employment they should be fully instructed in, and required to observe, the appropriate safety precautions and general and personal hygiene measures. Pre-employment and periodic in-employment medical examinations are also necessary.

Workers should be provided with, and use when necessary, appropriate personal protection equipment such as respirators, working clothes made of closely woven fabric, waterproof footwear, goggles, and protective helmets. (See also the corresponding section in the chapter Silicon compounds.)

REFERENCES

Altynbekov, B. E. *et al.* (1986) *Gig. Truda Prof. Zabol.*, No. 9, 47—48 (in Russian).

Andreas, M. O. *et al.* (1984) *Sci. Total Environ.*, 36, 79—80.

Arkhipova, O. G. *et al.* (1988) *Gig. Truda Prof. Zabol.*, No. 9, 24—27 (in Russian).

Balychev, Yu. V. (1982) *Gig. Truda Prof. Zabol.*, No. 1, 48—50 (in Russian).

Barknecht, Th. S. (1983) *Progr. Energy Combust. Sci.*, 9, 199—237.

Belobragina, E. V. & Yelnichnykh, L. N. (1978) *Gig. Truda Prof. Zabol.*, No. 3, 31—34 (in Russian).

Bennet, J. G. *et al.* (1979) *Brit. J. Ind. Med.*, 36, 206—210.

Berinia, D. Zh. & Lapinia, I. M. (1980) In: *Zagryazneniye prirodnykh sred vybrosami avtotransporta* [Environmental Pollution by Automobile Exhausts]. Riga, pp. 7—15 (in Russian).

Bezrukavnikova, L. M. & Gladkova, E. V. (1987) *Gig. San.*, No. 4, 33—36 (in Russian).

Bezzub, S. L. *et al.* (1988) *Gig. Truda Prof. Zabol.*, No. 9, 27—28 (in Russian).

Borisenkova, R. V. *et al.* (1983) *Gig. San.*, No. 5, 13—16 (in Russian).

Boyland, E. (1983) In: *Soot in Combustion Systems and its Toxic Properties. Proceedings of a NATO Workshop.* New York & London, pp. 13—22.

Bykhovskaya, I. A. *et al.* (1978) *Gig. Truda Prof. Zabol.*, No. 10, 34—39 (in Russian).

Chernov, A. D. (1979) *Gig. Truda Prof. Zabol.*, No. 1, 15—18 (in Russian).

Clarke, A. D. *et al.* (1984) *Sci. Total Environ.*, 36, 97—102.

Chylek, R. *et al.* (1984) *Sci. Total Environ.*, 36, 117—120.

Dechoux, J. & Pivateau, C. (1979) *Arch. Mal. Profes.*, 40, 25—41.

Dinkelis, S. S. *et al.* (1981) *Zdravookhraneniye Kazakhstana* [Health Care in Kazakhstan], No. 3, 44—47 (in Russian).

Dymova, E. G. *et al.* (1988) *Gig. Truda Prof. Zabol.*, No. 7, 19—21 (in Russian).

Einbrodt, H. (1976) *Staub-Reinhaltung Luft*, 36, 122—126.

Filippov, V. V. *et al.* (1971) *Gig. Truda Prof. Zabol.*, No. 10, 58—60 (in Russian).

Gerasimov, V. N. & Shatalova, N. A. (1967) *Gig. Truda Prof. Zabol.*, No. 7, 56—58 (in Russian).

Gladkova, E. V. & Abdyrakhmanova, A. A. (1984) *Gig. Truda prof. Zabol.*, No. 8, 31—34 (in Russian).

Gladkova, E. V. *et al.* (1988) *Gig. Truda Prof. Zabol.*, No. 8, 21—24 (in Russian).

Gridneva, N. V. *et al.* (1982) *Gig. Truda Prof. Zabol.*, No. 4, 19—22 (in Russian).

Gurvich, E. B. (1988) *Gig. Truda Prof. Zabol.*, No. 7, 16—18 (in Russian).

Gvozdeva, L. L. (1985) *Gig. Truda Prof. Zabol.*, No. 7, 12—16 (in Russian).

Heintzenberg, J. & Winkler, P. (1984) *Sci. Total Environ.*, 36, 27—38.

Heise, E. R. *et al.* (1979) *Amer. Rev. Respir. Dis.*, 119, 903—908.

Hurley, J. F. *et al.* (1982) *Brit. J. Ind. Med.*, 39, 120—127.

Kaganovich, D. I. & Kuznetsova, V. A. (1968) In: *Gigiyena* [Hygiene]. Novosibirsk, pp. 140—148 (in Russian).

Kandus, J. (1976) *Pracovni Lekarstvi*, 28 (No. 3), 75—80.

Karaboicheva, P. & Velikov, B. (1984)   *Stomatologiya* (Bulgaria), No. **4**, 18–23 (in Bulgarian).

Kukharenko, T. A. & Dinkelis, S. S. (1982) *Ugol* [Coal], No. **8**, 47–49 (in Russian).

Lavreneva, G. V. *et al.* (1988) *Vestnik Otolaringol.*, No. **3**, 67–68 (in Russian).

Lawther, P. J. (1983) In: *Encyclopaedia of Occupational Health and Safety.* International Labour Office, Geneva, p. 2090.

Liubomudrov, V. E. (1965) *Zabolevaniya legkikh u shakhterov-ugolshchikov* [Pulmonary Diseases in Coal Miners]. Kiev (in Russian).

Liubomudrov, V. E. (1969) *Gig. Truda Prof. Zabol.*, No. **3**, 20–22 (in Russian).

Mulleri, K.-M. & Werth, G. (1978) *Prax. Klin. Pneumol.*, **32**, 231–237.

Mumford, J. & Lewitas, J. (1983) *Toxicol. Lett.*, **18** (Suppl. 1), 132–134.

Novakov, T. (1984) *Sci. Total Environ.*, **36**, 1–10.

Oleru, U. G. *et al.* (1983) *Environ. Res.*, **30**, 161–168.

Orgen, J. A. & Charlson, R. J. (1983) *Tellus* (Stockholm), **35B**, 241–254.

Paustovskaya, V. V. *et al.* (1972) *Vrach. Delo*, No. **1**, 131–134 (in Russian).

Pichkhadze, G. M. (1985) *Gig. Truda Prof. Zabol.*, No. **2**, 36–39 (in Russian).

Puxbaum, H. & Baumann, H. (1984) *Sci. Total Environ.*, **36**, 47–52.

Pylev, L. N. (1974) *Gig. Truda Prof. Zabol.*, No. **4**, 52–53 (in Russian).

Risquez-Iribarren, R. (1983) In: *Encyclopaedia of Occupational Health and Safety.* International Labour Office, Geneva, pp. 390–392.

Rublevsky, V. P. (1983) *Atomnye elektrostantsii* [Atomic Power Stations], No. **5**, 226–227 (in Russian).

Seal, R. *et al.* (1986) *Thorax*, **41**, 531–537.

Shabad, L. M. & Dikun, P. P. (1959) *Zagryazneniye atmosfernogo vozdukha kantse-rogennym veshchestvom — 3,4-benzpirenom* [Atmospheric Pollution by the Cancerogenic Substance 3,4-Benzpyrene]. Leningrad (in Russian).

Shabad, L. M. *et al.* (1972) *Gig. Truda Prof. Zabol.*, No. **1**, 9–12 (in Russian).

Sorenson, J. *et al.* (1974) *Amer. Ind. Hyg. Assoc.-J.* **35**, 93–98.

Sorinson, N. S. (1957) In: *Materialy po voprosam toksikologii i kliniki professionalnykh zabolevaniy* [Industrial Toxicology and Clinical Features of Occupational Diseases], No. **8**. Gorky, pp. 54–56 (in Russian).

Szymczykiwicz, K. (1983) *Med. Pracy*, **34** (No. 1), 21–34.

Talakin, Yu. S. (1988) *Gig. Truda Prof. Zabol.*, No. **12**, 57–58 (in Russian).

Trijonis, J. (1984) *Sci. Total Environ.*, **36**, 131–140.

Vadkovskaya, I. K. & Lukashev, K. P. (1977) *Geokhimicheskie osnovy okhrany biosfery* [Protection of the Biosphere: Geochemical Principles]. Nauka i Tekhnika, Minsk (in Russian).

Velichkovsky, B. T. *et al.* (1979) *Gig. Truda Prof. Zabol.*, No. **5**, 1–5 (in Russian).

Vinarik, E. M. & Novikova, L. A. (1976) *Gig. Truda Prof. Zabol.*, No. **7**, 30–33 (in Russian).

Vorontsova, E. L. *et al.* (1982) In: *Borba s silikozom* [Silicosis Control], No. **11**. Moscow, pp. 106–114 (in Russian).

Zehnder, A. J. (1982) In: Hutzinger, O. (ed.) *The Handbook of Environmental Chemistry*, Vol. 1, Part B. Springer, Berlin, pp. 83–110.

Zorn, H. (1983) In: *Encyclopaedia of Occupational Health and Safety.* International Labour Office, Geneva, pp. 502–504.

Oxides of carbon

CARBON MONOXIDE (CARBON(II) OXIDE)

Physical and chemical properties

Carbon monoxide (CO) is a colorless, odorless, and tasteless gas that is slightly lighter than air and only slightly soluble in water. It is almost negligibly absorbed by activated carbon. In air it burns to CO_2 with a blue flame and with the evolution of much heat. Its flammability limits in mixture with air are 12.5–74.2%. A mixture of two parts of CO and one part of O_2 by volume will explode on ignition.

Carbon monoxide is chemically inert at low temperatures. At high temperatures and in the presence of a catalyst it becomes reactive, for example combines with chlorine to form carbonyl chloride (phosgene) and with metals to give metal carbonyls. It does not react with water, acids, or alkalies. In the presence of a mixture of magnesium and copper oxides, it oxidizes to CO_2 at room temperature. It can act as a strong reducing agent. See also Appendix.

Natural occurrence and environmental levels

Carbon monoxide is a constant component of the earth's atmosphere where its natural levels range from 0.01 to 0.9 mg/m^3. A large difference has been found between the atmospheric CO concentration in the Northern Hemisphere (0.13 ppm) and that in the Southern Hemisphere (0.04 ppm) [Spedding]. Its residence time in the atmosphere is believed to be ~0.2 year [WHO]. It enters the atmosphere as a result of incomplete combustion of organic matter and by being released from microorganisms, plants, animals, and man. Surface layers of the ocean were calculated to emit annually up to 220×10^6 tonnes of CO released by plankton and by blue–green, red, and other algae. The concentration of CO in the ocean surface water is relatively constant over all the oceans in both hemispheres, averaging 6×10^{-8} cm^3 CO per cm^3 water [Spedding]. Among other natural sources of CO in the atmosphere are forest and steppe fires, volcanoes, and marsh gases. A substantial source may be the oxidation of methane in the troposphere and also its biological oxidation. Some CO is formed in the upper atmosphere (above 75 km) by the photodissociation of CO_2 and some is produced in the clouds during electric storms and deposited on the earth's surface in rainwater. In addition, CO is formed during the germination of seeds and the growth of seedlings.

In man and animals, carbon monoxide is produced as a by-product of heme catabolism and lipid peroxidation.

The major mechanisms for removal of CO from the atmosphere are its atmospheric oxidation and its uptake by soil, vegetation, and inland fresh waters.

The carbon monoxide generated at the earth's surface migrates to the troposphere and stratosphere where it is oxidized to CO_2 (in particular, by reacting with the hydroxyl (OH) radical with formation of formaldehyde), accelerates both the oxidation of nitric acid to nitrogen dioxide and the rate of ozone formation, and interacts with various atmospheric pollutants. Much CO is metabolized by fungi and microorganisms present in the soil, which oxidize it to CO_2 and utilize to synthesize hydrogen sulfide and methane. Some plant species, both unicellular and higher, also remove CO from the atmosphere by oxidation to CO_2. Desert soils have been shown to take up CO at the lowest rates and tropical soils, at the highest. The uptake by cultivated soils occurs at lower rates than by uncultivated soils, probably because there is less organic matter in their surface layers. Animals and humans, too, can oxidize CO to CO_2, but only on a small scale, and this oxidation plays virtually no role in regulating the CO balance in nature. Some CO is removed from the atmosphere by inland waters. Appreciable amounts of CO are contained in rainwater, and its run-off into rivers and lakes accounts for additional removal [Tiunov & Kustov; WHO]. It has been suggested that the oceans are not only sources of CO but can also take it up from the atmosphere, thus acting as its sinks.

Production and uses

Carbon monoxide is produced when organic material such as coal, paper, oil, gasoline, gas, explosives, or any other carbonaceous material is burned in a limited supply of air or oxygen. It is also produced upon interaction of CO_2 with red-hot coal and during methane conversion in the presence of a catalyst. On an industrial scale, it is usually produced by the partial oxidation of hydrocarbon gases from natural gases or by gasification of coal or coke.

Carbon monoxide is one of the major starting materials in the modern industry of organic synthesis [Zorn *et al.*]. It is used for reducing metals from their oxides; for obtaining metal carbonyls, carbonyl chloride, carbonyl sulfide, aromatic aldehydes, formamide, formic acid, hexahydroxybenzene, aluminum chloride, and methanol; in reactions of carbonylation (in which it interacts with unsaturated organic compounds) and hydroformylation; and as a reducing agent in metallurgy. From a mixture of CO and H_2, synthetic gasoline and synthol (a mixture of carboxylic acids, alcohols, aldehydes, ketones, and hydrocarbons) can be obtained. Several industrial gases that are used for heating boilers and furnaces and driving gas engines contain carbon monoxide. Thus, water gas contains ~40% of CO, blast-furnace gas ~30%, producer gas 25%, and coal gas ~5%.

Man-made sources of emission into the environment

Global emissions of carbon monoxide due to human activities have been estimated to be between 350×10^6 to 600×10^6 tonnes per annum. The main domestic sources of CO are gas or coal stoves and heaters, faulty gas pipes, fires, and tobacco smoke. During a 2-hour operation of three gas burners in a kitchen of 16 m^3, for example, the CO concentration in the indoor air increased 11-fold to $37-40$ mg/m^3, with a resultant twofold elevation of carboxyhemoglobin (COHb) levels in the blood of inhabitants. CO poisoning is the cause of about 50% of all deaths during fires, and the blood of fire victims who died often contained more than 60% of COHb. Tobacco smoke contains on average $0.5-1.0$% of CO (the mainstream smoke of cigarettes may contain >4%), and tobacco smoking is considered as the principal source of nonoccupational exposure to CO [Zorn *et al.*].

The major industrial sources are the emissions of combustion products of carbon-containing compounds used as fuels in heat- and electricity-generating plants, boilers, furnaces, internal combustion engines, and also emissions from a variety of industrial establishments where the indoor air contains elevated CO concentrations. CO is a component of many gas mixtures (Table 1).

In coal mines, CO concentrations near the drill holes used for shotfiring attained 350 mg/m^3 5 to 10 min after the explosion. In opencut mining, concentrations ranged from 10 mg/m^3 at the surface to 80 mg/m^3 at 175 m below the ground level. CO also forms during surface oxidation of coal on coal-feeding lines. Coal dusts contain 0.1% to 3.9% of CO. A source of CO near mines may be smoldering rock dumps (e.g. concentrations exceeding the permissible level were recorded at distances up to 500 m from these). CO may be released from coal-containing dust when it is handled (particularly if the dust contains soot), from coal bunkers, from coal holds of ships, and during the crushing, processing, and storage of certain materials such as coke or slag.

Table 1. Carbon monoxide levels in various gases

Gases	CO (%)
Illuminating gases:	
Air gas	≈100
Coal Gas	4–11
Coke-oven gas	7
Oxygen gas	95–97
Semiwater gas	32–36
Shale gas	17
Vent gas	≈27
Water gas	37–39
Producer gases:	
From coal or coke	27
From lignite	29
Peat gas	20[a]
Wood gas	28
Oil gas from paraffin	9
Industrial fuel gases:	
From gas generators	12–50
From smoldering brown coal	10.3
Synthetic gases:	
For CO-based syntheses	85
For hydrogenation	0.4
Gases evolving in iron and steel works:	
Bessemer gas	25
Blast-furnace gas	30
From cupolas	13–15[b]
From mold boxes	18
From molten cast iron	3.4
Gunpowder gases	up to 50
Gases produced upon explosion of:	
Charcoal gunpowder	3–9
Dynamite	34
Melinite	61
Nitrocellulose	42.7
Picric acid	61
Trinitrotoluene	57

[a] Purified (without CO_2).
[b] The CO concentration may attain 24–29 mg/m^3 100–300 m away from the cupola.

In the chemical industry, significant sources are catalytic cracking units in petroleum refineries; distillation of coal and wood; manufacture of carbide, ammonia, formalin, and soda; hydrogenation of fats; synthesis of hydrocarbons, phosgene, methanol, formic and oxalic acids, methane, and some other compounds; CO may also be emitted during the production and processing of synthetic fibers. Elevated CO concentrations are generated near lime kilns, in brick-making and cement plants, in blast furnaces, coke ovens, and in the ceramics industry. Metallurgical works with an annual output of 1 million tonnes of steel were estimated each to emit 320 to 400 tonnes of CO daily [Ilkun]. CO is also emitted into the air during the fermentation of leaf tobacco, in bakeries, and from certain copying machines.

Further sources of CO release into the ambient air are the production of graphite electrodes and carbon black and gas welding using carbon dioxide. In fact, any industrial process or operation where incomplete combustion of carbonaceous material takes place is a potential source of environmental CO pollution. However, more than a half of all man-made CO emissions to the atmosphere are accounted for by the exhaust gases of internal combustion engines. It has been calculated that, on average, a gasoline-powered automobile covering a distance of 15 000 km per year, will add 530 kg of CO to the atmosphere and remove 4350 kg of O_2 over that period [Ilkun]. The CO concentration in the ambient air due to motor vehicle traffic depends on several variables, including engine characteristics (capacity, type of fuel, use of emission control devices), the number of operating vehicles, speed of traffic and gradient of speed, ambient temperature (as it affects the operating efficiency of the engine), meteorological conditions, and geometry of locality [WHO].

Depending on the engine and the mode of its operation, the amount of CO emitted per kg of burned gasoline varies from 150 to 800 g and the CO content in the exhaust gas, from 1% to 12%. As this gas contains in addition up to 200 different chemical substances, the combined toxic effect of these and CO is far greater than that of CO alone. Carbon monoxide poisoning of drivers is not an uncommon cause of road traffic accidents [Tiunov & Kustov].

The CO content of motor vehicle exhausts increases on congested or poorly paved roads and on sloping stretches of the roadway. CO concentrations in the ambient air over crossings may be 2.5–4 times higher than between them, and such increased levels have been shown to extend to distances of 500 m from the crossing along the highway and up to 150 m on both sides of it. In summer, CO accumulates in the green zones in front of houses on both sides of highways and also in closed, poorly ventilated yards. The air over highways was found to be effectively cleared from exhaust-derived CO at wind velocities above 3 m/s. Average CO concentrations over main streets of cities, for example, ranged from 6 to 57 mg/m^3 and the highest concentrations, from 290 to 600 mg/m^3, reaching up to 600 mg/m^3 in tunnels. Sources of large CO emissions may be garages with forced ventilation, especially multi-story parking garages (concentrations of

2000 mg/m^3 were recorded near ventilation gratings); in ground-level garages, CO concentrations reached $170-200 \text{ mg/m}^3$ during rush hours, while in underground garages the average concentration was 112 mg/m^3 and the highest, 570 mg/m^3 [Tiunov & Kustov].

Exhaust gases of properly operating diesel engines contain on average 0.1% CO. Diesel engine exhausts from locomotives and sea-going ships contained $70-93$ and $70-80 \text{ mg CO/m}^3$, respectively. CO was found to be the main contributor to the total mass of emissions from airplanes, with 80-90% of this being emitted while the plane is on the runway (before takeoff and after landing). There is evidence that CO may rank first among causes of poisoning in aircraft crews, and it has been reported that 19% of aircrashes were associated with CO poisoning [Bushtuyeva & Feldman; Tiunov & Kustov].

Toxicity

General considerations
Carbon monoxide is the most widely encountered gas of all gases known to have toxic effects upon humans and animals, and is recognized as the most common cause of human poisoning, both in industry and in homes.

CO exerts its toxic effects primarily by interrupting the normal oxygen supply to the body tissues as a result of readily combining with hemoglobin to form carboxyhemoglobin (COHb) because the affinity of hemoglobin for CO is much ($\geqslant 200$ times) higher than its affinity for oxygen.

In the tissues, CO also binds to myoglobin to form carboxymyoglobin, which impairs muscle metabolism, especially in the heart. Although the affinity of mammalian myoglobins for CO is lower than that of hemoglobin, it is still $25-50$ times as high as for oxygen; in severe poisoning, about 25% of myoglobin can be CO-bound and the carboxymyoglobin level in the muscles may reach 35%. CO can also combine with cytochrome, cytochrome oxidase, with the reduced form of peroxidase (to yield compounds resembling carboxyhemoglobin), and with catalase, and it inhibits tyrosinase and succinate dehydrogenase activities in the liver, heart, and brain [Zorn *et al.*]. In a number of cases, death from acute CO poisoning occurred in the presence of relatively low blood levels of COHb (45-55%). The development of chronic CO poisoning does not necessarily involve anoxemia. CO intensifies glycogen breakdown in the liver and impairs glucose utilization, with a consequent elevation of sugar in the blood and cerebrospinal fluid and its appearance in the urine. Acute poisoning results in increased levels of total catecholamines in brain tissue and of lactic acid in blood with the development of lactacidemia, and in elevated activity of glycolytic enzymes in erythrocytes. Moreover, CO impairs phosphorus metabolism; strongly excites carotid chemoreceptors; interferes with nitrogen metabolism, leading to azotemia, alterations in levels of plasma proteins, and reduced blood levels of vitamin B_6 and acetylcholinesterase activity; affects lipid peroxidation; inhibits lipase activity and disturbs lipid metabolism with the result that lipids rise in blood plasma and more cholesterol is deposited in the vessel walls. Hyperkaliemia may also occur, as a result of impaired water and salt metabolism [Aleksandrov *et al.*; Gadaskina *et al.*, 1961; Gutierrez *et al.*; Perret; Tiunov & Kustov; Tiunov *et al.*].

Carbon monoxide lowers the ventricular fibrillation threshold and depresses cardiac activity, increases blood coagulability and vessel wall permeability, and stimulates erythropoiesis by causing the kidneys to release more eythropoietin [Pankow & Ponsold]. It is capable of potentiating the hypertensive effect of food high in sodium [Shiotsuka *et al.*]. According to some authors, CO can also act on the myocardium directly [Chen & McGrath].

There is considerable individual variation in the sensitivity to carbon monoxide poisoning, both acute and chronic. Particularly sensitive are children, pregnant women, heavy smokers and alcohol drinkers, and also individuals with any condition that impairs oxygen supply to vital organs such as the heart, brain, and kidneys. Those with severe anemia may die even on exposure to relatively low CO concentrations.

Acute exposure

Animals

Experiments on animals have provided valuable information about the effects of carbon monoxide. Most animals die when COHb levels exceed 70%, while levels above 50% are often associated with damage to organs including the brain and the heart. Animals exposed to very high CO concentrations die within 1 min. At lower concentrations, the most conspicuous symptoms of poisoning are anxiety, respiratory distress, convulsions, loss of sensitivity, titanic muscular contractions, pupillary dilatation, and bulging of the eyes. Acute toxicity data collected by Tiunov and Kustov from various sources are presented in Tables 2 and 3. The 4-hour LC_{50} values were reported to be 2790 mg/m^3 for mice, 2070 mg/m^3 for rats, 6550 mg/m^3 for guinea-pigs; the lowest concentration lethal to rabbits and dogs after 46 min of exposure was 4580 mg/m^3 [Sax].

In mice, acute poisoning that led to convulsions was found to have caused acidosis and falls in glycogen, glucose, and DNA in the brain. Acutely poisoned rats had elevated levels of epinephrine and norepinephrine in brain tissue and of dopamine in the hypothalamus and frontal region of the cerebral cortex. Increased permeability of the blood-brain barrier, impaired glucose utilization, hyperglycemia, and glucosuria were also recorded, as were a number of biochemical changes, including elevated activities of serum transaminases, raised blood levels of lactic acid, altered peroxidation processes (a depression of superoxide dismutase activity in the blood and its rise in the liver), lowered xanthinoxidase activity in the liver, and enhanced steroidogenesis in the adrenal cortex. Also in rats, 4-hour exposure to 800 mg/m^3 disorganized the sleep-and-wakefulness rhythm, while 15-min exposure to $12\,500\ \mathrm{mg/m^3}$ resulted in considerable behavioral changes and impairment of short-term memory in the surviving animals [Nemby-Schmidt *et al.*; Schellenberger; Shamarin & Yurasova; Tiunov & Arutyunian; Tiunov & Kustov; Tiunov *et al.*].

Table 2. Percentage mortality in mice and rats exposed to various carbon monoxide concentrations

Mice: exposure time = 120 min		Rats: exposure time = 60 min	
CO concentration $(\mathrm{mg/m^3})$	Mortality (%)	CO concentration $(\mathrm{mg/m^3})$	Mortality (%)
5000	100	7000	100
4400	90	6700	60
4000	80	5750	40
3500	70	5200	15
2680	60	4700	3
2230	50	3500	0
1900	20		
1780	0		

Source: Tiunov & Kustov.

In guinea-pigs, acute CO exposure caused damage to the genital system and led to enhanced thyroid activity. A feature of acute poisoning in rabbits was increased platelet aggregation and drastic rises in blood cholesterol even after short-term exposures. A characteristic finding in cats was damage to subcortical centers. In exposed dogs with COHb concentrations above 26%, the myocardial blood flow was increased about 5-fold while the subendocardial blood supply was diminished; as the COHb level rose to 50–70%, the blood pressure and heart rate increased, the circulating blood volume diminished, and the blood flow in abdominal vessels fell sharply. A rise in COHb to 33% and collapse were observed in monkeys inhaling CO in a concentration of $1145\ \mathrm{mg/m^3}$ for 30 min.

Autopsy of acutely poisoned animals revealed numerous changes in various organs and systems. There were congestion and hemorrhages in all organs; necroses, cellular infiltration and fatty degeneration in the myocardium (predominantly in the left ventricular wall and papillary muscles); fatty, hydropic, and albuminoid degeneration and decreased glycogen content in the liver; peribronchial lymphocytic infiltrates and foci of bronchopneumonia in the lungs; nerve-cell degeneration and diffuse proliferation or nodular accumulations of glial cells in the brain; lipid depletion in the glomerular and, to a lesser extent, the fascicular zone of the adrenal cortex; disappearance of colloid and diminished follicular size in the thyroid.

Table 3. Toxicity of carbon monoxide for laboratory animals

Animals	CO concentration (mg/m³)	Exposure time (min)	COHb level (%)	Response
Rats	40	240		Increased plasma level of nonheme iron
	88	20		Reduced pO_2 in muscle
	176	20	3.3	Reduced pO_2 in muscle and brain
	800	240		Disorganized rhythm of sleep and wakefulness
	1600	180		42% mortality
	1900	240		50% mortality
	2000	11–12	52	Assumption of lateral position (narcosis)
	2400	240		84% mortality
	4100	50		100% mortality
	7700	30		100% mortality
	9000	15		No deaths
	13500	15		50% mortality
	18000	15		100% mortality
Cats	50–900	60		Impaired conditioned reflexes
	1150	150		No deaths
	2000	6	16	Early signs of poisoning
		20	38	Lateral position (narcosis)
	4600–5800	49–90		Deaths
	500–6000	30		Deaths
	5700	45		Deaths
	11500	15		Deaths
	34400–45800	3–4		Deaths
Dogs	2000	30–50	38	Lateral position (narcosis)
	2900	60–80		No deaths
	23000	20		Deaths
Guinea-pigs	2000	12–15	22	Early signs of poisoning
	2000	30–50	50	Lateral position (narcosis)
	4600	135		Deaths
	10300	60		Deaths

Table 3. (cont'd)

Animals	CO concentration (mg/m^3)	Exposure time (min)	COHb level (%)	Response
Rabbits	100–200	120		Augmented flexor reflex
	2000	60	35	Pulmonary edema
	6600–24000	10–30		Pulmonary edema
	17200	40		Deaths
	18000	90		Deaths
	20 000	60		Deaths
	66000	120–180		Deaths

Source: Tiunov & Kustov.

Threshold CO concentrations required to impair conditioned-reflex activity and the central nervous system's ability to summate subthreshold impulses in mice, were determined to be 2500 mg/m^3 with 5-min exposure and 66 mg/m^3 with 240 min exposure. That carbon monoxide is a highly hazardous poison is indicated by its narrow acute action zone when the exposure time is short and the concentration is high. (The acute action zone is a measure of acute toxicity and is defined by the ratio of the LC_{50} of a substance to its threshold concentration.) At longer exposure times and lower concentrations, the acute action zone progressively increases. For example, the acute action zone of CO for rats was calculated to be 4 with 15-min exposure and 47 with 4-h exposure, indicating that CO goes to the class of moderately hazardous substances on prolonged exposure to its relatively low concentrations.

Man

The acute toxicity of carbon monoxide, particularly at high concentrations, has long been recognized and is well documented. A mildly poisoned person remains conscious but may experience headache, drowsiness, nausea, and (occasionally) vomiting. Moderate poisoning often results in loss of consciousness for a short period; after regaining consciousness, the victim feels general weakness for some time and may experience lapses of memory, motor disturbances, and convulsions. Severe poisoning is marked by a longer-lasting unconscious state (>2 h), clonic or tonic convulsions, and involuntary urination and defecation. Those suffering prolonged unconsciousness may sustain permanent damage to the brain, the nervous system, and the heart [Zorn et al.]. The lowest lethal concentration for an adult with 30-min exposure has been reported to be 4580 mg CO/m^3 [Sax]. If the exposure is massive, loss of consciousness may take place almost instantaneously, with few or no premonitory signs and symptoms. Breathing concentrations of 10 000–40 000 ppm leads to death within minutes [Kurppa & Rantanen].

The blood from persons in a comatose state or dying from acute poisoning is usually found to contain COHb in concentrations of at least 50%, but fatal cases also occurred at lower COHb levels [Zorn *et al.*]. Despite the fairly large individual variation in the susceptibility to CO poisoning, which has been ascribed to genetic and a number of other factors, the blood level of COHb can serve as a rough guide to the severity of poisoning. Data on COHb levels in human blood in relation to CO concentrations in the air and to exposure time are given in Table 4, while Table 5 indicates conditions of CO exposure during physical activity under which COHb concentrations would not exceed 5%, which is considered as the highest permissible level.

Table 4. Carboxyhemoglobin levels in human blood (%) as a function of carbon monoxide concentration and exposure time

CO concentration	Exposure time (hours)			
(mg/m³)	1	2	4	>4
5000	75	—	—	—
2000	55–60	—	—	—
1350	28	45	60	70
1200	26	42	55	65
880	22	35	50	60
750	20	33	—	—
660	16	27	40	53
550	14	25	35	50
450	12	20	30	45
390	10	17	24	40
280	8	15	20	30
220	6	10	14	20
200	—	—	—	22–30
100	—	—	—	17

Source: Tiunov & Kustov.

A person exposed to medium CO concentrations (≤1000 mg/m³) is likely to experience tightness across the forehead, strong pains in the temples and forehead, vertigo, noises in the ears, flushing and burning of the skin, tremor, general weakness, fear, thirst, rapid pulse, pulsation of temporal arteries, shortness of breath, nausea and vomiting. With further exposure, if the victim remains conscious, he or she may be unable to leave the danger zone in good time because of the torpor, severe weakness, and apathy (or even a feeling of pleasant langour) that will have developed. Mental confusion and a state of drunkenness then ensue. The body temperature may rise to 38–40°C. In typical cases, consciousness is lost and spontaneous urination and defecation may occur. The coma often lasts for 1–2 days. Rarely, severely poisoned individuals remained conscious until death, but convulsions (sometimes tetanus) were usually observed.

Table 5. Conditions of exposure to carbon monoxide that would prevent carboxyhemoglobin levels exceeding 5% in non-smoking occupational groups performing light and heavy physical work

Concentrations		Exposure time not to be exceeded		Concentrations that would produce 5% COHb	
ppm	mg/m^3	Light work	Heavy work	Light work	Heavy work
200	230	15 min	—	298	—
100	115	30 min	15 min	157	193
75	86	60 min	30 min	87	105
50	55	90 min	60 min	64	62
35	40	4 h	2 h	37	41
25	29	8 h	8 h	31	30

Source: WHO.

The dyspnea may persist for many hours or even several days and terminate in death from respiratory arrest. The acute intoxication sometimes ran a biphasic course, with the initial symptoms of poisoning resolving within 2–3 weeks only to be succeeded by a sudden deterioration of the patient's condition involving severe neuropsychic disturbances (disorientation, apathy, abnormal behavior). The prognosis in such cases is grave [Kobayashi et al.]. Very occasionally, the short-term state of unconsciousness was followed by a rapid and apparently complete recovery.

Critical organs in CO poisoning are the brain and the heart, both of which are dependent on uninterrupted oxygen supply. A single acute exposure is particularly damaging to the central nervous system. As the anoxemia develops the victim loses the reasoning power, sustains a loss of cerebration, and fails to coordinate his movements; an early sign of progressive anoxemia is loss of pain sensitivity (e. g., even those who remain conscious may not notice burns). Following the cessation of anoxemia, pain sensitivity returns slowly and memory remains defective — sometimes to the extent that the persons cannot recognize even his close relatives or friends and (more often) recollect the circumstances of poisoning. Electroencephalographic changes were detectable 24 h post-exposure, on average.

In cases of early death, no morphological changes in the central nervous system were detected, whereas in those of later death, brain edema, areas of demyelination, and focal lesions in the corpus striatum, globus pallidus, and cerebellum were seen; at the cellular level, the greatest damage was suffered by neocortical cells (3rd, 5th, and 6th layers) and pyramidal cells of the hippocampal fields h_1 and h_{3-5} [Mitsuyama & Takamatsu; Siesjo].

Among common sequelae of acute poisoning are headache and dizziness. In severe cases, recurrent syncopes, encephalopathy, profound stupor, and coma may occur some time after the apparent recovery. Psychoses have been observed occasionally (CO poisoning may aggravate latent psychic disease). Severe nervous illnesses with signs of extrapyramidal insufficiency and parkinsonism have been described [Bezrodnykh & Turusina]. Very young children (aged 2–3 years) are apt to develop psychomotor instability and lose the previously learned habits. In older children (aged 10–15 years), speech disturbances, spatial disorientation, deterioration of short- and long-term memory, and depression or excitation have been reported [Klees et al.].

Long-term effects of acute poisoning may include choreoid hyperkinesias in young persons and depression, dementia, amnesia, and progressive cachexia in elderly individuals. The causes of nervous system damage are the impairment of cerebral circulation, the direct toxic action of CO on nerve tissue, and increased permeability of the blood–brain barrier. Damage to the peripheral nervous system is manifested in motor, sensory, and trophic disturbances, which sometimes persist for years; paralysis, poly- or mononeuritis, neurodystrophic arthritis, radiculitis, paresthesias, and increased sensitivity or, on the contrary, insensitivity of some skin areas are common sequelae.

Among sensory organs, the eyes suffer most. Even a single mild poisoning reduces color and night vision and the accuracy and speed with which visual space perception occurs; double vision, narrowing of the visual fields, abnormal pupillary reaction, paralysis of ocular muscles, nystagmus, and glaucoma have all been reported as possible consequences of such poisoning [Langeweg; Mamatsashvili; Zorn et al.]. Cases of temporary (for 6–7 days) cortical blindness (with preservation of the pupillary reaction to light) developing several days after acute poisoning have been described; the electroencephalograms taken in these cases showed abnormalities predominantly in the occipital lobes [Quattrocolo et al.; Werner et al.]. Hearing impairment and dysfunction of the vestibular system have also been observed.

Cutaneous sequelae of acute poisoning include trophic disturbances (resulting, e. g., in ulcers or in necrosis of sebaceous glands), hemorrhages, fragility and even detachment of nails, and graying and loss of scalp hair. Loosening of the gums and loss of healthy teeth may occur. The skin and mucous membranes become cherry red if the poisoning is severe.

Consistent findings in cases of acute carbon monoxide poisoning are functional cardiovascular disturbances such as labile pulse, tachycardia, extrasystoles, and transient deterioration of myocardial conductivity. There may be myocarditis and myocardial infarction (the latter sometimes developed after an acute intoxication sustained without any apparent consequences); gangrene of the extremities may develop because of increased capillary fragility [Labunsky & Struchenevskaya].

Hematological changes seen in mild and moderate poisoning mainly include leukocytosis, erythrocytosis, and increased hemoglobin. Anemia, lymphopenia, and eosinopenia tend to develop in more severe cases, as well as degenerative changes in bone marrow cells; coagulability of the blood also decreases.

Lesions in muscles and joints (necrosis, arthritis) and in respiratory organs (nasal hemorrhages, inflammations in the upper respiratory tract, and, several hours (usually 10–15) after the acute episode, pneumonia and pulmonary edema) have also occurred.

Among other sequelae of acute poisoning described in the literature are diseases and hyperactivity of the thyroid, spleen enlargement (particularly in elderly or asthenic individuals), adrenal dysfunction, renal lesions (nephrosis and focal nephritis), gastrointestinal disorders, reduced resistance to infection, and a number of metabolic abnormalities. The latter include weight loss, elevated blood levels of sugar, lactic acid, acetone bodies, cholesterol, and urea; impaired nitrogen metabolism; depressed transaminase activities in the brain, liver, and kidney together with their elevation to very high levels in blood serum; disrupted porphyrin metabolism; decompensated metabolic acidosis; reduced alkaline reserve; disordered water–salt metabolism; presence of sugar in urine [Tiunov & Kustov; Vyskočil].

Toxic CO concentrations in relation to manifestations of poisoning resulting from single exposures are indicated in Table 6.

Table 6. Toxic carbon monoxide concentrations and signs and symptoms of poisoning in human beings (data of various authors)

CO concentration (mg/m^3)	Length of exposure (min)	COHb level (%)	Signs or symptoms of poisoning
6	25	—	Weakened color discrimination and light sensitivity
11–12	300	3.8	Reduced accuracy in appreciating time intervals
13	7–8	1.6	Alterations in biochemical and physiological parameters
31	180	—	Reduced accuracy of visual space perception and weakened night vision
33	360–480	3.5–4	Lowered oxygen level in blood
50–60	120	—	Weakened hearing; alterations in the EEG
	300	—	Worsened performance of psychological tests

Table 6. (cont'd)

CO concentration (mg/m^3)	Length of exposure (min)	COHb level (%)	Signs or symptoms of poisoning
55	60–180	2.1–3.8	—
	300	—	Increased latency in the perception of signals
	360	5	Increased threshold of visual perception
57	—	2.5–3	Angina attacks on exertion in patients with coronary heart disease
60	300	4.7	—
	360	—	Changes in the EEG and heart rate
80–110	210–300	7–10	Reduced rate of visual perception; worsened performance of psychological and psychomotor tests; loss of dexterity in the performance of minor movements; decreased capacity for analytical thinking
220	60–180	—	Slight pain in the forehead region
220	3	14	Reduced working capacity
230	360	16–20	Pain in the forehead and a feeling of pressure against it (these symptoms both rapidly disappeared in fresh air); dilated skin vessels; reduced physical working capacity
230–340	300–360	23–30	Headache and a feeling of pulsation at the temples; dizziness
340	240	22–24	As above
	300	26–27	As above
345	30	11.9	Headache in some individuals
440	60	12	—
120	20		Overt signs of poisoning
	240	30	Vomiting, headache, palpitation
440–460	60	15–19	Pain in the forehead and occipital regions; palpitation
	120	21–29	As above
	180	—	As above
460–690	240–300	36–40	Severe headache, weakness, vertigo and mist before the eyes; nausea and vomiting; collapse
550	60	14	Headache, palpitation
600–700	15	—	Reduced mental and physical working capacity
660	60	16	First symptoms of poisoning; only light physical work could be performed
	120	27	Nausea; overt symptoms of poisoning
	240	40	Nausea and vomiting; loss of consciousness may occur

Table 6. (cont'd)

CO con-centration (mg/m³)	Length of exposure (min)	COHb level (%)	Signs or symptoms of poisoning
800–1000	5	—	Headache; first signs of intoxication; reduced working capacity
	20–30	—	Headache, general muscular weakness, nausea; partial loss of working capacity in some
800–1150	180–240	47–53	Severe headache, weakness, vertigo, mist before the eyes, nausea, vomiting, quick pulse; collapse
880	45	—	Headache, vertigo, nausea
	60	22	Symptoms of intoxication appeared
	120	35	Nausea, vomiting, complete loss of working capacity; loss of consciousness or collapse may occur
	240	50	Convulsions and coma. Death may occur
1100	120	40	Strongly marked picture of poisoning
1260	90–180	55–60	Increased breathing and pulse rates; coma interrupted by convulsions; Cheyne-Stokes respiration
1300	60	28	Nausea, palpitation, headache
	120	45	Severe poisoning
1350	30–90	—	Palpitation, slight staggering, shortness of breath during light muscular work, deterioration of vision and hearing
	120	—	Pulsating headache; mental confusion
1760	20	—	Headache, vertigo, nausea
	120	—	Loss of consciousness or collapse
1800–2300	60–90	61–64	Loss of consciousness or collapse; reduced respiration and cardiac activity. Death may occur
2000	15–35	50	Coma and convulsions
3000	60	—	Death
2300–3400	30–45	64–68	Loss of consciousness or collapse. Death may occur
3520	5–10	—	Headache and dizziness
	30	—	Vomiting and loss of consciousness
3400–5700	20–30	68–73	Weak pulse; slowed or arrested respiration and death
5000	17	—	Loss of consciousness; convulsions
5700	5–10	—	Death
5700–11500	2–5	73–76	Death
14080	1–3	—	Vomiting, loss of consciousness; death

Repeated and chronic exposures

Animals

Inhalation by rats and rabbits of CO at $100-200$ mg/m^3 several hours per day for $20-30$ days led to a number of blood changes, including pronounced erythrocytosis and reticulocytosis, increases in hematocrit and hemoglobin, reduced erythrocyte resistance to hemolysis, decreases in catalase activity and platelet counts, and intensified glycolysis. Hypertrophy and hyperfunction of the adrenal cortex and thyroid and, in male mice, alterations in the structure and function of sex organs were also observed. The metabolism of biogenic amines in the brain was impaired. Cardiomegaly (predominantly right ventricular hypertrophy) was noted, with an increase in cardiac output, and degenerative/dystrophic changes were seen in the myocardium. In rabbits, daily exposure to 110 mg CO/m^3 over 12 months led to hypercholesterolemia, while exposure to 170 mg/m^3 for 45 days accelerated the development of atherosclerosis in animals fed a hypercholesterolemic diet [Bishop & Penney; Davidson & Penney; Fay et al.; Penney et al.].

Prolonged inhalation of lower CO concentrations primarily resulted in damage to the central nervous system. Exposure of mice and rats to $10-30$ mg/m^3 for $4-6$ h/day over 10 to 15 weeks was accompanied and followed by growth retardation, decreased motor activity and working capacity, abnormal central nervous system functioning, leukocytosis and erythrocytosis, cardiovascular impairment, disturbances in carbohydrate, lipid, and porphyrin metabolism, reduced immunobiological reactivity, and disordered functions of the adrenal cortex and thyroid. The level of reduced glutathione was also altered, and the total content of sulfhydryl groups was increased. A 2-month exposure of rats to CO at 14.5 mg/m^3, which raised to the COHb level to 9.9%, produced a state of stress, reflected in depressed norepinephrine levels in the hypothalamus and thyroxine levels in the blood together with elevated corticosterone levels in the blood and of catecholamines in the adrenals [Vyskočil et al.].

After 30 days of continuous exposure to CO concentrations of $49-53$ mg/m^3, rats showed decreased body-weight gains, increased hemoglobin levels and erythrocyte counts, and abnormal succinate dehydrogenase and cytochrome oxidase activities in the liver. In rats similarly exposed to $100-300$ mg/m^3, such changes were observed after $1-2$ weeks [Abidin & Kustov]. Continuous exposure of cynomolgus monkeys to a CO concentration of 115 mg/m^3 resulted in electrocardiographic effects in the myocardium of both normal animals and those with myocardial infarction [De Bias et al., 1973].

Man

The diagnosis of chronic occupational carbon monoxide poisoning should be based on the work history, clinical presentation, and evidence of elevated COHb in the blood. Chronic poisoning is to be distinguished from the effects of repeated episodes of slight acute poisoning. Chronic poisoning occurred, for example, in individuals who had been continuously breathing air containing 10–50 mg CO/m^3 (COHb concentrations in their blood ranged from 3% to 13%).

Chronically poisoned persons present with a diversity of symptoms, of which the major ones are considered to be physical and psychic asthenia, headache, and vertigo. First symptoms usually appear 2 to 3 months after the start of exposure and may include any of the following: noises in the head and headache (particularly during work and in the morning), dizziness, a feeling of being intoxicated, increased fatiguability, weakened memory and reduced ability to concentrate (decrement of vigilance), apathy and labile mood, tinnitus and increased sensitivity to auditory stimuli, nausea, weight loss, lack of appetite, low alcohol tolerance, diarrhea, insomnia at night and drowsiness in daytime, pallor, grayish color of the skin, obsessive fears, pains in the cardiac and epigastric regions, arthralgia and neuralgia, sweating, and frequent urination; syncopes also occur in some cases. There is considerable individual variation in the occurrence of symptoms [Zorn et al.].

Nervous system abnormalities likely to develop in the early stages of intoxication are persistent bright-red dermatographism, trembling of the extremities, pyramidal disturbances, impaired cutaneous sensitivity, listlessness or complete absence of pupillary reactions, neuritis or polyneuritis. Speech disturbances are also possible. Subsequently, cerebrovascular and diencephalic crises may occur, as well as increased sweating of the hands and muscular atrophy; premature graying and loss of scalp hair have also been observed in some cases. Paresis (particularly of the facial nerve), encephalopathy, and convulsive seizures are not uncommon in severe poisoning. The central nervous system disturbances occasionally resembled parkinsonism. A twilight state with uncontrollable behavioral acts, amnesia with failure to recall recent events, and psychic disturbances may develop.

Chronic poisoning is associated with more severe cardiovascular disorders than acute episodes, especially in persons engaged in physical work activity; arrhythmia, rapid and labile pulse, extrasystoles, angina-like attacks, and unstable blood pressure with a tendency to hypotension are all common [Petry]. Cardiac abnormalities are usually detected 1 to 1.5 years after the termination of exposure. A frequent finding is increased capillary fragility in various organs; coronary thrombosis may occur. It has been suggested that chronic CO poisoning may aggravate preexisting atherosclerosis.

Frequent hematological abnormalities are increased hemaglobin and hematocrit and a left shift in the blood picture; less often, elevated platelet counts and reticulocytosis were noted. At later stages, anemia is common (mainly due to inadequate functioning of hematopoietic organs), as are various changes in the blood, including increased levels of pyruvic acid, calcium, phosphorus, β-globulins, total lipid, β-lipoproteins, cholesterol, fibrinogen, and catalase and acetylcholinesterase activities along with a reduction in blood sugar. The blood serum content of nonheme iron is also often elevated, as well as levels of protoporphyrin and coproporphyrin and of vitamin C in the urine [Tiunov & Kustov].

Reported gastrointestinal disturbances include loss of appetite, heartburn, nausea and vomiting, diarrhea, gastritis and colitis. Liver function tests were also abnormal occasionally.

Frequent ocular effects of CO are reduced visual acuity, diplopia, and narrowing of the visual fields; inadequate convergence and accommodation, poor dark adaptation, changes in the fundus oculi, and weakness of ocular muscles were also sometimes observed. Other sensory effects include impaired hearing, chronic otitis media, chronic inflammation of the auditory nerve, dysfunction of the vestibular system, and reduced sense of smell. Thyroid function is also commonly affected (with a tendency to hyperthyroidism), as is adrenal function. Men often showed weakened sexual activity, while menstrual disturbances, reduced libido, and miscarriages occurred in some women.

Chronic CO poisoning decreases resistance to infections, particularly to tuberculosis and pustular skin diseases.

Individuals repeatedly exposed to moderate concentrations of CO possibly become adapted to withstand its adverse effects, and the mechanisms of adaptation are believed to be similar to those inducing tolerance to hypoxia at high altitudes [Kurppa & Rantanen]. Workers who had been continually exposed to CO concentrations of $250-300$ mg/m^3, developed tolerance to this gas and felt better when exposed to its higher concentrations than did those exposed for the first time, and they could also endure the exposure for longer periods [Devyatka; Kovnatsky; Strelchuk]. However, the acquired increased resistance disappeared shortly after the cessation of occupational exposure.

The principal source of nonoccupational exposure to CO is tobacco smoke. The CO concentration in the mainstream smoke of cigarettes is approximately 4% by volume, and it has been estimated that the cigarette smoker may be exposed to a CO concentration of $460-575$ mg/m^3 ($400-500$ ppm) for the period needed to smoke a cigarette (~ 6 min); heavy cigarette smokers may have COHb levels as high as $15-17\%$ [WHO]. The average blood levels of COHb were reported to be about $3-4\%$ in moderate smokers and 6% in heavy smokers as compared to 2% in nonsmokers [Butt et al.]. The CO present in tobacco smoke is the major risk factor for vascular damage in smokers; it causes hypoxia in arterial walls and inhibits lipolytic enzymes.

Effects on reproduction

Animals
Animal experiments have shown that the level of COHb in fetal blood depends on that in maternal blood and on the duration of maternal exposure. Fetal blood

is saturated with CO much more slowly than maternal blood, but the transplacental release of CO also occurs far more slowly than through the lungs [Colmant & Wever]. As a result, the fetal COHb level may be 10–15% higher than the maternal [Longo]. When female mice and rats where maintained throughout pregnancy in an atmosphere containing 103–229 mg CO/m^3, fetal COHb levels rose to 15–27%. In such animals, the number of pups per litter was less than in the controls and the newborns had lower body-weights, greater heart weights, and lower hemoglobin levels and erythrocytes counts. These differences between the two groups of newborns were increasing during the first three weeks of their life. Periodic exposures of female rats to high CO concentrations during the second half of pregnancy resulted in increased post-implantation mortality; the pups born to these rats exhibited reduced learning capacity, abnormal development of noradrenergic and seroninergic systems of the brain, and defective catecholamine metabolism [Abbatiello & Mohrmann; Storm & Fechter; Tachi & Aoyama].

In female guinea-pigs exposed to CO from day 20 of pregnancy onward (550 mg/m³ for 10 min over 20 days or 480 mg/m³ for 60 min over 30 days), many abortions, low viability of newborn pups, and paralysis of the hind limbs in many of them were noted. Continuous 10-week exposure of female rabbits to 100–200 mg CO/m^3 (which raised their blood levels of COHb up to 15%), reduced the conception rate, litter size, and the body weight and length of the newborns; it also resulted in increased prenatal and postnatal death rates. A single short-term (76–150 min) exposure of female cats to a CO concentration of only 0.2–0.3% during late pregnancy was reported to have led to impaired brain development in their offspring [Okeda *et al.*].

Man

The effects of acute CO poisoning in pregnant women are more severe than in nonpregnant ones. Intrauterine or early postnatal mortality following acute maternal poisoning has been high (~65%). Fetal death may occur after acute (single or multiple) maternal poisoning even if the mother has not had any apparent untoward effects. Developmental abnormalities and deformities in fetuses and newborns and retardation of postnatal psychomotor development were observed in 10–15% of offspring from mothers who had sustained acute CO poisoning during pregnancy; lesions of fetal extrapyramidal pathways following maternal exposure to CO in the 8th month of pregnancy have been described. On the other hand, no adverse effects were detectable in some newborns whose mothers had been acutely poisoned with CO in the second trimester of pregnancy [Barois *et al.*; Longo].

Chronic exposure to CO may result from maternal smoking or be caused by environmental pollution. Women who have been long exposed to CO, particularly through smoking, are much more likely to have health problems during pregnancy. COHb concentrations in the blood of mothers and their fetuses during the third trimester of pregnancy reached 4.6% as compared to 0.4% in non-smoking mothers [Davies et al.]. Much higher COHb values (up to 14% in mothers and 9.8% in fetuses) have also been reported [WHO]. It has been demonstrated in several studies that babies delivered by mothers who smoke weigh less than those delivered by nonsmoking mothers. Babies born to smoking mothers are predisposed to cardiovascular abnormalities both because of their congenitally augmented susceptibility to adverse effects of smoking and because of the thinning out of their coronary walls.

Sex- and age-related differences in susceptibility to carbon monoxide

Animals
Male mice and rats are more susceptible to acute CO poisoning than females. Thus, male rats died earlier and in greater numbers than females under the same exposure conditions. Also, the 2-hour LC_{50} for male mice was 4150 mg/m^3 as compared to 5360 mg/m^3 for female mice. Castration of male mice and rats made them more resistant to CO, whereas castration of females did not. Young and neonatal male mammals are, as a rule, less sensitive to acute CO exposure than adult males [Tiunov & Kustov].

Man
Human females are also more resistant to toxic effects of CO than males. Blood levels of COHb in men were higher and the elimination half-time longer than in women exposed to the same CO concentration in the ambient air [Eyiam-Berdyev].

Cases of domestic poisoning by CO where the parents died whereas their infants survived have been reported [Scaroina et al.].

Species differences in susceptibility to carbon monoxide
These differences are due to differences between species both in hemoglobin affinity for CO and in CO absorption rate which depend in turn on metabolic rates, body-weight, heart rate, and breathing frequency. Laboratory mammals can be classified in order of decreasing CO susceptibility (and of decreasing rate of CO uptake into the blood) as follows: mice, hamsters, rats, cats, dogs, guinea-pigs, rabbits [Aleksandrov].

Combined effects of carbon monoxide and other occupational factors
Simultaneous inhalation of sulfur dioxide or nitrogen dioxide and CO (5000 mg/m^3) exacerbated the toxic effect of the latter: 50% of mice so exposed died after 25 and 13 min, respectively, whereas those exposed to CO alone at the same concentration died after 33.7 min [Datsenko et al.].

In humans, the combined effects of CO (300 mg/m^3) and high ambient temperature ($30-40°C$) were described as additive (body temperature elevation, rise in COHb levels, and reduced physical and mental working capacity were recorded) and those of these two factors plus physical activity (400 kg-m/min), as more than additive. Potentiation of effects was also observed in individuals performing heavy physical work while being exposed to CO (300 mg/m³) and CO_2 (3–5%). According to Sedov *et al.*, the longest time a person in resting state may be allowed to occupy a hermetically sealed space with a CO concentration of 300 mg/m³ is 120 min at an air temperature of 20°C, 90 min at 30°, and 60 min at 40°; the occupancy period should not exceed 60 min at 30°C and 30 min at 40° if the CO concentration is 800 mg/m³.

Absorption, distribution, and elimination

Carbon monoxide enters the body by inhalation exclusively. In man and higher animals, it can also be produced endogenously, mainly as a result of heme catabolism; another significant source may be lipid peroxidation.

Absorption of CO from the lungs into the bloodstream depends on several factors, including its concentration in the inspired air, duration of exposure, oxygen content in the inspired air, diffusion capacity of the lungs, characteristics of the pulmonary circulation, and pulmonary ventilation rate.

In the blood, CO combines with hemoglobin, displacing oxygen from oxyhemoglobin (O_2Hb) to form carboxyhemoglobin (COHb) by virtue of the fact that the affinity of hemoglobin for CO is many times higher than its affinity for oxygen — roughly 240 times higher for human hemoglobin and 130–550 times for hemoglobins of various vertebral animals; this explains in part the observed species differences in CO sensitivity.

Equilibrium between the concentration of CO in the air and that of COHb in the blood is reached in 6–11 h in adults and more rapidly in children [Zorn]. COHb dissociates very slowly — up to 3600 times more slowly than does O_2Hb.

In body tissues, CO also binds to other iron-containing substances such as myoglobin, cytochromes, cytochrome oxidase, and catalase.

In acute poisoning, about 13% of CO was found to combine with myoglobin in the myocardium and skeletal muscle. Some CO (up to 25–30% in chronic poisoning) can bind to nonheme iron whose levels in blood plasma rise in subacute (repeated) and chronic intoxications. Carbon monoxide may also be present in the liver, kidney, brain, and cerebrospinal fluid. Immediately after the cessation of exposure, its levels in the brain, myocardium, and skeletal muscle of female rats were lower by factors of 30–50, 40–50, and 3–4, respectively, than those in the blood [Sokal *et al.*]. About 0.1% to 3.3% of the CO entering the body is oxidized to CO_2; the oxidation is catalyzed by cytochrome oxidase whose activity is in turn inhibited by CO.

Carbon monoxide is eliminated largely or exclusively by exhalation. The elimination is rapid at first and becomes slower with time; in humans after acute poisoning, for example, 60–70% was cleared during the first hour and excretion continued over the next 3–4 h. In some cases of chronic poisoning, its blood level remains elevated for a long time and its elimination from the body is markedly retarded. Some (slight) excretion may also take place via the gastrointestinal tract and kidneys. The average biological half-life of carbon monoxide in man is about 5 hours. In small animals the elimination rate is higher than in larger ones.

Determination of COHb in the blood is of diagnostic value. The average normal COHb level in human blood is considered to be between 5% and 6%. If it has risen to 10–15%, chronic poisoning may be suspected. In cases of lethal poisoning, COHb levels in blood samples taken from cardiac cavities were higher than in those from vessels [Babakhanian *et al.*].

Hygienic standards

In the USSR, the following occupational exposure limits (MAC_{wz} in mg/m^3) have been adopted for carbon monoxide depending on the exposure time: 20 during the whole workday (~8 h), 50 for 60-min exposure, 100 for 30-min exposure, and 200 for 15-min exposure. At least 2 h must elapse before workers who have been thus exposed may be allowed to resume work under conditions of elevated CO concentrations. The values shown above should be halved for those doing heavy physical work. For short exposures, particularly in emergency situations, the following maximum permissible CO levels have been recommended (mg/m^3): 600 for 10 min, 400 for 15 min, and 300 for 30 min [Tiunov & Kustov]. Where both CO and oxides of nitrogen are present in the air, Tiunov & Kustov recommend decreasing the upper limit 1.5-fold for CO and 3-fold for the oxides. For combined exposures to CO and hydrogen sulfide, their respective MACs should be reduced (e.g. by factors of 1.5 and 3) since the toxic effects of these gases have been shown to be additive and, moreover, an enhancement of CO toxicity has been observed in some cases [Melnichenko].

Carbon monoxide is placed in hazard class 4 and its content in the air is to be monitored automatically. For the atmospheric air of residential areas, carbon monoxide exposure limits are set at 5.0 mg/m^3 (MAC_{hm}, i.e. the concentration that should not be exceeded for any length of time) and at 3.0 mg/m^3 (MAC_{ad}, i.e. the average daily concentration).

In the USA, threshold limit values for carbon monoxide have been set at 55.0 mg/m^3 (TWA) and 458 mg/m^3 (STEL).

Methods of determination

Commonly used methods for the routine estimation of carbon monoxide in *air* are the continuous analysis method based on nondispersive absorption spectroscopy, the method of semicontinuous analysis using gas chromatographic techniques, and a semiquantitative technique employing detector tubes [Peregud; WHO].

Chromatographic methods frequently employed in the USSR include the following: gas-adsorption or gas-liquid chromatography with an argon ionization detector (detection limit 1 µg) or with a katharometer and flame ionization detector with a differential system of gas supply (detection limit 8 μg) [*Guidelines for the Determination of Noxious Substances in Air*, 1979]; a more advanced method of gas-adsorption chromatography using a katharometer (detection limit 0.5 μg; measurable concentration range, 5–100 mg/m^3; measurement error, ≈10%) [*Guidelines for the Determination of Noxious Substances in Air*, 1981a]; a procedure in which carbon monoxide is first converted to methane and this is then determined in a flame ionization detector (detection limit 0.0005 μg) [*ibid*]; and a method based on the use of reactive gas chromatography with conversion of carbon monoxide to methane and detection of the latter with a flame ionization detector (detection limit 0.0002 μg; measurable concentration range, 0.1–300 mg/m^3; measurement error, ±25%) [*Guidelines for the Determination of Noxious Substances in Air*, 1981c].

Titrimetric gas analyzers may overestimate carbon monoxide levels in the air because they measure not only carbon monoxide but also some of the gaseous hydrocarbons unless these are removed. An optical-acoustic gas analyzer giving more accurate results has been described [Burenin; Pozhidayev & Elistratova]. For monitoring carbon monoxide levels in the air over long stretches (up to 1 km) of highways, a CO concentration meter based on the use of a diode laser has been developed [Astakhov *et al.*]. Barikhin described a gas-chromatographic technique for measuring carbon dioxide in air in the presence of several other inorganic gases.

For carbon monoxide determination in *blood* samples, photochemical methods [Sawicki & Gibson; Zorn *et al.*] and a method based on the reaction of carbon monoxide with palladium chloride [Tiunov & Kustov] may be used. Blood samples may be preserved for several days prior to analysis if kept cold (at 4°C). According to Ocak *et al.*, the carbon monoxide level in cadaveric blood contained in a tightly closed receptacle remains stable for 4 months at temperatures of 4 to 20°C. For the determination of blood carboxyhemoglobin, a variety of methods are available, including qualitative color reactions and colorimetric, spectrophotometric, and gas-chromatographic techniques [Baretta *et al.*; Gadaskina *et al.*, 1975; Tiunov & Kustov; WHO]. Radziszewski & Guillerm have described an infrared-spectrometric method for determining carboxyhemoglobin in microamounts of blood (measurable CO concentration range, 0.5–100%).

Measures to control exposure

Technical control measures should be applied to the maximum possible extent to reduce CO concentrations in the air and the possibility of its accidental escape. For each particular job situation carrying a hazard of CO exposure, a list of such measures should be worked out, with due consideration to the specific work conditions. Wherever possible, the sources of CO should be enclosed in an airtight manner to rule out any possibility of its release into the air.

Continual assessment of CO concentrations in the air is important. Where appropriate, automatic monitoring equipment should be installed, linked to an alarm system that is actuated whenever the CO concentration exceeds a preset ceiling level.

Good general dilution ventilation and local exhaust ventilation are also important. Ventilation should be designed to accommodate the largest levels of emission anticipated. All industrial heating devices should be checked at regular intervals to ensure that no unexpected CO emission occurs. When it is necessary to enter a space that may contain carbon monoxide, the precautions applicable to entry into confined spaces should be taken (e.g. the presence of a second person just outside the space to render assistance if necessary). Containers with CO and worksites with risk of high-level exposure should be provided with labels and warning signs. All workers employed in areas where a hazard of CO being present in high concentrations in the ambient air exists, should be instructed about its toxic properties, symptoms of poisoning, and appropriate emergency procedures [Kurppa & Rantanen].

Any person working under conditions of CO exposure is subject to preplacement and subsequent periodic medical examinations, with special attention to the cardiovascular and nervous system as well as to diseases or conditions that may be exacerbated by the hypoxic action of CO.

Those exposed to CO concentrations above the permissible level should use respiratory protection equipment — for example filter-type gas masks for short-term work (90–150 min) in an atmosphere containing no less than 18% O_2 and 0.5% CO by volume. At high CO concentrations (such as those obtaining in fires or in mines during accidents), supplied-oxygen respirators with closed circuit must be used.

Of special importance in protecting the general population from CO exposure is control over the emission of automobile exhaust gases of which CO is a major component (along with hydrocarbons and nitrogen oxides). The main control measures here consist in rational town planning; improving internal combustion engine designs and developing new types of engines; setting up limits for the CO level in automobile exhausts (e.g. by adopting motor-vehicle emission control legislation); more extensive use of electricity-driven vehicles; use of fuel additives that reduce the formation of toxic substances; and decontaminating the exhaust gases by means of neutralizers.

Emergency treatment in poisoning

A person acutely poisoned with carbon monoxide should be immediately removed from the contaminated atmosphere to fresh air in a recumbent position (even if he can walk unaided). If this cannot be done quickly, further entry of the monoxide into the victim's body should be stopped (e.g. by putting on an air-supplied respirator). All clothing interfering with breathing should be taken off and the patient placed in a comfortable position, assured complete rest, and kept warm (cooling is hazardous, but hot bags or bottles have to be used with caution since the poisoned persons are susceptible to burns, bruises, and trophic lesions). Oxygen should be administered as soon as possible. The duration of oxygen therapy depends on the severity of intoxication. If the respiration is not markedly inhibited, oxygen may be alternated with carbogen (oxygen for 40–60 min, carbogen for 15–20 min), but the latter must be used cautiously and only in the absence of respiratory distress. If respiration fails to be stimulated, oxygen alone should be used. Artificial respiration must be resorted to if indicated.

If the poisoning is mild, the patient may be given coffee or strong tea and allowed to smell a piece of cotton or gauze impregnated with ammonia water. Nausea and vomiting can be controlled by oral administration of a 0.5 novocaine (synonym: procaine) solution in teaspoons. Camphor (1–2 ml of 20% solution), caffeine (1 ml of 10% solution), and cordiamin [synonym: nikethamide] (1 ml of 0.5% solution) are given subcutaneously for stimulating central nervous system and respiratory functions.

In severe poisoning, hyperbaric oxygen therapy should be administered for 40–90 min as an emergency and be repeated if necessary. During the first few hours, Cytochrome C (an enzyme preparation derived from bovine myocardium) (15–60 mg), ascorbic acid (20 ml of 5% solution), glucose (20–50 ml of 40% solution), novocaine (50 ml of 2% solution), and pyridoxine (1–2 ml of 5% solution) are given in subcutaneous injections. If the patient is very excited and/or has developed cerebral edema, a lytic cocktail is administered in repeated intramuscular injections; in addition, 200 ml of a hypertonic glucose solution may be given by intravenous drip together with 10 units of insulin subcutaneously. Convulsions are treated by barbamyl [synonym: amobarbital sodium] intravenously (5–10 ml of 10% solution) or by chloral hydrate enema (100 ml of 2% solution).

For correcting respiratory abnormalities, a 2.4% euphylline solution is injected repeatedly (10 ml per injection); collapse is treated by parenteral administration of mesaton [synonym: phenylephrine hydrochloride] (0.5 ml of 1% solution), ephidrine (1 ml of 5% solution), or epinephrine (1 ml of 0.1% solution). For preventing cerebral edema in cases of long-lasting coma, use is made of such methods as hypothermia of the head (cooling with ice), forced osmotic diuresis (without water load), repeated lumbar punctures with removal of cerebrospinal fluid (10–15 ml), and hydrocortisone therapy. If respiratory arrest has occurred and no pulse can be felt, cardiac massage must be done.

As antidotal agents, preparations of reduced iron (such as Fercoven and Cytochrome C) are used. Broad-spectrum antibiotics are indicated. All emergency measures have to be continued until the respiration and circulation return to normal. Further treatment of patients with severe or moderate poisoning should be carried out in a hospital or similar inpatient facility.

REFERENCES

Abbatiello, E. R. & Mohrmann, K. (1979) *Clin. Toxicol.*, **14**, 401–406.
Abidin, B. M. & Kustov, V. V. (1974) *Kosmich. Biol. Med.* [Space Biology and Medicine], No. 4, 17–21 (in Russian).
Aleksandrov, N. P. (1973) *Gig. San.*, No. **11**, 92–95 (in Russian).
Aleksandrov, V. N. *et al.* (1973) *Voyenno-Med. Zh.* [Journal of Military Medicine], No. 1, 66–70 (in Russian).
Astakhov, V. I. *et al.* (1982) In: *Monitoring fonovogo zagryazneniya prirodnykh sred* [Monitoring of the Background Environmental Pollution], No. 1. Leningrad, pp. 175–180 (in Russian).
Babakhanian, R. V. *et al.* (1986) *Biol. Zh. Armenii* [The Biological Journal of Armenia], **39**, (No. 7), 635–637 (in Russian).
Baretta, E. D. *et al.* (1978) *Amer. Ind. Hyg. Assoc. J.*, **39**, 202–209.
Barikhin, S. Ya. (1981) In: Shitskova, A. P. (ed) *Sovremennye metody opredeleniya toksicheskikh veshchestv v atmosfernom vozdukhe* [Modern Methods of Determining Toxic Substances in Atmospheric Air]. Moscow, pp. 61–62 (in Russian).
Barois, A. *et al.* (1980) *Intensive Care Medicine*, **6**, 57–61.
Bezrodnykh, A. A. & Turusina, T. A. (1972) *Sov. Med.*, No. **3**, 139–140 (in Russian).
Bishop, P. & Penney, D. (1978) *J. Environ. Pathol. Toxicol.*, **2**, 407–415.
Burenin, N. S. (1984) In: *Atmosfernaya diffuziya i zagryazneniye vozdukha. Trudy Glavnoi geofizicheskoi observatorii* [Atmospheric Diffusion and Pollution of the Air. Transactions of the Main Geophysical Observatory], No. 479. Leningrad, pp. 109–115 (in Russian).
Bushtuyeva, K. A. & Feldman, Yu. G. (1976) In: *Rukovodstvo po gigiyene atmosfernogo vozdukha* [A Manual of Air Pollution Control]. Meditsina, Moscow, pp. 168–202 (in Russian).
Butt, J. *et al.* (1974) *Ann. Occup. Hyg.*, **17**, 57–63.
Chen, K. & McGrath, J. (1985) *Toxicol. Appl. Pharmacol.*, **81**, 363–370.
Chubar, S. V. (1970) *Vrach. Delo*, No. **6**, 121–122 (in Russian).
Colmant, H. & Wever, H. (1962) *Arch. Psychiat. Neurol.*, **204**, 271–287.
Datsenko, I. I. *et al.* (1986) In: *Problemy okhrany zdorovya naseleniya i zashchity okruzhayushchei sredy ot khimicheskikh vrednykh faktorov* [Protection of Public Health and the Environment from Noxious Chemical Agents]. Rostov-on-Don, pp. 125–127 (in Russian).
Davies, J. *et al.* (1979) *Brit. Med. J.*, No. **6186**, 355–356.
Davidson, S. & Penney, D. (1988) *Arch. Toxicol.*, **61**, 306–313.
De Bias, D. *et al.* (1973) In: *Arch. Environ. Health*, **27**, 161–167.
Devyatka, D. G. (1957) In: *Khronicheskiye oksiuglerodnye intoksikatsii. Sbornik nauchnykh rabot Lvovskogo meditsinskogo instituta* [Chronic Carbon Monoxide Intoxications. Transactions of the Lvov Medical Institute], Vol. 16. Lvov, pp. 113–122 (in Russian).
Eyiam-Berdyev, A. K. (1973) In: *Trudy Kharkovskogo meditsinskogo instituta* [Transactions of the Kharkov Medical Institute], Vol. 110. Kharkov, pp. 185–196 (in Russian).
Fay, E. *et al.* (1980) *Acta Physiol. Acad. Sci. Hungar.*, **56**, 27–30.
Gadaskina, I. D. *et al.* (1961) *Gig. Truda Prof. Zabol.*, No. **11**, 13–18 (in Russian).
Gadaskina, I. D. *et al.* (1975) *Opredeleniye promyshlennykh neorganicheskikh yadov v organisme* [In vivo Determination of Industrial Inorganic Poisons]. Leningrad (in Russian).
Guidelines for the Determination of Noxious Substances in Air (1979) [Metodicheskiye ukazaniya po opredeleniyu vrednykh veschestv v vozdukhe], No. 15. Publication of the USSR Ministry of Health, Moscow, pp. 32–41 (in Russian).
Guidelines for the Determination of Noxious Substances in Air (1981a) [Metodicheskiye ukazaniya po opredeleniyu vrednykh veshchestv v vozdukhe], No. 17. Publication of the USSR Ministry of Health, Moscow, pp. 64–68 (in Russian).

Guidelines for the Determination of Noxious Substances in Welding Aerosols (the solid phase and gases) (1981b) [Metodicheskiye ukazaniya po metodam opredeleniya vrednykh veshchestv v svarochnom aerozole (tverdaya faza i gazy]. Moscow, pp. 58—60 (in Russian).

Guidelines for the Determination of Noxious Substances in Air (1981c) [Metodicheskiye ukazaniya po opredeleniyu vrednykh veshchestv v vozdukhe], No. 19. Publication of the USSR Ministry of Health, Moscow, pp. 32—41 (in Russian).

Gutierrez, G et al. (1985) *J. Appl. Physiol.*, **58**, 558—563.

Ilkun, G. M. (1978) *Zagryazniteli atmosfery i rasteniy* [Pollutants of the Atmosphere and Plants]. Naukova Dumka, Kiev (in Russian).

Klees, M. et al. (1985) *Sci. Total Environ.*, 44, 165—176.

Kobayashi, K. et al. (1984) *Eur. Neurol.*, 23, 34—43.

Kovnatsky, M. A. (1961) *Gig. Truda Prof. Zabol.*, No. 10, 25—28 (in Russian).

Kurppa, K. & Rantanen, J. (1983) In: *Encyclopaedia of Occupational Health and Safety.* International Labour Office, Geneva, pp. 395—399.

Kustov, V. V. et al. (1987) *Gig. Truda Prof. Zabol.*, No. 4, 34—36 (in Russian).

Labunsky, V. Yu. & Struchenevskaya, I. L. (1984) *Kardiologiya*, No. 6, 106—108 (in Russian).

Langeweg, F. (1989) Chapters 3, 7, and 9 in: Bilthoven. National Institute of Public Health and Environmental protection: *Concern for Tomorrow—A National Environmental Survey, 1985 2010.*

Longo, L. (1977) *Amer. J. Obstet. Gynecol.*, 129, 69—103.

Mamatsashvili, M. I. (1967) *Gig. San.*, No. 8, 62—63 (in Russian).

Melnichenko, R. K. (1968) *Vrach. Delo*, No. 7, 87—89 (in Russian).

Mitsuyama, Y & Takamatsu, I. (1974) *Brain Nerve*, 20, 1111—1118.

Nemby-Schmidt, B. et al. (1980) *Neurotoxicology*, 1, 540—553.

Ocak, A. et al. (1985) *J. Anal. Toxicol.*, 9, 202—206.

Okeda, R et al. (1986) *Acta Neuropathol.*, 69, 244—252.

Pankow, D. & Ponsold, W. (1981) *Ztschr. Ges. Hyg.*, 27, 208—214.

Penney, D et al. (1979) *Toxicol. Appl. Pharmacol.*, 50, 213—218.

Peregud, E. A. (1976) *Khimicheskiy analiz vozdukha (novye i usovershenstvovannye metody* [Chemical Analysis of the Air (New and Improved Methods)]. Khimiya, Leningrad (in Russian).

Perret, C. (1973) *Schweiz. Med. Wochenschr.*, 103, 1161—1166.

Petry, H. (1960) *Arch. Gewerbepathol.*, 18, 22—36.

Pozhidayev, V. M. & Elistratova, L. A. (1983) *Gig. San.*, No. 1, 73—74 (in Russian).

Quattrocolo, G et al. (1987) *Ital. J. Neurol. Sci.*, 8, 57—58.

Radziszewski, E. & Guillerm, R. (1984) *J. Toxicol. Med.*, 4, 305—316.

Sawicki, Ch. & Gibson, Q. (1979) *Anal. Biochem.*, 94, 440—449.

Sax, N. I. (ed) (1984) *Dangerous Properties of Industrial Materials.* Van Nostrand Reinhold Co., New York.

Scaroina, F. et al. (1987) *Boll. Soc. Ital. Biol. Sper.*, 63, 939—946.

Schellenberger, M. (1981) *Neurotoxicology*, 2, 431—441.

Sedov, A. V. et al. (1985) *Gig. San.*, No. 9, 9—11 (in Russian).

Sedov, A. V. et al. (1985) *Gig. San.*, No. 12, 17—19 (in Russian).

Shamarin, A. A. & Yurasova, O. L. (1981) *Gig. San.*, No. 10, 33—35 (in Russian).

Shiotsuka, R. et al. (1984) *Toxicol. Appl. Pharmacol.*, 76, 225—233.

Siesjo, B. (1985) *J. Toxicol. Clin. Toxicol.*, 23, 267—280.

Sokal, J. A. et al. (1984) *Arch. Toxicol.*, 56, 106—108.

Spedding, D. J. (1979) In: Bockris, J. O'M. (ed.) *Environmental Chemistry.* Plenum Press, New York, pp. 213—242.

Storm, J. & Fechter, L. (1985) *Toxicol. Appl. Pharmacol.*, 81, 139—146.

Strelchuk, I. V. (1970) *Intoksikatsionnye psikhozy* [Intoxication Psychoses]. Moscow (in Russian).

Tachi, N. & Aoyama, M. (1983) *Bull. Environ. Contam. Toxicol.*, 31, 85—92.

Tiunov, L. A. & Kustov, V. V. (1980) *Toksikologiya okisi ugleroda* [The Toxicology of Carbon Monoxide]. Meditsina, Moscow (in Russian).

Tiunov, L. A. & Arutyunian, S. I. (1985) *Gig. Truda Prof. Zabol.*, No. **10**, 34–37 (in Russian).

Tiunov, L. A. *et al.* (1987) In: Shitskova, A. P. (ed.) *Prognozirovaniye toksichnosti i opasnosti khimicheskikh soyedineniy* [Predicting the Toxicities and Hazards of Chemical Substances]. Moscow, pp. 72–78 (in Russian).

Vyskočil, Y. (1957) *Int. Arch. Gewerbepathol. Gewerbehyg.*, **15**, 457–472.

Vyskočil, A. *et al.* (1986) *J. Appl. Toxicol.*, **6**, 443–446.

Werner, B. *et al.* (1985) *J. Toxicol. Clin. Toxicol.*, **23**, 249–265.

WHO (1979) *Environmental Health Criteria 13: Carbon Monoxide.* World Health Organization, Geneva.

Zorn, H. (1986) *Schriftenr. Ver. Wasserboden Lufthyg.*, No. **9**, 177–179.

Zorn, H. *et al.* (1988) In: Seiler, H. G. *et al.* (eds) *Handbook on Toxicity of Inorganic Compounds.* Marcel Dekker, New York, pp. 183–214.

CARBON DIOXIDE (CARBON(IV) OXIDE; CARBONIC ACID GAS)

Physical and chemical properties

Carbon dioxide (CO_2) is a colorless and odorless gas with a faint acid taste that condenses into a liquid under a pressure of about 0.6 MPa at room temperature. When intensely cooled at atmospheric pressure, it solidifies into a white snow-like mass (dry ice) that sublimes at -78.5°C directly into a gas. The solubility of CO_2 in water is not high, and the dissolution is accompanied by its partial reaction with the water to form carbonic acid (H_2CO_3). Its solubility coefficient in human blood serum is 0.58.

Carbon dioxide is rather inert chemically at ordinary temperatures, but reacts vigorously with strong bases to give carbonates. At high temperatures it is reduced to carbon monoxide by iron, zinc, and some other metals and by coal. It is reduced to methane by hydrogen in the presence of nickel at 350–400°C or of copper(II) oxide at 200°C. In reacting with calcium at red heat, it gives calcium carbide and calcium oxide. At such temperatures it gives urea with ammonia and yields sulfur and carbon monoxide with hydrogen sulfide in the presence of copper. It combines with, and carboxylates, many organic compounds. See also Appendix.

Natural occurrence and environmental levels

Carbon dioxide makes up, by mass, about 0.03% of the atmosphere which contains approximately 2.3×10^{12} tonnes of this compound, in concentrations normally varying from 0.03% to 0.06%.

Carbon dioxide is emitted into the atmosphere in volcanic gases, during the weathering of rocks, from mineral sources (e.g. the waters of mineral springs), from the fermenting, putrefying, or decaying organic substances, as a result of microbiological processes in the soil, and during the respiration of plants, animals, and man. It has been estimated that the fermentation and mineralization of organic matter together with the respiration of terrestrial and aquatic plant organisms contribute ~56% of the CO_2 entering the atmosphere, the activities of soil microorganisms contribute ~38%, and the respiration of animals and human beings, ~1.6% [Ilkun].

The oceans contain nearly 60 times as much CO_2 as the atmosphere ($\sim 1.3 \times 10^{14}$ tonnes). At a wind velocity of 10 m/s, the rates of the ascending flux of CO_2 from the ocean to the atmosphere were 0.3–0.35 mg/s per m^2 of the water area [Vadkovskaya & Lukashev]. The rate of CO_2 absorption by the ocean is limited by its slow diffusion from the surface water layers (down to 200 m) to the deep waters. Because it is more readily soluble in cold than warm water, at high latitudes more CO_2 is absorbed from the atmosphere by water and less is evolved from water into the atmosphere than at or near the equator. The ocean thus acts not unlike a pump in driving CO_2 from colder to warmer regions where the CO_2 pressure in the atmosphere is somewhat higher. The common range of CO_2 concentrations in surface oceanic water was reported to be 290–410 ppm [Budyko].

Approximately 1×10^{11} tonnes of CO_2 are in continuous circulation between the atmosphere and the oceans, The complete renewal of CO_2 in the surface and deep oceanic waters takes 5–25 years and 200–1000 years respectively, while the time required for its complete renewal in the atmosphere ranges from 300 to 500 years. The earth's crust is estimated to contain 20×10^{17} tonnes of CO_2 in the form of calcium carbonate. The estimated amount of CO_2 taken up by plants worldwide from the air in the course of photosynthesis is approximately 16×10^{10} tonnes per annum.

The major reservoirs of CO_2 are the stratosphere and troposphere (in the latter, CO_2 is uniformly distributed up to a height of 7 km) and deep layers of the oceans. Its temporary reservoir is the biosphere. Annually, the stratosphere has been estimated to receive 2.5×10^{16} tonnes of CO_2, the troposphere 3.1×10^{16} tonnes, and the mixed layer of the ocean 2.0×10^{16} tonnes. The CO_2 residence times in these three reservoirs are estimated to be 40 years, 2 years, and 1 year, respectively [Budyko].

Carbon dioxide plays an important role in regulating the influx of gamma, X-ray, ultraviolet, and infrared radiations to the earth, and it also decreases the thermal radiation from the latter.

Production

Carbon dioxide in the gaseous form is produced when carbonaceous substances burn in an excess of air or oxygen. It is a product of fermentation processes, synthetic ammonia production, limestone calcination, the reaction of sulfuric acid with dolomite, and numerous other reactions used in industrial chemistry.

Industrially, gaseous CO_2 is obtained by calcining carbonate of magnesium (as dolomite) or calcium (limestone, marble, chalk); from flue gases (which contain 10–18% CO_2); from gases generated in lime kilns (these gases contain 32–41% CO_2); from gases produced in the fermentation of molasses or grain (the gases contain 60 to 99.8% CO_2); or by burning coke in an excess of air. The liquid and solid forms of CO_2 are obtained from gaseous CO_2 after its purification.

Uses

In the gaseous or liquefied form, carbon dioxide is used in the production of sugar, in the carbonation of beverages (such as soft drinks), in fire extinguishers, in mining operations, in the chemical industry (for the manufacture of sodium hydrogen carbonate, ammonium carbonate and hydrogen carbonate, urea, and hydroxycarboxylic acids, in shielded-arc welding, as an addition to the atmosphere of greenhouses to increase the growth of plants, as a power source (as in spray painting and for inflating life rafts), and in medicine (in mixture with air or oxygen) to simulate respiration. In the solid form (dry ice or carbon dioxide snow), it is used for refrigeration and cooling in the production of ice cream, meat products, and frozen foods and in medicine as an escharotic agent to destroy certain skin lesions.

In its solid or gaseous form, carbon dioxide is utilized to produce an inert atmosphere for certain reactions and in vessels or plants where there would otherwise be a risk of explosion [Matheson].

Man-made sources of emission into the environment

Major industrial sources of the carbon dioxide present in the air are fuel combustion, geothermal electric power stations, putrefaction and fermentation processes, mining activities, and sugar refineries (where it is released from saturators). Carbon dioxide is also released during synthetic ammonium production, in the reaction of sulfuric acid with dolomite, and in many other reactions in industrial chemistry. In addition, it forms in various confined spaces such as tunnels, wells, cellars, caissons, etc. Its industrial emissions worldwide are estimated at about 5×10^9 tonnes annually, of which about 50% remains in the atmosphere and the rest enters the biosphere in the process of photosynthesis or is absorbed by the world's ocean.

The carbon dioxide content of the atmosphere has been rising with the expansion of industrial activities and has increased by about one-fifth since the burning of fossil fuels began on a large scale. Its estimated average atmospheric concentration rose from 290 ppm in 1950 to 310 ppm in 1960 and to 340 ppm in 1980; for the year 2000, a further rise to 380–390 ppm has been forecast. The progressive elevation of CO_2 concentration in the atmosphere has raised concern about the possible climatic changes that might take place in future as a result of the so-called greenhouse effect [Freedman; Langeweg; Neal; Zorn et al.]. It has been calculated that a rise of CO_2 to 600 ppm would entail an increase by 4–5°C in the mean annual temperature on the Earth, and that the increase would be greater in the north of the Northern Hemisphere than elsewhere. If so, then rapid melting of the Arctic ice should occur, with a consequent rise by 5–7 m in the levels of the world's ocean [Lockwood; Myers]. Actually, however, no warming of the climate has been observable. Although the world temperature increased by 0.5–0.6°C between the end of the 19th century and 1940, since then a reverse trend had been in evidence despite the rising CO_2 content in the atmosphere. Thus, the mean annual temperature in the Northern Hemisphere has decreased by 0.35°C and, moreover, ice caps in the Arctic and in Greenland have been growing while the levels of world's ocean has even decreased [Budyko].

Toxicity

General aspects

Carbon dioxide irritates skin and mucous membranes, stimulates the respiratory centers in relatively low concentrations while inhibiting them in very high ones, and can act as a narcotic. Its high content in the air, which is usually associated with depletion of oxygen supply, leads to rapid death. CO_2 is a central vasoconstrictor and a local vasodilator, causes acidosis, increases blood levels of epinephrine and norepinephrine, decreases those of amino acids, and inhibits enzymatic activity in tissues [Nahas]. Animals are, in general, less sensitive to CO_2 than humans.

Animals

Acute toxicity

Mice breathing air containing 20% CO_2 rapidly developed pulmonary edema. In rats exposed to a CO_2 concentration of 21%, impairment of myocardial metabolism and atrioventricular conduction and slowed heart rate were observed after 1 h; exposure to a level of 11% for 2–4 h resulted in a 15–25% fall in total oxygen consumption and in reticulocytosis. A gaseous mixture containing about 30% CO_2 and a normal oxygen concentration produced a short-term excitation succeeded by inhibition and a state of narcosis; it also raised blood pressure, impaired cardiac activity, and caused dilation of coronary vessels. A brief exposure of rats to a CO_2 concentration of 70% was accompanied by arrhythmia and a sharp rise in blood potassium [Gramenitsky *et al.*; Schaefer].

At an O_2 content of no less than 16%, the highest CO_2 concentration tolerated by rats was about 15% or 23% if it was raised to the latter level very slowly (over several days). In rabbits breathing air containing 5% CO_2, a rise and then a fall in erythrocytes and hemoglobin during 5-hour exposure were recorded, as were leukocytosis with a left shift and decreases in cytochrome oxidase and succinate dehydrogenase activities [Zheludkova]. In dogs, inhalation of CO_2 at 5–10% was accompanied by increases in blood pressure and heart rate, excitation of the sympathetic nervous system, and by narrowing of coronary vessels initially and their dilation later in the course of exposure. Concentrations above 7% led to elevated intraocular pressure in cats and dogs [Sazonov]. Slowed heart rate and respiratory arrhythmia due to increased vagal tone occurred in dogs breathing air containing 10–15% CO_2. Levels in excess of 60% were rapidly lethal to most experimental animals. Survival times were longer when the CO_2 concentration was raised to a high level gradually [Gramenitsky *et al.*].

Evidence for teratogenity of carbon dioxide was obtained following a single exposure of animals to its high concentrations during pregnancy [Sax].

Subacute and chronic toxicity

Rabbits exposed to CO_2 concentrations of 3–5% showed no deviations from normal in their blood after 3 days but developed reticulocytosis after 7 days. In dogs, guinea-pigs, and rabbits inhaling air with elevated CO_2 levels for a number of days, degenerative changes were found in the lungs, liver, kidneys, and brain [Zorn *et al.*]. Guinea-pigs sacrificed after 48 h of exposure to a CO_2 concentration of 16% had degenerative changes of the same severity as did rabbits exposed to 4.5% CO_2 for 13 days.

In an atmosphere high in CO_2, animals survived longer if the O_2 content was low (11–14%) than they did if it was normal, although the narcotic effect of CO_2 in the presence of O_2 deficiency was more strongly marked. When animals that had been exposed to high CO_2 levels (22–23%) for a long time were allowed to breathe fresh air, they developed attacks of tonic convulsions which proved fatal for some of them.

The possibility of habituation to CO_2 exposure was suggested by the findings in rats and guinea-pigs that pH of the blood, body temperature, and metabolic rate were reduced in the first 3–7 days of their residence in a CO_2-rich atmosphere but returned to normal thereafter [Schaefer].

There is some evidence that teratogenic and embryotoxic effects may occur in animals as a result of chronic exposure to carbon dioxide.

Man

Acute toxicity

Cases of intoxication with carbon dioxide have occurred in a variety of indoor situations, for example during cleaning of fermentation vats or tanks, sewer or water pipes, and wine cisterns or casks; in underground yeast vats of yeast plants; and in rooms where dry ice or carbonated mineral water were handled. There are descriptions of fatal cases of poisoning that took place during the descent into a cellar containing rotten or germinated potatoes, in manholes (where the air contained 11–24% CO_2), in cesspools (hydrogen sulfide, ammonia, and other compounds were also present in the air), and in tobacco storage sites.

Very high CO_2 concentrations (up to 60%) can be generated in potato pits, granaries, silo towers, and in confined spaces where vegetables (notably onion or pickled vegetables) are stored. Potential sources of CO_2 poisoning are such processes as acetone production from fermented corn and the manufacture of brick and lime. Air rich in CO_2 may have to be breathed for prolonged periods by those working in gas masks or diving suits. Expired air contains about 4% CO_2 and its concentration in confined spaces therefore rises with increasing intensity of the physical work being done there. Carbon dioxide does not usually give any warning of its presence, and a person may unwittingly enter a confined space and be overcome before becoming aware of the danger.

Breathing the air containing $0.25-1.0\%$ of CO_2 is accompanied by functional alterations in external respiration and blood circulation. At CO_2 concentrations of $2.5-5\%$, headache, irritation of the upper respiratory tract, a feeling of warmth in the chest, increased pulmonary ventilation (as a result of more frequent and deeper respirations), increased heart rate, and blood pressure elevation supervene [Breslav & Salatsinskaya; Yeliseyeva]. At 7% or higher CO_2 levels, sweating, noises in the ears, palpitation, and dizziness are added; excitation, vomiting, reduction in body temperature, visual disturbances (abnormal accommodation and dark adaptation, photophobia), and manifestations of toxic effects on the brain and spinal cord may also occur. At $4.5-6.5\%$ CO_2 levels, the ability to think and short-term memory were not affected if the exposure period was short (up to 20 min), whereas longer exposures (60–80 min) were accompanied by a retardation of mental processes. The concentration of 10.5% could be tolerated by some volunteers for 0.5 min at most [Freedman & Sevel; Malikman et al.; Sayers et al.].

In tightly enclosed spaces, morbid phenomena result from both the excess of CO_2 and the lack of O_2. Added danger in such environments comes from the upset of bodily thermoregulation as a result of the elevation of air temperature and humidity. When volunteers were requested to remain for 4 h in a hermetically sealed room where the CO_2 concentration was progressively rising from 0.48% to 4.7% as the O_2 concentration declined from 20.6% to 15.8%, some of them complained of stuffiness and headache at the end of that period and all experienced a fall in body temperature, increased breathing rate, and increased or decreased pulse rate. Some volunteers complained of headache and weakness on the next day.

In persons exposed to CO_2 in a completely enclosed space where its concentrations were gradually rising to 5.5% and those of O_2 were declining to 14.5%, a number of changes were observed by the end of the 10-hour exposure period, including a drastic increase in pulmonary ventilation rate (up to 30–35 L/min), a 50% increase in oxygen consumption, an acid shift of the blood pH value, altered pulse rate, elevated blood pressure (particularly diastolic), reduced body temperature (by $0.5°C$), and diminished physical and mental working capacity. The condition of individuals so exposed was aggravated by raising the CO_2 concentration to the same level at a faster rate. Strong headache and general weakness persisted in some for 12–72 h post-exposure.

Cases of asphyxation occurring at various CO_2 concentrations in the presence of O_2 deficiency have been described, the lowest concentration being 3.0% and the highest O_2 content, 13.6%. At very high CO_2 concentrations, death ensued as a result of respiratory arrest (in a matter of seconds at 20%) and was usually preceded by only slight, if any, convulsions; the heart continued to work for some time after the cessation of breathing. Autopsy revealed hemorrhages to mucous membranes in some cases. Livores mortis were of a bluish-red color. In cases of successful resuscitation efforts, consciousness usually returned in several minutes.

Consequences of acute poisoning during the first 24 h may include headache, general weakness, thirst, and a peculiar feeling of oppression in the chest. After severe poisoning, blood-stained sputum, fever, and inflammation of the bronchi and lungs were observed.

There is a wide individual variation in sensitivity to carbon dioxide exposure.

Chronic toxicity

The occurrence of chronic carbon dioxide poisoning has not been conclusively demonstrated. Persons working in fermenting cellars tolerated CO_2 concentrations of 0.5–2% without showing evidence of disease; when exposed to somewhat higher concentrations, they only experienced headache and labored breathing while climbing stairs. Similar observations have been made in miners. Persons continuously exposed for 1 month to an atmosphere containing 1% CO_2 did not notice any deleterious effects on their general health or physical well-being while those exposed to 2% for the same period only showed some decrease in working capacity [Zharov *et al.*].

Hygienic standards

In the USSR, no MACs for carbon dioxide have been established, but its concentration in the air of coal and ozokerite mines should not exceed 0.5%. Also, it has been recommended that the average daily and short-term carbon dioxide concentrations in the air of residential and public buildings should not exceed 0.05% and 0.1%, respectively [Yeliseyeva].

In the USA, threshold limit values for carbon dioxide have been set at 9000 mg/m^3 (TWA) and 54 000 mg/m^3 (STEL).

Methods of determination

Carbon dioxide in air can be measured by chromatographic and other methods. One commonly used method is gas-adsorption chromatography using a katharometer (detection limit, 1.2 µg; measurement error, ±12%). Automatic gas analyzers are available [Kozlitin; Nikitin & Novikov]. Analytical methods for determination of carbon dioxide have been reviewed by Zorn *et al.*

Measures to control exposure

A general and effective measure for preventing contamination of the workroom atmosphere with CO_2 gas is the provision of adequate exhaust ventilation.

Hazardous exposures are most likely to occur in wells, pits, cellars, vats and tanks, tunnels, mines, silos, and other similar confined or semiconfined spaces. Such spaces should be monitored for CO_2 (and other gases) and, if possible, be well ventilated before and during work there. Special air-purifying agents that absorb CO_2 and release O_2 may also be used. A worker entering a confined space where a high CO_2 concentration is believed to have built up, should wear a filter-type respirator and be accompanied by another, similarly equipped person. Such respirators should also be used during work in places where CO_2 concentrations exceed the MAC level. In plants where dry ice or liquid carbon dioxide are produced, the health and safety regulations laid down for the respective plants should be strictly observed.

REFERENCES

Breslav, I. S. & Salatsinskaya, E. N. (1967) *Fiziol. Zh. SSSR*, **53**, 957—963 (in Russian).
Budyko, M. I. (1983) In: *Marksizm-leninizm i globalnye problemy sovremennosti* [Marxism-Leninism and Global Problems of the Present Time]. Moscow, pp. 192-207 (in Russian).
Freedman, B. (1989) *Environmental Ecology—the Impacts of Pollution and Other Stresses on Ecosystem Structure and Function*. Academic Press, New York, pp. 9—52.
Freedman, A. & Sevel, D. (1966) *Arch. Ophthalmol.*, **76**, 59—65.
Gramenitsky, P. M. *et al* . (1978) *Byul. Eksper. Biol. Med.*, No. 9, 285—287 (in Russian).
Ilkun, G. M. (1978) *Zagryazniteli atmosfery i rasteniy* [Pollutants of the Atmosphere and Plants]. Naukova Dumka, Kiev (in Russian).
Kozlitin, A. A. *et al.* (1978) In: *Problemy kontrolya i zashchity atmosfery ot zagryazneniya. Respublikanskiy mezhvedomstvennyi sbornik* [Atmospheric Pollution: Monitoring and Protection. A Collection of Interdepartmental Papers], No. **4**, pp. 54—58 (in Russian).
Langeweg, F. (1989) Chapters 3, 7, and 9 in: Bilthoven. National Institute of Public Health and Environmental Protection: *Concern For Tomorrow—A National Environment Survey 1985—2010*.
Lockwood, J. (1982) *Nature*, **299**, (No. 5880), 203.
Malikman, I. I. *et al.* (1971) *Kosm. Biol. Aviakosm. Med.*, No. **5**, 17—22 (in Russian).
Matheson, D. (1983) In: *Encyclopaedia of Occupational Health and Safety*. International Labour Office, Geneva, pp. 392—393.
Myers, N. (1983) *J. Soc. Biol. Struct.*, **6**, 17—28.
Nahas, G. (1974) In: *Carbon Dioxide and Metabolic Regulation*. Springer, New York, pp. 107—116.
Neal, P. (1989) *The Greenhouse Effect and the Ozone Layer*. Batsford, London.
Nikitin, D. P. & Novikov, Yu. V. (1980) In: *Okruzhayushchaya sreda i chelovek* [Man and the Environment]. Moscow, pp. 177—181 (in Russian).
Sax, N. I. (ed.) (1984) In: *Dangerous Properties of Industrial Materials*. Van Nostrand Reinhold Co., New York.
Sayers, J *et al.* (1987) *J. Appl. Physiol.*, **63**, 25—30.
Sazonov, S. Ya. (1965) *Fiziol. Zh. SSR*, No. **9**, 1057—1065 (in Russian).
Schaefer, K. (1974) In: *Carbon Dioxide and Metabolic Regulation*. Springer, New York, pp. 253—265.
Vadkovskaya, I. K. & Lukashev, K. P. (1977) *Geokhimicheskie osnovy okhrany biosfery* [Protection of the Biosphere: Geochemical Principles]. Nauka i Tekhnika, Minsk (in Russian).

Yeliseyeva, S. V. (1966) In: Ryazanov, V. A. & Goldberg, I. S. (eds.) *Biologicheskoye deistviye i gigiyenicheskoye znacheniye atmosfernykh zagryazneniy* [Biological Actions and Health Implications of Atmospheric Pollution]. Meditsina, Moscow, pp. 7–27 (in Russian).

Zharov, S. G. *et al.* (1963) In: *Aviatsionnaya i kosmicheskaya meditsina* [Aviation and Space Medicine]. Meditsina, Moscow, pp. 182–185 (in Russian).

Zheludkova, T. N. (1967) *Gig. Truda Prof. Zabol.*, No. 2, 52–55 (in Russian).

Zorn, H. R. *et al.* (1988) In: Seiler, H. G. *et al.* (eds) *Handbook on Toxicity of Inorganic Compounds.* Marcel Dekker, New York, pp. 183–214.

Hydrogen cyanide and its derivatives

HYDROGEN CYANIDE (FORMONITRILE, HYDROCYANIC ACID, PRUSSIC ACID)

Physical and chemical properties

Hydrogen cyanide (HCN) is a readily flammable, colorless, very mobile volatile liquid or gas that has an odor of bitter almonds. In the presence of ammonia or trace alkalies, it darkens due to the formation of its polymerization products. In alkaline media and also when its aqueous solutions are boiled, hydrogen cyanide undergoes hydrolysis to give ammonium formate. It forms complex compounds. See also Appendix.

Production

Industrially, hydrogen cyanide is obtained by one of the following methods: (1) the action of dilute acids on cyanides; (2) dry distillation of molasses stillage, with the formation of methylamines that decompose at 1000–1100°C to give HCN; (3) absorption from coke-oven gas; (4) contact oxidation of methane and ammonia; (5) oxidation of methane and ammonia by electric discharge; (6) from ammonia and carbon monoxide by their direct reaction in the presence of a catalyst or by the formamide process (dehydration of the formamide resulting from the reaction of these two compounds in an alkaline alcoholic solution; (7) pyrogenetic decomposition (electro- or thermocracking of hydrocarbons with nitrogen or ammonia.

Uses

Hydrogen cyanide is used in many organic syntheses, for example in the production of monomers for nitrile resins, of synthetic fibers and plastics, organic glass, methylamide, formamide derivatives, hydroxy nitriles, cyanogen bromide, and cyanogen chloride. It has also been employed in electroplating, in mining, and as a rodenticide and insecticide.

Man-made sources of emission into the environment

Anthropogenic sources that may emit hydrogen cyanide (as well as sodium and potassium cyanides) into the ambient air include a wide variety of processes and industries. These are the production of benzene, toluene, and xylene; coke plants; hydrogenation of coal; electroplating and electroforming; combustion of celluloid and wool; heating of polymeric compositions (nylon, polyacrylonitrile, polyurethane, carbamide and melamine plastics); incomplete combustion or dry distillation of nitrogenous organic compounds and the production of cyanides from these; cyanidation of steel; manufacture of potassium ferricyanide (red prussiate of potash) and its use for dyeing and mordanting textile fabrics (HCN is also contained In waste waters from these processes); production of thiocyanates and oxalic acid; reactions of concentrated nitric or sulfuric acid with proteins; miscellaneous metallurgical processes (hardening and liquid cementation of metals, flotation of lead-zinc sulfide ores, briquetting of ferrosilicon or ferromanganese, etc.).

Cyanides were found in amounts of 0.03–0.3 g per 100 m^3 in blast-furnace gases and 2.7–9 mg per liter in wastewaters from gas scrubbing processes. Hydrogen cyanide is contained in tobacco smoke (10 g of tobacco were reported to yield 2 ml HCN), and it also may be produced by bacterial action in fermenting juice [Hüdepohl; Morikawa; Plisov; Wadden].

Hydrogen cyanide and other cyanides are likely to be present in industrial waste waters from mines, ore-concentrating factories, electroplating facilities, gas works, and coke plants. Such waters also often contain complex cyanides of metals that partly dissociate with the formation of the CN^- ion. When cyanide-containing water is treated with active chlorine in the presence of acid, the cyanides decompose giving rise to cyanogen chloride, a highly toxic gaseous compound slightly soluble in water.

The highest hydrogen cyanide concentration that did not interfere with biochemical processes in water was found to be 50 mg/L [Grushko].

Toxicity

General considerations

Hydrogen cyanide rapidly causes asphyxia by blocking respiratory enzymes and impairing tissue respiration [Zorn et al.; Way et al.]. This is also the mode of action by all other cyanides that are capable of splitting off HCN and yielding the CN⁻ ion [Abel].

In acute HCN poisoning, which affects most severely the respiratory and vasomotor centers, the initial deep breathing and blood pressure rise are succeeded by respiratory paralysis and a sharp blood pressure fall. Cyanides inhibit oxidative phosphorylation and energy metabolism in nerve cells, and they also inhibit the enzymes catalyzing the biotransformation of several amino acids including histidine, tryptophan, and tyrosine. Of major importance is the inhibition of cytochrome c oxidase, which is the terminal oxidase in the mitochondrial respiratory chain and is present in all cells. Oxygen starvation is the most critical early disorder in acute poisoning, and a good indication of the drastically reduced capacity of tissues to utilize oxygen is that venous blood becomes arterial in color. Later, degenerative changes may develop in the central nervous system.

In chronic poisoning, of great importance is the inhibition of thyroid hormone production caused by the thiocyanates (rhodanides) formed as a result of the hydrogen cyanide or its salts reacting with sulfhydryl-containing proteins (e.g. glutathione).

The susceptibility of warm-blooded animals to acute cyanide poisoning is dependent on the current level of oxygen consumption by the body. If the level is low as, for example, during hibernation, the susceptibility is markedly decreased because the body temperature is reduced and the resistance of the animal to hypoxia in general is enhanced.

Hydrogen cyanide and its salts can be absorbed through intact skin, particularly in the liquid state.

Acute exposure

Animals

Toxic doses and concentrations of hydrogen cyanide are shown in Table 1.

Symptoms and signs in lethal poisoning are similar in all species of laboratory animals. The initial increase in breathing frequency is followed by respiratory distress and then by respiratory arrest for a while, accompanied by convulsions and paralysis, followed in turn by resumption of respiratory movements, secondary respiratory arrest, and death. Cats, dogs, and monkeys usually vomit. The heart stops 5 to 10 min after the cessation of breathing.

The blockade of myocardial oxidase results in acute cardiac failure. The tonus of the parasympathetic nervous system increases with a concurrent elevation of myocardial sensitivity to vagal influences; the overall tonus of the sympathetic nervous system falls dramatically. The blood level of glutathione progressively rises, as does the secretion of hypertensive hormones and of epinephrine. If the animal does not die, symptoms of central nervous system damage are observed.

Table 1. Acute toxicity of hydrogen cyanide for animals

Species	LC_{50} (mg/m^3)	LD_{50} (mg/kg)	Minimal lethal dose (mg/kg)
Mouse	362 (5 min)	3.7 (oral)	3 (S.C.)
		2.7 (I.M.)	
		1.0 (I.V.)	
		2.99 (I.P.)	
Rat	542 (5 min)	0.8 (I.V.)	10 (oral)
Guinea-pig	2112 (1 min)		0.1 (S.C.)
Rabbit	980 (1 min)	1.11 (I.M.)	4 (oral)
		2.5 (S.C.)	
		0.66 (I.V.)	
		1.57 (I.P.)	
Cat	1221 (1 min)	0.81 (I.V.)	1.11 (S.C.)
Dog	616 (1 min)	1.34-1.43 (I.V.)	4 (oral)
			1.7 (S.C.)
Pig	1740 (1 min)		
Monkey	1616 (1 min)	1.3 (I.V.)	

Source: Sax.

On autopsy of rabbits, cats, and dogs that died within the first few minutes of poisoning, no changes were noticed except for a brighter than normal color of liquid blood and an odor of bitter almonds. Animals that succumbed somewhat later (after 5–15 min) were seen to have a congested and edematous trachea, a foamy (occasionally bloody) fluid in the tracheal cavity, hemorrhages under the pleura and less often also in the lungs themselves and under the endocardium and pericardium. The brains of surviving animals presented a picture of general diffuse nerve cell damage and symmetrical necrotic areas in the cerebral cortex.

The cerebral cortex is the structure primarily affected in animals exposed to cyanides in low to medium doses. Such animals develop general inhibition and a long-lasting state of ready exhaustibility of the nervous system. If the cerebral cortex was in a state of protective inhibition — for example in an animal under narcosis — the animal survived even after a high (lethal) dose. On the other hand, exposure to low hydrogen cyanide concentrations may be fatal if the ambient air contains carbon monoxide. Prior administration of ethanol slows the development of intoxication and prolongs survival.

Hydrogen cyanide can produce a lethal effect by being absorbed through intact skin or mucous membranes. Dogs died after pure HCN was applied to 5 cm^2 of their skin. Absorption of HCN through undamaged skin was also responsible for the death of cats when their skin was exposed to HCN concentrations of 550–15 000 mg/m^3 of air. Skin absorption is enhanced at high ambient temperatures [Minkina].

Man

Lethal concentrations of hydrogen cyanide have been assessed at 200 mg/m^3 for 10-min exposure, 150 mg/m^3 for 30 min, and 120 mg/m^3 for 60 min; the minimal lethal doses have been calculated to be 0.57 mg/kg *per os* and 1 mg/kg subcutaneously or intravenously [Sax]. Responses to various HCN concentrations are listed in Table 2.

Table 2. Toxic concentrations of hydrogen cyanide for man

Concentration (mg/m^3)	Response
1	Threshold concentration for odor perception
5–20	Headache and vertigo in some individuals
20–50	Headache, nausea, vomiting, and palpitation with prolonged (for hours) exposure
50–60	After 30–60 min exposure, toxic effects observed in some individuals several hours later but not in others at any time
100	May be fatal; death usually occurs within 60 min of exposure
120–150	Death after 30–60 min of exposure
200	Death after 10 min of exposure
300	In the resting state can be tolerated for 2 min without feeling dizzy
400	In the resting state can be tolerated for 1.5 min without feeling dizzy
550	Can be tolerated for 1 min without serious consequences
7000–12 000	For those wearing a gas mask, even 5 min of exposure is dangerous because of poisoning through the skin
22 000	Even those wearing a gas mask experience vertigo, weakness, and palpitation after 8–10 min of exposure and are unable to work during subsequent 2 or 3 days

Source: Zdrojewicz *et al.*

Inhalation of HCN at high concentration results in an almost instantaneous loss of consciousness, followed by respiratory and then cardiac arrest. In acute poisoning by lower concentrations, several stages may be distinguished. The initial stage is characterized by clinical symptoms such as a feeling of irritation in the throat, a burning bitter taste in the mouth, salivation, numbness in the mouth and pharynx, reddening of the conjunctivae, muscular weakness, staggering, difficulty in speaking, vertigo, acute headache, nausea and vomiting, urges to defecate; somewhat increased breathing rate at first and slower deep breathing later; ingress of blood into the head and palpitation. These symptoms will clear rapidly after moving into fresh air. The next, dyspneic stage is marked by progressive general weakness, pain and a feeling of tightness in the cardiac region; slowed pulse; slow deep breathing and marked dyspnea (sometimes accompanied by occasional short inspirations and long expirations); nausea and vomiting; pupil dilatation and exophthalmos. The third, convulsive stage is marked by a feeling of distress, intensifying dyspnea, loss of consciousness, severe convulsions (most often tetanic), and cramps of the masticatory muscles with biting of the tongue. The final stage is one of paralysis or asphyxia: there is a complete loss of sensation and reflexes, involuntary urination and defecation occur, respirations become rare and shallow, and death ensues.

At necropsy of fatal cases, typical signs of asphyxia can be observed such as liquid blood, congested meninges, hemorrhages, and sharply increased permeability of vessel walls. The blood and livores mortis usually appear bright red. Degenerative changes may be seen in nerve cells of the brain and cloudy swelling in cells of the liver, kidney, and heart. The brain and lungs smell of bitter almonds. Decomposition of the cadaver proceeds slowly.

The sequelae of poisoning may include a rapid pulse, increased physical and mental fatiguability, psychic and neural disorders with an extrapyramidal syndrome, weakened memory, and polyneuritis [Manz]. There is wide interindividual variation in the susceptibility to HCN poisoning; persons with affections of the organs that are targets for it, are likely to be more susceptible than others. No habituation occurs; rather, the susceptibility increases with further exposure.

Repeated and chronic exposures

Animals
Once in the body, hydrogen cyanide is rapidly decomposed and does not accumulate, but its toxic effects become progressively more severe on repeated (subacute) exposures. Thus, severe diseases and even deaths occurred among animals so exposed to HCN or its salts. In rabbits and dogs, complete flaccid paralysis of the hind limbs, loss of reflexes, and urinary and fecal incontinence were observed, followed by deaths 24 to 60 h after the onset of paralysis. Histological studies showed areas of malacia in various brain regions along with degenerative changes in peripheral nerves and in cells of the anterior spinal cord horns.

Chronic poisoning of dogs with cyanides resulted in neuroses and dystrophic dermal lesions that disappeared after the cessation of exposure. Dogs receiving small and apparently harmless doses developed persistent and profound changes in cerebral cortex activity as well as heightened sensitivity to these compounds that lasted up to 15 months after the exposure period. Changes in higher nervous activity preceded any other toxic effects. Individual sensitivity depends to a large extent on the type of this activity; thus, chronic exposure to low doses produced neurosis only in dogs with a weak type.

A relatively early sign of chronic poisoning is elevated serum acetylcholinesterase activity [Sinitsyn]. Chronic cyanide poisoning, unlike repeated oxygen starvation, does not lead to increased erythrocyte production. The chronic toxicity of cyanides is greatly augmented by alcohol, even in small doses.

As shown by animal experiments, chronic concurrent exposure to lead and cyanides or to arsenic and cyanides (combinations found in wastewaters from metallurgical and ore-concentrating works) at the MAC level adopted for each of these xenobiotics, results in simple summation of their toxic effects [Novakova & Dinoyeva; Wodichenska].

Man

Reported symptoms of chronic intoxication with hydrogen cyanide and its compounds include headache, weakness, rapid fatiguability, general malaise, motor incoordination, sweating, hyperemia of the upper part of the trunk, increased irritability, nausea, dyspepsia, tachycardia, and pains in the epigastric region, extremities, and heart. On detailed examination, the following abnormalities were revealed in various combinations: neurological disorders (a neuroasthenic, asthenic, or asthenic-vegetative syndrome that progressed with increasing length of exposure), vasoneurosis, nystagmus, cardiac enlargement and dull cardiac sounds, low blood pressure, frequent urination, proteinuria, thyroid enlargement; a number of blood changes including reduced sugar and increased lactic acid levels, depressed cholinesterase activity, altered hematocrit, and lowered phagocytic activity of leukocytes. All patients showed a morbid reaction to alcohol [Lutai *et al.*]. Paralysis and muscular atrophy have also developed occasionally.

On prolonged exposure to low concentrations of hydrogen cyanide or its salts, their metabolic products, thiocyanates, are not completely eliminated in the urine and will accumulate in the body. Because of their thyrostatic activity and ability to inhibit the peroxidase involved in the biosynthesis of thyroid hormones, the thiocyanates may produce hypothyroidism. Goiter and reduced glucocorticoid production by the adrenal cortex have also been reported [Goncharova *et al.*]. Diminished visual acuity as a result of retrobulbar neuritis of the optic nerve may occur [Vogel *et al.*].

In early phases of chronic poisoning, decreased blood pressure, myocardial dystrophy, and impaired myocardial contractility have been observed, as well as, occasionally, blocked branches of the bundle of His, depressed cardiac activity, and increased respiratory rate. Thyroid enlargement, altered acidity of the gastric juice, and slight body-weight losses have been observed in some cases.

Zdrojewicz *et al.* reported the presence of hypothyroidism in 10% of the electroplaters they examined. Chronic poisoning manifested by gastrointestinal disturbances, abdominal pain, constipation, and psychic alterations occurred among workers engaged in galvanoplastic operations. As a result of long-continued entry of cyanides into the body, even in microgram quantities, their metabolites (thiocyanates) may accumulate in the mucus of the uterine cervix to cause damage to spermatozoids of a sufficient degree for preventing conception. Cyanides also impair the transporting function of the placenta [Baumeister *et al.*; Kelman].

Observations in factories where hydrogen cyanide concentrations in the workroom air ranged from tenths of a milligram to several milligrams per m^3, failed to reveal any obvious signs of poisoning in the workers. However, in other workers exposed to low HCN concentrations for very long periods (up to 24 years), a number of abnormalities were detected, including pulmonary emphysema, cardiac enlargement, bradycardia, hypotension, erythrocyte hemolysis, morphological changes in the blood, hyperglycemia, liver enlargement, and mild pyramidal and cerebellar disturbances [Langauer-Lewowicka].

The risk of intoxication is increased by smoking. Cigarette smokers have been reported to contain 0.33 μmol/L of the CN$^-$ ion in the blood, 0.68 μmol/L in the erythrocytes, and 0.33 μmol/L in the plasma as compared to 0.13, 0.24, and 0.02 μmol/L, respectively, in nonsmokers [Lunguist *et al.*, 1988]. (The smoke from one cigarette contains 150–300 μg of CN$^-$.) The cyanides present in tobacco smoke have been incriminated as the cause of visual impairment and retrobulbar neuritis seen in smokers [Knapik *et al.*; Vogel *et al.*].

Consumption of cyanide-containing fruits and vegetables, especially by persons whose diet is deficient in sulfur-containing amino acids which are necessary for metabolizing cyanides in the normal way, may result in ataxic polyneuropathy, myelopathy, and impaired hearing and vision; the blood of such persons has a high content of thiocyanates [*Bulletin of the World Health Organization;* Cliff *et al.*; Eruman; Kläui *et al.*; Wilson]. Poisoning may be caused by the hydrogen cyanide formed in the body from the glycosides contained in flax (linamarine), in leaves of the tree *Laurocerasus* Roem (laurocerasin), and in white clover. It also may be due to the eating of large number of kernels from bitter almonds, peaches, plums, apricots, sweet cherries, apples, or some other members of the Rosaceae family, since the glycoside amygdalin contained in such kernels is decomposed by enzymes in the body to yield HCN. Apricot, peach and apple kernels have been found to contain cyanides at average levels of 2.92, 2.6, and 0.61 mg/g, respectively; cases of cyanide poisoning have been reported from the use of the amygdalin-containing drug Laetrile in the treatment of malignant neoplasms [Holzbecher *el al.*].

Dermal and ocular effects

Hydrogen cyanide is a mild skin irritant and causes itching and the formation of vesicles resembling those seen in burns. It also produces local irritation on entering the eyes.

Absorption, distribution, and elimination

Hydrogen cyanide and its salts take only seconds to be absorbed into the blood from the gastrointestinal tract or the lungs. Absorption through the skin occurs more slowly.

Concentrations of cyanide ions in the brain, myocardium, and muscles were higher than in any other organ or tissue regardless of the administration route and animal species; in the liver, they were higher after oral or intraperitoneal administration and lower after absorption by inhalation or via the skin. Whole blood contained more cyanide ions than did serum [Aleksandrov et al.; Ballantyne, 1980].

The major metabolites of hydrogen cyanide are poisonous thiocyanates. After its administration in a large dose, the production and elimination rates of thiocyanates were highest at 24–48 h. The cyanide undergoes partial oxidation to carbon dioxide via cyanic acid and also combines with cystine to form iminothiazolidine-4-carboxylic acid. Hydrogen cyanide was detected (at 0.67 nmol/L) in the air expired by healthy non-smoking men — a finding which was attributed to its formation from thiocyanate [Lunguist et al., 1985].

Hygienic standards

In the USSR, three hygienic standards have been adopted for hydrogen cyanide and its salts (as HCN) in workplace air, atmospheric air of resident areas, and water sources, respectively: $MAC_{wz} = 0.3$ mg/m^3 (hazard class 1; the content of HCN gas in the air is subject to automatic monitoring), $MAC_{ad} = 0.01$ mg/m^3, and $MAC_w = 0.1$ mg/L (based on sanitary and toxicological criteria).

In the USA, a TLV-C of 11.00 mg/m^3 has been adopted for hydrogen cyanide and a TLV-TWA of 5.0 mg/m^3 for the cyanides (with the 'skin' notation to indicate the possibility of absorption through skin, mucous membranes, and eyes).

Methods of determination

Two commonly used photometric methods for quantitative determination of hydrogen cyanide in *air* are the following: (1) a method in which hydrogen cyanide is reacted with bromine and the resulting cyanogen bromide is reacted with pyridine and aniline (detection limits: 0.1 µg in the solution volume analyzed and 0.1 mg/m^3 of air) [*Guidelines for the Determination of Noxious Substances in Air*, 1981a]; (2) a method in which cyanide ions are reacted with chloramine-T and the resulting cyanogen chloride is reacted with pyridine and barbituric acid to form a colored product (detection limits: 0.1 µg in the solution volume analyzed and 0.15 mg/m^3 of air; measureable concentration range, $0.15-1.5 \text{ mg/m}^3$; total measurement error, $\pm 25\%$); thiocyanates may interfere with the assay [*Guidelines for the Determination of Noxious Substances in Air*, 1981a]. A gas-chromatographic procedure for hydrogen cyanide determination in the presence of acrylonitrile using an electron-capture detector has been described by Pertsovsky (sensitivity 0.005 mg/m^3). Johnson & Isom have reported a radioisotopic technique for measuring hydrogen cyanide in expired air. One method for measuring hydrogen cyanide and black cyanide is based on conversion of the cyanide into cyanogen chloride and the determination of the latter by the formation of a polymethine dye (detection limits: 0.24 μg in the solution volume analyzed and 0.12 mg/m^3 of air; measurement error $\pm 20\%$; measurable concentration range $0.12-2.4 \text{ mg/m}^3$) [*Guidelines for the Determination of Noxious Substances in Air*, 1981b]. Qualitative methods for the determination of cyanides have been described by Peregud.

For the determination of cyanides in *food*, a colorimetric method using pyridine and barbituric acid has been proposed [Hönig *et al.*]. Spectrometric determination of cyanides after their isolation from samples of *soil* and *refuse* at pH = 7, pH = 4, and pH = 2, based on the use of pyridine and barbituric acid, has been described by Horvath *et al.*; the method is suitable for analysis of industrial wastes and of soils with varying levels of humus (detection limit ≈$0.06 \text{ mg CN}^-/\text{kg}$).

For quantitative determination of cyanides in *biological media*, methods of gas chromatography and mass spectrometry [Thompson & Anderson] and fluorimetry [Ganjeloo *et al.*] have been proposed, as has been a spectrophotometric method for cyanides in blood and urine samples, based on the conversion of organic and inorganic CN^- compounds to NaCN and then to a colored complex [Johnson & Williams]. A rapid method for measuring cyanide levels in blood has been reported by Holzbecher & Ellenberger (measurement error 3–9%). Groff *et al.* have reported a rapid method for the direct determination of free CN^- ions in blood plasma and of total CN^- in whole blood (measurement error 2.3%). A method for qualitative determination of cyanides is based on the formation of ferric ferrocyanide (Berlin blue) [Gadaskina *et al.*]. Procedures for determining cyanides in water are described in *Analytical Chemistry of Industrial Waste Waters* [Lurie]. Determination of cyanide and cyanogenic compounds in biological systems is discussed by Brimer.

Measures to control exposure

At the workplace

The provision of safe working conditions in places of employment where cyanides may be present is often a difficult task because of their high toxicity and because the safety and health measures need to be highly specific and take full account of the special conditions prevailing at the particular workplaces where dangerous exposure can occur.

Exposure should be primarily controlled by technical measures. Complete enclosure of the process equipment should be aimed at wherever feasible. Systematic monitoring of the workroom air for cyanides using instruments with automatic warning of their dangerous concentrations is mandatory. Careful attention should be always given to proper ventilation.

Workers handling cyanides are subject to a preplacement and subsequent periodic medical examinations, with special attention to the nervous and cardiovascular systems. It is essential that all personnel be trained to recognize the odor of hydrogen cyanide and apply proper first-aid measures in an emergency. First-aid kits should be provided and conveniently located.

For *personal protection*, a set of personal protective devices should be available and used properly when necessary. For respiratory protection, filter-type industrial gas masks or respirators should be used (as soon as even a faint odor of hydrogen cyanide is felt, it is necessary to leave the contaminated area immediately and replace the cartridge). For eye protection, chemical-safety goggles should be worn. Where skin can be exposed, protective clothing, including impervious hand protection such as gloves should be provided. Gloves should meet the requirements laid down for the specific jobs (rubber, capron, perchlorovinyl, or other types of gloves). There should be a ban on taking food and keeping personal effects in the working areas. Hand washing before meals and smoking is obligatory. It is recommended that before work the hands be anointed with a liquid consisting of ammonia water (20 parts), glycerin (100 parts), 96% ethanol (30 parts), boric acid (5 parts), and water (25 parts); after work, the hands should be treated with a 0.2% potassium permanganate solution or with hydrogen peroxide, the mouth thoroughly gargled, and a shower taken. Working clothes contaminated with hydrogen cyanide should be cleaned and decontaminated by keeping them for 2 h in a 2% soda solution containing 2% ferrous sulfate.

Protection of the general environment
Efforts must be primarily directed at ensuring effective purification of industrial wastewaters and, ideally, their compete recycling. The removal of cyanides from wastewater is based on their capacity for oxidation, and the best result can be obtained by treatment with active chlorine: this will destroy the cyanides completely in alkaline medium. If the pH is on the acid side and the dose of chlorine is too low, an excess of very volatile toxic cyanogen chloride may form.

Emergency treatment in poisoning
The victim should be moved to fresh air and allowed to inhale the vapor of amyl nitrite from a crushed ampoule (or a piece of cotton or gauze) held close to his nose; the contaminated clothes must be removed.

For antidotal therapy, nitrites, aminophenols, methylene blue, thiosulfates, and organic cobalt compounds have been used. In more severe cases, intravenous therapy is required. For it, 10–15 ml of a 2% sodium nitrite solution or 50 ml of chromosmon (1% methylene blue solution in 25% glucose solution) is slowly infused intravenously (at a rate of 2.5 to 5.0 ml/min), followed 3–5 min later by 50 ml of a 30% sodium thiosulfate solution, also intravenously at 2.5–5.0 ml/min. If indicated, the intravenous sodium nitrite and thiosulfate therapy is to be repeated at half dose two or three times at 10-minute intervals. Concurrently, other therapeutic measures are carried out as required, such as oxygen inhalation, subcutaneous injections of ephedrine and cordiamine [synonym: nikethamide] (arterial blood pressure monitoring is important) and repeated intravenous injections of a 40% glucose solution (20–40 mg) with 5% thiamine (1 ml), 2.5% niacin (1 ml), and 5% ascorbic acid (5 ml). If the pulse is weak and the blood pressure is low, 40 ml of a 1.5% cobalt edetate solution should be administered additionally by vein and the injection repeated at half dose in 2 or 3 min if the pressure has not risen. If respiration is depressed or has stopped, 0.5 ml of a 1% lobeline solution or 0.5–1.0 ml of cytiton (0.15% cytisine solution) is given intravenously.

If the patient is not breathing, which is likely in other than mild cases, artificial respiration (often prolonged) is necessary as the first step. After the patient has regained consciousness and breathes more or less normally, he must be transported to a hospital for further treatment.

SODIUM AND POTASSIUM CYANIDES

Physical and chemical properties
Sodium and potassium cyanides (NaCN and KCN) are solids with a weak odor of bitter almonds. In air, they decompose in the presence of moisture, releasing hydrogen cyanide. Their solutions have and alkaline reaction. See also Appendix.

Production
These salts can be produced by several methods, for example by neutralization of hydrogen cyanide with alkali or carbonate, fixation of nitrogen with a mixture of coal and soda or potash, reduction of the cyanates obtained from ammonium and carbonates, passage of ammonia over fused sodium or potassium (the resulting sodium and potassium amides yield the cyanides on heating with coal), or by absorption of hydrogen cyanide with an alkaline solution from gases that have formed during coal coking or dry distillation of stillage.

Uses
Sodium and potassium cyanides are used in the extraction of gold or silver from their ores, in galvanoplastic gilding or silvering, in gilding with the use of amalgams, for cleaning golden articles and precious stones, in the soldering and

liquid cementation of metals, in the case-hardening of steel, in copper-plating, bronzing, or zincing of metals; in the dressing of complex ores, for removing silver nitrate spots; in the silvering of mirrors, in photography, in lithographic printing, in the production of various cyanides and pharmaceutical preparations. They have also been used for the fumigation of fruit trees, ships, railway cars, warehouses, etc.

Man-made sources of emission into the environment

These are essentially the same as described above for hydrogen cyanide.

Toxicity

Fish

Toxic and lethal concentrations of KCN and NaCN for some fish species are shown in Table 3.

Table 3. Toxicity of potassium and sodium cyanides for some fish

Fish	KCN concentration (mg/L)		NaCN concentration (mg/L)	
	Toxic	Lethal	Toxic	Lethal
Crucian carp	0.04−0.12	0.1−0.3		
Trout	0.07 (74 h)			0.05 (124 h)
Loach		0.09		
Minnow			0.3	0.4 (1 h)
Perch				0.15 (96 h)

Source: Grushko.

Animals and man

The general toxicity of these compounds to warm-blooded animals is similar to that of hydrogen cyanide (q.v.).

Acute exposure

Animals. Median and minimal lethal doses of sodium and potassium cyanides on acute exposures are shown in Table 4. The main symptoms of poisoning were excitation, frequent breathing, convulsions and ataxia; lactate and oxypurine levels rose within minutes post-administration. Death resulted from respiratory arrest. The minimal blood concentration of cyanide ions at which death occurred was 2.5 µg/ml [Egekeze & Oeheme; Peterson & Cohen; Sax; Steven *et al.*].

Table 4. Acute toxicity of potassium and sodium cyanides for laboratory animals

Species	KCN LD_{50} (mg/kg)	KCN Minimal lethal dose (mg/kg)	NaCN LD_{50} (mg/kg)	NaCN Minimal lethal dose (mg/kg)
Mouse	8.5 (oral) 6.0–10.0 (S.C.) 6.7–7.9 (I.P.)	2.5 (I.V.)	4.9–5.4 (I.P.)	10.0 (S.C.)
Rat	6.0–10.0 (oral) 9.0 (S.C.) 4.0 (I.P.)	8.0 (I.M.) 2.5 (I.V.)	4.7–6.4 (oral) 4.3 (I.P.)	
Guinea-pig		8.0 (I.P.) 8.0 (S.C.)	5.8 (S.C.)	
Rabbit	5.0 (oral)			2.2 (S.C.)
Dog		5.0 (I.V.)		6.0 (S.C.) 1.3 (I.V.)

Source: Sax

Cats given an intravenous injection of NaCN in doses of 2.0 to 2.2 mg/kg died, while those injected it at 1.7–1.8 mg/kg all survived [Kvasenko; Smith *et al.*]. Sheep injected intravenously with KCN at 10 mg CN^-/kg died in 11 to 26 min; in such sheep CN^- concentrations reached 33 µg% in the blood, 124 µg% in the cerebrospinal fluid, 89 µg% in the caudate nucleus of the brain, and 99 µg% in the white matter of the brain [Ballantyne, 1975]. Reduced cytochrome oxidase activity in the brain and liver of mice was noted 3 min after an intraperitoneal injection of KCN at 10 mg/kg; NaCN given to rats by this route at 20 mg/kg raised L-dopa levels in the striate body of the brain [Cassel *et al.*].

A very fine sodium cyanide aerosol generated in the melting of cyanides proved to be more toxic that hydrogen cyanide, probably because of the associated formation of carbon monoxide. Toxic effects of this aerosol are indicated in Table 5.

Table 5. Toxicity of a fine sodium cyanide aerosol to mice

Effects	Concentration in air (mg/m^3)	Exposure time (min) before onset of indicated effects
Mucosal irritation,	40–90	25–40
salivation, excitation,	100–140	3–15
breathlessness	200–300	almost zero
Convulsions and	50–100	20–75
state of narcosis	150–200	26–42
	200–300	6–25
Death	150–170	2–76
	400–550	10–20

Source: Minkina.

On golden hamsters, evidence for fetotoxicity of NaCN has been obtained: decreased fetal size and weight and increased fetal mortality were observed for females exposed to this compound on days 6–9 of pregnancy; their offspring showed developmental abnormalities in the central nervous system [Doherty et al.].

Man. In occupational environments, poisoning may result from exposure to dusts, splashes, or mists of cyanides by inhalation (e. g. during the cleaning of golden articles or maintenance work at galvanizing baths), ingestion (e. g. accidental swallowing of these compounds when eating in the working area), and by percutaneous absorption, particularly if the skin integrity has been impaired (e. g. when the skin has small wounds or is affected by disease). A worker died after a hot mass containing 80% KCN was splashed into his face. The major hazard from NaCN and KCN comes from the release of hydrogen cyanide, which proceeds in a very violent manner in the presence of even weak acid.

Because of their good solubility in water, NaCN and KCN are more potent poisons than other cyanides. Fatal dosages by oral ingestion have been reported to be 2.2–2.8 mg/kg for NaCN and 2.86 mg/kg for KCN [Sax], but vary considerably depending on whether or not food is present in the stomach; sometimes much higher doses were tolerated. Delayed manifestations of acute poisoning occurred when the stomach was filled with food, and in some of such cases severe lesions of the air passages were seen at autopsy, whereas the characteristic signs of oral poisoning and the specific odor of bitter almonds were absent. An eruption of jelly-like consistency on the soles of a man who died on day 4 after poisoning was described [Janiszewska & Kostanecki; Sax].

Repeated exposure

Animals. Elevation of acetylcholinesterase activity was recorded in the cerebral cortex, hippocampus, and midbrain of rats after three subcutaneous injections of sodium cyanide in single doses of 0.4 mg/kg. Repeated dietary feeding of potassium cyanide to rats at 0.3 mg/kg over 2 weeks led to body-weight losses and to falls in hemoglobin, protein, and thyroxine in the blood with concurrent increases in the blood levels of thyocyanates; some animals developed intestinal tumors and all showed impaired reproductive performance [Iwasoyo & Iramain].

Chronic exposure

Animals. Demyelinization and degenerative changes developed in the cerebral cortex, hippocampus, and cerebellum of rats injected with potassium cyanide subcutaneously in doses of 0.5 mg/kg at weakly intervals over 22 weeks.

Man. Typical symptoms of chronic poisoning are headaches, weakness, distorted perception of taste and smell, nausea, and dyspnea; other reported symptoms and signs include abdominal and cardiac pains, weight loss, salivation, lacrimation, hyperemia of the throat, impaired accommodation, diminished potency and libido, neurocirculatory asthenia, anemia (or, on the contrary, increased quantity of hemoglobin), cyanmethemoglobinemia, leukopenia with a shift to the left, lymphocytosis, and elevated transaminases. Damage to the liver and myocardium, thyroid enlargement, impaired hearing and vision, and nervous system involvement including paralysis or (during exacerbations) mental disorders have been described as well [El-Ghawabi *et al.*; Jaroschka & Kreepp]. The cyanides contained in tobacco smoke have been implicated as the probable cause of decreased body-weight and body length in infants born to smoking mothers [Doherty *et al.*].

Effects from dermal exposure

Animals. Both NaCN and KCN are absorbed through undamaged skin. For rabbits, the LD_{50} of NaCN on skin application was 300 mg/kg. When one ear of a rabbit was kept immersed in a 1% KCN solution, the rabbit survived for only 60 min [Kolpakov & Prokhorenkov; Prokhorenkov & Kolpakov].

Man. Workers handling potassium cyanide solutions frequently suffer from dermatoses, mainly of allergic nature. Urticaria, dermatitis, eczema on the backs of fingers and wrists and at the radiocarpal and cubital articulations are all common and sometimes combine with the presence of deep and slowly clearing ulcers on the fingers. Ulcers tend to develop in abraded (if only slightly) areas. Powdered potassium cyanide elicits persistent eruptions that are prone to recur.

The onset of skin lesions is usually marked by itching. Thinning of the nail plates and subungual keratosis may occur. Cold weather is more favorable for the development of dermatoses than warm weather [Kolpakov & Prokhorenkov; Prokhorenkov & Kolpakov].

Absorption, distribution, and elimination

The absorption, distribution, and elimination of sodium and potassium cyanides are similar to those of hydrogen cyanide (q. v.). Levels of cyanide ions in the liver of rats after a single intragastric administration of potassium cyanide are shown in Table 6.

Table 6. Levels of cyanide ions in the blood and liver of rats after single oral (garage) administration of potassium cyanide

Dose (mg/kg)	Observation Time (min)	CN^- concentration Blood (µg/ml)	Liver (µg/g)
6	6–10	2.5	0.74
	60	2.41	0.41
10	6–10	3.95	1.98
	60	4.03	1.82
14	6–10	4.48	4.01
	60	4.16	1.3

Source: Egekeze & Oeheme.

In dogs, the elimination half-time of potassium cyanide after its intravenous injection at 0.82 mg/kg was 23 min; half of the remaining amount was removed over the next 5–6 h. Its blood concentration returned to the initial level (5 µg/100 ml) at 23 h post-injection [Bright & Marrs].

The urine of nonsmoking persons contained cyanide concentrations of about 6.7 µg%, while that of smokers contained 17.4 µg%. Excretion is largely (83–89%) with the urine; about 4% is removed in the expired air [Okoh].

Hygienic standards

See the corresponding section in *Hydrogen cyanide*.

Methods of determination

Methods for determining cyanides in water have been described by Lurie. A gas-chromotographic procedure for their determination in soil and plants has been proposed by Okuno *et al.* For other methods, see *Hydrogen cyanide*.

Measures to control exposure at the workplace

These measures are generally similar to those described above for hydrogen cyanide. Technical control measures should be primarily directed at segregating and enclosing the sources of toxic substances, mechanizing and automating technological processes, and eliminating any direct contact with cyanide salts.

Careful attention to proper ventilation is necessary. In workplaces where skin can be exposed to cyanide salts, protective clothing, including impermeable hand protection, must be provided. Those working with these salts should be instructed that contact of the latter with acids, even dilute ones, will result in the release of hydrogen cyanide (workers should be trained to recognize its odor). All containers with cyanide salts should be kept covered or in an exhausted hood when not in use [Hardy & Boylen]. It is important to remove promptly from the skin the dust or solutions containing cyanides.

First aid in poisoning

In moderate poisoning by swallowed cyanide-containing dust, the stomach should be washed with a 3% hydrogen peroxide solution or, preferably, a 0.5% sodium thiosulfate or 0.2% potassium permanganate solution. If the poisoning is more severe, the victim should be first allowed to inhale the vapor of amyl nitrite from a crushed ampoule containing this liquid. The patient should then drink, every 15 min, 1 tablespoon of a mixture of ferrous sulfate and magnesia powder in water (1 teaspoon per glass of water). In the event of skin contact, the sodium or potassium cyanide solution should be promptly flushed off the skin with large amounts of water, followed by the application of a wet dressing impregnated with boric acid solution or ointment.

CALCIUM CYANAMIDE

Physical and chemical properties

Calcium cyanamide ($CaCN_2$) is a finely divided powder that is snow-white when pure and grayish-black (owing to carbon impurities) when of technical grade. It is hygroscopic and decomposes in water. It does not dissolve in alcohol. In calcium cyanamide dusts, 60–70% of particles are 2 to 15 μm in diameter and 10–15% are 0.3–0.5 μm. Technical-grade calcium cyanamide contains about 60% $CaCN_2$ and also C (9–13%), CaO (18–23%), and CaCl, SiO_2, and FeS (5%); it may also contain unreacted CaC_2.

On storage, calcium cyanamide is decomposed by the action of moisture and carbon dioxide contained in the air, with the formation of various products, but not of cyanides — even after many months. See also Appendix.

Production and uses

Calcium cyanamide can be obtained by passing dry nitrogen over calcium carbide at 1000–1100°C or at lower temperatures if a suitable catalyst is added. It is mainly used in agriculture (as a fertilizer, a weed killer, and a defoliant for cotton plants), in the production of cyanides, urea, dicyanodiamide, and guanidine or its derivatives, and as a component of powders used for hardening metals.

Man-made sources of emission into the environment

The main sources are processes in which calcium cyanamide is produced or is used to obtain other nitrogen compounds or in agriculture, particularly for defoliation. In soil, it is rather rapidly decomposed by the action of moisture with the formation of urea and ammonium carbonate; in water it decomposes with the release of free cyanamide [Melnikov et al.].

Toxicity

General aspects

The toxicity of calcium cyanamide is mainly due to the action of two compounds formed upon its decomposition: lime, which is a strong local irritant of the skin and mucous membranes, and free cyanamide (NH_2CN), which exerts systemic effects, particularly on the vasomotor center and also on the respiratory center and on the blood [Zorn et al.].

Toxic effects of the free cyanamide are usually very short-lived because of its rapid decomposition in the body. It blocks the oxidation of alcohol, glutathione, dehydrases, and dehydrogenases either directly or via the formation of intermediate products. Exposure to calcium cyanamide can give rise to an asthenic-vegetative syndrome combining with dysfunction of endocrine organs and impaired basal metabolism.

Calcium cyanamide is a cumulative poison [Billewicz-Stankievicz & Pawlowski]. Poisoning usually occurs by the inhalation route. Contact with the skin frequently results in a skin disease which may be rather severe.

Effects in animals and man

Acute exposure

Animals. The oral LD_{50}s of calcium cyanamide have been determined at 335–385 mg/kg for mice, 250 mg/kg for rats, 220 mg/kg for guinea-pigs, and 350 mg/kg for rabbits. (Higher oral LD_{50} values — 844 mg/kg for mice and 738 mg/kg for rats — were reported by Markarian.) Acute poisoning was marked by an initial strong motor excitation succeeded by general inhibition, intermittent breathing, dyspnea, motor incoordination, and tremor. The highest oral dose tolerated by mice was 200 mg/kg. The LC_{50} on inhalation exposure to $CaCN_2$ aerosols was 220 mg/m^3 for mice and 251 mg/m^3 for rats [Bakhritdinov; Khalmetov].

Man. Exposure to calcium cyanamide leads to pronounced flushing of the skin on the face, neck, and upper part of the chest and back. The flushed skin usually has a bluish tinge but sometimes resembles that seen in scarlet fever. The conjunctivae, soft palate, and pharyngeal mucosa also turn reddish. There is no fever, but slight chills and particularly a sensation of cold in the hands have been frequently observed. Breathing is usually frequent and very deep. The pulse rate is above normal while the blood pressure may be slightly depressed [Zorn *et al.*]. Morphological changes in the blood have not been reported. The victim is likely to complain of palpitation, sinking of the heart, and a feeling of fear. The attack lasts for 1.5 to 2 h and usually regresses without appreciable sequelae, although in some cases organic nervous disease did develop, manifested in weakness in the limbs followed sometimes by transient focal myelitis, polyneuritis, and even paralysis. Choreiform arm movements due to involvement of the basal ganglia and cerebellum have been described. No fatalities have occurred.

There is a wide individual variation in the sensitivity to calcium cyanamide but individuals completely resistant to this compound do not exist; the sensitivity is increased markedly by the simultaneous intake of alcohol [Zorn *et al.*].

The acute toxicity of calcium cyanamide is low, the oral lethal dose for an adult being 40–50 g [Sollmann].

Repeated exposure

Animals. Mice and rats inhaling calcium cyanamide dust at concentrations of 50–500 mg/m^3 for 1 h daily over 24 days developed general emaciation and degenerative changes in the liver, gut, and spleen; most animals died before day 24. $CaCN_2$ crystals were found in their bronchi, but pathological changes in the air passages and lungs were relatively mild. Rabbits inhaling $CaCN_2$ dust showed more or less severe changes in the respiratory tract mucous membranes and the lungs. Cats experienced lacrimation, salivation, and severe weight losses, became lethargic and at times excited and dyspneic; autopsy showed degenerative changes in the central nervous system, inflammatory lesions in the upper airways, pneumoconiosis, acute inflammatory hyperplasia of lymph nodes, and signs of damage to the spleen and adrenals. The onset of death was accelerated by the intake of alcohol during the exposure period; such cats developed very severe inflammatory changes in their lungs and upper airways and contained thiocyanates in their urine. The main findings in rats given $CaCN_2$ by oral gavage at 25 mg/kg repeatedly (10–20 doses) were excitation, dyspepsia, large weight losses, tremor, difficult breathing, and reduced levels of sulfhydryls, leukocytes, erythrocytes, hemoglobin, and catalase and cholinesterase in the blood [Bakhritdinov].

Chronic exposure

Animals. In a 4-month inhalation study on rats, phasic changes in the central nervous system function and in the blood levels of sulfhydryls were observed, together with increases in blood levels of pyruvic and lactic acids and decreases in those of sugar, lactate dehydrogenase, and alkaline phosphatase; in addition, disordered mineral metabolism and chromosomal aberrations were detected [Markarian].

Man. Initial manifestations of chronic intoxication by calcium cyanamide dust may include headache, dizziness, rapid onset of fatigue, nervous excitability and irritability, poor appetite, disturbed sleep, sweating, heartburn, epigastric pain, gastrointestinal disorders, dryness and pallor of the skin (in the presence of raised erythrocyte counts), body-weight loss, weakness, myalgia, tendency of the gums to bleed readily, and dental caries.

The irritant action of $CaCN_2$ on the respiratory mucous membranes is manifested in frequent nasal discharges, chronic catarrhs on the pharynx, larynx, and trachea, and bronchitis accompanied by a feeling of tightness in the chest. Pneumonitis may also develop. The sensitivity to alcohol is greatly increased. After prolonged occupational exposure (>5 years), progressive weight lose, cardiac pain, memory deterioration, inflammation and ulceration of the gums, multiple caries, and severe lesions of the nasal mucosa, sometimes with perforation of the nasal septum, have been noted. Not uncommon conditions are said to be asthmoid bronchitis or bronchial asthma, chronic gastritis and hepatitis, hypotension, myocardial dystrophy, slowed or frequent pulse, and impaired sexual, thyroid and adrenal functions [Airapetian, 1970]. Schiele *et al.*, in contrast, have reported reduced resistance to alcohol as the only change they could observe in workers exposed to CaCN concentrations up to 8.3 mg/m^3.

Dermal and ocular effects

Man. Contact with calcium cyanamide makes the skin dry and very itchy and results in a dermatitic reaction that in mild cases is of short duration and terminates in desquamation of the skin in small sheets. The skin may appear reddened and edematous, and urticaria has been observed occasionally. In more severe cases, acute or subacute eczema of a papular or vesicular type may develop. Skin diseases tend to appear after a few days of work with $CaCN_2$ and may involve the interdigital folds, armpits, corners of the mouth, and margins of the nostrils; in those wearing defective footwear or working barefoot, the feet are affected (usually in agricultural laborers). $CaCN_2$ is particularly destructive to already damaged skin, and its cauterizing action is enhanced by moisture (e.g. rain, dew, profuse perspiration). Contact with the eyes may cause purulent conjunctivitis, keratitis and, in more severe cases, corneal ulceration and opacity [Airapetian; Yelizarov].

Transformation in the body
Calcium cyanamide converts to cyanamide which then reacts with the sulfhydryl groups of amino acids and proteins to form isothiourea derivatives.

Hygienic standards
In the USSR, the occupational exposure limit (MAC_{wz}) for calcium cyanamide has been set at 1.0 mg/m^3 (as aerosol, hazard class 2) and its MAC in water sources, at 1.0 mg/m^3 (based on organoleptic criteria; hazard class 3). In addition, a temporary MAC (TSEL) of 0.02 mg/m^3 has been adopted for calcium cyanamide in the atmospheric air of residential areas. No calcium cyanamide residues should be present in foodstuffs.

In the USA, a TLV-TWA of 5.0 mg/m^3 has been set for this compound.

Measures to control exposure at the workplace
The main technical control measures to protect the health of those working with calcium cyanamide should be directed at dust control. Where this cannot be assured, the skin, eyes, and respiratory tract should be protected by dustproof overalls, a hood fastened tightly round the neck, appropriate eyewear, respiratory protective equipment (dust masks), and gloves. Skin and face protection can be improved by the use of waterproof barrier creams; emergency eye-flushing installations should be provided [Bartalini].

All workers must be made aware of the hazards associated with this compound, including the potentiating effect of alcohol on its toxicity. Before employment and systematically during the period of employment, they should be instructed in, and required to observe, the corresponding safety precautions and general and personal hygiene measures. Preemployment and periodic health examinations are also necessary.

Emergency treatment in poisoning
Complete rest is important. The manifestations of general toxicity usually clear without any treatment. If the eyes are affected, they should be irrigated with clean water or physiological saline for 10–30 min, after which a 1% novocain [synonym: procaine] solution or a 0.5% dicain [synonym: tetracaine] solution with 0.1% epinephrine should be dropped into each eye (1 or 2 drops). The washing procedure should be repeated several times within 24 hours. In the event of an eye burn, safety glasses have to be worn; hydrocortisone ointment, synthomycin emulsion, or sodium sulfacetamide (in ointment or solution) should be applied if indicated. If there is acute dermatitis, dressings impregnated with

lead water or Burow's solution are applied to the affected area, followed by the application of an indifferent paste or ointment for the night, after which zinc ointment may be applied. The skin areas with mild burns are painted with solutions of gentian violet, methylene blue, or brilliant green. Treatment for pyodermas consists in the application of dressings impregnated with a streptocid [synonym: sulfanilamide] or synthomycin emulsion, followed by penicillin ointment or furacilin [synonym: nitrofuran] ointment. For the prevention and treatment of microtraumas, iodine tincture, rivanol (2-ethoxy-6,9-diaminoacridine lactate), and bactericidal adhesive tape are used.

BLACK CYANIDE

Properties
Black cyanide is a solid substance of black to brown color with a faint odor of bitter almonds. It releases hydrogen cyanide in the presence of moisture.

Production and use
Black cyanide is prepared by fusing calcium cyanamide with coal and soda or common salt. It is composed of $NaCN$ (42–47%), CaO or $CaCl_2$ (up to 50%), and the unreacted CaC_2 and $CaCN_2$. It is used in the hydrometallurgy of noble metals, for cyanidation of steels, in the manufacture of sodium ferrocyanide, as a seed fumigant, and for disinsectization and deratization.

Determination
See the section on hydrogen cyanide.

Toxicity
Black cyanide produces toxic effects similar to those of hydrogen cyanide.

Animals.
Cats exposed to black cyanide dust by inhalation at concentrations of 20–35 mg/m^3 for 2.5 h daily (the air contained hydrogen cyanide at 15 mg/m^3) all died within 17 to 128 days. Acute effects were mainly observed in the first few days of exposure and included vomiting, sluggishness, weakness, sometimes excitation, hypersalivation, lacrimation, and difficult breathing. Later, the animals developed severe anemia and degenerative changes in the central nervous system and internal organs.

Man.
Acute poisoning with black cyanide is marked by headache, dizziness, nausea and vomiting, and cardiovascular abnormalities; in severe cases, convulsions and loss of consciousness followed by death due to respiratory arrest may occur.

Reported symptoms and signs of chronic exposure include loss of appetite, pain in the epigastric and right hypochondrial regions, multiple dental caries, catarrh of the upper airways, bronchitis, chronic gastritis, myocardial lesions, increased capillary permeability, augmented basal metabolism, neuritis, radiculitis, and impaired renal, thyroid, and sexual functions. The minimal lethal dose is presumed to be between 0.20 and 0.25 g [Airapetian, 1961; Mirgorod; Zhivatova].

Local effects include dryness of the skin, dermatitis, eczema, and ulceration of the nasal mucosa with possible perforation of the nasal septum.

Measures to control exposure and first aid in poisoning

Exposure control measures and first-aid treatment are essentially the same as those described above for calcium cyanamide and hydrogen cyanide.

REFERENCES

Abel P. A. (1989) *Water Pollution Biology*. Ellis Horwood, Chichester.

Airapetian, M. A. (1961) In: *Trudy Yubileinogo plenuma Uchenogo soveta, posvyashchennogo 40-letiyu ustanovleniya Sovetskoi vlasti v Armenii* [Transactions of a Plenary Meeting of the Medical Research Council Devoted to the 40th Anniversary of the Establishment of Soviet Power in Armenia], Vol. 3. Yerevan, pp. 179−183 (in Russian).

Airapetian, M. A. (1970) In: *Trudy klinicheskikh otdeleniy NII gigiyeny truda i professionalnykh zabolevaniy* [Transactions of the Clinical Departments of the Institute of Occupational Health], Vol. 1. Yerevan, pp. 46−50 (in Russian).

Aleksandrov, N. P. *et al.* (1979) *Farmakol. Toksikol.*, No. 6, 670−674 (in Russian).

Bakhritdinov, Sh. S. (1975) In: *Problemy gigiyeny i organizatsii zdravookhraneniya v Uzbekistane* [Current Issues of Hygiene and Health Care Organization in Uzbekistan], No. 3. Tashkent, pp. 161−164 (in Russian).

Ballantyne, B. (1975) *J. Forens. Sci. Soc.*, **15**, 51−56.

Ballantyne, B. (1980) *Toxicol. Lett.*, **18** (Suppl. 1), 71−73.

Bartalini, E. (1983) In: *Encyclopaedia of Occupational Health and Safety*. International Labour Office, Geneva, pp. 360−361.

Baumeister, R. G. *et al.* (1975) *Arzneimittel-Forsch.*, **25**, 1056−1064.

Billewicz-Stankiewicz, J. & Pawlowski, L. (1961) *Med. Pracy*, No. **5**, 469−479.

Bright, J. & Marrs, T. (1988) *Hum. Toxicol.*, **7**, 183−186.

Brimer, L. (1988) In: *Cyanide Compounds in Biology — CIBA Foundation Symposium 140*. John Wiley & Sons, Chichester, pp. 177−200.

Bulletin of the World Health Organization (1984), **62**, 477−492.

Cassel, G. *et al.* (1985) *Acta Neurol. Scand.*, **72**, 804.

Cliff, J. *et al.* (1986) *Acta Endocrinol.*, **113**, 523−528.

Doherty, P. *et al.* (1982) *Toxicol. Appl. Pharmacol.*, **64**, 456−464.

Egekeze, J. & Oeheme, F. (1979) *Toxicol. Lett.*, **3**, 243−247.

El-Ghawabi, S. *et al.* (1975) *Brit. J. Ind. Med.*, **32**, 215−219.

Eruman, A. (1986) *Lancet*, No. **8478**, 441−442

Gadaskina, I. D. *et al.* (1975) *Opredeleniye promyshlennykh neorganicheskikh yadov v organizme* [Determination of Inorganic Industrial Poisons in Biological Media]. Meditsina, Moscow (in Russian).

Ganjeloo, A. *et al.* (1980) *Toxicol. Appl. Pharmacol.*, **55**, 103−107.

Goncharova, L. N. *et al.* (1984) *Gig. Truda Prof. Zabol.*, No. **5**, 23−25 (in Russian).

Groff, W. *et al.* (1985) *J. Toxicol. Clin. Toxicol.*, **23**, 133−163.

Grushko, Ya. M. (1979) *Vrednye neorganicheskiye soyedineniya v promyshlennykh stochnykh vodakh* [Harmful Inorganic Compounds in Industrial Waste Waters]. Khimiya, Leningrad (in Russian).

Guidelines for the Determination of Noxious Substances in Air (1981a) [Metodicheskiye ukazaniya po opredeleniyu vrednykh veshchestv v vozdukhe], No. 17. Moscow (in Russian).

Guidelines for the Determination of Noxious Substances in Air (1981b) [Metodicheskiye ukazaniya po opredeleniyu vrednykh veshchestv v vozdukhe], No. 19. Moscow (in Russian).

Hardy, H. L. & Boylen, G. W. (1983) In: *Encyclopaedia of Occupational Health and Safety*. International Labour Office, Geneva, pp. 574−577.

Holzbecher, M. *et al.* (1984) *J. Toxicol. Clin. Toxicol.*, 22, 341−347.

Holzbecher, M. & Ellenberger, H. (1985) *J. Anal. Toxicol.*, 9, 251−253.

Hönig, D. N. *et al.* (1983) *J. Agric. Food Chem.*, 31, 272−275.

Horvath, A. *et al.* (1982) *Ztschr. Ges. Hyg. Grenzgeb.*, 28, 101−105.

Hüdepohl, M. (1983) *Ztschr. Allgemeinmed.*, 59, 953−954.

Iwasoyo, J. O. & Iramain, C. A. (1980) *Toxicol. Lett.*, 6, 1−3.

Janiszewska, M. & Kostanecki, W. (1970) *Prz. Dermatol.*, 57, 755−757.

Jaroschka, R. & Kreepp, A. (1966) *Arch. Gewebepathol.*, 22, 202−207.

Johnson, J. & Isom, G. (1985) *J. Anal. Toxicol.*, 9, 112−115.

Johnson, D. & Williams, H. (1985) *Anal. Lett.*, 18, 855−869.

Kelman, B. J. (1979) *Fed. Proc.*, 38, 2246−2250.

Khalmetov, R. Kh. (1975) In: *Problemy gigiyeny in organizatsii zdravookhraneniya v Uzbekistane* [Current Issues of Hygiene and Health Care Organization in Uzbekistan], No. 3. Tashkent, pp. 198−200 (in Russian).

Kläui, H. *et al.* (1984) *Schweiz. Med. Wschr.*, 114, 983−989.

Knapik, Z. *et al.* (1981) *Gig. Truda Prof. Zabol.*, No. 11, 55−57 (in Russian).

Kolpakov, F. I. & Prokhorenkov, V. I. (1978) *Vestn. Dermatol. Venerol.*, No. 7, 76−79 (in Russian).

Kvasenko, O. Ya. (1962) *Farmakol. Toksikol.*, No. 6, 742−749 (in Russian).

Langauer−Lewowicka, H. (1964) *Med. Pracy*, 15, 259−264.

Lunguist, P. *et al.* (1985) *Clin. Chem.*, 31, 591−595.

Lunguist, P. *et al.* (1988) *Arch. Toxicol.*, 61, 270−274.

Lurie, Yu. Yu. (1984) *Analiticheskaya khimiya promyshlennykh stochnykh vod* [Analytical Chemistry of Industrial Waste Waters]. Khimiya, Moscow (in Russian).

Lutai, A. V. *et al.* (1975) *Gig. Truda Prof. Zabol.*, No. 4, 33−35 (in Russian).

Manz, A. (1968) *Zbl. Arbeitsmed. Arbeitsschüts.*, 18, 167−172 .

Markarian, K. L. (1986) In: *Problemy okhrany zdorovya naseleniya i zashchity okruzhayushchei sredy ot khimicheskikh vrednykh faktorov. Tezisy dokladov 1 Vsesoyuznogo syezda toksikologov* [Protection of Public Health and the Environment from Noxious Chemical Agents. Abstracts of papers presented to the First All-Union Congress of Toxicologists]. Rostov-on-Don, pp. 197−198 (in Russian).

Melnikov, N. N. *et al.* (1985) *Spravochnik po pestitsidam* [Pesticides (A Handbook)]. Khimiya, Moscow (in Russian).

Minkina, N. A. (1977) In: Lazarev, N. V. & Gadaskina, I. D. (eds) *Vrednye veshchestva v promyshlennosti* [Harmful Substances in Industry], Vol. 3. Khimia, Leningrad, pp. 260−265 (in Russian).

Mirgorod, A. V. (1966) In: *Materialy nauchnoi sessii po voprosam gigiyeny truda i professionalnoi patologii v khimicheskoi i gornorudnoi promyshlennosti* [Proceedings of a Scientific Meeting on Occupational Health in Chemical and Mining Industries]. Yerevan, pp. 45−48 (in Russian).

Morikawa, T. (1978) *J. Combust. Toxicol.*, 5, 315−330.

Novakova, S. & Dinoyeva, S. (1972) *Gig. San.*, No. 3, 89−93 (in Russian).

Okoh, P. (1983) *Toxicol. Appl. Pharmacol.*, 70, 335−339.

Okuno, Y. *et al.* (1979) *Bull. Environ. Contam. Toxicol.*, 22, 386−390.

Peregud, E. A. *et al.* (1970) *Bystrye metody opredeleniya vrednykh veshchestv v vozdukhe* [Rapid Methods for Determining Harmful Substances in the Air]. Khimiya, Moscow (in Russian).

Pertsovsky, A. L. (1988) *Gig. San.*, No. 9, 48—50 (in Russian).

Peterson, J. & Cohen, St. (1985) *Toxicol. Appl. Pharmacol.*, 81, 265—273.

Plisov, G. A. (1981) In: *Novye metody gigiyenicheskogo kontrolya za primeneniyem polimerov v narodnom khozyaistve. Tezisy dokladov III Vsesoyuznogo soveshchaniya* [New Methods for Monitoring Health Effects of Industrially Used Polymers. Abstracts of papers presented to the Third All-Union Conference]. Kiev, pp. 68—70 (in Russian).

Prokhorenkov, V. I. & Kolpakov, F. I. (1978) *Gig. Truda Prof. Zabol.*, No. 12, 44—45 (in Russian).

Sax, N. I. (ed.) (1984) *Dangerous Properties of Industrial Materials.* Van Nostrand Reinhold Co., New York.

Schiele, R. *et al.* (1981) *Zbl. Bakteriol.*, 173, 13—28.

Sinitsyn, S. N. (1961) *Farmakol. Toksikol.*, No. 5, 540—541 (in Russian).

Smith, A. *et al.* (1963) *Nature*, 200 (No. 4902), 179—181.

Sollmann, T. (1957) *A Manual of Pharmacology.* Saunders, Philadelphia.

Steven, J. *et al.* (1986) *Toxicol. Appl. Pharmacol.*, 82, 40—44.

Thompson, I. & Anderson, R. A. (1980) *J. Chromatogr.*, 188, 357—362.

Vogel, St. N. *et al.* (1981) *Clin. Toxicol.*, 18, 367—383.

Wadden, R. A. (1976) *Environ. Health Perspect.*, 14, 201—206.

Way, J. L. *et al.* (1988) In: *Cyanide Compounds in Biology — CIBA Foundation Symposium 140.* John Wiley & Sons, Chichester, pp. 232—243.

Wilson, J. (1988) *Hum. Toxicol.*, 7, 47—49

Wodichenska, C.S. (1970) *Gig. San.*, No. 10, 73—78 (in Russian).

Yelizarov, G. P. (1965) In: *Aktualnye voprosy professionalnoi dermatologii* [Current Problems in Occupational Dermatology]. Moscow, pp. 200—207 (in Russian)

Zdrojewicz, Z. *et al.* (1985) *Wiad. Lek.*, 38, 1351—1356.

Zhivatova, A. P. (1963) *Sud.-Med. Zapiski* [Journal of Forensic Medicine], No. 4, 41—46 (in Russian).

Zorn, H. R. *et al.* (1988) Chapter 16 *Carbon* in: Seiler, H. G. *et al.* (eds) *Handbook on Toxicity of Inorganic Compounds.* Marcel Dekker, New York, pp. 183—214.

Sodium and potassium cyanates

PROPERTIES

Sodium and potassium cyanates (NaOCN and KOCN) are hydrolyzed in moist air with the formation of sodium carbonate or potassium carbonate and ammonia. They give cyanic acid in reacting with mineral acids and form biuret in reacting with urea. See also Appendix.

PRODUCTION AND USES

Sodium and potassium cyanates are prepared by oxidation of the respective cyanides with air oxygen in the presence of nickel.

These cyanates are mainly used in organic synthesis to obtain urea and its derivatives, semicarbazide, and some other substances, and in the production of protective coatings for metals.

TOXICITY

Sodium and potassium cyanates do not display the toxicity typical for cyanides. They impair oxidative phosphorylation and elicit methemoglobin formation.

Acute toxicity to animals

The intraperitoneal LD_{50} of sodium cyanate was calculated to be 260 mg/kg for mice and 310 mg/kg for rats. The LD_{50} of potassium cyanate for mice by this route was 320 mg/kg; with oral administration (by gavage), it was determined at 841 mg/kg for mice and 1500 mg/kg for rats [Alter *et al.*; Cerami *et al.*; Haut *et al.*; Sax; Smith]. Two hours after an intraperitoneal injection of NaOCN at 410 mg/kg, reduced oxyhemoglobin levels and elevated methemoglobin levels were recorded in mice. When given to mice and rats in doses that caused partial mortality, the cyanates first exerted a sedative effect, followed by increased motor activity, tremor, and convulsions.

Repeated exposure of animals

Toxicity data obtained for rats and guinea-pigs with repeated oral administration of sodium cyanate are presented in Table 1.

Table 1. Toxicity of sodium cyanate for rats and guinea-pigs with repeated oral administration (by gavage)

Animals	Daily dose (mg/kg)	No. of doses (given once daily)	Effects
Rats	50	8	Mild adynamia
	200	10	Anorexia; gross adynamia; paralysis of hind paws; some animals died
	400	3	50% of animals died
Guinea-pigs	100	3	Adynamia and paralysis of hind paws in some animals
	220	3	Same effects as above but in all animals

Guinea-pigs given NaOCN in daily oral doses of 220 mg/kg for 3 days had decreased blood levels of glucose and glucose-6-phosphate dehydrogenase along with elevated transaminase activities; in addition, glycogen was greatly increased in their livers. Autopsy showed dystrophic changes in the liver, muscles, and spinal cord as well as lesions in peripheral nerves.

Intraperitoneal administration of sodium cyanate to rats at 60–80 mg/kg for 2 to 3 weeks resulted in reduced body weight and elevated reticulocyte, erythrocyte, and iron levels in the blood; 25% of the animals died [Alter *et al.*; Haut *et al.*].

Chronic toxicity

Animals

Intraperitoneal administration of potassium cyanate to mice at 32 mg/kg 5 times per week for 5 months led to a slight rise in the blood level of hemoglobin. In dogs, sodium cyanate added to the diet at 30 to 170 mg/kg body weight for 12 to 39 months caused weight losses, signs of disorders in the nervous and digestive systems, sclerotic changes in the crystalline lens (some dogs developed cataract), and corneal lesions; in young animals these changes all developed earlier and were more strongly marked than in old ones [Kern *et al.*]. The addition of NaOCN to the diet of monkeys at 100 mg/kg for 15 months, did not lead to any appreciable abnormalities. Monkeys given this cyanate subcutaneously at 35 mg/kg for 2 to 4 months developed demyelination in the pyramidal tract of the spinal cord, while those administered it intramuscularly at 40 mg/kg for 2 months experienced spastic paralysis in all extremities; at autopsy, multiple lesions in various parts of the central and peripheral nervous systems were found [Tellez *et al.*].

Man

Patients receiving sodium cyanate for sickle-cell anemia developed bilateral posterior subcapsular cataract. In young patients, signs of cataract appeared after a 6-month course of treatment (30 mg/kg NaOCN daily) but cleared up in some 6 months after its termination. In elderly patients, sclerotic changes in the lens were observed 30 to 36 months after such treatment. The lenticular abnormalities were attributable to the specific action of cyanates on protein structures [Nicholson *et al.*].

ABSORPTION AND ELIMINATION

Sodium cyanate readily passes through the blood-brain barrier and is rapidly eliminated from the body. In rats given it at 100 mg/kg intraperitoneally, CN^- ions were present in equal concentrations in the blood and cerebrospinal fluid at 50 min postinjection but only in trace amounts at 24 h [Leong *et al.*].

REFERENCES

Alter, B. *et al.* (1974) *Blood*, **43**, 69−77.
Cerami, A. et. al. (1973) *J. Pharmacol. Exp. Ther.*, **185**, 653−666.
Haut, M. *et al.* (1975) *J. Lab. Clin. Med.*, **85**, 140−150.
Kern, H. et. al. (1977) *J. Pharmacol. Exp. Ther.*, **200**, 10−16.
Leong, J. K. *et al.* (1974) *J. Pharmacol. Exp. Ther.*, **191**, 60−67.
Nicholson, D. *et al.* (1976) *Arch. Ophthalmol.*, **94**, 927−930.

Sax, N. I. (ed.) (1984) *Dangerous Properties of Industrial Materials.* Van Nostrand
Reinhold Co., New York.
Smith, A. (1973) *Proc. Soc. Exp. Biol. Med.*, **142**, 1041–1044.
Tellez, J. *et al.* (1979) *Acta Neuropathol.*, **47**, 75–79.

Thiocyanic acid and its salts (thiocyanates)

THIOCYANIC ACID

Properties

Thiocyanic acid (HSCN) is an unstable liquid acid that occurs as a white crystalline mass at temperatures below its melting point (-110°C). At -90°C the liquid acid begins to polymerize, forming white and then colored solid products. At 5°C the polymer melts and decomposes. Liquid thiocyanic acid is highly volatile and decomposes in air into hydrogen cyanide and persulfocyanic acid. Its dilute (<5%) aqueous solutions are stable and have properties of a strong acid. More concentrated (>5%) solutions decompose at room temperature.

Production

Thiocyanic acid can be obtained by heating dehydrated potassium thiocyanate with potassium hydrogen sulfate in a flow of hydrogen *in vacuo*.

Toxicity

Thiocyanic acid is irritating to mucous membranes. No cases of poisoning have been described. Unpurified acid may contain hydrogen cyanide or dicyanogen.

THIOCYANATES (OF SODIUM [NaSCN], POTASSIUM [KSCN], AND AMMONIUM [NH$_4$SCN])

Properties

These thiocyanates occur as colorless crystals when pure. Often, however, they have a faint pink color because of the iron present as an impurity. They form cyanides on fusion with alkalies.

Production

Sodium and ammonium thiocyanates are commonly obtained from waste solutions that result from the purification of water gas or coke-oven gas. Sodium thiocyanate can also be made by reaction of sodium cyanide with sulfur or sodium polysulfide, while ammonium thiocyanate can be obtained by reacting ammonia with carbon disulfide. Potassium thiocyanate is usually prepared by fusing potassium cyanide with sulfur.

Uses

Thiocyanates are used in analytical chemistry; in the pharmaceutical industry; as disinfectants; in the production of plastics; to make polythiocyanoalkanes with insectofungicidal and bactericidal properties. Ammonium and potassium thiocyanates are employed in the production of rhodanine and thiocarbamide, in dyeing and printing textiles, and in photography; potassium thiocyanate is also used to obtain other thiocyanates. Sodium thiocyanate is a defoliant and a weed killer.

Man-made sources of emission into the environment

The main anthropogenic sources are wastewaters from coking, ore-concentrating, and metallurgical plants. Thiocyanates may also be released during the thermal treatment of brown coals and the manufacture of fertilizers.

Toxicity

The acute toxicity of thiocyanates is low as compared to that of cyanides. Characteristic findings in subchronic and chronic poisoning are depressed thyroid function and lowered blood levels of thyroid hormones, and these effects are thought to be largely responsible for many of the signs and symptoms observed on repeated or chronic exposure to these compounds [Ivanov et al.; Kostovetsky et al.]. Ammonium thiocyanate can be absorbed through undamaged skin and exerts sensitizing effects. It caused allergic reactions in rats and guinea-pigs on inhalation exposure, after intracutaneous injection, and when applied to their skin [Savchenko].

Acute exposure

Animals

Median lethal and highest tolerated oral doses of thiocyanates are shown in Table 1.

Table 1. Median lethal and highest tolerated doses of thiocyanates with oral administration (data from various sources)

Compound	Animals	LD_{50} (mg/kg)	Highest tolerated dose (mg/kg)
Ammonium thiocyanate	Mice	720–750	
	Guinea-pigs	500	
Potassium thiocyanate	Mice	590	
Sodium thiocyanate:			
crystalline	Mice	360–380	200–250
	Rats, male	1180	875
	Rats, male	232	100
crushed	Mice	809	
	Sheep	520	
	Hens	837	

Acute poisoning is characterized by breathlessness, rales, motor incoordination, narrowing of the pupils, convulsions, diarrhea, and unsteady gait with falls, followed by blood pressure elevation, impairment of cardiac activity, and paresis of the hind paws; 20 to 20 min post-administration, the animals usually assumed a lateral position (narcosis). Death occurred within 48 h. Subacutely poisoned animals exhibited erythropenia and leukocytosis. Following the administration of finely crushed sodium thiocyanate, elevated transaminase activities and depressed activities of erythrocytic acetylcholinesterase and blood serum cholinesterases were observed; blood levels of sugar and residual nitrogen were also elevated. Highest concentrations of thiocyanates were recorded in the blood at 24 h post-administration. Autopsy demonstrated pulmonary edema and mucosal lesions in the small gut, liver, kidneys, and spleen [Grin et al.; Ivanov et al.; Kostovetsky et al.].

The threshold of acute toxicity for mice exposed to ammonium thiocyanate by inhalation was determined at 86 mg/m^3 [Grin et al.].

Man

Acute psychosis (resembling schizophrenia with disorientation and hallucinations) and acute gastritis developed 3 to 8 h after an oral intake of 30 g potassium thiocyanate; psychic disturbances cleared in 3 days [David & Miketukova].

Repeated exposure

Animals

Rats and rabbits given sodium or potassium thiocyanate repeatedly in a dose of 0.5 mg/kg by gavage for about a month, exhibited reductions in the blood level of cholesterol and markedly abnormal liver and kidney functions. Post-mortem examination showed dystrophic changes in the liver, kidneys. heart, and gastric and intestinal mucosas, atrophy of splenic follicles, and perivascular edema and congestion in the brain [Ivanov et al.; Kostovetsky et al.].

Chronic exposure

Animals

In rats, prominent findings in the course of their exposure to ammonium thiocyanate by inhalation at 20.7 mg/m^3 4 h/day over 4 months included abnormal behavioral responses, impaired oxidation-reduction processes, and leukocyte lysis. At autopsy, pneumonia, dystrophic changes in the liver, and lesions in the brain and thyroid were observed. The threshold concentration for chronic toxicity was determined at 5 mg/m^3, while the concentration of 0.5 mg/m^3 was described as 'ineffective' [Grin et al.; Savchenko].

Man

Nonspecific resistance of the body has been reported to be appreciably reduced in workers occupationally exposed to ammonium thiocyanate for a period of about 3 years (workroom air contained concentrations of 0.3 to 12.0 NH_4SCN/m^3, and thiocyanates were detected in washings from the hands of these workers) [Savchenko].

Blood serum samples from adult nonsmokers had a mean thiocyanate (SCN$^-$ ion) concentration of 43.3 μmol/L, whereas those from smokers contained significantly higher concentrations (70 to 145 μmol/L) [Pechacek et al.; Robertson et al.]. The thiocyanate content of normal human saliva averaged 0.01%.

Ocular and dermal effects

Ammonium thiocyanate irritated the eyes, but not the skin, of rabbits. Allergic dermatitis developed in the hands of workers following prolonged contact with this compound.

Distribution and elimination

At 24 h after an intravenous injection into rats, ammonium thiocyanate was more or less uniformly distributed in their bodies, although higher concentrations than elsewhere were recorded for the thyroid and kidneys. Clearance from the blood occurred with a half-time of 36 h. With the urine, 46% of the injected amount was excreted during the first 24 h and another 40% over the next 24 h. For complete removal about 5 days were required. After the termination of inhalation exposure, ammonium thiocyanate concentration in the blood was rising during the first 60 min only [Ivanov et al.].

Hygienic standards

In the USSR, the occupational exposure limit for thiocyanates (MAC_{wz}) has been set at 50 mg/m^3 and their MAC in water sources, at 0.1 mg/L (based on sanitary and toxicological criteria). The thiocyanates have been placed in hazard class 2.

Measures to control exposure

These measures are essentially the same as those described above for hydrogen cyanide. The removal of thiocyanates from waste waters is based on the ability of these compounds to undergo oxidation. The more commonly used oxidizing agents are liquid chlorine, chlorinated lime (bleaching powder), and calcium or sodium hypochlorite; ozone is also used occasionally.

Emergency treatment in poisoning

This is virtually the same as in hydrogen cyanide poisoning (q.v.).

REFERENCES

David, A. & Miketukova, V. (1967) Arch. Toxicol., 23, 68—72.
Goncharuk, E. I. (1977) Sanitarnaya okhrana pochvy ot zagryazneniya khimicheskimi veshchestvami [Sanitary Measures to Protect Soil from Chemical Pollution]. Zdorovye, Kiev (in Russian).
Grin, N. V. et al. (1987) Gig. San., No. 8, 75—76 (in Russian).
Ivanov, A. G. et al. (1975) In: Farmakologiya i toksikologiya novykh produktov khimicheskogo sinteza [Pharmacology and Toxicology of New Synthetic Chemicals]. Minsk, pp. 167—169 (in Russian).
Ivanova, L. A. et al. (1988) Gig. Truda Prof. Zabol., No. 10, 52—53 (in Russian).
Kostovetsky, Ya. M. et al. (1967) In: Promyshlennye zagryazneniya vodoyomov [Industrial Pollution of Water Bodies], Vol. 5. Meditsina, Moscow, pp. 170—185 (in Russian).
Pechacek, T. F. et all. (1985) JAMA, 254, 3330—3332.
Robertson, A. S. et al. (1987) Brit. J. Ind. Med., 44, 351—354.
Savchenko, M. V. (1987) Gig. San., No. 11, 29—32 (in Russian).

Silicon

IDENTITY AND PHYSICOCHEMICAL PROPERTIES

Silicon (Si) is a nonmetallic element in group IV of the periodic table, with the atomic number 14. Its natural isotopes are ^{28}Si (92.27%), ^{29}Si (4.68%), and ^{30}Si (3.05%).

Silicon is a semiconductor. It is chemically inert at low temperatures but reacts with many substances at elevated temperatures; it is particularly active in fused form. Its usual oxidation state in compounds is +4. Reacting with oxygen at temperatures above 400°C, it yields the silicon oxide SiO_2 that gives rise, on hydration, to weak silicic acids difficultly soluble in water. It reduces the oxides of many elements. It is resistant to acids with the exception of hydrofluoric acid in mixture with nitric acid or with potassium chlorate or nitrate. Its interaction with alkali solutions results in the liberation of hydrogen and the formation of silicates of the corresponding alkali metals. It forms halides of the general formula $SiHal_4$ in reacting with fluorine (at 20°C), chlorine (300°C), bromine or iodine (500°C), as well as with hydrogen fluoride (at 20°C) and hydrogen bromide and hydrogen iodide (400–500°C). Its reaction with halogens can also yield halides of the general formula Si_nHal_{2n+2}. It reacts with sulfur at 600°C to give the sulfide SiS_2 and with nitrogen at 1000°C yielding the nitride Si_3N_4 (which is resistant to hydrofluoric acid and to melted alkalis and metals), and it also reacts with phosphorus, arsenic, and carbon. It forms silicides when dissolved in melts of alkali metals, alkaline-earth metals, lanthanides, group VIII metals, titanium, zirconium, chromium, molybdenum, tungsten, and manganese. See also Appendix.

NATURAL OCCURRENCE

Silicon is the most abundant element in the earth's crust after oxygen, constituting, on average, 29.5% of the lithosphere by weight. Approximately 12% of the latter is made up of silica (silicon dioxide SiO_2) which occurs in several polymorphic modifications as the mineral quartz and its varieties and in other silica-containing minerals whose total number exceeds 400.

Elemental (free) silicon does not occur naturally. Average silicon contents are (by weight) 33% in soil, $5 \times 10^{-5}\%$ in seawater, 0.15% in plants, and $1 \times 10^{-5}\%$ in animal organisms.

PRODUCTION

In the laboratory, silicon is obtained by reduction of silicon tetrachloride with zinc vapor at high temperature, by electrolysis of fused sodium hexafluorosilicate, or by decomposition of silicon halides. Industrially it is made most commonly by reducing silicon dioxide (quartz) with coke in an electric arc furnace.

USES

Silicon is mainly used to make crystalline detectors, semiconductors, and photoresistors; in high-temperature silit (silicon carbide) rods; for the production of organosilicon compounds; for deoxidizing metals; and for making alloys such as ferrosilicon, silicon copper, and silicon bronze. Metallurgical silicon is used as a starting material for silicon resins, elastomers, and oils.

TOXICITY OF SILICON AND ITS COMPOUNDS: GENERAL ASPECTS

The toxicity of silicon and of nearly all of its compounds is generally low. Silicon-containing dusts deposited in the airways tend to cause slowly developing pathological changes similar to those seen in chronic catarrh of the upper respiratory tract, chronic bronchitis, and pneumoconiosis. However, the rate of progression and the severity of these changes may vary markedly depending on the compound, and in some cases the disease process (notably pneumoconiosis) elicited by a particular compound shows distinctive qualitative features.

The soluble sodium and potassium silicates (q.v.) have more strongly marked toxic properties than other compounds; of major concern are the skin lesions they cause. Silicon halides such as silicon tetrachloride ($SiCl_4$) and tetrafluoride (SiF_4) are gases that produce appreciable irritant effects on mucous membranes. The chronic toxicity of silicon tetrachloride and of silicofluorides is mainly determined by the fluorine of these compounds and is very similar to that of hydrogen fluoride and hydrofluoric salts, respectively. Moreover, it is to be appreciated that highly poisonous substances may arise from impurities present in some silicon compounds (examples are hydrogen phosphide and hydrogen arsenide produced from impurities contained in ferrosilicon).

Six to 12 months after intratracheal administration of elemental silicon dust into the lungs of rats or rabbits, accumulations of the dust with scanty cellular elements and with minimal development of fibrous connective tissue were noted, along with somewhat thickened interalveolar septa and signs of bronchitis [Arkhangelskaya; Kulikov *et al.*; Velichkovsky *et al.*]. In addition, minor dystrophic changes in parenchymal organs were observed [Velichkovsky *et al.*], as was the presence of slight hyperglobulinemia, although without any elevation of gamma globulin which is a typical response to silicon dioxide exposure [Arkhangelskaya]; these effects were attributable to the action of the silicic acid that had formed in the animals to be absorbed through the lungs, given that increased urinary excretion of this acid was recorded in the dust-exposed rats [Velichkovsky *et al.*].

How silicon dust acts on humans has not been described. During electrothermal smelting of silicon, the risk of silicosis can arise from contamination of the breathing zone with condensation aerosols of the silicon dioxide (q.v.) that is produced via the readily sublimable silicon monoxide.

REFERENCES

Arkhangelskaya, L. N. (1967) In: *Novye dannye po toksikologii redkikh metallov i ikh soedineniy* [New Data on the Toxicology of Rare Metals and their Compounds]. Meditsina, Moscow, pp. 181–188 (in Russian).
Kulikov, V. G. *et al.* (1988) In: *Professionalnye bolezni pylevoi etiologii* [Occupational Diseases Caused by Dust]. Publication of the Erisman Research Institute of Hygiene, Moscow, Vol. 2, pp. 10–14 (in Russian).
Velichkovsky, B. T. *et al.* (1960) In: *Sbornic rabot po silikozu* [Silicosis: A collection of papers]. Publication of the Uralian Branch of the USSR Academy of Sciences, No. 2, pp. 171–184 (in Russian).

Silicon compounds
(silicon carbide and silicon dioxide)

SILICON CARBIDE

Physical and chemical properties
These are indicated in the Appendix.

Production
Silicon carbide (SiC) is usually made in an electric furnace by fusing together quartz sand or quartzite and anthracite or petroleum coke at 1920–2000°C with the addition of common salt, wood sawdust, and 'returns' of amorphous silicon carbide.

Uses
Silicon carbide is used as an abrasive for grinding metals with low tensile strength, stones, glass, ebonite, and other materials; in heating elements, heat-resistant thermocouples, semiconductors, and electrodes; in the manufacture of highly refractory and acid-resistant articles; and in dentistry as an abrasive agent.

Toxicity
Silicon carbide dust administered to rats in a chronic inhalation experiment or intratracheally in an amount of 50 mg led to pronounced bronchitis and displayed weak fibrogenic activity. Among workers occupationally exposed to silicon carbide dusts for a long time, chronic bronchitis, relatively benign

pneumoconiosis (which progressed slowly and was rarely complicated by tuberculosis), or both were observed. Similar respiratory changes have been described in metal workers engaged in grinding operations that involved exposure to a mixed dust containing, in addition to silicon carbide and some other abrasive materials, the metal being ground and components of the binder, including free silica [Ashbel *et al.*; Latushkina, 1959, 1960a & b].

Hygienic standards

In the USSR, an occupational exposure limit (MAC_{wz}) of 6.0 mg/m^3 has been set for silicon carbide (carborundum) as well as silicon nitride (as aerosols, hazard class 4), with the notation that these compounds are fibrogenic. In the USA, the TLV-TWA and TLV-STEL values for silicon carbide (and silicon) are 10 mg/m^3 and 20 mg/m^3, respectively.

Measures to control exposure at the workplace

Effective control of exposure in silicon carbide manufacture can be achieved through the use of: traveling hearth furnaces; mechanization for such operations as sorting, grading, and discarding of the materials; wetting of cooled fused blocks and lump material before their sorting and grading and before each stage of primary (coarse) crushing; wet processes for fine crushing and hydraulic classification of powders; mechanizing the separation and packaging of powders and the cleaning of sieves; and pneumatic conveyers. Technical control measures to be applied for reducing exposure to dust during grinding include the use of improved methods for forging and stamping metals so that the more dusty abrasive operations are excluded; wet grinding; and hooding of abrasive wheels with provision being made for dust aspiration.

Persons employed in the production or use of carborundum or other abrasives should undergo periodic medical examinations. Regular sojourn of the workers in prophylactoria is desirable for disease-preventing purposes.

Personal protection

Workers on dust-generating jobs must use personal protection devices. For respiratory protection, disposable respirators should be worn. Dust from work clothes should be removed in special chambers. Only wet cloths or vacuum cleaners may be used to clean up workrooms.

SILICON DIOXIDE (SILICA, SILICON OXIDE)

Aerosils; chalcedony [min.]; coesite; cristobalite [min.]; diatomaceous earth (diatomite); fuller's earth; infusorial earth; kieselguhr [min]; opal [min.]; quartz [min.]; quartz glass (vitreous silica); silica filler (highly dispersed amorphous silica); silica gel; silicic anhydride; stishovite [min.]; tridymite [min.]; tripoli [min.]

Physical and chemical properties

Silicon dioxide (SiO_2), or silica, is a hard mineral polymer [$(SiO_2)_x$] that occurs naturally in three crystalline modifications (as quartz, tridymite, and cristobalite), in cryptocrystalline (e.g. chalcedony) and amorphous (e.g. opal) forms, and in combined forms in silicates. Fused SiO_2 hardened in the amorphous state is known as quartz glass (vitreous or fused silica).

In crystalline silicon dioxide, the silicon and oxygen atoms are arranged in a definite regular pattern throughout the crystal. On mechanical disintegration, various surface defects are produced in the crystalline lattice, including the formation of a layer of highly disrupted structure ('pseudoamorphous layer'). Crystalline silicon dioxide has piezoelectric and semiconductor properties. The free valencies in incomplete silicon-oxygen tetrahedra on the surface of silicon particles hydrate in water and in water vapor-containing air with the formation of silanol groups $\equiv Si(OH)$ and $= Si(OH)_2$, and they can also form bonds with metal atoms and with organic radicals. Metal atoms can occur in the silica crystalline lattice as minor impurities which, however, may have important biologic implications.

When SiO_2 is 'solubilized' in water, the surface layer of the silicon-oxygen tetrahedra undergoes hydration, and molecules of the resulting monosilicic acid go into solution where they polymerize. The pseudoamorphous layer hydrates more rapidly than the undamaged crystalline layers, with the result that there is an initial phase of relatively fast silicon dioxide 'dissolution'; no such phase is observed with particles that have been pretreated with alkali or hydrofluoric acid which destroy the pseudoamorphous layer in the first place.

The 'solubility' of SiO_2 in aqueous media declines with decreasing pH (e.g. it is 1120 mg/L at pH 10.6 but only 140 mg/L at pH 1.0). The solubility of SiO_2 powder is greater the larger the specific surface of the powder. Amorphous silicas are more soluble than crystalline ones; the solubility is decreased by impurities that are in contact with the surface of silica particles (e.g. by aluminum, iron, and other nonalkali metals) and also by mineral components (as coal ash) and a number of organic compounds (e.g. alkylchlorosilanes). Hydrofluoric acid can destroy SiO_2 completely, with the formation of silicofluoric acid (H_2SiF_6). On heating, SiO_2 also reacts with phosphoric acid; other acids do not practically attack it. Heating SiO_2 with metal hydroxides, oxides, or carbonates or with salts of strong acids will yield silicates. SiO_2 dissolves in strong organic bases such as ethylenediamine. See also Appendix.

Natural occurrence

Silicon dioxide (silica) is found predominantly in crystalline forms as various quartzes (rock crystal, vein quartz, quartz sand, sandstone, quartzite, chalcedony, etc.) and in most rocks. Crystalline silica is the most widely distributed of all minerals.

Fossil amorphous silicas—kieselguhr (infusorial earth), diatomaceous earth (diatomite), and fuller's earth — are derived from deposits of ancient aquatic organisms that possessed silicified shells or skeletons and usually contain oxides of aluminum (up to 14%) and of some other metals. The natural gel of amorphous SiO_2, opal, may be of different origin, for example be deposited from thermal waters or derived from the weathering of silicate rocks.

Uses

Silicon dioxide is extensively used to make glass (e.g. quartz glass and silicate glass), ceramics (e.g. porcelain, faience, and dinas [refractory material]), abrasive and concrete materials, and silica bricks. As quartz, it is employed in radio and ultrasonic equipment. Infusorial earth finds chief applications as a filler, as a vehicle for catalysts, and as a filtering, thermal insulation, or abrasive material; for these purposes, use is often made of calcined diatomite in which a variable proportion (depending on the calcining conditions) of SiO_2 is present in the crystalline form of cristobalite. The dried and calcined artificial gel of amorphous SiO_2 (silica gel) is used as an adsorbent and catalyst carrier. Certain varieties of chemically pure amorphous silica (so-called aerosils) are utilized as fillers for lacquers, plastics, and rubbers. To impart special properties (e.g. hydrophobicity) to aerosils, the surface of their particles is modified by dimethylchlorosilane or other compounds.

Possible sources of occupational exposure

Silicon dioxide is a constituent of dusts that pollute the working area during the extraction and primary processing of most minerals as well as in many manufacturing industries (e.g. metallurgical, machine-building, and construction materials industries). Finely dispersed condensation aerosols of amorphous SiO_2 are likely to be emitted into the air of workplaces where technical-grade silicon or siliceous ferroalloys are produced by electrothermal processes, high-silicon steels are obtained by electric-arc smelting, heavy-coated electrodes are used for electric welding, or quartz-blowing or other high-temperature manufacturing processes are employed.

Mechanism of silica dust deposition and clearance in the respiratory tract

As dusty air passes in the airways during inspiration and expiration, it is cleared of suspended dust particles through processes of inertial impaction (which mainly clears particles more than 10 µm in diameter in the nasal passages and nasopharynx), sedimentation (particles up to decimal fractions of a micrometer in diameter all over the thracheobronchial tree), and collision with the mucous membrane in the course of chaotic Brownian movement (still smaller particles mainly in the distal lung regions, but also in the nasal passages). The particulate matter deposited in respiratory organs may exceed 90% of the exhaled amount by weight [Feoktistov], but its quantity varies greatly with particle size, shape, and density. Moreover, the number of deposited particles is different for different airway regions. Generally, the number and size of particles penetrating and deposited in the airways decreases in the downward direction. Also, the diameter of particles that are predominantly deposited at a given airway level decreases somewhat with growing particle density. For non-spherical (filamentous, needle-like, or scaly) particles, the dependence of deposition on particle size may be very different from that for spherical particles, but such non-spherical particle shapes are not characteristic of free silica (see also the section on asbestos). As a result of Brownian motion, particles less than 0.4–0.3 μm in diameter are deposited in much greater amounts than are larger particles in the submicron range.

Particles that have penetrated into the most distal airway region (which is referred to as the pulmonary or alveolar region and usually taken to comprise the area below the boundary between the upper and middle thirds of the terminal bronchiole), play a special role in causing pulmonary lesions because this so-called fine or respirable dust is retained in the lungs for long periods of time. To measure the mass concentration of this dust, a separator of larger particles is placed before the dust-collecting filter. According to the widely accepted view, such a separator should be designed to pass the following percentages of spherical particles (specific gravity 1.0): 95% of 1.5-μm particles, 75% of 3.5-μm, 50% of 5.0-μm, and 0% of 7.1-μm or larger particles. Adverse effects, however, are exerted by dust particles deposited at any level of the respiratory tract (see, e.g., the section *Silicosis and dust bronchitis*), and the measurement of respirable dust concentrations is therefore useful only when the total concentration of airborne dust is also considered.

For predicting the harmful effects of inhaled particulate matter, mathematical models of fractional dust deposition in various parts of the human respiratory tract are frequently used. Of these, the most extensively used one is a three-compartment model of the International Commission on Radiological Protection (ICRP) [Brain & Valberg], but experimental findings [e.g. Stahlhofen *et al.*] indicate that the model needs to be improved.

When pulmonary ventilation increases (as during physical work), so does not only the volume of dusty air entering the airway, but also the percentage of dust particles deposited therein because deeper breaths are taken. The particles primarily deposited on the nasopharyngeal and tracheobronchial mucous membranes — i.e. at levels not lower than the upper third of the terminal bronchiole — are carried in mucus upward to the pharynx at a rate that increases with decreasing distance from the latter. This clearance is effected by specialized ciliated cells of the epithelial mucosa ('mucociliary escalator'). Dust particles enter the gastrointestinal tract in swallowed saliva and food.

Particles that are primarily deposited in the alveolar region are phagocytized or enter lymphatic pathways of the interstitial lung tissue whence a proportion of them may be carried to the bronchial mucosa by lymph flow to be eliminated subsequently by the mucociliary clearance mechanism. Some of the particles in the lung interstitium (which is not in direct contact with air) are retained there for a long time (or even permanently) while some are transferred to lymph nodes, but this cannot be considered as being a true bronchopulmonary clearance since silicon dioxide can exert its damaging effects in lymph nodes also. In contrast, most of the dust particles phagocytized in the alveolar region are removed from the body. The major role in this clearance is played by alveolar macrophages, but an important contribution to the phagocytosis of SiO_2 particles is also made by polymorphonuclear (neutrophilic) leukocytes that compensate for the loss of alveolar macrophages which are readily damaged by these particles; the autoregulation of this compensatory mechanism is effected through macrophage breakdown products [Katsnelson & Privalova; Privalova et al.]. Cells that have ingested dust particles (coniophages) are transported to the airways together with non-phagocytized particles under the action of physical forces generated by respiratory movements; the coniophages are then eliminated in mucus, as are free particles. When an overreaction of alveolar phagocytosis occurs in response to a very large dust load, alveolar clearance is impaired, especially if exposure to some other phagocytosis-stimulating agent (e.g. cold) is added [Bykhovsky & Komovnikov; Gross et al., 1966; Katsnelson, 1972].

The efficiency with which dust is cleared depends both on its properties, which determines in turn whether the mucociliary or the phagocytizing mechanism will predominantly operate, and on how well these mechanisms function themselves. The functional state of these mechanisms varies widely, which is one of the main reasons for the observed individual variation in the susceptibility to dust diseases. Dust particles are eliminated from the respiratory tract the more rapidly, the larger they are and, consequently, the shorter the distance from the site of their predominant deposition to the nasopharynx and the higher the rate of mucus movement [Holma]. Accordingly, the less finely divided fractions of primarily deposited dust, which usually make up most of the dust load, are removed relatively rapidly, whereas the remaining dust is eliminated slower. In small laboratory animals, lung clearance from dust takes much less time than in humans. In the latter, the elimination half-time of dust particles amounts to days for the free alveolar surface and to months for the interstitial lung tissue [Bailey et al.; Brain]. Dusts that are eliminated from the alveolar region at particularly slow rates are those causing severe damage to coniophages and thus disrupting the phagocytic mechanism of clearance; such cytotoxicity is possessed by silica.

Prolonged exposure to quartz dust by inhalation results in progressive deterioration of macrophage-mediated clearance, and yet the ratio of lung clearance rates remains virtually unchanged owing to a parallel increase in the compensatory contribution of neutrophil-mediated phagocytosis [Privalova et al.]; as a result, the accumulation of quartz dust in the lungs reaches a plateau.

The cytotoxicity of dust particles has also been associated with their preferentially extracellular location in the interstitial lung tissue, one consequence of which is their transfer by lymph flow to intra- and extrapulmonary lymphoid tissues [Gross & Hatch]. For this reason, other conditions being equal, silica dust accumulates in the lungs and especially in the tracheobronchial, mediastinal, and more remote lymph nodes in greater amounts than do many other dusts [Dolgner et al.; Katsnelson & Babushkina; Strecker]. Dissolution is not an important factor in lung clearance from most mineral particulates, including disintegration aerosols of crystalline silica, but it does contribute appreciably to the liberation of lung tissue from particles of highly disperse condensation aerosols of amorphous silica [Glömme & Swensson; Velichkovsky]. The disintegration aerosols of amorphous silica (e.g., of quartz glass or silica gel), although highly cytotoxic, are cleared from the lungs more efficiently than are quartz aerosols, which is also explained by the greater solubility of the former aerosols [Podgaiko et al.].

Thus, long-lasting dust retention in lung tissue results both from the penetration of dust particles into the interstitial lung tissue (where they can be phagocytized) and from alveolar plugging. It has been estimated that prolonged occupational exposure to a dusty atmosphere results in approximately 1% to 2% of the primarily deposited dust being retained in the lungs [Gross], which may amount to several grams or even dozens of grams. Because particles of different sizes are eliminated with different efficiencies, the retained dust markedly differs in size distribution both from the total airborne dust and from its respirable fraction. For instance, small particles of 0.5–2 μm in diameter constituted by weight most of the dusts recovered from the lungs of mine workers who had died with silicosis [Nagelschmidt]. Although a 6-μm particle weighed 16 times more than a 1.5-μm particle, the total mass of 6-μm particles was 10 times less than that of 1.6-μm ones, whereas the contribution of larger particles (10–20 μm) to the total dust mass was negligible. As reported by Leiteritz et al., the dust extracted from coal miners' lungs contained (by weight) 3 to 4 times more particles in the 0.35–1.12 μm range and 1.9 times more in the 1.12–2.00 μm range than did the airborne dusts of coal mines, while containing much less particles of larger diameters. In a similar vein, no dust particles more that 7.2 μm in size were found in rats following chronic inhalation of coal dusts in which this fraction accounted for up to 16% of the airborne dust weight [Cartright and Skidmore].

Toxicity: general considerations

Following pulmonary deposition of relatively readily soluble fine particles of amorphous silica, the silicic acid that forms can produce some systemic effects after being absorbed through the lung tissue (in particular, it may affect the liver) [Velichkovsky]. Usually however, the primary pathologic changes occur at sites where silica dust particles are deposited (respiratory tract mucosa and alveoli), eliminated (respiratory and gastrointestinal tract mucosas), or retained (pulmonary parenchyma and lymph nodes), whereas various systemic manifestations of silica toxicity (metabolic disturbances, alterations in reactivity, immunopathologic phenomena) develop secondarily. The typical disease caused by inhaled silica-containing dusts is silicosis, the main feature of which is progressive pulmonary fibrosis produced by these sclerogenous dusts as they accumulate in the lungs. In silicosis the fibrotic reaction of the lungs to the retained dust (dust pneumosclerosis) is more severe than it is in other pneumoconioses caused by mineral dusts, although the silicosis-inducing potency of silica-containing dusts varies widely.

The most important single determinant of the silicosis hazard presented by a dust is its percentage silica content which, generally speaking, is directly related to its biologic aggressiveness (particularly fibrogenicity), but this relation can be disturbed to varying degrees depending on the nature and amount of admixtures present in the crystalline lattice and the surface of the silica particles and on the type of mechanical, chemical, and thermal treatments the dust-generating material has been subjected to [Beck et al.; Robock; Robock & Klosterkötter]. It should also be appreciated that the silicosis-producing potential of a silica dust is determined not only by its fibrogenicity but also by its capacity for prolonged retention in the lungs; because of this, more fibrogenic but rapidly eliminated varieties of amorphous silica may eventually prove not so damaging as less fibrogenic ones [Katsnelson et al., 1984; Podgaiko et al.]. The fibrogenicity of artificial amorphous silica powders such as aerosils varies with the method of their production, their fineness, their particle size, and the chemical modification undergone by their surfaces; the changes induced by aerosils in the lungs often regress, principally as a result of the elimination of their solubilized particles [Klostekötter; Schepers et al.; Timchenko]. Brieger & Gross [1967] disproved the contention that coesite is only slightly fibrogenic; the only form of silica found to be biologically inert is the mineral stishovite which differs from other silica modifications by having an octahedral rather than tetrahedral structure. Freshly ground silicas are more fibrogenic than 'old' dust [Kuselá & Holusă].

The silicosis-inducing potency of silicas rises with a decrease in the size of their particles, though only to a certain limit because the disintegration of particles is accompanied by increasing disruption of their crystalline structure [Aronova et al.; Velichkovsky]. The most hazardous substances in disintegration aerosols actually present in occupational environments are considered to be silica particles $1-2$ μm in diameter.

The property of silica underlying both its high fibrogenicity and the high capacity of its particles to be retained in the lungs and lymph nodes is its cytotoxicity to macrophages. As demonstrated in numerous experimental studies, there is a direct relationship between the cytotoxicity of various silica forms (and of silica-containing dusts) and their fibrogenicity. Any treatment that decreases the cytotoxicity of silica dust or makes the body less susceptible to its damaging action will also slow down the development of silicotic pneumosclerosis.

It has been shown in in *vitro* studies that macrophages injured by quartz release a humoral 'fibrogenic factor' causing cultured fibroblasts to synthesize more collagen—a fibrous connective tissue protein whose excess is a feature of silicosis; such a factor was not produced by macrophages following their physical breakage. A similar fibrogenic factor has been isolated from silicotic rat lungs; this is a protein of molecular mass ~16000 differing from other macrophagic proteins by its amino acid sequence [Kulonen *et al.*]. The damage of macrophages by silica particles also 'disinhibits' the general protein-synthesizing activity of fibroblasts which is normally regulated by macrophagic RNase [Kulonen *et al.*].

The mechanism of silica cytotoxicity and the reasons for differences between the cytotoxicities of different silica dusts have not been fully elucidated. It has been suggested that the cytotoxicity of any mineral particles is closely associated with the presence of chemically active sites on their surface—with the presence of silanol groups (primarily free ones) in the case of silica particles [e.g. Velichkovsky; Velichkovsky & Katsnelson]. The surface of such particles interacts with cell components to form various chemical linkages with them [Lomonosov *et al.*], predominantly hydrogen bonds [Velichkovsky]. The interaction impairs selective permeability of the cell membrane with a consequent disruption of cellular metabolic processes, which causes hydrolytic enzymes to diffuse out of lysosomes to the cytoplasm, and this is thought to result in cell death [e.g. Allison].

In addition to the fibrogenic factor of protein nature, macrophage breakdown by silica also elicits the formation of a lipid factor that stimulates the maturation of new macrophages of pulmonary and extracellular origin, causes macrophages and neutrophils to be recruited into the airways, and activates the phagocytic, immunologic, and other macrophage functions [Katsnelson & Privalova; Privalova].

Special attention has been given in several studies to elucidating the role the immune reactivity of the body might play in silicosis, since immunopathologic reactions are known to be an important factor in its pathogenesis in phases that follow the death of macrophages. The silicotic process has a number of clinical, morphologic, serologic, and biochemical characteristics that make it similar to autoimmune diseases [see, e.g., Velichkovsky & Katsnelson]. There is some evidence that specific autoantigens are produced as a result of silica coming in contact with macrophages, although the view prevails that the stimulation of immune responses by macrophage breakdown products is predominantly nonspecific, being associated with excessive activation of the primary immune interaction between macrophages and T lymphocytes [Pernis & Vigliani].

Experimental silicosis in animals

Experimental studies of silicosis make wide use of various biologic models raging from microsomal, mitochondrial, and erythrocytic suspensions (to examine how SiO_2 acts on membranes) through macrophage and macrophage-like cell cultures (to evaluate its cytotoxic activity) to whole animal models. The animal species most often utilized is the rat, but rabbits, guinea-pigs, hamsters, cats, dogs, monkeys, and sheep have also been used. Apart from chronic exposure to silica dusts by inhalation in chambers, animals may be exposed to these dusts by the intratracheal route in the form of suspension (for rats, the usual dose is 50 mg of dust in 1 ml of physiologic saline), intraperitoneally (the animals are then observed for a brief period to assess phagocytic responses or over a longer period to induce silicotic fibrosis of the omentum), intravenously (to induce silicotic changes in the liver), or subcutaneously.

The so-called silicotic nodule — a hallmark of silicosis — is a connective-tissue structure whose essential features are the same irrespective of the organs in which it develops. The pathologic changes occurring in animal lungs in response to silica dust particles administered intratracheally and particularly by inhalation are by and large similar to those found in silicosis-affected human lungs [Dvizhkov].

The range of dust concentrations that have produced silicotic changes in animal lungs is fairly broad (several mg to hundreds of mg per m^3). The concentrations used most commonly range from 30 to 100 mg/m^3 and the exposure period, from several weeks to 12–18 months (5–6 h daily by inhalation).

Both the severity of silicosis and the rate of its development in animals of the same species may vary widely even when the exposure conditions are the same. This is explained by the use of nonstandard silica samples (whose fibrogenic potency, as noted above, may differ greatly) and by the variability of responsiveness in laboratory animals due to genetic factors or/and uncontrolled environmental variables. For example, significant interstrain differences in susceptibility to the fibrogenic action of silica have been shown among inbred rats and mice [Callis et al.; Governa et al.]. Nevertheless, both the qualitative characteristics and the sequence of pulmonary changes induced by silica dust are always reproducible.

The main early effects of silica are local thickening of interalveolar septa which contain diffusely located dust-laden macrophages as well as free-lying dust particles, and hyperplasia of intrapulmonary lymphoid structures. Somewhat later, silicotic nodules are formed, which present initially as roundish structures composed of coniophages, fibroblasts, and histiocytes; in addition, lymphocytes and plasma cells can be seen more peripherally within the nodule. These plasma cells (and also plasma cells appearing in large numbers in the lung tissue, regional lymph nodes, and in the thymus, spleen, and other organs) are believed to be responsible for the production of autoantibodies involved in the pathogenesis of silicosis [Ambrosi; Engelbrecht *et al.*; Ragolskaya]. A characteristic feature of the rapidly proliferating pulmonary lymphocyte population is the preponderance of helper T cells which are consistently found in abundance in lung tissue and in bronchoalveolar lavage fluid [Struhar *et al.*]. A number of histochemical changes in both macrophages and fibroblasts have also been described, including those testifying to diffusion of lysosomal enzymes and to inhibition of the mitochondrial oxidation-reduction enzyme system [Raikhlin & Shnaidman].

Silicotic cellular nodules can form in the sites of former alveoli as their lumens become filled with coniophages, in interalveolar septa, or along the path by which dust particles are transported in lymph flow. The nodules gradually enlarge and neighboring nodules not infrequently coalesce into more or less large conglomerates. Concurrently, the nodules undergo sclerotization. Later, after its complete collagenation, the nodule may also undergo hyaline degeneration; the hyaline of silicotic nodules is more similar in composition to amyloid or fibrinoid than, for example, to hyalinized scar tissue, and this is taken as further evidence that immunopathology plays an important role in the pathogenesis of silicosis. Central necrosis of the nodules is sometimes also observed.

Even a completely collagenated and hyalinized nodule usually remains surrounded at the periphery by cellular elements that are typical for the early phases of its formation. Also, cellular nodules in the early phases of formation tend to be present in the lung in addition to mature collagenized nodules — even after a single exposure to silica dust. This phenomenon, which can serve as a model for the consistently observed progression of human silicosis after cessation of dust exposure, is due to the fact that dust particles, including those 'immured' by collagen fibers in the nodule, can be carried by coniophages from the nodule's center to its periphery or to more remote locations in the lung [Heppleston; Heppleston & Morris]. Such mobilization of dust particles may possibly be facilitated by edema arising for some reason or other [Gross & de Treville]. Yet the sclerotization of nodules imparts a degree of stability to the silicotic process, as is indicated by several pieces of experimental evidence. For example, severe progression of 'atypical silicosis' occurred in Syrian hamsters whose lungs showed diffuse consolidation of alveolar tissue with the stroma that, though argyrophilic, remained uncollagenated [Gross *et al.*, 1967]. Similarly, severe alveolar proteinosis or lipoproteinosis developed in guinea-pigs, hamsters, conventional rats, and particularly in so-called specific pathogen-free rats in the absence of sclerotic nodules in their lungs despite massive exposure to silica dust [Friedberg & Schiller; Gross & de Treville, 1968; Heppleston *et al.*].

In animals with ordinary experimental silicosis, the mass of silica in the lungs is always seen to decrease progressively after termination of dust exposure, because dust particles are not only carried from one lung region to another but are also eliminated from the lungs via the bronchi (by coniophages) and in lymph flow. The later pathway of elimination is conducive to the formation of new silicotic nodules in regional and more remote lymph nodes and also in the liver, bone marrow, spleen, and kidneys as a result of dust particles not retained in the lymph nodes being transferred in lymph to the blood and hence to these organs. The clearance of silica particles retained in the lungs usually proceeds more slowly than the progression of silicotic pneumosclerosis, and the nodules do not normally undergo regression. On the other hand, pneumosclerosis was clearly observed to regress in experiments where pulmonary clearance of silica was quite rapid—for instance, in the case of highly dispersed condensation aerosols or of aerosols from a colloidal silicic acid solution [Podgaiko *et al.*] or when the cytotoxic action of the silica accumulated in the lungs could be blocked by administered poly-2-vinylpyridine-N-oxide [Aronova & Katsnelson].

Silicotic nodules at various stages of development are often classified according to King & Belt, as follows: Stage 0—a cellular nodule without signs of fibrosis; Stage 1—one with rare argyrophilic fibrils; Stage 2—one with a dense network of argyrophilic fibers, possibly including a small number of collagen fibers; Stage 3—a largely collagenated nodule with small numbers of cellular elements; Stage 4—a completely collagenated nodule; Stage 5—coalescent collagenated nodules undergoing hyaline degeneration.

In addition to nodules, silica-containing lungs always show a degree of diffuse fibrosis (septal, perivascular, and/or peribronchial), accompanied by development of emphysema and atelectasis, deformation of the bronchial tree, and obliteration of vessels. It may well be that these pathologic processes rather than the nodular lesions are mainly responsible for the functional respiratory and circulatory abnormalities typical of silicosis.

The silica level in dust determines not only the severity of pneumosclerosis but also in large measure whether nodular or diffuse sclerotic lesions will predominate. When the level is low, the typical features of nodules such as round shape, clear delimitation from surrounding tissue, extensive growth, tendency to coalesce, and progressive collagenation, are less well defined. The lower the silica level and the smaller the pulmonary dust load, the more difficult it is to induce combined coniotic and tuberculous lesions in the animal [Gross *et al.*, 1961]. Animal models of silicotuberculosis are usually obtained through chronic exposure or intratracheal introduction of quartz dust combined with inoculation of live or killed tubercle bacilli, in particular organisms with weakened virulence [Gross *et al.*, 1961]. Lesions resembling those seen in silicotuberculosis were also produced experimentally using mycobacteria other than *Mycobacterium tuberculosis* [Tacquet *et al.*], as well as after intratracheal administration of tuberculin together with quartz dust in guinea-pigs presensitized by subcutaneous injection of tubercle bacilli [Gross *et al.*, 1961].

In animal models simulating the action of silica-containing dusts on human lungs, much attention has also been given to cytologic and biochemical changes in bronchoalveolar lavage fluid. The most characteristic (and an early) response to the deposition of cytotoxic dust particles has been found to be large increases in the number of alveolar macrophages (AM) and especially neutrophilic leukocytes (NL) in the lavage fluid; the NL/AM ratio becomes higher as the silicosis progresses and may serve as a useful index of dust cytotoxicity and also (when the capacity of the dust for retention is taken into account) as a predictor of dust fibrogenicity [Katsnelson et al., 1984], since this ratio depends on the mass of macrophage breakdown products [Katsnelson & Privalova]. For rapid evaluation of the cytotoxicity possessed by dust particles, many in vitro tests are available and used—for example the triphenyltetrazolium reduction test, trypan-blue exclusion test, lipid peroxidation enhacnement test, and so on.

Human silicosis

Accumulation of silica in the lungs and tracheobronchial lymph nodes of individuals working in a dusty atmosphere is accompanied by pathologic changes that are similar in principle to those described above for animals. These changes sooner or later modify the roentgenographic appearance of thoracic organs. Not uncommonly, a number of functional abnormalities as well as of biochemical and immunologic alterations indicative of silicosis are detectable even during the so-called preroentgenologic phases of the disease. Such early signs, however, are so inconstant and unspecific that diagnosis of silicosis (and other pneumoconioses) is virtually impossible unless roentgenographic signs of nodular or diffuse pneumosclerosis have also appeared. Their onset is therefore usually taken as that of the disease itself.

The time taken by silicosis to develop after the start of exposure may vary widely from several months (which is rare nowadays) to dozens of years depending on fibrogenicity of the dust (primarily its silica content), dust level in the workplace atmosphere, concomitant action of certain other factors (heavy physical work, cool microclimate, irritant gases, carbon monoxide) that may increase the silicosis hazard of the work being performed, and, last but not least, susceptibility of the organism to silica. Interindividual variation in susceptibility is large and explains why, under identical dust exposure conditions, some individuals become silicotic after only a very few years of employment, others do not contract silicosis until after many years, while still others never show any overt symptoms or signs of the disease either during long-term employment or after retirement. One reason for such variability is that the efficiency of pulmonary clearance mechanisms may vary from person to person. Another factor, and a very important one, is interindividual variation in general and immune reactivity. Other important intrinsic factors are genetic predisposition, age, and sex. All these factors interact in a complex way and are influenced in turn by many environmental variables. The correct identification of the set of factors that determine, for example, the occurrence of silicosis in only a proportion of workers under relatively low-level dust exposure is not possible unless the health-related conditions of life and work are taken into consideration. Their identification in real situations where effects of many factors are closely intertwined and interdependent, requires multifactorial analysis. Such studies [Katsnelson *et al.*, 1986] have shown, for instance, that factors contributing to individual vulnerability to silicosis include young age at onset of silica dust exposure, bronchitis, smoking, alcohol abuse, and certain dietary habits and lifestyle features; genetic predisposition has also been proved to play a role.

In recent decades, although silicosis has been diagnosed earlier than it used to be in the past, in industrially developed countries the average period of exposure before its first signs can be detected has become much longer owing to improved (less dusty) working conditions, and this has also led to reduced silicosis prevalence. Silicosis that has taken considerable less than the average time to develop is usually spoken of as 'early' or 'acute', while the term 'late silicosis' is reserved for cases where the disease appears long after the work in a dust-laden atmosphere had ended; the latter cases reflect the progressive nature of silicosis (see below), but in those who have not reached a roentgenologically recognizable phase after many years of work in a dusty environment.

The clinical classification of silicosis adopted and still frequently used in the USSR (and some other countries) is based on radiologic features of this disease supplemented by general clinical and functional assessments. In it, three forms of silicotic fibrosis are distinguished according to lung patterns: nodular, interstitial (diffuse sclerotic), and mixed. Moreover, three stages are recognized depending on the magnitude and profusion of fibrotic changes.

Stage I:
A symmetrical bilaterally deformed and accentuated lung pattern that is most pronounced in the middle and lower lobes and comprises fine reticular irregularly shaped opacities and a small number of rounded ('nodular') shadows reaching up to 1.5–2.0 mm in diameter; in the diffuse sclerotic form of silicosis, the latter shadows are indetectable with the conventional roentgenologic techniques. A more or less strongly marked thickening of the interlobar pleura is often seen; the hilar shadows are somewhat dilated.

Stage II:
Pathologic shadows are larger and more numerous than in Stage I, and they extend over all lung fields although without visible coalescence of the nodular opacities. In the nodular form of silicosis, many well-defined irregularly rounded shadows up to 3–5 mm in diameter are seen. In the interstitial form, the lung pattern is strongly accentuated, predominantly in the lower parts of the lung fields; the shadows of vessels and bronchi are more deformed than in Stage I: nodular shadows are few in number and not more than 1–2 mm in diameter or are absent altogether. The hili appear thickened and dense, with uneven outlines. Interlobar pleural thickening and pleurodiaphragmatic adhesions are

usually present; less often, pleuropericardial adhesions are seen. Signs of pulmonary emphysema are nearly always present.

Stage III:

Silicosis in this stage is marked by coalescence of the nodular and reticular opacities into larger shadows (termed 'nodes') 1–2 cm in diameter (sometimes into massive tumor-like shadows up to 10 cm in diameter) with uneven contours, most often bilateral and located predominantly in the upper and middle lung regions. Elsewhere in the lung, nonconfluent nodular lesions are seen in the nodular form of silicosis. The hili are invariably thickened, dense, and deformed; the basal regions and apices of the lungs are emphysematous (bullous emphysema is present in places). Extended pleural adhesions are observed in most cases.

It is a common practice to supplement the three stages of silicosis described above by transitional stages (Stage I–II and Stage II–III); in addition, when the roentgenographic picture seen in a silica-exposed worker is abnormal to a degree that does not yet justify the diagnosis of Stage I silicosis, the condition is not infrequently referred to as 'suspected silicosis', or Stage 0–I.

This classification is currently used in various combinations with the *International Classification of Radiographs of Pneumoconioses* adopted in 1971 by the International Labour Organization (ILO) and the International Union Against Cancer (UICC). The ILO/UICC classification is designed to codify the qualitative and quantitative features of radiologic appearances associated with silicosis, asbestosis, and other pneumoconioses. In it, small rounded opacities are designated, depending on their size, by the letters *p*, *q*, or *r* and small irregular opacities, by *s*, *t*, or *u*. The profusion of rounded and irregular opacities (i.e. their concentration in the affected lung zones) is designated by consecutive codes from 0/0 (which corresponds to an absolutely normal lung pattern) to 4/4 (the categories 0/1 and 1/0 roughly correspond to the 'suspected silicosis' mentioned above); in the short version of this classification, only the codes 0, 1, 2, and 4 are used to designate the profusion of opacities. In addition, the zones into which the opacities occur are recorded (each lung is divided into thirds—upper, middle, and flower zones). If one or more opacities >1 cm in the greatest diameter are present, only the letters *A*, *B*, or *C* are used depending on the sum of greatest diameters or on the combined area of the opacities. Special symbols are employed to indicate the nature, location, and extent of pleural changes as well as certain additional features of the radiograph; in particular, since the classification is a purely radiologic one, no distinction is made between silicosis and silicotuberculosis, but if opacities suggestive of tuberculosis are present, an additional symbol is introduced in the coding system. For more objective evaluation of radiographic appearances, each radiograph should be preferably read independently by several experts. The classification is supplemented by a set of standard radiographs to facilitate correct coding.
In 1980, the 1971 Classification was revised—a few amendments were introduced to clarify some ambiguities in the text and to facilitate further its use. The revised ILO classification is accompanied by a new set of standard radiographs.

The general clinical symptoms and signs in uncomplicated silicosis are scarce, particularly in Stage I. The disease may remain asymptomatic for a long time; the most frequent and earliest complaint is shortness of breath, often even on slight exertion. Other common symptoms are chest pain associated with pleural changes, and cough whose nature and severity mainly depend on the presence and degree of chronic 'dust' bronchitis. The changes detectable on percussion and auscultation of the chest are slight, even when the radiographic picture of silicosis is very well defined. Physical signs of pulmonary emphysema (especially basal emphysema) are often present, however; if tumor-like lesions have developed, dull percussion sounds may be heard over them. The presence and extent of dry rales also depend on the severity of chronic bronchitis and its exacerbations, although such rales are present, on average, in only 10–20% of patients.

Functional respiratory abnormalities in silicosis tend to be both of restrictive and obstructive types and are devoid of specific features, reflecting the severity of pneumosclerosis, emphysema, and chronic bronchitis in equal measure. In some cases they attain the degree of respiratory insufficiency and lead to arterial hypoxemia, although dyspnea in silicotic patients is often also observed in the presence of normal blood oxygenation because it may be due not only to disturbances of pulmonary ventilation and alveolocapillary gas diffusion, but may result from disordered regulation of respiration.

Polycardiographic and ballistocardiographic studies reveal — usually already in the early stage of silicosis — cardiac changes attributable principally to overloading of the right ventricle, but it is only in more advanced cases that these changes may be severe enough to resemble those seen in cases of a pronounced cardiopulmonary syndrome.

Among the other systems and organs affected in silicosis, the most significant changes are observed in the upper respiratory tract (subatrophic or atrophic chronic inflammatory processes) and digestive organs whose mucous membranes are directly exposed to the damaging action of dust particles eliminated from the respiratory tract through the pharynx and the esophagus; increased prevalence of gastrointestinal disorders is common in a number of dusty occupations.

A characteristic feature of silicosis is a lack of correlation between the severity of radiologic changes, on the one hand, and the clinical and functional manifestations of the disease on the other. Nevertheless, it is generally agreed that when sufficiently large groups of patients are compared, both the average severity and the frequency of clinical and functional symptoms and signs increase with each subsequent stage of radiographically defined silicosis.

Workers exposed to silica are at an increased risk of tuberculosis which is still regarded as the most serious complication of silicosis. In cases of silicosis complicated by tuberculosis as well as in those diagnosed as silicotuberculosis in the first place, the clinical picture is supplemented by signs and symptoms absent in simple (or pure) silicosis [Goldelman]. The earliest of these are usually subfebrility, somewhat accelerated erythrocyte sedimentation rate and mild leukocytosis; complaints of weakness, rapid onset of fatigue, and sweating; and the presence on the chest X-ray film of somewhat enlarged and polymorphic nodular opacities (chiefly in the subclavian region) and of deformed and dense hilar shadows [Goldelman]. Occasionally, the sole radiographic manifestation of tuberculous complication is enlarged paratracheal and bifurcational lymph nodes, with eggshell calcification in some cases; the diagnosis of silicotuberculosis can then be confirmed by biopsy of intrathoracic lymph nodes [Senkevich *et al.*; Shaklein *et al.*].

The following types of silicosis associated with tuberculous infection have been identified: (a) silicotuberculous bronchadenitis, either isolated or combining with silicosis or silicotuberculosis; (b) silicosis at any stage combining either with the early manifestations of tuberculosis mentioned above or with focal, infiltrative, hematogenously disseminated, or fibrous-cavernous varieties of pulmonary tuberculosis; (c) silicotuberculoma; and (d) massive progressive silicotuberculosis in which the elements of silicotic and tuberculous genesis cannot be clearly separated from each other.

The phases of silicosis-complicating tuberculosis are the same as those of isolated pulmonary tuberculosis but their clinical manifestations and time course may be rather different. Both abacillary (closed) and bacillary (open) forms are encountered, but in most cases there is so-called oligobacillary silicotuberculosis which is more difficult to diagnose. The infective agent is usually *Mycobacterium tuberculosis*, although other ('atypical') mycobacteria (e.g. *M. kansasii* and *M. scrofulaceum*) may also be responsible; in such cases the disease is clinically similar to silicotuberculosis but even more severe and the prognosis is worse.

Silicosis and tuberculosis may develop in the lungs independently or in close topographic relation. A pure silicosis can proceed to its most severe form without any tuberculosis. On the other hand, silicosis facilitates the development of tuberculosis and the latter frequently combines with advanced forms of silicosis. There is also epidemiologic evidence that silicotuberculosis is most often not a typical complication of preexistent simple silicosis, but a special combined disease, the predisposition to which depends on a number of environmental and genetic factors [Goldelman *et al.*].

Both simple silicosis and silicotuberculosis are characterized by a progressive course. Even when all exposure to silica dust has ceased, a more or less rapid progression of radiographic signs of silicosis (to the next classification stage in some cases) had been observed, with the addition of progressive tuberculosis and the deterioration of pulmonary function. The probability and rate of disease progression depend both on individual characteristics of the patient and on a set of environmental factors. The prognosis is, as a rule, least favorable in cases of 'early' silicosis. Occupational factors promoting progression include high silica content in the dust, high dust concentrations in the air during silicosis development, late removal of the diseased person from the dusty environment and especially his continued or resumed work in such environment, and heavy physical work. The course of silicotuberculosis largely depends on the form of tuberculous complications and also of course on the timeliness and appropriateness of antituberculosis treatment. In general, however, silicotuberculosis is a more severe and disabling disease than either uncomplicated silicosis or tuberculosis in individuals occupationally unexposed to silica. Factors increasing the risk of death are the same that aggravate the disease. Other conditions being equal, the risk increases with each subsequent stage of silicosis and is particularly high in silicotuberculosis. The latter ranked first among notifiable causes of death for patients who had been first diagnosed as having uncomplicated Stage I silicosis on which tuberculosis was later superimposed. When effective dust control measures had been implemented, newly detected silicotic patients were found to run a significantly lower risk of premature death than did those who continued working in a dusty environment [Mokronosova et al.].

Silicosis is also associated with increased mortality from cardiopulmonary insufficiency. Moreover, lung cancer as well as cancers of some other locations (particularly gastrointestinal cancer) have also been shown to be more frequent causes of death in silicotic persons as compared to the general population; cancer mortality rates are also significantly elevated among those exposed to various non-carcinogenic dusts, though usually only in men, and this suggests a potentiating effect of smoking [Katsnelson & Mokronosova]; this matter has been the subject for debate [Goldsmith et al.; IARC].

Silicosis and dust bronchitis

The damage caused to the respiratory mucosa by deposited dust gradually leads to chronic inflammation; an important part in its development is played by respiratory microflora. The peribronchial sclerosis, which is typical of silicosis and which is accompanied by bronchial deformation, together with the alterations in physical properties of mucus as a result of silica action on goblet cells, impair the normal transport of mucus with dust particles and pathogenic micro-organisms, thereby aggravating the endobronchitic process. For these reasons, the condition referred to as 'dust bronchitis' is usually a combined infectious and dust disease in which one or the other of the two components may predominate. However, chronic bronchitis may also arise on the basis of recurrent respiratory infections of the catarrh type, i.e. without substantial involvement of exogenous stimulants such as dust; on the other hand, important exogenous contributors to chronic bronchitis are of non-occupational nature (primarily smoking). But heightened prevalence of chronic bronchitis in occupational groups exposed to various industrial dusts is a well established fact, and in such groups this disease has therefore been classified in the USSR as occupational, being termed 'chronic dust bronchitis'. The differential diagnosis is based on certain clinical features (such as gradual development and less sputum than in chronic infectious bronchitis) and on a consideration of the nature and concentration of dust, the length of exposure to it, and some other occupational factors such as combined exposure to dust and irritant gases, cold, or heavy physical work. It is important to know whether the dust contains caustic alkalies or other irritant components and allergens. Epidemiologic data indicate that exposure to dusts high in silica is associated with increased prevalence of chronic bronchitis. Although it is rather common for silicosis to combine with chronic bronchitis, there have been many cases of silicosis not accompanied by either clinical or endoscopic signs of bronchitis (especially among workers exposed to relatively low dust concentrations). Conversely, many cases of definite chronic bronchitis have been recorded among workers without roentgenographic signs of silicosis after long-term employment on dusty jobs; in such cases, however, the risk of silicosis developing in future is significantly higher [Katsnelson et al., 1986].

Mixed dusts

In addition to silica, most industrial dusts carrying a risk of silicosis contain various silicates (they are usually measured together with silica as 'total silica'), metal oxides, coal, or other components. The nature and severity of deleterious effects caused by such 'dust mixes' is determined at least by four factors: (i) the amount and form of SiO_2 they contain; (ii) the biologic aggressiveness (toxicity, fibrogenicity) of dust components other than silica; (iii) the impact of these other dust components on silica fibrogenicity; (iv) the possible impact of silica on the other dust components. As a rule, any admixture makes the silica less fibrogenic by contaminating the surface of its particles or through chemical interaction with their silanol groups. Such an effect is produced, for example, by oxides of iron [Gross et al., 1961], some compounds of aluminum [Engelbrecht et al.] and copper [Katsnelson & Babushkina], and coal dusts [Dolgner et al.]. On the other hand, any of these admixtures may themselves exert a more or less marked fibrogenic effect as they accumulate in the lungs, and this effect may in turn be enhanced by silica.

It should also be borne in mind that because of differences in their particle shapes and sizes and in their specific weights, the individual components of a mixed dust may partly separate from one another already in the course of their primary deposition in the lung shortly after inhalation. And since the components also differ in cytotoxicity, they are separated further during the subsequent processes of pulmonary clearance and macrophage-mediated intrapulmonary transfer of their particles [Policard et al.]. For these reasons, typical silicotic fibrotic nodules have frequently been observed to combine in the same lung with a focal cellular reaction to another dust — for example a predominantly iron dust in the case of pneumoconiosis induced by mixed silica-iron oxide dust [Dvizhkov].

Suppression of silica cytotoxicity and fibrogenicity

The search for agents capable of protecting macrophages from the cytotoxic action of silica offers one of the most promising approaches to the suppression of its fibrogenic activity. Agents particularly effective in this respect in experimental trials have proved to be, on the one hand, poly-2-vinylpyridine-N-oxide (polyvinoxide for short) and some other polymers containing positively charged nitrogen [Aronova et al.; Holt; Marchisio et al.; Katsnelson et al., 1967, 1989; Schlipköter, 1969] and, on the other hand, certain active metabolites, notably glutamic acid and sodium glutamate, that are able to enhance macrophage resistance to the damaging action of quartz [Katsnelson et al., 1984, 1989; Morozova et al.]. Parenteral administration of polyvinoxide resulted in decreased retention of silica in the lungs and lymph nodes of test animals and in marked inhibition of the fibrotic process. Similar effects were produced by sodium glutamate given per os. Encouraging results have also come from clinical trials of polyvinoxide [Schlipköter, 1969; Zislin et al.].

Hygienic standards

Occupational exposure limits adopted in the USSR for silicon dioxide (silica) and silica-containing substances are presented in Table 1.

In the USA, the threshold limit values have been established for crystalline silicas (quartz, cristobalite, tridymite, fused silica, and infusorial earth) and amorphous silica. For *quartz*, the TLV is expressed in mppcf (million of particles per cubic foot of air based on samples collected using an impinger and counted by a light-field technique) and given by the formula 300/(% quartz + 10), where the percentage of quartz is the amount determined from airborne samples (except in those instances in which other methods have been shown to be applicable). The TLV for *respirable dust* is expressed in mg/m³ and given by the formula 10 mg/m³/(% respirable quartz + 2), if 10 mg/m³ contains <1% quartz; otherwise, the corresponding formula for quartz is used. The TLV for 'total dust' (both respirable and nonrespirable) is given by the formula: 30 mg/m³/(%quartz + 3). For *cristobalite*, the TLV is equal to one half the value calculated from the particle count or mass formulas for quartz. For *tridymite*, the TLV also equals one half the value calculated from the formulas for quartz. For *fused silica*, the same value as for quartz is used. For *infusorial earth* (tripoli), the formula for respirable quartz dust is used. Finally, the TLV for *amorphous silica* is set at 6 mg/m³ (total dust of all particle sizes).

Table 1. Exposure limits for silicon dioxide

	Workroom air		Atmospheric air				
	MAC_{wz} (mg/m^3)	Aggregative state	MAC_{hm} (mg/m^3)	MAC_{ad} (mg/m^3)	$TSEL_{aa}$ (mg/m^3)	Hazard class	Note
Quartz glass	1.0	Aerosol	–	–	–	3	F[a]
Silicon-copper alloy	4.0	Aerosol	–	–	–	4	F
Silicon dioxide, amorphous: condensation aerosol containing:							
10–60% SiO$_2$	2.0	Aerosol	–	–	0.02	4	F
>60% SiO$_2$	1.0	Aerosol	–	–	–	3	F
Silicon dioxide, crystalline: dust containing:							
2–10% SiO$_2$[b]	4.0	Aerosol	–	–	–	4	F
10–70% SiO$_2$[c]	2.0	Aerosol	0.3	0.1	–	3	F
>70% SiO$_2$[d]	1.0	Aerosol	0.15	0.05	–	3	F

[a] F - predominantly fibrogenic activity.
[b] Kukruze oil shales, copper-sulfide ores, carbon and coal dusts, clay, etc.
[c] Granite, chamotte, raw mica, coal mine dust, etc.
[d] E.g., quartzite and Dinas bricks.

Methods of determination

In air
In accordance with the approaches to hygienic standard setting adopted in the USSR, the usual practice has been to determine the percentage content of free silica in airborne dust rather than measuring silicon concentration in the air directly. A given volume of air is drawn through a pre-weighed filter, the mass of the dust on the filter is measured, and the filter is ashed in a platinum crucible at 450–500°C (at 650–700°C in the case of coal dust). The ash residue is then boiled in a 1:1 mixture of diluted nitric and hydrochloric acids to remove the metal oxides and acid-soluble silicates. For the determination of crystalline silica, the amorphous silica is removed by treating the residue additionally with borofluorohydric acid for 1 h at 70°C. Petrographic and x-ray diffraction analyses are usually carried out where the crystalline modifications of silica need to be identified in addition to chemical determination. For separate

determination of crystalline and amorphous silica in industrial dusts, infrared spectroscopy has been used with much success [Radulescu & Borda].

The method most commonly used in the USSR for determining small quantities of free silica in dust samples is the one proposed by Polezhayev, in which free silica is selectively fused with a mixture of sodium hydrogen carbonate and sodium chloride (or of potassium hydrogen carbonate and potassium chloride) in the 1 : 1 ratio, the resulting sodium (or potassium) silicate is dissolved, and the silicon is detected by the formation of silicomolybdic heteropoly acid. To make the assay more sensitive, this acid is reduced to a blue complex compound, usually with ascorbic acid in the presence of tartaric acid at pH 2 [Dobrovolskaya *et al.*; Peregud & Kozlova].

In biologic materials

The method most frequently employed is based on the reaction with ammonium molybdate, resulting in the formation of colored silicomolybdic acid. The reduction of the yellow complex to a blue one after preliminary removal of phosphorus with magnesial mixture underlies a method for determining silicic acid in urine [Ivanov & Rozenberg; Lastykina *et al.*]. Spectrographic procedures for measuring silicon in blood, urine, and other media have also been used. For a brief review of analytical methods, see Berman and Friedberg & Schiller.

Measures for preventing exposure at the workplace

Preventive measures are particularly important as there is no effective treatment for the pulmonary lesions of silicosis. Such measures must be primarily of engineering nature and be directed at suppressing dust formation and precluding the dust that has formed from reaching the breathing zone of workers. It is usually only in emergency situations and during breakdown of the normal dust control measures that the use of personal respiratory protection is justified — for example the wearing of dust masks or insulating hoods with air supply. The workers should also be supplied with goggles and protective clothing.

Many techniques and a wide range of equipment have been devised to suppress dust at workplaces and keep its concentrations below maximum permissible levels. The suppression of dust by health engineering measures appropriate for the particular industry and workplace should be rigidly enforced and any residual dust should be controlled by proper ventilation. Respirable dust levels and the free silica content of dust should be monitored regularly.

In the mining industry, dust concentrations cannot be stably maintained at acceptable (below MAC) levels unless a whole range of control measures have been implemented such as dust collection by wet or dry methods in drilling operations; use of hydraulic (water) stemming, mist sprayers, and water curtains in shotfiring; dust suppression during the use of rock-winning and drivage machinery by wetting the material won; and proper ventilation of the working with removal of dust from the ventilation air. The efficiency of dust suppression, particularly during drilling, can be improved by various methods, for example by adding surface-active agents to the water; these agents will accelerate the wetting process and improve dust binding. In coal mines, it is common practice to spray water from spray nozzles at the coal-cutting mechanisms of combines and at other coal-getting machines as well as at skip loading points and conveyer delivery points; an effective means of suppressing dust even before it is released is infusion of water into the coal seam under high pressure.

In processing industries, apart from the introduction of dust-free or less dusty processes, the most widely used method of dust control where the work situation makes the release of much dust unavoidable consists in isolating the source of dust as much as possible from the ambient air, with removal of the dust-laden air from the enclosed spaces by exhaust ventilation and its purification. The air is then usually released into the atmosphere.

In addition to technical measures for preventing silicosis (and other pneumoconioses) and dust bronchitis, an important role is played by medical prevention. This mainly consists in preplacement and periodic medical examinations that should include a medical history of the employee and a physical examination with special attention to the respiratory system. A full-size chest film should also be taken and basic lung function tests be carried out, including measurement of the vital capacity and forced expiratory volume. The frequency of periodic examinations depends on the level of dust exposure. If control measures are satisfactory, the examination could be repeated at 3-year intervals. Regular medical examinations enable silicosis to be detected at an early stage and thus can prevent many cases of advanced or complicated silicosis. When the first signs of silicosis appear, the worker should be immediately withdrawn from further exposure [Goldelman; WHO]. In view of the wide interindividual variation in susceptibility to the harmful effects of silica, it has been recommended that special tests be devised for assessing the susceptibility of workers during the preplacement examination [Katsnelson et al., 1989].

It is important to prevent tuberculosis in silicotic patients. A specific means of preventing silicotuberculosis is considered to be vaccination and chemo-prophylaxis, both in silicotic patients and in healthy workers exposed to silicogenous dusts [Goldelman].

For preventive purposes, inhalation of poly-2-vinylpyridine-N-oxide solutions has been proposed, mainly on account of their beneficial effect on lung clearance from silica [Schlipköter, 1986]. It has also been proposed that trials of glutamic acid be carried out on exposed workers since this acid has proved effective against silicosis in experimental animals [Katsnelson et al., 1984; Morozova et al.]. A system of measures (referred to collectively as 'biologic prophylaxis') aimed at increasing the resistance of individuals and of the working population as a whole to silica dust has been proposed [Katsnelson et al., 1989].

Persons suffering from bronchopulmonary disorders should be transfered to jobs where there is little or no risk of silicosis. In the USSR, persons under the age of 18 are not permitted to be employed in occupations involving silicosis hazards.

REFERENCES

Allison, A. C. (1970) *Adv. Sci.*, **27**, 137–140.

Ambrosi, L. (1966) *Med. Lav.*, **57**, 249–256.

Aronova, G. V. & Katsnelson, B. A. (1977) *Gig. Truda prof. Zabol.*, No. **6**, 49–50 (in Russian).

Aronova, G. V. *et al.* (1983) *Farmakol. Toksicol.*, No. **4**, 89–92 (in Russian).

Ashbel, S. I. *et al.* (1973) *Voprosy gigieny truda i profpatologii rabochikh-shlifovshchikov* [Occupational Health Services for Workers Engaged in Grinding]. Volgo-Vyatsk Book Publishers, Gorky (in Russian).

Bailey, M. R. *et al.* (1985) *J. Aerosol Sci.*, **16**, 293–295 & 295–303.

Beck, E. H. *et al.* (1973) *Staub*, **33**, 3–7.

Berman, E. (1980) *Toxic Metals and their Analysis*. Heyden, London.

Brain, J. D. (1985) In: Fishman, A. P. & Fisher, A. B. (eds) *Handbook of Physiology. Section 3. The Respiratory System. Vol. 1: Circulation and Nonrespiratory Functions.* American Physiological Society, Bethesda, pp. 447–471.

Brain, J. D. & Valberg, P. A. (1974) *Arch. Environ. Health*, **28**, 1–11.

Brieger, H. & Gross, P. (1966) *Arch. Environ. Health*, **13**, 38–43.

Brieger, H. & Gross, P. (1967) *Arch. Environ. Health*, **15**, 751–757.

Bykhovsky, A. V. & Komovnikov, G. S. (1970) In: *Patogenez pnevmokoniozov* [Pathogenesis of Pneumoconioses]. Publication of the Sverdlovsk Research Institute of Occupational Health, pp. 341–361 (in Russian).

Callis *et al.* (1985) *J. Lab. Clin. Invest.*, **105**, 547–553.

Cartright, J. & Skidmore, J. W. (1964) *Ann. Occup. Hyg.*, **7**, 151–167.

Devulder, B. *et al* (1972) *Lille Méd.*, **17**, 1133–1143.

Dolgner *et al.* (1965) *Zeitschr. Arbeitsmed.*, **3**, 26–30.

Dobrovolskaya, V. V. *et al.* (1958) *Opredelenie svobodnoi dvuokisi kremniya v gornykh porodakh i rudnichnoi pyli* [Determination of Free Silicon Dioxide in Rocks and Mine Dust]. Publication of the USSR Academy of Sciences, Moscow (in Russian).

Dvizhkov, P. P. (1985) *Pnevmokoniozy* [Pneumoconioses]. Medistsina, Moscow (in Russian).

Engelbrecht, F. M. *et al.* (1972) *S. Afr. Med. J.*, **46**, 462–464.

Feoktistov, G. S. (1968) *Gig. San.*, No. **3**, 21–26 (in Russian).

Friedberg, K. D. & Schiller, E. (1988) Chapter 54 *Silicon* in: Seiler, H. G. *et al.* (eds) *Handbook on Toxicity of Inorganic Compounds*. Marcel Dekker, New York, pp. 595–617.

Glömme, J. & Swensson, A. (1957) *Acta Med. Scand.*, **158**, 387–393.

Goldelman, A. G. (1967) *Klinika, lechenie i profilaktika silikotuberkuleza* [Clinical Features, Treatment, and Prevention of Silicotuberculosis]. Sverdlovsk (in Russian).

Goldelman, A. G. *et al.* (1989) *Gig. Truda Prof. Zabol.*, No. **8**, 10–13 (in Russian).

Goldsmith, D. F. *et al.* (eds) (1986) *Silica, Silicosis and Cancer. Controversy in Occupational Medicine*. Praeger Publishers, New York.

Governa, M. *et al.* (1967) *Lavor Med.*, **21**, 225–230.

Gross, P. (1964) *Health Phys.*, **10**, 995–112.

Gross, P. & Hatch, Th. (1962) *Intern. Arch. Gewerbepathol. Gewerbehyg.*, **19**, 660–666.

Gross, P. & de Treville, R. T. P. (1968) *Arch. Pathol.*, **86**, 251–261.

Gross, P. & de Treville, R. T. P. (1972) *Amer. Rev. Respir. Dis.*, **106**, 684–691.

Gross, P. *et al.* (1961) *Amer. Rev. Respir. Dis.*, **83**, 510–527.

Gross, P. *et al.* (1966) *Amer. Rev. Respir. Dis.*, **94**, 10–19.

Gross, P. *et al.* (1967) *Arch. Pathol.*, **84**, 87–94.

Heppleston, A. G. (1984) *Environ. Health Perspect.*, **55**, 11–128.
Heppleston, A. G. & Morris, T. G. (1965) *Amer. J. Pathol.*, **46**, 945–958.
Heppleston, A. G. *et al.* (1970) *J. Pathol.*, **101**, 293–307.
Holma, B. (1967) *Lung Clearance of Mono- and Di-disperse Aerosols Determined by Profile Scanning and Whole-body Counting.* Emil Kihlström Trycken, Stockholm.
Holt, P. F. (1971) *Brit. J. Ind. Med.*, **28**, 72–77.
IARC (1987) *Monographs on the Evaluation of the Carcinogenic Risk of Chemicals to Humans*, Vol. **42**: *Silica and some Silicates.* International Agency for Research on Cancer, Lyons.
Ivanov, V.I . & Rozenberg, N. A. (1960) *Gig. Truda Prof. Zabol.*, No. **4**, 39–42 (in Russian).
Katsnelson, B. A. (1972) In: *Kombinirovannoye deistvie khimicheskikh i fizicheskikh faktorov proizvodstvennoi sredy* [Joint Action of Chemical and Physical Environmental Agents]. Publication of the Sverdlovsk Research Institute of Occupational Health, pp. 10–19 (in Russian).
Katsnelson, B. A. & Babushkina, L. G. (1968) *Gig. San.*, No. **3**, 108–110 (in Russian).
Katsnelson, B. A. & Mokronosova, C. A. (1979) *J. Occup. Med.*, **21**, 15–20.
Katsnelson, B. A. & Privalova, L. I. (1984) *Environ Health Perspect.*, **55**, 313–325.
Katsnelson, B. A. *et al.* (1967) *Gig. Truda Prof. Zabol.*, No. **6**. 35–39 (in Russian).
Katsnelson, B. A. *et al.* (1984) *Med. Lav.*, **75**, 450–461.
Katsnelson, B. A. *et al.* (1984) *Arbeitsmed. Sozialmed. Präventivmed.*, **19**, 153–156.
Katsnelson, B. A. *et al.* (1986) *Environ. Health Perspect.*, **68**, 175–185.
Katsnelson, B. A. *et al.* (1989) *Environ. Health Perspect.*, **82**, 311–321.
King, E. J. & Belt, T. (1938) *Physiol. Rev.*, **18**, 329–356.
Klostekötter, W. (1968) *Arch. Hyg. Bakteriol.*, **158**, 7–22.
Kulonen, E. *et al.* (1983) *Environ. Health Perspect.*, **51**, 119–124.
Kyselá, B. & Holusá, R. (1971) *Pracov. Lěk.*, **23**, 317–320.
Lastykina, K. O. *et al.* (1986) *Gig. San.*, No. **11**, 42–44 (in Russian).
Latushkina, V. B. (1959) In: *Sbornik nauchnykh rabot institutov okhrany truda* [A Collection of Research Papers by Workers of Institutes for Occupational Safety], Vol. 1. Profizdat, Moscow, pp. 37–43 (in Russian).
Latushkina, V. B. (1960a) In: *Sbornik nauchno-issledovatelskikh rabot* [A Collection of Research Papers]. Profizdat, Moscow, pp. 53–71 (in Russian).
Latushkina, V. B. (1960b) *Gig. Truda Prof. Zabol.*, No. **7**, 49–52 (in Russian).
Leiteritz, H. *et al.* (1966) In: *Inhaled Particles and Vapours*, Vol. II. Pergamon Press, London, pp. 381–390.
Lomonosov, S. A. *et al.* (1970) In: *Patogenez pnevmokoniozov* [Pathogenesis of Pneumoconioses]. Publication of the Sverdlovsk Research Institute of Occupational Health, Sverdlovsk, pp. 55–60 (in Russian).
Marchisio, M. A. *et al.* (1970) *Securitas*, **55**, 101–153.
Mokronosova, K. A. *et al.* (1972) *Gig. Truda Prof. Zabol.*, No. **5**, 16–19 (in Russian).
Morozova, K. I. *et al.* (1984) *Brit. K. Ind. Med.*, **41**, 518–525.
Nagelschmidt, G. (1965) *Amer. Ind. Hyg. Assoc. J.*, **26**, 1–7.
Peregud, E. A. & Kozlova, N. P. (1954) *Zh. Analit. Khimii*, No. **1**, 47–50 (in Russian).
Pernis, B. & Vigliani, E. C. (1982) *Amer. J. Ind. Med.*, **3**, 113–137.
Podgaiko, G. A. *et al.* (1982) In: *Professionalnye bolezni pylevoi etiologii* [Occupational Diseases Caused by Dust]. Publication of the Erisman Research Institute of Hygiene, Moscow, pp. 93–100 (in Russian).
Policard, A. *et al.* (1964) *Arch. Malad. Profess.*, **25**, 293–298.
Polezhayev, N. G. (1957) *Gig. San.*, No. **11**, 91–94 (in Russian).
Privalova, L. I. (1986) *Med. Lav.*, **77**, 613–621.
Privalova, L. I. *et al.* (1987) *Brit. J. Ind. Med.*, **44**, 228–235.
Ragolskaya, F. Sl (1966) In: *Sbornik rabot po silikozu* [A Collection of Papers on Silicosis]. Middle Uralian Book Publishers, Vol. 4, pp. 205–209 (in Russian).
Radulescu, D. N. & Borda, M. (1973) *Z. Ges. Hyg. Grenzgeb.*, **19**, 493–496.
Raikhlin, N. T. & Shnaidman, I. M. (1970) *Gistokhimiya soedinitelnoi tkani pri silikoze* [Histochemistry of Connective Tissue in Silicosis]. Meditsina, Moscow (in Russian).

Robock, K. (1974) *Beitr. Silikose-Forsch.*, **26**, 11−262.
Robock, K. & Klosterkötter, W. (1973) *Staub*, **33**, 60−63.
Schepers, G. W. H. *et al.* (1957) *Arch. Ind. Health*, **16** (No. 2), 125−146; (No. 3), 200−204; (No. 4), 280−301; (No. 5), 363−379; (No. 6), 499−513.
Schlipköter, H.-W. (1966) *Zentralbl. Arbeitsmed. Arbeitsschutz*, **16**, 221−226.
Schlipköter, H.-W. (1969) *Canad. Mining J.*, **90**, 72.
Schlipköter, H.-W. (1986) In: *Umwelthygiene, Jahresbericht 1985*, Vol. 18. Verlag S.W. Alberts, pp. 159−173.
Senkevich, N. A. *et al.* (1970) *Probl. Tuberk.*, No. 3, 61−65 (in Russian).
Shaklein, I. A. *et al.* (1970) *Probl. Tuberk.*, No 3, 56−60 (in Russian).
Stahlhofen, W. *et al.* (1983) *J. Aerosol Sci.*, **14**, 181 188.
Strecker, F. J. (1967) In: *Inhaled Particles*, Vol. 2. Oxford, pp. 141−153.
Struhar, D. *et al.* (1989) *Amer. Rev. Respir. Dis.*, **139**, 28−32.
Tacquet, A. *et al.* (1972) *Lille Méd.*, **17**, 1121−1132.
Timchenko, A. N. (1973) *Gig. Truda Prof. Zabol.*, No. 8, 47−49 (in Russian).
Velichkovsky, B. T. (1980) *Fibrogennye pyli* [Fibrogenic Dusts]. Volgo-Vyatsk Book Publishers, Gorky (in Russian).
Velichkovsky, B. T. & Katsnelson, B. A. (1964) *Etiologiya i patogenez silicoza* [Etiology and Pathogenesis of Silicosis]. Meditsina, Moscow (in Russian).
Velichkovsky, B. T. *et al.* (1970) In: *Borba s silikozom* [Silicosis Control], Vol. 8. Nauka, Moscow, pp. 191−198 (in Russian).
WHO (1986) In: *Early Detection of Occupational Diseases*. World Health Organization, Geneva, pp. 9−25.
Zizlin, D. *et al.* (1985) *Gig. Truda Prof. Zabol.*, No. 11 . 17−19 (in Russian).

Silicic acids and their salts (silicates)

I. Silicic acids and general considerations regarding silicates

IDENTITY

The silicic acids have a general formula $x SiO_2 \cdot y H_2O$. For metasilicic acid (H_2SiO_3), $x = y = 1$, while for orthosilicic acid (H_4SiO_4), $x = 1$ and $y = 2$. For the various polysilicic acids (of which many naturally occurring silicates are derivatives) $x > 1$. So-called simple silicates are salts of particular silicic acids in which the hydrogen has been completely or partially replaced by a metal(s). Complex silicates are salts of acids having the general formula $x R_2O_3 \cdot y SiO_2 \cdot z H_2O$; most commonly, $R = Al$ and the salts of such acids are known as aluminosilicates.

PHYSICAL AND CHEMICAL PROPERTIES

Silicic acids are weakly acid substances. They are almost never found in ionic molecular forms, and the degree of their polymerization in colloidal solutions depends on the concentration and pH of the latter (acid media promote polymerization) and increases as the solution ages.

Silicates are built up of a fundamental structural unit, the silicon-oxygen tetrahedron $(SiO_4)^{4-}$. They vary greatly in their crystalline structure but may be classified into several major types according to the arrangement of the tetrahedra (individual tetrahedra, rings, infinite chains, infinite sheets, etc.). On hydration, silanol groups appear on the fracture surface of silicates (as they may on the SiO_2 surface). With the exception of the simple silicates of alkali metals, all silicates and aluminosilicates are virtually insoluble in water, although they give off certain amounts of metal ions and silicic acid to it. Silicates have more or less high melting points. See also Appendix and the section *Silicates*.

NATURAL OCCURRENCE

Feldspars, i.e. the aluminosilicates of potassium (orthoclase), sodium (albite), and calcium (anorthite), make up together more than a half of the earth's crust by weight. Their weathering products are silicon dioxide in the form of quartz (crystalline SiO_2) and kaolinite ($Al_2O_3 \cdot 2SiO_2 \cdot 2H_2O$). The latter constitutes the basis of common clays and in a purer state sometimes forms deposits of white clay (kaolin). See also the section *Silicates*.

PRODUCTION

Silicic acids are obtainable in the form of colloidal solutions by exposing silicon dioxide (q.v.) to water; much more concentrated solutions can be obtained by treating a sodium silicate solution (water glass) with dilute sulfuric acid on an ion exchange resin. See also the section *Silicates*.

USES AND OCCUPATIONAL EXPOSURE

Colloidal silicic acid solutions in concentrations of the order of 200–250 g/L (silica sol) are used to provide protective or (when mixed with milled quartz glass, zircon, or other mineral grains) refractory coatings (e.g. for tubes and steel molds).

Among the natural silicates, of the greatest industrial importance are asbestos (q.v.), talc (q.v.), and olivines (q.v.), while the most important man-made silicates are glasses. The composition of ordinary glass can be represented by the formula $Na_2O \cdot CaO \cdot SiO_2$. By suitably varying this basic composition, in particular by partially substituting other elements for the sodium, calcium, and silicon, special kinds of glass can be obtained with properties required for particular applications.

Of the natural aluminosilicates, industrial use is made especially of clays, micas (q.v.), nepheline (q.v.), and zeolites (q.v.).

Occupational exposure to silicate-containing dusts frequently occurs in many industries because natural silicates are the ores of various metals (lithium, beryllium, nickel, rare metals) and are widely used on account of their own valuable properties (asbestos, talc, micas, clays, gem stones) or as raw materials for the production of refractories and other artificial silicates and as building materials (e.g. silicate-containing rocks such as granite).

Artificial silicates (synthetic asbestos and mica, vitreous mineral fibers, special-purpose glasses) are manufactured as commercial products. In addition, many silicates similar in composition to silicate minerals (mullite, forsterite, volcanic glass, etc.) become constituents of various refractory materials, cements, metallurgical slags, and building materials — mainly as a result of high-temperature manufacturing processes. See also the section *Silicates*.

TOXICITY

Animals

The lungs of rats administered a colloidal silicic acid solution intratracheally developed inflammatory changes that progressed to focal or diffuse sclerosis, but no typical nodules seen in experimental silicosis (q.v.) were observed [Belobragina]. Some manifestations of the systemic toxic action of silica, notably after the administration of relatively readily soluble particles of its condensation aerosols, appear to be associated with absorption of silicic acid. The latter is excreted in increased amounts with the urine in both humans and animals with silicosis and also following the introduction of silicates or elemental silicon into the lungs. With increasing polymerization of the colloidal silicic acid solution, its toxicity was found to increase initially and decrease subsequently as the polymerization proceeds [Siehoff & Antweiler].

In chronic inhalation studies on rats, exposure to an atomized colloidal silicic acid solution (i.e. silica sol), induced typical silicotic nodules (see *Silicon dioxide*) much more rapidly than did inhaled quartz dust, although the quartz-induced silicosis was more marked in the long term [Podgaiko *et al.*]. A characteristic feature of the silicosis caused by silica sol is its substantial regression after cessation of exposure. Mixtures of silica sol and zircon were less potent in inducing silicosis than silica sol alone, whereas those of silica sol and quartz glass were more potent.

Man

Effects of silicic acid (in the form of silica sol) on man have not been described, but the systemic toxicity of silicates is generally very low. When their dust particles are deposited and retained in the lungs, local lesions of the pneumoconiosis and chronic dust bronchitis types develop, accompanied by systemic manifestations similar to those seen in these conditions (see *Silicon dioxide*). Since the fibrotic response of the lung (dust pneumosclerosis) is a major consequence of dust deposition in this organ, the relative pneumoconiosis-inducing potencies of silicates are often assessed in terms of their relative fibrogenicity. The severity of fibrogenic effects, however, does not correlate with the amount of silicic acid given off into solution.

There is evidence indicating that the fibrogenicity of certain silicates depends on their crystalline structure [Velichkovsky & Katsnelson], but the fibrogenicity of any silicate is lower than that of free silica. Accordingly, the pneumoconioses caused be silicates (silicatoses) run a more favorable course than silicoses. In particular, they progress more slowly even in individuals exposed to much higher concentrations of their dusts, and are much less frequently complicated by tuberculosis. In other respects, the roentgenologic appearance, clinical picture, and course of silicatoses are generally similar to those of silicosis in its interstitial form [Kovnatsky], as is the staging of silicatoses (see *Silicon dioxide*). However, the notion that silicatoses are clinically benign conditions does not apply to all silicatoses and is certainly inapplicable to asbestosis (see the section on asbestos). Moreover, in some silicatoses, even those marked by relatively slow progression, functional abnormalities in the respiratory and circulatory systems may be even more pronounced than in cases of typical silicosis [Kovnatsky]. This fact may be explained, on the one hand, by the occurrence of peribronchial and perivascular changes in the diffuse sclerotic form of pneumoconiosis caused by silicates and, on the other, by the still frequent presence of dusts at fairly high concentrations in industries where silicates are produced or used because less attention may be given in these industries to dust control measures as compared to those where the silicosis hazard is known to be high; as a result, the deposition of silicate dust in increased amounts promotes the development of chronic bronchitis which aggravates functional abnormalities in patients with silicatosis.

It should be appreciated that many silicate dusts, especially those from weathered aluminosilicate rocks, contain a certain percentage of free silica that may add substantially to their fibrogenic potential and so alter to some extent the nature of the pneumoconiosis (which is sometimes referred to as 'silicosilicatosis'). Pneumoconiosis in workers engaged in chamotte production is not infrequently complicated by tuberculosis and shows a marked propensity to progression, being often even more severe in terms of functional derangements than the silicosis in workers producing dinas refractories containing as much as 90% or more of free silica.

The fibrogenicity of certain silicate building materials is enhanced by firing because this increases both the content of free silica and the specific surface of the dust particles. Thus, ceramsite (claydite) is more fibrogenic than the Cambrian clay from which it is obtained; expanded perlite is more fibrogenic than the parent rock which is largely composed of volcanic glass; and expanded vermiculite is more fibrogenic than the crude vermiculite which is a hydromica [Oksova *et al.*]. See also the section *Silicates*.

HYGIENIC STANDARDS

In the USSR, an occupational exposure limit (MAC_{wz}) of 1 mg/m^3 has been adopted for disintegration aerosols of colloidal silicic acid solutions (silica sols) as well as those of silica sols containing ground quartz glass. For the condensation aerosols of silica sols containing zircon, the MAC_{wz} is set at 2 mg/m^3, and this is also the MAC_{wz} for dusts consisting largely of silicates and containing >10% free silica (dusts of chamotte and raw mica, mixed clay — chamotte — graphite dusts, etc.). See also the section *Silicates*.

DETERMINATION IN AIR

The indentification of individual silicates present in dust is a rather difficult matter (see, e.g., *Asbestos*) and their content is often estimated indirectly from the difference between the amounts of total and free SiO_2. For the determination of total SiO_2, the sample is fused with soda, potash, or their mixture, thereby transferring both the free SiO_2 and that within the silicates to soluble sodium and potassium silicates. These are then determined in solution by the formation of a yellow complex using ammonium molybdate; to make the assay more sensitive, the yellow complex is reduced to a blue one with ascorbic acid.

II. Silicates

Asbestos; zeolites; talc; micas; olivines; nepheline; man-made silicate fibers; silicate-containing and silicate building materials; soluble glass; ultramarine

ASBESTOS

'Asbestos' is a collective name given to a large number of silicates that are fibrous varieties of certain natural or man-made minerals.

Natural asbestos

Asbestos is widely distributed in the lithosphere and is found in many soils. Important asbestos minerals are contained in two mineral groups — amphibole and serpentine, both groups being hydrated silicates with complex crystal structures. Four classes of amphibole-group minerals have been distinguished [Andreyev] : alkaline (crocidolite, riebeckite, rhodusite, magnesiarfvedsonite), alkali-containing (richterite), calcium (tremolite, actinolite), and magnesial or magnesium-iron (amosite, cummingtonite, anthophyllite). Of the major industrial significance are anthophyllite, crocidolite (or riebeckite which is closely related to the latter in properties), and, outside the USSR, amosite. The serpentine-group minerals include alpha- and beta-chrysotiles (mountain flax) and the rarely encountered picrolite. Chrysotile is the most abundant and economically important form of asbestos. A quarter of a century ago, approximately 95% of all asbestos mined in the world was chrysotile asbestos [Fedoseyev *et al.*].

Man-made asbestos

Asbestos is usually prepared artificially by crystallization of silicates from their fluorine-containing melts at normal atmospheric pressure or under hydrothermal conditions at 300–550°C and very high pressures (up to 1000 atmospheres). In addition to analogues of natural asbestos minerals, various fibrous silicates differring in composition and properties from natural asbestos varieties have been obtained. Artificial asbestos have been classified into fluorine amphiboles, hydroxyl amphiboles, and chrysotile.

Structure and properties

Naturally occurring varieties of asbestos are silicates of various metals such as magnesium (MgO makes up, by weight, about 40% of chrysotile and 4% to 30% of amphibole asbestos of various types), iron (which is most abundant in amosite), calcium (especially in tremolite and actinolite), and sodium (especially in crocidolite); admixtures of other cations may be present.

Silica is not usually present in the free state but may be released in an amorphous form or as cristobalite on heating the asbestos to temperatures that cause its decomposition. The content of crystallization water ranges from 0.2–3.5% (tremolite, actinolite) to 12–14% (chrysotile).

The crystalline lattice of amphiboles comprises continuous double chains of anions of the composition Si_4O_{11}, cross-linked by bridging cations of the corresponding metals (the principal cations being magnesium, iron, calcium, and sodium) that form, together with the oxygen and hydroxyl, a band of so-called brucite structure (brucite being a fibrous mineral consisting of magnesium hydroxide).

The breakage (both parting and longitudinal splitting) of amphiboles occurs along defined crystallographic planes. The amphibole structure is such as to allow great latitude in cation replacement, and consequently the chemical composition and physical properties of different amphibole asbestos types vary widely [Langer & Nolan].

Chrysotile is a sheet silicate composed of planar-lined silicon—oxygen tetrahedra with an overlying layer of brucite. These silicon—brucite sheets are slightly warped because of a structural mismatch, as a result of which chrysotile fibrils are tubes in the form of a rolled scroll. The fibrils tend to be long and curved and to form bundles [WHO, 1986].

Asbestos fiber bundles can split lengthwise into very fine fibrils. In the case of amosite and crocidolite the fibrils are straight rods about 0.4 and 0.2 μm in diameter, respectively. The fibrils of chrysolite are about 0.16 μm in diameter.

Asbestos fibers are characterized by high tensile strength and flexibility. They do not alter their structure and properties at relatively high temperatures, are poor heat conductors, resist attack by acids (with the exception of chrysotile) and alkalies, and show high adsorptive capacities.

Uses

Well over 1000 uses for asbestos have been described, and there are wide variations in the pattern of its use in various countries. In some countries, for example, the production and application of some asbestos products has been discontinued (in part, at least) because of serious health risks associated with their production [WHO, 1986].

The greatest use of asbestos fibers occurs in the manufacture of composites such as asbestos cement (used for building purposes and for various pipes), friction materials (brake linings, clutch facings, etc.), fire-resistant insulation boards, millboard and paper, reinforced plastics, vinyl tiles and sheets, jointings and packings. Asbestos products are also widely used in conjunction with water as insulating plasters, cement, or spray mixtures.

Asbestos can be spun into yarn and woven into cloth, and the resulting textile products can be used for further processing into friction materials, packings, and laminates, or they may find direct applications such as insulation cloth, protective clothing, fire protection, and electric insulation.

Man-made sources of exposure

The main industrial sources of exposure to asbestos fibers, principally in airborne dusts, are the mining and milling of asbestos and the production, transportation, handling, use, and disposal of various asbestos or asbestos-containing products (e.g. brake shoes, thermal insulation materials, floor tiles, cement articles). The percentage of fibrous asbestos particles in the airborne dust may be particularly high in the asbestos textile industry (which is the dustiest of all asbestos-manufacturing activities) and in areas where finished asbestos products are packed or unpacked. Special problems are created by asbestos fibers entering the atmosphere from weathered asbestos-cement building materials and from worn-out braking shoes and also by those introduced into water from asbestos-cement piping and asbestos-containing minerals or in industrial effluents.

Health effects and their mechanisms

On direct contact with skin, asbestos fibers can penetrate the epidermis which may respond by hyperkeratosis and cellular proliferation with giant-cell formation resulting in the appearance of asbestos 'warts'. These are seen most commonly among asbestos textile workers and tend to occur on the flexor surfaces of extremities. In connection with the frequent presence of asbestos fibers in drinking water and many beverages, attention was called some time ago to the experimentally demonstrated ability of chrysotile to penetrate into the bloodstream from the stomach and be deposited in many organs [Pontefract & Cunningham]. However, by far the greatest health risks arise from the inhalation of asbestos dust that, when deposited in the airways, may lead to asbestosis — the severest variety of silicatosis closely associated with lung cancer and pleural mesothelioma.

The deposition, clearance, and retention of asbestos particles differ substantially from these processes observed with nonfibrous (spherical or granular) particles — largely because the sedimentation of fibers does not follow Stokes' law; in addition, fibers (particularly those >5 μm in length) are more likely than spherical particles to be deposited in bronchi and bronchioles by interception, mainly at bifurcations [Gross, 1981; Timbrell; Timbrell & Skidmore]. Respirable fibers, i.e. those capable of greater penetration and deposition in distal portions of the lung, are considered to be fibers not more than 3 μm in diameter, although they may reach up to 100 μm (or even more) in length; in general, however, the deposition efficiency of fibers decreases with increases in both their diameter and their length.

The deposition efficiency of various asbestos fibers also depends on their ability to curl up and form coils. For example, the deposition of the curly chrysotile fibers in the bronchial airway was found to be about 5 times greater than that observed for the straighter amphibole fibers [Timbrell]. This property also appears to determine in large measure the ability of fibers to enter the pleural cavity in lymph flow — an ability that was found to be most marked in the case of durable and relatively thick crocidolite fibers [Pott & Friedrichs].

There is considerable evidence that both the degree of primary deposition of dust in the lungs and its clearance rate from them may be different for the same asbestos and for different varieties of asbestos; other conditions being equal, amphibole asbestos dust usually accumulates in the lungs in greater quantities than chrysotile dust [Wagner & Skidmore]. However, since chrysotile fibers are capable of splitting longitudinally into much finer fibrils, the number of its particles in the lungs may be much greater than that of amphibole particles.

The fate of fibers deposited on surfaces within the lungs depends on both the site of deposition and the characteristics of the fibers. Fibers deposited in the tracheobronchial airways can be swallowed. In the nonciliated airspaces below the terminal bronchioles fibers are cleared much more slowly from their deposition sites by various less effective mechanisms such as translocation and disintegration.

Longer fibers are cleared less readily than shorter ones. When rats were exposed for 6 weeks to various asbestos dusts, the clearance rate of chrysotile was higher by a factor of 3 than that for amosite and crocidolite and, moreover, the retention of chrysotile, as measured a few days after the end of the exposure period, was only about one-third that of the amphiboles [Wagner & Skidmore].

Inhalation of long amosite fibers led to considerably more severe changes in the lungs of animals than that of its short fibers [Timbrell & Skidmore]. This is just one example of numerous studies in which relatively long asbestos fibers have been demonstrated to be more hazardous, which may explain why a number of countries have set exposure limits only for fibers longer than 5 μm.

In rats chronically inhaling amosite dust, the retention of its particles in the respiratory tract resulted in obvious damage to the dust clearance mechanisms after a certain period, leading to a progressive increase in the mass of dust in the lungs [Bolton et al.]. Mathematical analysis of epidemiologic data also indicates that once it has accumulated in human lungs to a certain level, the asbestos dust virtually fails to be cleared [Finkelstein]. Breakdown of the pulmonary clearance mechanism favored the development of 'asbestos cancer' in rats [Gross *et al.*].

How exactly asbestos exerts its fibrogenic action has not been elucidated. There are several facts militating against the hypothesis that the fibrogenicity of asbestos is due to a traumatizing action of its longer fibers on lung tissue. *First*, fibrogenic activity is possessed by both long and short chrysotile fibers [Kogan, 1970] and even by the nonfibrous magnesium silicate serpentine which has a similar chemical structure [Kogan, 1975]; *second*, the fibrogenic potential of asbestos particles coated by bakelite, rubber, or cement is low [Kogan, 1975]; *third*, traumatizing action is unlikely to be exerted by particles that are predominantly located intracellularly and/or surrounded by a special capsule (asbestos bodies — see below). Predominantly intracellular location is characteristic even of the longest asbestos particles among those accumulating in the lungs of experimental animals [Davis, 1965–1981; Holt & Young]. On the other hand, such particles cannot be effectively phagocytized: they pierce, as it were, the cell membrane of macrophages causing the release of their enzymes that are possibly damaging to the tissue — a mechanism to which some investigators attach much importance, particularly in considering the blastomogenic action of asbestos [Maroudas *et al.*]. A further argument against the 'traumatization hypotheses' is the existence of a well-defined parallelism between the fibrogenicity and cytotoxicity of asbestos. Thus, chrysotile is more fibrogenic than amphibole asbestos varieties, and it is also more damaging both to macrophage cell cultures and to erythrocyte suspensions. Both the fibrogenic and the cytotoxic potential of chrysotile can be substantially reduced by calcining it at 600–800°C [Gross, 1981; Kogan, 1970].

A proportion of longer fibers, especially amphiboles, retained in the lungs become coated with a deposit of gel-like iron-containing protein, forming so-called 'asbestos bodies'. These extracellularly located bodies arise during the intracellular residence of asbestos fibers or (in the case of very long fibers) from cytoplasmic fragments left on the fiber after an unsuccessful 'attempt' by the macrophage to engulf it [Davis *et al.*]. No etiologic significance has been ascribed to the formation of asbestos bodies, but their biologic role is likely to consist in isolating the aggressive surface of the asbestos fiber. Such bodies have been found in large numbers in individuals occupationally exposed to asbestos and also in the lungs of most adults who have lived in urban areas [WHO, 1986].

Experimental asbestosis

The biologic activity of chrysotile asbestos and, to a lesser extent, that of certain amphibole asbestos varieties has been investigated on animals (mainly rats and guinea-pigs) exposed to their dusts by the intratracheal or inhalation route [e.g., Davis, 1981; Kogan, 1975; Wagner; Wagner & Skidmore]. The main morphologic events observed in their lungs following asbestos dust deposition included phagocytosis of dust particles by macrophages; diffuse or focal proliferation of histiocytic and lymphoid elements, not infrequently with a giant-cell response; and the gradual formation of reticulin and then collagen fibers eventually resulting in pneumosclerosis, predominantly of a diffuse type.

These animal models, however, reproduce human asbestosis less well than do the animal models of silicosis (see *Silicon dioxide*), and do not give an adequate concept of its severity and progression. For example, no progression of pulmonary changes could be detected in rats over periods up to 24 months after relatively small quantities of chrysotile asbestos had been deposited in their lungs: the initial cellular response led to only a minimal focal collagen sclerosis, and the amount of asbestos in the lungs gradually decreased [Gross & de Treville]. Asbestos-exposed hamsters showed a progressive cellular reaction with a much less efficient dust clearance from the lungs than in rats but with the stroma always retaining its precollagen character [Gross & de Treville], and this pathologic process, too, is not representative of the human disease which is marked by collagen fiber formation. In chronic inhalation studies, rats contained less collagen in their lungs after 9 months of exposure to a fairly large concentration of chrysotile dust than they did after 3 months of the exposure period when the collagen content was significantly higher than in control animals [Kogan, 1970]. Such a time course, which is quite uncharacteristic for experimental silicosis, also exemplifies the failure of animal models to reproduce the progressive nature of pneumosclerosis seen in human asbestosis.

Human asbestosis

Inhalation of asbestos dust can cause asbestosis, changes in one or both surfaces of the pleura, bronchial carcinoma (lung cancer), mesothelioma of the pleura and peritoneum, and possibly cancer in other sites [Friedberg & Schiller].

It is not known for certain whether or not asbestos fiber type influences the frequency or severity of pulmonary fibrosis, although there is evidence that workers in the textile industry, in mining and milling, and in the manufacture of friction products are exposed to higher risks than other asbestos workers. The content of mineral fibers in the lungs of asbestos textile workers was reported to have ranged from 1 to 10 g/kg dry weight, as compared to levels of about 0.3 g/kg dry weight in the general population [Beattie & Knox].

The hallmark of asbestosis is a diffuse fibrosis (which is most marked in the middle and lower portions of the lung) with substantial dystrophic (degenerative) and sclerotic changes in bronchial and vessel walls leading to their deformation; emphysema is common. Asbestos bodies are often present in large numbers. A frequent and early sign of asbestosis well recognizable radiologically, is pleural thickening, either circumscribed (in the form of so-called plaques) or diffuse, with evidence of hyaline degeneration and (at a later stage) calcification. Distinctive fibrotic pleural lesions may also occur. These pleural changes are not uncommonly referred to together as 'pleural asbestosis' in contradistinction to 'parenchymal asbestosis'. Radiologic and clinical presentations of the latter are similar to those of silicosis, particularly in its diffuse fibrotic stage (see *Silicon dioxide*).

The principal radiographic signs of asbestoses are small irregular opacities with or without pleural thickening. The earliest changes usually occur at the lung bases with the appearance of small irregular (linear) opacities superimposed on the normal branching architechture of the lung. As the disease advances, the opacities become larger and more profuse, and shrinkage of the lung occurs with elevation of the diaphragm [Gilson]. The radiographic appearances are well illustrated in the set of standard films provided with the ILO 1980 *International Classification of Radiographs of Pneumoconioses* (see *Silicon dioxide*) which is recommended for use in making the diagnosis. As with silicosis, chest radiography remains the most important diagnostic tool.

Clinical symptoms and signs of asbestosis are similar to those of silicosis but manifestations of bronchitis, peribronchitis, and emphysema are somewhat more pronounced. Fine rales and crepitations may be found on physical examination as an early sign, and breathlessness on exertion may be the first symptom. Cough, with little sputum, may be present, especially in more advanced cases when distressing paroxysms often occur. Functional respiratory and circulatory abnormalities (the latter being mainly associated with an increased load on the left ventricle) often fail, as they do in silicosis, to correlate with the radiographically defined stage of asbestosis, although morphologic and consequently radiographic changes are usually more pronounced than in silicosis of the same stage, especially if pleural changes are present [Becklake *et al.*].

Information on the frequency with which asbestosis is complicated by tuberculosis is inconclusive. Some authors contend that there is a close association between the two, while others point to a substantial weakening of this association in recent decades. Kogan *et al.* [1971], for instance, reported almost a threefold excess of mortality from tuberculosis as compared to other causes of death among asbestos miners and millers who died in the period 1948–67. The relative risks of tuberculosis in asbestoses caused by different asbestos varieties have been given little study, but amphiboles are likely to be less hazardous in this respect than other asbestos minerals. For example, no tuberculosis was found in 25 cases of anthophyllite asbestosis detected among 100 individuals working in a very dusty environment [Bunimovich].

Asbestos and neoplasms

The principal asbestos-related hazards for man are two types of respiratory cancer: bronchial carcinoma (lung cancer) and mesothelioma which most often affects the pleural surfaces but may also occur in the peritoneum [Friedberg & Schiller].

Lung cancer has been variously reported to occur in 7.5% to 50% of patients with asbestosis. Cancers of all sites ranked first among causes of deaths in 90 individuals with chrysotile asbestosis [Kogan *et al.*, 1971] and, in contrast to silicosis, not only in males, but also in females. Exposure to asbestos also increases the risk of lung cancer in those who have not yet developed detectable signs of asbestosis [e.g. Ehrenreich *et al.*; McDonald *et al.*; Kogan *et al.*, 1972]. Occupational groups exposed to asbestos also show increased mortality rates due to cancers of other locations, primarily the digestive system. During the period 1967–71, for example, of 5119 insulation workers in the USA who had been exposed to asbestos for at least 20 years by that period, 408 died from cancers of all locations, including 191 from lung cancer, 15 from gastric cancer, 12 from esophageal cancer, and 23 from intestinal cancer — mortality rates that are 5.1, 2.7, 4.3, and 1.5 times higher, respectively, than the expected rates due to cancer affecting these organs [Ehrenreich *et al.*].

Cancer mortality depends on several factors. One of them is duration of exposure to asbestos (e.g., the excess mortality from lung cancer was only 3.1 among insulation workers with less than 20 years of exposure). Another is the level (intensity) of exposure determined by dustiness of the work environment (among English asbestos-textile workers exposed for more than 2 years and followed up for 17–31 years, the ratio of observed to expected deaths from all cancers was close to 1.0 for low to moderate exposures but 7.3 for lung and pleural cancers and 2.7 for cancers of other sites in cases of heavy exposure [Newhouse]. An important factor in determining mortality is also considered to be the period elapsing since first exposure [Finkelstein; Liddell & Hanley; Selikoff]. A major determinant of cancer mortality is the content of asbestos fibers in the dust. Chrysotile millers, for example, were found to experience a higher mortality than chrysotile miners [Kogan *et al.*, 1972].

The risk of lung cancer in asbestos workers is increased by cigarette smoking. Evidence was provided that the effects of asbestos and smoking on the risk of lung cancer might be multiplicative [Selikoff *et al.*, 1968], but more recent epidemiologic studies have indicated that the joint effect of these two carcinogenic agents is likely to be additive (or somewhat more than additive) rather than multiplicative [Berry *et al.*; Liddell *et al.*].

In addition to lung cancer, of considerable concern has been the occurrence of pleural or peritoneal mesotheliomas as a result of asbestos exposure. Indeed, a large majority of these highly malignant tumors (e.g. 65–85% [Hain *et al.*]) have been associated with a history of such exposure. Although a rare tumor in the general population, mesothelioma gives mortality rates in certain occupational groups comparable to those due to lung cancer [Ehrenreich *et al.*; Kogan & Berzin; McDonald *et al.*]. These tumors have been usually observed in individuals exposed to amphibole asbestos (mainly crocidolite), either alone or in amphibole-chrysotile mixtures; they seldom followed exposure to chrysotile asbestos only [Kogan *et al.*, 1971; McDonald *et al.*; Thomas *et al.*; WHO, 1986], although the tumor yield in rats was fairly high when chrysotile was administered intrapleurally. The higher mesothelioma incidence seen with amphiboles appears to be due to the heightened capacity of their particles for intrapleural penetration. Exposure to amphiboles is also frequently associated with non-neoplastic pathologic changes (so-called pleural plaques) in the absence of roentgenologic signs of pulmonary fibrosis, but these signs are added subsequently more often than in individuals without such changes [McMillan & Rossiter]. The risk of pleural mesothelioma, unlike that of lung cancer, has been shown not to depend on smoking [Berry *et al.*].

There is indirect experimental evidence that the relative carcinogenic potencies of different asbestos varieties depend on differences in the levels of certain trace elements contained in them, including such known carcinogens as, for example, cobalt, chromium, and nickel [Gross *et al.*]. Samples of different crude long-fibered asbestos types and those of airborne dusts collected after their milling were both found to have unequal contents of trace elements. Thus, chrysotile samples contained much more chromium, cobalt, and nickel and much less manganese and iron than did crocidolite and amosite samples which did not differ from one another in the content of the former three metals [Roy-Chowdhury *et al.*].

It has also been suggested that the carcinogenicity of asbestos may be associated with the traces of polycyclic aromatic hydrocarbons (including benz(a)pyrene) contained therein, which have been shown capable of being adsorbed onto the asbestos from the air or from some types of container [Harington & Roe; Pylev & Krivosheina]. Animal experiments provided evidence for increased carcinogenicity of benz(a)pyrene following its intratracheal administration with asbestos [Pylev]. The blastomogenic (and other) hazards of asbestos are reviewed in greater detail in *IARC Monographs on the Evaluation of Carcinogenic Risk of Chemicals to Humans* [IARC, 1977] and in *Asbestos and Other Natural Mineral Fibers* [WHO, 1986]. Information on carcinogenic activity of synthetic chrysotiles of various fiber sizes and chemical compositions can be found in the paper by Vasiliev *et al.* [1989].

Hygienic standards

Occupational exposure limits adopted in the USSR for asbestos and asbestos products are given in Table 1. In the USA, the following TLV-TWAs have adopted for asbestos minerals dusts: 0.2 fiber more than 5 μm in length per cm^3 of air for crocidolite, 0.5 fiber >5 μm/cm^3 for amosite, and 2 fibers >5 μm/cm for chrysotile and other forms of asbestos.

In East European countries, a maximum permissible concentration of 2.0 mg/m^3 has been set for chrysotile asbestos. In other countries, the following maximum permissible concentrations per cm^3 have been established for fibers >5 μm long: 1 for chrysotile in Canada and Switzerland; 0.2 for amosite and crocidolite and 2 for other amphibole asbestos forms in Great Britain; 1 for all asbestos forms in Germany.

Methods of determination

In air

The concentration of airborne asbestos particles is expressed either as fiber number concentration (i.e. the number of fibers per unit volume of air) or as mass concentration (i.e. the weight of particles per unit volume of air). A fiber is defined (especially for optical microscopy) as a particle having a length to diameter ratio (aspect ratio) $\geqslant 3$ and a length of $\geqslant 5$ µm; fibers with diameters $\leqslant 3$ µm are referred to as 'respirable' (the alveolar fraction) and those with diameters >3 µm as 'nonrespirable'.

Table 1. Exposure limits for asbestos and asbestos products

| | Workroom air | | | |
	MAC_{wz} (mg/m^3)	Aggregative state	Hazard class	Note
Asbestos, natural or man-made	2.0	Aerosol	3	Fibrogenic and carcinogenic
Asbestos rock dusts:				
containing >10% asbestos	2.0	Aerosol	3	Fibrogenic and
containing <10% asbestos	4.0	Aerosol	3	carcinogenic
Asbestos cements, including uncolored cements and colored ones containing $\leqslant 5\%$ MnO_2, $\leqslant 7\%$ CrO_3, and $\leqslant 10\%$ iron oxides	6.0	Aerosol	4	Fibrogenic
Asbestos bakelite and asbestos rubber	8.0	Aerosol	4	Fibrogenic

Fibers with diameters smaller than approximately 0.25 µm cannot be seen in the light microscope, and an electron microscope is necessary for counting and identifying them. Electron microscopes furnished with auxiliary equipment can give information on both the structure and elemental composition of asbestos fibers [WHO, 1986].

For the assessment of health risks from exposure to asbestos, the inspirable fraction as a whole is considered in measuring the concentration of airborne asbestos dust in some countries, while in the majority of countries only the alveolar fraction is measured. Fiber number concentrations are determined in most instances. In the USSR, however, total airborne mass concentrations are usually determined.

Chemical identification of individual silicates in asbestos-containing dust may be complicated by the fact that the asbestos and enclosing rock have similar or identical chemical compositions. Also, such methods of analysis as optical and electron microscopy, infrared spectrography, and X-ray diffraction analysis may be difficult to use because of the presence of other minerals or of particles of different size categories; moreover, different methods may give differing results.

Detailed description of the collection and preparation of samples and of various analytical procedures can be found in *Safety in the Use of Asbestos* [ILO].

In biologic tissues

For the recovery of asbestos and other mineral dusts from human tissues, a number of techniques have been developed, including wet chemistry methods (using, e.g., formamide, glacial acetic or other acids) and physical methods (e.g., ashing), for tissue destruction [WHO, 1986]. In addition, there is a carbon-extraction replication technique enabling asbestos fibers to be examined *in situ* in tissue slices by electron microscopic analysis [Langer *et al.*].

Determination of asbestos fibers in lung tissue may be of forensic-medical importance, for example in cases where exposure to asbestos is suspected of having been associated with death from cancer. For post-mortem examination of such cases, Ehrenreich *et al.* recommended the following sequence of procedures: ordinary optical microscopy of thin (5 μm) lung sections stained with hematoxylin and eosin, microscopy of spodograms (ash pictures) obtained after incineration of 175-μm sections, and electron microscopy of fine sections after their incineration and dusting them with carbon; in rare instances, electron diffraction and microchemical analysis using an electron probe may be also necessary. In may be difficult to identify asbestos in tissue because the fibers undergo fragmentation with time.

Measures to control exposure at the workplace

Suppression of airborne dust formation by engineering controls is the basis for preventing adverse health effects of asbestos in workplaces where it is mined, milled, or used. As in other industries where dust is a major health problem, the basic techniques are segregation and enclosure of dusty machinery, installation of local exhaust ventilation if openings in the equipment are unavoidable, application of moisture to materials, provision of adequate general ventilation, and filtration of the air. In recent decades, dust emissions in various branches of the asbestos industry have become much lower through the introduction of technical control measures.

In asbestos mills, unless dust emissions are well controlled, special care should be taken to ensure thorough cleaning of the workroom air (e.g. using electric or fabric filters); a very cost-effective method, particularly in regions with a cold climate, has proved to be one by which the air cleaned in hose filters is returned to the workroom atmosphere [Kogan, 1975]. Special attention should also be given to automating the packaging of end products and making this process as dust-free as possible. In this regard, substantial reductions of dust formation in the packaging departments of asbestos mills, as well as in the user industries, have been achieved through briquetting of milled asbestos.

Asbestos textile manufacturing remains the dustiest of all asbestos-manufacturing processes, although during the recent decades dust emissions have been substantially reduced thanks to technical control measures. An effective dust suppression measure is dumping of the fiber before spinning and weaving. A major reduction of dust emission in the first stage of textile production has resulted from the introduction of a wet process for forming the thread [Gilson].

Hazards are particularly high in manual handling of asbestos. Whenever possible, the asbestos should be made wet before working with it.

Substitution of asbestos by alternative safer materials should also be considered (legislative restrictions on the use of asbestos, especially for insulation, have been imposed in some countries).

Asbestos workers must be provided with appropriate protective clothing which should be laundered regularly. Respirators or dust masks have to be used where hazardous dust exposure cannot be avoided — as, for example, during the removal of old insulation and spraying new material.

Medical measures for preventing the adverse effects of asbestos dust are essentially the same as those described in the section on silicone dioxide (except for the use of poly-2-vinylpyridine-N-oxide). All workers should be informed of the nature of the hazard and of the possible methods of protection. They are subject to a pre-employment and subsequent periodic medical examinations which should include a chest x-ray and lung function tests. It is important to stress the need for a high 'index of suspicion' with regard to cancer during periodic examinations.

ZEOLITES

Identity
Zeolites comprise a group of various crystalline microporous aluminosilicates in which the primary building blocks are either silicon-oxygen tetrahedra or those in which Al^{3+} substitutes for Si^{4+} ion. The tetrahedra combine, linked together by oxygen bridges and cations, to yield ordered three-dimensional frameworks.

In addition to natural zeolites that vary with regard to the Si to Al ratio, there are also synthetic zeolites with predetermined ratios.

Physical and chemical properties
Distinction is made between high-silicon (the Si:Al ratio >10) and low-silicon (<6) zeolites. The former are highly resistant to heat and acids, while the latter are less resistant but are characterized by higher sorptive capacities that are determined both by their high specific surface areas (which may be more than 700 m^2/g) and by their cationic composition.

Zeolites are soluble in acid media and can pass into them — mainly the silicon, whose concentration was found to have reached 16% over 30 days [Tkachev *et al.*]. In some deposits, zeolites contain admixtures of lead, mercury, cadmium, iron, and/or arsenic (in concentrations of the order of $10^{-3}-10^{-5}\%$) [Tkachev *et al.*] and also adsorbed benz(a)pyrene [Pylev *et al.*].

Natural occurrence

Approximately 40 natural zeolites are known. Together with feldspars, clays, and silicon dioxide minerals, they are the most widely distributed rock constituents. The most abundant zeolites are ferrierite, phillipsite, mordenite, chabazite, clinoptilolite, and erionite, the latter four being the only zeolites having fibrous structure. In the USSR, about 60 zeolite deposits have been discovered, and the total projected reserves of clinoptilolite, mordenite, and phillipsite are around 3.5 billion tonnes.

Uses

Zeolites are used for drying, purifying, and desulfurizing petroleum products; as sorbents in the production of hydrogen, ammonia, unsaturated and aromatic hydrocarbons; for the removal of sulfur dioxide from industrial emissions into the atmosphere and of chemical and radioactive contaminants from wastewaters; in the manufacture of sulfuric acid; in agriculture as mineral additives to feeds, as deodorants of livestock stables, and for improving the quality of soil; in drinking-water purification.

Toxicity

The presence of admixtures mentioned above may limit the use of zeolites in agriculture and for water treatment, but special attention has been called to the possible carcinogenic and fibrogenic activity of certain zeolites.

Animals

A zeolitized tufa from the Liutog deposit in Russia administered in single oral doses up to 44.4 mg/kg did not cause any deaths among mice or young (broiler) chickens, nor were any adverse effects detectable in these animals during or after long-term ingestion of this mineral in the diet at levels of 3 to 25% by weight [Vedmitsky *et al.*]. The lungs of rats given zeolites intratracheally in the form of suspension presented a typical histologic picture of experimental silicatosis whose severity differed with different zeolites.

The fibrogenicity of zeolites appears to be mainly due to their cytotoxicity which is displayed by both natural and synthetic zeolites. The cytotoxic activity of most zeolite samples from 11 deposits in Siberia and the Far East was similar to that of chrysotile asbestos [Velichkovsky *et al.*], while some clinoptilolite samples closely resembled the highly cytotoxic quartz DQ_{12} in this respect [Korkina *et al.*]. That zeolites are cytotoxic is also evidenced by their high hemolytic activity [Deyeva *et al.*; Domnin *et al.*].

Erionite samples, particularly those from certain Turkish, North American, and Transcaucasian deposits, have been shown to be highly carcinogenic. In a chronic inhalation study where rats were exposed to eritonite from several sources at 10 mg/m^3 (7 h/day, 5 days/week, over 12 months), the incidence of mesothelioma in the surviving animals was as high as 96.4% [Wagner *et al.*]. Erionite has also proved to be extremely potent (more potent than chrysotile) in the induction of mesothelial tumors following intrapleural injection [Lipkin; Maltoni *et al.*; Ozelmi *et al.*; Pylev *et al.*; Spurny; Suzuki]. For example, the incidence of mesotheliomas was 100% in rats that had received erionite from Oregon and 95% in those given erionite collected near the village of Karain in Anatolia, Turkey [Wagner *et al.*]. A high incidence of malignant tumors was also observed among mice given erionite intraperitoneally [Suzuki].

A high mesothelioma-inducing capacity following injection by the intrapleural and intraperitoneal routes has been demonstrated for clinoptilolite and mordenite from Bulgarian sources, whereas the blastomogenic activity of Georgian (USSR) clinoptilolite was low [Pylev *et al.*]. No tumors were observed in mice given a mixture of fibrous and nonfibrous mordenite [Suzuki & Kohyama].

Some zeolites (notably that from the Pegas deposit in the USSR) displayed a mutagenic activity comparable to that of chrysotile, as indicated by micronuclear tests and by their ability to induce chromosomal aberrations [Domnin *et al.*; Snegireva *et al.*].

Man

No cases of pneumoconiosis have been reported, but the prevalence of chronic bronchitis was high among workers exposed to dusts of Bulgarian clinoptilolite [Nikolova, unpublished dissertation]. After a very high incidence of mesothelial tumors in some remote Anatolian villages (Turkey) was first reported by Baris in 1975, intensive environmental and epidemiologic studies have been undertaken. In the village of Karain, with a population of less than 600 in 1977, 42 cases of malignant mesothelioma were recorded during the previous 8 years, while in Tuzkoy, a large village of 2729 inhabitants 5 km away from Karain, at least 27 cases occurred in the period 1978–1980. Both sexes were equally affected and at relatively young age. These cases were all attributable to environmental exposure, from infancy, to fine erionite fibers that were contained in dusts from the local soil and were identified in the lung tissue of patients [Artvinli & Baris; Baris *et al.*, 1978]. A detailed review of experimental and epidemiologic data can be found in the IARC (1987) publication. Eritonite has been classified as a confirmed animal and human carcinogen.

Hygienic standard

An occupational exposure limit (MAC_{wz}) of 2 mg/m^3 has been set in the USSR for all natural and artificial zeolites.

TALC

Identity

Talc is a hydrous magnesium silicate of the approximate composition $Mg_3Si_4O_{10}(OH)_2$; iron or, less often, nickel may substitute for some magnesium ions. In practice the term 'talc' is applied to a group of multicomponent mineral mixtures (particularly those of silicates) containing various percentages of the talc proper.

Properties

True talc is a sheet silicate composed of continuous double layers of silicon-oxygen tetrahedra with a brucite layer interposed between them. Such triple sheets are linked up by weak Van der Waals bonds which explains the ability of soapstone (massive talc) to split into fine leaflets or flakes. For this reason, talc-containing dusts characteristically contain a more or less high proportion of foliated or flaky particles. Moreover, such dusts sometimes contain needle-like particles of tremolite (an accessory mineral found in talc deposits, though rarely in the USSR) or fibrous particles of amphibole asbestos. The other accessory minerals, listed below, and also quartz that is occasionally encountered as an admixture, give rise to particles of irregular granular form. The color of talc powders varies from white to gray depending on their mineral composition.

Natural occurrence

Talc occurs in foliated, granular, or fibrous masses in veins or other rock bodies, although rarely in a 'pure' form (as steatite or 'noble' talc). Much more often it is encountered in talcites that contain >75% of talc along with various silicates (chlorite, serpentinite, tremolite, actinolite, enstatite), carbonates (calcite, dolomite, brewnnerite, magnesite), or ore minerals (magnetite, hematite, pyrite, etc.). A large proportion of the mined talc is contributed by the so-called talc ores which contain >40% of talc but usually less than >75% [Romanovich].

Uses

Talc is used in a finely ground form (except for the use of steatite in ornamental carving). The ground talcites are not, as a rule, concentrated; if they are, the end product is only partially freed from the associated minerals. Ground talc is used as a stabilizing filler in paints, rubber, and paper; as a dielectric in the ceramic and cable industries; to prevent the sticking together of surfaces (e.g. in sheet-rolling, in rubber articles, and in the production of filaments); as an insecticide carrier; as adsorbing and absorbing agents in cosmetics, toilet preparations, and in the pharmaceutical industry; as a dusting agent in medicine. Soapstone, which is obtainable in large blocks and lends itself to cutting and sawing, is useful in the fabrication of smelting furnaces for paper mills, electrical switchboards, table tops, sinks, and crayons. Talc-chlorite and talc-magnesite ores are employed as refractory mortars.

Toxicity

It is generally recognized that prolonged inhalation of significant amounts of talc dusts, especially those containing tremolite, serpentinite (serpentine), anthrophylite, or similar talcose minerals, can produce symptomatic pneumoconiosis. As indicated by animal experiments [e.g. Bardin & Kusevitsky; Schepers & Durkan] and numerous clinical observations [e.g. Alivasates *et al.*; Bardin; Jaques & Bernischke; Kovnatsky], chronic exposure to talc dusts of various mineral compositions induces predominantly a diffuse sclerotic form of pneumoconiosis known as talcosis. At present, talcosis usually runs the course of a relatively benign silicatosis and, in particular, is rarely seen to proceed to an advanced stage (II or III) or to be complicated by tuberculosis — even in industries where advanced and complicated forms of talcosis were common in the past [Boitsov & Bakhireva; Vagina *et al.*].

The main symptoms and signs of talcosis are productive cough, progressive shortness of breath, diminished breath sounds, limited chest expansion, diffuse rales, and sometimes clubbing of the finger tips [Kleinfeld, 1983]. In cases where talcosis and asbestosis present similar roentgenologic appearances in their early phases, functional respiratory and cardiovascular disturbances are less marked in talcosis [Vagina]. The reported cases of 'acute talcosis' developing in workers after 1 to 5 years of talc exposure [Alivasates *et al.*] as well as the view previously taken by some investigators that talcosis is a disease similar to asbestosis (in severity, the rate of progression, and the tendency to be complicated by tuberculosis), can probably be explained by very high exposures to talc dust in the past. The occurrence of such cases could not be attributed to any qualitative characteristics of the responsible dust, such as admixtures of quartz [Alivasates *et al.*] or fibrous particles [Takahashi *et al.*]. However, postmortem studies in humans [Einbrodt *et al.*] and animal experiments have provided clear evidence that the presence of quartz in talc dust did increase fibrogenicity of the latter, and the term 'silicotalcosis' has therefore been proposed to designate the silicatosis cased by such dust [Spiegel].

There is also a view that pure talc dust is biologically 'inert' and that no hazard of talcosis exists unless the dust also contains needle-like tremolite particles [Schepers]. This view, however, has not received support from clinical or experimental studies. Thus, cases of talcosis have occurred among workers of talc factories exposed to talc that was not admixed with tremolite. In an experimental study carried out to assess quantitatively the fibrogenicity of various dusts (those of talc, tremolite asbestos, nonasbestiform tremolite, quartz, and mixtures consisting of talc and 20% of one of these three minerals), talc was found to be more fibrogenic than any of these mineral admixtures with the exception of quartz; the mixtures of talc and tremolite or tremolite asbestos were less fibrogenic than the pure talc dust [Boitsov]. On the other hand, symptomatic pneumoconiosis was observed in six workers exposed, for an average of 23 years, to talc dust admixed with tremolite and anthrophyllite and a small amount of free silica [Kleinfeld, 1983].

Exposure to industrial talc dust high in tremolite and anthrophyllite has been associated with increased carcinogenic hazards. Mortality from lung cancer and pleural mesothelioma among talc workers was reported to be 4 times that in a control population, although definite casual relationships between exposure to talc dusts and lung cancer have not been established [Kleinfeld, 1983; Kleinfeld *et al.*]. Increased risks of lung cancer, and also of cancer in other locations, is likely to exist also in talc factories in the absence of tremolite or amphibole asbestos [Boitsov; Boitsov & Katsnelson; Katsnelson & Mokronosova], but even in factories with very high concentrations of airborne talc dust (which were not uncommon in the past) these risks were significantly lower than in asbestos factories with much lower dust concentrations. When effective dust control measures were applied in a factory processing tremolite talc, the oncogenic risk was reduced virtually to zero [Kleinfeld, 1973].

In experiments where various types of asbestos varieties or talc containing 50% tremolite were implanted intrapleurally into hamsters, mesotheliomas developed only in animals receiving asbestos [Smith]. It has been suggested that the high mortality from gastric cancer seen in Japan could be associated with the consumption of rice that had been polished using tremolite-containing talc [Merliss].

Recently, a group of experts has concluded that only talcs containing asbestiform fibrous minerals present a real carcinogenic hazard to human beings, and that the available epidemiologic evidence regarding the carcinogenic risks of talc dusts that do not contain such minerals is inconclusive [IARC, 1987]. However, the validity of this conclusion appears questionable.

Hygienic standards

In the USSR, occupational exposure limits (MAC_{wz}) have been set at 4 mg/m^3 for talc-rock and talc dusts containing <10% quartz and at 2 mg/m^3 for those with a quartz content of >10%. In the USA, a TLV-TWA of 2 mg/m^3 has been adopted for respirable dust of non-asbestiform talc; for asbestiform (fibrous) talc, the threshold limit is set at 2 fibers/cm^3 for fibers >5 μm in length.

Measures to control exposure

These are similar to those described above for silicon dioxide and asbestos and should focus on dust control. Special attention is to be given to the provision of adequate exhaust ventilation, particularly in grinding and packing operations. Workers exposed to talc dusts should use respiratory protection devices and undergo periodic medical examinations.

MICAS

Identity

Micas comprise a group of aluminosilicates having a layer structure and readily separable into very thin smooth-surfaced leaves or flakes. The basic crystalline structure of micas is made up of silicon-oxygen tetrahedra in which one Si atom is replaced by an Al atom and which form double layers joined together into flat packets by brucite $[Mg(OH)_2]$ or hydrargillite $[Al(OH)_3]$ layers; between the packets, various cations (e.g. K, Mg, Fe, Li, Al) are located; accordingly, micas are classified into several categories including magnesium-iron (biotite, lepidomelane, sometimes phlogopite), lithium (lepidolite, zinnwaldite), and aluminum (muscovite, paragonite) varieties. As a result of weathering, micas lose their alkalies and are enriched with water, transforming into hydromicas; an example of hydromica is vermiculite, which is a weathering product of biotite.

Physical properties

Micas are heat resistant (with the exception of hydromicas) and have high specific resistance. Their thin sheets are usually transparent.

Natural occurrence

Micas are constituents of granites, syenites, diorites, gneisses, and some sedimentary rocks. Of industrial importance are principally deposits of large plates of muscovite, phlogopite, and biotite and also large accumulations of lithium micas and of vermiculite.

Uses

Muscovite and phlogopite are indispensable materials for electric insulation in electric, radio, and aircraft equipment. As electric insulators, use is also made of micanites (leaflets of bonded mica splittings), mica foils (similar leaflets bonded to paper), and mica tape paper (layers of split mica to which paper is glued on both sides). Lithium micas are used as raw materials for the production of lithium and some special glasses. Vermiculite expanded by heating is widely used as a heat insulating material, especially in the building industry.

Mica powder is used in the manufacture of electric cables, pneumatic tyres, welding electrodes, bituminized cardboard, and flameproof insulators. Man-made micas are also used.

Toxicity

Animals

In animals administered a fine muscovite dust intratracheally, the following evidence for pneumoconiotic activity of this dust was obtained: the appearance of abundant reticulin fibers and later (after 290 days) also of collagen fibers in the dust foci; well-defined transport of mica particles to mediastinal lymph nodes; and macrophagic hemosiderosis, which is regarded as a manifestation of hemolytic action of the dust. However, no completely collagenated nodules typical of silicosis were observed, even after a long time [Kaw & Zaidi]. In other studies, with muscovite and phlogopite, the authors came to the conclusion that these two minerals are both fibrogenic, though to a much lesser extent than free silica [Pushkina, 1960; Sedov et al.]. Expanded vermiculite was found to be appreciably more fibrogenic than crude vermiculites, and nodules resembling siliotic nodules were seen in the lungs of some exposed animals [Oksova et al.].

Man

Mica miners may be exposed to a wide variety of dusts including quartz, feldspar, and various silicates; in particular, many cases of silicosis have been observed in India, and the advanced stage of silicosis may obscure the specific effect of mica. When mica was prepared by dry grinding, the handling and packaging operations were very dusty, but this hazard has been substantially reduced by the introduction of wet grinding and by mechanization of work processes [Meiklejohn]. In one study, the content of free silica in the airborne dust during primary mica processing did not exceed 2.5% and no free silica was detectable in dust samples taken in the departments where the material was processed further [Sedov et al.].

Mica pneumoconiosis in persons exposed to mica dusts without any admixture of quartz or containing only small amounts of the latter has been described by a number of authors [e.g. Freidberg & Schiller; Heimann et al.; Pushkina; Sedov et al.], but this disease may be regarded as a relatively benign silicatosis. In one plant, for example, signs of pneumoconiosis were only detectable in workers after 20–25 years of employment involving mica exposure; of the 50 individuals diagnosed as having pneumoconiosis, 44 had disease of stage I and only two had coniotuberculosis [Sedov et al.]. At autopsy, 'mica bodies' resembling 'asbestos bodies' were observed.

Cericite (potash mica) — a variety of muscovite found in many ores — was once thought to be an important factor in the causation of silicosis and other pneumoconioses which were considered to have resulted from combined exposure to the dust and radiation because of the presence of the K^{40} isotope in the cericite dust [Flügge]. It has been shown, however, that even a cericite-rich mine dust deposited in the lungs of workers increased only slightly the content of K^{40} in the lung tissue [Vincent et al.]. That cericite is unlikely to present a serious silicosis hazard is also indicated by the very moderate fibrogenicity of muscovite and other potassium-containing micas. The fibrogenicity of quartz–cericite mine dusts correlates with their radioactivity, though only in the case of dusts with enhanced radioactivity due to admixtures of elements from the uranium series; even the most fibrogenic of such dusts proved to be less potent than a nonradioactive dust with a high (94%) quartz content [Engelbrecht *et al.*].

In addition to pneumoconiosis, cases of chronic bronchitis and occupational dermatosis among mica workers have been described [Sedov *et al.*].

Hygienic standards

In the USSR, occupational exposure limits (MAC_{wz}) have been set at 4 mg/m^3 for muscovite and phlogopite dusts and at 2 mg/m^3 for crude silica dusts containing $\geqslant 10\%$ free silica. In the USA, a TLV of 30 million particles per cubic foot has been adopted for mica dust.

OLIVINES

Identity

Olivines are magnesium, iron, and manganese orthosilicates or their isomorphous mixtures. In addition to the olivine proper [$(Mg,Fe)_2SiO_4$], this group includes forsterite (Mg_2SiO_4), fayalite (Fe_2SiO_4), knebelite [$(Fe,Mn)_2SiO_4$], monticellite ($Ca\,MgSiO_4$), tephroite (Mn_2SiO_4), and some other minerals. The crystalline lattice of olivine is built up of isolated silicon–oxygen tetrahedra and metal ions surrounded by six oxygen ions.

Natural occurrence

Olivines are constituents of ultrabasic and basic rocks (olivinites, dunites, diabases, basalts, etc.) with which deposits of several metals may be associated.

Uses

Olivinites and dunites are used as raw materials for the manufacture of forsterite refractories. The transparent olivine varieties chrysolite and peridot are used as gem stones. Minerals of the olivine group are components of open-hearth slags (fayalite, monticellite), blast-furnace slags (monticellite), ferromanganese smelting slags (tephroite), and of iron-ore agglomerates (fayalite).

Toxicity

As suggested by experimental studies reported by Borshchevsky, the olivine proper is one of the least fibrogenic silicates. This author also described autopsy findings consistent with moderate diffuse pulmonary sclerosis in a worker who had been exposed to high concentrations of olivine dusts over 14 years. Other silicates in the olivine group are also characterized by a low degree of aggressiveness, one indication of which is the greatly diminished cytotoxicity of chrysolite following its conversion to forsterite as a result of calcining. The fibrogenicity of forsterite refractories is also low, although it is higher than that of other magnesia refractories, apparently because they contain an admixture of free silica [Katsnelson et al.]. In a similar vein, the relatively low fibrogenic potency of fayalite-containing dusts of iron-ore agglomerates was found to increase appreciably if they contained even a small amount of free silica [Katsnelson].

Clinical features of the disease caused by dust inhalation have been described only for olivine proper. This disease (so-called olivinosis) is relatively benign, and no severe cases were observed even in a group of workers with olivinosis resulting from massive exposure (hundreds to thousands of milligrams of olivine dust per m^3 of air); the disease progressed very slowly and was seldom complicated by tuberculosis [Kovnatsky]. The severe functional abnormalities sometimes noted in olivine workers appear to have been associated with the chronic dust bronchitis caused by exposure to very high dust concentrations; similarly, the strongly marked roentgenologic manifestations of pneumoconiosis occasionally seen in such workers were due to high iron levels in the dust (45–50% as Feo).

Hygienic standards

In the USSR, occupational exposure limits (MAC_{wz}) are set at 6.0 mg/m^3 for olivine as such and for forsterite (as aerosols, hazard class 4, with the notation 'fibrogenic') and at 4.0 mg/m^3 for olivine-containing iron and nickel agglomerates (aerosols, hazard class 4).

NEPHELINE

Identity and occurrence

Nepheline is an aluminosilicate of sodium and potassium [$(Na,K)_2 \cdot Al_2O_3 \cdot 2SiO_2$] which is the major constituent of nepheline syenites and urtites as well as of apatite-nepheline and some other rocks. The principal impurities are CaO, Fe_2O_3, Cl, and H_2O.

Uses

Nepheline is used for the production of alumina, soda, potash, silica gel, Portland cement, glass, ultramarine, and some other substances. It also finds use as a fertilizer.

Toxicity

Experimental nepheline pneumoconiosis is characterized by moderate sclerosis and some irritant effects of the dust [Borshchevsky; Domnin & Solomina; Gasparian]. (See also the section on silicic acids; Barrie & Gosselin; Kovnatsky.)

Hygienic standard

In the USSR, an occupational exposure limit (MAC_{wz}) of 6.0 mg /m^3 has been adopted for nepheline concentrates and for syenite (as aerosols, hazard class 4, with the notation 'fibrogenic').

MAN-MADE SILICATE FIBERS

Identity

Man-made silicate fibers, which outside the USSR are usually referred to as man-made mineral fibers (MMMF), are amorphous silicates or aluminosilicates manufactured from glass, slag, natural rock, or other minerals. Most of them have a vitreous structure and are therefore called man-made vitreous fibers.

The nomenclature of MMMF is not consistent or well defined. Thus their categories and materials are variously named with reference to use, structure, raw material, or manufacturing process, and many trade names are also in use. However, MMMF have been classified into four broad groups: continuous filaments, insulation wools (including rock/slag wool, and glass wool), refractory (including ceramic) fibers, and special-purpose fibers [WHO, 1988].

Properties

Man-made mineral fibers possess high mechanical strength, chemical and heat resistance, and (depending on the fiber type) excellent properties as electric, thermal, acoustic, and/or vibration insulations. The most important physical property of these fibers is that, in contrast to naturally occurring fibers such as asbestos, they cannot fracture longitudinally but may break transversely into shorter segments.

Chemical resistance of MMMF and their solubility in aqueous and physiologic solutions vary according to their chemical composition and fiber size distribution. The solubility increases with increasing alkali content, and fine fibers degrade more rapidly than coarse ones. Finer fibers also have a lower thermal conductivity. Melting points of MMMF have been reported to range from 1700 to 2600°C and their densities, from about 2.5 to 4 g/cm^3 [WHO, 1988].

Although amorphous, aluminosilicate fibers undergo progressive crystallization, with the possible formation of mullite or cristobalite after a long period of service at high temperature.

Production

Continuous filaments and special-purpose fibers are made from glass. Insulation wools are made from glass, rock (rock wool) or slag (slag wool), and the two latter kinds of wool are also referred to together as mineral wool. Refractory fibers are produced most commonly from kaolin clay or from oxides of silicon, aluminum, or other metals.

Continuous filaments are made by mechanical drawing of the molten material through small openings in a drum or spinner rotating at high speed. Wools (entangled masses of fibers in contrast to the more ordered fibers in continuous filaments) are produced by fiberizing a steam of molten material by centrifugal forces, by blowing jets of steam or some other energy carrier, or by a combination of these processes. Refractory fibers are obtained by the mechanical blowing and/or hot gas blowing processes, while special-purpose (extremely fine) fibers are made by a flame attenuation (flame-blowing) process which consists essentially of mechanical drawing of molten glass rods followed by further attenuation of the fibers in a stream of hot gas emerging from the exhaust nozzle of an internal combustion burner.

MMMF products usually contain a binder and an oil for dust suppression. As binders, a variety of chemicals are used, including among others organic or inorganic synthetic resins (e.g. melamine formaldehyde, phenol formaldehyde, and urea formaldehyde resins).

MMMF are manufactured to controlled nominal diameters. These usually range 6–15 µm for continuous filaments, 2–9 µm for insulation wools, 1.2–3.5 µm for refractory fibers, and 0.1–3 µm for special-purpose fibers [WHO, 1988].

The manufacture of MMMF is a rapidly developing industry, and the number of MMMF products of varying chemical compositions, structures, and properties is continuously growing. The estimated global production of MMMF rose from about 4.5 million tonnes in 1973 to between 6 and 6.6 million tonnes in 1985 [WHO, 1988].

Uses

Continuous filaments (glass fibers) are used for reinforcing cement and resinous materials, paper, and rubber products; in textiles (e.g. for draperies and curtains); and in electric insulation. Wool materials (glass, rock, and slag wool) are mainly used in thermal or acoustic insulation (they are usually compressed, e.g., into boards, blankets, or sheets, or are bagged as loose wool for blowing or pouring into structural spaces). Refractory fibers are chiefly used for high-temperature insulation (e.g. thermal blankets for industrial furnaces). Finally, special-purpose fibers are designed for specialty applications such as high-efficiency filter papers or high-performance insulation for aircraft and space vehicles [WHO, 1988].

Toxicity

Deposition of MMMF on the skin and in the conjunctival sac has been reported to result in high prevalence of skin itching, occupational dermatoses, and conjunctivitis in industrially exposed groups [e.g. Sadkovskaya & Matsak; Swensson *et al.*]. Fibrous glass and rock wool fibers (mainly those greater than 4.5–5 µm in diameter) cause mechanical irritation of the skin characterized by a fine, punctate, itching erythema, which often disappears with continuous exposure [WHO, 1988]. Skin conditions have been the major health problem among workers handling glass fibers [Hill].

Alveolar deposition of MMMF depends primarily of their size, and decreases with their decreasing length. Short MMMF (<5 µm in length) are efficiency cleared by alveolar macrophages, whereas for fibers longer than 10 µm this form of clearance appears to be much less effective. In general, longer fibers and fibers of fine diameter dissolve more readily in lung tissue than do shorter or coarse fibers of the same type. The translocation of fibers to other organs from the lungs is limited [WHO, 1988]. (See also the section on asbestos for a discussion of the deposition and retention of fibrous particles in the distal lung regions.)

Reported effects of fiber retention in the deep airways are contradictory and range from complete absence of any pathological effects [e.g. Gross, 1976, 1982] to a full-fledged pneumoconiosis [Skomarokhin *et al.*].

In the majority of experimental inhalation studies, conducted on a range of animal species exposed to glass fibers of various types in concentrations up to 100 mg/m^3 for periods up to 2 years, little evidence of pulmonary fibrosis has been obtained. The severity of the tissue reaction to glass fibers and to glass wool was much less than that for equal masses of chrysotile or crocidolite asbestos. Moreover, no statistically significant increases in lung tumors have been found in inhalation studies on animals exposed to glass fibers (including glass microfibers) or rock wool. Both fibrosis and malignant tumors were, however, observed in animals of several species following intratracheal, intrapleural, or intraperitoneal administration of various types of MMMF. In particular, mesotheliomas were induced in rats by intrapleurally or intraperitoneally injected fibers, the highest tumor-inducing potency being shown by fibers less than 0.5 µm in diameter and more than 10 µm in length [Davis, 1981; Pott *et al.*; Stanton & Layard].

Reports in the earlier literature describe cases of acute irritation of the upper respiratory tract as well as more serious conditions such as bronchiectasis, pneumonia, chronic bronchitis, and asthma attributed to occupational exposure to MMMF; however, exposure to MMMF in most of these cases was probably incidental rather than causal, since such conditions have not been observed in more recent epidemiologic studies [WHO, 1988]. Chronic inflammatory processes in the upper respiratory tract have also been reported [Pushkina, 1965; Rozanov *et al.*; Sidenko & Bodyako].

There does not appear to be any reliable evidence that pleural or peritoneal mesotheliomas are associated with occupational exposure to MMMF. However, an excess of mortality due to lung cancer has been observed in large epidemiologic studies on rock wool/slag wool production workers, though not in those on glass wool or continuous filament workers. Epidemiological studies have also demonstrated an increased risk of lung and gastric cancer in MMMF workers [Enterline *et al*; Shannon *et al.*; Saracci *et al.*; WHO, 1988]. Fibrosis is not regarded as a human hazard with these fibers. On the basis of experimental evidence, glass, basalt, and slag wool and also ceramic fibers have been classified as possible human carcinogens [IARC, 1988].

In general, airborne MMMF concentrations present in workplaces with modern control technology are low. There have been isolated reports of respiratory symptoms and dermatitis associated with exposure to MMMF in the home and office environments [WHO, 1988].

The fibrogenicity of vitreous aluminosilicate fibers has been shown to increase as a result of crystallization in the course of their use at high temperatures [Skomarokhin].

Hygienic standards

In the USSR, an occupational exposure limit (MAC_{wz}) of 4 mg/m^3 has been adopted for all man-made silicate and aluminosilicate fibers of vitreous structure (as aerosols, hazard class 3, with the notation 'fibrogenic'). In the USA, a TLV-TWA of 10 mg/m^3 has been set for mineral wool fiber.

Measures to control exposure at the workplace

A particularly important general measure is provision of adequate local exhaust ventilation in areas where substantial amounts of dust are emitted. Glass fiber should be wetted before its application for heat and sound insulation purposes. Workers should be made aware of the hazards (especially skin and eye irritation) associated with MMMF, and encouraged to observe good hygienic practices and to report any sings of skin irritation at an early stage when treatment is usually easy and effective.

Good sanitary facilities are important, and workers handling MMMF are advised to rinse their hands and arms in cold or lukewarm water before applying soap.

Workers should be supplied with protective overalls and gloves which must be laundered separately from other clothes to prevent their contamination [Hill]. If necessary, disposable individual respirators should be used for respiratory protection. Wearing of personal protective equipment such as gloves, goggles, face screens, respirators, and protective clothing is necessary when working in confined spaces. Efforts should be made to mechanize MMMF production processes to the maximum possible extent. (Further details regarding preventive measures can be found in Hill and Sadkovskaya & Matsak.)

SILICATE-CONTAINING AND SILICATE BUILDING MATERIALS

Many silicate rocks (e.g. asbestos, pumice, tufa, marl, opoka, perlite, vermiculite, schungite, pegmatite) are extensively used as building materials per se or in various composite materials (often after heat treatment). Extensive use is also made of man-made silicates. The latter include, in particular, glasses (e.g. window glass, glass blocks, and glass tiles) and cements (e.g. portland cement, portland-slag cement, and pozzolan). Various bricks (e.g. clay bricks, silica bricks, and firebricks), ceramics (e.g. tiles and tubes), and expanded materials (e.g. vermiculite, ceramsite, and perlite) are made by firing aluminosilicates or other silicate rocks. On the basis of cements, a diversity of concretes are prepared (e.g. gas, foam, asphalt, gypsum, perlite, vermiculite, and slag concretes, heavy concrete, etc), as are dry gypsum plaster and asbestos cement. The firing is usually carried out at temperatures above 1000°C and various additives are included in the mixture (e.g. dyes and accelerators).

Physical and chemical properties
A number of silicate-containing building materials have a certain percentage of free silica—e.g. portland cement contains 0.8–3.0% (sometimes up to 9%), while gas concrete and foam concrete contain 8.3–24.0%. During the firing process, the content of free silica may rise as a result of both the loss of crystallization water and the partial decomposition of the (alumino)silicate. See also *Silicic acids and general considerations regarding silicates*.

Toxicity
The toxicity of dusts generated by building materials is determined, in the main, by silicates (which generally make up the bulk of the dust) and may increase with increasing level of free silica in the building material concerned.

Animals
Experimental silicatosis developed in animals following insufflation into their lungs or various dusts such as those from perlite and vermiculite (including their expanded forms), ceramsite, gas concrete, or foam concrete [Oksova et al.], ceramsite concrete [Rumyantsev & Kochetkova], glass tiles, glass blocks, or pegmatite [Blinova et al.], building bricks [Snegs et al.], refractory bricks [Katsnelson et al.], and asbestos cement [Dedkova et al.]. It has been shown that fired materials may be responsible for more strongly marked sclerotic changes in the lungs than their unfired counterparts.

Man
Among workers exposed to elevated dust concentrations in factories manufacturing building materials, cases of silicatosis as well as of chronic dust

bronchitis have been recorded from time to time. Some of those exposed to cement dust developed bronchial asthma, possibly under the influence of chromium(VI) contained in the dust.

Measures to control exposure at the workplace

Detailed measures are set out in health and safety regulations pertaining to particular building materials industries or to the building and civil engineering sector. As with silicates in general, prevention of dust formation is the basis of all technical control measures in workplaces where silicates are produced or used. Much dust is generated by crushing and grinding equipment, and high-level exposure can occur particularly during its maintenance, loading and unloading or when dusty materials are transported. Enclosing dust-generating equipment, applying local exhaust ventilation where openings are unavoidable, and moving dusty materials by mechanical means rather than manually are therefore important. If large quantities of airborne dust form in a workroom, general ventilation (even if well organized) will reduce exposure to only a limited extent.

Workers engaged in manual operations in the production and application of building materials should be provided with protective clothing; eye, hand, and foot protection are often required. Work clothes should be regularly freed from dust and laundered. Depending on the concentration and toxicity of silicate-containing dusts to which workers are exposed, pre-employment and regular medical examinations may be required. It is highly desirable that the rest rooms or prophylactoria be provided with facilities for administering inhalations and other measures to prevent deterioration of respiratory function in exposed workers. Further details concerning preventive measures can be found in *Occupational Hygiene in the Building Materials Industry* [Sadkovskaya & Matsak].

SOLUBLE GLASS

Identity

Soluble glass is an alkaline metal silicate of the general formula $Me_2O \cdot nSiO_2$, where Me = Na or K. The sodium salt is more common and its saturated aqueous solution is called water glass (or liquid glass). The molar ratio of SiO_2 to Na_2O varies from 1 to 4.2.

Physical and chemical properties

Soluble glass usually occurs as a gel-like or glassy mass of greenish-gray color. It is hydrolyzable and has a strongly alkaline reaction so that both free sodium hydroxide and colloidal silicic acid are present in a sodium silicate solution. It can be decomposed even by weak acids and by carbon dioxide.

Production and use

Soluble glass is obtained by fusing sand or kieselguhr with sodium sulfate or soda or with potash (potash water glass). It is used as a flotation agent; in acid-resistant cements and putties, as a builder for soaps and synthetic detergents; as a protective coating and a fireproofing agent in the textile industry; and in the manufacture of silicate paints, silica gel, matches, paper, etc.

Toxicity

Spatters or mist of water glass are irritating to the upper airways, and spatters may cause burns if they contact the eyes. The skin of workers handling soluble glass may develop lichenoid thickenings, particularly on the hands, and occasionally also ulcerations. The systemic toxicity of sodium silicates appears to have been studied only in rabbits in which they caused anemia.

Measures to control exposure

These consist in avoiding direct contact of soluble glass with the skin, and washing the hands with water or acetic acid solution. An indifferent ointment or cream may be recommended for hand protection.

ULTRAMARINE

The term 'ultramarine' is used to designate a blue pigment formerly prepared by powdering lapis lazuli (a semiprecious stone consisting essentially of the mineral lazurite) as well as a brilliant blue pigment (also called French blue) of similar composition and of variable shades which is prepared artificially; at the present time, only the latter is of practical importance, and it has no definite formula.

Physical and chemical properties

Man-made ultramarine is a fine blue powder of fairly variable casts (most commonly reddish or greenish). It is resistant to light, atmospheric precipitation, alkalies, and soaps. In does not dissolve in water, alcohol, or ether. Nor is it attacked by alkalies but can be readily decomposed and decolorized by weak acids with the release of H_2S and SO_2. It is also decolorized when heated in air.

Production and uses

Ultramarine is usually prepared by powdering the product resulting from the calcining, without access of air, of a mixture of powdered kaolin, sulfur, and soda or of kaolin, sodium sulfate, and charcoal.

Ultramarine is used in the dyeing industry; as a pigment in paints and printing inks (e.g. for wallpaper and cotton printing); in the manufacture of colored paper; for blueing paper, starch, sugar, and soap and in laundry blueing.

Toxicity

Animals

Rabbits administered ultramarine by gavage in amounts of 5 to 10 g daily became lethargic, refused to eat, and were rapidly losing weight. Decreased erythrocyte counts and marked proteinuria were recorded. Deaths occurred at days 45–47. Autopsy demonstrated degenerative changes in the kidneys.

Man

Workers inhaling ultramarine dust complained of appetite deterioration and belching; their gastric juice was found to have reduced acidity.

REFERENCES

Alivasates, C. *et al.* (1955) *Brit. J. Ind. Med.*, **12**, 43.

Andreyev, Yu. K. (1963) *Trudy Instituta geologii rudnykh mestorozhdeniy, petrografii, mineralogii i geokhimii AN USSR* [Transactions of the Institute for the Geology of Ore Deposits, Petrography, Mineralogy, and Geochemistry under the USSR Academy of Sciences), No. **95**, p. 61 (in Russian).

Artvinli, M. & Baris, Y. I. (1977) *J. Natl Cancer Inst.*, **63**, 17–27.

Bardin, N. S. (1958) In: *Voprosy gigieny truda, profpatologii i promyshlennoi toksikologii* [Labor Hygiene, Occupational Diseases, and Industrial Toxicology]. Publication of the Sverdlovsk Research Institute of Occupational Health, Sverdlovsk, Vol. 2, pp. 434–438 (in Russian).

Bardin N. S. & Kusevitsky, I. A. *Ibid*, pp. 439–455 (in Russian).

Baris, Y.I. (1975) *Hacettepe Bull. Med. Surg.*, **8**, 165–168.

Baris, Y. I. *et al.* (1978) *Thorax*, **33**, 181–192.

Barrie, H. & Gosselin, J. (1960) *Arch. Environ. Health*, **1**, 109–117.

Beattie, J. & Knox, J. F. (1961) In: Davis, C.N. (ed.) *Inhaled Particles and Vapours*. Pergamon Press, pp. 419–433.

Becklake, M. E. *et al.* (1970) *Bull. Physio-Pathol. Respir.*, **6**, 637–667.

Belobragina, G. V. (1959) In: *Voprosy gigieny truda, profpatologii i promyshlennoi toksikologii* [Labor Hygiene, Occupational Diseases, and Industrial Toxicology]. Publication of the Sverdlovsk Research Institute of Occupational Health, Vol. 5. Sverdlovsk, pp. 70–88 (in Russian).

Berry, G. *et al.* (1985) *Brit. J. Ind. Med.*, **42**, 12–18.

Blinova, T. S. *et al.* (1972) In: *Patomorfologiya silikatozov* [Pathomorphology of Silicatoses]. Transactions of the Leningrad Institute for Continuing Education of Physicians, Leningrad, No. **115**, pp. 61–92 & 100–104 (in Russian).

Boitsov, V.V. (1978) [*The Dust Factor in Talc Production and its Health Effects*]. Dissertation, Sverdlovsk Research Institute of Occupational Health (in Russian).

Boitsov, V.V. & Bakhireva, I.D. (1977) In: *Professionalnye bolezni pylevoi etiologii* [Occupational Diseases Caused by Dust]. Publication of the Erisman Research Institute of Hygiene, Moscow, No. 4, pp. 25–29 (in Russian).

Boitsov, V.V. & Katsnelson, B. A. (1977) *Ibid*, pp. 18–24.

Bolton, R. E. *et al.* (1983) *Brit. J. Ind. Med.*, **40**, 264–272.

Borshchevsky, Yu.M. (1964) In: *Borba s silikozom* [Silicosis Control]. Nauka, Moscow, Vol. 6, pp. 246–255 (in Russian).

Bunimovich, G. I. (1967) In: *Borba s silikozom* [Silicosis Control]. Nauka, Moscow, Vol. 7, pp. 320–322 (in Russian).

Davis, J. M. G. (1965) *Ann. N.Y. Acad. Sci.*, **32**, 98–111.

Davis, J. M. G. (1967) *Brit. J. Exper. Pathol.*, **48**, 379–385.

Davis, J. M. G. (1972) *Brit. J. Exper. Pathol.*, **53**, 652–658.
Davis, J. M. G. (1981) *Ann. Occup. Hyg.*, **24**, 227–230.
Davis, J. M. G. *et al.* (1970) *Arch. Pathol.*, **89**, 364–373.
Dedkova, L. E. *et al.* (1972) In: *Pathomorfologiya silikatozov* [Patomorphology of Silicatoses]. Transactions of the Leningrad Insitute for Continuing Education of Physician, Leningrad, No. **115**, pp. 128–133 (in Russian).
Deyeva, I. B. *et al.* (1988) In: *Teoreticheskie i prikladnye problemy vnedreniya prirodhykh tseolitov v narodnom khozyaistve RSFSR* [Industrial Uses of Natural Zeolites in the Russian Federation: Theoretical and Applied Aspects]. Kemerovo, pp. 90–91 (in Russian).
Domnin, S. G. & Solomina, S. N. (1972) In: *Voprosy gigieny truda i professoinalnoi patologii v metallurgii* [Occupational Health in Metallurgy]. Publication of the Erisman Research Institute of Hygiene, Moscow, pp. 132–142 (in Russian).
Domnin, S. G. *et al.* (1978) In: *Professionalnye bolezni pylevoi etiologii* [Occupational Diseases Caused by Dust]. Publication of the Erisman Research Institute of Hygiene, No. **5**, pp. 98–105 (in Russian).
Ehrenreich, T. *et al.* (1973) *Arch. Malad. Profess.*, **34**, 189–204.
Einbrodt, H. J. *et al.* (1965) *Arch. Hyg. Bakteriol.*, **149**, 408–412.
Engelbrecht, F. M. *et al.* (1960) *Ann. Occup. Hyg.*, **2**, 257–266.
Enterline, P. E. *et al.* (1983) *Amer. Rev. Respir. Dis.*, **128**, 1–7.
Fedoseyev, A. D. *et al.* (1966) *Voloknistye silikaty. Prirodnye i sinteticheskie asbesty* [Fibrous Silicates. Natural and Man-made Asbestoses]. Nauka, Moscow (in Russian).
Finkelstein, M. M. (1985) *Brit. J. Ind. Med.*, **42**, 319–325.
Flügge, G. (1958) *Glüchauf*, **94**, 185–187.
Friedberg, K. D. & Schiller, E. (1988) Chapter 54 *Silicon* in: Seiler H. G. *et al.* (eds) *Handbook on Toxicity of Inorganic Compounds*. Marcel Dekker, New York, pp. 595–617.
Gasparian, A. A. (1968) In: *Materialy 3-ey itogovoi nauchnoi konferentsii po voprosam gigieny truda* [Proceedings of the 3rd Scientific Meeting on Occupational Hygiene]. Publication of the Yerevan Research Institute of Industrial Hygiene and Occupational Diseases, pp. 128–130 (in Russian).
Gilson, J. C. (1983) In: *Encyclopaedia of Occupational Health and Safety*. International Labour Office, Geneva, pp. 185–191.
Gross, P. (1976) *Arch. Environ. Health*, **32**, 101–107.
Gross, P. (1981) *Amer. Ind. Hyg. Assoc. J.*, **42**, 449–452.
Gross, P. (1982) *Internat. Arch. Occup. Environ. Health*, **50**, 103–112.
Gross, P. & de Treville, R. T. P. (1967) *Arch. Environ. Health*, **15**, 638–639.
Gross, P. *et al.* (1967) *Arch. Environ. Health*, **15**, 343–355.
Hain, E. *et al.* (1973) Staub, **33**, 51–56.
Harington, J. S. & Roe, F. (1965) *Ann. N.Y. Acad. Sci.*, **132**, 439–451.
Heimann, H. *et al.* (1953) *Arch. Ind. Hyg. Occup. Med.*, **8**, 531–532.
Hill, G.W. (1983) In: *Encyclopaedia of Occupational Health and Safety*. International Labour Office, Geneva, pp. 852–855.
Holt, P. F. & Young, D. K. (1967) *J. Pathol. Bacteriol.*, **93**, 696–699.
IARC *Monographs on the Evaluation of Carcinogenic Risk of Chemicals to Man*. Vol. **14**: *Asbestos* (1977). International Agency for Research on Cancer, Lyons.
IARC *Monographs on the Evaluation of Carcinogenic Risk of Chemicals to Humans*. Vol. **42**: *Silica and Some Silicates* (1987). International Agency for Research on Cancer, Lyons.
IARC *Monographs on the Evaluation of Carcinogenic Risk of Chemicals to Humans*. Vol. **43**: *Man-made Mineral Fibers and Radon* (1988). International Agency for Research on Cancer, Lyons.
ILO (1984) *Safety in the Use of Asbestos*. International Labour Office, Geneva.
Jaques, W. & Bernischke, K. (1952) *Arch. Ind. Health*, **5**, 451–463.
Katsnelson, B. A. (1957) *Gig. Truda Prof. Zabol.*, No. **2**, 24–30 (in Russian).
Katsnelson, B. A. & Mokronosova, K. A. (1979) *J. Occup. Med.*, **21**, 15–20.
Katsnelson, B. A. *et al.* (1964) *Gig. San.*, No. **12**, 30–38 (in Russian).

Kaw, J. L. & Zaidi, S. H. (1973) *J. Exper. Pathol.*, **8**, 224–231.

Kleinfeld, M. (1973) *Proceedings of a Symposium on Talc.* Washington, D. C., USA Bureau of Mines Information Circular No. 8639, pp. 4–11.

Kleinfeld, M. (1983) In: *Encyclopaedia of Occupational Health and Safety.* International Labour Office, Geneva, pp. 2141–2142.

Kleinfeld, M. *et al.* (1967) *Arch. Environ. Health,* **14**, 662–667.

Kogan, F. M. (1970) In: *Pathogenez pnevmokoniozov* [Patogenesis of Pneumoconioses]. Publication of the Sverdlovsk Research Institute of Industrial Hygiene and Occupational Diseases, Sverdlovsk, pp. 16–34 (in Russian).

Kogan, F. M. (1975) *Asbestosoderzhashchie pyli i mery preduprezhdeniya ikh vrednogo vliyaniya na zdorovie rabotayushchikh* [Asbestos-containing Dusts and Measures to Prevent their Adverse Health Effects on Workers]. Middle-Uralian Book Publishers, Sverlovsk (in Russian).

Kogan, F. M. & Berzin, S. A. (1973) In: *Epidemiologiya i genez raka zheludka* [Epidemiology and Genesis of Gastric Cancer]. Vilnius, pp. 114–116 (in Russian).

Kogan, F. M. *et al.* (1966) *Gig. San.,* No. **8**, 28–33 (in Russian).

Kogan, F. M. *et al.* (1971) *Gig. Truda Prof. Zabol.,* No. **4**, 43–46 (in Russian).

Kogan, F.M. *et al.* (1972) *Gig. San.,* No. **7**, 29–32 (in Russian) .

Korkina, K. T. *et al.* (1988) In: *Teoreticheskie i prikladnye problemy vnedreniya prirodnykh tseolitov v narodnom khozyaistve RSFSR* [Industrial Uses of Natural Zeolites in the Russian Federation: Theoretical and Applied Aspects]. Kemerovo, pp. 49–50 (in Russian).

Kovnatsky, M. A. (1971) *Silikatozy* [Silicatoses]. Meditsina, Leningrad (in Russian).

Langer, A. M. & Nolan, R. P. (1985) In: Beck, T. & Bignon, J. (eds) *Proceedings of the 3rd International Workshop on In Vitro Testing of Mineral Dusts, October, 1984.* Springer-Verlag, Berlin.

Langer, A. M. *et al.* (1972). *J. Histochem. Cytochem.,* **20**, 735–740.

Liddell, F. D. K. & Hanley, J. A. (1985) *Brit. J. Ind. Med.,* **42**, 389–396.

Liddell, F. D. K. *et al.* (1983) *Paper presented to the Third International Symposium on Epidemiology in Occupational Health,* Singapore, September 1983.

Lipkin, L. E. (1980) *Environ. Health Perspect.,* **34**, 91–102.

Maltoni, C. *et al.* (1981) *Acta Oncol.,* **2**, 137–142.

Maroudas, N. G. *et al.* (1973) *Lancet,* **1**, 807–809.

McDonald, J. C. (1984) *Ann. Acad. Med. Singapore,* **13**, 345–352.

McDonald, A. D. *et al.* (1983) *Brit. J. Ind. Med.,* **40**, 361–367.

McMillan, G. H. G. & Rossiter, C. E. (1982) *Brit. J. Ind. Med.,* **39**. 54–59.

Meiklejohn A. (1983) In: *Encyclopaedia of Occupational Health and Safety.* International Labour Office, Geneva, pp. 1358–1359.

Merliss, R. R. (1971) *Science,* **173**, 1141–1142.

Newhouse, M. L. (1969) *Brit. J. Ind. Med.,* **26**, 294–301.

Oksova, E. E. *et al.* (1972) In: *Patomorfologiya silikatozov* [Pathomorphology of Silicatoses]. Transactions of the Leningrad Institute for Advanced Medical Studies, Leningrad, No. **115**, pp. 32–52 (in Russian).

Ozelmi, M. *et al.* (1985) *Brit. J. Ind. Med.,* **42**, 746–749.

Podgaiko, G. A. *et al.* (1982) In: *Professionalnye bolezni pylevoi etiologii* [Occupational Diseases Caused by Dust]. Publication of the Erisman Research Institute of Hygiene, Moscow, pp. 68–77 (in Russian).

Pontefract, R. D. & Cunningham, H. M. (1973) *Nature* (London), **243**, 252–253.

Pott, F. & Friedrichs, K.-H. (1972) *Naturwissenschaft,* **59**, 318.

Pott, F. *et al* (1980) In: Wagner, J. C. (ed.) *Biological Effects of Mineral Fibers.* International Agency for Research on Cancer, Vol. 1., pp. 337–342 (IARC Scientific Publications No. 30), Lions.

Pushkina, I. K. (1965) *Gig. Truda Prof. Zabol.,* No. **6**, 28–31 (in Russian).

Pylev, L. N. (1972) *Vopr. Onkol.,* No **6**, 40–46 (in Russian).

Pylev, L. N. & Krivosheina, L. V. (1972) In: Shabad, L. M. (ed.) *Voprosy profilaktiki zagryazneniya okruzhayushchei cheloveka sredy kantserogennymi veshchestvami* [Prevention of Environmental Pollution by Cancerogenic Substances]. Valgus Publishers, Tallinn, pp. 46—51 (in Russian).

Pylev, L. N. *et al.* (1986) *Gig. Truda Prof. Zabol.*, No. 5, 29—33 & No. 6, 33—37 (in Russian).

Romanovich, I. F. (1974) *Talk* [Talc]. Nedra, Moscow (in Russian).

Roy-Chowdhury, A. K. *et al.* (1973) *Arch. Environ. Health*, 26, 253—255.

Rozanov, L. S. *et al.* (1952) *Gig. San.*, No. 3, 14—18 (in Russian).

Rumyantsev, G. I. & Kochetkova, T. A. (1966) In: *Voprosy gigieny truda i profpatologii* [Occupational Health: Selected Topics]. Meditsina, Moscow, pp. 260—268 (in Russian).

Sadkovskaya, N. I. & Matsak, V. G. (1972) *Gigiena truda v promyshlennosti stroitelnykh materialov* [Occupational Hygiene in the Building Materials Industry]. Meditsina, Moscow (in Russian).

Saracci, R. *et al.* (1984) *Brit. J. Ind. Med.*, 41, 425—436.

Schepers, G. W. H. (1973) *Proceedings of a Symposium on Talc*. Washington, D. C., USA Bureau of Mines Information Circular No. 8639, pp. 49—76.

Schepers, G. W. H. & Durkan, T. (1955) *Arch. Ind. Health*, 12, 317—328.

Sedov, K. R. *et al.* (1969) *Slyudyanoi pnevmokonioz* [Mica Pneumoconiosis]. Meditsina, Moscow (in Russian).

Selikoff, I. J. (1977) In: *Origins of Human Cancer*. Cold Spring Harbour, N.Y., pp. 1765—1784.

Selikoff, I. J. *et al.* (1968) *JAMA*, 204, 106—112.

Shannon, H. S. *et al.* (1984) *Brit. J. Ind. Med.*, 41, 35—38.

Sidenko, A. G. & Bodyako, V. S. (1965) *Gig. Truda Prof. Zabol.*, No. 9, 48 (in Russian).

Siehoff, F. & Antweiler, H. (1956) *Arch. Gewerbepathol. Gewerbehyg.*, 15, 158—170.

Skomarokhin, A. F. (1985) *Pylevoi faktor v proizvodstve i ispolzovanii iskusstvennykh mineralnykh volokon i yego deistviye na organizm* [The Dust Factor in the Production and Use of Man-made Mineral Fibers and its Health Effects]. Author's synopsis of his dissertation. Institute for Advanced Medical Studies, Leningrad (in Russian).

Skomarokhin, A. F. *et al.* (1986) *Gig. Truda Prof. Zabol.*, No. 2, 1—7 (in Russian).

Smith, E. (1973) *Proceedings of a Symposium of Talc*. Washington, D. C., USA Bureau of Mines Information Circular No. 8639, pp. 43—48.

Snegireva, T. V. *et al.* (1988) In: *Teoreticheskie i prikladnye problemy vnedreniya prirodnykh tseolitov v narodnom khozyaistve RSFSR* [Industrial Uses of Natural Zeolites in the Russian Federation: Theoretical and Applied Aspects]. Kemerovo, pp. 112—116 (in Russian).

Snegs, R. N. *et al.* (1972) In: *Patomorfologiya silikatozov* [Pathomorphology of Silicatoses]. Transactions of the Leningrad Institute for Continuing Education of Physicians, Leningrad, No. 115, pp. 92—100 (in Russian).

Spiegel, R. H. (1973) *Proceeding of a Symposium of Talc*. Washington, D. C., USA Bureau of Mines Information Circular No. 8639, pp. 97—102.

Spurny, K. R. (1983) *Sci. Total Environ.*, 30, 147—166.

Stanton, M. F. & Layard, M. (1977) *J. Natl Cancer Inst.*, 58, 587—603.

Suzuki, J. (1982) *Environ. Res.*, 27, 433—445.

Suzuki, J. & Kohyama, N. (1984) *Environ. Res.*, 35, 277—292.

Swensson, A. *et al.* (1981) *Arbete och Hälsa*, 26, 31—32.

Takahashi, A. *et al.* (1958) *Bull. Osaka Ind. Res. Inst.*, 9, 242—248.

Thomas, H. F. *et al.* (1982) *Brit. J. Ind. Med.* 39, 273—276.

Timbrell, V. (1976) In: *Occupational Exposure to Fibrous Glass. Proceedings of a Symposium, College Park, Maryland, 26-27 June 1974*. Washington D. C., US Department of Health, Education and Welfare, pp. 33—50.

Timbrell, V. & Skidmore (1971) In: Walton, W. H. (ed.) *Inhaled Particles. III. Proceedings of an International Symposium, London, 14—23 September, 1970*. Old Woking, Surrey, Unwin, Vol. 1, pp. 49—56.

Tkachev, P. G. *et al.* (1988) In: *Teoreticheskie i prikladnye problemy vnedreniya prirodnykh tseolitov v narodnom khozyaistve RSFSR* [Industrial Uses of Natural Zeolites in the Russian Federation: Theoretical and Applied Aspects]. Kemerovo, pp. 116—118 (in Russian).

Vagina, E. F. *et. al.* (1984) In: *Professionalnye bolezni pylevoi etiologii* [Occupational Diseases Caused by Dust]. Publication of the Erisman Research Institute of Hygiene, Moscow, pp. 75—79 (in Russian).

Vagina, E. F. (1980) In: *Professionalnye bolezni pylevoi etiologii* [Occupational Diseases Caused by Dust]. Publication of the Erisman Research Institute of Hygiene, Moscow, pp. 31—36 (in Russian).

Vasiliev, L. A. *et al.* (1989) *Eksper. Onkol.*, No. 4. 26—29 (in Russian).

Vedmitsky, V. A. *et al.* (1988) In: *Teoreticheskie i prikladnye problemy vnedreniya prirodnykh tseolitov v narodnom khozyaistve RSFSR* [Industrial Uses of Natural Zeolites in the Russian Federation: Theoretical and Applied Aspects]. Kemerovo, pp. 88—90 (in Russian).

Velichkovsky, B. T. *et al.* (1988) *Ibid*, pp. 116—118.

Velichkovsky, B. T. & Katsnelson. B. A. (1964) *Etiologia i patogenez silikoza* [Etiology and Pathogenesis of Silicosis], Meditsina, Moscow (in Russian).

Vincent, D. *et al.* (1958) *Beitr. Silikose-Forsch.*, **54**, 15—29.

Wagner, J. C. (1982) In: *Biological Effects of Man-made Mineral Fibers. Proceedings of a WHO/IARC Conference, Copenhagen, Denmark, 20—22 April 1982.* Copenhagen, World Health Organization, Regional Office for Europe, Vol. 2, pp. 209—233.

Wagner, J. C. & Skidmore, J.W. (1965) *Ann. N.Y. Acad. Sci.*, **132**, 77—86.

Wagner, J. C. *et al.* (1985). *Brit. J. Cancer*, **51**, 727—730.

WHO (1986) *Environmental Health Criteria 53: Asbestos and Other Natural Mineral Fibers.* World Health Organization, Geneva.

WHO (1988) *Environmental Health Criteria 77: Man-made Mineral Fibers.* World Health Organization, Geneva.

Germanium and its compounds

Germanium(IV) chloride (germanic chloride, germanium tetrachloride);
g. hydride (germanium tetrahydride, germane, monogermane); **g.(II) oxide**
(germanous oxide); **g.(IV) oxide** (germanic oxide, germanium dioxide)

IDENTITY AND PHYSICOCHEMICAL PROPERTIES OF THE ELEMENT

Germanium (Ge) is an element in group IV of the periodic table with the atomic number 32. Its natural isotopes are ^{70}Ge (20.55%), ^{72}Ge (27.35%), ^{73}Ge (7.78%), ^{74}Ge (36.50%), and ^{76}Ge (7.86%).

Germanium is brittle and will not lend itself either to cold or hot plastic working up to 550°C, but becomes ductile at higher temperatures. It can be sawn.

Germanium is resistant to the action of air, water, H_2SO_4, and HCl. In compounds it is present in the oxidation state +2 (germanium(II), or germanous, compounds) and +4 (germanium(IV), or germanic, compounds). It is oxidized by oxygen to the oxides GeO and GeO_2. In reacting with strong alkalies, it forms meta- and orthogermanates that are strong reducing agents. It readily combines with halogens to form halides of the $GeHal_4$ type, and forms alloys with many metals. Germanium is a semiconductor. See also Appendix.

NATURAL OCCURRENCE AND ENVIRONMENTAL LEVELS

Germanium is always found in combination with other elements, never in the free state. The more common, although rare, germanium minerals are sulfides such as germanite [Cu_3(Ge, Fe, Zn, etc.) (S, As)$_4$], argyrodite (Ag_8GeS_6), and confildite $(Ag_8(Sn,Ge)S_6)$. The bulk of germanium is widely distributed in a large number of rocks and other minerals including sulphide ores of nonferrous metals, iron ores, some oxide minerals (chromite, magnetite, rutile, etc.), in granites, diabases, and basalts. It is found in nearly all silicates and in some oil deposits. Its relatively high concentrations (20–280 mg/kg ash weigh) in different types of coal have often been reported.

The clarke of germanium is estimated at $(1.4-1.5) \times 10^{-4}\%$ in the earth's crust and at $1.3 \times 10^{-4}\%$ in the granite layer of the continental crust [Dobrovolsky].

The lithosphere has an average germanium content of $1.4 \times 10^{-4}\%$ (clays and shales contain $2 \times 10^{-4}\%$ and sandstones, $3 \times 10^{-4}\%$) [Vadkovskaya & Lukashev]. During weathering, germanium is partly mobilized, but is then readily fixed to clay minerals, iron oxides, and organic matter [Kabata-Pendias & Pendias].

The oceans contain an estimated 82.2 million tonnes of dissolved germanium, mainly as $Ge(OH)_3^-$, at average concentrations of 0.06 µg/L in water and $17 \times 10^{-8}\%$ in total salts; clay muds have an average content of $2 \times 10^{-4}\%$ and calcareous muds $0.2 \times 10^{-4}\%$, i.e. nonbiogenic sediments contain the metal at much higher levels [Dobrovolsky].

In surface soils of the USA, the abundance of germanium is rather uniform and averages 1.1 mg/kg dry weight [Kabata-Pendias & Pendias].

The distribution of germanium in the environment (as well as in foods and living organisms) is described by Fishr et al.

PRODUCTION

Germanium is mainly obtained from by-products generated in the processing of nonferrous metal ores containing 0.001 to 0.1% of this element, or from the ash of coals, the dust of gas generators, or the wastes of coking plants. The resulting concentrate contains 2 to 10% germanium which is then extracted in several steps (e.g. chlorination, hydrolysis, and reduction). Highly purified metal can be obtained by special methods such as multiple-zone melting and fractional distillation.

USES

Germanium is widely used in electronics (for diodes, triodes, transistors, crystal detectors, amplifiers, and rectifiers) and in many alloys with metals (germanides), notably vanadium and niobium which have relatively high temperatures of transition to a state of superconductivity. It is also used in the manufacture of special optical glasses and in infrared detectors. In addition, it finds application in the chemical and machine-building industries and in the production of ceramics and enamels. Monocrystalline germanium is employed in dosimeters and in instruments for measuring the strength of constant and variable magnetic fields.

Germanium(IV) chloride is an intermediate in the production of germanium(IV) oxide and organogermanium compounds. Germanium(IV) oxide is used in the manufacture of optical glasses and in cathodes for electronic tubes.

Other uses of germanium and its compounds are mentioned in Fisher *et al.* and Gerber.

MAN-MADE SOURCES OF EMISSION INTO THE ENVIRONMENT

Considerable and increasing quantities of germanium are released into the environment from coal burning. Thus, an estimated 3.9 thousand tonnes (or 26 g/km^2 of land area) entered the atmosphere from this source in 1970 and 8.3 thousand tonnes (or 55 g/km^2 of land area) in 1980; for the year 2000, 13.1 thousand tonnes (or 91 g/km^2 of land area) has been forecast [Dobrovolsky].

In the manufacture of germanium monocrystals, the workroom air contained germanium(IV) chloride (GeCl$_4$) and its hydrolysis products (GeO$_2$, HCl) at concentrations up to 70 mg/m^3 (corresponding to about 7 mg Ge/m^3). The main sources of soil pollution with germanium are sewage sludges and organic fertilizers [Kabata-Pendias & Pendias].

TOXICITY

Acute/single exposure

Animals

The oral LD$_{50}$ of **germanium(IV) oxide** for mice was 1.25 g/kg; after this dose deaths occurred within 1 to 12 days and were preceded by adynamia and diarrhea. Autopsy revealed congested internal organs, edematous and thickened interalveolar septa in the lungs, proliferation of reticuloendothelial elements, and cellular alterations in the liver and spleen [Antoniuk]. With intraperitoneal administration, lethal doses of the oxide for rats ranged from 0.6 to 1.2 g/kg. The LC$_{50}$ of **germanium(IV) chloride** for mice was 44 mg/L with 2 h exposure by inhalation. In animals thus exposed, a short period of excitation was followed by general inhibition, convulsions, and signs of irritation in the upper airways and eyes. Pathomorphologic examination demonstrated congested internal organs, subpleural hemorrhages, necrosis of the tracheal mucosa, pulmonary edema, and necrobiotic changes in hepatic cells and in the epithelium of convoluted renal tubules [Kurliandsky *et al.*].

The 2-hour LC$_{50}$ of **germanium hydride** (monogermane) for mice was determined at 1380 mg/m^3. The clinical picture of poisoning was marked by dyspnea, convulsions, and coma; pathomorphologic examination showed congested internal organs, hemorrhages, nerve cell vacuolation, and degenerative changes in the cells of parenchymal organs. Similar pathomorphologic findings were made in guinea-pigs after a single 4-h exposure to germanium hydride concentrations of 260–1400 mg/m^3 [Batkina; Boyarchuk].

Thickened interalveolar septa and hyperplastic peribronchial lymph nodes were seen in rats 7 months after a single intratracheal administration of **metallic germanium** dust (with an admixture of iron) in a dose of 500, 750, or 1000 mg/kg.

Man

No signs of poisoning were observed in individuals who had ingested up to 1.4 g of metallic germanium in fractional doses of 0.1 to 0.2 g.

Repeated exposure

Animals

Mild irritation of the respiratory mucous membranes and conjunctiva developed in mice in the course of 40-day inhalation exposure to **germanium(IV) chloride** at concentrations of 0.5–0.7 mg/L (2 h/day, 5 days/week). After this exposure period, focal thickening of alveolar septa, bronchitis, and interstitial pneumonia were seen in the surviving animals. Rats inhaling **germane** at 0.013–0.17 mg/L for 4 h daily over 30 days did not show any visible signs of intoxication, but traces of occasional hemorrhages with gliotic nodules were discovered in their brains at autopsy. When rats were given **germanium(IV) oxide** in a dose of 17 mg/kg body weight in their drinking water, half of them died within 4 weeks.

Chronic exposure

Animals

Rats given **germanium(IV) oxide** for 100 days in a daily dose of 10.0 mg/kg by oral gavage showed reductions in nervous system excitability, oxygen consumption, and hemoglobin levels. In rats exposed by inhalation to **germanium(IV) chloride** concentrations of 0.003–0.007 mg/L for 7 months, the prominent findings were shifts in the differential blood count, leukocytosis, reduced capacity for hippuric acid formation, diminished oxygen consumption, and increased excitability of the higher autonomic centers [Kurliandsky *et al.*]. An increased incidence, compared to controls, of fatty degeneration of the liver occurred among rats, but not mice, on lifelong exposure to germanium in drinking water at 5 mg/L [Vouk].

In a long-term toxicity study with continuous low-level exposure of rats to germanium(IV) oxide by inhalation, the following concentrations of the oxide were considered as threshold levels: 0.55 mg/m^3 for shortening the time of sperm motility in male rats; 0.4 mg/m^3 for lowering erythrocyte count; 2.1 mg/m^3 for producing a decrease in the summated threshold index; and 3.8 mg/m^3 for reducing the number of sulfhydryl groups [Akbarov].

Dermal and ocular effects

Ulceration and necrosis of the skin were observed in rabbits after 2-h exposure to germanium(IV) chloride. When 0.05–0.1 ml of germanium(IV) chloride solution was placed in their conjunctival sac, severe irritation of the conjunctiva and prolonged blindness ensued. Local effects of the chloride are likely to result from the combined action of this compound and its hydrolysis products [Kurliandsky et al.]. Germanium(IV) oxide did not irritate the skin, but caused conjunctival irritation.

ABSORPTION, DISTRIBUTION, AND ELIMINATION

The daily human diet contains, on average, about 1.5 mg of germanium which is readily absorbed and completely eliminated from the body (1.4 mg with the urine and 0.1 mg with the feces). Germanium was detected in nearly all of 125 analyzed samples of foods and bevarages, and some samples (canned tuna, tomato juice, and beans) contained it in concentrations of 3 to 5 ppm [Whanger]. Germanium concentrations in grains ranged from 0.09 to 0.7 ppm (on a fresh weight basis) [Kabata-Pendias & Pendias].

In man, germanium levels average (mg/kg) 0.2 in blood, 0.0009 in lymph nodes, 0.003 in muscle, 0.04 in liver, 0.09 in lung tissue, 0.1 in brain, 0.5 in testis, 9.0 in kidney.

Germanium(IV) oxide was rapidly absorbed from the gastrointestinal tract of rats after oral gavage (77.4% in 4 h, 96.4% in 8 h). When rats were exposed to elemental germanium dust by inhalation, 52% of its particles (mean size 1.4 μm) were cleared from their lungs in 24 h and only 18% remained there 7 days after exposure. The clearance of germanium(IV) oxide (mean particle size 0.4 μm) was more rapid (79% within 24 h) [Vouk].

Absorbed germanium was reported to be distributed between erythrocytes and plasma in a 2:3 ratio and not to combine with plasma proteins [Rosenfeld]. In rabbits and dogs injected with germanium(IV) oxide, the highest germanium concentrations were found in the kidney, liver, spleen, gastrointestinal tract, and bone [Vouk]. In general, germanium is widely distributed in the body and does not appear to be selectively retained in any tissue, although the spleen has been reported as the main organ of its accumulation in man [Linder].

HYGIENIC STANDARDS

Exposure limits adopted in the USSR for germanium and its hydride and oxides are shown in Table 1. In the USA, a TLV-TWA of 0.63 has been set for germanium tetrahydride.

Table 1. Exposure limits for germanium and its compounds

	Workroom air		Atmospheric air	
	MAC_{wz} (mg/m^3)	Aggregative state	MAC_{ad} (mg/m^3)	Hazard class
Germanium	2.0	Aerosol	–	3
Germanium hydride	5.0/1.8*	Vapor	–	3
Germanium (II) oxide	2.0	Aerosol	–	3
Germanium (IV) oxide	–	–	0.04	3

* 5.0 mg/m^3 is the concentration not to be exceeded at any time, while 1.8 mg/m^3 is the average maximum allowable concentration per shift.

METHODS OF DETERMINATION

Photometric methods for determining germanium and its hydride (germane) in *air* are based on the oxidation of germanium with nitric acid followed by measurement of the colored substance formed in the reaction of the oxidation product with phenylfluorone, or on the reaction of germanium or its hydride with phenylfluorone in a hydrochloric acid medium (sensitivity, 0.05 mg/m^3 for germanium and 2 mg/m^3 for germane) [*Guidelines for the Determination of Noxious Substances in Air; Peregud*]. In *biologic materials*, germanium can be determined by methods of atomic absorption spectroscopy (sensitivity up to 0.015 mg/L), spectrophotometry with phenylfluorone (sensitivity 0.1–0.5 mg/L; relative standard deviation, 5–10%), and emission spectrography (detection limit ~1 μg) [Vouk]. (See also Fisher *et al.* and Gerber.)

MEASURES TO CONTROL EXPOSURE

Technical control measures *at the workplace* should be primarily aimed at preventing contamination of the air by dusts and fumes of germanium and its compounds during their production or use. This can be best assured by tight enclosure of the processes and equipment.

Adequate local exhaust ventilation is essential in operations such as dry cutting and grinding of germanium crystals, and local ventilation should also be provided near melting furnaces during the manufacture of semiconductors as well as during the cleaning of the furnaces. Exhaust ventilation is also required in premises used for the chlorination, rectification, and hydrolysis of germanium (IV) chloride. Appliances, connections, and fittings in such premises should be made of corrosion-resistant materials. The process of manufacturing and alloying germanium monocrystals should be carried out in a vacuum [Mogilevskaya].

For *personal protection*, workers should be supplied with acid-proof clothing and footwear. Respirators must be used during the cleaning of germanium-contaminated equipment. Strict observance of personal hygiene is important.

Measures for minimizing the release of germanium into the *general atmosphere* include the use of gas instead of coal in heating systems, the development of more efficient methods of fuel combustion, and thorough purification of gases and dusts emitted from heat and electric power plants.

REFERENCES

Akbarov, A. A. (1981) *Gig. San.*, No. **12**, 57−58 (in Russian).

Antoniuk, O. K. (1975) *Gig. San.*, No. **8**, 98−99 (in Russian).

Batkina, I. B. (1966) *Gig. San.*, No. **12**, 18−22 (in Russian).

Boyarchuk, I. F. (1968) *Gigiyena truda v proizvodstve slozhnykh mineralnykh udobreniy* [Occupational Hygiene in the Manufacture of Composite Mineral Fertilizers]. Meditsina, Moscow (in Russian).

Dobrovolsky, V. V. (1983) *Geografiya mikroelementov. Globalnoe rasseyanie* [Geography of Trace Elements. Global Dispersion]. Mysl, Moscow (in Russian).

Fisher, B. R. *et al.*, (1991) Chapter II *Germanium* in: Merian, E. (ed.) *Metals and their Compounds in the Environment—Occurrence, Analysis and Biological Relevance*. VCH, New York, pp. 921−929.

Gerber, G. B. (1988) Chapter 24 *Germanium* in: Seiler, H. G. *et al.* (eds) *Handbook on Toxicity of Inorganic Compounds*. Marcel Dekker, New York.

Goldman, Z. I. (1960) *Gig. Truda Prof. Zabol.*, No. **10**, 30−35. (in Russian).

Guidelines for the Determination of Noxious Substances in Air (1979) [Metodicheskiye ukazaniya po opredeleniyu vrednykh veshchestv v vozdukhe], No. 13. Publication of the USSR Ministry of Health, Moscow (in Russian).

Kabata-Pendias, A. & Pendias, H. (1986) *Trace Elements in Soils and Plants*. CRC Press, Boca Raton (Fl.).

Kurliandsky, B. A. *et al.* (1968) *Gig. Truda Prof. Zabol.*, No. **5**, 51−53. (in Russian).

Linder, M. C. (1985) In: Linder M. C. (ed.) *Nutritional Biochemistry and Metabolism.* Elsevier, New York & Amsterdam, pp. 151−197.

Mogilevskaya, O. Ja. (1983) In: *Encyclopaedia of Occupational Health and Safety*. International Labour Office, Geneva, pp. 964−965.

Peregud, E. A. (1976) *Khimicheskiy analiz vozdukha (novye i usovershenstvovannye, metody* [Chemical Analysis of the Air (New and Improved Methods)]. Khimiya, Leningrad (in Russian).

Rosenfeld, G. (1954) *Arch. Biochem. Biophys.*, **48**, 84−94.

Vadkovskaya, I. K. & Lukashev, K. P. (1977) *Geokhimicheskie osnovy okhrany biosfery* [Protection of the Biosphere: Geochemical Principles]. Nauka i Tekhnika, Minsk (in Russian).

Vouk, V. (1979) In: Friberg, L. *et al.* (eds) *Handbook on the Toxicology of Metals.* Elsevier, Amsterdam, pp. 421—428.

Whanger, P. D. (1982) In: Hatchcock, J. N. (ed.) *Nutritional Toxicology*, Vol. 1. Academic Press, New York, pp. 163—208.

Tin and its compounds

Tin(II) compounds and stannites (stannous compounds): **chloride; chloride dihydrate; fluoride; orthophosphate; oleate; oxide; sulfate; sodium pentachlorostannite; sodium pentafluorostannite**
Tin(IV) compounds (stannic compounds and stannates): **hydride, oxide** (tin dioxide; cassiterite (α) or tinstone [min.]); **sodium chlorostannate; sodium hexahydroxostannate**

IDENTITY AND PHYSICOCHEMICAL PROPERTIES OF THE ELEMENT

Tin (Sn) is a metallic element in Group IV of the periodic table, with the atomic number 50. It has 10 isotopes: ^{112}Sn, $^{114-120}Sn$, ^{122}Sn, and ^{124}Sn, the last being slightly radioactive. The most widely distributed isotope is ^{120}Sn (33%).

Tin is a heavy soft lustrous metal that is malleable and ductile at ordinary temperatures. It exists in three allotropic modifications: α, β, and γ. It slowly tarnishes in air. In compounds it is present in the oxidation state +2 (tin(II), or stannous, compounds) and in +4 (tin(IV), or stannic, compounds), the latter state being more common. A film of oxides that forms on the metal protects it from further oxidation. Tin is resistant to chemical agents under ordinary conditions. When melted, it burns in oxygen to the oxide SnO_2. Tin readily forms the fluoride SnF_4 with F_2 and the chloride $SnCl_4$ with Cl_2. On heating with sulfur it forms the sulfide SnS. When alloys of tin are reacted with magnesium and acids, the hydride SnH_4 results, which gradually decomposes on storage. Tin does not interact with carbon or silicon. It gives tin(II) salts with diluted acids. It reduces hot concentrated sulfuric acid to SO_2 with formation of the sulfate

$Sn(SO_4)_2$. It forms stannites (salts of the stannous acid H_2SnO_2) in concentrated alkalies and stannates (salts of the stannic acid H_2SnO_3) in the presence of oxidizing agents. Tin(II) salts are usually readily soluble in water, whereas the phosphates $Sn_3(PO_4)_2$ and $Sn(HPO_4)_2$ are virtually insoluble. In solution, tin(II) salts are easily oxidized, and the oxidation proceeds more rapidly in the light than in the dark. Tin(IV) salts are readily hydrolyzable. An important property of tin is its ability to form alloys with many other metals. See also Appendix.

NATURAL OCCURRENCE AND ENVIRONMENTAL LEVELS

Of the greatest commercial importance among tin minerals is cassiterite (tinstone) which usually contains nearly 79% of tin. The tin content of stannite is 27% and that of nordenskioldine (a calcium tin borate) is 43%. The impurities found in tin ores include sulfur and a number of metals (Cu, Pb, Zn, Li, Be, Ta, etc.). The clarke of tin is estimated at $(2.0-2.5) \times 10^{-4}\%$ in the earth's crust and at $2.7 \times 10^{-4}\%$ in the granite layer of the continental crust [Dobrovolsky].

Elevated levels of tin (6–10 mg/kg) occur in argillaceous sediments and low levels (0.35–0.50 mg/kg) in ultrabasic and calcareous rocks. The mobility of tin during weathering is strongly pH-dependent; in particular, Sn^{2+}, being a potent reducing agent, can be present only in media that are acid and reducing [Kabata-Pendias & Pendias].

The oceans have been estimated to contain 27.4 million tonnes of dissolved tin, at an average concentration of 0.020 µg/L. Concentrations in river water average 0.5 µg/L, and the estimated global run-off of tin from rivers amounts to 19 000 tonnes annually ($K_w = 1.56$) [Dobrovolsky]. In aqueous media, inorganic tin undergoes methylation with the formation of various methylated species capable of being bioaccumulated [Craig].

In soil, a common range of tin concentrations is considered to be 1 to 11 mg/kg dry weight. Standard soil samples were found to contain 4.5 mg Sn/kg dry weight; peats contained from 50–100 mg/kg ash weight. In surface soils of the USA, mean tin levels ranged from 0.6 mg/kg dry weight (soils on glacial till and drift) to 1.7 mg/kg (alluvial soils) [Kabata-Pendias & Pendias].

Continental vegetation contains an estimated 630 000 tonnes of tin, at average fresh weight, dry weight, and ash weight concentrations of $0.1 \times 10^{-4}\%$, $0.25 \times 10^{-4}\%$, and $5 \times 10^{-4}\%$, respectively. The amount of tin taken up annually through the increment in phytomass over the global land area is estimated at 69 000 tonnes, or 0.46 kg/km^2 (the $K_b = 1.85$) [Dobrovolsky]. Tin levels in mammalian tissues range from $n \times 10^{-3}\%$ to $n \times 10^{-5}\%$ dry weight [Kovalsky]. The common range of tin concentrations in plants is 20 to 30 mg/kg ash weight.

One step in tin movement through the food chain to higher trophic levels involves algae, some of which, such as *Ankistrodesmus falcatus*, are able to accumulate the metal to a level equalling 10% of their dry cell weight — i. e. in concentrations that exceed several thousandfold its concentration in the medium. In various algal species of the *Enteromorpha* and *Sargassum* genera, inorganic tin compounds undergo biodegradation (biomethylation) [Ischii; Wong *et al.*]. Inorganic tin is also biomethylated by some *Pseudomonas* bacteria which can take up the metal to a level of up to 18% of the cell mass [Ischii; Wong *et al.*, 1982, 1984]. (More detailed information on biomethylation can be found in Ridley *et al.* and Wood *et al.*) Plants growing in areas of tin mineralization accumulated the metal to levels as high as 300 mg/kg ash weight. The best tin accumulators were found to be sedges and mosses. Sugar beets grown near a chemical factory contained about 1000 mg Sn/kg dry weight, while vegetation in the vicinity of a tin smelter contained twice as much [Kabata-Pendias & Pendias].

Tin is relatively seldom detected in air and when detected, it is generally in low concentrations (≈ 0.01 μg/m^3), except in the proximity of some industrial sources. In the USA, tin concentrations ranged from 0.003 to 0.3 μg/m^3 in >60% of 754 samples taken in 22 cities; the lowest values were found in rural areas (tin was indetectable in >50% of samples from three urban and three rural sites) [WHO]. In the Heidelberg area in Germany, levels around 100 ng/m^3 were reported [Piscator]. Continental aerosols may be relatively rich in tin: values for the coefficient of aerosol concentration (defined as the ratio of the level of a metal in the solid phase of an aerosol to its level in the granite layer of the continental earth's crust) lie between 50 and 100.

PRODUCTION

The primary tin-containing material is first concentrated by flotation or gravity flotation. The concentrates are roasted to remove sulfur and then smelted in electric or flame furnaces to obtain the so-called black tin which contains 94–98% metallic tin and several impurities (Cu, Pb, Fe, As, Sb, Bi). This tin is filtered and then refined in various ways. A considerable proportion (10–20%) of the tin produced is secondary tin recovered from scrap (tinplate clippings from container manufacture, solder, bronze rejects, various alloys lacquers, used tin containers, etc.). The total world production of tin in 1975 was 236 000 tonnes, of which about 92% was produced as primary tin [WHO]. For the year 2000, 300 000 tonnes has been forecast [Kabata-Pendias & Pendias].

USES

Tin is mainly used in industries producing tinplate, solder, babbitt for bearings, brasses and bronzes, pewter, printer's alloy (type metal), plastics, and tin-containing chemicals. The largest single use of tin is in the production of tin-plated steel either by hot-dipping or by continuous electroplating. Tinplate is extensively used in food and beverage packaging and in aerosol containers. Tin is used as solder in transportation, machinery, electrical, plumbing, and heating trades and industries. In lead-free solder alloys, tin is alloyed with silver, antimony, zinc, or indium to obtain materials with special properties such as increased strength or corrosion resistance. The largest quantities of solder are utilized in car radiators, air conditioners, heat exchangers, plumbing and sheet metal joining, container seaming, electronic equipment, and computers. Special alloys containing tin include dental amalgams (silver — mercury alloys), titanium alloys for aircraft, and zirconium alloys for nuclear reactors. Inorganic tin compounds are employed in a variety of processes (e.g. for the strengthening of glass, in the dyeing and printing of textiles, as catalysts, as stabilizers in perfumes and soaps, etc.).

MAN-MADE SOURCES OF EMISSION INTO THE ENVIRONMENT

Anthropogenic sources of tin in the environment include industrial effluents and atmospheric emissions from tin-mining and tin-processing operations, various metallurgical and chemical processes, and miscellaneous industrial activities involving the use of tin or its compounds.

Emissions of highly dispersive aerosols from a large tin-producing plant were found to have high proportions (34–47%) of tin and its inorganic compounds (SnO_2, up to 90%; SnO, 5–6%; Sn, 2.3%) [Grin et al.]. Effluents from certain nonferrous metal works contained several hundred milligrams (up to 530) of tin per liter, while the precipitation-soluble forms of tin deposited daily in the area of some electric power stations amounted to 0.015 mg/km^2 and accounted for 11 to 44% of the total tin fallout depending on the distance from the station; within a radius of 5 km from the latter, the quantity of tin entering the environment each year was as high as 1 kg/km^2 [Popov & Konyusheva]. Around the cities of Zaparozhye and Donetsk in the Ukraine, the highest recorded tin concentrations in soil from emissions by iron and steel plants were in excess of 400 mg/kg [Makhonko]. At a distance of 700 m from an industrial facility, tin levels in the air were 3.8–4.4 μg/m^3 compared with 100 ng/m^3 at most in a non-industrial area [Doncheva et al.]. In Japan, tin concentrations of up to 640 μg/m^3 were found in emissions from some electric furnaces; at a distance of 700 m from these, atmospheric tin concentrations still ranged from 3.6 to 4.4 μg/m^3. In the USA, tin levels in air samples from an industrial site in Boston attained 0.8 μg/m^3. Peak concentrations of tin were found in particles 1 to 3 μm in diameter.

Tin is also released into the environment from the burning of coal and oil in which it was reported to be present at average levels of 2.0 and 0.01 mg/kg dry weight, respectively [WHO]. According to Dobrovolsky, the dry substance of coals contains 1.2×10^{-4}% of tin and the ashes of coals and oils, 15×10^{-4} and

(20–500) × 10⁻⁴%, respectively. The quantity of tin emitted into the environment from coal combustion has been estimated at 1600 tonnes (or 11 g/km² of land area) in 1970 and twice as much in 1980; for the year 2000, 5200 tonnes (35 g/km²) has been forecast [Dobrovolsky]. Sorokina *et al.* found for an industrial city that the amounts of tin entering the environment in atmospheric precipitation, solid waste, and waste water were 0.445 mg/day, 118 mg/day, and 2.70 mg/day per inhabitant, respectively; thus, the main source of environmental pollution with tin was solid waste; about 40% of the tin contained in it was accounted for by domestic waste.

Soil may be polluted with tin from tin-containing mineral fertilizers and pesticides. In some areas of England, for instance, tin levels above 1000 mg/kg were observed in fertilizer-treated soils compared with 75 mg/kg in those free from fertilizer. Tin levels ranged from 0.2 to 15.0 mg/kg in phosphorites and 3 to 4 mg/kg in phosphate fertilizers [Kabata-Pendias & Pendias].

In water- and soil-polluting industrial effluents, tin concentrations typically range from 40 to 700 mg/kg, the mean value being 160 mg/kg. The tin entering soil from the re-use and land disposal of industrial and domestic wastes, the incineration products of domestic garbage high in tin, and the application of fertilizers prepared from municipal/domestic refuse, can increase the natural soil content of the metal by tens of percents [Popov & Konyusheva; Sorokina *et al.*]. Tin contents in agricultural sources of soil contamination with trace elements were reported to be as follows (mg/kg dry weight) : sewage sludges, 40–700; phosphate fertilizers, 3–19; limestones, 0.5–4.0; nitrogen fertilizers (principally ammonium sulfate), 1.4–16.0; organic fertilizers (manure), 3.8 [Kabata-Pendias & Pendias].

Considerable health risks may be presented by processed foods and drinks containing elevated levels of tin because of corrosion and leaching out of the metal from plain unlacquered cans or from the foil used for packaging. The corrosion of tinplate in food containers such as cans depends on several factors including the type of food product, the duration and temperature of storage, the acidity of the product, and the amount of air present in the headspace of the can. The corrosion is accelerated by oxidizing agents (e.g. nitrates and ferric and cupric salts), anticyanin pigments, methylamine, sulfur dioxide, and other sulfur compounds, and is retarded by tin salts in solution, sugars, and colloids such as gelatin. A general reduction in corrosion and leaching has been achieved through the introduction of lacquered cans and the crimping of can tops, which minimizes the direct contact of food with tin. Some foods are unsuitable for packing in unlacquered cans because they promote corrosion. For example, nitrate can accumulate in tomato fruit grown under conditions of high temperature, high levels of nitrogen fertilization, and low light intensity. Such tomatoes can bring about considerable detinning of internal surfaces in plain (unlacquered) cans [WHO].

The transfer of tin into canned food is augmented by high storage temperatures; thus, an increase of 1°C in temperature raised the tin content of the food by 2 mg/kg per month of storage. The tin content of a canned food rapidly increases after opening the can. In pineapples stored at +8°C in unlacquered cans, for example, an increase from 50–77 to 260–300 mg/kg occurred in 72 h after the cans had been opened. Cow's milk in glass bottles contained tin in a mean concentration of only 0.0078 mg/L compared with 16 mg/L in evaporated milk taken from unlacquered cans and diluted to the equivalent volume of bottled cow's milk, while the milk from lacquered cans had less than 5 mg/L. The highest concentration of tin found in condensed milk stored for less than 4 weeks was 40 mg/L, and the concentration did not increase much upon further storage for 5 months but concentrations as high as 160 mg/L were recorded after 2 years of storage [WHO].

Tin concentrations ranging from 2 to 200 mg/kg were detected by Gelfand *et al.*, after 2 to 3 months of storage, in various canned foodstuffs such as tomato (containing up to 60% of dry substance); infant and dietetic foods; fruit jams, compotes, and marinades; snacks; and soups.

The distribution of tin in the environment (as well as in foods and living organisms) has been reviewed by Bulten & Meinema.

TOXICITY

Plants

Excessive or toxic tin levels in leaves of many plant species have been reported by most authors to be 60 mg/kg upward. On the criterion of phytotoxicity, the maximum permissible concentration of tin in surface soil is considered to be <50 mg/kg dry weight [Kabata-Pendias & Pendias]. Nutrient solutions containing more than 40 mg Sn/L are toxic to plant seeds [Bulten & Meinema].

Aquatic organisms

Young eel survived for an average of 2.8 h and 50 h in water containing **tin(II) chloride** at concentrations of 6.0 mg/L and 1.2 mg/L, respectively. To daphnids, this compound was lethal at 2.5 mg/L [Grushko]. Exposure to **tin(II) chloride dihydrate** at 1 g/L (or 530 mg Sn/L) was lethal for all fish tested (survival times varied from 4 min to 2 h) [Metelev *et al.*].

Animals and man

Single or short-term exposure

Animals

Threshold concentrations for acute effects in rats on 4 hour inhalation exposure were as follows (mg/m^3): **sodium hexahydroxostannate**, 6.5; **tin(II) sulfate**, 6.0; **tin(II) oxide**, 9.6; **tin(IV) oxide**, 21.8. LD$_{50}$ values of some tin compounds for mice and rats are given in Tables 1 and 2. **Tin(IV) hydride** is a strong convulsive poison. Exposure of mice by inhalation to its concentration of 2.0 mg/L of air caused convulsions in 7 min and death in 12 min. Guinea-pigs died within 15 min after being exposed to this concentration for 15 min by the same route.

Table 1. Median lethal doses (LD$_{50}$, mg/kg) of some tin compounds administered orally (by gavage)

	Animals	Salt	Metal
Tin sulfate	Rats	2205.0±295.0	1218.0±163.0
	Mice	2145.0±252.6	1184.0±138.5
Sodium hexahydrox-ostannate	Rats	4350.0±277.0	1934.4±123.2
	Mice	2685.0±288.6	1194.0±128.3
Tin(II, IV) oxide	Rats/mice	No deaths after	10 000 mg/kg

Source: Grin *et al.*

Table 2. Median lethal doses (LD$_{50}$, mg/kg) of sodium pentafluorostannite and tin(II) chloride dihydrate

Route	Sodium pentafluorostannite			Tin(II) chloride dihydrate
	Male mice	Male rats	Female rats	Male rats
Intravenous	18.9	12.9	12.9	29.3
Intraperitoneal	80.9	75.4	65.0	258.4
Oral: fasted	—	223.1	218.7	2274.6
fed	592.9	573.1	—	3190.1

Source: WHO.

In rats administered **sodium pentafluorostannite** or **tin(II) chloride dihydrate** by oral gavage in a single dose at the LD$_{50}$ level, ataxia, muscular weakness, and depression of the central nervous system were observed. A single intraperitoneal injection of the stannite at 35 mg/kg body weight or of the chloride at 44.4 mg/kg resulted in extensive necrosis of epithelial cells in the kidneys, mainly involving the proximal tubules. Similar changes along with a reduction in hemoglobin were noted in rats after their daily oral treatment with the stannite at 175 mg/kg for 15 days; the dose of 20 mg/kg had no effect on the hemoglobin level. Daily oral administration at 100 and 175 mg/kg for 30 days resulted in a dose-related retardation of growth in rats. A single subcutaneous injection of rats with tin(II) chloride dihydrate in doses from 5.6 to 56.4 mg/kg led, in 16 h, to a threefold increase in heme oxygenase activity in the liver and a 20- to 30-fold increase in heme-oxidizing activity in the kidneys [WHO]. Rats given tin as **tin(II) chloride, orthophosphate,** or **sulfate** in the diet at concentrations of 3 to 10 g/kg for 4 weeks exhibited decreased food intake, growth retardation, edema and atrophy of the pancreas, and reductions in hemoglobin levels, hematocrit, erythrocyte counts, and serum iron concentration; dietary supplements of iron had a strong protective effect [De Groot *et al.*]. Intratracheal administration of **metallic tin** dust to rats in an amount of 50 mg per animal was well tolerated and no pulmonary fibrosis was seen to develop within 1 year of exposure [Robertson].

Man

Cases of acute intoxication have mainly resulted from the ingestion of tin in food. Poisoned individuals experienced persistent headache, dizziness, stomach ache, nausea, vomiting, diarrhea, retention of urine, photophobia, rapid weight loss, and occasionally pareses and psychic disturbances. In cases of lethal poisoning, death occurred in a state of coma in the presence of cardiac and pulmonary abnormalities [Berman].

The lowest dose of tin causing clinical manifestations of poisoning such as nausea, vomiting and diarrhea, is usually 250 mg/kg of food. In a questionnaire study of 85 individuals most of whom had developed symptoms of poisoning after eating or drinking nothing else except tin-contaminated canned peaches, tin was found in concentrations of 413–597 mg/kg in the peaches and 298–405 mg/L in the syrup. Symptoms of poisoning occurred in 74 out of the 78 persons who had ingested tin in an amount of 100 mg or more, and also in two out of seven who had consumed only 50 mg [Piscator]. Symptoms and signs of acute gastroenteritis were recorded after the consumption of canned peaches and tomato juice containing tin in mean concentrations of 563 mg/kg and 400 mg/L, respectively [Winship]. Case studies of human intoxications have been reviewed by Bulten & Meinema and Pressel.

Chronic exposure

Animals

For rats continuously exposed to **tin(II) sulfate** by inhalation for 4 months, the threshold concentration was estimated to be 0.35 mg/m^3 for effects on hemoglobin, urea, sulfhydryl groups, and blood acetylcholinesterase and 0.09 mg/m^3 for effects on erythrocytes, total protein, chlorides, Ca, Na, K, β-lipoproteins, total lipids, triglycerides, cholesterol, catalase activity, alanine and aspartate transaminases, lactate dehydrogenase, and leukocyte bioluminescence; the no-observed-effect concentration was 0.04 mg/m^3. The compound also displayed embryotoxic and teratogenic activity. For rats exposed as indicated above, the threshold concentration for effects on relative ovarian weight, pre- and post-implantation mortality, and biochemical changes in amniotic fluid was

0.29 mg/m^3, while that for effects on the overall embryonal mortality, ossification of embryonal skeleton, and other related indices was 0.13 mg/m^3; the concentration of 0.045 mg/m^3 produced no embryotoxic or teratogenic effects. Male rats exposed to 0.09±0.07 mg/m^3 showed reduced sperm counts, while those exposed to 0.35±0.027 mg/m^3 had spermatozoa with reduced osmotic and acidic resistance. On histologic examination, desquamation of spermatogenic epithelium in the seminiferous tubules, increased numbers of abnormal spermatogonia, and signs of reduced spermatogenesis were observed. A tin sulfate concentration of 0.04 mg/m^3 has been recommended as the MAC for ambient air in communities. It has also been recommended that the 24-h (time-averaged) concentration of inorganic tin compounds in the ambient air be set at 0.02 mg/m^3 (as tin) [Bessmertnyi et al.; Grin et al.].

Tin(II) chloride or tin(II) sulfate added to the diet of rats at 10 g/kg for 4 and 13 weeks, led to homogenization of liver cell cytoplasm and hyperplasia of the bile ducts. After 15 weeks' administration, changes in the liver were more marked, and the chloride, in addition, caused testicular degeneration and a spongy state of the white substance of the brain. Similar, though smaller, alterations in the liver were observed in rats administered tin(II) chloride or tin(II) orthophosphate at 3 g/kg of diet. Tin(II) chloride in a dose of 5 mg/L given to rats in drinking water from weaning until their natural death, increased the incidence of fatty hepatic degeneration in female rats and produced vacuolar changes in the renal tubules of animals of both sexes; it also reduced the lifespan of female (but not male) rats [WHO].

In mice receiving tin in the form of sodium chlorostannate at 1 g/L in drinking water or in the form of tin(II) oleate at 5 g/kg in the diet, no adverse effects were observed. In pregnant rats fed sodium pentafluorostannite, sodium pentachlorostannite, and tin(II) fluoride in the diet (at levels corresponding to tin concentrations of 125, 250, and 500 mg/kg of diet) and killed on day 20 of gestation, no teratogenic effects were noted; concentrations of the metal in the fetuses were about 1 mg/kg compared with 0.65 mg/kg in control fetuses, which indicated that the transplacental transfer of tin was low [Theuer et al.]. Reliable data concerning the carcinogenicity or mutagenicity of inorganic tin compounds are lacking [WHO].

In rabbits administered tin at 0.1–3.0 mg/kg or at 6.0 mg/kg, disturbances of lipid and cholesterol metabolism, reduced lipolytic activity of the blood, and, after the larger dose, an abnormal ratio of protein fractions were recorded. Similar doses of tin(II) chloride lowered the serum activity of ceruloplasmin and produced a hypoglycemic effect. These changes were associated with inhibited function of the adrenal cortex and thyroid.

A specific adverse effect of tin is considered to be that on calcium metabolism. In rats given tin in a dose of 3 mg/kg, high levels of calcium in the renal cortex and of tin in renal cortical microsomal fractions along with reduced calcium levels in blood serum and in the epiphyses of femoral bones were observed, as was inhibition of calcium transport in the intestinal mucous membrane. In totally thyroid- and parathyroidectomized rats, however, calcium levels in the femoral bones remained unchanged although the serum level of tin was reduced. The diminished content of calcium in femoral bones has been attributed to its mobilization to the bloodstream under the action of parathyroid hormone and is regarded as a specific manifestation of tin toxicity [Popov & Konyusheva].

Man

Tin concentrations in the air near foundary furnaces ranged from 8.6 to 14.9 mg/m^3 [Piscator]. Workers may be exposed to aerosols formed during the smelting of tin and consisting mainly of tin(IV) oxide.

Occupational exposure to tin fumes or dusts by inhalation does not cause acute poisoning but can lead to stannosis — a form of pneumoconiosis characterized by deposition of tin and connective tissue development in the lungs, and by a relatively benign slow course without evidence of pulmonary dysfunction and with little or no effect on the working capacity. Radiologically, small dense shadows (somewhat denser that in silicosis) can be seen in the lungs. The initial finding is an increase of bronchovascular markings followed by hilar thickening; later, well-defined nodular elements may appear. Radiologic signs of stannosis have been observed after 3 to 5 years of exposure to tin fumes or dusts.

More than 200 cases of stannosis have been described. Stannosis developed, for example, in a workman who had been bagging, for 15 years, a material containing 96.5% of tin(IV) oxide and small amounts of Al, Fe, and Na but not Si; in a worker who attended a furnace in which metallic tin was burnt to produce tin(IV) oxide; and in 10 workers engaged in the reduction of tin ore concentrates to metal and exposed to dust with a high tin content. These 10 workers did not have any disability and although radiologic changes in their lungs were strongly marked, the vital capacity, maximal breathing and resting minute volume, and respiratory reserve remained within normal limits; the results of urine and blood examinations were also normal [WHO].

In a man who died 18 years after his last exposure to tin oxide fumes and who had stannosis without pulmonary dysfunction, 1100 mg Sn/kg wet weight was found in the lungs at autopsy (which is more than 1000 times the normal value); no reaction of the lung tissue to such a high tissue concentration had taken place, however [Piscator].

Cases of chronic bronchitis with initial signs of emphysema, moderate respiratory insufficiency, and other signs of pneumoconiosis occurred among workers of tin plants, particularly smelters who had been exposed to tin dust concentrations of 10 to 150 mg/m^3 for 6 to 8 years [Izraelson; Vorotnikova & Kuzmina].

Dermal and ocular effects

The application of 1% **tin(II) chloride** and 0.25% **tin(II) fluoride** solutions to intact skin of rabbits did not result in any injury to the skin [WHO]. Tin powder is moderately irritating to the eyes and respiratory tract.

ABSORPTION, DISTRIBUTION, AND ELIMINATION

Most of the tin enters the human body in food; much less enters from air. Tin levels ranged 0.19–0.65 mg/kg in potatoes and beets, up to 18.7 mg/kg in sunflowers and peas, and 0.1–1.0 mg/kg in meats, breads, fats, and miscellaneous edible plants.

In the USA, common hard, common soft, and durum wheats contained tin in concentrations of 5.6, 7.9, and 6.8 mg/kg, respectively, while samples of the flour and semolina prepared from these wheats, contained it in concentrations of 4.1, 3.7, and 6.0 mg/kg, respectively. In Bulgaria (around Sofia), tin concentrations in wheat, corn, beans, potatoes, tomatoes, cabbage, carrots, spinach, lettuce, onions, apples, and peaches ranged from 0.02 to 1.02 mg/kg [WHO]. A diet composed largely of fresh meats, cereals, and vegetables would contain about 1 mg of tin/kg, whereas a diet that includes considerable quantities of canned vegetables and fish, may contain some 40 mg of the metal [Grushko]. The mean total daily intake of tin in food has been variously estimated to be between 187 µg and 8.8 mg.

The absorption of ingested tin and its inorganic compounds from the gastrointestinal tract is less than 5% according to most studies, although values up to 20% have been reported. When rats were administered a tin(II) compound (citrate or fluoride) at 20 mg/kg body weight, only 2.8% of the ingested dose was absorbed; even less (0.6%) of the dose of the same compounds in the tetravalent form was absorbed [WHO].

With both oral and parenteral administration of tin to animals, the highest tin levels occurred in the kidneys, liver, and bone; the main site of its deposition was the bone. Two days after oral or intravenous administration, tin levels in the blood of rats were extremely low, and the metal was detected in erythrocytes only. Experimental studies on tin distribution in the body indicate that the metal is not rapidly oxidized or reduced in the process of absorption and systemic transportation [WHO].

Adult human tissues and organs were estimated to contain tin in the following amounts (mg/kg wet weight): muscle, 0.1; bone, 0.5; brain, 0.26–0.34; lung, 0.45–0.8, kidney, 0.2; liver, 0.4–0.6; spleen, 0.2; lymph nodes, 2.1; blood, 0.007–0.01 [Berman]. Human hair was reported to contain 0.007 µg Sn/g dry weight on average [Revich].

The distribution of tin in human tissues and organs varies considerably with age, geographic location, and other factors. For instance, all samples of kidney, liver, and lung tissue from Africa, Asia, and Europe had lower mean and median tin contents than those from the USA (the differences were least in the lung samples). Tin has been rarely detected in the tissues of stillborn infants, which indicates that it does not easily cross the placental barrier. With increasing age, tin was found to accumulate in the lungs, but not in other organs [WHO]. Most of the tin present in the lungs is thought to be derived from air.

The main route for excretion of absorbed tin is the kidney; a small fraction is excreted into the bile. In adult humans, the mean rate of urinary excretion was reported to be 23.4 µg/24 h (16.6 µg/L of urine). In rats given a single oral dose of tin(II) or tin(IV) citrate or fluoride at 20 mg/kg (as the metal), about 50% of the absorbed tin was excreted within 48 h. Studies on the elimination of ^{113}Sn(II) from animals of five species showed that, on average, 19.8 (mice) to 62.8% (dogs) of the intravenously injected dose was excreted in the urine and 5.6 to 10.6% in the feces within 3 days. In mice, rats, monkeys, and dogs, the biologic half-time of tin in the skeleton after intravenous or intraperitoneal injections of ^{113}Sn(II) was 90–100 days [Piscator; WHO].

HYGIENIC STANDARDS

Exposure limits adopted for tin and its inorganic compounds in the USSR are given in Table 3. For canned foods (condensed milk, vegetables, fruits, berries, juices, meat, poultry, and fish), the MAC has been set at 200.0 mg/kg.

In the USA, threshold limits values have been set at 2.0 mg/m^3 (TWA) and 4.0 mg/m^3 (STEL) for metallic tin and for tin oxide and inorganic tin compounds (as Sn).

Table 3. Exposure limits for tin and its compounds

	Workroom air		Atmospheric air			
	MAC_{wz} (mg/m^3)	Aggregative state	MAC_{hm} (mg.m^3)	MAC_{ad} (mg/m^3)	$TSEL_{aa}$ (mg/m^3)	Hazard class
Tin (and various inorganic compounds)	4.0	Aerosol	—	—	—	—
Tin(II) chloride	—	—	0.05	0.05	—	3
Tin(II) & tin(IV) oxide	4.0	Aerosol	—	—	0.05	—
Tin(II) sulfate	—	—	—	—	0.05	—

METHODS OF DETERMINATION

A wide variety of analytical methods for determining tin at low concentrations are available (for reviews, see Bulten & Meinema; Pressel; WHO). Some of the more commonly used methods are the following. In *air*: emission spectroscopy and neutron activation analysis; in *food*: atomic absorption spectroscopy, emission spectroscopy, and spectrography; in *biologic samples*: atomic absorption spectroscopy and neutron analysis; in *soil*: fluorimetry; in *water*: anodic stripping voltametry [Ilyashenko; Portretnyi *et al.*; Shau *et al.*; Winship]. In *wastewaters*: photometry with rhodanide and methyl violet, spectrography with hollow electrode, and atomic absorption spectroscopy (the detection limits of the these three methods are 0.002 mg/L, 1.2 µg/sample, and 10.0 mg/L, respectively) [*Guidelines for the Determination of Toxic and Very Toxic Substances...*].

MEASURES TO CONTROL EXPOSURE

At the workplace

Although most of the operations associated with the extraction and treatment of tin ore are wet processes, tin dust and oxide fume may escape during bagging of concentrates, in ore rooms, and during smelting operations, as well as during the periodic cleaning of the filters used to remove particulate matter from smelter furnace flue gas.

Safer working conditions can be assured by introducing changes in the manufacturing processes and equipment to reduce the pollution of workroom air by fumes and aerosols of tin and its compounds. Open surfaces associated with flotation processes should be reduced to a minimum, manual operations should be mechanized to the maximum possible extent, and automated lines monitored and controlled from a distance should be installed.

In the hydrometallurgical, reagent, sulfur, and furnace departments, hydrants (for flushing eyes with water) and emergency showers should by provided not more than 25 m away from permanent working sites in conjunction with sound alarms for summoning medical personnel in case of need. Food must be taken in separate rooms or in canteens. All those working with tin or its compounds should undergo preplacement and subsequent periodic medical examinations.

For *personal protection*, workers exposed to dusts of tin or its compounds should wear respirators. If the air contains tin hydride, all work operations must be performed in gas masks. Before leaving the plant, thorough washing of the hands, showering, and changing of the working clothes is mandatory, as is dust removal from the latter.

Protection of the general environment

Protective measures include provisions for reducing atmospheric emissions and the discharge of effluents from tin-mining and tin-processing plants and for ensuring strict adherence to the existing regulations on the recovery and disposal of industrial and domestic wastes. High priority should be given to the setting of health standards for environmental media, notably water and soil in view of the fact that tin compounds of heightened toxicity can be produced naturally in these media through biochemical processes of tin methylation.

For the prevention of food poisoning in the general population, the introduction of lacquered cans and the crimping of can tops is highly recommended, at this will minimize the direct contact of food with tinplate and bring about a general reduction in the corrosion and leaching of the metal.

REFERENCES

Berman, E. (1980) *Toxic Metals and their Analysis*. Heyden, London.
Bessmertnyi, A. N. *et al.* (188) *Gig. San.*, No. 9, 62–64 (in Russian).
Bulten, E. J. & Meinema, H. A. (1991) Chapter II. 30. *Tin* in: Merian, E. (ed.) *Metals and their Compounds in the Environment—Occurrence, Analysis and Biological Relevance*. VCH, New York, pp. 1243–1257.
Craig, P. J. (1986) In: Berhard, M. *et al.* (eds) *The Importance of Chemical Speciation in Environmental Processes*. Springer Verlag, New York, p. 443.
Dobrovolsky, V. V. (1983) *Geografiya mikroelementov. Globalnoe rasseyanie* [Geography of Trace Elements. Global Dispersion]. Moscow (in Russian).
Doncheva, A. V. *et al.* (1982) *Khimia v selskom khoziaistve* [Chemistry in Agriculture], No. 3, 8–10 (in Russian).
Gelfand, S. Yu. *et al.* (1984) *Vopr. Pitaniya*, No. 6. pp. 65–66 (in Russian).
Grin, N. V. *et al.* (1989) In: *Gigiena naselennykh mest* [Hygiene in Populated Areas], No. 28. Moscow, pp. 80–83 (in Russian).
de Groot *et al.* (1973) *Food Cosmet. Technol.*, **11**, 19–30.
Grushko, Ya. M. (1972) *Yadovitye metally i ikh neorganicheskie soedineniya v promyshlennykh stochnykh vodakh* [Toxic Metals and their Inorganic Compounds in Industrial Waste Waters]. Meditsina, Moscow (in Russian).
Guidelines for the Determination of Toxic and Very Toxic Substances Present in Low Concentration in Various Media and of their Migration (in various forms) in the Environment [Metodicheskie ukazaniya po opredeleniyu nizkikh kontsentratsiy toksichnykh i osobo toksichnykh veshchestv v razlichnykh sredakh i stepeni migratsii (v razlichnykh formakh) etikh veshchestv v okruzhayushchei srede] (1986). Publication of the USSR Ministry of Health, Moscow (in Russian).
Iliyashenko, L. A. (1978) *Gig. San.*, No. 12, 71–73 (in Russian).
Ishii, T. (1982) *Bull. Jap. Soc. Sci. Fish.*, **48**, 1609–1615.
Izraelson, Z. I. (ed.) (1963) *Toksikologiya redkikh metallov* [Toxicology of Rare Metals]. Medgiz, Moscow (in Russian).
Kabata-Pendias, A. & Pendias, H. (1986) *Trace Elements in Soils and Plants*. CRC Press, Boca Raton (Fl).
Kovalsky, V. V. (1974) *Geokhimicheskaya ekologiya* [Geochemical Ecology]. Nauka, Moscow (in Russian).
Makhonko, E. P. (1981) *Zagriaznenie atmosphery, pochvy i prirodnykh vod* [Pollution of the Atmosphere, Soils, and Natural Waters]. Leningrad (in Russian).
Metelev, V. V. *et al.* (1971) *Spravochnik po pestitsidam* [Pesticides: A Reference Handbook]. Moscow (in Russian).

Piscator, M. (1979) In: Friberg, L. *et al.* (eds) *Handbook on the Toxicology of Metals.* Elsevier, Amsterdam, pp. 613—626.

Popov, Yu. P. & Konyusheva, L. V. (1983) *Gig, San.*, No. 9, 55—57 (in Russian).

Portretnyi, V. P. *et al.* (1973) *Zh. Analit. Khimii*, 28, 437—439 (in Russian).

Pressel, G. (1988) Chapter 64 *Tin* in: Seiler, H.G. *et al.* (eds) *Handbook on Toxicity of Inorganic Compounds.* Marcel Dekker, New York, pp. 697—703.

Revich, B. A. (1983) In: *Metally. Gigienicheskie aspekty otsenki i ozdorovleniya okruzhayushchei sredy* [Metals: Hygienic Aspects of Evaluation and Environmental Health Promotion]. Moscow, pp. 73—80 (in Russian).

Ridley, W. P. *et al.* (1977) *Science*, 197, 329—332.

Robertson, A. J. (1960) In: King, E. J. & Fletcher, C. M. (eds) *Symposium on Industrial Pulmonary Diseases.* Boston, Little-Brown, pp. 168—184.

Shau, V. K. *et al.* (1982) *Anal. Chem.*, 54, 246—249.

Sorokina, E. P. *et al.* (1984) In: *Geochimicheskie issledovaniya dlya tselei ekologicheskoi otsenki urbanizirovannykh territoriy* [Geochemical Studies for Ecological Evaluation of Urbanized Areas]. Published by the Institute of Geography, USSR Academy of Sciences, under Project MAB-II, No. 2, pp. 36—59 (in Russian).

Theuer, R. C. *et al.* (1971) *J. Nutr.*, 101, 525—532.

Vorotnikova, A. I. & Kuzmina, O. I. (1973) *Gig. Truda Prof. Zabol.*, No. 11, 43—44 (in Russian).

WHO (1980) *Environmental Health Criteria 15: Tin and Organotin Compounds (A preliminary review).* World Health Organization, Geneva.

Winship, K. A. (1988) *Adverse Drug React. Acute Poison. Rev.*, 7, 19—38.

Wong, P. T. S. *et al.* (1984) *Can. J. Fish. Aquat. Sci.*, 41, 1570—1574.

Wong, P. T. S. *et al.* (1982) *Can. J. Fish. Aquat. Sci.*, 39, 483—488.

Wood, J. M. *et al.* (1978) *Fed. Proc.*, 37, 16—21.

Lead and its compounds

Lead acetate; l. azide; l.(II) bromide chloride; l. carbonate (cerussite [min.]);
l.(II) chloride (cotunnite [min.]); l.(II) chromate (chrome yellow, crocoite
[min.]); l. hydrogen orthoarsenate (acid lead orthoarsenate, disubstituted);
l. hydrogen orthoarsenite (acid lead orthoarsenite, disubstituted); l. metasilicate
(lead silicate); l. nitrate; l. orthoarsenate; l. orthophosphate; l.(II) oxide (lead
monoxide, litharge; massicot(β) [min.]); l.(II, IV) oxide (trilead tetroxide, red
lead, minium); l.(IV) oxide (lead dioxide); l.(II) sulfate (anglesite(α) [min.]);
l. sulfide (galena [min.], lead glance [min.]); l. zirconate-titanate
(l. metazirconate-metatitanate)
Tetralead dihydroxide dioxide sulfate (basic lead sulfate)
Trilead dihydroxide dicarbonate (lead carbonate-lead hydroxide (2/1), basic lead
carbonate, white lead)

IDENTITY AND PHYSICOCHEMICAL PROPERTIES
OF THE ELEMENT

Lead (Pb) is a metallic element in group IV of the periodic table, with the atomic
number 82. Its natural stable isotopes are ^{204}Pb (1.48%), ^{206}Pb (23.6%), ^{207}Pb
(22.6%), and ^{208}Pb (52.3%); the naturally occurring radioactive isotopes are
^{209}Pb (half-life = 3.3 h), ^{210}Pb (23.3 years), ^{211}Pb, ^{212}Pb, and ^{214}Pb.

Lead is a heavy malleable and ductile plastic metal that is softer than any
other heavy metal and much softer than aluminum, and can be cut with a knife.
It can be readily rolled into sheets but cannot be drawn out into a wire because
of its low strength.

In compounds lead is present in the oxidation state +2 (lead(II), or plumbous, compounds) or, much less commonly, +4 (lead(IV), or plumbic, compounds). In air at ordinary temperatures, the surface of lead oxidizes readily and is then very resistant to corrosion. Molten lead oxidizes to the lead(II) oxide PbO, then to the lead(II, IV) oxide Pb_2O_3, and finally to the lead(II, IV) oxide Pb_3O_4. On heating, lead directly combines with halogens, sulfur, and tellurium. It does not react with dilute sulfuric acid, dilute hydrochloric acid, or hydrofluoric acid because of a protective layer of the sulfate $PbSO_4$, chloride $PbCl_2$, or fluoride PbF_2 forms on its surface. It, however, reacts wits dilute nitric acid. It forms alloys with many metals. See also Appendix.

NATURAL OCCURRENCE AND ENVIRONMENTAL LEVELS

The clarke of lead in the earth's crust is estimated at $(12.5-16) \times 10^{-4}\%$. The major lead-containing minerals are galena (lead glance) PbS, anglesite $PbSO_4$, and cerussite $PbCO_3$. The total global reserves of lead are estimated at around 100 million tonnes, with most of this amount occurring in the form of sulfate. Annually, up to 210 thousand tonnes of lead has been estimated to enter the environment from the various natural sources such as silicate dusts of soils, volcanic smoke, evaporation from forests, forest fires, sea salt aerosols, meteoric and meteoritic dust, These emissions give rise to natural background levels ranging from 2×10^{-9} to 5×10^{-4} μg/m^3. Lead levels rise with time in all environmental media [Rovinsky et al.].

The oceans contain an estimated 41.1 million tonnes of dissolved lead, at an average concentration of 0.03 μg/L. (Average lead concentrations were reported to be 0.01–0.03 μg/L in deep water and 0.3 μg/L in surface water [Chow & Patterson].) Large quantities of lead are contained in ferromanganese oxide deposits of the oceans. In the Pacific, the amount of lead scavenged by these deposits has been calculated to be 5.4 thousand tonnes per annum. The global mean lead content in lakes and rivers has been estimated at 1–10 μg/L [WHO, 1977], and the amount entering the oceans with river discharges, at 37 thousand tonnes per annum $(K_w = 0.52)$ [Dobrovolsky]. Analyses of ground water demonstrated lead concentrations varying from 1 to 60 μg/L [WHO, 1977]. Lead levels in river, lake, and marine sediments have been reviewed by Huynh-Ngoc et al.

In soils, lead concentrations are of the order of its crustal abundance because of the stability of this element in the geochemical cycle [Brooks]. Although it is difficult to separate background lead levels in soils from those due to human activity, the average natural concentrations are considered to range from 1 to 27 mg/kg and are highest in the upper 15 cm of the soil profile due to the high lead-absorbing capacity of the humus horizon [Ewers & Schlipköter]. The major contribution to lead fixation in soil is made by organic matter. The principal mechanism of fixation is coordination bonding of the metal by groups with a free electron pair. Lead adsorption by humus and the strength of the bonds formed increase with increasing alkalinity of the soil medium. The intensity of lead fixation is strongly influenced by clay minerals present in the soil [Bolshakov *et al.*]. Significantly increased lead concentrations in surface soils have been found in inner cities, near busy highways and near lead works [Ewers & Schlipköter].

The earth's biomass, most of which is made up by land vegetation, has been estimated to contain up to 50 million tonnes of lead and is a major factor in its migration and distribution in the environmental media. Continental vegetation has been estimated to contain 6.3 million tonnes of lead, at average wet weight, dry weight, and ash weight concentrations of $1.0 \times 10^{-4}\%$, $2.5 \times 10^{-4}\%$, and $50 \times 10^{-4}\%$, respectively, and its estimated amount taken up annually worldwide through the increment in terrestrial phytomass is 431 thousand tonnes, or 2.87 kg/km^2 ($K_b = 3.73$) [Dobrovolsky].

For the ambient air, average lead concentrations of 1.1 μg/m^3 for urban and 0.21 μg/m^3 for nonurban areas have been reported [WHO, 1977]. Concentrations in air above the north-central Pacific Ocean and south Indian Ocean are of the order of 0.001 μg/m^3 [Tsuchiya].

Average lead contents in organisms were reported to be (mg/100 g dry weight): marine algae, 0.84; land plants, 0.27; marine animals, 0.05; land animals, 0.2; bacteria, 0.6–1.0.

Normal lead concentrations were estimated to be (mg/kg dry weight) 2.5 in leaves and twigs of woody plants and 0.1–1.0 in vegetables and cereals; the usual concentrations in pasture grasses were found to be about 1.0 mg/kg dry weight [WHO, 1977]. Some blue algae are capable of accumulating lead by a factor of more than 100 000 [WHO, 1977], and some plankton species by a factor of 12 000 [Mullins]. Coniferous plants and mosses are also efficient lead concentrators [Kovalsky]. See also *Environmental Health Criteria* [WHO, 1989].

PRODUCTION

Lead is commonly obtained by roasting lead(II) sulfide followed by reduction of the resulting lead(II) oxide with coke at 1500°C; lead of very high purity (99.99%) is made by electrolytic refining..

Lead(II) azide is obtained by reaction of sodium azide and lead(II) nitrate solutions. **Lead(II) carbonate** is prepared by bubbling carbon dioxide through an aqueous solution of lead(II) acetate, while **trilead dihydroxide dicarbonate** is made by bubbling carbon dioxide and air through a boiling aqueous solution of lead(II) carbonate. **Lead(II) chloride** is chiefly obtained either by dissolving lead(II) oxide or trilead dihydroxide dicarbonate $2PbCO_3 \cdot Pb(OH)_2$ in hydrochloric acid or by dissolving granulated lead in nitric acid followed by precipitation with hydrochloric acid. **Lead(II) chromate** is obtained as a product of the exchange reaction between lead(II) nitrate and potassium chromate. **Lead(II) nitrate** can be obtained by dissolving lead, lead(II) oxide, or trilead dihydroxide dicarbonate in a hot dilute nitric acid. **Lead(II) orthoarsenate** is made by reaction of lead(II) acetate with sodium orthoarsenate or by electrolysis of the latter on a lead anode. **Lead(II) hydrogen orthoarsenate** and **hydrogen orthoarsenite** are by-products in the smelting of nonferrous metals. **Lead(II) oxide** is obtained by oxidation of lead in air at 600°C; by boiling lead(II) hydroxide with a sodium hydroxide solution; or in the process of removing impurities from lead in metallurgy. **Lead(II,IV) oxide** is made by heating lead(II) oxide in the presence of air at 400–500°C and **lead(IV) oxide**, by decomposing lead(II,IV) oxide in nitric acid, by electrolytic oxidation of lead(II) salts, or by the action of strong oxidizing agents on lead salt solutions.

USES

The manufacture of electric storage batteries is responsible for the largest consumption of lead [WHO, 1977]; thus, 40% of all lead consumed is used for the production of lead-acid batteries [Ewers & Schlipköter]. Lead is also used in cable production; in the chemical industry to obtain lead pigments and tetraethyllead and tetramethyllead (the production of these alkylleads, which are used as antiknocking agents in gasoline, accounts for approximately 10% of the world lead consumption [Ewers & Schlipköter]); in the chemical equipment industry; in the building and construction industries for roofing, wall cladding, and sound insulation; in solders, bearing metals, and type metal; for shielding against gamma radiation; as a component of many alloys; and in a variety of minor applications. Lead sheets, cable sheathings, solder, ammunition, bearing alloys, type metal, tubes, weights and ballast, and low melting alloys account for 20% of the lead used [Ewers & Shlipköter].

Many lead compounds (e.g. oxides, carbonate, chloride, chromate, trilead dihydroxide dicarbonate) are used in the manufacture of paints including white lead.

Lead(II) oxide is used in making plates for electric storage batteries, in compounding rubber, in glass, glazes, and vitreous enamels, and in the manufacture of drying oil and of other lead compounds. **Lead(II,IV) oxide** is used in storage-battery plates, in glass and ceramics, in printing inks, in putties, as a paint pigment, and for the production of **lead(IV) oxide** which finds use as an oxidizing agent in the dye and chemical industries and in the manufacture of matches.

Lead acetate is employed as a mordant in dyeing and printing, as a drier for paints, and in the manufacture of other lead salts; **l. azide**, as a detonating agent; **l.(II) chromate**, in pigments and as an oxidizing agent; **l. metasilicate**, in the preparation of frits for glazes and as a stabilizer in the manufacture of plastics; **l. nitrate**, in pyrotechnics and to obtain metallic lead and lead compounds; **l. orthoarsenate**, as a pesticide. **L.(II) sulfate** and **tetralead dihydroxide dioxide sulfate** (basic lead sulfate) are added to lacquers and paints to improve the weatherproofness of the coatings, while the sulfate is also used for the production of metallic lead and lead compounds. L. **sulfide** is used to make photoresistors. L. **zirconate-titanate** (a solid solution containing 56–67% lead, 20–22% zirconium, 11–12% titanium, and ~16% strontium) is used to make piezoelectric cells. For recycling of lead, see Ewers & Schlipköter.

MAN-MADE SOURCES OF EMISSION INTO THE ENVIRONMENT

The major sources of lead in the environment that are of significance to human health arise from the industrial and other uses of lead. The most important source of lead pollution is emissions from motor car exhausts because antiknock agents such as tetraethyllead and tetramethyllead are added to gasoline supplies. Other important and widespread sources are lead mining, smelting, and refining, incineration and discharge of industrial and municipal wastes, miscellaneous high-temperature manufacturing processes, and combustion of fossil fuels. It has been estimated that no less than 260 000 tonnes of lead are deposited on the earth's surface from exhaust gases and 90 000 tonnes from metallurgical industries. The quantity of lead emitted into the environment from coal burning has been estimated at 13 000 tonnes (or 33 g/km^2 of land area) in 1970 and twice as much in 1980; for the year 2000, 43 000 tonnes (290 g/km^2) has been forecast [Dobrovolsky]. The control of lead pollution has become one of the most important problems facing mankind.

Human activities are responsible for the release of between 430 and 650 thousand tonnes of lead into the world's ocean annually. The recent upper layers of oceans contain approximately five times as much lead as the lower ancient layers which have not been exposed to appreciable anthropogenic activity. Highest concentrations were found in the surface 10-μm film of ocean water—up to 2920 μg/L as compared to a maximum of 13 μg/L at a depth of 50 cm. Calculations have shown that the concentration of lead in the surface waters of the North Atlantic Ocean has risen from 0.01 to 0.07 μg/L during the last few decades since the introduction of lead in gasoline [Mullins]. Particular hazards to the hydrosphere are created by wastewaters from metallurgical, metal-processing, machine-building, chemical, pharmaceutical, photographic materials, and match industries. Lead concentrations in wastewaters sometimes reached about 8200 mg/L [Grushko, 1979] (lead in wastewaters is reviewed by Ewers & Schlipköter and Huynh-Ngoc et al.). Raw sludges in the primary settling tanks of biologic wastewater-treatment plants contained lead at levels up to 13.5 mg/kg [Nainshtein]. The presence of lead in the initial links of the food chain results in its accumulation from wastewaters by certain bacteria that are in turn consumed by oligochaete worms of the Tucidae family. Powerful lead accumulators are gastropods: some of them contained close to 7 mg Pb/kg dry weight in an aquatic medium that contained only 3.5×10^{-4} mg Pb/L [Zhulidov et al.].

One may consider, as a first approximation, that the input of lead to the biosphere from anthropogenic sources is probably of the same order as from natural sources, although it is believed by some that the environmental pollution from the former sources exceeds by a factor of 100 that from the latter. In major industrial centers, lead concentrations in the ambient air may rise up to 40 $\mu g/m^3$, i.e. exceed by hundreds of times the natural background. Annually, lead levels in air may increase by 2 $\mu g/m^3$ in large cities, 1 $\mu g/m^3$ in small towns, and by 0.5 $\mu g/m^3$ in rural areas. According to one estimate, the content of lead in the human environment has increased by four orders of magnitude over the period of human evolution [Eisinger].

The transport and distribution of lead from both stationary and mobile sources is mainly via air. Large quantities are also discharged into soil and water, but lead tends to localize near the points of such discharges because of low solubility of the lead compounds that are formed upon contact with soil and water. Lead discharged into the air over areas of high traffic density falls out mainly within their immediate vicinity, although the fraction that remains airborne (about 20%) is widely dispersed. Depending on their size, airborne suspended particles may have a long residence time in the atmosphere. Lead can be removed from the atmosphere either by dry or wet deposition. Residence time for small particles is of the order of several days and is influenced by rainfall. Despite its widespread dispersion, with consequent dilution, lead may build up at points extremely remote from human activity, as is evidenced, for example, by lead levels in the ice strata of Greenland. Up to 1750, the lead content of these was about 20 μg per tonne. As a result of the Industrial Revolution, this value had increased to 50 $\mu g/$tonne by 1860. With the explosive growth of the automobile industry, the level rose sharply to 120 $\mu g/$tonne in 1950 and to 210 $\mu g/$tonne by 1965 [Murozumi et al.].

The main sources of lead entry into soils are the atmosphere, inadequately purified wastewater and sewage sludge, fertilizers, and particularly automobile exhausts. A conspicuous feature of lead from the latter source is its tendency to concentrate in soils and vegetation along a relatively narrow strip on both sides of major highways. The exhaust gases of gasoline engines yield lead chloride, bromide, oxide, fluoride, and other particles, at least 20% of which settle in the immediate proximity of the highway. When lead was measured in the ash of grasses bordering two major highways in Denver, Colorado, values up to 1000 ppm were obtained for samples nearest the road; the influence of lead could be detected up to 200 m from the road [Cannon & Bowles]. In close proximity to highways, grasses were found to contain lead at about 100 mg/kg (background level, 10 mg/kg) in Sweden and 12–19 mg/kg in Japan and the USSR (background level, 3–4 mg/kg). Within a big parking site and at different distances from it, soil, grasses and insects (beetles of various species) contained the following lead concentrations (mg/10 kg dry weight): soil, 72 within that site, 30 at 600 m, 11 at 1500 m, and 9 at 5000 m; grasses, 65, 31, 10, and 5, respectively; insects, 31–131, 21–34, 16–25, and 15–25, respectively [Makarova et al.; Zhulidov & Yemets].

Traffic is also a highly serious pollutant of urban air. In Los Angeles, for instance, the monthly mean lead concentration near traffic was as high as 6.4 $\mu g/m^3$ as compared to the general ambient air level of 2–4 $\mu g/m^3$ in that city [WHO, 1977]. Appreciably elevated blood lead levels occurred in workers of filling stations (329±0.72 $\mu g/L$ vs 143±0.72 $\mu g/L$ in the control).

Considerable health hazards are presented by lead-containing street dust, particularly for children. For example, 5 to 50 mg of lead was found to have entered their bodies from this source over 30 min of intensive physical activity on a playground [Dinerman et al.].

A relationship has been found between traffic density, proximity to the highway, engine acceleration, wind direction, and the amount of lead in the air. Particle size studies revealed that 65% of the lead between 10 m and 550 m from the roadway consisted of particles under 2 μ diameter, and 85% of particles under 4 μ diameter [Daines et al.]. These authors also found that high lead pollution extended usually to about 80 m from the main highway. It has been stated that dwelling houses, vegetable plots, and wells should preferably be located at least 100 m away from motorways [Nikiforova].

Substantial sources of pollution in local areas are lead deposits, mining activities, smelters and other metallurgical works, and cement factories. In one area of lead deposits, the soil contained lead at 14–27 mg/kg dry weight and water at 0.008–0.03 mg/L; as a result, elevated lead levels were detected in 12 food crops — for example (mg/100 g dry weight) 0.248 in onion (vs 0.121 in control samples), 0.228 in potatoes (vs 0.0609), 0.117 in beet (vs 0.08), 0.129 in corn (vs 0.004), 0.242 in tomatoes (vs 0.125) [Agafonova & Zakirova]. In Derbyshire, England, extensive abandoned lead workings dating back to Roman times gave rise to a very considerable accumulation of lead in soils (up to 7000 ppm) [Brooks].

A source of very substantial local pollution are smelters. Their influence on the surrounding air and soil depends in large measure on the height of the stack, the trapping devices in the stack, the topography, and other local characteristics [Ewers & Schlipköter]. The emissions can cover a considerable area. In one study, for instance, anomalous lead values were detected in soil and grass up to 10 km from a smelting works [Burkitt et al.]; air and water pollution can extend up to 100 km away from smelters. Lead in soil near a lead mining area in Idaho reached 20 000 mg/kg [Tsuchiya]. In the area adjoining a lead smelter, lead concentrations in the ambient air regularly ranged from 0.62 to 0.95 $\mu g/m^3$ and occasionally peaked up to 12 $\mu g/m^3$. Appreciable contamination may occur in fruits and vegetables grown in areas exposed to smelter emissions. When lead in the total diet of peasants living near a smelter was determined, the daily ingestion of lead with food was found to be 670–2640 mg [Kerin].

Solid emissions from one metallurgical works contained 250 to 440 mg of lead per kg and raised its concentrations in the nearby soil up to 1.8 mg/kg [Garmash]. In the vicinity of such works, leaves of trees contained 69–290 mg Pb/kg against 0.11 mg/kg in a control area. Lead concentrations as high as these may be lethal even to such powerful plants as chestnuts and oaks.

A serious source of lead pollution is cement production using metallurgical dross and pyrite cinders; atmospheric emissions from this source reached 0.0028 mg/m^3. There are reasons to belive that cement factories may well be comparable to metallurgical works and motor vehicles as environmental pollutants.

Two special sources of lead pollution should be mentioned — improperly glazed earthenware containers and lead-based paints. Such containers have been responsible for numerous cases of lead poisoning due to the release of soluble lead from their glazed surfaces. Flaking lead-based paints have long been considered as a major source of excessive lead intake by children of preschool age because of their habit of licking, chewing, or actually eating foreign objects. It has been estimated that 50% of paint is removed from surfaces protected by lead pigments in a period of 7 years before repainting [Patterson]. In one study, 75% of children with elevated lead concentrations in the blood ($\geqslant$60 µg/100 ml) lived in houses with at least one painted surface [Greenfield et al.]. Children may also be exposed to lead from colored newsprint or lead-painted toys.

Further sources of lead pollution worthy of mention are lead pipes and fittings in the plumbing of domestic water supplies, illicit whisky, certain wines, lead solder in the seams and caps of cans, and tobacco. Lead content of tobacco has been reported to vary from 3 to 12 µg per cigarette, resulting in the inhalation of 1.2 to 4.6 µg lead per 20 smoked cigarettes [Tsuchiya].

Attention has been called to soil pollution in areas where geese were hunted on a massive scale: the number of lead shot in one such area exceeded 100 000 per square hectare. A fairly common source of lead has been reported to be the crumbs and dust of aging leaded rubber in X-ray rooms [Usoltsev et al.].

In occupational environments, the highest exposure is experienced by workers engaged in mining, smelting (primary and secondary), and various manufacturing processes where lead is used, the major pathway of exposure being inhalation. Lead concentrations in smelters and storage battery factories have often exceeded 1 000 µg/m^3. In many countries, however, considerable success has been achieved in reducing lead exposure. (For a review of occupational exposure to lead, see Louekari et al.)

Good bioindicators of the urban atmospheric pollution are found in certain domestic animals and plants. From one study, lead levels of 339, 333, and 349 µg/L were reported for, respectively, the blood, plasma, and erythrocytes of city dogs as compared to 213, 255, and 163 µg/L, respectively, in dogs living in a rural area [Gilli *et al.*]. The hair of dogs accumulated lead to 1.89 µg/g [Nemenko *et al.*]. Sheep grazing at distances of 3–50 m from a highway with a traffic density of 5000 vehicles/day, contained appreciably elevated lead levels in their wool [Ward & Brooks]. In dolphins and large marine fishes, tissue levels of lead were found to increase from one year to another and it has been suggested that its levels in their muscles may serve as indicators of pollution of the marine environment by lead in the long term [Honda *et al.*].

The maximum permissible lead levels in soils and sewage sludges, established several years ago in the East European countries and the USSR, are 1 000 mg/kg for sludge and 100 mg/kg for soil (recommended levels are 750 mg/kg for sludge and 50 mg/kg for soil) [Leschber]. In the USA, limits of 5 mg Pb/L and 20 mg Pb/L, respectively, have been set for waters used for long-term and short-term irrigation [Grushko, 1979]. Lead levels in the food chain are reviewed by Ewers & Schlipköter.

TOXICITY

Microorganisms

Soil microflora show a strong tendency to accumulate lead at a high rate, and its elevated concentrations in the soil make soil microbiocenoses less abundant and decrease the maneralization of organic nitrogen in the soil and its enzymatic and total biologic activities. The most sensitive tests for biologic activity of soil have been reported to be those for the quantity of free amino acids and enzymatic activity in the soil; nitrification processes have proved to be less sensitive to lead [Bolshakov *et al.*].

The toxicity of lead on soil microflora depends of soil type; in a chernozem soil, for example, its toxicity will be neutralized more rapidly than in a soddy podzol soil. Soil microorganisms particularly sensitive to lead are the actinomycetes and bacteria assimilating organic nitrogen, and they may therefore serve as indicators of soil pollution by this metal; the most resistant microorganisms are fungi and bacilli [Bulavko].

Lead concentrations ⩾0.07 mg/L interfered with the biologic purification of wastewater, while those ⩾0.1 mg/L adversely affected the activated sludge from sewage treatment plants [Grushko, 1972, 1979].

Inorganic lead compounds are, in general, less toxic to microorganisms than organic ones such as trialkyl- and tetralkyllead compounds. A major factor influencing lead toxicity to aquatic organisms is the free ionic concentration, which affects lead availability to organisms. The toxicity of inorganic lead salts is strongly dependent on environmental conditions such as water hardness, pH, and salinity [WHO, 1989].

The toxic threshold of lead nitrate for the bacterium *Escherichia coli*, related to cell numbers produced, was 1.3 mg /L. The nitrate had little effect on the growth of the human skin bacterium *Micrococcus luteus* at a concentration of 0.6 mg/L over 48 h [WHO, 1989].

Land plants

Lead is mainly accumulated by vegetative organs of plants, namely leaves, stalks, roots, and tubers (e.g. in crop plants such as potato, beet, radish, carrot, wheat, pea, and tomato), whereas its content in the seeds and fruits remains virtually constant. Translocation of lead from soil to plants depends on the composition and pH of the soil, the degree to which the metal is bound to soil complexes, and on its ability to pass through the root membrane [Dinerman *et al.*]. The tendency of inorganic lead to form highly insoluble salts and complexes with various anions, together with its tight binding to soils, drastically reduces its availability to terrestrial plants via the roots. Translocation of the lead ion in plants is generally limited and most bound lead remains at root or leaf surfaces. For this reason, high lead concentrations (100 to 1 000 mg/kg soil) were usually required to cause visible toxic effects on photosynthesis, growth, or other parameters. Thus, lead is only likely to affect plants at very high environmental concentrations [WHO, 1989].

Lead levels in soil that decrease the height of plants by 5–10% are considered as toxic and were found to be 50 ppm for oat and clover plants. In experimental plots containing 500–2500 mg Pb/kg soil, the yields of radish, lettuce, and onion were lower by 23–50%, 40–68%, and 23–74%, respectively, than in the control plots with normal lead levels [Pollak & Fisher]. The lowest lead concentrations in soil adversely affecting the growth of herbage and trees were 364 mg/g and 1600 mg/g, respectively [Dinerman, *et al.*]. Among plants growing in a soil containing about 8 mg Pb/kg, the smallest amounts of this metal ($\leqslant$0.5 mg/kg) were accumulated by legumes and the greatest amounts by turnip leaves ($\leqslant$16.2 mg/kg) and pumpkin fruits; in most other plants, lead levels did not exceed 3 mg/kg [Grigorieva & Khramova]. At soil lead levels of 50–300 mg/kg, lead concentrations in the edible parts of vegetables exceeded the permissible limits.

Aquatic organisms

The toxicity of lead-contaminated water to fish varies considerably, depending on the availability and uptake of the lead ion [Prosi]. Factors affecting the availability are water hardness, pH, salinity, and organic matter, while lead uptake is affected by the presence of other cations and the oxygen content of the water. The toxicity of inorganic lead may be 10 to 100 times lower than that of organic lead. The 96-hour LC_{50}s for most fish vary between 1 and 25 mg/L in soft water and between 440 and 540 mg/L in hard water. Juvenile stages are generally more sensitive than adults, whereas eggs are less sensitive because lead is absorbed onto the egg surface and excluded from the embryo [WHO, 1989].

In soft water, manifestations of intoxication in most fish species appeared at 0.1–0.4 mg/L. **Lead(II) nitrate** caused mortality among stickleback at 0.1 mg/L, inhibited growth in tadpoles at 1.6 mg/L, and killed daphnids within 24 h at 5 mg/L. **Lead(II) chloride** was lethal to some whitefish at 0.58 mg/L and to daphnids at about 0.5 mg/L. **Lead(II) sulfate** was highly toxic to minnow and crucian carp at 2.5 mg/L. Fish are capable of detecting and avoiding lead [Grushko, 1979; Metelev *et al.*]. Digestive and tissue enzymes were found to be inhibited in freshwater fishes whose muscles contained lead at concentrations of 0.04–5 mg/g [Sastry & Gupta].

Birds

Metallic lead is not toxic to birds exept at very high dosages when administered in powder form, but is highly toxic when given as lead shot; thus, ingestion of a single pellet of lead shot can be fatal for some birds. Since birds have often been found in the wild with large numbers of lead shot in the gizzard (20 shot was not unusual), lead shot is a major hazard to bird species feeding on river margins and in fields where much shot has accumulated. Lead salts are only toxic at high dietary doses (>100 mg/kg). Although a variety of effects at high dose levels have been reported, most can be considered as primary effects on food consumption (anorexia and weight loss) [WHO, 1989]. In the poisoned birds lead shot is found in the gizzard, and lead levels are elevated in the liver, kidneys, and bones. Poisoning occurred in a falcon that had preyed upon mallard ducks containing lead in their bodies [Benson *et al.*].

Experimental animals and man

General considerations

Lead belongs to the large group of moderately toxic metals that can cause acute and especially chronic poisoning with a great variety of clinical manifestations. Lead affects particularly the central and peripheral nervous systems, the hematopoietic, gastrointestinal, and renal systems, vessels, blood, and protein synthesis, and it can also exert genetic, gonadotoxic, reproductive, and embryotoxic effects [Roshchin; Tsuchiya]. There is also some evidence that it may be teratogenic [Ewers & Schlipköter; Fergusson; Goyer, 1988; Shepard].

The classification adopted in the USSR distinguishes between the following four forms of lead poisoning (saturnism) : carrier state (excess lead in the urine or a lead line on the gum margin without clinical manifestations of intoxication); mild poisoning (blood changes such as reticulocytosis and basophilic stippling of erythrocytes; porphyrinuria; asthenic–vegetative syndrome [a combination of asthenia with vegetative disturbances]); moderate poisoning (anemia, toxic hepatitis, lead colic, pronounced asthenic–vegetative syndrome); severe poisoning (progressive anemia; lead colic, lead encephalopathy, paralysis) [Artamonova].

The effects of lead on the nervous system vary with the duration and intensity of exposure and may be manifested in altered behavior, depression, loss of memory and ability to concentrate, electroencephalographic changes, motor dysfunction, impaired speech and writing, muscular rigidity, and peripheral neuropathy.

Lead effects on the hematopoietic system are primarily due to interference with the activities of the enzymes δ-aminolevulinic acid dehydratase (ALAD), which catalyzes the formation of porphobilinogen from δ-aminolevulinic acid (ALA), and heme synthetase, which incorporates iron into protoporphyrin IX [Ewer & Schlipköter; Fergusson; Goyer, 1988; Shestakov; Tsuchiya]. Hematopoietic changes appear before any clinical signs and symptoms of poisoning, and they also provide quantitative information on the intensity of lead exposure.

The effects on the gastrointestinal system are most commonly manifested clinically in intestinal colic, which is usually preceded, and almost invariably accompanied, by constipation.

There is evidence that lead may adversely affect the genetic apparatus of the cell [Fomenko *et al.*]. Increased mortality from lung cancer among lead miners has been reported [Axelson & Sundell]. Lead acetate has been classified among substances possessing carcinogenic properties and presenting particular hazards as pollutants of atmospheric air, drinking water, and food.

An important role in lead intoxication is played by its competitive relationships with other metals, including iron [Fergusson; Goyer, 1988]. It has been shown in many studies that lead inhibits the incorporation of iron into protoporphyrin, which results in increased iron excretion in the urine. As lead accumulates in the body, blood levels of copper, manganese, aluminum, and zinc rise, with concurrent increases in the urinary excretion of these metals. Of major importance is the competition of lead with calcium, since lead has a strong tendency to accumulate in bone and replace calcium (94–95% of the total lead present in the human body was calculated to be in the bones [Barry & Mossman]).

Lead and its compounds are capable of exerting direct effects on the living cell. An important factor in the pathogenesis of lead poisoning has been reported to be the damage inflicted by lead compounds to lysosomal membranes; lead ions have been shown capable of causing destructive changes in erythrocyte membranes, blocking sulfhydryl groups, disrupting lipoprotein complexes in tissues, damaging mitochondria, and impairing oxidative phosphorylation [Teras & Kahn].

Most inorganic lead compounds have similar modes of action, and the differences between their toxicities are mainly due to those in their solubility in the body fluids, in particular gastric juice. Hydrochloric acid in the stomach and carbonic acid in the lungs may convert insoluble compounds into soluble and readily absorbable products. The greatest toxicity is exhibited by basic lead carbonate (white lead) and lead(II) oxide. Special modes of action are displayed by lead compounds containing a toxic anion, such as the orthoarsenates, chromates, and azide [Fairhall et al.]. In terms of their decreasing toxicity to guinea-pigs, lead compounds were arranged in the following order: nitrate > chloride > oxide > carbonate > orthophosphate [Tartler].

Animals most sensitive to lead and its compounds are dogs and horses; cats and rabbits are moderately sensitive; guinea-pigs, rams, and goats show low sensitivity, while rats, mice, and birds are fairly resistant [Cantarow & Trumper].

Lead poisoning in cattle and transfer of lead to milk have been examined by Oskarsson et al.

Exposure by inhalation

Animals

In a 6-month study, rats inhaling **lead sulfide** at 0.048 mg/m^3 for 6 h/day accumulated lead in the bones, showed signs of impaired higher nervous activity, and developed morphologic changes in cerebral cortical cells; the concentration of 0.013 mg/m^3 was without detectable effects. Exposure of rats to dusts of **lead oxides** at 0.01 mg/m^3 under similar conditions (6 h/day for 6 months) produced a 10-fold increase in the lead content of the bones and altered conditioned reflexes, the activity of several enzymes, and the histologic structure of the brain and spinal cord; in rabbits so exposed, impairment of porphyrin metabolism was also observed. The oxides did not affect conditioned reflexes in rats at 0.0013 mg/m^3 and porphyrin metabolism in rabbits at 0.0039 mg/m^3 under the exposure conditions used [Talakin, 1978, 1979].

In an inhalation toxicity study where rats were continuously exposed to **lead** for 78 days, the lowest effective (threshold) concentration was 0.0024 mg/m^3, as judged by impaired motor chronaxie of antagonistic muscles, reductions in cholinesterase activity and sulfhydryl groups in whole blood, and increased counts of reticulocytes and erythrocytes with basophil granules. Other threshold concentrations of the acetate were calculated to be (mg/m^3) 0.016 for inhibition of sperm motility and 0.010, 0.012, and 0.022 for effects on the summated threshold index, the blood content of sulfhydryl groups, and the erythrocyte count, respectively [Kamildzhanov].

Daily inhalation exposures of rats to **lead metasilicate** dust for 3 to 12 months caused blood changes typical of lead poisoning, blockade of sulfhydryl groups, and dystrophic changes in internal organs at the 2.2 mg/m^3 level, inhibited immunobiologic reactivity without producing detectable clinical or morphologic changes at ≈1.0 mg/m^3, and were virtually harmless at 0.1 mg/m^3 [Lagutin].

Rats and monkeys inhaling a **lead(II) bromide chloride** aerosol at 0.6 mg/m^3 for periods of 1 to 2 years exhibited chronic nephritis, impaired porphyrin metabolism, and diminished osmotic resistance of erythrocytes. Female rats chronically exposed to dusts of **tetralead dihydroxide dioxide sulfate** at 1 mg/m^3

(as Pb) and then mated to unexposed males produced less viable offspring than did control animals.

Narrowing of the nasal cavity by mucosal edema and infiltration, necrosis and desquamation of the upper respiratory tract epithelium, and considerably elevated levels of lead in the laryngeal cartilages, tracheal rings, and the intima of pharyngeal blood vessels were observed in rabbits that had been continuously maintained for prolonged periods in a lead smelter where concentrations of airborne lead dust ranged from 0.00092 to 0.10667 mg/m^3.

Rats administered a glass-ceramic cement dust intratracheally in the amount of 50 mg three times at monthly intervals developed blood changes typical of lead intoxication without any evidence of pulmonary fibrosis [Fedoseyeva et al.].

Man
The main manifestations of lead intoxication are the appearance of a lead line near the gum margin, particularly at the anterior teeth (although the line may not appear even in cases of severe intoxication), a sallow skin, reticulocytosis (>10%), the presence of basophilic stippled erythrocytes (>15 per 10 000 cells), and an elevation of δ-aminolevulinic acid in blood (>2 mg%), of porphyrins in urine (>50–60 μg/L) [Rashevskaya et al.], and of lead in both blood and urine. A fairly consistent early warning of potentially more serious effects likely to occur with further exposure is gastrointestinal colic. Taken separately, the above-mentioned symptoms and signs are not pathognomonic of lead poisoning and the correct diagnosis can only be made when all or some of them are present together. The lead line and lead in the urine in the absence of other symptoms (e.g. a disturbance in porphyrin metabolism) are taken to be signs of a lead carrier state.

In chronic lead poisoning, the following main syndromes or effects on specific organs and systems have been observed.

Nervous system. Asthenic syndrome. This syndrome usually develops in the early stages of intoxication and results from dysfunction of the central nervous system. Its chief symptoms are headache, vertigo, easy fatiguability, irritability, sleep disturbances, and memory deterioration. Diminished sensory perception, muscular hypotension, and excessive sweating are also common [Rashevskaya et al.].

Encephalopathy. This tends to start acutely and is characterized by such symptoms as headache, vertigo, sleep disturbances, epileptic fits, transient disorders of speech and vision, spastic paralysis, clouding of consciousness, and coma.

Conditions such as 'lead depression' and 'lead mania' have also been described, as have attacks of acute excitation, hallucinations, mental confusion, and other manifestations. Acute encephalopathies have often been fatal, while chronic encephalopathies are marked by a more favorable course and outcome.

Included under the 'encephalopathy syndrome' is lead meningitis accompanied by fever, intracranial pressure elevation, and a characteristic lymphocytosis in the cerebrospinal fluid [Rashevskaya *et al.*]. In cases of chronic progressive lead intoxication, an early cerebral atherosclerosis may develop. Autopsy of individuals who died with the latter condition revealed flattened gyri and diffuse dystrophic changes in the brain with sclerotic shrinkage and breakdown of the nerve cells. Computer tomograms taken in three males aged 54–57 years who had worked in lead smelters for more than 30 years and contained lead at concentrations of 54–72 μg/100 ml in their blood serum, demonstrated calcium deposits in the subcortical area of the cerebrum, basal ganglia, and cerebellar hemispheres; neurologic abnormalities included dementia, reduced visual acuity, peripheral neuropathy, syncopal states, nystagmus, rapid onset of fatigue, and pain in the back [Reyes *et al.*].

Motor disturbances. These may be manifested as polyneuritis that mainly affects the extensor muscles of the wrist and of fingers. A typical disease of the peripheral nervous system is lead paralysis, especially in the most active muscles, including paralysis of the radical nerve with wristdrop; the afferent nerves are not affected; there is no loss of sensation and no pain. In severe cases, the lower extremities (extensors of the feet) may be involved, with the development of footdrop due to paralysis of the perineal nerve.

The effects of lead on motor neurons may not be manifested clinically in any way or there may be only a decrease in the strength of the hand grip and abnormal excitability of nerves and muscles in the upper extremities. Marked bioelectric changes in muscles may combine with the complete absence of clinically manifest motor dysfunction [Okhnyanskaya & Komarova]. A prominent feature of chronic lead poisoning may be reduced motor conduction velocity (especially of arm nerves), which has been observed at blood lead levels of 200–700 μg/L. Early motor disturbances include tremor of the hands, legs, upper eyelids, and tongue, nystagmoid movements of the eyeballs, and paresis of the laryngeal muscles [Atchabarov].

Polyneuritis with sensory changes has been seen most frequently with latent or moderate lead intoxications. Such patients complained of aching limbs, pain along the course of nerves on palpation, and rapid fatiguability of limb muscles; vasomotor disturbances, manifested as cyanosis of the limbs, temperature asymmetry, and sweating, were common. The sciatic nerve was affected more frequently than any other. The proximal portions of peripheral nerves are generally more vulnerable than their distal portions which are usually affected in severe cases only. As the intoxication progresses, the pathologic process tends to spread not so much down the affected nerve(s) as to involve other nerves, leading to polyradiculoneuritis [Atchabarov].

Hematopoietic system. Among the early stages of lead intoxication are reticulocytosis, anisocytosis, and microcytosis. Subsequently, oligochromic or, less commonly, normochromic anemia develops; characteristic features are basophilic stippling of erythrocytes and monocytosis [Rashevskaya *et al.*]. Intravital examination of bone marrow samples often revealed signs of bone marrow irritaion, increased numbers of nucleated erythroblastic elements, and the presence of erythroblasts with damaged mitochondria and ribosomes, and it is these changes that are actually responsible for the formation of basophilic stippled erythrocytes. As indicated by cytochemical studies, the basophilic granules of these cells may consist exclusively of RNA. The primary toxic effect of lead on the bone marrow thus entails the release of abnormal erythrocytes into the bloodstream, and these cells then undergo hemolysis as a secondary phenomenon, which is confirmed by reductions in their osmotic and acidic resistance and their shortened life-spans [Sokolov *et al.*]. One of the main causes of lead anemia is the impairment of porphyrin metabolism which underlies hemoglobin synthesis [Aldanazarov]. In the erythrocyte, lead directly combines with the hemoglobin to form insoluble complexes, followed by the release of potassium from this cell and deterioration of its mechanical stability [Moore]. The major effects of lead on hematopoiesis that are readily measurable in man are those on the rate of excretion of aminolevulinic acid and coproporphyrin.

Gastrointestinal tract. Characteristic symptoms of lead intoxication are an unpleasant taste in the mouth, poor appetite, nausea, heartburn, belching, intense intermittent abdominal pain (lead colic), gingivitis, and pigmentation of the gums and teeth. Among chronically poisoned persons, increased prevalences of dental caries and periodontosis have been observed [Rashevskaya *et al.*]. Decreased pancreatic function and hypersalivation with low amylolytic activity of the saliva may occur [Okshina; Timoshina *et al.*].

Lead colic usually begins suddenly in chronically intoxicated individuals and is characterized by a triad of symptoms: severe spasmodic abdominal pains, constipation unrelievable by laxatives (although diarrhea may occur), and a rise in arterial pressure. Other possible symptoms and signs include chills followed by fever (up to 38°C), nausea, vomiting, and presence of protein, erythrocytes and casts in the urine; oliguria (anuria occasionally) and sinus bradycardia may be added. The pathogenesis of colic has been associated with damage to the solar plexus and higher autonomic centers as well as to autonomic nerves in the intestinal wall directly. Some individuals with severe lead intoxication accompained by gastrointestinal colic developed atrophic gastritis involving enhanced motor and evacuative functions of the stomach [Lobanova *et al.*]. Colic is most commonly encountered in industrially exposed workers, often at relatively low exposure levels.

Liver. Liver damage is manifested in elevated values of serum transaminases and in γ- and β-hyperglobulinemia. Lead can act directly on the oxidative systems and structure of hepatic cell mitochondria and affect the intralobular circulation in an indirect manner.

Kidney. Lead can cause spasm of renal vessels and damage the renal glomeruli and tubules, resulting in epithelial and then interstitial nephropathy which is seen with prolonged lead exposure. Biopsy in cases of nephropathy showed focal fibrosis, degeneration of the tubular epithelium, and increased permeability of the basement membrane to proteins. Renal functional impairment is mainly manifested in reduced urea and creatinine clearance, proteinuria, aminoaciduria, and glucosuria [Dynnik; Liubchenko *et al.*]. The nephropathy may be associated with hypertension and gout.

Cardiovascular system. Lead-intoxicated persons may complain of cardiac pain: arrythmia, tachycardia or bradycardia, dull sounds, and systolic murmurs at the cardiac apex may be present. Electrocardiographic abnormalities have also been observed. The effects of lead on the cardiovascular system are reviewed (Ewers & Schlipköter, 1991).

The toxic action of lead on the heart has been associated with impairment of the extracardiac regulation of cardiac activity and with inhibition of enzyme metabolism in the myocardium. It is thought that lead effects on the heart are not primary ones, and that myocardial lesions can only occur in very severe poisoning as a result of impaired coronary circulation. Capillary vasoneurosis may precede the appearance of the major signs of saturnism.

In instrument-making and storage battery factories, the prevalence of hypertension increased with increasing duration of lead exposure [Cramer & Dahlberg; Kaplan]. In general, however, hypertension is not unduly prevalent among workers exposed to this metal [WHO, 1977] (a slight hypertension due to vascular contraction has not infrequently been observed in cases of acute lead poisoning [Goyer, 1988]).

One long-term effect of lead intoxication has been reported to be connective tissue restructuring in the aorta, capillaries, and myocardium similar to that seen when this tissue is undergoing accelerated aging [Sanotsky *et al.*].

Metabolic, endocrine, and reproductive effects. A major metabolic effect of lead is the impairment of porphyrin metabolism. Porphyrinuria and especially elevation of δ-aminolevulinic acid are held to be the cardinal signs of lead poisoning. In workers occupationally exposed to lead, the urinary level of iron rose, on average, from 342 to 460 μg% and the urinary excretion of lead increased 3- to 6-fold over 5 years of exposure [Bikezina]. Lead alters the energetics of cells. Lead intoxication may result in impaired protein, lipid, and carbohydrate metabolism, vitamin B_1 deficiency, drastically reduced nicotinic acid levels in the blood and urine, and lowered blood concentration of vitamin B_{12}, as well as in a deficiency of vitamin C which probably promotes lead deposition in the form of insoluble lead ascorbate [Rashevskaya et al.].

Lead may also impair thyroid and adrenal functions [Ewers & Schlipköter; Fergusson; Monaenkova]. The impairment of thyroid function was manifested in altered iodine accumulation in this gland and in reduced thyroxine secretion by it. Workers exposed to lead for a long time showed reduced activity of the pancreatic insular apparatus and of the hypophyseoadrenal system [Liubchenko et al.; Talakin].

Menstrual disturbances (most often hypomenorrhea) and high rates of anovulatory cycles with luteal abnormality were observed among women aged 21–40 years after 6 to 20 years of work in lead trades [Panova}.

Lead effects on the reproductive function have been manifested in increased rates of premature delivery, miscarriages, and fetal death [Goyer, 1988]. Children born to lead-exposed mothers tend to show decreased growth rates and high mortality. It has been estimated that lead concentrations in the maternal blood that are hazardous to the fetus and particularly the embryo are $\geqslant 30$ µg/L, and that such concentrations are likely to occur in women exposed to lead concentrations in the ambient air of ~5 µg/m³. In children, blood lead levels between 250 and 550 µg/L were associated with behavioral deficiencies and higher levels ($\geqslant 600$ µg/L), with debility [Bridbord].

It has been shown that both lead poisoning and moderately increased lead absorption may decrease fertility in men. Direct gonadal effects of lead such as asthenospermia, oligospermia, and teratospermia and reduced secretory function of the prostate and seminal vesicles were observed in men with blood lead concentrations of 410–745 µg/L who worked in nonferrous metals plants, printing establishments, or storage battery-manufacturing factories [Cullen et al.; Fomenko et al.; Goyer, 1988]. Depressed serum testosterone levels were recorded in some male lead workers; it has been suggested that lead first acts on the testes directly, and that this is followed by involvement of the hypothalamohypophyseal system [Radamilans et al.]. (See also Ewers & Schlipköter.)

Ocular effects. Periarteritis and lead retinopathies, congestion and atrophy of the optic papilae, and narrowing of the visual fields have been described [Atchabarov]; adverse effects of lead on the visual fields have been considered as one of the early and most frequent signs of lead intoxication [Vints]. Paralysis of the oculomotor nerves may also occur.

Other effects. Many workers in lead industries had atrophic changes in the nose, larynx, and pharynx [Basamygina et al.]; on roentgenologic examination, structural abnormalities in bones and occasional large foci of resorption in the epiphyses and metaphyses of the femoral bones were seen [Grinberg et al.]. In children attending creches and kindergartens located 150–300 m away from a lead-processing plant, signs of focal sclerosis were present in calcification sites of the long bones.

Chronic lead poisoning may also result in depressed natural immunity, as is evidenced by the reductions observed in the phagocytic activity and osmotic resistance of leukocytes, in complement, agglutinin, and hemolysin titers, in bactericidal properties of plasma and saliva, and in antimicrobial activity of the oral mucosa [Rakhimova]. Weakened mitogenic activity of T lymphocytes was recorded in workers of a lead storehouse [Alomran & Shleamoon]. Allergic diseases (mainly dermatoses) and, less often, enteropathies were seen in about 20% of the workmen and other employees of a lead plant [Iskakov]. Many lead workers with periods of exposure up to 5 years showed sharply increased histamine levels and lowered histaminase activity in their blood [Talakin].

Toxic concentrations. According to reports in the literature reviewed by Atchabarov, a single exposure to lead at concentrations of $271-795$ mg/m^3 was fatal, exposure to $9.9-11.4$ mg/m^3 led to poisoning in $1-16$ days and to severe intoxication (which could be fatal) in $4-9$ months, and that to $0.7-1.7$ mg/m^3 resulted in signs of intoxication several days to months later; concentrations of 0.07 to 0.14 mg/m^3 were described as hazardous; concentrations between 0.011 and 0.04 mg/m^3 caused functional alterations in the higher nervous activity after 6 months of periodic exposure and signs of poisoning after 8 years of regular exposure.

Examination of workers in storage battery factories revealed chronic poisoning in 177 individuals exposed to lead at 0.15 mg/m^3 [Dressen *et al.*]. In an instrument-making factory where manual soldering operations using lead were performed and its concentrations in the workroom air ranged from 0.004 to 0.009 mg/m^3, an increased prevalence of hypertension that correlated with the duration of exposure was recorded [Kaplan]. Predominantly sensory disturbances similar to those seen in polyneuritis developed in workers after long-term exposure to lead at $0.00002-0.00005$ mg/L of air [Shraiber & Mosevich].

Children aged 2 to 11 years living in areas with mean annual lead concentrations in the ambient air between 0.0004 and 0.002 mg/m^3 showed reduced scores in intelligence tests and were found to have lead levels of $320-430$ μg/L in the blood as well as a fourfold higher lead content in the deciduous teeth and more than a 200-fold higher lead content in the surface of the hand skin as compared to their peers from control areas [Roels *et al.*; Winneke *et al.*, 1982, 1983].

Exposure by other routes

Animals

In mice, the intraperitoneal LD_{50}s were calculated to be 217 mg/kg for **lead(II) oxide**, 291 mg/kg for **lead(IV) oxide**, 600 mg/kg for **lead(II) sulfate**, and 17,7 g/kg for a glass-ceramic cement containing **lead(II,IV) oxide**; the intraperitoneal LD_{50} of lead(II) sulfate for rats was 282 mg/kg. In mice, threshold intraperitoneal doses of lead(II) oxide and lead(IV) oxide for effects on body weight and blood levels of sulfhydryl groups were 36.1 mg/kg and 48.5 mg/kg, respectively [Liublina & Dvorkin].

Orally administered **lead acetate** and **lead nitrate** caused considerable damage to the nervous tissue of rats, rabbits, and dogs [Michaelson; Okhnyanskaya & Komarova; Tolgskaya]. In rabbits, a marked reduction in the erythropoietic activity of the bone marrow was observed. In rats and rabbits, cytochemical and electron-microscopic studies demonstrated vacuolated erythrocytes and platelets and signs of impaired nucleoprotein metabolism in bone marrow cells. Dogs developed atrophic and necrotic lesions in the gastrointestinal tract; in some animals, erosions and ulcers were observed, and histologic studies revealed impaired innervaion of the digestive tract glands, decreased numbers of neurons in the ganglia, and gross changes in the myalinated fibers [Shevchenko & Tolgskaya]. In the kidney, the most vulnerable structure is the epithelium of proximal tubules where the greatest morphologic changes were seen; in particular, inclusion bodies composed of lead-protein complexes appeared in epithelial cell nuclei. These intranuclear inclusion bodies probably represent a mechanism for sequestration of lead, given that they have a high and specific affinity for lead and were found to contain about 90% of the lead present in the kidney [Goyer 1971; Sun et al.].

Lead carbonate fed to rabbits and cats in the diet caused loss of teeth by destroying the dental alveoli [Shevchenko & Tolgskaya].

Lead metasilicate induced blood changes in rats, guinea-pigs and rabbits not only on inhalation exposure but also when applied to their skin [Lagutin].

Lead acetate produced ocular changes in rabbits when fed to them in the diet at concentrations of 750 to 2250 ppm for 12 months: rabbits receiving medium to high concentrations in this range developed changes in neuronal elements of the retina, while those receiving low concentrations, corresponding to environmental levels of the acetate in industrial cities, showed changes in the pigmented epithelium; similar dose-dependent ocular effects were detected among rabbits after a single intraperitoneal injection of lead carbonate in the 20–80 mg/kg dose range [Hennekes et al.].

In rats repeatedly exposed to **lead** by the oral route in doses that did not cause any alterations in the general condition, blood, or nervous system of the animals (0.002, 0.02, or 0.2 mg/kg six times on alternate days), a number of gonadotoxic effects were observed. In male rats, these included increased weights of the testes and prostate, hypertrophy of the latter gland, impaired spermatogenesis, diminished sperm motility, the appearance of abnormal spermatozoa, and decreased RNA synthesis in the spermatozoa. Findings in females included disturbances of the estrous cycle, atrophy of the ovarian cortex, and degenerative changes in the ova [Golubovich & Gnevkovskaya; Tarabayeva; Yegorov *et al.*].

Miscarriages, stillbirths, and defective offspring were frequent among rats exposed to lead through the diet before mating and during pregnancy. When only male rats were fed lead, a 15% decrease in litter size, a 12% decrease in the birth weight of live pups, and a 18% decrease in their survival rate were recorded; when only females were fed, the corresponding decreases were greater (26%, 19%, and 41%, respectively), and they were still greater when both the males and females were exposed to lead (35%, 29%, and 67%) [Stowe & Goyer]. Marked effects on rat pups were exerted by the lead transmitted in the maternal milk; as compared to the controls, such pups contained 10 to 45 times more lead in the internal organs, had less myalin in the brain and smaller axons, and showed delayed development of the nervous system [Heyxmancik *et al.*; Krigman *et al.*].

Prolonged low-level exposure of rats to lead through food and water (in doses of 0.0015–0.05 mg/kg body weight) increased the percentage of chromosomal aberrations in bone marrow cells, decreased the mitotic index, induced subclinical neurologic alterations, and depressed the activity of Na^+- and K^+-dependent adenosinetriphosphatase in erythrocytes [Krasovsky *et al.*; Litvinov *et al.*; Teras & Kahn].

An excess of lead in soil and in plant feedstuffs may present health hazards for domestic, agricultural, and wild animals [Ewers & Schlipköter]. Sources of poisoning may be, for example, pastures and watering places contaminated by emissions from lead mines or smelters or by lead-containing automobile exhausts. Acute poisoning may result from the eating of leaded paint, linoleum, or plaster. The main symptoms and signs of such poisoning were loss of appetite, lethargy, anemia, diarrhea, and cramps. Osweiler *et al.* detected 80 cases of lead poisoning among grazing cows; in animals that died, lead levels reached 780 µg/L in the blood, 442.5 mg/L in the rumen's contents, 29.7 mg/kg in the liver, and 57.7 mg/kg in the kidney. According to these authors, an unequivocal diagnosis of lead poisoning cannot be made if the lead content of the liver, as determined by biopsy, is less than 10 mg/kg.

Milk with lead levels of 0.05 to 0.15 ppm obtained from cows grazing in the vicinity of a lead mine was reported to be unsafe for human consumption [Bahat & Koshnamaehan].

The minimal daily oral dosages of lead that caused poisoning in cattle were usually 5 to 7 mg/kg, but could be lower depending on the nature of feed and the time and method of administration. Horses are more sensitive to lead: they died after about 2 months when ingesting lead at 2 mg/kg daily; characteristic signs of poisoning were paralysis of the pharyngeal muscles and highly labored respiration [Aronsen]. Pigs are fairly resistant to lead: only slight signs of intoxication were evident in pigs maintained on drinking water containing 290 μg Pb/L [Lassen & Buck]. Sheep dogs poisoned through eating lead-containing paint flakes in their kennels failed to guard sheep and became excited and cowardly; in some dogs, the picture of poisoning resembled that seen in canine distemper, and the final diagnosis could only be made after measuring lead in their blood.

Man

Reported lethal oral doses of lead usually range from 155 to 455 mg/kg. Cases of acute and chronic poisoning resulting from ingestion of lead in food or water in substantial quantities (>5 mg/kg) have mainly occurred in the domestic environment.

Under certain circumstances, the concentration of lead in drinking water can become very high and cause poisoning. For example, two cases of severe clinical poisoning attributable to a municipal water supply that contained lead at 2.6 mg/L were described [Gajdos & Gajdos-Török]. A group poisoning involving 14 persons who had been ingesting as much as about 100 mg of lead daily in their drinking water for many years was reported from the USA (cited in Grushko [1979]). In Great Britain, some 5 million people were reported to live in homes with lead pipes in the plumbing; as a consequence, lead concentrations up to 8500 μg/L were acquired by tap water (which exceeded more than 15-fold the limit of 50 μg/L set by WHO), and many inhabitants had blood lead levels between 500 and 1500 μg/L. Attempts have been made to diminish the plumbosolvency of water in lead pipes by increasing its pH and hardness [Akers & Fellows].

A fairly common source of domestic poisoning by lead is handcrafted glazed earthenware, since improper glazing is conducive to the leaching of lead, particularly if such earthen vessels contain acidic liquids, and cases of both fatal and nonfatal poisoning have resulted from their use. For example, two cases (one fatal) of poisoning with apple juice that had been stored in a glazed earthen vessel for 3 days and contained 1300 μg Pb/L were described [Klein et al.]. Sixty people were poisoned (with four fatalities) by jam stored in glazed pots and containing as much as 120 mg Pb/L. Other, similar examples are cases of acute poisoning with kvas (a national Russian beverage usually prepared from bread) kept in an earthen jar for 5 days, with pickled cucumbers, and numerous cases of poisoning with canned cowberries from improperly glazed pottery. Acid products kept in such vessels may acquire extremely high lead concentrations. For instance, values up to 48.8 mg/L were recorded for sauerkraut, 29.0 mg/L for pickled tomatoes, 48.9 mg/L for pickled cucumbers, 59.3 for fruit paste (these values by far exceed the permissible limit of 7 mg/L set in the USA for the leaching of lead from glazed ceramic articles). The amount of leached lead depends on the duration of storage, the temperature of the medium, and how often the vessel has been used [Henderson *et al.*].

Another not uncommon source of lead poisoning is illicitly distilled whisky or homebrew produced in homemade stills containing lead components or lead solder. Poisoning may also result from the consumption of milk products stored in flasks containing lead solder or of orange juice stored in an open lead-soldered can. In all such cases the victims complained of abdominal cramps, constipation, general weakness, dizziness, and pain in the limbs and lumbar region; characteristic findings on examination were pallor of the skin which appeared sallow, yellowish coloration of the sclerae, a gray line on the gum margin, and a painful liver [Girard *et al.*; Henderson *et al.*; Novoselsky; Raskovalov].

A source of lead poisoning, especially in Oriental countries, has been facial cosmetics. Some infants were poisoned by the milk of their mothers who had used a lead-bearing cosmetic paint (the milk contained up to 74 µg Pb/L). Further (very rare) instances of poisoning that may be mentioned are a case of poisoning in a child who had swallowed a 'pearl' whose brilliant surface contained 18.6% Pb, and a typical picture of acute lead intoxication developed by persons 2 to 3 months after receiving a blind gunshot wound [Galkin & Patrikeyev].

Effects of dermal exposure

Inorganic lead compounds do not irritate the skin. Absorption through undamaged skin may occur, particularly upon contact with organic lead compounds such as tetraethyllead used as an antiknock additive in gasoline [Rastogi & Clausen; Shimaitis & Tsiunene].

Effects of lead hydrogen orthoarsenate and lead hydrogen orthoarsenite in animals

The oral LC_{50} of lead hydrogen orthoarsenate (which is used as an insecticide) was determined at 1500 mg/kg for mice and 3100 mg/kg for rats, and its intraperitoneal LD_{50} for mice at 128 mg/kg. The corresponding LD_{50}s of lead hydrogen orthoarsenite (used for pest control in agriculture) were 150 mg/kg (oral for mice), 190 mg/kg (oral for rats), and 74 mg/kg (intraperitoneal for mice).

The acute effects of these two compounds were similar and included dyspnea, tremor, diarrhea, hypothermia, hypodynamia and, as found at autopsy, dystrophic changes and vascular lesions in parenchymatous organs. The threshold oral doses for acute toxicity in mice, as gauged by the appearance of hypothermia and diarrhea, were 370 mg/kg for the orthoarsenate and 20 mg/kg for the orthoarsenite. Mice repeatedly exposed by the oral route to the orthoarsenate at 220 mg/kg or to the orthoarsenite at 20 mg/kg over 11 days developed abnormalities in porhpyrin metabolism, hypochloric anemia, gastritis, enterocolitis, hepatitis, and nephronephritis.

The orthoarsenate irritated the conjunctivae of rabbit eyes, and both compounds were absorbed through the intact skin of mice. In mice, the LD_{50} of the orthoarsenite on dermal application was determined at 1700 mg/kg [Davydov et al.].

Effects of lead zirconate-titanate

Animals

Acute/single exposure. Lead zirconate-titanate (LZT) composed of 66.6% Pb, 20.0% Zr, 11.8% Ti, and 1.6% Sr (the most used brand of this compound) did not cause any deaths among rats and guinea-pigs when administered by inhalation at 1.5 g/m^3 for 4 h or either intraperitoneally or orally in single doses of up to 15 g/kg. The threshold dose increasing the urinary excretion of δ-aminolevulinic acid in rats was of the same order as that of lead(II) oxide (PbO) and equalled 150 mg/kg (90 mg/kg as Pb). The threshold concentration required to produce acute effects was 104 mg/m^3 for a rat kept in an individual exposure chamber so designed as to rule out the licking off of this compound by the animal from its coat, and 20 mg/m^3 for a rat that was unrestrained in its movements in the chamber. Slightly marked fibrotic effects were observed in rats administered this compound intratracheally [Zhislin et al.].

Chronic exposure. In a 4 month inhalation toxicity study, rats exposed to a LZT aerosol at 25.6 m/m^3, began to excrete increased amounts of δ-aminolevulinic acids and coproporphyrin in the urine after 2 weeks of exposure; by the 4th week, their urinary concentration had increased 2- to 3.4-fold. The concentration of 6.8 mg/m^3 was described as the threshold level and that of 1.5 mg/m^3, as ineffective. Rats inhaling the aerosol in high concentrations (~ 300 mg/m^3) developed abnormalities in porphyrin metabolism and blood morphology typical of lead poisoning; the histologic structure of their internal organs was damaged [Zhislin et al.].

Effects of local application. In rats, LZT applied to the shaved skin of the back produced a 2- to 3-fold rise in the urinary excretion of δ-aminolevulinic acid and dystrophic changes in parenchymal organs and in the testes. No local irritant effects were seen [Zhislin et al.].

Man

Increased rates of respiratory disease were observed among workers exposed to dusts containing lead zirconate-titanate in concentrations up to 0.4 mg/m^3 (average values, 0.1–0.19 mg/m^3); no specific biochemical changes were recorded in their blood or urine. Both the aggressiveness and cumulative potential of lead in this compound appear to be much weakened [Zhislin et al.].

Biologic indicators of human exposure

There is a good positive correlation between the concentration of lead in the ambient air and its level in biological media such as blood and urine [Ewers & Schlipköter; Kowal et al.]. It has been calculated that an average increase of 5 µg Pb/L blood corresponds to an average increase of 1 µg/m^3 (40 h per week average exposure) of lead in air up to a lead blood level of 500 µg/L. On this basis, a range of 30–60 µg/m^3 has been recommended for lead in air [Tsuchiya]. Blood lead levels have been found to correlate, in turn, with the absorbed dose and the severity of clinical poisoning [Irwig et al.].

The lead level in blood is accepted as the epidemiologic index of choice. δ-Aminolevulinic acid dehydratase (ALAD) activity estimates in erythrocytes may be equally useful, but only at lead blood levels below 600 µg/L. A health-based biologic exposure limit of 400 µg/L for blood lead is recommended for males and for females of reproductive age. The blood lead level in females in the reproductive age range should be kept as low as possible and, to protect the fetus, should not exceed 300 µg/L [WHO, 1980]. The limit of 300 µg/L has also been proposed for children [Waldron].

In urine, values ranging from 150 to 200 µg/L may be considered as borderline, and an excess of 200 µg/L as indicative of harmful exposure. It has been estimated that the air lead concentrations of 0.1–0.15 mg Pb/m^3 correspond to 120–140 µg Pb/L in urine if exposure is stable over a period of a few months or more [Tsuchiya].

Tests for δ-aminolevulinic acid and coproporphyrin estimations in urine are widely used since they are simple, avoid the possibility of external contamination, and may provide a picture of the internal exposure [Adamovich; Berlin; Waldron; WHO, 1977, 1980].

In general, blood lead and urine lead values provide a direct and reliable quantitative indication of lead exposure, but only indirect evidence of lead effects.

The hematopoietic system shows effects at lower blood lead levels than any other system. The effects are, in order of sensitivity: inhibition of erythrocyte ALAD, elevation of erythrocyte protoporphyrin IX, rise in urinary δ-aminolevulinic acid and coproporhyrin excretion, inhibition of erythrocyte sodium-potassium adenosine triphosphatase, and fall in hemoglobin level. A fall in hemoglobin is clearly an indication of adverse effects. The no-detected-effect level for this fall is a lead blood concentration equivalent to 50 µg/100 ml in adults and 40 µg/100 ml in children [WHO, 1977].

Reticulocytosis was found to appear at blood lead concentrations of 0.097–0.19 µmol%, basophilic stippled erythrocytes at 0.29 µmol%, and a complete symptom complex of lead anemia at 0.48 µmol% [Gribova et al.]. There is evidence for the existence of still earlier and specific signs of lead intoxication; these include enhanced formation of anti-erythrocytic auto-antibodies [Digoyeva & Berezov], elevation of arginase activity in the blood (even in dry blood spots) [Fukumoto *et al.*], and alteration in the zincprotoporphyrin content of erythrocytes.

ABSORPTION, DISTRIBUTION, AND ELIMINATION IN MAN

Lead and its compounds normally enter the body by ingestion (with food and water) and inhalation. The inhalation of airborne lead contributes relatively little to the total intake except in certain occupational and polluted urban environments. Absorption through the skin is of importance only in the case of organic lead compounds.

Dietary intake of lead varies with eating habits and the lead content of water sources, but the majority of estimates from various countries suggest that the daily oral lead intake ranges from approximately 100 μg to more than 500 μg (most studies showed the lead intake from dietary sources to be 200–300 μg/day) [WHO, 1977], although figures as low as about 20 μg have been quoted, and daily intakes between 75 and 120 μg were reported for children aged 6 months to 2 years. People living in the vicinity of some lead smelters were found to have particularly high intakes (up to 2640 μg) [Goyer; Shtenberg & Zayeva]. Daily doses considered to be toxic are 2 to 3 mg for adults and about 1 mg for children [Goyer, 1971]. According to Caplun *et al.*, the average daily lead intake with food items was 400–500 μg in Belgium and Italy, 200 μg in France, 100 μg in Great Britain, and 30 μg in Switzerland. However, the large differences in daily intake reported for different countries may be due to inadequacies or differences in sampling and in the analytical methods used. Lead intake via drinking water does not usually exceed 50 μg/day, the mean value being close to 20 μg/day [Tsuchiya].

Reported lead concentrations in various items and categories of food are also highly variable. Schroeder *et al.* found, for example, that the concentration range was (mg/kg) 0–1.5 for condiments, 0.2–25 for fish and seafood, 0–0.37 for meat and eggs, 0–1.39 for grains, and 0–1.3 for vegetables. Skurikhin has reported mean lead contents (μg/100g) of 45 for fish and 15 for meat (edible parts), 20 for bread, 5 for milk, 20 for vegetables, 15 for fruits (edible parts). In another study, tissues of crab, fish (cod), rabbit, and pig had mean lead values (μg/100 g) of 17, 90, 104, and 115, respectively, while apples, nuts, and coffee beans had values of 8, 23, and 30, respectively; tea infusions contained 2 to 120 μg/L (depending on the pH and hardness of the water) and cigarettes contained from 1.5 to 5 μg each [Oudart *et al.*]. A substantial source of lead for some people may be wine (values of 130 to about 300 μg/L were recorded). Lead concentrations in human milk have been reported to range from 0.006 to 0.58 μg/ml [Dillon *et al.*], and comparable values have been found for cow's milk. A significant contribution to the lead content of processed milk and of other food products can be made by lead solder used in the seams and caps of cans [WHO, 1977].

Only about 5–10% of ingested lead is absorbed through the gastrointestinal tract, and about 30–60% of the inhaled lead is retained in the lungs depending on the particle size of airborne lead. Particles smaller than 10 μg can be retained in the lungs, whereas larger particles tend to be deposited in the upper respiratory tract, from where they are transported to the nasopharynx by mucociliary movement and swallowed. It has been suggested that the maximal tolerable daily dose of inhaled lead is 60 μg; workers in a storage-battery manufacturing plant were found to inhale from 1000 to 8600 μg of lead per shift [Liubchenko et al.].

Conclusive evidence that lead is an essential trace element is lacking, although some authors speak of a definite daily requirement for lead [Dubina & Leonov]. It has long been thought that a certain background lead level exists in the body, and the level rises as the environment becomes increasingly polluted from man-made sources. The contribution of different sources to human lead exposure has been discussed by Ewers & Schlipköter.

The absorbed lead is transported to other organs and tissues in the blood, mostly ($\approx 95\%$) by erythrocytes, in the form of organic complexes with proteins. In the general population, lead levels in blood are influenced by a number of demographic, social and local factors and by individual habits. Lead has special affinity for bone where over 90% of the absorbed lead may be firmly fixed. The skeleton is thus the major repository that reflects the long-term accumulative human exposure to this metal. The body pools of lead other than blood may be divided into two types: hard tissues (bones, hair, nails, teeth) and soft tissues (bone marrow, nervous system, kidneys, liver). If is believed that only lead in soft tissues is directly toxic. Lead in hard tissues remains tightly bound to them and is toxic only when it is leached out into the blood under particular circumstances.

Lead readily passes the placental barrier, and its concentration in the blood of newborns is similar to that of their mothers [Ewers & Schlipköter].

The total body burden of lead increases from birth to old age and has been reported to be approximately 100–400 mg in an average (70-kg) man [Tsuchiya] (it is usually lower in women). Blood lead levels reported from most studies for occupationally unexposed (rural and urban) populations fall into the range of between 10 and 25 μg/100 ml. Average lead levels in organs were found to be as follows (mg/100 g wet tissue): myocardium and skeletal muscle, 0.01; aorta, 0.2; liver, 0.04–0.26; kidney, 0.027; stomach, 0.022; intestine, 0.023. The lead content of $2.15 \times 10^{-4}\%$ found in the epiphyses of femoral bones from cadavers of rural people, has been proposed as the natural background level for the present human generation [Dubrovina et al.; Shvaikova]. Lead concentrations in bones of men and women aged over 16 years ranged from 0.9 to 0.34 mg/100 g wet weight [Tsuchiya]; lead in bone, but not in soft tissues, builds up throughout life [Drasch & Ott]. Appreciable amounts occur in the hair and teeth [Ewers & Schlipköter; Paterson et al.]. For example, the hair of rural people contained up to 20 μg/g, that of city dwellers up to 60 μg/g, and that of lead smelter workers from 95 to 250 μg/g. A correlation exists between lead levels in the hair and blood; thus, the concentration of 7 μg/g in hair has been found to correspond to 600 μg/L in blood. There is also a correlation between hair lead and clinical manifestation of lead poisoning [Revich]. The evolution of lead content in teeth may be a useful index of the body burden of heavy metals in relation to the degree of industrialization. Thus, the dentine of Egyptian mummies was reported to contain 9.7±10.7 μg/g, that of 12th-century Peruvian Indians 13.6±19.8 μg/g, and that of modern Eskimos in Alaska and of modern Philadelphia inhabitants, 56.0±30.1 μg/g and 188.3±37.9 μg/g, respectively [Shapiro et al.].

Only 4–10% of ingested lead is absorbed from the gastrointestinal tract (the fraction absorbed may be higher in children), the rest being eliminated in feces. Absorbed lead is excreted in urine (75–80%) and, to a lesser extent (~15%), via the gastrointestinal tract; other excretion routes are hair, sweat, nails, exfoliated skin, and milk [WHO, 1986].

Because of the distribution of lead between soft and hard tissues, its biologic half-life is difficult to establish but the clearance of one-half of the body burden of lead would certainly require a number of years [WHO, 1977].

HYGIENIC STANDARDS

Exposure limits adopted for lead and its compounds in the USSR are shown in Table 1. In addition, it has been recommended to set occupational exposure limits for lead hydrogen orthoarsenate at 0.03 mg/m^3 and for lead hydrogen orthoarsenite at 0.02 mg/m^3 (as As), with the notation that both compounds can be absorbed through undamaged skin [Davydov et al.].

In the USA, TLV-TWA values have been set for inorganic lead dusts and fumes at 0.15 mg/m^3 (as Pb), for lead arsenate also at 0.15 mg/m^3 (as $Pb_3(AsO_4)_2$), and for lead chromate at 0.05 mg/m^3 as Pb and 0.012 mg/m^3 as Cr, this compound being classified as a possible human carcinogen.

Table 1. Exposure limits for lead and its compounds

	Workroom air		Atmospheric air	Water sources	Soil	
	MAC_{wz} (mg/m^3)	Aggregative state	MAC_{ad} (mg/m^3)	MAC_w (mg/L)	MAC_s (mg/kg)	Hazard class
Lead and its inorganic compounds (as Pb)	$0.01/0.005^a$	Aerosol	0.0003	0.03^b	32.0^c	1
Lead sulfide (as Pb)	—	—	0.0017	—	—	1

[a] 0.01 mg/m^3 is the concentration not to be exceeded at any time, while 0.005 mg/m^3 is the average maximum allowable concentration per shift.
[b] Based on sanitary and toxicologic criteria.
[c] Based on the general sanitary index of harmfulness that characterizes the impact of a compound on the self-purification capacity of the soil and on the soil microbiocenosis.

The maximum permissible levels adopted for foods in the USSR are shown in Table 2. Regulations concerning lead in gasoline, ambient air, soil, compost, sewage sludge, drinking water, food, beverages, and in the workplace are summarized in Ewers & Schlipköter.

METHODS OF DETERMINATION

In air. One method for estimating the lead content of air is based on measuring the turbidity of a solution formed upon the interaction of Pb^{2+} with potassium chromate; for comparison, a standard scale is used (sensitivity 1 µg in the analytical volume). The oldest and best known of the general methods for lead analysis in air and various materials are those based on the formation of a red complex by lead with dithizone (diphenylthiocarbazone), and numerous specific procedures based on spectrophotometric determination of the lead dithizonate have been developed. In a commonly used method based on the reaction of lead with dithizone, the colored dithizonate dissolved in chloroform or carbon tetrachloride is determined colorimetrically (sensitivity 2 µg in the analytical volume) [Peregud]. A procedure involving amperometric titration of lead acetate solution by an aqueous solution of ammonium molybdate with the formation of an insoluble compound is also used (with this method, soluble, insoluble, and total lead can be determined, the minimal measurable quantity of lead being 10^{-5} mg in a 10 ml sample [*Specifications for Methods of Determining Noxious Substances in Air*]. Nekrasova & Tashi have described a procedure using alternating-current polarography for lead determination, based on the reduction of its ions on a mercury-dropping electrode in the presence of a mixture of acetic acid and ammonium acetate solutions (detection limit 0.04 µg/ml).

Table 2. MACs for lead in foods (mg/kg)

Cereal grains, groats, flour, and pastries	0.5
Bran	1.0
Sugar, candies, and halva	1.0
Bread	0.3
Milk and sour milk products	0.1
Condensed milk, canned	0.3
Butter, margarines, and cooking fats	0.1
Vegetable oils	0.1
Cheese, curds, and casein	0.3
Vegetables, fresh or canned (in glass jars)	0.5
Fruits, fresh or canned	0.4
Mushrooms, fresh	0.5
Spices	5.0
Tea	10.0
Meat and poultry, fresh or canned	0.5
Sausages	0.5
Eggs	0.3
Fish, fresh or canned	1.0
Mollusks and crustaceans (shellfish)	10.0
Mineral waters	0.1
Beverages, based on infusions or essences	0.3
Beer, wine, and other alcoholic beverages	0.3
Infant foods:	
milk-based	0.05
grain- and milk-based	0.1
vagetable- or fruit-based	0.3
meat- or poultry-based	0.3
fish-based	0.5

In water and food. One method recommended for lead determination in water and food samples is based on the concentration of lead and its extraction by chloroform as a complex with diethyldithiocarbamate and 8-hydroxyquinoline followed by spectrometry (sensitivity limits 4–80 µg) [Novikov *et al.*]. In *aqueous* and *model media*, lead can be determined by chromatography in a thin adsorbent layer; at the basis of this method is the formation of a complex between lead cations and sodium diethyldithiocarbamate (sensitivity 0.01 mg/L) [Katayeva]. A potentiometric method for determining lead in water using a lead-selective electrode has been described [Dracheva *et al.*].

Methods for measuring lead in *plant materials* have been described by Katayeva & Arkhipenko and Sokolova *et al.*

In *urine* and *blood* samples, lead can be determined using dithizone after mineralization of the sample [Gadaskina *et al.*]. Pavlovskaya *et al.* described a rapid method for polarographic determination of lead in microgram amounts in blood samples without preliminary mineralization.

In recent years, anodic stripping voltametry and especially atomic absorption spectroscopy (spectrophotometry) have been gaining increasing popularity for lead analysis. Nondestructive methods such as neutron activation analysis and X-ray fluorescence have also been used to some extent. A review of the various analytical methods for lead (as well as of methods for measurement of some biochemical affects of lead) can be found in *Environmental Health Criteria 1: Lead* [WHO, 1977]; for a recent review of methods for measuring lead in environmental and biologic samples, see Ewers & Schlipköter. The advantages of atomic absorption spectrophotometry over other methods of lead determination in environmental samples and biologic materials have been discussed by Kuntsevich *et al.*

MEASURES TO CONTROL EXPOSURE

At the workplace

Hazardous exposure to lead occurs in many industries, and if healthy and safe working conditions are to be assured, the relevant standards, regulations, and guidelines pertaining to the respective industries where lead and its compounds are produced or used must be strictly observed.

The main hazards in lead smelting are the lead dust produced during crushing and dry grinding operations, and lead fumes and oxides encountered in sintering, blast-furnace reduction, and refining of this metal. Radical general measures of a technical nature for dust control include tight enclosure and proper venting of the dust-generating equipment, use of automata and automatic lines (particularly in the production of glass-ceramic cement), remote operation of the technological processes, and substitution of less toxic substances for lead compounds in as far as this is possible.

Workers who are to be exposed to lead should receive a pre-employment medical examination followed by regular medical examinations during the period of employment, with special attention to the hematopoietic, nervous, and renal systems. Pregnant women must be transferred to jobs not associated with exposure to lead (and other toxic substances). Persons under 18 years of age should not be employed in workplaces where lead is produced or used. All workers must of course be instructed in technical safety and personal hygiene measures.

It is recommended to supply workers with acidic milk products (e.g. kefir and yoghurt; 0.5 L/day) and pectin (2 g/day in the form of fruit juices or beverages enriched with this substance). Drinking water for workers in soldering operations should be supplied through waterspouts installed near the working area but in no case within it.

There should be a ban on entering the lunchroom or canteen in working clothes as well as on keeping any clothes and personal effects in the workrooms. Changing rooms with separate accommodation for outdoor clothing and working clothes should be provided; the latter must not be allowed to be taken home. Eating, drinking, and smoking in the working areas should be prohibited.

Good housekeeping is important. It is essential that the rooms and the plant associated with lead processes be kept clean through regular cleaning by a wet process or by vacuum cleaners, never by dry methods. Where, in spite of these precautions, workers may still be exposed to lead, respiratory protective equipment should be provided and properly maintained.

Workers who are exposed to lead in any of its forms must wear personal protective clothing, which should be washed or renewed at least once a week. Protective clothing made of certain man-made fibres retains much less dust than cotton overalls and should be used where the conditions of work render it possible; turn-ups, pleats and pockets in which lead dust may collect should be avoided [Zielhuis].

Protection of the general environment
Since lead is an important pollutant of the environment and may have adverse effects on living organisms, every effort should be made to prevent the atmosphere, soils, bodies of water, and foods from being polluted with lead and its compounds. Atmospheric pollution can be abated through such measures as a changeover to diesel fuel (which is free of tetraethyllead) and unleaded petrol, the development of electric cars, the use of gas in heating systems, and the improvement of fuel-combustion processes.

Water bodies can be prevented from pollution in the most radical way by constructing recirculating water-supply systems whereby the discharge of industrial wastewaters to the outside will be completely excluded. Valuable products contained in mine waters should be recovered. Wastewater may only be discharged into water streams after effective purification. Solid industrial wastes containing lead should be decontaminated in special installations.

High priority is to be given to the setting and observance of hygienic (health) standards for lead in environmental media, including soils and foods.

FIRST AID IN POISONING

The first measure to be taken in any case of lead poisoning is removal of the victim from the source of exposure. Lead—both free and that deposited in the bones—can be cleared from the body by using a chelating agent such as calcium disodium edetate ($CaNa_2$—EDTA), pentacin [synonym: calcium trisodium pentetate], D-penicillamine, or unithiol (sodium 2,3-dimercapto-1-propane sulfonate) using the dose schedules accepted for lead poisoning with these drugs. Good results are obtainable with succimer (*meso*-2,3-dimercaptosuccinic acid), a dithiol drug. If signs of encephalopathy are present, 10 — 12 intravenous injections of 10 ml of a 25% magnesium sulfate solutions should be given. Epileptiform seizures can be controlled by enemas containing 50 — 100 ml of a 2% chloral hydrate solution; soporifics may be helpful. For the relief of lead colic, warm baths or hot bottles applied to the abdomen, atropine and morphine in subcutaneous injections, magnesium sulfate with glucose, potassium bromide intravenously, and dibasol [synonym: bendazole hydrochloride] orally are indicated.

REFERENCES

Adamovich, G. G. *et al.*(1981) *Gig. San.*, No. 11, 75—76 (in Russian).

Agafonova, L. L. & Zakirova, V. S. (1966) *Meditsinskiy Zhurnal Uzbekistana* [Uzbek Medical Journal], No. 8, 9—12 (in Russian).

Akers, C. & Fellows, R. (1979) *Environ. Health*, 87, 148—152.

Aldanazarov, A. T. (1974) *Izmeneniya sistemy krovi pri saturnizme* [Changes in the Circulatory System in Saturnism]. Alma-Ata (in Russian).

Alomran, A. & Shleamoon, M. (1988) *J. Biol. Res* ., 19, 575—580.

Aronsen, A. (1971) *J. Wash. Acad. Sci.*, 61, 110—113.

Artamonova, V. G. (1981) *Neotlozhnaya pomoshch pri professionalnykh intoksikatsiyakh* [Emergency Care in Occupational Intoxications]. Meditsina, Leningrad (in Russian).

Atchabarov, V. A. (1966) *Porazheniye nervnoi sistemy pri svintsovoi intoksikatsii* [Affections of the Nervous System in Lead Intoxication]. Alma-Ata (in Russian).

Axelson, O. & Sundell, L. (1978) *J. Work. Environ. Health.*, 4, 46—52.

Bahat, R. & Koshnamaehan (1980) *Bull. Environ. Contam. Toxicol.*, 25, 142—145.

Barry, P. & Mossman, D. (1970) *Brit. J. Ind. Med.*, 27, 339—351.

Basamygina, I. Ya. (1968) In: *Voprosy gigiyeny i profilaktiki v ugolnoi, gornorudnoi i metallurgicheskoi promyshlennosti* [Hygienic and Preventive Measures in the Coal, Mining, and Metallurgical Industries]. Alma-Ata (in Russian).

Benson, W. *et al.* (1974) *Bull. Environ. Contam. Toxicol.*, 11, 105—108.

Berlin, A. (1987) In: *Encyclopaedia of Occupational Health and Safety.* International Labour Office, Geneva, pp. 1206—1209.

Bikezina, V. G. (1970) In: *Aktualnye problemy profpatologii* [Current Problems of Occupational Pathology]. Kiev, pp. 149—154 (in Russian).

Bolshakov, V. A. et. al. (1978) *Zagryazneniye pochv i rastitelnosti tyazhelymi metallami* [Pollution of Soils and Plants by Heavy Metalls]. Moscow (in Russian).

Bridbord, K. (1978) *Prev. Med.*, 7, (No. 3), 311—321.

Brooks, R. (1979) In: Bockris, J. O'M. (ed.) *Environmental Chemistry.* Plenum Press, New York, pp. 429—476.

Bulavko, G. I. (1982) *Izvestiya SO Akad. Nauk SSSR. Ser. Biol.* [Herald of the Siberian Division of the USSR Academy of Sciences. Biology Series], No. 5, 79—85 (in Russian).

Burkitt, A. et. at. (1972) *Nature*, 238, 327.

Cannon, H. & Bowles, J. (1962) *Science*, 137, 765.

Cantarow, A. & Trumper, M. (1944) *Lead Poisoning.* Williams & Wilkins Co., Baltimore.

Caplun, E. et. al. (1984) *Endeavour*, 8, 135—144.

Chow, T. & Patterson, C. (1966) *Earth Planet Sci. Lett.*, 1, 397—400.

Cramer, K. & Dahlberg, L. (1966) *Brit. J. Ind. Med.*, 23, 101—104.

Cullen, M. *et al.* (1984) *Arch. Environ. Health*, 39, 431—440.

Daines, R. *et al.* (1970) *Environ. Sci. Technol.*, 4, 318.

Davydov, V. I. *et al.* (1986) *Gig. San.*, No. 1, 74 (in Russian).

Digoyeva, M. D. & Berezov, T. T. (1981) *Vopr. Med. Khimii*, No. 2, 223–226 (in Russian).

Dillon, H. *et al.* (1974) *Amer. J. Dis. Child.*, 128, 491–492.

Dinerman, A. A. *et al.* (1978) In: *Svinets v okruzhayushchei srede* [Lead in the Environment]. Moscow, pp. 36–45 (in Russian).

Dobrovolsky, V. V. (1983) *Geografiya mikroelementov. Globalnoe rasseyanie* [Geogrphy of Trace Elements. Global Dispersion]. Mysl, Moscow (in Russian).

Dracheva, L. V. *et al.* (1984) *Gig. San.*, No. 4, 39–42 (in Russian).

Drasch, G. A. & Ott, J. (1988) *Sci. Total Environ.*, 68, 61–69.

Dressen, W. *et al.* (1941) *Publ. Health Bull.* (Wash.), No. 262.

Dubina, T. L. & Leonov, V. A. (1968) *Usp. Sovr. Biol.*, No. 3, 453–470 (in Russian).

Dubrovina Z. V. *et al.* (1967) *Gig. San.*, No. 4, 43–46 (in Russian).

Dynnik, V. I. (1964) *Gig. Truda Prof. Zabol.*, No. 9, 57–59 (in Russian).

Eisinger, J. (1979) *Quart. Rev. Biophys.*, 11, 439–466.

Ewers, U. & Schlipköter, H. W. (1991) Chapter II, 16 *Lead* in: Merian, E. (ed.) *Metals and Their Compounds in the Environment—Occurrence, Analysis and Biological Relevance*. VCH, New York, pp. 971–1014.

Fairhall, L. *et al.* (1943) *Publ. Health Repts*, 56, 607–611.

Fedoseyeva, N. M. *et al.* (1982) *Gig. Truda Prof. Zabol.*, No. 8, 51–53 (in Russian).

Fergusson, J. E. (1990) *The Heavy Elements: Chemistry, Environmental Impact and Health Effects*. Pergamon Press, New York.

Fomenko, V. N. *et al.* (1982) *Gig. Truda Prof. Zabol.*, No. 10, 38–41 (in Russian).

Fukumoto, K. *et al.* (1983) *Brit. J. Ind. Med.*, 40, 106–110.

Gadaskina, I. D. *et al.* (1975) *Opredeleniye promyshlennykh neorganicheskikh yadov v organizme* [In vivo Determination of Inorganic Industrial Poisons]. Meditsina, Leningrad (in Russian).

Gajdos, A. & Gajdos-Török, M. (1973) In: *Proceedings of the International Symposium on Environmental Health Aspects of Lead, Amsterdam, 2–6 October, 1972*. Commission of the European Communities, Luxemburg, pp. 501–505.

Galkin V. M. & Patrikeyev, Yu. V. (1972) *Klin. Med.*, No. 7, 145–147 (in Russian).

Garmash, G. A. (1983) In: *Khimiya v selskom khozyaistve* [Chemistry in Agriculture], No. 10, pp. 45–48 (in Russian).

Gilli, G. *et al.* (1975) *Jg. Mod.*, 68, 195–201.

Girard, R. *et al.* (1973) *Arch. Malad. Profess.*, 34, 563–565.

Golubovich, E. Ya. & Gnevkovskaya, T. V. (1967) In: *Toksikologiya novykh promyshlennykh khimicheskikh veshchestv* [Toxicology of New Industrial Chemicals], No. 9. Moscow, pp. 86–91 (in Russian).

Goyer, R. (1971) *Amer. J. Pathol.*, 64, 167–182.

Goyer R. A. (1988) Chapter 31 *Lead* in: Seiler, H.G. *et al.* (eds.) *Handbook on Toxicity of Inorganic Compounds*. Marcel Dekker New York, pp. 359–382.

Greenfield, S. *et al.* (1973) In: *Environmental Health Aspects of Lead*. Commission of European Communities Directorate General for Dissemination of Knowledge, Centre for Information and Documentation (CID), Luxemburg, pp. 19–27.

Gribova, I. A. *et al.* (1983) *Gig. Truda Prof. Zabol.*, No. 2, 22–25 (in Russian).

Grigorieva, T. & Khramova, S. (1978) In: *Svinets v okruzhayushchei srede* [Lead in the Environment]. Moscow, pp. 22–25 (in Russian).

Grinberg, A. V. *et al.* (1970) *Vestn. Rentgenol.*, No. 6, 11–17 (in Russian).

Grushko, Ya. M. (1972) *Yadovitye metally i ikh neorganicheskiye soyedineniya v promyshlennykh stochnykh vodakh* [Poisonous Metals and their Inorganic Compounds in Industrial Waste Waters]. Meditsina, Moscow (in Russian).

Grushko, Ya. M. (1979) *Vrednye neorganicheskiye soyedineniya v promyshlennykh stochnykh vodakh* [Harmful Inorganic Compounds in Industrial Waste Waters]. Khimiya, Leningrad (in Russian).

Haar, G. (1975) In: *Lead*. Stutthart, pp. 76–94.

Henderson, R. *et al.* (1979) *Bull. Environ. Contam. Toxicol.*, 21, 102.

Hennekes, R. *et al.* (1987) *Fortschr. Ophthalmol.*, 84, 374–376.

Heyxmancik, M. *et al.* (1982) *J. Toxicol. Environ. Health*, 9, 77–86.

588   Lead and its compounds

Honda, K. *et al.* (1982) *Agric. Biol. Chem.*, **46**, 3011–3021.
Huynh-Ngoc, L. *et al.* (1988) *J. Toxicol. Environ. Chem.*, **17**, 223–236.
Irvig, L. *et al.* (1978) *Lancet*, No. **8079**, 4–7.
Iskakov, T. K. (1984) In: *Trudy NII krayevoi patologii Kazakhskoi SSR* [Transactions of the Research Institute for Regional Diseases of the Kazakh SSR]. Alma-Ata, pp. 201–204 (in Russian).
Kamildzhanov, A. K. (1983) *Gig. San.*, No. **6**, 79–81 (in Russian).
Kaplan, L. M. (1973) *Zdravookhraneniye Belorussii* [Health Care in Byelorussia], No. **4**, 37–39 (in Russian).
Katayeva, S. E. (1982) *Gig. San.*, No. **5**, 60–61 (in Russian).
Katayeva, S. E. & Arkhipenko, L. S. (1989) *Gig. San.*, No. **2**, 52–53 (in Russian).
Kerin, Z. (1972) *Protectio Vitae*, **71**, 22–23.
Klein, M. *et al.* (1970) *New Engl. J. Med.*, **283**, 669–672.
Kovalsky, V. V. (1974) *Geokhimicheskaya ekologiya* [Geochemical Ecology]. Nauka, Moscow (in Russian).
Kowal, W. A. *et al.* (1988) In: Merian, E. *et al.* (eds.) *Carcinogenic and Mutagenic Metal Compounds.* 3. Gordon & Breach Science Publishers, New York, pp. 135–142.
Krasovsky, G. N. *et al.* (1984) *Gig. San.*, No. **3**, 15–17 (in Russian).
Krigman, U. *et al.* (1974) *J. Neuropathol. Exper. Neurol.*, **33**, 58–73.
Kuntsevich, I. E. *et al.* (1984) *Zdravookhraneniye Belorussii*, No. **7**, 41–42 (in Russian).
Lagutin, A. A. (1970) *Gig. San.*, No. **10**, 90–91 (in Russian).
Lassen, E. & Buck, W. (1979) *Amer. J. Vet. Res.*, **40**, 1359–1364.
Leschber, R. (1983) *Zbl. Bact. Hyg.*, **178**, 186–193.
Litvinov, N. N. *et al.* (1983) *Gig. San.*, No. **9**, 22–24 (in Russian).
Liubchenko, P. N. *et al.* (1989) *Gig. Truda Prof. Zabol.*, No. **3**, 7–9 (in Russian).
Liublina, E. I. & Dvorkin, E. A. (1983) In: Dudarev, A. Ya. (ed.) *Gigiyenicheskaya toksikologiya metallov* [Hygienic Toxicology of Metals]. Moscow, pp. 25–29 (in Russian).
Lobanova, E. A. *et al.* (1988) *Gig. Truda Prof. Zabol.*, No. **9**, 42–43 (in Russian).
Louekari, K., *et al.* (1989) *Sci. Total Environ.*, **84**, 1–12.
Makarova, A. I. *et al.* (1983) *Gig. San.*, No. **7**, 63–64 (in Russian).
Metelev, V. V. *et al.* (1971) *Vodnaya toksikologiya* [Aquatic Toxicology]. Kolos, Moscow (in Russian).
Michaelson, J. (1973) *Toxicol. Appl. Pharmacol.*, **26**, 539–548.
Monaenkova, A. M. (1957) *Gig. Truda Prof. Zabol.*, No. **2**, 44–48 (in Russian).
Moore, M. (1988) *Sci. Total Environ.*, **71**, 419–431.
Mullins, T. (1979) In: Bockris, J. O'M. (ed.) *Environmental Chemistry.* Plenum Press, New York, pp. 331–400.
Murozumi, M. *et al.* (1969) *Geochim. Cosmochim. Acta*, **33**, 1247.
Nainshtein, S. Ya. (1975) *Aktualnye voprosy gigiyeny pochvy* [Priorities in the Field of Soil Hygiene]. Shtiintsa, Kishinev (in Russian).
Nekrasova, S. V. & Tashi, P. F. (1988) *Gig. San.*, No. **5**, 48–49 (in Russian).
Nemenko, B. A. *et al.* (1981) *Gig. San.*, No. **1**, 74–75 (in Russian).
Nikiforova, E. M. (1981) In: *Tekhnogennye potoki veshchestv v landshaftakh i sostoyaniye ekosistem* [Anthropogenic Flows of Substances in Landscapes and the State of Ecosystems]. Moscow, pp. 220–229 (in Russian).
Novikov, Yu. V. *et al.* (1981) *Metody opredeleniya vrednykh veshchestv v vode vodoyemov* [Methods for Measuring Noxious Substances in the Water of Water Bodies]. Meditsina, Moscow (in Russian).
Novoselsky, I. I. (1971) *Vopr. Pitaniya*, No. **2**, 88–89 (in Russian).
Okhnyanskaya, L. G. & Komarova, A. A. (1970) *Elektromiografiya v klinike professionalnykh zabolevaniy* [Electromyography in the Clinical Management of Occupational Diseases]. Meditsina, Moscow (in Russian).
Okshina, L. N. (1977) In: *Trudy NII krayevoi patologii Kazakhskoi SSR* [Transactions of the Research Institute for Regional Diseases of the Kazakh SSR], Vol. 29, Part 2. Alma-Ata, pp. 28–34 (in Russian).
Oskarsson, A. *et al.* (1992) *Sci. Total Environ.*, **111**, 83–94.

References 589

Osweiler, G. *et al.* (1973) *Clin. Toxicol.*, **6**, 367—376.
Oudart, N. *et al.* (1976) *Eur. J. Toxicol. Environ. Hyg.*, **9**, 423—428.
Panova, Z. [1972) *Akush. Ginek.* [Obstetrics & Gynecology] (Bulgaria), No. **6**, 473—478 (in Bulgarian).
Paterson, L. J. *et al.* (1988) *Sci. Total Environ.*, **74**, 219—233.
Patterson, C. (1965) *Arch. Environ. Health*, **11**, 344—363.
Pavlovskaya, N. A. *et al.* (1982) *Lab. Delo*, No. **1**, 26—29 (in Russian).
Peregud, E. A. (1976) *Khimicheskiy analiz vozdukha (novye i usovershenstvovannye metody)* [Chemical Analysis of Air (New and Improved Methods)]. Khimiya, Leningrad (in Russian).
Pollak, G. F. & Fisher, A. M. (1971) In: *Trudy NII krayevoi patologii Kazakhskoi SSR* [Transactions of the Research Institute for Regional Diseases of the Kazakh SSR], Alma-Ata, pp. 113—128 (in Russian).
Prosi, F. (1989) *Sci. Total Environ.*, **79**, 157—170.
Radamilans, M. *et al.* (1988) *Hum. Toxicol.*, **7**, 125—128.
Rakhimova, M. T. (1970) In: *Trudy NII krayevoi patologii Kazakhskoi SSR* [Transactions of the Research Institute for Regional Diseases of the Kazakh SSR], Vol. **19**, Alma-Ata, pp. 239—241 (in Russian).
Rashevskaya, A. M. *et al.* (1973) In: *Professionalnye bolezni* [Occupational Diseases]. Meditsina, Moscow, pp. 111—257 (in Russian).
Raskovalov, M. G. (1962) In: *Trudy Sverdlovskogo meditsinskogo instituta* [Transactions of Sverdlovsk Medical Institute], No. **38**. Sverdlovsk, pp. 181—182 (in Russian).
Rastogi, S. & Clausen, J. (1976) *Toxicology*, **6**, 371—376.
Revich, B. A. (1980) In: *Gigiyena okruzhayushchei sredy. Ekspress-informatsiya* [Environmental Hygiene: Current Awareness Information], No. **9** (in Russian).
Reyes, P. *et al.* (1986) *Amer. J. Roentgenol.*, **146**, 267—270.
Roels, H. *et al.* (1980) *Environ. Res.*, **22**, 81—94.
Roshchin, A. V. (1977) *Gig. Truda Prof. Zabol.*, No. **11**, 28—35 (in Russian).
Rovinsky, F. Ya. *et al.* (1982) In: *Monitoring fonovogo zagryazneniya prorodnykh sred* [Montoring the Background Pollution of Environmental Media], No. **1**. Leningrad, pp. 14—35 (in Russian).
Sandstead, H. *et al.* (1970) *Clin. Res.*, **18**, 76 (abstract).
Sanotsky, I. V. *et al.* (1978) In: *Svinets v okruzhayushchei srede* [Lead in the Environment]. Moscow, pp. 35—47 (in Russian).
Sastry, K. & Gupta, P. (1978) *Environ. Res.*, **22**, 472—479.
Sastry, K. & Gupta, P. (1979) *Toxicol. Lett.*, No. **3**, 145—150.
Schroeder, H. *et al.* (1961) *J. Chron. Dis.*, **14**, 408—425.
Shapiro, J. *et al.* (1975) *Arch. Environ. Health*, **30**, 483—486.
Shepard, T. (1986) *Catalogue of Teratogenic Agents*. Johns Hopkins University Press, Baltimore.
Shestakov, N. M. (1972) In: *Trudy Ryazanskogo meditsinskogo instituta* [Transactions of Ryazan Medical Institute], Vol. **43**. Ryazan, pp. 12—16 (in Russian).
Shevchenko, V. I. & Tolgskaya, M.S. (1976) *Gig. Truda Prof. Zabol.*, No. **10**, 35—38 (in Russian).
Shimaitis, R. S. & Tsiunene, E. P. (1977) *Gig. Truda Prof. Zabol.*, No. **12**, 50—53 (in Russian).
Shraiber, L. B. & Mosevich, P. N. (1963) *Med. Zh. Uzbekistana* [Medical Journal of Uzbekistan], No. **7**, 35—39 (in Russian).
Shtenberg, A. I. & Zayeva, G. N. (1978) *Vestn. Akad. Med. Nauk SSSR*, No. **3**, 78—82 (in Russian).
Shvaikova, M. D. (1975) *Toksikologicheskaya khimiya* [Toxicological Chemistry]. Meditsina, Moscow (in Russian).
Skurikhin, I. M. (1981) *Vopr. Pitaniya*, No. **2**, 10—16 (in Russian).
Sokolov, V.V. *et al.* (1966) *Gig. Truda Prof. Zabol.*, No. **8**, 31—36 (in Russian).
Sokolova, G. N. *et al.* (1989) *Gig. San.*, No. **5**, 52—54 (in Russian).

Specifications for Methods of Determining Noxious Substances in Air (1973) [Tekhnicheskiye usloviya na metody opredeleniya vrednykh veshchestv v vozdukhe], No. 9. Publication of the USSR Ministry of Health, Moscow (in Russian).

Stowe, H. & Goyer, R. (1971) *Fertil. Steril.*, **22**, 755–760.

Sun, C. *et al.* (1966) *Arch. Pathol.*, **82**, 156–163.

Talakin, Yu. N. (1978) *Gig. Truda Prof. Zabol.*, No. 6, 18–21 (in Russian).

Talakin, Yu. N. (1979) *Gig. San.*, No. **9**, 17–19 (in Russian).

Tarabayeva, G. I. (1961) *Deistviye svintsa na organizm i lechebno-profilakticheskiye meropriyatiya* [Health Effects of Lead, and Therapeutic and Preventive Measures]. Alma-Ata (in Russian).

Tartler, J. (1941) *Arch. Hyg.*, **125**, 273–279.

Teras, L. E. & Kahn, Ch. A. (1966) *Vopr. Med. Khimii*, No. 1, 41–45 (in Russian).

Timoshina, I. V. *et al.* (1985) *Terapevtich. Arkhiv*, No. 2, 91–95 (in Russian).

Tolgskaya, M. S. (1964) In: *Toksikologiya novykh promyshlennykh khimicheskikh veshchestv* [Toxicology of New Industrial Chemicals], No. **6**. Meditsina, Moscow, pp. 128–144 (in Russian).

Tsuchiya, K. (1979) In: Friberg, L. *et al.* (eds) *Handbook on the Toxicology of Metals.* Elsevier, Amsterdam, pp. 451–484.

Usoltsev, V. I. *et al.* (1978) *Gig. San.*, No. **12**, 87–89 (in Russian).

Vints, L. A. (1975) *Vestn. Oftalmol.*, No. 1, 74–75 (in Russian).

Waldron, H. (1974) *Arch. Environ. Health*, **29**, 271–273.

Ward, W. & Brooks, R. (1979) *Bull. Environ. Contam. Toxicol.*, **21**, 403–408.

Winneke, G. *et al.* (1982) *Intern. Arch. Occup. Environ. Health*, **51**, 169–183.

Winneke, G. *et al.* (1983) *Intern. Arch. Occup. Environ. Health*, **51**, 231–252.

WHO (1977) *Environmental Health Criteria 3: Lead.* World Health Organization, Geneva.

WHO (1980) *Technical Report Series 647: Recommended Health-based Limits in Occupational Exposure to Heavy Metals.* World Health Organization, Geneva.

WHO (1986) *Early Detection of Occupational Diseases.* World Health Organization, Geneva.

WHO (1989) *Environmental Health Criteria 85: Lead—Environmental Aspects.* World Health Organization, Geneva.

Yegorov, G. M. *et al.* (1966) In: *Toksikologiya novykh promyshlennykh khimicheskikh veshchestv* [Toxicology of New Industrial Chemicals], No. **6**. Moscow, pp. 33–41 (in Russian).

Zhislin, L. E. *et al.* (1978) *Gig. Truda Prof. Zabol.*, No. **5**, 31–34 (in Russian).

Zhulidov, A. V. *et al.* (1980) *Dokl. Akad. Nauk SSSR*, No. 4, 1018–1020 (in Russian).

Zhulidov, A. V. & Yemets, V. M. (1979) *Dokl. Akad. Nauk SSSR*, No. 6, 1515–1516 (in Russian).

Zielhuis, R. (1983) In: *Encyclopaedia of Occupational Health and Safety.* International Labour Office, Geneva, pp. 1200–1205.

Titanium and its compounds

Titanium carbide; t.(IV) chloride (titanium tetrachloride); **t. diboride; t. hydride; t. nitride** (osbornite [min.]); **t.(IV) oxide** (titanium dioxide, titanium white, anatase [min.], brookite [min.], rutile [min.]); **t. oxide dichloride** (titanium oxychloride); **t. phosphate; t. silicide; t. sulfide**
Barium metatitanate; barium tetratitanate; barium titanate-zirconate; calcium metatitanate (perovskite [min.]); **lead metatitanate; metatitanic acid** (β-titanic acid); **potassium octatitanate**

IDENTITY AND PHYSICOCHEMICAL PROPERTIES
OF THE ELEMENT

Titanium (Ti) is a metallic element in group IV of the periodic table, with the atomic number 22. Its natural isotopes are ^{46}Ti (7.95%), ^{47}Ti (7.75%), ^{48}Ti (73.45%), ^{49}Ti (5.51%), and ^{50}Ti (5.34%).

Titanium exists in two modifications, α and β. It has a lower density (4.507 g/cm^3) and is more resistant to corrosion, more ductile, and stronger than most other metals. The plasticity of titanium is greatly reduced by O_2, N_2, H_2, and C dissolved in it. At very low temperatures it becomes stronger while remaining ductile. Titanium filings and powders are pyrophoric, and molten titanium burns in air.

In compounds, by far the most common oxidation state of titanium is +4 (titanic compounds), but titanium(III) (titanous) and titanium(II) compounds are also known. Its high resistance to corrosion by many agents, including atmospheric air and sea water, is due to the formation of a strong surface film of titanium(IV) oxide. Titanium is moderately reactive, and its reactivity increases with rising temperature. See also Appendix.

NATURAL OCCURRENCE AND ENVIRONMENTAL LEVELS

The clarke of titanium is estimated at 0.45–0.57% in the earth's crust and 0.33% in the granite layer of the continental crust by weight [Dobrovolsky]. It is widely distributed in nature and occurs in many minerals but only a few of them are of industrial significance. The major titanium-bearing minerals are ilmenite (~53% of Ti), rutile, anatase, brookite, and leucoxene (≤100% TiO_2), perovskite (~59% TiO_2), and titanite (~39% TiO_2).

The oceans contain at estimated 1370 million tonnes of dissolved titanium, mainly as $Ti(OH)_4$, at an average concentration of 1.0 μg/L [Dobrovolsky; Henderson] but values up to 9 μg/L have been reported for seawater [Berlin & Nordman]; its residence time in ocean water is estimated to be 1.3×10^4 years [Henderson] and its estimated amount taken up annually by ferromanganese oxide deposits of the Pacific is 40 000 tonnes [Dobrovolsky]. In river water, titanium is present at an average concentration of 3.0 μg/L and its total amount discharged by rivers worldwide is estimated to be 110 000 tonnes per year (K_w=0.01) [Dobrovolsky], most (~98%) of this amount being associated with mechanical particles suspended in water.

The titanium content of surface soils commonly ranges from 0.1 to 0.9%. In surface soils of the USSR, its contents range (on a dry weight basis) from 0.45% in chernozems to 0.1% in podzolic and sandy soils [Lukashev & Petukhova], while in surface soils of the USA average reported levels are 0.14% in organic light soils, 0.36 in clay, clay loamy, and forest soils, and 0.53% in soils over volcanic rocks [Kabata-Pendias & Pendias].

Titanium levels in plants vary widely within the range of 0.15 to 80 mg/kg dry weight [Kabata-Pendias & Pendias]. The metal is poorly absorbed by plants (e.g. herbage was found to contain only 1×10^{-5}% of its amount present in the soil). Continental vegetation contains a total of 81 million tonnes of titanium, at average fresh weight, dry weight, and ash weight concentrations of 13.0×10^{-4}%, 32.4×10^{-4}%, and 650×10^{-4}%, respectively, and its estimated amount taken up annually worldwide through the increment in phytomass is 5606 thousand tonnes, or 37.4 kg/km² (K_b=0.20) [Dobrovolsky].

Titanium concentrations in urban air are mostly below 0.1 μg/m³ and are still lower in rural air [WHO]. Its atmospheric concentrations were found to be (ng/m³): 10 over the Shetland Islands, 22–210 over Germany (West), 2.6 over Norway, 5–690 over Japan, and 10–230 over North America [Kabata-Pendias & Pendias].

PRODUCTION

Titaniferous ores are processed by various methods, such as gravity concentration, flotation, or electromagnetic separation, to obtain concentrates containing titanium(IV) oxide (TiO_2). The concentrates are treated to produce industrial grade TiO_2 or slag containing up to 80% of this oxide. These are the raw materials for the production of titanium(IV) chloride ($TiCl_4$) which is the basis for the industrial production of metallic titanium. In the most commonly used process, the slag is chlorinated in the presence of carbon, the resulting titanium(IV) chloride is freed from impurities by distillation and rectification methods and then reduced with magnesium metal in an inert atmosphere. This yields a titanium sponge which is melted in an arc or vacuum furnace and the molten metal is cast into ingots. High-purity titanium is obtained by electrolytic refining. Titanium metal powder is usually produced by reaction of the metal with hydrogen; the resulting brittle titanium hydride is crushed and heated in a vacuum to remove the hydrogen.

USES

Titanium compares favourably with other structural materials in such properties as corrosion resistance and lightness combined with high strength, and it is therefore extensively used, as a pure metal and especially in alloys, in the aerospace industry. Thus, about 95% the titanium metal consumed in the USA is for aerospace applications, including aircraft and space craft (titanium, e.g., makes up 34% of the F-15 fighter and 22% of the B-1 bomber [Kelto et al.]). Titanium and its alloys and compounds are also used in ship-building, powder metallurgy, in the rubber, leather, paper pulp, textile, electric equipment, and some other industries. Titanium is employed in high-temperature and cryogenic technologies; as a coating for other materials to impart them corrosion resistance; as a component of paints, lacquers, enamels, and artificial pearls. It is also used to make abrasive materials and various tools. The technical-grade titanium used in industry contains admixtures of O, N, Fe, C, and Si that increase its strength but reduce its ductility. It is used in pipes, fittings, pumps, reservoirs for chemical reactors and in other types of chemical equipment operating in aggressive media. Equipment made of titanium is utilized in the hydrometallurgy of nonferrous metals. Platinized titanium anodes are employed in electroplating with gold, platinum, copper, silver, and other metals.

Titanium alloys with iron known as ferrotitaniums (2–50% Ti) serve as alloying additives and deoxidizers in the metallurgy of high-grade steels and special alloys. The high strength and harmlessness of titanium make it eminently suited for use in surgical implant materials and prostheses.

Titanium compounds are used as moderators in nuclear power plants. Their other miscellaneous applications include the production of floor and wall coverings, rubber tyres, porcelain enamels, inks, artificial leather, oilcloth, upholstery materials, and other coated fabrics.

Titanium oxide (TiO_2) — by far the most important titanium compound — is extensively used as a white pigment in surface coatings such as paints, lacquers, and enamels. It also finds use in paper coatings or as a paper filler; in the plastics industry; in ceramic capacitors and electromechanical transducers; in welding-rod coatings; in the production of glass fibers; in food industries and in medicine. **Titanium tetrachloride** is primarily an intermediate in the production of titanium metal and pigments and is also used for the low-pressure polymerization of ethylene, propylene, and other hydrocarbons, and as the intermediate material for the production of organic titanium compounds. **Titanium carbide** and **titanium nitride** are employed in powder metallurgy.

The total world production of titanium concentrates in 1979 was estimated at about 4.60 million tonnes. It has been calculated that the world demand for titanium in the year 2000 will range between 2.1 and 4.5 million tonnes [WHO].

MAN-MADE SOURCES OF EMISSION INTO THE ENVIRONMENT

The major anthropogenic sources of titanium in the general environment are the combustion of fossil fuels, mainly coal and oil, and the incineration of titanium-containing wastes. Titanium levels in coal and oil were reported to average 500 mg/kg and 0.1 mg/kg, respectively [WHO]. As a result of coal burning, 2080 thousand tonnes of titanium, or 14 g/km^2 of land area, were estimated to have been emitted into the environment in 1970 and twice as much in 1980 (4400 thousand tonnes, or 29 g/km^2); for 2000, about 7000 thousand tonnes, or 47.5 g/km^2, has been forecast [Dobrovolsky].

In urban air, concentrations above 1.0 µg/m^3 were recorded, especially in industrialized areas, though they are below 0.1 µg/m^3 in most urban environments. The atmosphere of urban and other areas may be polluted with titanium through the incineration of titanium-containing materials such as paper, inks, plastics, and painted wood. The titanium released may eventually enter the soil. Problems of soil, air and water pollution are created by titanium-containing wastes generated in large quantities during the mining and concentration of titanium and the production of titanium(IV) oxide and titanium pigments.

Soils in the vicinity of power and incineration plants may contain tens of milligrams of titanium per kg. The upper 2 cm of soil around a coal-burning power plant, for example, contained, on average, 92 mg Ti/kg as compared with a background level of 56 mg Ti/kg [Klein & Russell].

Although titanium is poorly absorbed and retained by plant (and animal) tissues, food crops may contain elevated concentrations of the metal in localized areas as a result of soil contamination. Some algae have been found capable of accumulating titanium by factors up to 10 000 and to have the potential of introducing it into the food chain [Schroeder *et al.*]. An indicator of soil contamination by titanium may be its level in herbage samples.

Occupational exposure occurs during the mining and processing of titanium minerals and the production of the metal, its compounds and alloys. Nearly all exposures are to dusts, but some exposure to fume or vapor is experienced by workers handling titanium(IV) chloride. In the processing of raw materials, considerable concentrations of titanium dust were generated in such operations as crushing, grinding, mixing and sieving. In crushing rooms, for example, concentrations amounting to 30–50 mg/m^3 were reported [Kokorev *et al.*]. Much higher concentrations of titanium dusts (up to 500 mg/m^3) were recorded in the breathing zone of workers employed in the manufacture of titanium hydride [Shkurko & Brakhnova]. In the production of titanium alloys, dust concentrations in the air of work premises ranged from 20.3 to 40.2 mg/m^3 during the sieving of titanium carbide and reached 22 mg/m^3 in the process of titanium carbonization [Mezentseva *et al.*].

TOXICITY

Plants
High titanium concentrations in the soil may have adverse effects on plants. In sunflower plants, the rate of photosynthesis was inhibited by 50% when their leaves contained the metal at about 63 mg/kg dry weight. For barley, the critical level (defined as the concentration of a contaminant reducing the harvest by at least 10% on a dry weight basis) was 20 mg Ti/kg [Stepanova]. Chlorotic and necrotic spots occurred on bush bean leaves that contained about 200 mg Ti/kg dry weight [Kabata-Pendias & Pendias].

Animals and man

Acute/single exposure

Animals
For rats, the intraperitoneal LD$_{50}$ of **lead metatitanate** was 2.0 g/kg and its oral LD$_{50}$ >12.0 g/kg. The acute toxicity of titanium compounds with barium (e.g. **barium titanate**, **barium tetratitanate**, and **barium titanate-zirconate)** is determined by that of barium [Nechayeva *et al.*].

In rats given a single injection of **titanium(IV) oxide** intraperitoneally at 139–156 mg/kg or intravenously at 250 mg/kg, no alterations in the external appearance or behavior were noted (the oxide behaved as an inert substance). Observations made 2 months after an intratracheal administration of titanium(IV) oxide to rats at 20 mg/animal and 3 months after its administration by this route to rabbits at 400 mg/animal, did not demonstrate any specific responses. Intrapharygeal administration of a suspension containing 50 mg of titanium(IV) oxide dust did not cause any increase in proline hydroxylase levels (an indicator of increased collagen synthesis in lung tissue) [WHO].

The **carbide, diboride, hydride,** or **nitride** of titanium administered to rats intratracheally caused dystrophic changes in the liver. In addition, the nitride produced swelling of the tubular renal epithelium and exerted a weak fibrogenic effect seen after 6 months, while the hydride caused dystrophic changes in the myocardium and kidneys, impaired protein metabolism, and increased the hydroxyproline content of the lungs. A single intragastric administration of titanium phosphate in oil (30% solution) to rats elicited signs of intoxication (mainly sluggishness and abdominal distension) that disappeared within 24 h; there were no deaths. Rapid deaths in the presence of respiratory distress were observed among mice inhaling aerosols of **titanium(IV) chloride** hydrolysis products (including hydrogen chloride); at autopsy, edemas and hemorrhages were present in the lungs. In studies with 2-h inhalation exposure to concentrations of the order of 0.1–3.0 mg/L, dose-dependent death rates were observed that were considerably higher than those caused by the inhalation of hydrogen chloride alone in equivalent concentrations [Yakusheva & Shnaider]. The higher toxicity was attributed to the adsorption of hydrogen chloride on particles of hydrated titanium oxide which then penetrated to the deeper regions of the lungs not reached by the highly soluble hydrogen chloride.

The LC_{50} of **titanium(IV) cloride** on 2-h inhalation exposure was 0.1 mg/L for mice and 0.4 mg/L for rats [Izmerov et al.].

Man
Accidentally inhaled fumes of metatitanic acid and **titanium oxide dichloride** caused marked congestion of mucous membranes in the pharynx, vocal cords, and trachea; much later, scar formation and laryngeal stenosis were observed [Heimendinger & Klotz].

Chronic exposure

Animals
After disks made of **titanium metal** were implanted into the muscle tissue of dogs and left in place for 7 months, no skin irritation or skin absorption were noted, nor was the normal process of wound healing affected; the metal was encapsulated by fibrous tissue [Beder & Eade]. A similar lack of response to titanium was shown by the bone tissue of dogs 120 to 180 days following the fixation of their fractured bones with titanium plates [Beder et al.].

In guinea-pigs, intratracheal instillation of **barium tetratitanate** (suspended in saline) in three doses at weekly intervals for a total dose of 75 mg did not elicit any fibrotic reactions over 12 months of the observation period [Pratt et al.]. In contrast, rats, hamsters, and guinea-pigs that had been inhaling needle-like fibers of **potassium octatitanate** in doses of 2.9 to 41.8×10^6 fibers per liter 6 h daily for 3 months, developed dose-dependent fibrosis 15–24 months later [Lee *et al.*, 1981]. Fibrosis also occurred in rats administered titanium phosphate fibers intratracheally [Gross *et al.*].

Feeding of **titanium(IV) oxide** to guinea-pigs (at 0.6 g/day), rabbits (3 g/day), cats (3 g/day), and a dog (9 g/day) for 390 days produced no adverse effects in any of the animals [Lehman & Herget].

Rats exposed to **titanium(IV) chloride** by inhalation at 10 mg/m^3 6 h/day, 5 days/week over 24 months developed acute or chronic rhinitis and tracheitis as well as abnormalities in the lungs, such as alveolar proteolysis and hyperplasia of the alveolar epithelium and tracheobronchial lymph nodes [Lee *et al.*, 1986].

Man

In factories manufacturing **titanium(IV) chloride,** many workers were found to have pathological changes in the respiratory tract (hyperemia, mucosal thinning, toxic bronchitis), which were believed to have been caused by titanium(IV) chloride [Kokorev *et al.*].

Workers employed in plants producing titanium sponge were exposed to fumes of **metallic titanium, titanium(IV) chloride,** chlorine, and magnesium chloride [Feigin, 1983]. As reported by this author, 91 episodes of illness with 757 days off work were recorded per year per 100 workers thus employed, against 71 episodes and 648 days off in a matched control group. About a half of all episodes in titanium workers involved the upper and/or deep airways, including hypertrophic, subatrophic, and atrophic changes in the nose, pharynx, and larynx with alterations in respiratory function of the nose, in olfactory sensitivity, and in motor function of the ciliated nasal epithelium.

The incidence of upper respiratory tract disorders among workers exposed to high levels of airborne **titanium metal** or **titanium(IV) oxide** dust was nearly the same as among those exposed to toxic fumes and gases of **titanium(IV) chloride** and its hydrolysis products. The bronchopulmonary diseases induced by metallic titanium or titanium(IV) oxide dust were diagnosed either as dust bronchitis (in 70% of cases) with frequent subsequent development of bronchogenic pneumosclerosis, or (in 30% of cases) as primary pneumosclerosis (primary interstitial pulmonary fibrosis). Titanium pneumosclerosis was described as a relatively benign condition that does not tend to progress rapidly and is not accompanied by substantial cardiopulmonary insufficiency [Feigin, 1983], although there is radiological evidence that workers who have handled titanium(IV) oxide for long periods develop changes similar to those seen in silicosis [Mogilevskaya].

Workers engaged in the major processes of titanium production have been found to have also pathologic changes in the central nervous and cardiovascular systems, neurmuscular abnormalities, and impaired thermoregulation. In young workers exposed to **titanium carbide** concentrations reaching 42 mg/m^3 during titanium carbonization, signs of mucosal hypertrophy in the nasopharynx, larynx, and trachea and predisposition to arterial hypertension and liver dysfunction after less than 5 years of employment were observed and an increased prevalence of chronic bronchits was recorded. In older workers, exposed for periods up to 15 years, a number of abnormalities were detected singly or in various combinations, including myocardial dystrophy, vegetovascular dystonia (vasoneurosis), hemodynamic disturbances, chronic bronchitis with bronchiectasis, and, less commonly, diffuse pneumoconiosis and pulmonary emphysema, impaired pulmonary respiratory function, and hepatic and renal disorders [Feigin, 1988].

Dermal and ocular effects

Animals
When a suspension of **titanium phosphate** was applied to the intact rat skin daily for 3 weeks, no signs of irritation were noted at the application site. Nor was any irritation observed in rabbit eyes exposed to a dust containig 10% of this compound [Yakusheva & Shnaider]. **Titanium(IV) chloride** caused purulent conjunctivitis and corneal opacity in rabbit eyes.

Man
Hot liquid **titanium(IV) chloride** splashed on the skin causes irritation and burns that heal with scar formation. Even short contact of the conjunctiva with this compound leads to purulent conjunctivitis and keratitis [Feigin, 1983].

ABSORPTION, DISTRIBUTION, AND ELIMINATION

In urban environments outside occupational settings, the daily intake of titanium depends on several factors (geographic area, dietary habits, etc.) and may range from 300–450 µg to 2 mg; about 99% of this amount enters the body in food and water. The daily intake from urban air has been reported to range from 1 to 5 µg [Berlin & Nordman]. The daily intake calculated for four American cities was approximately 3.8 µg assuming that 20 m^3 of air per day was respired, the maximum titanium concentration in the air being 0.19 µg/m^3. The daily intake of titanium from drinking water is usually low, probably less than 5 µg [WHO].

The mean titanium contents in plant foods were as follows (mg/kg dry weight): wheat grains, 0.9; corn grains, 2.0; snap bean pods, 3.2; onion bulbs, 1.6; apple fruits, 0.18; orange fruits, 0.15; edible parts of various plants, 0.2–80. On a fresh weight basis, the mean levels (mg/kg) were <0.2 in bean pods, <0.3 in lettuce leaves, <0.7 in cabbage leaves, <0.5 in carrots and cucumbers, <0.004 in apples, 0.1 in oranges [Kabata-Pendias & Pendias].

Corn oil, corn-oil margarine, butter, milled grains, and lettuce had titanium levels ranging from 1.76 to 2.42 mg/kg wet weight [Schroeder *et al.*], while shrimps, garlic, and ground black pepper had mean contents of 7.4, 6.3, and 16.9 ppm [Whanger]. Wheat flour from the USA and Japan contained 0.41 and 0.99 mg/kg (wet weight), respectively [Berlin & Nordman].

In man, titanium levels in different organs have been found to vary widely, but the highest levels consistently occurred in the lungs. For individual tissues, the following levels have been reported (mg/kg): blood, 0.02–0.15; brain, 0.8; liver, 1.3; kidney, 1.3; lung, 3.7; muscle tissue, 0.2 [Berman; WHO].

After a single exposure of rats to titanium dioxide particles by inhalation in concentrations of 15 or 100 mg/m^3, 40–45% of the deposited particles were cleared from the lungs in 25 days. Following exposure to 15 mg/m^3, 0.7% of the particles occurred in the hilar lymph nodes, indicating penetration of particles from the alveoli into the lymphatic system and their partial clearance by the lymphatic route. After the inhalation of 100 mg/m^3, the clearance rate was drastically reduced. Experiments on rats suggest that one-third of the inspired titanium may be retained in the lungs, and that this metal can also enter the lungs from the blood. Titanium dioxide was detected in the lymphatic system of workers employed in the processing of titanium dioxide pigments [WHO].

After an intravenous injection of rats with titanium dioxide at 250 mg/kg, about 70% of the injected dose was detected in the liver at 5 min post-administration. The highest concentrations were recorded in the liver and spleen after 6 h; after 24 h, the concentration was highest in the celiac lymph nodes, which filter lymph from the liver. These lymph nodes still had the highest concentration one year after the injection [Berlin & Norman; WHO].

Titanium is poorly absorbed from the gastrointestinal tract. It was calculated that only about 3% of the titanium entering the human body with the diet would be absorbed. When 5 g of titanium dioxide were administered to five adult male volunteers on three consecutive days, no significant increase in the urinary content of titanium was observed. Titanium can pass the blood-brain barrier, and it can also pass the placenta, as is indicated by its presence in the tissues of newborn infants. Most of the ingested titanium is eliminated unchanged in the feces. After an intraperitoneal or intravenous injection of mice with ^{44}Ti, its mean biological half-life was found to be 640 days. In man, its biological half-life was calculated to be 320 days, although there is evidence that the whole body retention of titanium in man may be longer than that found for mice [WHO].

HYGIENIC STANDARDS

Exposure limits adopted for titanium compounds in the USSR are shown in Table 1. In the USA, a TLA–TWA value of 10 mg/m^3 has been set for total dust of titanium dioxide containing no asbestos and <1% crystalline silica.

Table 1. Exposure limits for titanium and its compounds

	Workroom air		Atmospheric air	Water sources		
	MAC_{wz} (mg/m^3)	Aggregative state	$TSEL_{aa}$ (mg/m^3)	MAC_w (mg/L)	Hazard class	Note
Titanium and its dioxide	10.0	Aerosol	0.5	0.1[*]	4	Fibrogenic
Titanium nitride and silicide	4.0	Aerosol	—	—	—	Fibrogenic
Titanium sulfide and disulfide	6.0	Aerosol	—	—	3	

[*] Based on the general sanitary index of harmfulness that characterizes the impact of a compound on the self-purification capacity of the soil and on the soil microbiocenosis.

METHODS OF DETERMINATION

For the determination of titanium in various media, a wide variety of analytical procedures have been used [Berman; Wenning & Kirsch; Whitehead; WHO]. For its determination in *air*, the use of X-ray fluorescence and atomic absorption spectroscopy has been widely reported. Neutron activation analysis and X-ray emission spectroscopy have also been employed, particularly for the estimation of titanium (and other trace elements) in aerosols [WHO]. For the determination of titanium and its oxide and chloride in air, colorimetric procedures have been developed, which measure the red-colored complex formed in the reaction of titanium ion with chromotropic acid (sensitivity, $1 mg/m^3$) [Peregud; *Specifications for Methods of Determining Noxious Substances in Air*]. Titanium can also be determined in air using a spectral method, in which titanium atoms are excited in an alternating-current arc followed by photography of the spectrum and measurement of the intensity of blackening of the titanium analytical line relative to background (detection limit, $0.004 mg/m^3$; measurement error, ±5%) [Muraviyeva].

For the determination of titanium (and other trace elements) in *water*, wide use is made of x-ray fluorescence, spark-source mass spectrophotometry, spectrography, and photometric techniques [WHO]. One method for measuring titanium in water samples is based on its interaction with the disodium salt of chromotropic acid in a weakly acid medium followed by measurement of the resulting red-colored titanium-chromotropate complex (sensitivity 1 μg/L) [Novikov et al.].

Titanium levels in various *foods* have been determined, in the main, by spectrographic and colorimetric techniques. For their determination in canned fruit and vegetable juices, use of emission spectrum analysis has been reported. Titanium in *biologic materials* has been commonly determined by photometric and spectroscopic methods [WHO].

MEASURES TO CONTROL EXPOSURE AT THE WORKPLACE

In factories where titanium and its compounds are produced, special stands fitted with suction hoods should be used to disassemble, clean, and reassemble the apparatus for vacuum distillation, rectification, and reduction of titanium(IV) chloride and to recover titanium sponge from the retorts. All operations involved in the transportation, installation, and removal of the apparatus and in the knocking out and crushing of titanium sponge should be mechanized. Enclosures and suction hoods should be integral parts of the process equipment (with provision being made for cleaning the suctioned air) and be so designed as to be readily accessible for maintenance and repair operations. The chlorinator should be equipped with an independent outlet to the system for disposal of acidic wastewaters. The removal of dust from the cleaning devices and its transportation should be mechanized and preclude the escape of dust to the environment. Processes such as unloading, sifting, and blending of dry concentrates should also be mechanized and exclude indoor air pollution by dusts and gases.

Provision of enclosures and exhaust ventilation is necessary to control dust emission in all plants where powdery metallic titanium, titanium carbide, or titanium(IV) oxide are processed.

Dust control is also the prime concern in factories where barium metatitanate, barium zirconate titanate, or calcium metatitanate are produced or used since these compounds occur in powder form. The basic technical measures for dust control include mechanization and automation of the technological processes; hermetic sealing and effective venting of the dust-generating equipment; avoidance, to the greatest possible extent, of manual operations when servicing individual pieces of the equipment; containment of dust emissions at source; and application of local exhaust ventilation at points where the equipment has to be opened [Nechayeva].

Workers exposed to dusts of titanium or its compounds should be provided with respirators. Those working with titanium(IV) chloride must be supplied with well-fitting gas masks. During work with a hot solution of this compound, impervious gloves and chemical safety goggles should be worn.

All workers should undergo a preplacement and subsequent regular (at least once a year) medical examinations, which should include, as a minimum, a chest X-ray, an electrocardiogram, and hematologic studies. Medical contraindications to work with titanium and its compounds are chronic respiratory, cardiovascular, and nervous diseases and pathologic conditions of the genital organs.

FIRST AID IN POISONING

In the event of skin contact with titanium halides, they should be immediately removed from the skin with a cotton pad or blotting paper and the affected area washed thoroughly with running water. The eyes, if contaminated, should also be thoroughly irrigated with running water or isotonic sodium chloride solution for several minutes while keeping the lids apart. If the poisoning is by ingestion, liberal quantities of a liquid such as milk (if possible, with several beaten-up raw eggs) should be given to the victim. Activated charcoal is indicated after vomiting [Ludewig & Lohs].

REFERENCES

Beder, O. E. & Eade, G. (1956) *Surgery*, **39** (1), 470–473.
Beder, O. E. *et al.* (1957) *Surgery*, **41**, 1012–1015.
Berlin, M. & Nordman, C. (1979) In: Friberg, L. *et al.* (eds) *Handbook on the Toxicology of Metals*. Elsevier, North-Holland, pp. 627–636.
Berman, E. (1980) *Toxic Metals and their Analysis*. Heyden, London.
Dobrovolsky, V. V. (1983) *Geografiya mikroelementov. Globalnoe rasseyanie* [Geography of Trace Elements. Global Dispersion]. Mysl, Moscow (in Russian).
Feigin, B. T. (1983) In: *Metally. Gigiyenicheskiye aspekty otsenki i ozdorovleniye okruzhayushchei sredy* [Metals. Hygienic Evaluation and Environmental Health Promotion]. Moscow, pp. 180–186 (in Russian).
Feigin, B. T. (1988) *Gig. Truda Prof. Zabol.*, No. 7, 30–33 (in Russian).
Gross, P. *et al.* (1977) *Arch. Pathol. Lab. Med.*, **101** (10), 550–554.
Heimendinger, E. & Klotz, G. (1956) *Arch. Otolaryng.*, **73**, 645 .
Izmerov, N. F. *et al.* (1982) *Toxicometric Parameters of Industrial Toxic Chemicals under Single Exposure*. Published by the USSR Commission for the United Nations Environmental Programme (UNEP), Moscow (in Russian).
Kabata-Pendias, A. & Pendias, H. (1986) *Trace Elements in Soils and Plants*. CRC Press, Boca Raton (Fl.).
Kelto, S. A. *et al.* (1985) In: *Poroshkovaya metallurgiya titanovykh splavov* [The Powder Metallurgy of Titanium Alloys]. Moscow, pp. 10–27 (in Russian).
Klein, D. H. & Russell, P. (1973) *Env. Sci. Technol.*, **7**, 357–359.
Kokorev, N. P. *et al.* *Gig. Truda Prof. Zabol.*, No. 10, 23–29 (in Russian).
Lee, K. P. *et al.* (1981) *Am. J. Pathol.*, **102** (3), 314–323.
Lee, K. P. *et al.* (1986) *Toxicol. Appl. Pharmacol.*, **83**, 30–45.
Lehman, K. B. & Herget, L. (1927) *Chem. Zig.*, **51**, 793–794.
Ludewig, R. & Lohs, K. (1981) *Akute Vergietungen*. VEB Gustav Fischer, Jena.
Lukashev, K. I. & Petukhova, N. N. (1974) *Pochvovedeniye*, No. **8**, 47 (in Russian).
Mezentseva, N. V. *et al.* (1963) In: *Toksikologiya redkikh metallov* [Toxicology of Rare Metals]. Medgiz, Moscow, pp. 58–71 (in Russian).
Mogilevskaya, O. Ya. (1983) In: *Encyclopaedia of Occupational Health and Safety*. International Labour Office, Geneva, pp. 2179–2181.
Muraviyeva, S. I. (ed.) (1982) *Sanitarno-khimicheskiy kontrol vozdukha promyshlennykh predpriyatiy* [Physical and Chemical Monitoring of Air Quality in Industrial Workplaces]. Meditsina, Moscow (in Russian).
Nechayeva, E. N. (1981) In: *Professionalnaya patologiya pri vozdeistvii metallov* [Occupational Diseases Caused by Exposure to Metals], Moscow, pp. 84–87 (in Russian).
Nechayeva, E.N. *et al.* (1982) *Gig. Truda Prof. Zabol.*, No. **9**, 10–13 (in Russian).

Novikov, Yu. V. *et al.* (1981) *Metody opredeleniya vrednykh veshchestv v vode vodoyemov* [Methods for Measuring Noxious Substances in Water Bodies]. Meditsina, Moscow (in Russian).

Peregud, E. A. (1976) *Khimicheskiy analiz vosdukha (novye i usovershenstvovannye metody* [Chemical Analysis of the Air (New and Improved Methods)]. Khimiya, Leningrad (in Russian).

Pratt, P. C. *et al.* (1953) *Arch. Ind. Hyg. Occup. Med.*, **8**, 109—117.

Schroeder, H. A. *et al.* (1963) *J. Nutrit.*, **80**, 39—47.

Shkurko, G. A. & Brakhnova, I. T. (1973) *Poroshkovaya Metallurgiya* (Kiev), No. **8**, 100—102 (in Russian).

Specifications for Methods of Determining Noxious Substances in Air (1965) [Tekhnicheskiye usloviya na metody opredeleniya vrednykh veshchestv v vozdukhe], No. 4, Publication of the USSR Ministry of Health, Moscow (in Russian).

Stepanova, M. D. (1982) In: *Khimicheskiye elementy v sisteme pochva-rasteniye* [Chemical Elements in the Soil-Plant System]. Novosibirsk, pp. 92—105 (in Russian).

Wennig, R. & Kirsch, N. (1988) In: Seiler, H. G. *et al.* (eds) *Handbook on Toxicity of Inorganic Compounds*. Marcel Dekker, New York, pp. 705—714.

Whanger, P. D. (1982) In: Hatchcock, J. N. (ed.) *Nutritional Toxicology*, Vol. 1. Academic Press, New York, pp. 163—208.

Whitehead, J. (1991) In: Merian, E. (ed.) *Metals and their Compounds in the Environment—Occurrence, Analysis and Biological Relevance*. VCH, New York, pp. 1261—1267.

WHO (1982) *Environmental Health Criteria 24: Titanium*. World Health Organization, Geneva.

Yakusheva, N. A. & Shnaider, L. E. (1982) *Gig. San.*, No. **10**, 81—82 (in Russian).

Zirconium and its compounds

Zirconium carbide; z. carbonate; z.(IV) chloride (zirconium tetrachloride);
z. diboride; z. dichloride oxide (zirconium oxychloride, zirconyl chloride);
z. dichloride oxide octahydrate (zirconium oxychloride octahydrate, zirconyl
chloride octahydrate); z. dinitrate oxide hexahydrate (zirconium oxynitrate
hexahydrate, zirconyl nitrate hexahydrate); z.(IV) fluoride (zirconium
tetrafluoride); z.(IV) hydroxide; z. nitride; z. orthosilicate (zircon [min.]);
z. oxide (zirconium dioxide, zirconia; baddeleyite (α) [min.]); z. sulfate;
z. sulfate tetrahydrate
Potassium hexafluorozirconate; sodium metazirconate

IDENTITY AND PHYSICOCHEMICAL PROPERTIES
OF THE ELEMENT

Zirconium (Zr) is a metallic element in group IV of the periodic table, with the
atomic number 40. Its natural isotopes are ^{90}Zr (51.46%), ^{91}Zr (11.23%), ^{92}Zr
(17.11%), ^{94}Zr (17.40%), and ^{96}Zr (2.80%).

Zirconium exists in two modifications, α and β. In the pure state it is a strong
ductile metal that readily lends itself to hot and cold working by rolling, forging,
and stamping. The presence, even in small amount, of dissolved O_2, N_2, H_2, C,
or their compounds in zirconium, makes the latter rather brittle.

Zirconium is oxidized only at high temperatures (200–800°C), but may ignite
in air at ordinary temperature when in powder form. It is resistant to water and
water vapor at temperatures up to 300°C. It is also resistant to dilute acids and

alkalies, and is the only metal resistant to ammonia-containing alkalies. It reacts with nitric acid and aqua regia at >100°C. Its usual oxidation state in compounds is +4 but it occurs in the +2 or +3 state in some of its halides.

Zirconium oxide exists in three modifications and zirconium(IV) fluoride in two. See also Appendix.

NATURAL OCCURRENCE AND ENVIRONMENTAL LEVELS

Zirconium is contained in 27 minerals, the most important of which are baddeleyite (ZrO_2) and zircon ($Zr(SiO_4)_2$).

The clarke of zirconium has been estimated at $(165-170) \times 10^{-4}\%$ in the earth's crust and $170 \times 10^{-4}\%$ in the granite layer of the continental crust [Dobrovolsky].

The oceans contain an estimated 34 million tonnes of zirconium, mainly as $Zr(OH)_4$, at average concentrations of $0.026-0.03$ µg/L in water and $0.00070 \times 10^{-4}\%$ in total salts [Dobrovolsky; Henderson]. The amount of zirconium taken up annually by ferromanganese oxide deposits of the Pacific is estimated at 3.8 thousand tonnes and its average abundance in these, at $630 \times 10^{-4}\%$, i.e. is about 4 times greater than its crustal abundance. Its average estimated concentration in non-biogenic dispersed particles (clay muds) of the world's ocean is $150 \times 10^{-4}\%$, while that in biogenic particles (calcareous muds) is $20 \times 10^{-4}\%$. In river water, zirconium is contained at an average concentration of 2.6 µg/L, and its global run-off from rivers is estimated to be 96 000 tonnes annually ($K_w = 0.13$) [Dobrovolsky].

The content of zirconium in soils was reported to average $3 \times 10^{-2}\%$ ($2 \times 10^{-2}\%$ in shales and clays) [Vadkovskaya & Lukashev]. No substantial variation in its abundance in soils of different types is observed, but lower levels are found in soils on glacial moraines and higher levels in residual soils derived from zirconium-rich rocks. The average content calculated for Australian soils was 350 mg/kg, while that for soils in the USA was 224 mg/kg. In the USA, mean zirconium concentrations in various surface soils were as follows (mg/kg): 140 in alluvial soils, 240 in soils over limestones and calcareous rocks, 200–278 in garden soils, and 330 in light desert soils. Zirconium levels in food plants vary from 0.005 to 2.6 mg/kg dry weight [Kabata-Pendias & Pendias].

Continental vegetation contains an estimated 18.8 million tonnes of zirconium, at average fresh weight, dry weight, and ash weight concentrations of $3 \times 10^{-4}\%$, $7.5 \times 10^{-4}\%$, and $150 \times 10^{-4}\%$, respectively. The amount of zirconium taken up annually worldwide through the increment in phytomass amounts to ≈ 1300 thousand tonnes, or 8.62 kg/km^2 of land area ($K_b = 0.88$) [Dobrovolsky].

PRODUCTION

The principal industrial source of zirconium is the mineral zircon. Zircon ores are concentrated by gravitational methods and the concentrates are refined by magnetic or electrostatic separation. Zirconium compounds are commonly decomposed by treatment with alkali, chlorination, or fusion with potassium hexafluorosilicate. The zirconium dichloride oxide, zirconium sulfate, and hexafluorozirconate thus obtained are crystallized or hydrolytically precipitated and then calcined to zirconium oxide. Since zirconium compounds extracted from ores always contain an admixture of hafnium, zirconium is separated from this impurity by fractional crystallization, ion exchange, or other methods. Metallic zirconium is obtained in a powder or spongy form by reduction of the zirconium chloride, zirconium oxide, or potassium hexafluorozirconate.

USES

Zirconium-based alloys free of hafnium have found wide use in the nuclear power industry for the cores of thermal-neutron reactors (in fuel-element jackets, channels, fuel assemblies, and core lattices).

Zirconium-based alloys may also contain Nb, Sn, Fe, Cr, Ni, Co, and Mo. Zirconium is a constituent of alloys based on Mg, Ti, Ni, Mo, Nb, and other metals used as structural materials in rockets and various aircraft and in the windings of superconducting magnets. Zirconium is also employed in piezoceramic materials and in chemical engineering as a corrosion-resistant material. Zirconium additives are used for deoxidizing and desulfurizing steel. Powdered zirconium is used in pyrotechnics, tracer bullets, and detonators.

Zirconium oxide (zirconia) or **zircon** are basic components of special refractory compositions used in steel-casting and aluminum industries, in the smelting of platinum, palladium, and other metals, as linings for high-temperature furnaces or to provide high-temperature insulation. **Zirconium sulfate** is used as a tanning agent in the leather industry.

Other uses of zirconium are described in Deknudt and Schaller.

The world production of zirconium was 250 000 tonnes in 1974 and is expected to attain 500 000 tonnes in 2000 [Kabata-Pendias & Pendias].

MAN-MADE SOURCES OF EMISSION INTO THE ENVIRONMENT

Major anthropogenic sources of zirconium in the environment are emissions from coal- and petroleum-burning power plants. Its average concentrations in coal and coal ash are $70 \times 10^{-4}\%$ and $480 \times 10^{-4}\%$, respectively, while its common concentration range in oil ash is $50{-}500 \times 10^{-4}\%$. From coal burning alone, a

total of 91 000 tonnes of zirconium, or 607 g/km^2 of land area, was estimated to
have been released worldwide in 1970 and 192 500 tonnes, or 1283 g/km^2, in
1980; for the year 2000, 305 000 tonnes, or 2035 g/km^2, has been predicted
[Dobrovolsky]. In an industrial city, the main source of environmental pollution
with zirconium is usually solid wastes; according to one estimate, this source
accounted for 46 mg Zr per person per day *vs* 14 mg from atmospheric
precipitation. Most of the zirconium deposited in atmospheric precipitation is
contributed by emissions from electric power plants and the dust from building
materials. Zirconium carbide dust reached a concentration of 190 mg/m^3 in the
air of work premises when electric furnaces were being charged with this
compound [Brakhnova].

Some plants from the Leguminosae family growing in a zirconium-enriched
soil accumulated this metal to 20–70 mg/kg on an ash weight basis [Gough *et
al.*].

In sewage sludges, phosphate fertilizers, and organic fertilizers applied to soil,
zirconium was found in concentrations of 5–90, 50, and 5.5 mg/kg dry weight,
respectively [Kabata-Pendias & Pendias].

TOXICITY

Fish

On 96-hour exposure, **zirconium sulfate** was toxic to minnow at a concentration
of 14 mg/L in soft water and 115 mg/L in hard water (calculated as Zr). The
corresponding toxic concentrations of **zirconium dichloride oxide** 18 mg/L and
240 mg/L for minnow and 15 mg/L and 270 mg/L for black bass [Grushko].

Animals and man

Acute exposure

Animals

LD$_{50}$ values of zirconium compounds are shown in Table 1. Inhalation exposure
to **zirconium dichloride oxide** in concentrations of 0.5–0.8 mg/L was lethal to
rats within 30 to 40 min.

In animals that died from acute poisoning by **zirconium(IV) chloride**, the
severest changes were found in the esophagus (epithelial necrosis, congestion,
submucosal edema), gastric and intestinal mucosas, convoluted renal tubules
(granular and vacuolar degeneration), and liver (fatty degeneration and necrotic
areas). In mice acutely exposed to **potassium hexafluorozirconate** by oral gavage,
the poisoning was manifested by restlessness at first, followed shortly by
sluggishness, ataxia, clonic convulsions, adynamia, abdominal distension,
nonresponsiveness to external stimuli, and death within 60 min.

Histopathological examination revealed cyanotic and collapsed lungs with hemorrhages, congested abdominal organs, albuminous and fatty degeneration with necrotic areas in the liver, albuminous degeneration of the convoluted tubular epithelium in the kidneys, and microabscesses in the spleen. It has been pointed out that this compound should be classed among markedly toxic industrial poisons [Mogilevskaya; Shalganova; Zhilova & Kasparov].

Table 1. Median lethal doses of zirconium compounds

Compound	Animal	Route of administration	LD_{50} (mg/kg)
Zirconium carbonate	Rat	Oral	>10 000[*]
(hydrated)		Intraperitoneal	>1500[*]
Zirconium (IV) chloride	Rat	Oral	1688[**]
		Intraperitoneal	662.5[**]
	Mouse	Intraperitoneal	400[**]
Zirconium dichloride oxide	Rat	Oral	990[*]
octahydrate		Intraperitoneal	113[*]
Zirconium dinitrate oxide	Rat	Oral	853[*]
hexahydrate		Intraperitoneal	426[*]
Zirconium (IV) hydroxide	Rat	Oral	10 000[**]
Zirconium sulfate	Rat	Oral	1253[*]
tetrahydrate		Intraperitoneal	63[*]
Potassium hexafluorozirconate	Rat	Oral	250[**]
	Mouse	Oral	97[**]
	Guinea-pig	Oral	200[**]
Sodium metazirconate	Rat	Oral	2290[*]
		Intraperitoneal	939[*]

Sources: [*]Stokinger (expressed as Zr); [**]Shalganova and Zhilova & Kasparov.

Rats administered 50 mg of **zirconium carbide** or **diboride** intratracheally were found, 6 to 8 months later, to have developed a chronic productive interstitial process with emphysema and mild fibrosis in their lungs. At 1 and 3 months after the intratracheal administration of zirconium diboride, their livers were seen to undergo fatty and parenchymatous degeneration that became more severe at 6 months when the central areas of hepatocytes had very conspicuous necrobiotic foci containing necrotic masses; there was little or no evidence of regeneration. Epithelial degeneration with necrotic areas was also noted in the convoluted renal tubules [Brakhnova].

Chronic exposure

Animals
In rabbits exposed daily to **zirconium(IV) chloride** by oral gavage at 100 mg/kg for 4 months, the only notable finding during that period was a decreasing tendency shown by the prothrombin consumption index. However, histopathologic examination disclosed moderate edema of the gastric and small intestinal submucosas, lymphohistiocytic infiltration of hepatic structures, and epithelial swelling in the convoluted renal tubules. Stronger toxic effects were produced by zirconium(IV) chloride in rats given this compound at 100 mg/kg by the same route every third day for 2 months.

In rats exposed to **potassium hexafluorozirconate** by inhalation for 2 h/day, 6 days/week over 5.5 months, growth retardation was observed, and histopathological examination revealed thickened, distended, or ruptured interalveolar septa, extensive peribronchial accumulation of lymphoid elements, destructive changes in the bronchial tree, and metaplasia and detachment of the bronchial epithelium in places. It should be noted that a significant contribution to the chronic as well as acute effects of the chloride and hexafluorozirconate are made by the chlorine (with formation of hydrochloric acid on hydrolysis) and the fluorine.

Metallic zirconium and its insoluble compounds are physiologically inert.

Man
Allergic skin reactions occurred in individuals who were regular users of lotions or deodorants containing zirconium salts.

Workers of zirconium-producing plants complained of cardiac pain, weakness, and headache, and some had chronic disorders of the upper respiratory tract, leukocyte abnormalities, erythropenia, reduced hemoglobin levels, pulmonary fibrosis, or pulmonary granulomas [Gough et al.]. The effects of zirconium on plants, animals and humans have been summarized [Schaller, 1991]. Pulmonary granuloma has been reported to develop in zirconium workers following exposure to zirconium lactate [Deknudt].

Dermal and ocular effects
Zirconium(IV) chloride was very irritating to the skin and conjunctiva of rabbits on local application, causing edema and reddening of the skin and conjunctivitis of 24 h in duration. **Potassium hexafluorozirconate** applied to intact rabbit skin produced profound necrosis of the soft tissues.

ABSORPTION, DISTRIBUTION, AND ELIMINATION IN MAN

Normally, zirconium enters the human body predominantly in food, the average daily dietary dose being 4.2 mg. Its levels in foods were reported to be as follows

610 Zirconium and its compounds

(ppm) : wheat, 2.8; rice, 3.0; beans, 4.0; oats, 6.6; red peppers, 9.0; tea, 12.0; mutton, 4.0; butter, 6.0; vegetable oils, 3.0–6.0 [Whanger]. Zirconium levels in a number of plant foodstuffs are indicated in Table 2.

The adult (70 kg) body has been estimated to contain 250 to 420 mg of zirconium, of which 67% is present in the fatty tissue, 2.5% in the blood, and the rest in the skeleton, aorta, liver, brain, kidneys, and some other organs [Linder; Schaller; Whanger].

Table 2. Zirconium levels in plant foods

Plant	Tissue	Zirconium (mg/kg)		
		Wet weight	Dry weight	Ash weight
Cereal	Grains	0.08–10	0.02–1	8–1033
Corn	Grains	—	—	<20
Bean	Pods	<0.13	2.6	—
Lettuce	Leaves	0.41–0.62	0.56	4
Cabbage	Leaves	<0.4	—	<20
Carrot	Roots	<0.32	—	<20
Potato	Tubers	—	0.5	12
Onion	Bulbs	0.45–0.84	—	<20
Tomato	Fruits	0–1.79	—	<20
Apple	Fruits	0.31	—	<20
Orange	Fruits	0.05	—	<20
Grape	Raisins	—	1.5	—
Peanut	Seed	—	2.3	—

Source: Kabata-Pendias & Pendias.

In the blood, zirconium commonly occurs in concentrations of 0.012–0.026 $\mu g/g$, its main depot being erythrocytes. The following concentrations have been reported as average values ($\mu g/g$) : brain, 2.04; kidney, 5.39; liver, 9.19; lung, 7.49; muscle, 2.63; fatty tissue, 18.7. The principal route of excretion is into the bile [Linder; Schaller]. Absorption from the gastrointestinal tract is very slight (about 0.01% of the ingested amount). Elimination in the urine accounts for less than 1% of the daily intake [Berman].

HYGIENIC STANDARDS

Exposure limits adopted for zirconium and its compounds in the USSR are shown in Table 3. In the USA, threshold limit values for zirconium compounds have been set 5 mg/m³ (TWA) and 10 mg/m³ (STEL).

METHODS OF DETERMINATION

Zirconium in *air* can be determined by a photometric method, based on its reaction with pyrocatechol violet (sensitivity 0.2 mg/m³; measurable concentration range, 5–40 µg in the photometric volume) [Peregud], and by a colorimetric method in which tetravalent zirconium ion reacts with arsenazo III to form a complex compound and the resulting color change of the solution from pink to violet to blue is measured colorimetrically (sensitivity 0.5 µg Zr in the analytical volume) [*Specifications for Methods of Determining Noxious Substances in Air*].

Table 3. Exposure limits for zirconium and its compounds

	Workroom air			
	MAC_wz (mg/m³)	Aggregative state	Hazard class	Note
Zirconium, metallic	6.0	Aerosol	3	
Fluorozirconates	1.0	Aerosol	2	
Zircon	6.0	Aerosol	4	Fibrogenic
Zirconium carbide	6.0	Aerosol	4	Fibrogenic
Zirconium dioxide	6.0	Aerosol	4	Fibrogenic
Zirconium nitride	4.0	Aerosol	3	Fibrogenic

For the determination of zirconium in *water*, a photometric method has been developed, in which an orange complex formed by zirconium with a specific reagent (picramine-epsilon) is subjected to photometry (detection limit 0.1 mg/L) [Goryacheva & Yershova]. For a review of methods for zirconium determination, see *Toxic Metals and their Analysis* [Berman] (see also Deknudt and Schaller).

MEASURES TO CONTROL OCCUPATIONAL EXPOSURE

The main technical measures for restricting exposure in the processing of zirconium and its compounds in powder form include mechanization of manual operations, sealing and venting of the dust-generating equipment, and provision of adequate ventilation in the workrooms. Any spilled powder should be cleaned immediately with water. Powdered materials should be transported in the wet state. Workers should be meticulous to keep their clothing free of powder and dust.

It is necessary to install local exhaust ventilation in places where hydrofluoric and hydrochloric acids are formed and released during hydrolysis of zirconium salts. Contact of workers with zirconium chlorides and fluorides should be kept to a minimum.

In the production and use of barium metazirconate and lead zirconate titanate, the control measures described for barium (q.v.) and lead (q.v.) and their compounds apply.

Those working with zirconium or its compounds are subject to preplacement and subsequent periodic medical examinations.

REFERENCES

Berman, E. (1980) *Toxic Metals and their Analysis*. Heyden, London.

Brakhnova, I. T. (1971) *Toskichnost poroshkov metallov i ikh soyedineniy* [Toxicity of Metals and their Compounds in Powder Form]. Naukova Dumka, Kiev (in Russian).

Deknudt, G. (1988) Chapter 72 *Zirconium* in: Seiler, H. G. *et al.* (eds.) *Handbook on Toxicity of Inorganic Compounds*. Marcel Dekker, New York, pp. 801–804.

Dobrovolsky, V. V. (1983) *Geografiya mikroelementov. Globalnoe rasseyanie* [Geography of trace elements. Global Dispersion]. Mysl, Moscow (in Russian).

Goryacheva, N. A. & Yershova, K. P. (1980) *Gig. San.*, No. 6, 72 (in Russian).

Gough, L. P. *et al.* (1979) *Element Concentrations Toxic to Plants, Animals, and Man.* Government Printing Office, Washington, DC.

Grushko, Ya. M. (1979) *Vrednye neorganicheskiye soyedineniya v promyshlennykh stochnykh vodakh* [Harmful Inorganic Compounds in Industrial Waste Waters]. Khimiya, Leningrad (in Russian).

Henderson, P. (1982) *Inorganic Geochemistry*. Pergamon Press, Oxford.

Kabata-Pendias, A. & Pendias, H. (1986) *Trace Elements in Soils and Plants*. CRC Press, Boca Raton (Fl.).

Linder, M. C. (1985) In: Linder, M. C. (ed.) *Nutritional Biochemistry and Metabolism*. Elsevier, Amsterdam, pp. 151–197.

Mogilevskaya, O. Ya. (1967) In: Izraelson, Z. I. (ed.) *Novye dannye po toksikologii redkikh metallov i ikh soyedineniy* [New Data on the Toxicology of Rare Metals and Their Compounds]. Meditsina, Moscow, pp. 160–165 (in Russian).

Peregud, E. A. (1976) *Khimicheskiy analiz vozdukha (novye i usovershenstovovannye metody)* [Chemical Analysis of Air (New and Improved Methods]. Khimiya, Leningrad (in Russian).

Schaller, K.-H. (1991) Chapter II. 37 *Zirconium* in: Merian, E. (ed.) *Metals and their Compounds in the Environment—Occurrence, Analysis and Biological Relevance*. VCH, New York, pp. 1343–1347.

Shalganova, I. V. (1967) In: *Novye dannye po toksikologii redkikh metallov i ikh soyedineniy* [New Data on the Toxicology of Rare Metals and Their Compounds]. Meditsina, Moscow, pp. 105–115 (in Russian).

Specifications for Methods of Determining Noxious Substances in Air (1968) [Tekhnicheskiye usloviya na metody opredeleniya vrednykh veshchestv v vozdukhe], No. 5. Publication of the USSR Ministry of Heatlh, Moscow (in Russian).

Stokinger, H. E. (1963) In: Patty, F. A. (ed.) *Industrial Hygiene and Toxicology*, Vol. II: *Toxicology*. Interscience Publishers, New York, pp. 1188–1194.

Vadkovskaya, I. K. & Lukashev, K. P. (1977) *Geokhimicheskiye osnovy okhrany biosfery* [Protection of the Biosphere: Geochemical Principles]. Nauka i Tekhnika, Minsk (in Russian).

Whanger, P. D. (1982) In: Hatchcock, J. N. (ed.) *Nutritional Toxicology*, Vol. 1. Academic Press, New York, pp. 163–208.

Zhilova, N. A. & Kasparov, A. A. (1967) In: *Novye dannye po toksikologii redkikh metallov i ikh soyedineniy* [New Data on the Toxicology of Rare Metals and Their Compounds]. Meditsina, Moscow, pp. 98–105 (in Russian).

Hafnium and its compounds

Hafnium carbide; h. chloride (h. tetrachloride); **h. diacetate oxide** (h. oxyacetate, hafnyl acetate); **h. dichloride oxide octahydrate** (h. oxychloride octahydrate, hafnyl chloride octahydrate); **h. dinitrate oxide** (h. oxynitrate, hafnyl nitrate); **h. hydride; h. nitride; h. oxide** (h. dioxide, hafnia)

IDENTITY AND PHYSICOCHEMICAL PROPERTIES
OF THE ELEMENT

Hafnium (Hf) is a metallic element in group IV of the periodic table, with the atomic number 72. Its natural isotopes are ^{174}Hf (0.163%), ^{176}Hf (5.21%), ^{177}Hf (18.56%), ^{178}Hf (27.10%), ^{179}Hf (13.75%), and ^{180}Hf (35.22%).

Hafnium is a hard and ductile metal readily amenable to cold and hot working (rolling, forging, stamping). It exists in two crystalline polymeric modifications.

The usual oxidation state of hafnium in compounds is +4, although it can exhibit the +2 and +3 states under special circumstances (hafnium(II) and hafnium(III) compounds). Its chemical properties closely resemble those of zirconium. It readily absorbs gases. In air, it becomes coated with a film of oxide. The oxide also forms when hafnium is heated with oxygen. On heating with halogens, halides of the HfHal$_4$ type are produced. At high temperatures, it reacts with nitrogen to form the nitride HfN and with carbon to form the carbide HfC. See also Appendix.

NATURAL OCCURRENCE AND ENVIRONMENTAL LEVELS

Hafnium has no minerals of its own and is usually found associated with zirconium. Its clarke has been estimated at $(1.0-3.0) \times 10^{-4}\%$ in the earth's crust and at $3.5 \times 10^{-4}\%$ in the granite layer of the continental crust [Dobrovolsky].

The common range of hafnium concentrations in rocks is 0.1–10 mg/kg. Soils have an average hafnium content of $6 \times 10^{-4}\%$. Its content in Bulgarian and Canadian soils ranged from 1.8 to 18.7 and 1.8 to 10 mg/kg, respectively, and averaged 5 mg/kg in standard soil samples from Great Britain [Kabata-Pendias & Pendias].

Sea and surface river waters contain hafnium at average concentrations of 0.007 μg/L and ~0.004 μg/L, respectively. Rain and snow waters were reported to have mean concentrations of 0.06 and 0.125 μg/L, respectively [Henderson; Vadkovskaya & Lukashev].

Hafnium concentrations measured in the atmospheric air over several regions of countries were as follows (ng/m^3): Greenland, 40–60; Germany (West), 300; Japan, 18–590; Central, North, and South America, 60–70, 0.5–290; and 40–760, respectively [Kabata-Pendias & Pendias]. In the Upper Volga River region of the USSR, the amount of hafnium deposited with atmospheric precipitation was estimated at 0.008 kg/m^2 per year [Vadkovskaya & Lukashev].

In plants, the common range of hafnium was given as 0.01 to 0.4 mg/kg dry weight; in food plants, the range was from 0.6 to 1.1 μg/kg fresh weight, but the element was not always detectable [Kabata-Pendias & Pendias].

PRODUCTION

Hafnium compounds are usually separated at the final stage in the production of zirconium from its ore. Metallic hafnium is obtained by reduction of hafnium chloride or oxide with magnesium or sodium.

USES

Hafnium and its compounds are used in neutron screens and as control rod material for nuclear reactors; in cathodes, getters, and electrical contacts; in the aircraft and rocket building industries; in optical glasses, heat-resistant or gastight materials, and special crucibles for the melting of high-melting metals; in metallurgy; in industrial catalytic processes; in the textile industry; in the production of gunpowder and explosives; in filaments for incandescent lamps; and in medicine.

In 1974, about 1.3 tonnes of hafnium was produced worldwide.

MAN-MADE SOURCES OF EMISSION INTO THE ENVIRONMENT

During the mining of zirconium ores, some hafnium is released into the ambient air as a constituent of multicomponent dusts. In workrooms where zirconium was processed and refined or its compounds were produced, airborne dusts contained hafnium carbide and nitride at levels up to 75 and 165 mg/m^3, respectively; over 70% of the dust particles were fairly small in size (<3 μm) and only 3–6% were 9 μm or larger [Zhislin and Ovetskaya].

TOXICITY TO ANIMALS

Acute/single exposure

Acute toxicity data for three hafnium compounds administered intraperitoneally are given in Tables 1 and 2. The oral LD$_{50}$ of **hafnium chloride** for rats was calculated to be 2000 mg/kg.

Table 1. Median lethal doses of hafnium compounds after intraperitoneal injection

Compound	Animal	LD$_{50}$ (mg/kg)
Diacetate oxide	Rat	758
Dichloride oxide	Rat	850
	Mouse	732
Dinitrate oxide	Rat	560

Source: Ovetskaya.

Table 2. Mean survival times (days) of rats following a single intraperitoneal injection of hafnium compounds

Dose (mg/kg)	Diacetate oxide	Dichloride oxide	Dinitrate oxide
500	—	—	5.5 (4.7)
750	—	5.7 (5.5)	3.8 (2.0)
1000	3.3 (1.8)*	5.3 (3.0)	2.5 (1.7)
1250	2.3 (1.5)	2.0 (2.4)	2.5 (1.0)

* Figures in parentheses are the mean survival times for the corresponding zirconium salts after a single injection by the same route.

Source: Ovetskaya.

After oral administration of hafnium compounds to rats by gavage, coagulation necrosis without marked inflammatory changes developed in the stomach, while after intratracheal administration, coagulation necrosis of the bronchial mucosa and pulmonary edema were seen. Following dermal application, the major changes occurred in the epidermis and included superficial necrosis and a weak inflammatory response.

One, three, and six months after an intratracheal administration of **hafnium hydride** to rats in an amount of 50 mg, slight fibrogenic and systemic toxic effects of this compound were in evidence. Hydroxyproline levels in the lungs were elevated, and sulfhydryl groups in the blood and internal organs were reduced. Histopathological examination demonstrated thickened interalveolar septa in the lungs, fatty degeneration in the liver, swollen epithelium of convoluted tubules and swollen endothelium of glomerular capillaries in the kidneys, and hyperplastic intrafollicular tissue in the spleen [Brakhnova & Shkurko].

Multiple exposure

After oral administration of **hafnium chloride** by gavage to rats at 400 mg/kg ($1/5$ of the LD_{50}) for 3 weeks, slightly marked manifestations of general toxicity were observed, such as altered protein metabolism and granular degeneration in the liver and kidneys; there were also clear signs of local irritant action (necrotic and inflammatory lesions in the small intestine) [Spiridonova].

Long-term exposure

Rats exposed to a suspension of **hafnium carbide** or **nitride** dust by inhalation at 10.5–10.8 mg/L for 5 h/day over 6 or 9.5 months had increased relative weights of the lungs and adrenals. The lungs had thickened interalveolar septa and showed moderate emphysematosis, vascular anormalities (erythrostasis), chronic interstitial pneumonia with pronounced fibrosis, disseminated purulent bronchitis, and bronchiectasis. Circulatory disturbances and granular degeneration were evident in the liver. Superficial purulent cysts were present in the liver and, in some animals, the lungs. The main findings in the myocardium were eosinophilia and granular degeneration of muscle fibers. The most conspicuous abnormality in the kidneys was granular dystrophy of the convoluted tubular epithelium. The splenic reticuloendothelium appeared congested and hyperplastic [Zhislin & Ovetskaya].

ABSORPTION, DISTRIBUTION, AND ELIMINATION IN ANIMALS

The activity of ^{181}Hf injected into rats intravenously as a constituent of hafnium-sodium oxyphenyl acetate was highest in the spleen, lower in the liver, bones, and kidneys, and low in the adrenals, thyroid, pancreas, salivary glands, and testes. In the blood, [181]Hf was detectable during 4 days post-injection and most of it (95%) occurred in the plasma. About 7% of the injected dose was eliminated over a period of 16 days; the rate of urinary excretion was 2.4 times higher than that of fecal excretion.

No complete elimination of radioactive hafnium oxide occurred in rats even 3 months after its intravenous injection [Riedel *et al.*]. Hafnium chloride was reported to undergo hydrolysis in body fluids within 30 min with the formation of hydrochloric acid; the rate of hydrolysis, however, varied from one medium to another, being slowest in the stomach [Spiridonova].

HYGIENIC STANDARDS

In the USSR, a TSEL value (temporary MAC) of $5 \, mg/m^3$ has been recommended for hafnium carbide and hafnium nitride in workroom air [Zhislin & Ovetskaya]. In the USA, threshold limits values for hafnium have been set at $0.5 \, mg/m^3$ (TWA) and $1.5 \, mg/m^3$ (STEL).

METHODS OF DETERMINATION

Methods of hafnium determination have been reviewed by Mukharji. Spectrophotometric methods are based on the reaction of hafnium with 1-(2-thiazolylazo)-2-naphthol [Kalyanaraman *et al.*] or with 4-(2-pyridylazo) resorcinol [Kalyanaraman & Fukasawa]. See also Duverger-van Bogaert & Lambotte-Vandepaer.

MEASURES TO CONTROL OCCUPATIONAL EXPOSURE

The major adverse factors in the production and handling of hafnium and its compounds are their dusts and the volatile substances that form during the decomposition of these compounds. To ensure healthy working conditions, the guidelines established for zirconium production should be followed (see *Zirconium and its Compounds*). The processing and concentration of the crude ore as well as all processes using hafnium should be designed and ventilated to keep dust generation to a minimum. It is also important to reduce as much as possible the transportation of hafnium-containing materials. The powdered metal should be transported in the wet state (water is usually used for wetting). Any spilled powder should be cleaned up immediately with water. Workers should be meticulous in keeping the powder off their clothing.

Those working with hafnium and its compounds should undergo a preplacement and subsequent medical examinations.

For *personal protection*, respirators and goggles should be provided to protect the respiratory organs and eyes. Appropriate clothes, footwear, and gloves are recommended for skin protection.

REFERENCES

Brakhnova, I. T. & Shkurko, G. A. (1972) *Gig. San.*, No. **7**, 36—39 (in Russian).

Dobrovolsky, V. V. (1983) *Geografiya mikroelementov. Globalnoe rasseyanie* [Geography of Trace Elements. Global Dispersion]. Mysl, Moscow (in Russian).

Duverger-van Bogaert, M. & Lambotte-Vandepaer, L. (1988) Chapter 26 *Hafnium* in: Seiler, H. G. *et al.* (eds) *Handbook on Toxicity of Inorganic Compounds.* Marcel Dekker, New York, pp. 315—318.

Henderson, P. (1982) *Inorganic Geochemistry.* Pergamon Press, Oxford.

Kabata-Pendias, A. & Pendias, H. (1986) *Trace Elements in Soils and Plants.* CRC Press, Boca Raton (Fl.).

Kalyanaraman, S. & Fukasawa, T. (1983) *Anal. Chem.*, **55**, 2239—2241.

Kalyanaraman, S. *et al.* (1985) *Analyst*, **110**, 213—214.

Mukherji, A. K. (1970) *Analytical Chemistry of Zirconium and Hafnium.* Pergamon Press, New York.

Ovetskaya, N. M. (1969) In: *Aktualnye voprosy krayevoi gigiyeny i epidemiologii Donbassa* [Current Issues of Hygiene and Epidemiology in the Donbass Region]. Kiev, pp. 70—75 (in Russian).

Riedel, W. *et al.* (1979) *Environ. Res.*, **18**, 127—139.

Spiridonova, V. S. (1970) In: *Organizm i sreda. Chast 1. Materialy 6-i nauchnoi konferentsii gigiyenicheskikh kafedr 1-go MMI* [The Organism and the Environment. Part 1. Proceedings of the Sixth Workshop of the Departments of Hygiene of the First Moscow Medical Institute]. Moscow, pp. 97—99 (in Russian).

Vadkovskaya, I. K. & Lukashev, V. P. (1977) *Geokhimicheskiye osnovy okhrany biosfery* [Protection of the Biosphere: Geochemical Principles]. Nauka i Tekhnika, Minsk (in Russian).

Zhislin, L. A. & Ovetskaya, N. I. (1969) In: *Gigiyena Truda* [Occupational Hygiene] (Kiev), No. **1**, pp. 36—40 (in Russian).

Appendix*: Physical and chemical properties of the elements (Groups I–IV) and their compounds

Abbreviations:

alk.	=	alkalies
conc.	=	concentrated
cryst.	=	crystals or crystalline
decomp.	=	decomposes
dil.	=	dilute
expl.	=	explosive
hygr.	=	hygroscopic
insol.	=	insoluble
liq.	=	liquid
org. solv.	=	organic solvents
sol.	=	soluble
sl.sol.	=	slightly soluble
subl.	=	sublimes

Designations:

AcOH	=	acetic acid
AmOH	=	1-pentanol
EtOAc	=	ethyl acetate
EtOH	=	ethanol
Et_2O	=	diethyl ether
MeOH	=	methanol

---

* Compiled by A. V. Moskvin.

Name	Symbol/ formula	Atomic/ molecular mass	Color & crystal system	Density (g/cm³ at 20°C unless otherwise stated)
HYDROGEN AND ITS COMPOUNDS				
HYDROGEN	H_2	2.016	colorless gas, hexagonal	0.08988 g/L
Deuterium oxide	D_2O	20.03	colorless liq., hexagonal	1.1042 (25°)
Hydrogen peroxide	H_2O_2	34.1	colorless liq., tetragonal	1.45
LITHIUM AND ITS COMPOUNDS				
LITHIUM	Li	6.941	silvery-white, hexagonal(α), cubic(β)	0.534, 0.507 (200°) liq.
Lithium bromide	LiBr	86.85	colorless, cubic, hygr.	3.464 (25°)
carbonate	Li_2CO_3	73.89	colorless, monoclinic	2.11 (0°)
chloride	LiCl	42.39	colorless, cubic, hygr.	2.07 (25°)
dihydrogen phos- phate	LiH_2PO_4	103.93	colorless, cryst.	2.461

Melting point, °C	Boiling point, °C	Vapor pressure, kPa (at t°C)	Solubility, g/100 g (at t°C)	Relationship with bases and/or acids
-259.19	-252.87	1.33 (-261.3) 13.3 (-257.9) 53.3 (-254.6)	H_2O: 1.82 ml/100 g (20); EtOH: 6.92 ml/100 g (0)	insol. in acids & alk.
3.81	101.43	13.3 (54) 53.3 (84.8)	sol. in H_2O & EtOH; sl. sol. in Et_2O	—
-0.42	152	0.133 (14.6) 1.33 (49.9) 13.3 (97.9)	sol. in H_2O, EtOH, & Et_2O	—
180.5	1340	0.133×10^{-3} (430) 1.33×10^{-3} (502) 13.3×10^{-3} (592)	rects with H_2O & EtOH	reacts with acids & NH_3
552	1290	0.133 (748)	H_2O: 160(20), 266 (100); EtOH: 32.6 (0), 73(40); acetone: 18.2(20), 39.7 (60); sol. in MeOH & Et_2O	—
732	decomp.	—	H_2O: 1.33(20), 0.72 (100); insol. in EtOH & acetone	reacts with HNO_3, H_2SO_4, H_3PO_4, & $Ca(OH)_2$; insol. in NH_3
614	1382	0.133×10^{-3} (513) 0.0133 (677) 0.133 (789)	H_2O: 83.2(20), 128.8 (100); EtOH: 14.4 (0), 24.3(20), 25.4(40), 23.5(60); MeOH: 45.2(0), 44.1 (40); acetone: 1.2 (20), 0.61(50); pyridine: 7.8(15)	reacts with NH_3
>100	$-H_2O$	—	H_2O: 126(0)	

Name	Symbol/ formula	Atomic/ molecular mass	Color & crystal system	Density (g/cm^3 at 20°C unless otherwise stated)
LITHIUM COMPOUNDS (cont'd)				
hydride	LiH	7.95	colorless, cubic	0.776 (25°)
metaniobate	LiNbO$_3$	147.85	white powder	4.35
metatantalate	LiTaO$_3$	235.89	white powder	7.35
metaphosphate	LiPO3	85.91	powder	2.55
nitrate	LiNO$_3$	68.94	colorless, hexagonal, hygr.	2.36
nitrate trihydrate	LiNO$_3$·3H$_2$O	122.99	colorless, cryst., hygr.	—
sulfate	Li$_2$SO$_4$	109.94	colorless, monoclinic(α), cubic(β)	2.221
SODIUM AND ITS COMPOUNDS				
SODIUM	Na	22.98977	silvery, cubic	0.968
Sodium bromide	NaBr	102.90	colorless, cubic, hygr.	3.21

Melting point, °C	Boiling point, °C	Vapor pressure, kPa (at t°C)	Solubility, g/100 g (at t°C)	Relationship with bases and/or acids
691	decomp. >700	— —	reacts with H_2O & EtOH; sl. sol. in Et_2O	reacts with NH_3
1164	—	—	H_2O: 0.004(25), 0.009(75)	—
1650	—	—	sl. sol. in H_2O	
630	—	—	insol. in H_2O	
253	decomp. >600	—	H_2O: 70.0(20), 206 (70); pyridine: 38 (25); sol. in EtOH & acetone	—
$-H_2O$ at 29.9	—	H_2O: 302(20)		
860; $\alpha \rightarrow \beta$ 575	—	—	H_2O: 34.7(20), 30.6 (100); insol. in EtOH & acetone	—
97.9	886	1.33×10^{-4} (236.5) 1.33×10^{-3} (290) 0.0133 (356)	reacts with H_2O & EtOH; insol. in Et_2O	reacts with NH_3
747	1390	1.33×10^{-4} (554) 1.33×10^{-3} (621) 0.0133 (701)	H_2O: 90.8(20), 121.2(100); EtOH: 2.45(0); MeOH: 16.8 (20), 15.3(60); glycerol: 38.7(20); sl. sol. in acetone	—

Name	Symbol/ formula	Atomic/ molecular mass	Color & crystal system	Density (g/cm^3 at 20°C unless otherwise stated)
SODIUM COMPOUNDS (cont'd)				
carbonate	Na_2CO_3	105.99	colorless, monoclinic(α) hexagonal(β)	2.53
carbonate decahydrate	$Na_2CO_3 \cdot 10H_2O$	286.14	colorless, monoclinic	1.45 (17°)
chlorate	$NaClO_3$	106.44	colorless, cubic	2.49 (15°)
chloride	$NaCl$	58.44	colorless, cubic	2.165 (25°)
chlorite trihydrate	$NaClO_2 \cdot 3H_2O$	144.49	colorless, triclinic, hygr.	—
fluoride	NaF	41.99	colorless, cubic	2.79
hexafluoro- aluminate	Na_3AlF_6	209.94	colorless, monoclinic	2.90
hexahydroxo- stannate(IV)	$Na_2[Sn(OH)_6]$	266.67	colorless, hexagonal	—
hydrogen carbonate	$NaHCO_3$	84.01	colorless, monoclinic	2.16

Melting point, °C	Boiling point, °C	Vapor pressure, kPa (at t°C)	Solubility, g/100 g (at t°C)	Relationship with bases and/or acids
858; $\alpha \rightarrow \beta$ 480	decomp.	—	H_2O: 21.8(20), 44.7 (100); very sol. in glycerol; sl. sol. in EtOH; insol. in acetone & CS_2	reacts with acids
32.5	—	—	sol. in H_2O; insol. in EtOH	reacts with acids
263	—	—	H_2O: 95.9(20), 203.9(100); EtOH: 14.7(25); MeOH: 51.35(25); acetone: 51.8(25); sol. in glycerol	reacts with liq. NH_3
801	1490	1.33×10^{-4} (567) 1.33×10^{-3} (636) 0.0133 (718)	H_2O: 35.9(20), 39.4 (100); EtOH: 0.065 (20); MeOH: 1.31 (25); glycerol: 8.2 (25); insol. in acetone & Et_2O	reacts with H_2SO_4
$-3H_2O$ 37.4	decomp. >180	—	sol. in H_2O	—
966	≈1700	1.33×10^{-4} (761) 1.33×10^{-3} (843) 0.0133 (926)	H_2O: 4.28(20), 4.69 (80); EtOH: 0.095 (20); MeOH: 0.413 (20)	very sol. in HF
1010	—	—	H_2O: 0.06(20)	—
decomp. 140	—	—	H_2O: 61.3(15.5), 50(100); insol. in EtOH & acetone	reacts with acids, alk., & NH_4OH
decomp. >50	—	—	H_2O: 9.6(20), 24.3 (100); EtOH: 1.2 (15.5); glycerol: 7.9(20)	reacts with acids

Name	Symbol/ formula	Atomic/ molecular mass	Color & crystal system	Density (g/cm^3 at 20°C unless otherwise stated)
SODIUM COMPOUNDS (cont'd)				
hydroxide	NaOH	40.00	colorless, rhombic(α), monoclinic(β) cubic(γ), hygr.	2.13 (α)
iodide	NaI	149.89	colorless, cubic	3.665 (4°)
metasilicate	Na$_2$SiO$_3$	122.06	colorless, rhombic	2.61
nitrite	NaNO$_2$	69.00	colorless, rhombic	2.17 (0°)
orthophosphate	Na$_3$PO$_4$	163.94	colorless, tetragonal	2.54 (17.5°)
sulfate	Na$_2$SO$_4$	142.04	colorless, rhombic(α, β), hexagonal(γ)	2.70
sulfite	Na$_2$SO$_3$	126.04	colorless, hexagonal	2.63 (15°)
thiosulfate pentahydrate	Na$_2$S$_2$O$_3$· 5H$_2$O	248.19	colorless, monoclinic	1.73 (17°)

Melting point, °C	Boiling point, °C	Vapor pressure, kPa (at t°C)	Solubility, g/100 g (at t°C)	Relationship with bases and/or acids
322; $\alpha \rightarrow \beta$ 243; $\beta \rightarrow \gamma$ 297	1378	0.133 (738) 1.33 (895) 13.3 (1110)	H_2O: 108.7(20), 337 (100); very sol. in EtOH, MeOH, & glycerol; insol. in Et_2O & acetone	reacts with acids
661	1300	1.33 (882) 13.3 (1066) 53.3 (1221)	H_2O: 179.3(20), 302 (100); EtOH: 43.3 (25); MeOH: 78.0 (25); acetone: 30.0 (20)	—
1088	—	—	H_2O: 18.8(20), 160.6 (800); insol. in EtOH	—
284	decomp. >320	—	H_2O: 82.9(20), 160 (100); sol. in EtOH & pyridine	sol. in liq. NH_3
1510	—	—	H_2O: 14.5(25), 94.6 (100)	—
884; $\alpha \rightarrow \beta$ 185; $\beta \rightarrow \gamma$ 241	—	—	H_2O: 19.2(20), 42.3 (100); MeOH: 2.46(20), 1.84(50); EtOH: 0.44 (20); sol. in glycerol	—
911	—	—	H_2O: 26.1(20), 20.6 (100); sl. sol. in EtOH	reacts with acids
48.5	$-5H_2O$ 100	—	H_2O: 70.1(20), 245 (100); insol. in EtOH	reacts with acids

Name	Symbol/ formula	Atomic/ molecular mass	Color & crystal system	Density (g/cm^3 at 20°C unless otherwise stated)

POTASSIUM AND ITS COMPOUNDS

Name	Symbol/ formula	Atomic/ molecular mass	Color & crystal system	Density
POTASSIUM	K	39.098	silvery, cubic	0.86
Potassium carbonate	K_2CO_3	138.21	colorless, monoclinic(α), hexagonal(β), hygr.	2.43 (19°)
chloride	KCl	74.56	colorless, cubic	1.99
fluoride	KF	58.10	colorless, cubic, hygr.	2.48
hydroxide	KOH	56.11	colorless, monoclinic(α), cubic(β), hygr.	2.12 (25°)
nitrate	KNO_3	101.11	colorless, rhombic(α), hexagonal(β)	2.11 (16°)
orthophosphate	K_3PO_4	212.28	colorless, cubic, hygr.	2.56 (17°)
peroxodisulfate	$K_2S_2O_8$	270.32	colorless, cryst.	2.477
sulfate	K_2SO_4	174.27	colorless, rhombic(α), hexagonal(β)	2.66

Melting point, °C	Boiling point, °C	Vapor pressure, kPa (at t°C)	Solubility, g/100 g (at t°C)	Relationship with bases and/or acids
63.5	761	0.000133 (165) 0.00133 (212) 0.0133 (270)	reacts with H_2O & EtOH	reacts with acids
900; $\alpha \rightarrow \beta$ 420	decomp.	—	H_2O:111.0(20), 155.8(100); insol. in EtOH & acetone	reacts with acids
776	1430	0.000133 (568) 0.00133 (637) 0.0133 (720)	H2O: 34.4(20), 56.0 (100); EtOH: 0.03 (25); MeOH: 0.54(25); glycerol: 6.7(25); insol. in acetone	—
857	1505	0.000133 (637) 0.00133 (711) 0.0133 (800)	H_2O: 94.9(20), 150 (90); insol. in EtOH	—
405; $\alpha \rightarrow \beta$ 244	1320	0.133 (719) 1.33 (863) 13.3(1059)	H_2O: 112.4(20), 179.3(100); EtOH: 38.7(20); MeOH: 55 (28); insol. in Et_2O	reacts with acids
334.5; $\alpha \rightarrow \beta$ 129	decomp. >400	—	H_2O: 31.6(20), 243.6 (100); sol. in glycerol; insol. in EtOH & Et_2O	—
1640	—	—	H_2O: 98.5(20), 178.5(60); insol. in EtOH	—
$\rightarrow K_2S_2O_7$ $\approx$100	—	—	H_2O: 5.3(20)	reacts with acids
1069; $\alpha \rightarrow \beta$ 584	$\approx$1700	0.000133 (883) 0.00133 (980) 0.0133 (1100)	H_2O: 11.1(20), 24.1 (100); insol. in EtOH, MeOH, CS_2, & acetone	—

Name	Symbol/ formula	Atomic/ molecular mass	Color & crystal system	Density $(g/cm^3$ at 20°C unless otherwise stated)
POTASSIUM AND ITS COMPOUNDS (cont'd)				
magnesium sulfate hexahydrate	$K_2SO_4{\cdot}MgSO_4{\cdot} 6H_2O$	402.73	colorless, monoclinic	2.15
RUBIDIUM AND ITS COMPOUNDS				
RUBIDIUM	Rb	85.4678	silvery, cubic	1.532
Rubidium chloride	RbCl	120.92	colorless, cubic	2.76
hydroxide	RbOH	102.48	colorless, rhombic, hygr.	3.203 (11°)
iodide	RbI	212.37	colorless, cubic	3.55
sulfate	Rb_2SO_4	267.00	colorless, rhombic (α), hexagonal (β)	3.613

Melting point, °C	Boiling point, °C	Vapor pressure, kPa (at t°C)	Solubility, g/100 g (at t°C)	Relationship with bases and/or acids
decomp. ≈72	—	—	H_2O: 18.3(20), 43.8 (75)	—
39.3	690	0.000133 (129) 0.00133 (174) 0.0133 (229)	reacts with H_2O & EtOH	reacts with liq. NH_3
723	1390	0.000133 (536) 0.00133 (602) 0.0133 (682)	H_2O: 91.1(20), 138.9(100); EtOH: 0.08(25); acetone: 0.0002(18)	sol. in HCl
385	—	—	H_2O: 179(15), 946 (95); sol. in EtOH	—
656	1327	1.33 (866) 13.3 (1055) 53.3 (1208)	H_2O: 124.7(0), 281 (100); sol. in EtOH & acetone	reacts with conc. H_2SO_4 & conc. HNO_3; sol. in HCl, dil. H_2SO_4, dil. NHO_3, & alk.
1070; α→β 655	≈1700	—	H_2O: 48.2(20), 81.8(100); sl. sol. in EtOH	—

Name	Symbol/ formula	Atomic/ molecular mass	Color & crystal system	Density (g/cm^3 at 20°C unless otherwise stated)
CESIUM AND ITS COMPOUNDS				
CESIUM	Cs	132.9054	yellow, cubic	1.90
Cesium bromide	CsBr	212.81	colorless, cubic	4.44
carbonate	Cs$_2$CO$_3$	325.82	colorless, cryst., hygr.	—
chloride	CsCl	168.36	colorless, cubic, hygr.	3.97
dihydrogen orthoarsenate	CsH$_2$AsO$_4$	273.84	colorless, cryst.	—
hydroxide	CsOH	149.91	colorless, cryst, hygr.	3.675
iodide	CsI	259.81	colorless, hygr.	4.51 (20°)
nitrate	CsNO$_3$	194.91	colorless, hexagonal(α), cubic(β)	3.685
sulfate	Cs$_2$SO$_4$	361.87	colorless, hexagonal(β), rhombic(α)	4.243

Melting point, °C	Boiling point, °C	Vapor pressure, kPa (at t°C)	Solubility, g/100 g (at t°C)	Relationship with bases and/or acids
28.5	672	0.000133 (114) 0.00133 (156) 0.0133 (208)	reacts with H_2O & EtOH	reacts with acids
638	1290	0.000133 (492) 0.00133 (556) 0.0133 (633)	H_2O: 123.3 (25), 214 (80); sol. in EtOH	reacts with conc. H_2SO_4; sol. in HCl
decomp. 793	—	—	H_2O: 260.5 (15); very sol. in hot H_2O; EtOH: 11 (19); sol. in Et_2O	reacts with acids
645	1300	0.000133 (482) 0.00133 (546) 0.0133 (622)	H_2O: 186.5 (20), 270.5 (100); very sol. in EtOH	sol. in HCl; reacts with NHO_3
		—	H_2O: 0.275 (20)	—
343	subl. ≈ 400	—	H_2O: 385.6 (15), 303 (30); sol. in EtOH	reacts with acids
632	1280	0.000133 (470) 0.00133 (533) 0.0133 (610)	H_2O: 44.1 (0), 170.8 (75); sol. in EtOH	sol. in HCl, alk., & NH_4OH; reacts with conc. H_2SO_4 & conc. HNO_3
409; $\alpha \rightarrow \beta$ 154	decomp.	—	H_2O: 23.0 (20), 196.8 (100); sol. in acetone	—
1015; $\alpha \rightarrow \beta$ 667	—	—	H_2O: 178.7 (20), 220.3 (100); insol. in EtOH & acetone	—

Name	Symbol/ formula	Atomic/ molecular mass	Color & crystal system	Density (g/cm^3 at 20°C unless otherwise stated)
COPPER AND ITS COMPOUNDS				
COPPER	Cu	63.546	red, cubic	8.96
Copper(II) carbo- nate hydroxide	$CuCO_3$· $Cu(OH)_2$	221.10	dark green, monoclinic	4.0
(I) chloride	CuCl	98.99	colorless, cubic(α), hexagonal(β)	3.7 (α)
(II) chloride	$CuCl_2$	134.45	brown, monoclinic, hygr.	3.054
(II) chloride- copper(II) hydroxide- water(1/3/1)	$3Cu(OH)_2$· $CuCl_2 \cdot H_2O$	445.12	light-green powder	—
(II) hydroxide	$Cu(OH)_2$	97.55	blue gel or amorphous powder	3.368
(I) iodide	CuI	190.44	colorless, cubic(α,γ); hexagonal(β)	5.62
(II) nitrate trihydrate	$Cu(NO_3)_2$· $3H_2O$	241.60	blue, cryst., hygr.	2.32
(I) oxide	Cu_2O	143.08	red, cubic	6.0
(II) oxide	CuO	79.54	black, monoclinic	6.45

Melting point, °C	Boiling point, °C	Vapor pressure, kPa (at t°C)	Solubility, g/100 g (at t°C)	Relationship with bases and/or acids
1083	2540	0.000133 (1130) 0.00133 (1260) 0.0133 (1430)	insol. in H_2O	reacts with HNO_3 & hot conc. H_2SO_4
decomp. >200	—	—	insol. in cold H_2O; reacts with hot H_2O	reacts with acids & NH_4OH
430; $\alpha \rightarrow \beta$ 408	1212	0.133 (517) 1.33 (687) 13.3 (951)	sl.sol. in H_2O; insol. in Et_2O & acetone	reacts with HCl & NH_4OH
596	decomp. 993	—	H_2O: 74.5(20), 110.5 (100); EtOH: 50(20); MeOH: 58.6(20); sol. in Et_2O, acetone, & pyridine	sol. in liq. NH_3
-3H_2O 140	—	—	insol. in H_2O	—
decomp.	—	—	insol. in cold H_2O; reacts with hot H_2O	reacts with acids, conc. alk., & NH_4OH
600	1320	0.133 (606) 1.33 (780) 13.3 (1036)	H_2O: 0.00044(18); sol. in pyridine	—
114.5	—	—	H_2O: 124.7(20), 247.2 (100); EtOH: 100(12.5)	—
1242	1800	—	insol. in H_2O & EtOH	reacts with HCl & NH_4OH
decomp. >800	—	—	insol. in H_2O	reacts with acids

Name	Symbol/ formula	Atomic/ molecular mass	Color & crystal system	Density (g/cm^3 at 20°C unless otherwise stated)
COPPER AND ITS COMPOUNDS (cont'd)				
(II) sulfate	$CuSO_4$	159.60	colorless, rhombic	3.603
(II) sulfate pentahydrate	$CuSO_4 \cdot 5H_2O$	249.68	blue, triclinic	2.284
(I) sulfide	Cu_2S	159.14	black, rhombic (α), hexagonal (β), cubic (γ)	5.5—5.8 (α)
(II) sulfide	CuS	95.60	black, hexagonal	4.68

SILVER AND ITS COMPOUNDS

Name	Symbol/ formula	Atomic/ molecular mass	Color & crystal system	Density
SILVER	Ag	107.867	silvery, cubic	10.5
Silver bromide	AgBr	187.78	light yellow, cubic	6.473 (25°)
carbonate	Ag_2CO_3	275.75	light yellow, monoclinic	6.077
chloride	AgCl	143.32	colorless, cubic	5.56

Melting point, °C	Boiling point, °C	Vapor pressure, kPa (at t°C)	Solubility, g/100 g (at t°C)	Relationship with bases and/or acids
decomp.	—	—	H_2O: 20.5(20), 77.0 (100); MeOH: 1.04(18); insol. in EtOH	—
-2H$_2$O 100, -4H$_2$O 150	-5H$_2$O 250	—	H_2O: 35.6(20), 205 (100); sol. in MeOH; insol. in EtOH	—
1129; $\alpha \to \beta$ 103; $\beta \to \gamma$ 445	—	—	insol. in H_2O & EtOH	insol. in acids & alk.; reacts with HNO_3 & NH_4OH
decomp. >450	—	—	H_2O: 1×10^{-21}(20); insol. in EtOH	insol. in acids & alk.; reacts with HNO_3 & hot conc. H_2SO_4
960.5	2167	0.000133 (920) 0.00133 (1030) 0.0133 (1170)	insol. in H_2O	reacts with HNO_3 & conc. H_2SO_4; insol. in alk.
424	decomp. 1505	—	H_2O: 1.6×10^{-6}(25), 3.7×10^{-4}(100); insol. in EtOH	sol. in liq. NH_3; sl. sol. in NH_4OH
decomp. $\approx$200	—	—	H_2O: 0.0032(20), 0.05(100); insol. in EtOH	sol. in liq. NH_3 & NH_4OH
455	1550	0.000133 (609) 0.00133 (692) 0.0133 (791)	H_2O: 8.9×10^{-5}(20), 0.0021(100); sol. in pyridine	sol. in liq. NH_3 & NH_4OH

Name	Symbol/ formula	Atomic/ molecular mass	Color & crystal system	Density (g/cm^3 at 20°C unless otherwise stated)
SILVER AND ITS COMPOUNDS (cont'd)				
nitrate	AgNO$_3$	169.87	colorless, rhombic(α) hexagonal(β)	4.352 (19°)
oxide	Ag$_2$O	231.74	brown, cubic	7.14 (16.5°)
sulfide	Ag$_2$S	247.80	black or dark gray, monoclinic (α), cubic(β), cryst. (γ)	7.32
GOLD AND ITS COMPOUNDS				
GOLD	Au	196.9665	yellow, cubic	19.3
Gold(I) cyanide	AuCN	222.98	light yellow powder	7.12 (25°)
Potassium dicyanoaurate(I)	K[Au(CN)$_2$]	288.10	colorless, cryst.	3.452 (25°)

Melting point, °C	Boiling point, °C	Vapor pressure, kPa (at t°C)	Solubility, g/100 g (at t°C)	Relationship with bases and/or acids
209.7; $\alpha \to \beta$ 159.5	decomp. ≈ 300	—	H_2O: 222.5(20), 770 (100); MeOH: 3.6(20); EtOH: 2.12(20); acetone: 0.44(18); pyridine: 33.6(20)	—
decomp. >200	—	—	H_2O: 0.0013(20), 0.0053(80); insol. in EtOH	reacts with acids; sol. in NH_4OH
838; $\alpha \to \beta$ 177	—	—	insol. in H_2O	insol. in NH_4OH; reacts with HNO_3 & conc. H_2SO_4
1063.4	2880	0.000133 (1300) 0.00133 (1450) 0.0133 (1630)	insol. in H_2O	insol. in acids; reacts with aqua regia & hot H_2SeO_4
decomp.	—	—	insol. in H_2O & Et_2O	sol. in NH_4OH; insol. in alk.
decomp. >200	—	—	sol. in H_2O; sl.sol. in EtOH & glycerol; insol. in Et_2O	reacts with acids

Name	Symbol/ formula	Atomic/ molecular mass	Color & crystal system	Density $(g/cm^3$ at 20°C unless otherwise stated)
BERYLLIUM AND ITS COMPOUNDS				
BERYLLIUM	Be	9.0122	light gray, hexagonal(α), cubic(β)	1.85
Beryllium bromide	BeBr$_2$	168.83	colorless, rhombic, hygr.	3.465 (25°)
carbide	Be$_2$C	30.04	yellow, cubic	1.90 (15°)
carbonate tetrahydrate	BeCO$_3$·4 H$_2$O	141.08	colorless, hexagonal	—
chloride	BeCl$_2$	79.92	colorless, rhombic(α,β), hygr.	1.899 (25°)
fluoride	BeF$_2$	47.01	colorless, tetra-gonal(α), hexa-gonal(β), cubic(γ), or vitreous	1.99 (25°)
hydroxide	Be(OH)$_2$	43.03	colorless, tetra-gonal(α), rhom-bic(β), amorphous	1.926 (cryst.)
iodide	BeI$_2$	262.82	colorless, rhombic	4.325 (25°)
nitrate tetra-hydrate	Be(NO$_3$)$_2$·4H$_2$O	187.07	colorless, cryst., hygr.	—
orthophosphate	Be$_3$(PO$_4$)$_2$	216.979	colorless, cryst.	—

Melting point, °C	Boiling point, °C	Vapor pressure, kPa (at t°C)	Solubility, g/100 g (at t°C)	Relationship with bases and/or acids
1285; α→β 1275	2470	0.000133 (1080) 0.00133 (1210) 0.0133 (1360)	insol. in cold H_2O; reacts with hot H_2O	reacts with dil. acids & alk.
508	540	0.000133 (183) 0.00133 (217) 0.0133 (257)	sol. in H_2O, EtOH, Et_2O & pyridine; insol. in C_6H_6	—
decomp. >2100	—	0.000133 (1240) 0.00133 (1375)	reacts with H_2O	reacts with acids & alk.
decomp. 100	—	—	H_2O: 0.36(0)	reacts with acids
415; α→β 403	550	0.133 (303) 1.33 (351) 13.3 (409)	H_2O: 67.6(20), 77.3 (30); very sol. in EtOH, Et_2O,C_6H_6, & pyridine; sl.sol. in $CHCl_3$ & CS_2; insol. in acetone	—
800; α→β 130	1175	0.000133 (549) 0.00133 (614) 0.0133 (690)	very sol. in H_2O; sl. sol. in EtOH	—
decomp. 138	—	—	sl.sol. in H_2O	reacts with acids & alk.
490	530	0.133 (281) 1.33 (339) 13.3 (410)	reacts with H_2O; sol. in EtOH, Et_2O, & CS_2	—
61	→ BeO 320	—	H_2O: 107(20), 142(50); very sol. in EtOH	—
—	—	—	sl.sol. in H_2O	—

Name	Symbol/ formula	Atomic/ molecular mass	Color & crystal system	Density (g/cm^3 at 20°C unless otherwise stated)
BERYLLIUM COMPOUNDS (cont'd)				
oxide	BeO	25.01	colorless, hexagonal(α), tetragonal(β)	3.01
oxyfluoride	5BeF$_2$·2BeO	285.069	white mass	2.3 (15°)
sulfate	BeSO$_4$	105.07	colorless, tetragonal(α), rhombic(β), cubic(γ)	2.443

MAGNESIUM AND ITS COMPOUNDS

Name	Symbol/ formula	Atomic/ molecular mass	Color & crystal system	Density
MAGNESIUM	Mg	24.305	silvery, hexagonal	1.74
Magnesium carbonate	MgCO$_3$	84.32	colorless, hexagonal	2.96
chlorate hexahydrate	Mg(ClO$_3$)$_2$· 6H$_2$O	299.30	colorless, cryst.	1.80 (25°)
chloride	MgCl$_2$	95.22	colorless, hexagonal, hygr.	2.316
hydroxide	Mg(OH)$_2$	58.32	colorless, hexagonal	2.35−2.46
oxide	MgO	40.31	colorless, cubic	3.58 (25°)

Melting point, °C	Boiling point, °C	Vapor pressure, kPa (at t°C)	Solubility, g/100 g (at t°C)	Relationship with bases and/or acids
2580 α→β 2100	≈4100	0.000133 (2094) 0.00133 (2276) 0.0133 (2490)	H_2O: 0.00002(3)	reacts with conc. H_2SO_4 & fused alk.
—	—	—	sol. in H_2O	reacts with H_2SO_4
decomp. ≈1100; α→β 588; β→γ 639	—	—	H_2O: 40(20), 85.9 (100)	—
650	1095	0.000133 (380) 0.00133 (440) 0.0133 (514)	insol. in cold H_2O; reacts with hot H_2O	reacts with acids; insol. in alk.
decomp. >500	—	—	sl.sol. in cold H_2O; reacts with hot H_2O; insol. in AcOH	reacts with acids; insol. in NH_4OH
35	decomp. 120	—	H_2O: 130(18), 281 (93); sol. in EtOH	—
714	1370	0.000133 (576) 0.00133 (636) 0.0133 (706)	H_2O: 54.8(20), 73.0 (100); EtOH: 5.6(20), 15.9(60); MeOH: 16.0 (20), 20.4(60); sl. sol. in acetone	—
-H_2O 350	—	—	H_2O: 6.43×210^{-4}(25), 0.004(100)	reacts with acids
2825	3600	—	H_2O: 0.00062(0), 0.0086(30), insol. in EtOH	reacts with acids

Name	Symbol/ formula	Atomic/ molecular mass	Color & crystal system	Density (g/cm^3 at 20°C unless otherwise stated)
MAGNESIUM COMPOUNDS (cont'd)				
sulfate	$MgSO_4$	120.37	colorless, rhombic, hygr.	2.66
sulfate heptahydrate	$MgSO_4 \cdot 7H_2O$	246.48	colorless, rhombic	1.68
CALCIUM AND ITS COMPOUNDS				
CALCIUM	Ca	40.08	silvery, cubic (α, β)	1.54
Calcium carbide	CaC_2	64.10	colorless, tetragonal(α), cubic(β)	2.2
carbonate (aragonite)	$CaCO_3$	100.09	colorless, rhombic	2.93
carbonate (calcite)	$CaCO_3$	100.09	colorless, hexagonal	2.711 (25°)
chlorate dihydrate	$Ca(ClO_3)_2 \cdot 2H_2O$	243.01	light yellow, rhombic, monoclinic, hygr.	2.71
chloride	$CaCl_2$	110.99	colorless, rhombic, hygr.	2.51 (25°)
hydroxide	$Ca(OH)_2$	74.09	colorless, hexagonal	2.24

Melting point, °C	Boiling point, °C	Vapor pressure, kPa (at t°C)	Solubility, g/100 g (at t°C)	Relationship with bases and/or acids
1137	—	—	H_2O: 35.1(20), 50.2 (100); EtOH: 0.025 (15), 0.016(55); MeOH: 3.5(20); Et_2O: 1.16(18); sol. in glycerol; insol. in acetone	—
decomp. 54	—	—	H_2O: 107(20), 215(100); sol. in EtOH, MeOH, & glycerol	—
842; $\alpha \rightarrow \beta$ 443	1495	0.000133 (528) 0.00133 (605) 0.0133 (700)	reacts with H_2O; sl. sol. in EtOH; insol. in C_6H_6	reacts with acids
2160; $\alpha \rightarrow \beta$ 450	—	—	reacts with H_2O	reacts with acids
decomp. 825	—	—	insol. in H_2O	reacts with acids
decomp.	—	—	H_2O: 0.0065(20)	reacts with acids
-H_2O 100	—	—	H_2O: 196(20) [anhydrous]; sol. in EtOH & acetone	—
772	1960	—	H_2O: 74.5(20), 158(100); MeOH: 29.9(20), 38.5(40); EtOH: 25.8(20), 35.4(40); acetone: 0.01(20)	—
-H_2O 580	—	—	H_2O: 0.16(20), 0.072 (100); insol. in EtOH	reacts with acids

Name	Symbol/ formula	Atomic/ molecular mass	Color & crystal system	Density $(g/cm^3$ at 20°C unless otherwise stated)
CALCIUM COMPOUNDS (cont'd)				
hypochlorite trihydrate	$Ca(ClO)_2\cdot$ $3H_2O$	197.03	colorless, tetragonal	2.1
oxide	CaO	56.08	colorless, cubic	3.4
sulfate	$CaSO_4$	136.14	colorless, rhombic (α), hexagonal(β)	2.90–2.99 (α)
sulfate dihydrate	$CaSO_4\cdot2H_2O$	172.17	colorless, monoclinic	2.32
STRONTIUM AND ITS COMPOUNDS				
STRONTIUM	Sr	87.62	silvery, cubic$(\alpha,$ (γ), hexagonal(β)	2.63(α)
Strontium carbonate	$SrCO_3$	147.63	colorless, rhombic (α), hexagonal(β)	3.70
chloride	$SrCl_2$	158.53	colorless, cubic, hygr.	3.05
chromate	$SrCrO_4$	203.61	yellow, monoclinic	3.90 (15°)
fluoride	SrF_2	125.62	colorless, cubic	4.24
hydroxide	$Sr(OH)_2$	121.64	colorless, tetragonal	3.63
nitrate	$Sr(NO_3)_2$	211.63	colorless, cubic	2.99

Melting point, °C	Boiling point, °C	Vapor pressure, kPa (at t°C)	Solubility, g/100 g (at t°C)	Relationship with bases and/or acids
-3H$_2$O 60	—	—	reacts with hot H$_2$O	reacts with acids
≈2630	2850	—	reacts with H$_2$O	reacts with acids
1460	—	—	H$_2$O: 0.176(20), 0.162 (100); sol. in glycerol	reacts with acids
-1.5H$_2$O 128	-2H$_2$O 163	—	H$_2$O: 0.206(20), 0.212 (30); sol. in glycerol	reacts with acids
768; α→β 215; β→γ 605	1390	0.000133 (454.5) 0.00133 (528) 0.0133 (620)	reacts with H$_2$O & EtOH	reacts with acids; sol. in liq. NH$_3$
α→β 925	-CO$_2$ >1000	—	H$_2$O: 0.0011(18), 0.065(100)	reacts with acids
874	≈2040	0.000133 (865) 0.00133 (970) 0.0133 (1100)	H$_2$O: 53.1(20),102 100; sol. in EtOH, acetone, & glycerol; insol. in pyridine	sl.sol. in N$_2$H$_4$; insol. in NH$_3$
1283 (in O$_2$)	—	—	H$_2$O: 0.12(15), 3(100)	reacts with acids
1477	≈2460	0.000133 (1233) 0.00133 (1345) 0.0133 (1475)	H$_2$O: 0.012(20); sl. sol. in Et$_2$O, EtOH, & acetone	sol. in HF & hot HCl
535	decomp. 710	—	H$_2$O: 0.81(20), 27.9 (100); sol. in MeOH; insol. in acetone	reacts with acids
decomp. 645	—	—	H$_2$O: 70.4(20), 102 (100); sl. sol. in EtOH, MeOH, pyridine, & acetone	sl. sol. in conc. HNO$_3$; very sol. in NH$_3$

Name	Symbol/ formula	Atomic/ molecular mass	Color & crystal system	Density (g/cm^3 at 20°C unless otherwise stated)
STRONTIUM COMPOUNDS (cont'd)				
oxide	SrO	103.62	colorless, cubic	4.7
sulfate	SrSO$_4$	183.68	colorless, rhombic (α), hexagonal (β)	3.96

BARIUM AND ITS COMPOUNDS

Name	Symbol/ formula	Atomic/ molecular mass	Color & crystal system	Density
BARIUM	Ba	137.34	silvery, cubic	3.76
Barium carbonate	BaCO$_3$	197.35	colorless, rhombic (α), hexagonal (β), cubic (γ)	4.43
chloride	BaCl$_2$	208.25	colorless, rhombic (α), cubic (β)	3.92
fluoride	BaF$_2$	175.34	colorless, cubic	4.83
hydroxide	Ba(OH)$_2$	171.35	colorless, rhombic	4.5
nitrate	Ba(NO$_3$)$_2$	261.35	colorless, cubic	3.24 (23°)
oxide	BaO	153.34	colorless, cubic	5.72

Melting point, °C	Boiling point, °C	Vapor pressure, kPa (at t°C)	Solubility, g/100 g (at t°C)	Relationship with bases and/or acids
2650	≈3000	—	reacts with H_2O; sl. sol. in EtOH & MeOH; insol. in acetone & Et_2O	reacts with acids
1605; $\alpha \to \beta$ 1156	decomp.	—	H_2O: 0.0132(20), 0.0113(95); insol. in acetone & EtOH	sl. sol. in acids
727	1860	0.000133 (620) 0.00133 (725) 0.0133 (835)	reacts with H_2O	reacts with acids
$\alpha \to \beta$ 810; $\beta \to \gamma$ 960	decomp. 1400	—	H_2O: 0.002(20), 0.006 (100); insol. in EtOH	reacts with acids
960; $\alpha \to \beta$ 925	2050	0.00133 (910) 0.0133 (1030) 0.133 (1180)	H_2O: 36.2(20), 58.2 (100); insol. in EtOH	—
1370	2250	0.000133 (1065) 0.00133 (1176) 0.0133 (1307)	H_2O: 0.159(10), 0.162(30)	sol. in HF, HCl & HNO_3
408	decomp.	—	H_2O: 3.89(20), 101.4 (80); sl. sol. in acetone & EtOH	reacts with acids
decomp. 595	—	—	H_2O: 9.05(20), 34.2 (100); insol. in EtOH	—
2020	—	0.000133 (1395) 0.00133 (1552) 0.0133 (1740)	reacts with H_2O; sol. in EtOH; insol. in acetone	reacts with acids; insol. in NH_4OH

Name	Symbol/ formula	Atomic/ molecular mass	Color & crystal system	Density (g/cm^3 at 20°C unless otherwise stated)
BARIUM COMPOUNDS (cont'd)				
sulfate	BaSO$_4$	233.40	colorless, rhombic (α), cubic(β)	4.50 (15°)
sulfide	BaS	169.4	colorless, cubic	4.25 (15°)
ZINC AND ITS COMPOUNDS				
ZINC	Zn	65.38	silvery, hexagonal	7.133
Zinc carbonate	ZnCO$_3$	125.40	colorless, hexagonal	4.40
chloride	ZnCl$_2$	136.28	colorless, tetragonal, hygr.	2.91 (25°)
nitrate hexahydrate	Zn(NO$_3$)$_2$· 6H$_2$O	297.47	colorless, rhombic	2.07 (14°)
orthophosphate	Zn(PO$_4$)$_2$	386.05	colorless, monoclinic	4.00 (15°)
oxide	ZnO	81.37	colorless, hexagonal	5.7
selenide	ZnSe	144.33	light yellow, cubic	5.42 (15°)

Melting point, °C	Boiling point, °C	Vapor pressure, kPa (at t°C)	Solubility, g/100 g (at t°C)	Relationship with bases and/or acids
1680; $\alpha \to \beta$ 1150	—	—	H_2O: 0.0002(18), 0.00041(100)	reacts with fused alk. sol. in conc. H_2SO_4; insol. in dilute acids
1200	—	—	H_2O: 7.86(20),60.3 (100); insol. in EtOH	reacts with acids
419.5	906.2	0.000133 (247) 0.00133 (291) 0.0133 (343)	insol. in H_2O	reacts with acids & alk.
-CO_2 300	—	—	H_2O: 0.001(15); insol. in acetone	reacts with acids; insol. in NH_4OH
317	732	0.000133 (260) 0.00133 (301) 0.0133 (364)	H_2O: 367(20), 614 (100); EtOH: 100 (12.5); acetone: 43.5(18); pyridine: 2.6(2); very sol. in Et_2O	reacts with NH_4OH
36.4 -6H_2O 105	—	—	H_2O: 118.8(20), 871 (70); very sol. in EtOH	—
1060	—	—	insol. in H_2O & EtOH	sol. in acids
1975	—	—	H_2O: 0.00016(20); insol. in EtOH	reacts with acids & alk.; insol. in NH_4OH
1520	—	0.000133 (703) 0.00133 (788) 0.0133 (890)	insol. in H_2O	sol. in acids

Name	Symbol/ formula	Atomic/ molecular mass	Color & crystal system	Density (g/cm^3 at 20°C unless otherwise stated)
ZINC COMPOUNDS (cont'd)				
sulfate	$ZnSO_4$	161.43	colorless, rhombic	3.74
sulfate heptahydrate	$ZnSO_4 \cdot 7H_2O$	287.54	colorless, rhombic	1.96 (25°)
sulfide	ZnS	97.43	colorless, cubic (α), hexagonal(β)	4.09(α) 3.98−4.08 (β)
Trizinc diphosphide	Zn_3P_2	258.06	dark gray, tetragonal	4.55

CADMIUM AND ITS COMPOUNDS

Name	Symbol/ formula	Atomic/ molecular mass	Color & crystal system	Density (g/cm^3 at 20°C unless otherwise stated)
CADMIUM	Cd	112.40	gray, hexagonal	8.65
chloride	$CdCl_2$	183.31	colorless, hexagonal(α), cryst. (β), hygr.	4.047 (25°)
nitrate tetrahydrate	$Cd(NO_3)_2 \cdot 4H_2O$	308.47	colorless, rhombic, hygr.	2.455 (17°)
oxide	CdO	128.39	light brown, cubic	8.15

Melting point, °C	Boiling point, °C	Vapor pressure, kPa (at t°C)	Solubility, g/100 g (at t°C)	Relationship with bases and/or acids
decomp. >600	—	—	H_2O: 54.1(20), 67.2 (80); EtOH: 0.038 (15), 0.029(35); MeOH: 0.485(15), 0.408(35)	—
-H_2O 48	-7H_2O 280	—	H_2O: 165(20), 202 (100); sl. sol. in EtOH; insol. in acetone	—
1775; $\alpha \rightarrow \beta$ 1175	—	0.00133 (828) 0.0133 (923) 0.0133 (1036)	H_2O: 3.4×10^{-11}(18); insol. in AcOH	reacts with acids; insol. in alk.
1193	—	0.00133 (433) 0.00133 (493) 0.0133 (564)	reacts with H_2O; insol. in EtOH	—
321	766.5	0.000133 (210) 0.00133 (257) 0.0133 (308)	insol. in H_2O	reacts with acids; insol. in alk.
568; $\alpha \rightarrow \beta$ 472	964	0.000133 (364) 0.00133 (415) 0.0133 (475)	H_2O: 144.1(20), 146.9 (100); EtOH: 1.52(15); sol. in MeOH; insol. in Et_2O & acetone	—
59.4	132	—	H_2O: 149.4(20), 681 (100); insol. in EtOH & EtOAc	sol. in NH_4OH
—	—	0.000133 (707) 0.00133 (792) 0.0133 (893)	insol. in H_2O	reacts with acids; insol. in alk.

Name	Symbol/ formula	Atomic/ molecular mass	Color & crystal system	Density (g/cm^3 at 20°C unless otherwise stated)
CADMIUM COMPOUNDS (cont'd)				
sulfate	$CdSO_4$	208.46	colorless, rhombic (α,β), cryst.(γ)	4.691
sulfide	CdS	144.46	light yellow or orange-red, hexagonal(α), cubic(β)	4.82

MERCURY AND ITS COMPOUNDS

Name	Symbol/ formula	Atomic/ molecular mass	Color & crystal system	Density (g/cm^3 at 20°C unless otherwise stated)
MERCURY	Hg	200.59	silvery liq.; hexagonal	13.5461 14.139 when solid (-38.9°)
Mercury(I) acetate	$Hg_2(CH_3COO)_2$	519.27	colorless, cryst.	—
(II) acetate	$Hg(CH_3COO)_2$	318.67	colorless, powder	3.27
(II) amido-chloride	$HgNH_2Cl$	252.07	white powder	5.70
(II) bromide	$HgBr_2$	360.41	colorless, rhombic	6.109 (25°)
(I) chloride	Hg_2Cl_2	472.09	colorless, tetragonal	7.15

Melting point, °C	Boiling point, °C	Vapor pressure, kPa (at t°C)	Solubility, g/100 g (at t°C)	Relationship with bases and/or acids
1135; α→β 792 β→γ 859	—	—	H_2O: 76.4(20), 58.0 (100); insol. in EtOH, MeOH, & acetone	—
1475 (under pressure)	1380 subl.	0.000133 (604) 0.00133 (677) 0.0133 (762)	H_2O: 0.00013(18)	reacts with conc. acids
-38.89	356.66	0.000173 (20) 0.00179 (50)	insol. in H_2O 0.0123 (80)	reacts with aqua regia, HNO_3, & hot conc. H_2SO_4
decomp.	—	—	H_2O: 0.75(12); insol. in EtOH	sol. in H_2SO_4 & HNO_3
84.5	—	—	H_2O: 25(10), 100(100); sol. in EtOH & AcOH	—
—	—	—	H_2O: 0.14(20); reacts with hot H_2O; insol. in EtOH	reacts with acids
238	319	0.000133 (45.6) 0.00133 (71) 0.0133 (100.7)	H_2O: 0.55(20), 4.9 (100); EtOH: 27.3(0), 28.6(20), 34.0(40), 42.3(60); MeOH: 65.3 (20), 76.0(40); glycerol: 15.7(25); pyridine: 24(10), 39.6(30); sol. in acetone, C_6H_6 & CS_2	—
525 (under pressure)	383.7 subl.	0.133 (199) 1.33 (247) 13.3 (309)	H_2O: 0.0002(25), 0.001 (43); insol. in EtOH, Et_2O, & acetone	sol. in aqua regia & hot conc. HNO_3; sl.sol. in conc. HCl

Name	Symbol/ formula	Atomic/ molecular mass	Color & crystal system	Density (g/cm^3 at 20°C unless otherwise stated)
MERCURY COMPOUNDS (cont'd)				
(II) chloride	$HgCl_2$	271.50	colorless, rhombic (α), cryst. (β)	5.44 (25°)
(II) iodide	HgI_2	454.40	red, tetragonal (α); yellow, rhombic (β)	6.36 (α) 6.09 (β)
(I) nitrate dihydrate	$Hg_2(NO_3)_2 \cdot 2H_2O$	561.22	colorless, monoclinic	4.78
(II) nitrate semihydrate	$Hg(NO_3)_2 \cdot 0.5H_2O$	333.61	colorless, cryst., hygr.	4.39
(II) oxide	HgO	216.59	yellow or red, rhombic	11.14 (yellow) 11.08 (red)
(I) sulfate	Hg_2SO_4	497.25	colorless, monoclinic	7.56
(II) sulfate	$HgSO_4$	296.65	colorless, rhombic	6.47

Melting point, °C	Boiling point, °C	Vapor pressure, kPa (at t°C)	Solubility, g/100 g (at t°C)	Relationship with bases and/or acids
280; $\alpha \rightarrow \beta$ 155	302	0.000133 (46) 0.00133 (74.5) 0.0133 (104)	H_2O: 6.59(20), 58.3 (100); EtOH: 42.5(0), 47.1(20), 55.3(40); MeOH: 25.2(0), 51.5 (20), 141.6(40); pyridine: 15.1(0), 25.2 (20); sol. in Et_2O, acetone, C_6H_6,$CHCl_3$, dioxane, & AcOH	—
256; $\alpha \rightarrow \beta$ 127	354	0.000133 (63) 0.00133 (89) 0.0133 (119.3)	H_2O: 0.0061(25); EtOH: 2.29(25); MeOH: 3.16(19.5), 6.51(66); acetone: 2.1(25); sol. in Et_2O, C_6H_6, dioxane, pyridine, & $CHCl_3$	—
decomp. 70	—	—	reacts with H_2O; sol. in CS_2	reacts with conc. HNO_3 & alk.; sol. in dilute HNO_3
145	decomp.	—	very sol. in cold H_2O; reacts with hot H_2O; sol. in acetone; insol. in EtOH	sol. in dilute HNO_3 & NH_4OH
decomp. 400	—	—	H_2O: 0.0051(25) (yellow), 0.0049(25) (red), 0.041(100) (yellow), 0.0379(100) (red); insol. in EtOH, acetone, & Et_2O	reacts with acids; insol. in alk. & NH_4OH
decomp.	—	—	H_2O: 0.04(25), 0.09 (100)	sol. in H_2SO_4 & HNO_3
decomp. >500	—	—	reacts with H_2O; insol. in EtOH & acetone	sol. in acids; reacts with NH_4OH

Name	Symbol/ formula	Atomic/ molecular mass	Color & crystal system	Density (g/cm^3 at 20°C unless otherwise stated)
MERCURY COMPOUNDS (cont'd)				
(II) sulfide	HgS	232.66	black, cubic; red, trigonal	7.73 (cubic) 8.10 (trigonal)

BORON AND ITS COMPOUNDS

Name	Symbol/ formula	Atomic/ molecular mass	Color & crystal system	Density
BORON	B	10.811	brown, amorphous; dark gray, hexagonal	2.34
Boron fluoride	BF$_3$	67.81	colorless, gas	2.990 (g/L)
nitride	BN	24.82	colorless, hexagonal(α), cubic(β)	2.34
oxide	B$_2$O$_3$	69.62	vitreous or colorless hexagonal, hygr.	1.84 (vitreous) 2.46 (hexagonal)
Calcium hexaboride	CaB$_6$	104.95	black, cubic	2.33 (15°)
metaborate	Ca(BO$_2$)$_2$	125.70	colorless, rhombic	—
Chromium diboride	CrB$_2$	73.62	—	5.22
Decaborane(14)	B$_{10}$H$_{14}$	122.22	colorless, monoclinic	0.94 (25°)

Melting point, °C	Boiling point, °C	Vapor pressure, kPa (at t°C)	Solubility, g/100 g (at t°C)	Relationship with bases and/or acids
820 (under pressure); -HgS (cubic) 344	subl.	—	reacts with hot H_2O; insol. in cold H_2O & EtOH	reacts with conc. H_2SO_4 & conc. HNO_3; insol. in alk. & NH_4OH
2075	3700	0.000133 (1880) 0.00133 (2050) 0.0133 (2260)	insol. in H_2O, EtOH	insol. in alk.; reacts with HNO_3 & H_2SO_4
-128	-101	1.33 (-141.5) 13.3 (-123.5) 53.3 (-108.9)	H_2O: 106 ml(0); reacts with hot H_2O & EtOH; sol. in C_6H_6	—
decomp. ≈3000	—	—	insol. in cold H_2O, sl.sol. in hot H_2O	insol. in acids; reacts with hot alk.
450	2100	0.000133 (1071) 0.00133 (1186) 0.0133 (1323)	H_2O: 1.1(0),15.7 (100)	sol. in alk.
2235	—	—	insol. in H_2O	—
1162	—	—	H_2O: 0.13(100)	—
2200	—	—	—	—
99.5	213	0.133 (61.1) 1.33 (89.7) 13.3 (141.3)	reacts with H_2O; sol. in CS_2, EtOH, & Et_2O	—

Name	Symbol/ formula	Atomic/ molecular mass	Color & crystal system	Density (g/cm^3 at 20°C unless otherwise stated)
BORON COMPOUNDS (cont'd)				
Diborane (6)	B_2H_6	27.67	colorless gas; monoclinic	0.577 (-183°) (solid); 0.447 (-112°) (liq.)
Lead (II) meta- borate hydrate	$Pb(BO_2)_2 \cdot H_2O$	310.82	colorless, cryst.	5.598 anhydrous
Magnesium dodecaboride	MgB_{12}	154.03	—	2.46
Molybdenum boride	MoB	106.75	tetragonal (α), rhombic (β)	8.65
pentaboride	Mo_2B_5	245.93	rhombic	8.01
Orthoboric acid	H_3BO_3	61.83	colorless, triclinic	1.435 (15°)
Pentaborane (9)	B_5H_9	63.13	colorless liq.; tetragonal	0.66 (0°)
Potassium pentaborate tetrahydrate	$KB_5O_8 \cdot 4H_2O$	293.21	colorless, rhombic	—
Sodium metaborate	$NaBO_2$	65.80	colorless, hexagonal	2.46
tetraborate decahydrate	$Na_2B_4O_7 \cdot 10H_2O$	381.38	colorless, monoclinic	1.69—1.72

Melting point, °C	Boiling point, °C	Vapor pressure, kPa (at t°C)	Solubility, g/100 g (at t°C)	Relationship with bases and/or acids
-166	-92.5	1.33 (-144) 13.3 (-121) 53.3 (-100)	reacts with H_2O	pyrophoric in air; sol. in conc. H_2SO_4
-H_2O 160	—	—	insol. in H_2O	—
2200	—	—	—	—
2560; $\alpha \rightarrow \beta$ 1950	—	—	—	—
2100	—	—	—	—
decomp. 171	—	—	H_2O: 4.87(0), 38(100); glycerol: 28(20); Et_2O: 0.078(20); EtOH: 5.56(20); sl. sol. in acetone	—
-48.8	60	1.33 (-31.6) 13.3 (8.8) 53.3 (40.4)	reacts with H_2O	pyrophoric in air
780	—	—	H_2O: 0.007(0); insol. in EtOH	reacts with HNO_3 & fused alk.
966	1434	—	H_2O: 25.4(20), 125.2 (100); insol. in EtOH & Et_2O	—
64	-10H_2O 380	—	H_2O: 2.12(0), 22.0 (50); sol. in glycerol; insol. in EtOH	—

Name	Symbol/ formula	Atomic/ molecular mass	Color & crystal system	Density (g/cm^3 at 20°C unless otherwise stated)
BORON COMPOUNDS (cont'd)				
Tetraboron carbide	B$_4$C	55.26	black, hexagonal	2.52
Titanium diboride	TiB$_2$	69.52	hexagonal	4.5

ALUMINUM AND ITS COMPOUNDS

Name	Symbol/ formula	Atomic/ molecular mass	Color & crystal system	Density (g/cm^3 at 20°C unless otherwise stated)
ALUMINUM	Al	26.98154	silvery, cubic	2.70
Aluminum chloride	AlCl$_3$	133.34	colorless, monoclinic(α) hexagonal(β)	2.44 (25°)
hydroxide	Al(OH)$_3$	78.00	colorless, monoclinic	2.424
nitrate nonahydrate	Al(NO$_3$)$_3$· 9H$_2$O	375.13	colorless, monoclinic, hygr.	—
nitride	AlN	40.99	colorless, hexagonal	3.26
oxide	Al$_2$O$_3$	101.96	colorless, hexagonal	3.96

Melting point, °C	Boiling point, °C	Vapor pressure, kPa (at t°C)	Solubility, g/100 g (at t°C)	Relationship with bases and/or acids
2350	3500	—	insol. in H_2O	insol. in acids; sol. in fused alk.
2900±80	—	—	insol. in H_2O	—
660.1	2520	0.000133 (1100) 0.00133 (1230) 0.0133 (1380)	insol. in H_2O & EtOH	reacts with HCl & H_2SO_4
192.6 (0.229 MPa); $\alpha \rightarrow \beta$ 138	subl. 180	0.000133 (41.2) 0.00133 (57.7) 0.0133 (76.2)	H_2O: 45.1(25), 46.5 (60); EtOH: 100 (12.5); $CHCl_3$: 0.72 (25); sol. in Et_2O & CCl_4; insol. in C_6H_6	—
$\rightarrow$ boehmite >150	—	—	insol. in H_2O	reacts with hot acids & with alk.
73.6	—	—	H_2O: 62.6(20), 159.7 (100); sol. in EtOH	
2200 (0.4 MPa)	subl. 2000	—	reacts with H_2O & EtOH	—
2050	2980	0.133 (2145) 1.33 (2380) 13.3 (2670)	insol. in H_2O	—

Name	Symbol/ formula	Atomic/ molecular mass	Color & crystal system	Density (g/cm^3 at 20°C unless otherwise stated)
ALUMINUM COMPOUNDS (cont'd)				
sulfate	$Al_2(SO_4)_3$	342.15	colorless, hexagonal	2.71
sulfate octadeca- hydrate	$Al_2(SO_4)_3 \cdot 18H_2O$	666.42	colorless, monoclinic	1.69 (17°)
sulfate-potassium sulfate-water (1/1/24)	$Al_2(SO_4)_3 \cdot K_2SO_4 \cdot 24H_2O$	348.78	colorless, cubic	1.76
Alunite	$K_2O \cdot 3Al_2O_3 \cdot 4SO_3 \cdot 6H_2O$	828.4	colorless	2.6—2.9
Bauxite	$Al_2O_3 \cdot 2H_2O$	137.99	red	2.9—3.5
Lanthanum ortho- aluminate- calcium meta- titanate(1/1)	$LaAlO_3 \cdot CaTiO_3$	349.78	white-pink or brown	4.7—4.8
Kaolin	$Al_2O_3 \cdot 2SiO_2 \cdot 2H_2O$	258.16	colorless	2.5—2.6
Mullite	$Al_6Si_2O_{13}$	426.05	colorless, rhombic	3.16

Melting point, °C	Boiling point, °C	Vapor pressure, kPa (at t°C)	Solubility, g/100 g (at t°C)	Relationship with bases and/or acids
decomp. >700	—	—	H_2O: 38.5(25), 89(100); sl. sol. in EtOH	—
decomp. 86.5	—	—	H_2O: 36.2(20) anhydrous; insol. in EtOH	—
92.5	—	—	H_2O: 5.9(20), 109(90); insol. in EtOH	—
$-H_2O$ 400-600	decomp. 700-1000	—	sl. sol. in H_2O	sl. sol. in acids
—	—	—	—	—
$-2H_2O$ 430—550	—	—	—	reacts with HF, hot conc. H_2SO_4, & fused alk.
>200	—	—	sl. sol. in hot H_2O	sl. sol. in dilute HCl; sol. in hot HCl+H_2SO_4+HF
1935	—	—	insol. in H_2O	insol. in acids

Name	Symbol/ formula	Atomic/ molecular mass	Color & crystal system	Density (g/cm^3 at 20°C unless otherwise stated)
GALLIUM AND ITS COMPOUNDS				
GALLIUM	Ga	69.72	silvery, rhombic	5.904
Gallium chloride	GaCl$_3$	176.08	colorless, cryst., hygr.	2.47 (25°)
hydroxide	Ga(OH)$_3$	120.74	white, gel	—
nitrate	Ga(NO$_3$)$_3$	255.73	colorless, cryst., hygr.	—
nitride	GaN	83.73	gray, hexagonal	6.1
oxide	Ga$_2$O$_3$	187.44	colorless, monoclinic(α) hexagonal(β)	6.48 (α) 5.88 (β)
phosphide	GaP	100.69	cubic	4.10
sulfate	Ga$_2$(SO$_4$)$_3$	427.62	colorless, hexagonal	—

Melting point, °C	Boiling point, °C	Vapor pressure, kPa (at t°C)	Solubility, g/100 g (at t°C)	Relationship with bases and/or acids
29.8	2295	0.000133 (940) 0.00133 (1050) 0.0133 (1190)	insol. in H_2O	reacts with acids & alk.
77.9±0.2	201.2	0.133 (48.2) 1.33 (78) 13.3 (132)	reacts with H_2O; sol. in EtOH; sl.sol. in petroleum ether	reacts with HCl, alk., & NH_4OH
decomp 420–440	—	—	H_2O: 7.6×10^{-9} (20)	reacts with acids, alk., & NH_4OH
decomp 200	—	—	very sol. in H_2O	—
subl. >800	—	—	insol. in H_2O	reacts with hot alk.
1740±25	—	—	insol. in H_2O	reacts with acids & alk.
1467	—	0.000133 (738) 0.00133 (815.5) 0.0133 (906)	insol. in H_2O	reacts with hot conc. H_2SO_4
decomp >500	—	—	sol. in H_2O & EtOH; insol. in Et_2O	—

Name	Symbol/ formula	Atomic/ molecular mass	Color & crystal system	Density (g/cm^3 at 20°C unless otherwise stated)
INDIUM AND ITS COMPOUNDS				
INDIUM	In	114.82	silvery, tetragonal	7.31
Indium antimonide	InSb	236.57	gray	—
arsenide	InAs	189.74	dark gray, cryst.	—
chloride	InCl$_3$	221.19	colorless, monoclinic, hygr.	3.45
hydroxide	In(OH)$_3$	165.84	colorless, cubic	—
nitrate trihydrate	In(NO$_3$)$_3$· 3H$_2$O	354.88	hygr.	—
nitride	InN	128.83	black, cryst.	—
oxide	In$_2$O$_3$	277.64	yellow, amorphous or cubic	7.179
sulfate nonahydrate	In$_2$(SO$_4$)$_3$· 9H$_2$O	679.96	colorless, hygr.	3.438

Melting point, °C	Boiling point, °C	Vapor pressure, kPa (at t°C)	Solubility, g/100 g (at t°C)	Relationship with bases and/or acids
156.4	2000	0.000133 (810) 0.00133 (915) 0.0133 (1045)	insol. in H_2O	reacts weakly with NaOH
546	—	0.000133 (546) 0.00133 (616) 0.0133 (700)	insol. in H_2O & EtOH	reacts with acids; insol. in alk.
943	—	—	insol. in H_2O & EtOH	reacts with acids
585 (under pressure)	subl.	0.000133 (243) 0.00133 (279) 0.0133 (319)	H_2O: 195(22), 374(80); sol. in EtOH; sl. sol. in Et_2O	—
$-H_2O$ <150	—	—	insol. in H_2O	reacts with acids; reacts weakly with NaOH; insol. in NH_4OH
$-2H_2O$ 100	decomp.	—	very sol. in H_2O; sol. in EtOH	—
—	—	0.000133 (468) 0.00133 (509) 0.0133 (554)	—	reacts with HCl
1910	—	—	insol. in H_2O	amorphous reacts with acids; cryst. insol. in acids
$-H_2O$ 300	decomp. >600	—	sol. in H_2O	—

Name	Symbol/ formula	Atomic/ molecular mass	Color & crystal system	Density (g/cm^3 at 20°C unless otherwise stated)
THALLIUM AND ITS COMPOUNDS				
THALLIUM	Tl	204.37	silvery, hexagonal(α), cubic(β)	11.85 (α)
Thallium(I) bromide	TlBr	284.28	light yellow, cubic	7.557 (17.3°)
(I) carbonate	Tl$_2$CO$_3$	468.75	colorless, monoclinic	7.16 360
(I) chloride	TlCl	239.82	colorless, cubic	7.00
(I) iodide	TlI	331.27	yellow, rhombic(α); red, cubic(β)	7.07 (α) 7.10 (β)
(I) oxide	Tl$_2$O	424.74	black, hexagonal, hygr.	9.52 (16°)
(III) oxide	Tl$_2$O$_3$	456.74	dark brown, amorphous or cubic	10.0 (cryst.)
(I) sulfate	Tl$_2$SO$_4$	504.80	colorless, rhombic (α), hexagonal(β)	6.67 (α)
RARE-EARTH ELEMENTS AND THEIR COMPOUNDS				
CERIUM	Ce	140.12	silvery, cubic(α, γ,δ), hexagonal(β)	6.77 (β)
Cerium (III) chloride	CeCl$_3$	246.48	colorless, hexagonal, hygr.	3.92 (0)

Melting point, °C	Boiling point, °C	Vapor pressure, kPa (at t°C)	Solubility, g/100 g (at t°C)	Relationship with bases and/or acids
304; $\alpha \to \beta$ 234	1475	0.000133 (535) 0.00133 (615) 0.0133 (714)	insol. in H_2O & EtOH	reacts with acids, insol. in alk.
460	824	0.000133 (268) 0.00133 (615) 0.0133 (368)	H_2O: 0.05(25), 0.25 (68); sol. in EtOH; insol. in acetone	insol. in HBr
269	$-CO_2$	—	H_2O: 5.23(18), 27.2 (100); insol. in EtOH, Et_2O, & acetone	—
431	820	0.000133 (205) 0.00133 (295) 0.0133 (355)	H_2O: 0.32(20), 2.38 (100); sol. in Et_2O & EtOH	reacts with HCl
441; $\alpha \to \beta$ 178	decomp. 883	0.000133 (272) 0.00133 (319) 0.0133 (377)	H_2O: 0.0064(20), 0.12 (100); sl. sol. in EtOH, acetone, & pyridine	sol. in HNO_3 & aqua regia
579	600 (0.001 MPa)	0.000133 (182.5) 0.00133 (216) 0.0133 (255)	reacts with H_2O; sol. in EtOH	reacts with acids
717 (O_2 pressure)	$\to Tl_2O$ 875	—	insol. in H_2O	reacts with acids; insol. in alk.
632; $\alpha \to \beta$ 500	—	—	H_2O: 4.87(20), 18.5 (100)	very sol. in H_2SO_4
804; $\alpha \to \beta$ -130, $\beta \to \delta$ ≈130, $\gamma \to \sigma$ 730	≈3450	0.000133 (1520) 0.00133 (1695) 0.0133 (1905)	insol. in cold H_2O; reacts with hot H_2O	reacts with acids
822	1650	—	sol. in cold H_2O, acetone, & EtOH; pyridine: 1.55(2); reacts with hot H_2O	—

Name	Symbol/ formula	Atomic/ molecular mass	Color & crystal system	Density (g/cm^3 at 20°C unless otherwise stated)
RARE-EARTH ELEMENTS AND THEIR COMPOUNDS (cont'd)				
(III) fluoride	CeF$_3$	197.12	colorless, hexagonal	6.16
hexaboride	CeB$_6$	204.89	blue, cubic	—
(III) oxide	Ce$_2$O$_3$	328.24	gray-green, cubic	6.86
(IV) oxide	CeO$_2$	172.12	light brown, cubic or colorless	7.3
(III) sulfide	Ce$_2$S$_3$	376.43 trigonal	red or brown, cubic	5.02
DYSPROSIUM	Dy	162.50	silvery, hexagonal (α), cubic(β)	8.559
Dysprosium chloride	DyCl$_3$	268.86	colorless, monoclinic	3.67 (0°)
oxide	Dy$_2$O$_3$	373.00	yellow, powder; cubic	7.81 (27°)
sulfide	Dy$_2$S$_3$	421.22	yellow, monoclinic	—
ERBIUM	Er	167.26	silvery, hexagonal	9.062
Erbium oxide	Er$_2$O$_3$	382.52	red-yellow or pink, cubic	8.640
EUROPIUM	Eu	151.96	silvery, cubic	5.245

Melting point, °C	Boiling point, °C	Vapor pressure, kPa (at t°C)	Solubility, g/100 g (at t°C)	Relationship with bases and/or acids
1432	2180	—	insol. in H_2O	—
2190	decomp.	—	insol. in H_2O	insol. in HCl
2180	—	—	insol. in cold H_2O; reacts with hot H_2O	pyrophoric; reacts with acids
2400	—	—	insol. in H_2O	reacts with H_2SO_4 & HCl
decomp.	—	—	insol. in cold H_2O; reacts with hot H_2O	reacts with acids
1409; $\alpha \to \beta$ 1384	≈ 2600	0.000133 (1015) 0.00133 (1130) 0.0133 (1270)	reacts with H_2O; insol. in EtOH	reacts with conc. acids
653	1539	—	sol. in H_2O & EtOH	reacts with alk.; sol. in HCl
2400	≈ 4300	—	insol. in H_2O & EtOH	reacts with acids
1470-1490	—	—	insol. in H_2O	reacts with acids
1525	2860	0.000133 (1140) 0.00133 (1270) 0.0133 (1430)	reacts with H_2O; insol. in org. solv.	reacts with acids; insol. in HF & H_3PO_4
>2200	—	—	H_2O: 5×10^{-4} (29)	reacts with acids
826	1560	0.000133 (530) 0.00133 (610) 0.0133 (710)	insol. in cold H_2O & in EtOH; reacts with hot H_2O	reacts with acids

Name	Symbol/ formula	Atomic/ molecular mass	Color & crystal system	Density (g/cm^3 at 20°C unless otherwise stated)
RARE-EARTH ELEMENTS AND THEIR COMPOUNDS (cont'd)				
Dieuropium dioxide sulfide	Eu_2O_2S	367.98	pink, trigonal	7.04 (calculated)
GADOLINIUM	Gd	157.25	silvery, hexagonal (α), cubic(β)	7.87
Gadolinium chloride	$GdCl_3$	263.59	colorless, rhombic (α), hexagonal(β)	4.52 (0°)
oxide	Gd_2O_3	262.50	colorless, hexagonal(α), monoclinic(β)	7.407 (15°)
HOLMIUM	Ho	164.93	silvery, hexagonal	8.80
Holmium oxide	Ho_2O_3	377.86	yellow, cubic	—
LANTHANUM	La	138.906	silvery-gray, hexagonal, cubic	6.162
Lanthanum chloride	$LaCl_3$	245.27	colorless, hexagonal, hygr.	3.842
hexaboride	LaB_6	203.78	purple, cubic	2.61
nitrate hexahydrate	$La(NO_3)_3 \cdot 6H_2O$	433.02	colorless, triclinic, hygr.	—
oxide	La_2O_3	325.82	colorless, hexagonal, cubic	6.51 (15°)

Melting point, °C	Boiling point, °C	Vapor pressure, kPa (at t°C)	Solubility, g/100 g (at t°C)	Relationship with bases and/or acids
2260	—	—	—	—
1312; α→β 1262	3280	0.000133 (1840) 0.00133 (2100) 0.0133 (2450)	reacts with H_2O; insol. in EtOH	reacts with acids
605; α→β 97	1600	0.000133 (686) 0.00133 (772) 0.0133 (875)	sol. in H_2O	—
2350; α→β 2200	—	—	insol. in H2O	sol. in acids
≈1500	≈2700	0.000133 (1100) 0.00133 (1230) 0.0133 (1390)	reacts with H_2O; insol. in EtOH	reacts with acids
—	—	—	insol. in H_2O & EtOH	reacts with acids
920	3450	0.000133 (1555) 0.00133 (1735) 0.0133 (1950)	reacts with H_2O; insol. in org. solv.	reacts with acids
862	1710	—	H_2O: 97.2(25), 170.3 (92); sol. in EtOH & pyridine; insol. in Et_2O, acetone, & C_6H_6	sol. in HCl; reacts with H_2SO_4, alk., & NH_4OH
2715	decomp.	—	insol. in H_2O	reacts with HCl
decomp. 43	—	—	H_2O: 113(25); sol. in EtOH & acetone	—
2280	≈4200	—	H_2O: 0.0004(29); reacts with hot H_2O; sol. in MeOH; insol. in acetone	reacts with acids

Name	Symbol/ formula	Atomic/ molecular mass	Color & crystal system	Density (g/cm^3 at 20°C unless otherwise stated)
RARE-EARTH ELEMENTS AND THEIR COMPOUNDS (cont'd)				
Lanthanum sulfate	$La_2(SO_4)_3$	566.00	colorless powder, hygr.	3.60 (15°)
LUTECIUM	Lu	174.97	silvery, hexagonal	9.849
Lutecium chloride	$LuCl_3$	281.33	colorless, monoclinic	3.98
oxide	Lu_2O_3	397.94	colorless, cubic	9.42
NEODYMIUM	Nd	144.24	light yellow, hexagonal(α), cubic(β)	7.01 (α) 6.80 (β)
Neodymium chloride	$NdCl_3$	250.60	pink—violet, hexagonal	4.134 (25°)
oxide	Nd_2O_3	336.48	light blue powder, cubic, hexagonal	7.24
PRASEODYMIUM	Pr	140.9077	light yellow, hexagonal(α), cubic(β)	6.77 (α) 6.44 (β)
Praseodymium chloride	$PrCl_3$	247.27	green-blue, hexagonal	4.02 (25°)
(III) oxide	Pr_2O_3	329.81	yellow-green, trigonal or amorphous powder	7.07
PROMETHIUM	Pm	144.913	silvery, hexagonal	7.26

Melting point, °C	Boiling point, °C	Vapor pressure, kPa (at t°C)	Solubility, g/100 g (at t°C)	Relationship with bases and/or acids
decomp. 1150	—	—	H_2O: 2.142(25), 0.69 (100); sl. sol. in EtOH; insol. in Et_2O	reacts with alk. & NH_4OH
1660	≈3410	0.000133 (1500) 0.00133 (1670) 0.0133 (1860)	reacts with H_2O; insol. in EtOH	reacts with acids
925	1420	—	very sol. in H_2O	—
2450	—	—	insol. in cold H_2O; reacts with hot H_2O	reacts with acids
1024; α→β 855	≈3080	0.000133 (1200) 0.00133 (1340) 0.0133 (1520)	reacts with H_2O	reacts with acids
760	1620	—	H_2O: 96.7(13), 140 (100); EtOH: 44.5 (20); insol. in Et_2O & $CHCl_3$	—
2320	—	—	H_2O: 0.0019(18), 0.003(75)	sol. in acids
932; α→β 796	≈3500	0.000133 (1325) 0.00133 (1510) 0.0133 (1700)	reacts with H_2O	reacts with acids; insol. in HF & H_3PO_4
786	1630	—	H_2O: 91.4(0), 141.6 (100); very sol. in EtOH; sol. in pyridine; insol. in Et_2O & $CHCl_3$	—
decomp.	—	—	H_2O: 0.00002(29)	reacts with acids
1170	≈3000	—	insol. in cold H_2O; reacts with hot H_2O; insol. in org. solv.	reacts with acids

Name	Symbol/ formula	Atomic/ molecular mass	Color & crystal system	Density (g/cm³ at 20°C unless otherwise stated)
RARE-EARTH ELEMENTS AND THEIR COMPOUNDS (cont'd)				
SAMARIUM	Sm	150.35	silvery, hexagonal (α), cubic(β)	7.536
Samarium oxide	Sm_2O_3	348.70	light yellow, cubic(α), monoclinic(β)	8.35
SCANDIUM	Sc	44.9559	silvery, hexagonal(α), cubic(β)	3.02 (25°)
Scandium oxide	Sc_2O_3	137.91	colorless, cubic	3.86
nitrate	$Sc(NO_3)_3$	230.97	colorless powder, hygr.	—
TERBIUM	Tb	158.9254	colorless, hexagonal(α) cubic(β)	8.253
THULIUM	Tm	168.9254	silvery, hexagonal	9.318
YTTERBIUM	Yb	173.04	silvery, cubic	6.953
Ytterbium oxide	Yb_2O_3	394.08	colorless, cubic	9.175
YTTRIUM	Y	88.9059	gray, hexagonal (α), cubic (β)	4.48
Yttrium chloride	YCl_3	195.26	colorless, monoclinic	2.8 (18°)

Melting point, °C	Boiling point, °C	Vapor pressure, kPa (at t°C)	Solubility, g/100 g (at t°C)	Relationship with bases and/or acids
1072; $\alpha \rightarrow \beta$ 917	≈ 1800	0.000133 (680) 0.00133 (760) 0.0133 (860)	reacts with H_2O; insol. in org. solv.	reacts with acids
2270; $\alpha \rightarrow \beta$ 875	—	—	insol. in H_2O	reacts with acids
1541; $\alpha \rightarrow \beta$ 1336	≈ 2850	0.000133 (1240) 0.00133 (1380) 0.0133 (1555)	reacts with H_2O; insol. in org. solv.	reacts with acids
≈ 2450	—	—	insol. in H_2O	sl. sol. in cold dil. acids; sol. in hot conc. acids
150	decomp.	—	sol. in H_2O	—
1360; $\alpha \rightarrow \beta$ 1290	≈ 3200	0.000133 (1380) 0.00133 (1540) 0.0133 (1740)	reacts with H_2O; insol. in org. solv.	reacts with acids
1545	1950	0.000133 (777) 0.00133 (870) 0.0133 (983)	reacts with H_2O	reacts with acids
824	1211	0.000133 (410) 0.000133 (471) 0.0133 (550)	reacts with H_2O; insol. in EtOH	reacts with acids
>2000	4300	—	insol. in cold H_2O; reacts with hot H_2O	reacts with acids
1528; $\alpha \rightarrow \beta$ 1482	3300	0.000133 (1480) 0.00133 (1650) 0.0133 (1855)	insol. in cold H_2O & in EtOH; reacts with hot H_2O	reacts with acids & hot KOH; insol. in HF
721	1482	—	H_2O: 73.6(0), 78.4 (80); EtOH: 60.1(15); pyridine: 60.6(15)	—

Name	Symbol/ formula	Atomic/ molecular mass	Color & crystal system	Density (g/cm^3 at 20°C unless otherwise stated)

RARE-EARTH ELEMENTS AND THEIR COMPOUNDS (cont'd)

diyttrium dioxide sulfide	Y_2O_2S	241.87	colorless, trigonal	4.86
fluoride	YF_3	145.90	rhombic (α), hexagonal (β)	4.01
oxide	Y_2O_3	225.81	colorless, cubic (α), hexagonal (β)	4.84 (α)
sulfide	Y_2S_3	274.00	yellow-gray, monoclinic	—

THORIUM AND ITS COMPOUNDS

THORIUM	Th	232.0381	silvery, cubic	11.7 (25°)
Thorium chloride	$ThCl_4$	373.85	colorless, rhombic (α), tetragonal (β), hygr.	4.59 (15°)
fluoride	ThF_4	308.03	colorless, monoclinic	6.32 (24°)
nitrate	$Th(NO_3)_4$	480.08	colorless, cryst.	—

Melting point, °C	Boiling point, °C	Vapor pressure, kPa (at t°C)	Solubility, g/100 g (at t°C)	Relationship with bases and/or acids
2120	—	—	insol. in H_2O & EtOH	reacts with conc. acids
1155; $\alpha \rightarrow \beta$ 1077	—	—	insol. in H_2O	—
2430; $\alpha \rightarrow \beta$ 2300	4300	—	H_2O: 0.00018 (29)	reacts with acids; insol. in alk.
≈ 2000	—	—	reacts with H_2O	reacts with acids
1750	≈ 4800	0.000133 (2180) 0.00133 (2420) 0.0133 (2710)	insol. in H_2O	insol. in alk.; reacts with hot HCl & with aqua regia; reacts slowly with HF, H_2SO_4, & HNO_3
770; $\alpha \rightarrow \beta$ 405	921	0.000133 (423) 0.00133 (474) 0.0133 (534)	very sol. in cold H_2O; reacts with hot H_2O; sol. in EtOH & Et_2O	sol. in acids
1110	≈ 1650	0.000133 (852) 0.00133 (932) 0.0133 (1025)	H_2O: 0.2(25)	insol. in HF; sol. in hot H_2SO_4 & in $HClO_4$
$\rightarrow ThO_2$ >400	—	—	H_2O: 190.7(20); reacts with hot H_2O; very sol. in EtOH, Et_2O, & acetone	—

Name	Symbol/ formula	Atomic/ molecular mass	Color & crystal system	Density (g/cm^3 at 20°C unless otherwise stated)
THORIUM COMPOUNDS (cont'd)				
oxide	ThO$_2$	264.04	colorless, cubic	9.7

URANIUM AND ITS COMPOUNDS

URANIUM	U	238.029	silvery, rhombic(α), tetragonal(β), cubic(γ)	19.04(α) (26°)
Uranium carbide	UC	250.04	gray-black, cubic	13.63
(IV) chloride	UCl$_4$	379.84	green, tetragonal(α) cubic(β), hygr.	4.87
(V) chloride	UCl$_5$	415.30	red-brown, monoclinic, hygr.	3.81
dicarbide	UC$_2$	262.04	light gray, tetragonal	11.28 (16°)
(IV) fluoride	UF$_4$	314.02	green, monoclinic (α), cryst.(β)	6.7−6.9
(VI) fluoride	UF$_6$	352.02	colorless, rhombic, hygr.	5.06

Melting point, °C	Boiling point, °C	Vapor pressure, kPa (at t°C)	Solubility, g/100 g (at t°C)	Relationship with bases and/or acids
≈3350	≈4400	0.000133 (2240) 0.00133 (2440) 0.0133 (2670)	insol. in H_2O	insol. in acids & alk.; reacts with HNO_3+HF
1134; α→β 668, β→γ 775	≈4200	0.000133 (1735) 0.00133 (1940) 0.0133 (2190)	reacts weakly with hot H_2O	insol. in alk.; reacts weakly with H_2SO_4, cold H_3PO_4, & HF; reacts with HCl & HNO_3
decomp. 2400	—	—	reacts with H_2O	reacts with acids
590; α→β 547	792	0.000133 (367) 0.00133 (409) 0.0133 (456)	reacts with H_2O; sol. in acetone, pyridine, & EtOAc; insol. in C_6H_6, $CHCl_3$, & Et_2O	—
decomp. 320	—	—	reacts with H_2O, acetone, Et_2O, & EtOH; sol. in CCl_4 & CS_2	—
2350	4320	—	reacts with H_2O; insol. in EtOH	reacts with acids & alk.
1036; α→β 845	≈1700	0.000133 (718) 0.00133 (788) 0.0133 (869)	H_2O: 0.01(25); reacts with hot H_2O	reacts with conc. acids & alk.
64.0 (0.14 MPa)	56.5 subl.	1.33 (-7) 13.3 (23) 53.3 (45)	reacts with H_2O, EtOH, Et_2O, & C_6H_6; very sol. in $C_2H_2Cl_4$; sl. sol. in CCl_4 & $CHCl_3$; insol. in CS_2	—

Name	Symbol/ formula	Atomic/ molecular mass	Color & crystal system	Density (g/cm^3 at 20°C unless otherwise stated)
URANIUM COMPOUNDS (cont'd)				
hydride	UH_3	241.06	black, cubic	11.0
(IV) oxide	UO_2	270.03	dark brown, cubic	10.95
(VI) oxide	UO_3	286.03	orange, rhombic(α), monoclinic(β,δ), cubic(σ), triclinic(ε), amorphous	8.34 (α) 8.02 (β)
(V)oxide— uranium(VI) oxide(1/1)	U_3O_8	842.09	olive green, rhombic	8.3 1300
peroxide dihydrate	$UO_4{\cdot}2H_2O$	338.06	light red, amorphous or rhombic	4.66 260
Ammonium diuranate	$(NH_4)_2U_2O_7$	624.13	yellow, cryst	—
Sodium diuranate	$Na_2U_2O_7$	634.06	orange-yellow, rhombic	—
Uranyl acetate dihydrate	$UO_2(CH_3COO)_2{\cdot}2H_2O$	424.12	yellow, rhombic	2.89 (15°)
carbonate	UO_2CO_3	330.04	light yellow, tetragonal	5.24
fluoride	UO_2F_2	308.03	light yellow, trigonal	5.8

Melting point, °C	Boiling point, °C	Vapor pressure, kPa (at t°C)	Solubility, g/100 g (at t°C)	Relationship with bases and/or acids
decomp. 300	—	—	reacts with H_2O	reacts with acids & alk.
≈2850 decomp..	—	0.00133 (2030) 0.0133 (2200) 0.133 (2400)	insol. in H_2O	reacts with conc. HNO_3, aqua regia, hot conc. H_2SO_4, & H_3PO_4
decomp. >500	—	—	insol. in H_2O	reacts with acids & alk.
→UO_2 1300	—	—	insol. in H_2O	reacts with HNO_3 & H_2SO_4
decomp. 260	—	—	H_2O: 0.0006(20), 0.008(90)	—
→UO_3 350	—	—	insol. in H_2O	reacts with conc. H_2SO_4 & conc. HNO_3
—	—	—	insol. in H_2O	—
-2H_2O 110	—	—	H_2O: 7.73(15); reacts with hot H_2O; very sol. in EtOH & Et_2O	reacts with acids
—	—	—	sl. sol. in H_2O; sol. in EtOH & Et_2O	—
—	—	—	H_2O: 64.4(0), 74.1 (100); sol. in EtOH; insol. in Et_2O & AmOH	—

Name	Symbol/ formula	Atomic/ molecular mass	Color & crystal system	Density (g/cm^3 at 20°C unless otherwise stated)
URANIUM COMPOUNDS (cont'd)				
Uranil nitrate dihydrate	$UO_2(NO_3)_2 \cdot 2H_2O$	430.07	yellow, monoclinic, hygr.	3.35
sulfate trihydrate	$UO_2SO_4 \cdot 3H_2O$	420.14	light yellow-green, cryst.	3.28 (16.5°)
CARBON AND ITS COMPOUNDS				
AMORPHOUS CARBON	C	12.01115	black, amorphous	1.80−2.10
DIAMOND	C	12.01115	colorless, cubic	3.515
GRAPHITE	C	12.01115	gray-black, hexagonal	2.265
Carbon dioxide (carbon(IV) oxide)	CO_2	44.01	colorless gas, cubic	1.977 g/L (0°), 1.56 (-79°) (solid); 1.101 (-37°) (liq.)
Carbon monoxide (carbon(II) oxide)	CO	28.01	colorless gas, cubic(α), hexagonal(β)	1.25 g/L (0°); 0.814 (-195°) (liq.)
Ammonium thiocyanate	NH_4SCN	76.12	colorless, monoclinic, hygr.	1.305
Calcium cyanamide	$CaCN_2$	80.10	colorless, trigonal	2.29

Melting point, °C	Boiling point, °C	Vapor pressure, kPa (at t°C)	Solubility, g/100 g (at t°C)	Relationship with bases and/or acids
184	—	—	H_2O: 119(20), 400 (80); sol. in EtOH, Et_2O, & acetone	—
decomp. 100	—	—	H_2O: 151.4(30), 237.8 (100); sol. in EtOH	—
→ graphite 1500-1600		—	insol. in H_2O	insol. in acids & alk.
→ graphite >1000 (in vacuum)		—	insol. in H_2O	insol. in acids & alk.
→ liq. carbon 3700 (>1.05 × 10^5 Pa)	subl. 3700	—	insol. in H_2O	insol. in acids & alk.
-56.6 (0.52 MPa)	subl. -78.5	1.33 (-119.6) 13.3 (-100.2) 53.3 (-86.0)	H_2O: 88 ml(20), 35.9 ml(75); sol. in EtOH MeOH, acetone, $CHCl_3$, CCl_4, & C_6H_6	reacts with alk.
-205; α→β -211.6	-191.5	1.33 (-215.3) 13.3 (-205.7) 53.3 (-196.3)	H_2O: 3.5 ml(0), 1.43 ml(80); sol. in EtOH	—
149.6	decomp. 170	—	H_2O: 170(20), 431(70); sol. in acetone & EtOH	sol. in NH_4OH
subl. 1200	—	—	reacts with H_2O; insol. in EtOH	—

Name	Symbol/ formula	Atomic/ molecular mass	Color & crystal system	Density (g/cm³ at 20°C unless otherwise stated)
CARBON COMPOUNDS (cont'd)				
Hydrogen cyanide	HCN	27.03	colorless gas or liq.	0.901 g/L; 0.688 (liq.)
Potassium cyanate	KOCN	81.12	colorless, tetragonal	2.06
cyanide	KCN	65.12	colorless, cubic, hygr.	1.56
thiocyanate	KSCN	97.18	colorless, rhombic (α), tetragonal (β,γ), hygr.	1.89 (14°)
Sodium cyanate	NaOCN	65.01	colorless, hexagonal	1.937
cyanide	NaCN	49.01	colorless, cubic, hygr.	1.596
thiocyanate	NaSCN	81.07	colorless, rhombic, hygr.	1.73

Melting point, °C	Boiling point, °C	Vapor pressure, kPa (at t°C)	Solubility, g/100 g (at t°C)	Relationship with bases and/or acids
-13.3	25.65	1.33 (-49.0) 13.3 (-18.6) 53.3 (9.5)	sol. in H_2O, EtOH, & Et_2O	reacts with alk.
decomp. >700	—	—	H_2O: 75(25); insol. in EtOH	—
620	—	—	H_2O: 71.6(25), 122 (100); EtOH: 0.88 (19.5); MeOH: 4.9 (19.5); glycerol: 32(15.5)	
177.0; $\alpha \to \beta$ 141.4, $\beta \to \gamma$ 175.0	decomp. 500	—	H_2O: 217(20), 673(99); EtOH: 20.75(22); sol. in acetone & AmOH	sol. in liq. NH_3
decomp. 700 (in vacuum)	—	—	sol. in cold H_2O; reacts with hot H_2O; insol. in EtOH & Et_2O	—
564	1497	0.00133 (586) 0.0133 (687) 0.133 (816)	H_2O: 58.2(20), 82.5 (35); sl. sol. in EtOH	sol. in NH_4OH
307.5	—	—	H_2O: 166(25), 225 (100); MeOH: 35(16); EtOH: 18.37(18.8); ethylenediamine: 93.5 (25); sol. in acetone	—

Name	Symbol/ formula	Atomic/ molecular mass	Color & crystal system	Density (g/cm^3 at 20°C unless otherwise stated)
SILICON AND ITS COMPOUNDS				
SILICON	Si	28.086	dark gray, cubic; brown, amorphous	28.086 (cryst.), 2.0 (amorphous)
Silicon carbide	SiC	40.10	colorless, cubic(α), hexagonal(β)	3.217
Silicon dioxide: cristobalite	SiO$_2$	60.08	colorless, tetragonal(α), cubic(β)	2.320
quartz	SiO$_2$	60.08	colorless, hexagonal(α,β)	2.650
tridymite	SiO$_2$	60.08	colorless, rhombic(α), hexagonal(β)	2.26 (25°)
Metasilicic acid	H$_2$SiO$_3$	78.10	colorless, amorphous	3.17
Orthosilicic acid	H$_4$SiO$_4$	96.11	colorless, amorphous	2.1–2.3

Melting point, °C	Boiling point, °C	Vapor pressure, kPa (at t°C)	Solubility, g/100 g (at t°C)	Relationship with bases and/or acids
1420	≈3300	0.0001 (1207) 0.01 (1467) 1.0 (1867)	insol. in H_2O & org. solv.	reacts with HNO_3+HF & (amorphous Si) with HF & KOH
>2830 decomp..	—	—	insol. in H_2O	insol. in acids & alk.; reacts with HNO_3+HF, & fused alk.
1730; α→β 242	—	—	insol. in H_2O	reacts with HF
≈1610; α→β 573; β-quartz→ β-tridymite 867	—	—	insol. in H_2O	insol. in alk.; reacts with HF
1680; α→β 117; β-tridymite→ β-cristobalite 1470	—	—	insol. in H_2O	reacts with HF
—	—	—	insol. in H_2O	reacts with alk.
—	—	—	insol. in H_2O	reacts with alk.

Name	Symbol/ formula	Atomic/ molecular mass	Color & crystal system	Density (g/cm^3 at 20°C unless otherwise stated)
GERMANIUM AND ITS COMPOUNDS				
GERMANIUM	Ge	72.59	light gray, cubic	5.323 (25°)
Germanium (IV) chloride	GeCl$_4$	214.40	colorless liq.	1.874 (25°)
(II) oxide	GeO	88.59	amorphous powder	—
(IV) oxide	GeO$_2$	104.59	colorless, tetragonal (α), hexagonal (β)	6.239 (α), 4.703 (β) (18°)
Germane	GeH$_4$	76.62	colorless gas	3.420 g/L

Melting point, °C	Boiling point, °C	Vapor pressure, kPa (at t°C)	Solubility, g/100 g (at t°C)	Relationship with bases and/or acids
937	2850	0.000133 (1270) 0.00133 (1415) 0.0133 (1590)	insol. in H_2O	insol. in alk., HCl, cold H_2SO_4, & HNO_3; reacts with hot H_2SO_4, aqua regia, & alk.+H_2O_2
-49.5	83.1	1.33 (-14.5) 13.3 (27.7) 53.3 (64.0)	reacts with H_2O; sol. in EtOH, Et_2O, CCl_4, & CS_2	—
subl. >700	—	0.133 (614) 1.33 (675) 13.3 (745)	sl. sol. in H_2O	insol. in alk.; reacts with acids
1086(α); 1116(β); $\alpha \rightarrow \beta$ 1049	—	—	H_2O: 0.43(20) (β), 1.0(100) (β); insol. in H_2O(α)	α: sl. sol. in NaOH, insol. in HCl & HF; β: sl. sol. in acids, reacts with alk.
-165.8	-88.5	1.33 (-145.6) 13.3 (-120.8) 53.3 (-100.5)	reacts with H_2O	reacts with alk.

Name	Symbol/ formula	Atomic/ molecular mass	Color & crystal system	Density (g/cm^3 at 20°C unless otherwise stated)
TIN AND ITS COMPOUNDS				
TIN	Sn	118.69	gray, cubic(α); silvery, tetragonal (β); silvery, hexagonal(γ)	5.8466 (α) 7.2984 (β) 6.52–6.56 (γ)
(II) chloride	SnCl$_2$	189.60	colorless, rhombic	3.95 (25°)
Tin(II) chloride dihydrate	SnCl$_2$·2H$_2$O	225.63	colorless, monoclinic	2.710 (15.5°)
(II) fluoride	SnF$_2$	156.69	colorless, monoclinic	—
(IV) hydride	SnH$_4$	122.72	colorless gas	—
(II) ortho- phosphate	Sn$_3$(PO$_4$)$_2$	546.01	white powder, amorphous	3.823 (17°)
(II) oxide	SnO	134.69	black, tetragonal	6.446 (0°)
(IV) oxide	SnO$_2$	150.69	colorless, tetragonal(α); cryst.(β)	7.01
(II) sulfate	SnSO$_4$	214.75	colorless, rhombic	—
Sodium pentafluoro- stannite(II)	NaSn$_2$F$_5$	355.36	colorless, tetragonal	4.24
Sodium hexa- hydroxostan- nate(IV)	Na$_2$[Sn(OH$_6$)]	266.73	colorless, hexagonal	—

Melting point, °C	Boiling point, °C	Vapor pressure, kPa (at t°C)	Solubility, g/100 g (at t°C)	Relationship with bases and/or acids
231.9 $\alpha \to \beta$ 14, $\beta \to \gamma$ 173	2620	0.000133 (1105) 0.00133 (1240) 0.0133 (1405)	insol. in H_2O & org. solv.	reacts with acids & hot conc. NaOH
247	670	0.133 (306); 1.33 (392); 13.3 (509)	reacts with H_2O; sol. in EtOH, Et_2O, acetone, pyridine, & EtOAc	—
37.7	decomp.	—	reacts with H_2O; sol. in EtOH, Et_2O, acetone, & AcOH	—
215	853±5	0.000133 (278) 0.00133 (326) 0.0133 (382)	H_2O: 63(25); insol. in Et_2O, EtOH, & $CHCl_3$	reacts with HF
-150	-52.6	1.33 (-118.8) 13.3 (-89.5) 53.3 (-63.8)	—	reacts with conc. acids & with H_2SO_4
—	—	—	insol. in H_2O	reacts with acids & alk.
$\to SnO_2$ at 550 (in air)	—	0.133 (803) 1.33 (961) 13.3 (1173)	insol. in H_2O	reacts with acids
2000; $\alpha \to \beta$ 425	≈2500	—	insol. in H_2O	insol. in acids; reacts with alk.
decomp. 360	—	—	H_2O: 19(20), 18.1 (100)	reacts with H_2SO_4
281	—	—	H_2O: 5.6(25), 11.7 (60); insol. in org. solv.	—
decomp 140	—	—	H_2O: 61.3(15.5), 50(100); insol. in EtOH & acetone	reacts with NH_4OH

Name	Symbol/ formula	Atomic/ molecular mass	Color & crystal system	Density (g/cm^3 at 20°C unless otherwise stated)
LEAD AND ITS COMPOUNDS				
LEAD	Pb	207.2	blue-gray, cubic	11.336
Lead azide	Pb(N$_3$)$_2$	291.23	colorless, rhombic	4.71 (α), 4.93 (β)
carbonate	PbCO$_3$	267.20	colorless, rhombic	6.56
chloride	PbCl$_2$	278.10	colorless, rhombic	5.85
(II) chromate	PbCrO$_4$	323.18	yellow, monoclinic	6.12 (15°)
hydrogen orthoarsenate	PbHAsO$_4$	347.12	colorless, monoclinic	5.79
orthoarsenite	PbHAsO$_3$	331.1	white powder	—
metasilicate	PbSiO$_3$	283.27	colorless, monoclinic	6.49
nitrate	Pb(NO$_3$)$_2$	331.20	colorless, cubic or monoclinic	4.53
(II) oxide	PbO	223.19	red, tetragonal (α); yellow, rhombic (β)	9.51 (α) 8.70 (β)
(II,IV) oxide	Pb$_3$O$_4$	685.57	red, tetragonal	8.79

Melting point, °C	Boiling point, °C	Vapor pressure, kPa (at t°C)	Solubility, g/100 g (at t°C)	Relationship with bases and/or acids
327.4	1745	0.000133 (628) 0.00133 (721) 0.0133 (837)	insol. in H_2O; reacts with AcOH	insol. in HF, dilute HCl, H_2SO_4, conc. HNO_3; reacts with dilute HNO_3, hot conc. H_2SO_4, & HCl
expl. 350	—	—	H_2O: 0.023(18), 0.09 (70)	—
decomp. >300	—	—	H_2O: 0.000011 (20); reacts with hot H_2O; insol. in EtOH	reacts with acids & alk.
495	953	0.000133 (383) 0.00133 (433) 0.0133 (492)	H_2O: 0.98(20), 3.25 (100); sol. in EtOH, pyridine, & glycerol	sol. in conc. alk.
844	—	—	H_2O: 5.8×10^{-6}	—
decomp. 200	—	—	sl. sol. in hot H_2O	—
780	—	—	H_2O: 0.006−0.012(20)	—
766	—	—	insol. in H_2O	reacts with acids
decomp. >200	—	—	H_2O: 52.2(20), 127.3 (100); EtOH: 0.04(20); MeOH: 1.42(25); pyridine: 5.46(25)	—
886; $\alpha \to \beta$ 540	1535	0.000133 (670) 0.00133 (745) 0.0133 (834)	H_2O: 0.0017(22) (α), 0.0023(22) (β)	reacts with acids & alk.
$\to$ PbO 550	—	—	insol. in H_2O	reacts with dilute acids

Name	Symbol/ formula	Atomic/ molecular mass	Color & crystal system	Density (g/cm^3 at 20°C unless otherwise stated)
LEAD COMPOUNDS (cont'd)				
(IV) oxide	PbO_2	239.19	brown-black, tetragonal(α), rhombic(β)	9.33 (β), 9.67(α)
orthoarsenate	$Pb_3(AsO_4)_2$	899.41	colorless, cryst.	7.30
orthophosphate	$Pb_3(PO_4)_2$	811.51	colorless, hexagonal	6.9—7.3
(II) sulfate	$PbSO_4$	303.25	colorless, rhombic(α), monoclinic(β)	6.35 —
(II) sulfide	PbS	239.25	gray-black, cubic	7.59
Trilead dihydroxide dicarbonate	$2PbCO_3 \cdot Pb(OH)_2$	775.60	colorless, hexagonal or amorphous powder	6.14
TITANIUM AND ITS COMPOUNDS				
TITANIUM	Ti	47.90	silvery, hexagonal (α), cubic (β)	4.505
Titanium diboride	TiB_2	69.52	gray, hexagonal	4.5
carbide	TiC	59.89	silvery, cubic	4.92
(IV) chloride	$TiCl_4$	189.71	colorless liq.	1.726

Melting point, °C	Boiling point, °C	Vapor pressure, kPa (at t°C)	Solubility, g/100 g (at t°C)	Relationship with bases and/or acids
decomp. >280(β), >229(α)	—	—	insol. in H_2O & EtOH	reacts with HCl & conc. H_2SO_4
decomp. 1042	—	—	sl. sol. in H_2O	reacts with acids
1014	—	—	H_2O: 0.000014(20)	—
decomp. 1170; $\alpha \to \beta$866	—	—	H_2O: 0.0045(25), 0.0057(50); insol. in EtOH	sol. in conc. HNO_3, HCl, & hot conc. H_2SO_4
1077	1281	0.000133 (589) 0.00133 (659) 0.0133 (740)	H_2O: 8×10^{-14}(20)	insol. in alk., dilute HCl, & H_2SO_4; reacts with HNO_3, conc. H_2SO_4, & HCl
decomp. 400	—	—	insol. in H_2O	—
1668; $\alpha \to \beta$ 882	$\approx$3330	0.000133 (1590) 0.00133 (1750) 0.0133 (1950)	insol. in cold H_2O & in org. solv.; reacts with hot H_2O	reacts with acids
2900	—	—	insol. in H_2O & EtOH	insol. in HCl; reacts with H_2SO_4 & HNO_3 +H_2O_2
2780 decomp..	—	—	insol. in H_2O	insol. in HCl & H_2SO_4; reacts with HNO_3+HF & fused alk.
-24.1	136.3	0.133 (-17.6) 1.33 (20.7) 13.3 (72.4)	reacts with H_2O	reacts with acids, alk., & NH_4OH

Name	Symbol/ formula	Atomic/ molecular mass	Color & crystal system	Density (g/cm^3 at 20°C unless otherwise stated)
TITANIUM COMPOUNDS (cont'd)				
hydride	TiH_2	49.92	gray, tetra-gonal or cubic	3.9 (12°)
nitride	TiN	61.91	yellow-brown, cubic	5.43
(IV) oxide: anatase	TiO_2	79.90	colorless, tetragonal	3.84
brookite	TiO_2	79.90	colorless, rhombic	4.17
rutile	TiO_2	79.90	yellow or red, tetragonal	4.26
silicide	$TiSi_2$	104.07	light gray, rhombic	4.02
Barium metatitanate	$BaTiO_3$	233.21	colorless, tetragonal	6.08
Calcium metatitanate	$CaTiO_3$	135.98	colorless, monoclinic	4.10
Lead metatitanate	$PbTiO_3$	303.09	yellow, rhombic or tetragonal	7.52
Metatitanic acid	H_2TiO_3	97.91	white powder, amorphous	—

Melting point, °C	Boiling point, °C	Vapor pressure, kPa (at t°C)	Solubility, g/100 g (at t°C)	Relationship with bases and/or acids
decomp. 400	—	—	—	—
2950	—	—	insol. in H_2O	insol. in hot HCl, HNO_3, & H_2SO_4; weakly reacts with hot HF+aqua regia & KOH
→ rutile 800—850	≈3000	—	insol. in H_2O	insol. in acids; reacts with alk., H_2SO_4, & HF
—	—	—	insol. in H_2O	reacts with alk. & H_2SO_4
1870	—	—	insol. in H_2O	insol. in acids; reacts with alk., H_2SO_4, & HF
1470	—	—	insol. in H_2O	reacts with HF
1625	—	—	insol. in EtOH	—
—	—	—	—	—
—	—	—	insol. in H_2O	—
—	—	—	insol. in H_2O	insol. in acids & conc. alk.; reacts with conc. H_2SO_4, & HF

Name	Symbol/ formula	Atomic/ molecular mass	Color & crystal system	Density (g/cm^3 at 20°C unless otherwise stated)
ZIRCONIUM AND ITS COMPOUNDS				
ZIRCONIUM	Zr	91.22	silvery, hexagonal(α), cubic(β)	6.45 (α)
Zirconium carbide	ZrC	103.23	gray, cubic	6.73
(IV)chloride	ZrCl$_4$	233.03	colorless, monoclinic, hygr.	2.803
diboride	ZrB$_2$	112.84	gray, hexagonal	6.09
dichloride oxide octahydrate	ZrOCl$_2$· 8H$_2$O	322.25	colorless, tetragonal	1.552
dinitrate oxide hexahydrate	ZrO(NO$_3$)$_2$· 6H$_2$O	339.32	colorless, cryst.	2.08
(IV)fluoride	ZrF$_4$	167.21	colorless, tetragonal(α), monoclinic(β)	4.43
(IV)hydroxide	Zr(OH)$_4$	159.25	colorless powder, amorphous	3.25

Melting point, °C	Boiling point, °C	Vapor pressure, kPa (at t°C)	Solubility, g/100 g (at t°C)	Relationship with bases and/or acids
1855; α→β 863	≈4340	0.000133 (2140) 0.00133 (2350) 0.0133 (2610)	insol. in H_2O & org. solv.	insol. in alk. & dil. acids; reacts with conc. H_2SO_4, HF, aqua regia, & fused alk.
3735	≈5100	—	insol. in H_2O	reacts with acids & fused alk.
437 (1.99 MPa)	subl. 333	0.133 (186) 1.33 (228) 13.3 (280)	reacts with H_2O; sol. in EtOH & Et_2O	reacts with acids & alk.
3200	—	—	insol. in EtOH	reacts with acids
-6H2O 150; -8H2O 210; → ZrO_2 400	—	—	H_2O: 60(20); reacts with hot H_2O; sol. in EtOH & Et_2O	reacts with hot conc. HCl
—	—	—	very sol. in H_2O; sol. in EtOH	reacts with acids
910 (0.11 MPa); α→β 690	subl. 600	0.000133 (489) 0.00133 (536) 0.0133 (590)	H_2O: 1.5(25), 1.39 (50)	sol. in HF; reacts with acids & alk.
-2H2O 500	—	—	sl. sol. in H_2O; insol. in EtOH	reacts with HF; insol. in alk.

Name	Symbol/ formula	Atomic/ molecular mass	Color & crystal system	Density (g/cm^3 at 20°C unless otherwise stated)
ZIRCONIUM AND ITS COMPOUNDS (cont'd)				
nitride	ZrN	105.23	yellow-green, cubic	7.09
orthosilicate	ZrSiO$_4$	183.31	colorless or red, tetragonal	4.56
oxide	ZrO$_2$	123.22	colorless, monoclinic (α), tetragonal (β), cubic (γ)	5.68 (α)
sulfate tetrahydrate	Zr(SO$_4$)$_2$· 4H$_2$O	355.40	colorless, rhombic	3.22 (16°)
Potassium hexa- fluorozirconate	K$_2$ZrF$_6$	283.41	colorless, monoclinic	3.48
Sodium metazirconate	Na$_2$ZrO$_3$	185.20	colorless, cryst	—
HAFNIUM AND ITS COMPOUNDS				
HAFNIUM	Hf	178.49	silvery, hexagonal (α), cubic (β)	13.31
0.94-carbide	HfC$_{0.94}$	189.78	gray, cubic	12.20

Melting point, °C	Boiling point, °C	Vapor pressure, kPa (at t°C)	Solubility, g/100 g (at t°C)	Relationship with bases and/or acids
2990	—	—	insol. in H_2O	reacts weakly with aqua regia, HNO_3+HF, & hot conc. acids
1680 decomp.	—	—	insol. in H_2O	insol. in acids, alk., & aqua regia
2700; $\alpha \to \beta$ 1175; $\beta \to \gamma$ 2350	≈4300	0.000133 (2400) 0.00133 (2610)	insol. in H_2O & EtOH	reacts with HF & conc. H_2SO_4
-3H_2O 100—160; -4H_2O 190—340; decomp.. >450	—	—	H_2O: 64(18),79(40); insol. in EtOH	sol. in H_2SO_4
600 decomp..	—	—	H_2O: 0.78(2), 25(100)	reacts with conc. alk.
decomp. 1500	—	—	—	reacts with acids
2230; $\alpha \to \beta$ 1740	≈4620	0.00133 (2220) 0.00133 (2450) 0.0133 (2720)	insol. in H_2O	reacts with conc. HF, aqua regia, & conc. H_2SO_4
3960	—	—	—	reacts with conc. H_2SO_4

Name	Symbol/ formula	Atomic/ molecular mass	Color & crystal system	Density (g/cm^3 at 20°C unless otherwise stated)
HAFNIUM COMPOUNDS (cont'd)				
chloride	HfCl$_4$	320.30	colorless, cubic, hygr.	—
dichloride oxide octahydrate	HfOCl$_2\cdot$ 8H$_2$O	409.52	colorless, tetragonal	—
nitride	HfN	192.50	dark brown, cubic	—
oxide	HfO$_2$	210.49	colorless, monoclinic	9.68

Melting point, °C	Boiling point, °C	Vapor pressure, kPa (at t°C)	Solubility, g/100 g (at t°C)	Relationship with bases and/or acids
432 (3.38 MPa)	subl. 315	0.000133 (103) 0.00133 (129.5) 0.0133 (160.5)	reacts with H_2O & EtOH	reacts with acids & alk.
-H_2O 65	decomp. >300	—	sol. in cold H_2O; reacts with hot H_2O; insol. in EtOH	reacts with acids & alk.
3300	—	—	—	—
2780	≈5400	—	insol. in H_2O hot conc.	reacts with aqua regia, H_2SO_4, HNO_3, HF, & fused alk.

Index of the elements and their compounds